NANOWIRES - IMPLEMENTATIONS AND APPLICATIONS

Edited by **Abbass Hashim**

Nanowires - Implementations and Applications
Edited by Abbass Hashim

Published by InTech
Janeza Trdine 9, 51000 Rijeka, Croatia

Publishing Process Manager Viktorija Zgela

Technical Editor Teodora Smiljanic

Cover Designer InTech Design Team

Additional hard copies can be obtained from orders@intechopen.com

Nanowires - Implementations and Applications, Edited by Abbass Hashim
p. cm.
ISBN 978-953-307-318-7

For EU product safety concerns:
IN TECH d.o.o., Prolaz Marije Krucifikse Kozulić 3, 51000 Rijeka, Croatia,
info@intechopen.com or visit our website at intechopen.com.

Meet the editor

Currently I am carrying out research in several areas of interest, mainly covering work on environmental pollution, nanotechnology, chemical and bio-sensors, semiconductor thin film device fabrication and characterisation. At the moment I have very strong interest in radiation environmental pollution and bacteriology treatment. The teams of researchers are working very hard to bring novel result in this field. I am also a member of team in charge for the supervision of Ph.D. students in the fields of development of silicon based planar waveguide sensor devices, study of inelastic electron tunnelling in planar tunnelling nanostructures for sensing applications and development of organotellurium(IV) compounds for semiconductor applications. I am a specialist in data analysis techniques and nanosurface structure. I have served as the editor for many books, been a member of the editorial board in science journals, have published many papers and hold many patents.

Contents

Preface

This potentially unique work offers various approaches on the implementation of nanowires. As it is widely known, nanotechnology presents the control of matter at the nanoscale and nanodimensions within few nanometers, whereas this exclusive phenomenon enables us to determine novel applications.

Nanowire technology has considerably improved and developed many technologies and industry sectors in the field of IT, communications, computer technology, medicine devices, sensors applications and many others.

The main advantage of using nanowire is the possiblity of adaptation of the essential structures of materials at the nanoscale to achieve the required properties while, at the same time, contributing to diversification of toolkits used in material science. Currently, nanowires are already in use in many computing, communication, and other electronics applications to provide faster, smaller, and more portable systems that can download and store even considerably large amounts of information.

The superiority of nanoscale wires lies in the fact that they create interesting solid state systems with extremely individual geometry, that offer great opportunity for further development in optoelectronic devices, computer chip and nanosensors application developments, along with several possibilities for studying exciting physical phenomena arising from carrier confinement and large surface/volume ratio.

This book presents an overview of recent and current nanowire application and implementation research worldwide. We examine methods of nanowire synthesis, types of materials used, and applications associated with nanowire research. Wide surveys of global activities in nanowire research are presented, as well.

The book is divided into five sections and the authors contributed to this book with 24 chapters.

The semiconductor nanowires (NWs) explore the nanoscale wire and are to play the critical role in future electronic and optoelectronic devices. In section one we reviewed recent and advanced research in growth, characterization, assembly and integration of synthesized, atomic scale semiconductor NWs. Four different chapters are enclosed in this section. Subsequently, the chapters are framed to discuss surface optical mode, novel properties associated with one-dimensional (1D) structures, methods of semiconductor NWs production via electrochemical synthesis and the last chapter is about the growth, doping and device applications.

Oxide nanowire is one of the most important types of nanowire application. There is a high demand in recent years to advance the reesearch of this type of nanowire. Section two is concerned with applications of oxide nanowire. The section contains four diverse chapters that are focused on application of the Kirkendall effect on morphology control of nanowires, gas sensors oxide nanowire formed by thermal evaporation, copper oxide and zinc oxide nanowire photovoltaic cells, and investigatation of ZnO nanowires and their application for solar cells.

Electronic, optoelectronic and other devices interfere in many areas of industry, from simple household piece of equipments and multimedia systems to communications, computing, control, medical instruments and engineering test instrumentations.

The progress of technology demands more powerful systems to enable new functions and enhance the performance of instruments. Semiconductor nanowires are, therefore, becoming a privileged class of materials that can control the growth of nanoscale and open up substantial opportunities for novel nanoscale photonic and electronic devices. The third section discusses a range of the most recent research of nanowire devices in seven chapters.

An important aspect of nanowires fabrication is to study the ability of preparing wires of an individual material on substrate with a tiny thickness. In particular, to use various substrates which are highly desirable and to study the optional methods. To achieve this goal we suggest section four, the section that discusses the fabrication of nanowire in six chapters.

The authors employed a variety of nanocharacterization techniques to understand and ultimately improve the nanowire properties. The techniques that have been adapted to study the nanostructures by using Scanning Microscopy, Raman Spectroscopy, Single Beam Optical Tweezers and Electron Multiplying Charges Coupled Device (EMCCD), have revealed magnificent details of nanowire structures. These details can be viewed in section five which contains three outstanding chapters.

We reviewed a broad range of nanowire application and implementation research and finally we think that we managed to implement the most valuable and useful chapters on nanowires. Furthermore, by building blocks of research papers available in this book, we have made the valuable knowledge accessible to researchers and discussed a range of electronic and optoelectronic nanodevices, as well as integrated device arrays that could allow diverse and exciting applications in the future.

Dr. Abbass Hashim
Material and Engineering Research Institute
Sheffield Hallam University
UK

Part 1

Semiconductor Nanowires

1

Surface Optical Mode in Semiconductor Nanowires

S. Dhara, Prasana Sahoo, A. K. Tyagi and Baldev Raj
Surface and Nanoscience Division,
Indira Gandhi Center for Atomic Research, Kalapkkam,
India

1. Introduction

Semiconducting nanowires represent interesting solid state systems with unique geometry offering great possibility for further development of optoelectronic devices and sensors applications with numerous possibilities for studying exciting physical phenomena arising from carrier confinement and the large surface-to-volume ratio (Hayden 2008, Pauzauskie & Yang, 2006; Yang et al., 2010; Lu & Leiber, 2007; Patolsky et al., 2007; Cao et al., 2009; Algra et al., 2008 & Wang et al., 2001). However, the growth of nanowires free of contaminants, controlled surface states and structural defects is still one of the key issues.

Studies comparing Raman scattering experiments of bulk and nanostructured materials have been reported in literature for several different kind of systems with extensive use as a primary characterization tool for study of phase and optical properties of nanowires and nanoparticles. (Zeng et al., 2006; Lin et al., 2003; Mahan et al., 2003; Adu et al., 2006; Hartschuh et al., 2003; Zardo et al., 2009; Cao et al., 2006; Shan et al., 2006; Pauzauskie et al., 2005; Xiong et al., 2006).With respect to one-dimensional (1-D) structures, several new phenomena have been reported to date, e.g., the high surface-to-volume ratio enabling the measurement of surface optical (SO) phonon modes (Gupta et al., 2003; Xiong et al., 2004; Spirkoska et al., 2008; Sahoo et al., 2009a; Sahoo et al., 2010; Adu et al., 2006). A increase in the scattered intensity is reported for nanoscale structures with respect to their bulk counterpart, describing the effect denominated as 'Raman antenna effect' (Xiong et al., 2004; Xiong et al., 2006; Cao et al., 2007). In addition, polarization dependent experiments on single carbon nanotubes and nanowires have shown that the physics behind Raman scattering of such 1-D nanostructures can differ significantly from the bulk (Frechette et al., 2006; Pauzauskie et al., 2005; Livneh et al., 2006; Cao et al., 2006). As a matter of fact, the highly anisotropic shape of the nanowires can lead to angular dependencies of the phonon modes which otherwise would not be expected from selection rules (Pauzauskie et al., 2005; Nobile et al., 2007). In general, Raman spectroscopy is an efficient tool to provide qualitative information from semiconductor by probing SO modes.

1.1 Surface optical phonons

SO phonons are similar in concept to bulk phonons, except that the atomic amplitudes are confined to the near-surface region of the material. The amplitudes of these phonons decay exponentially with distance measured away from the sample SO modes. In general, two different types of SO phonon modes, depending on the penetration depths, can be distinguished as macroscopic and microscopic ones. Macroscopic SO modes are classified as optical (e.g. Fuchs-Kliewer; Fuchs & Kliewer, 1965) and acoustic (e.g. Rayleigh) SO phonon frequencies. Acoustic SO modes in isotropic elastic media propagate along the surface or interface with displacements in the sagittal plane which is defined by the normal to the surface and the direction of propagation (Wallis, 1994). These modes decay exponentially into the bulk and the penetration depth is proportional to the wavelength in the long-wave limit. Contrary to an acoustic mode the Fuchs-Kliewer phonon is connected to a macroscopic electric field (Fuchs & Kliewer, 1965). The frequency of a Fuchs-Kliewer phonon can be determined by solving the Laplace equation for the electrostatic potential of an ionic crystal under appropriate electromagnetic boundary conditions. Then the frequency is found to lie in between the frequency of the transverse optical (TO) and longitudinal optical (LO) bulk phonon. The corresponding pictorial representation of the atomic motion in a phonon mode has been illustrated in figure 1 (Richter et al., 2000).

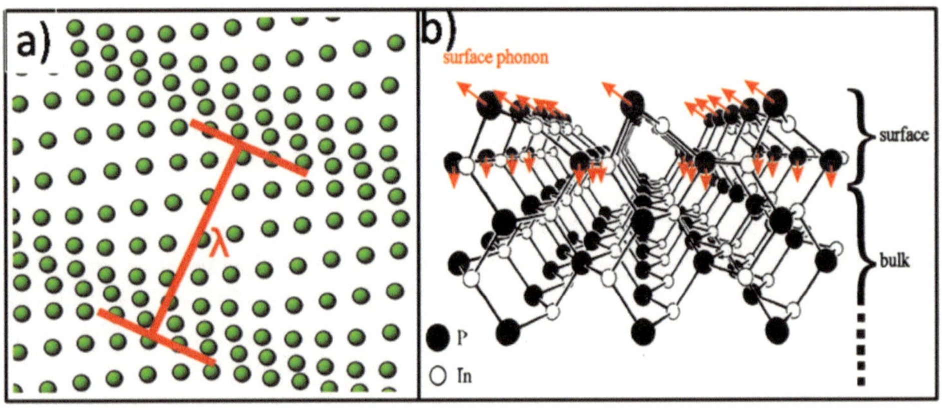

Fig. 1. Schematic representation of the a) atomic motion in a phonon mode (http://en.wikipedia.org/wiki/Surface_phonon) and b) SO phonon. The atomic displacements of SO phonons are confined to the top atomic layers. The atomic positions and the eigenvector in this example correspond to the higher frequency gap mode (270 cm^{-1}) of InP (110) at point $\bar{\Gamma}$ in the SO Brillouin Zone. (Richter, et al., 2000 Copyright © Xiamen University Press).

Experimental techniques for analyzing SO phonon properties have so far been based on SO phonon sensitive probes such as low-energy electrons and atoms, i.e. high resolution electron energy loss spectroscopy and He-atom scattering (Ibach & Mills, 1982; Benedek & Toennies, 1994). Due to the strong interaction of low-energy electrons and atoms with matter, the penetration depth of these probes is limited to the few outermost atomic layers giving rise to the sensitivity of SO modes in these techniques. Raman spectroscopy, in contrast, is based on the comparably weak interaction of photons with matter. Even under strong absorbing conditions the penetration depth of photons is no less than approximately fifty atomic layers and thus the Raman signal is bulk rather than SO mode related. However, the resolution and sensitivity of modern Raman equipment allows also for the observation of Raman signals generated by SO phonons (Gupta et al., 2003; Xiong et al., 2004). The reasons are, firstly the vibrational frequencies of SO phonons are different from those of the bulk, since the top layer is free and bonding atoms are missing. Secondly, the energies of electronic states corresponding to the surface are different from bulk states and thus the resonance condition of SO phonons is different from that of bulk phonons. These differences in SO and bulk phonon scattering are pronounced for a perturbed surface or for hetero-termination i.e., when a surface is terminated with atoms not being present in the bulk.

SO phonons are reported in polar semiconductors exhibiting a formal charge separation between cation and anion sub-lattices. These SO modes show dispersion, i.e. their frequency depends on the wave vector q measured along the surface. For any q vector, the SO mode frequency ω_{SO} lies between those of the $q = 0$, LO and TO phonons. The SO phonon dispersion depends both on the dielectric function of the semiconductor as well as on the material, e.g. air, oxide, or liquid in contact with the surface. Wherever there is an interface between different materials with different dielectric functions, generally, there will be interface modes. They are solutions to Maxwell's equations with appropriate boundary conditions (Sernelius, 2001). These modes, which are localized to the interface, can propagate along the interface.

Normally, SO modes are observed when the translational symmetry of the SO potential is broken. This break down of symmetry can activate a larger wave vector SO mode whose frequency is sufficiently separated from the other Raman-active optical phonons, to be observed experimentally. An interplay of a strong Fourier component of the perturbed SO potential acts as a source of wave vector supplied by the symmetry breaking (Sernelius 2001). This can be achieved in several ways, e.g., by surface roughness, or by formation of a grating along the surface. Another method for detecting SO modes involves placing a prism at the surface of the sample. The evanescent wave from the prism–sample interface can probe the surface and the desired SO mode (Falge et al., 1974)

Overall, Raman spectroscopy of nanostructures represents an extremely active and exciting field for the benefit of science and technology at the nanoscale. The arising new phenomena and technical possibilities not only open new vistas for the characterization of materials but also for the understanding of fundamental processes at nanoscale. In this chapter, we provide a review of SO modes on nanowires using Raman spectroscopy depicting overview of the appearance of new modes and their effect on geometry of samples.

2. Surface optical phonons from semiconductor nanowires

There are geometry and size-related phonons appearing for several different kind of systems when dealing generally with nanowires and nanoparticles, such as the SO and breathing modes arising from acoustic confinement. In case of nanowires, the geometry is one of the long filaments. The cross-sectional shape of nanowire matters for the SO phonon dispersion. It is usually observed that the TO and the LO modes have a position in energy close to that observed in bulk. When scaling down the size and the dimensionality of the structures, the position can change along with the appearance of new Raman modes due to breakdown of translational symmetry in the finite size. Moreover, effects related to the shape of the system can become significant. The existence of boundary conditions at the nanoscale gives rise to electric and polarization forces. The surface represents a new mechanical boundary, since the surface atoms are less bound and experience a different local field than that from the bulk. This has consequences even in the propagation of an optical phonon.

Several works have reported the presence SO modes in Raman spectra of semiconductor nanowires which have been assigned to SO phonons (Gupta et al., 2003; Shan et al., 2006; Sahoo et al., 2008a, 2010b; Sahoo et al., 2010; Lin et al., 2003; Zeng et al., 2006; Spirkoska et al.,

2008). The SO phonons are generated at the interface between different materials with different dielectric functions and propagate along the interface. The atoms involved in their propagation are those close to the surface, so that the amplitude of the oscillations decays exponentially with the distance from the surface. This mode is activated by breakdown of the translational symmetry of the SO potential, which in the case of the nanowire can be addressed to the presence of roughness, saw tooth faceting on the nanowire sidewall or to a diameter oscillation along the nanowire length. There are two characteristics which are distinctive of the SO modes and can therefore allow a reliable assignment of the mode: the dependence of the phonon frequency (1) on the dielectric constant of the medium surrounding the wires and (2) on the diameter or on the period of the diameter oscillation of the wires. Indeed, it was already observed that blue shift of the SO mode value increases with the dielectric constant of the surrounding optical medium and decreases with the nanowire diameter (Gupta et al., 2003; Sahoo et al., 2009). Furthermore, the frequency of the SO modes at the center of the Brillouin zone is located between those of the TO and the LO and also the long-wavelength (q = 0) SO modes are hard to identify experimentally, as they strongly overlap with the TO (q =0) phonon.

2.1 SO phonons in cylindrical geometry

The SO modes dispersion at the interface between a semiconductor and a dielectric material can be calculated taking into account of the geometrical constraint and surrounding dielectric medium, by imposing the condition:

$$\varepsilon(\omega) + \varepsilon_m = 0 \tag{1}$$

where $\varepsilon(\omega)$ the dielectric function of the semiconductor and ε_m is the dielectric constant of the medium. For a cylindrical interface, the SO mode frequency can be obtained by solving the following equation,

$$\varepsilon\,(\omega) + \epsilon_m f(qr) = 0 \tag{2}$$

where r is the radius of the cylinder, and

$$f(x) = \frac{I_0(x)K_1(x)}{I_1(x)K_0(x)}, \tag{3}$$

where I and K are Bessel functions and $x = qr$. For an interface between a dielectric medium and a semiconductor surface, on solving (1) with the dielectric function

$$\varepsilon(\omega) = \varepsilon_\infty + \frac{\varepsilon_0 - \varepsilon_\infty}{1 - \frac{\omega^2}{\omega_{TO}^2}} \tag{4}$$

for a semiconductor, the SO mode frequency is

$$\omega_{SO} = \omega_{TO}\sqrt{\frac{\varepsilon_0 + \varepsilon_m}{\varepsilon_\infty + \varepsilon_m}}, \tag{5}$$

where ε_0 and ε_∞ are the static and high-frequency dielectric constants, respectively. ω_{TO} is the TO mode frequency at zone center. This frequency is the asymptotic limit of the SO polariton, i.e. coupled SO phonons and photons.

$$\omega_{SO}^2 = \omega_{TO}^2 + \frac{\tilde{\omega}_p^2}{\varepsilon_\infty + \varepsilon_m f\,(x)},\ x = qr, \tag{6}$$

where $\widetilde{\omega}_p^2$ is the screened ion plasma frequency given by $\omega_{LO}^2 = \omega_{TO}^2 + \widetilde{\omega}_p^2/\varepsilon_\infty$, ω_{LO} is the LO mode frequency at zone center, and $d = 2r$ is the wire diameter. For infinite-diameter cylinders, i.e. $r \to \infty$, $f(x) \to 1$,

$$\omega_{SO}^2 = \omega_{TO}^2 + \frac{\varepsilon_\infty(\omega_{LO}^2 - \omega_{TO}^2)}{\varepsilon_\infty + \varepsilon_m} \Rightarrow \omega_{SO} = \omega_{TO}\sqrt{\frac{\varepsilon_0 + \varepsilon_m}{\varepsilon_\infty + \varepsilon_m}}, \tag{7}$$

This is exactly the SO phonon frequency of an infinite semiconductor flat surface expressed by (5). Therefore, equations (6) and (7) establishes the dependency of the SO phonon energy on the external medium and on the size of the wire, since the position of the SO phonon can be related to the dielectric constant of the surrounding medium as well as to the nanowire radius. Furthermore, values of q for the activation of the SO mode can be determined experimentally (Gupta et al., 2003). Instead, the line width of the SO mode has not been yet well understood. The effect of the position of the SO modes can be evidently observed by comparing semiconductor nanowires with various diameters.

2.1.1 GaP nanowires

The SO potential along the axis of the GaP nanowire is perturbed by a strong component with wave vector q (Gupta et al., 2003). Thus, inelastic light scattering via SO phonons with this average wave vector is particularly important. The wave vector q, responsible for the observation of the SO band, can be deduced by comparing the computed SO frequencies $\omega_{SO}(q)$ to those obtained experimentally.
Figure 2 shows the Raman spectra of the optical phonons for GaP nanowires with diameter of 50 nm in air ($\varepsilon_m = 1$), dichloromethane ($\varepsilon_m = 2$), and aniline ($\varepsilon_m = 2.56$).

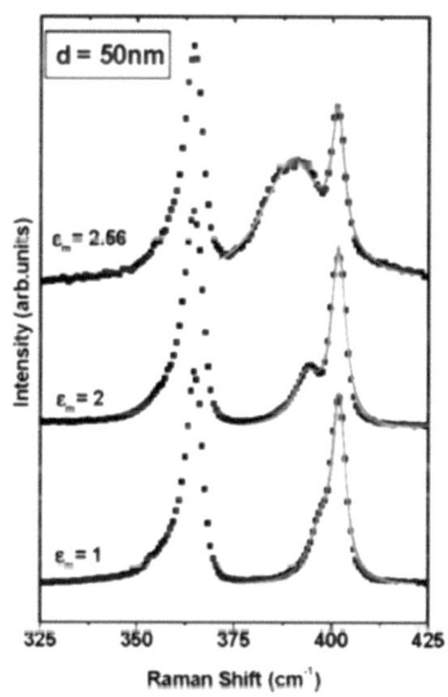

Fig. 2. Raman spectra of GaP nanowires with diameter d= 50 nm recorded in three different media with different dielectric constant (ε_m). The low, middle and high frequency bands are identified respectively with the TO, SO, and LO phonons. The solid lines represent Lorentzian line shapes used to fit the SO and LO band. (Gupta et al., 2003 Copyright © American Chemical Society; Applied for permission).

It is assumed that the LO and TO phonon branches are dispersion less as shown by horizontal lines. The nanowire diameter is given by $2r$, and the thin lines are the result of a Lorentzian line shape analysis. The decomposition of the band into individual Lorentzians is also shown. The lowest and highest frequency bands, respectively, are identified with the first-order modes of TO at 367 cm^{-1} and LO at 401 cm^{-1}. A third Raman band lies midway between the TO and LO bands (Fig. 2). This band is assigned to SO modes because of its sensitivity to the dielectric constant of an external medium (ε_m) e.g. air, dichloromethane, and aniline in contact with the nanowire. The SO band show red shift (Fig. 2), as the dielectric constant of the surrounding medium is increased. The peak positions of the LO and TO bands, however, do not change. Based on this observation, the intermediate frequency band has been unambiguously assigned to SO modes.

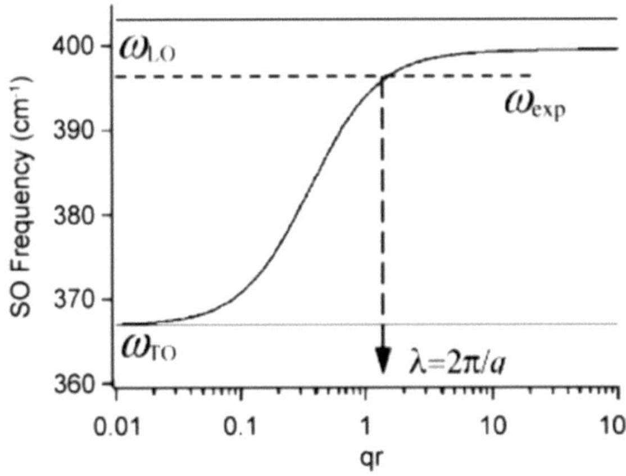

Fig. 3. Calculated dispersion of the SO mode for a GaP nanowire (ε_m =1). The arrow indicate the q= 0, LO and TO phonon frequencies. (Gupta et al., 2003 Copyright © American Chemical Society; Applied for permission).

From the calculated SO dispersion and the ω_{SO} obtained from experiments with the nanowires in various dielectric media (e.g. the crossing points between the solid curve and the dashed lines in Fig. 3), the wavelength $\lambda = 2\pi/q$ of the SO potential (perturbation) which is important for activating the SO mode scattering has been estimated. For cylindrical cross section GaP nanowires, $\lambda = 2\pi/q$ is ~ 40 nm. In case of d= 20 nm wire, the maximum dispersion takes place at q^* wire ~ 2 $\times 10^7$/m by using equation (6). The q^* value marking maximum dispersion is much larger than its counterpart for SO modes at a plane surface (i.e., $q_{plane} \sim \omega_{LO}/c$ =2.5× 10^5/m). Thus, it is clearly important to use the correct (cylindrical) geometry while describing the SO phonons in small diameter nanowires (Adu et al., 2006; Spirkoska et al., 2008).
It appears that values of the SO band frequency, as calculated for x= 4, correspond to the experimentally obtained values for the sample of d= 50 nm, while x= 1.5 values are close to those SO mode observed for the sample with a diameter, d= 20 nm. Approximately the same value of q = 1.5× 10^8/m is found for the two sets of wires d= 20 or 50 nm. Therefore, the length scale which is responsible for excitation of the SO mode is given by $2\pi/q$ = 40 nm. Since this length scale is similar in both cases, there must be an inherent length scale in the nanowires responsible for the activation and observation of the SO band. The symmetry

breakdown mechanism responsible for the observation of the SO band is due to the presence of the wire diameter modulation along the nanowire growth direction. This modulation can indeed be observed in TEM images (Fig. 4) where it has been shown that four different images with different wire diameters with modulation wavelength λ. The diameter modulation wavelengths, λ ($\approx$ 30-70 nm), for the small subset of nanowires (Fig. 4) is consistent with the length scale ($\approx$ 40 nm).

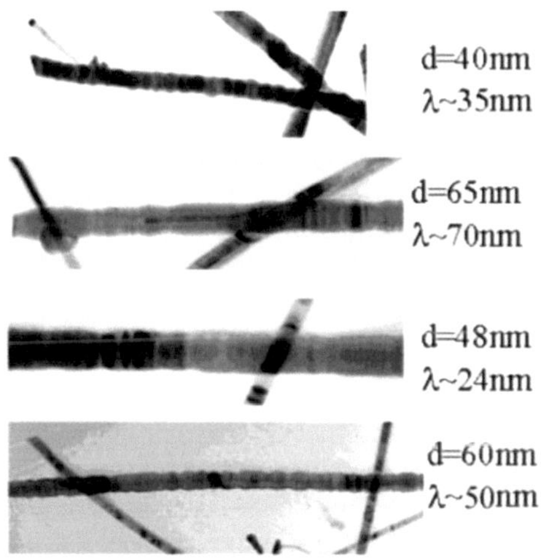

Fig. 4. Bright field TEM images of GaP nanowires showing nearly periodic modulation of wire diameter. The diameter and the modulation wavelength for each image appear to the right. Diameter oscillations have been reported to be more prominent in larger diameter nanowires. (Gupta et al., 2003 Copyright © American Chemical Society; Applied for permission).

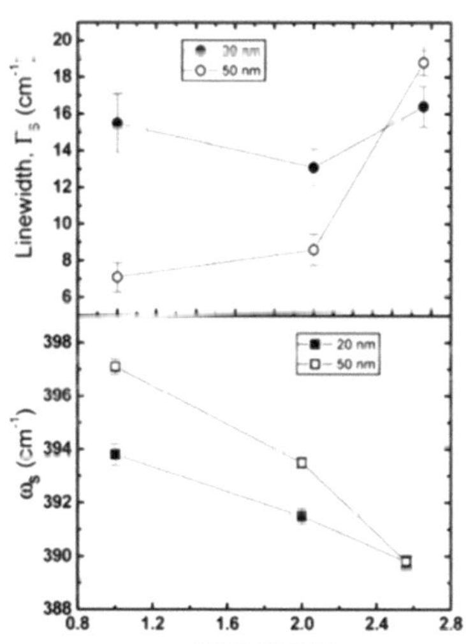

Fig. 5. Variation of the line shape parameters for the SO band (peak position and line width) as a function of the dielectric constant of the overlaying medium. (Gupta et al., 2003, Copyright © American Chemical Society; Applied for permission).

Again, it is already discussed that the dispersion for a SO mode calculated in a cylindrical geometry is crucial for estimating the wavelength of the important Fourier component of the SO potential that breaks the symmetry and "activates" the SO modes in a Raman scattering event. It is also possible that SO Raman mode from larger nanowire diameters ($d > 40$ nm) in the lognormal diameter distribution with a more prominent diameter modulation (Fig. 5) might dominate the SO spectrum for the nanowire ensemble. This assumption can be proved if SO mode frequency can be measured from individual wires with known diameter modulation wavelength.
The SO band position does not include any dynamics of the SO phonons, and therefore it is inadequate to describe the variation of the SO mode line width as a function of ε_m as shown in figure 5 (top). The line width measured in air for the SO band in the 20 nm diameter sample is twice of that observed in the 50 nm sample, even if the line widths of the LO mode in the two nanowire samples with different diameter distributions are comparable to within 10% in all overlaying media. This implies that the SO modes in the 20 nm nanowires have a lifetime that is half of the 50 nm nanowire batch. This may results due to the difference in the decay channel phase space available to the SO phonons for the two samples as the dielectric constant varies.

2.1.2 GaAs nanowires

Figure 6 represents scanning electron microscopy image of the GaAs nanowires, grown by molecular beam epitaxy (Spirkoska et al., 2008). The Raman spectra have been recorded in ensembles of GaAs nanowires having diameters 160 and 69 nm (Fig. 7a). The existence of a SO mode is confirmed by comparing the spectra of the nanowires in an environment of air and PMMA with higher value of dielectric constant. As expected, the SO mode shifts to lower wave numbers for smaller diameters, as it is observed in the spectra obtained for nanowires with an average diameter of 69 nm. The entire trend of the position as a function of the diameter is shown in figure 7b. The SO mode is barely observed as the position is very close to the LO phonon for the nanowire with the largest diameter. Continuous line indicates data for nanowires with a circular section. GaAs nanowires exhibit a hexagonal section explaining the discrepancy with the experimental data.

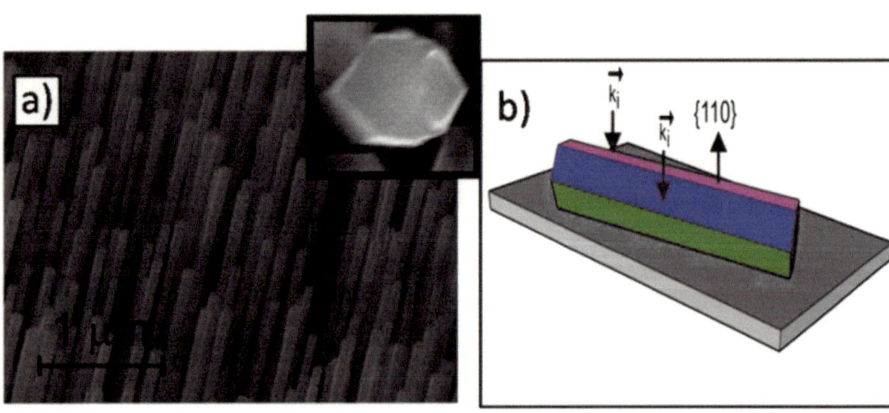

Fig. 6. a) Scanning electron micrograph from the as-grown GaAs nanowires. The hexagonal cross section of the nanowires is presented in the inset. b) Schematic drawing of the scattering geometry of the measurements after transferred of the GaAs nanowires onto Si substrate. (Spirkoska et al., 2008 Copyright © Institute of Physics.)

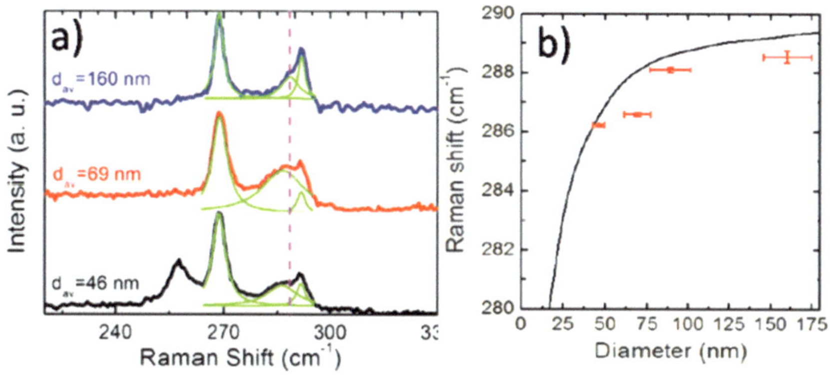

Fig. 7. a) Raman spectra of GaAs nanowire bundles with respectively an average diameter of 160 and 60 nm. The SO mode can be observed on the left of the LO mode in which the position of the SO phonon down-shifts with the decrease in diameter. b) Evolution of the SO phonon position as a function of the diameter of the nanowires. The line corresponds to the theoretical values expected for hexagonal cross section of GaAs nanowires. (Spirkoska et al., 2008 Copyright © Institute of Physics.)

2.1.3 GaN nanowires

In our latest study,(Sahoo et al., 2010) the SO phonon modes has been identified in GaN nanowires grown by vapor liquid solid (VLS) technique showing a strong correlation between the surface morphology and Raman phonon spectra. The SO phonons associated with A_1(TO) at 533 cm^{-1} and E_1(TO) at 560 cm^{-1} are calculated for GaN by taking the qr value of 1.07, which yields SO phonon frequencies (ω_{SO}) pertaining to A_1 (654 cm^{-1}) and E_1 (688 cm^{-1}) characters (Fig. 8). Hence, the reported peaks around 652 and 691 cm^{-1} has been accounted for SO phonon modes in GaN.

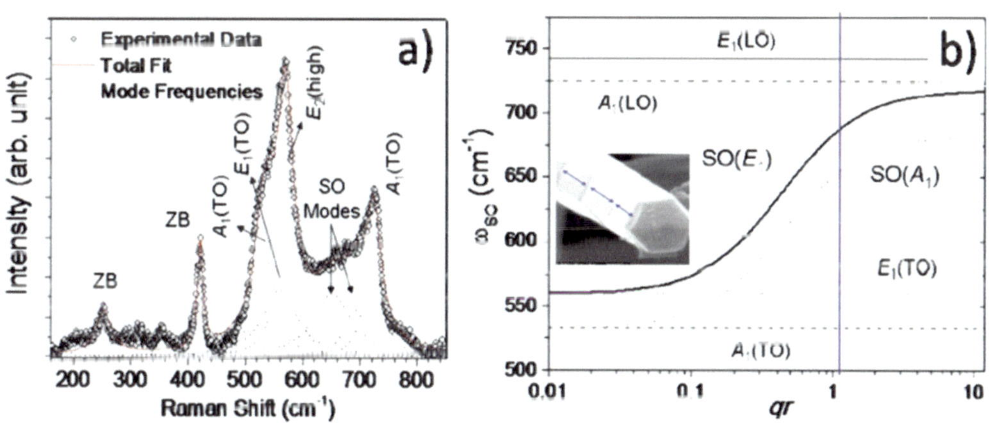

Fig. 8. a) Room-temperature Raman-scattering spectra of GaN nanowires showing the experimentally observed data and the Lorentzian fit for the individual peaks. b) Calculated SO phonon frequencies as a function of qr, full curve: SO(E_1), dashed curve: SO(A_1), horizontal lines are the LO and TO frequencies of E_1 and A_1 modes, as inscribed in the figure. Vertical dotted line is marked for qr = 1.07. Inset is a GaN nanowire with diameter modulation. (Sahoo et al., 2010 Copyright © Inderscience Enterprises Ltd.)

It can also be seen that the intensities of the observed SO phonons in the present study are comparable to other phonons. The magnitude of surface roughness determines the SO peak intensity owing to the breakdown of the translational symmetry of the SO potential. The wavelength $\lambda = 2\pi/q$ ($qr = 1.07$; $r\sim250$) corresponding to the perturbation of SO potential has been estimated as ~1500 nm for the GaN nanostructures (using the dispersion relations in Fig. 8b for typical nanostructure diameter of ~500 nm). Surface roughness with a modulation of ~150–300 nm is observed (indicated by arrows in the inset of Fig. 8b) with any integral multiple equivalent to the λ (10×150 nm = 5×300 nm = 1500 nm) will be sufficient to initiate breakdown of translational symmetry in contributing SO potential toward SO modes. The process makes the intensity of the SO mode comparable to that of the other phonons.

SO mode in a single GaN nanowire grown by molecular beam epitaxy has also been reported by Hsiao and co-workers (Hsiao et al., 2007) during angle dependent Raman spectroscopy measurement with the z axis pointing along the crystallographic c axis of the rod. The direction of different angles between the laser polarization and the rod length (c axis) are shown in figure 9 (a). The reference frame (x, y polarization) is rotated against the rod. The angle between the laser polarization and the z axis (x axis) is represented by θ (φ) (Fig. 9a). In the Porto notation, (Arguello et al., 2969) the representations of θ = 0° and θ = 90° scattering configurations with φ =90° are $x(z, z)\bar{x}$ and $x(y, y)\bar{x}$, respectively. The hexagonal geometry shape with vibrational modes are resolved in $x\,(x+y,\ x+y)\,\bar{x}$ scattering configuration to obtain all main Raman peaks which further demands for application as angle selective nanosensors. However, a new peak at 708.5 cm^{-1} have been ascribed to the SO mode (Fig. 9b) which is absent in GaN compact film (also shown in lower spectra of figure 9b). With high surface-to-volume ratio and the high perfection of the hexagonal surface with highly crystalline structure, the peak is assigned to the SO mode of GaN nanocrystals rather than the disorder-active Raman mode which refers to the effect of defects in crystal or a-GaN and agrees well with the calculated value.

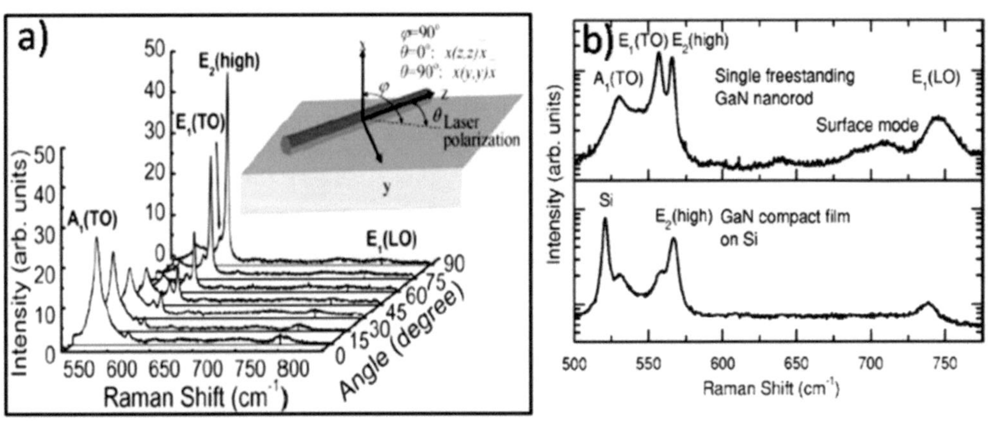

Fig. 9. a) Angle-dependent Raman spectroscopy measured at different angles between the laser polarization and the rod length (c axis). The inset shows a schematic diagram of the Raman scattering configurations measured on a single freestanding nanorod. b) Micro-Raman spectra measured from a single freestanding GaN nanorod and a compact GaN thin film. A prominent peak at 708.5 cm^{-1} is contributed from the SO vibrational mode of GaN nanorod. (Hsiao et al., 2007 Copyright © American Institute of Phys.)

2.1.4 InN nanowires

An optical mode of one-dimensional nanostructures of InN, such as nanowires and nanobelts grown by chemical vapor deposition, has been studied in our earliest report (Sahoo et al., 2008). Typical Raman spectra of nanowires and nanobelts are shown in figures 10(a) and (b), respectively and SO phonons modes have been reported taking into consideration of cylindrical geometry to both. The range of qr is determined from the dispersion relation of ω_{SO} as a function (Fig. 11a) by the spread in diameter (d~75 to 150 nm for nanowires and ~50 to 200 nm for nanobelts). For d=100 nm the value of qr = 1.18 is shown as a vertical dot-dashed line in figure 11. Based on the above agreement the modes around 528–560 cm^{-1} have been assigned to SO modes with A_1 and E_1 characters, respectively. A further confirmation has been clearly reported by recording the spectra in a higher dielectric medium (ε_m=2.24 for CCl_4) which shows inherently a red shift of 4 cm^{-1} to SO modes. The spread in SO phonon essentially arises due to the wide range of diameter and dimension of the nanostructures.

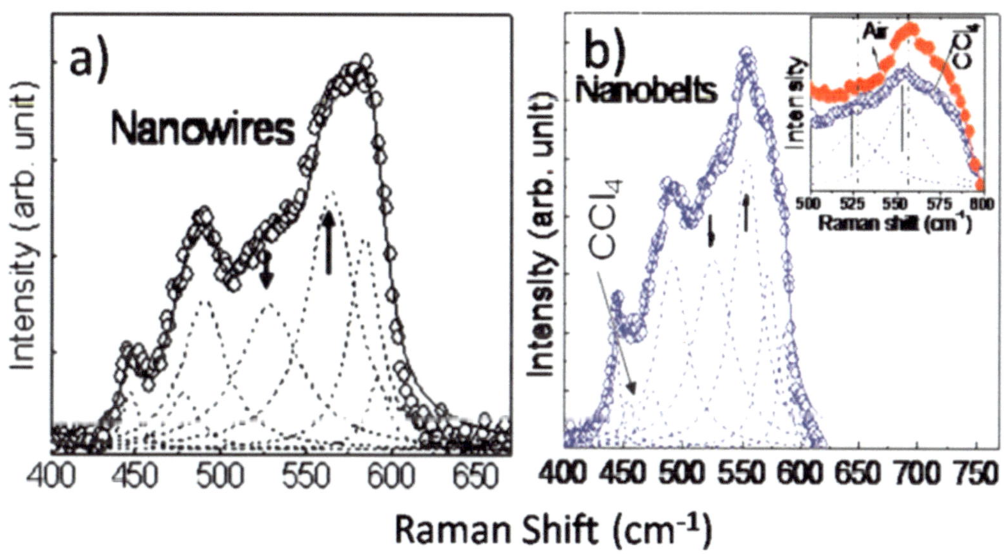

Fig. 10. Raman spectra for InN a) nanowires and b) nanobelts. Full curve: total fitted spectrum, dashed curves: individual fitted peaks. Peaks corresponding to SO mode are indicated by arrows. Raman spectra of InN nanobelts b) immersed in CCl_4. The inset shows SO modes under the expanded range of 500–600 cm^{-1}. (Sahoo et al., 2008 Copyright © American Institute of Physics.)

The wavelength $\lambda=2\pi/q$, corresponding to the SO potential, can be estimated as ~135 nm for nanowires and ~110 nm for nanobelts using the dispersion relations in figure 11 with maximum qr=1.65 for nanowires and 1.4 for nanobelts. Surface roughness with modulation of ~65 nm for nanowires (Fig. 11b) and ~55 nm for nanobelts in figure 11b have been observed. However, the intensity of SO modes found to be higher than that for the TO modes. Defects related to abrupt termination of translational symmetry in the nanostructures may reduce the intensities of fundamental phonon modes; on the other hand, rough surface and interface may result in intense SO modes.

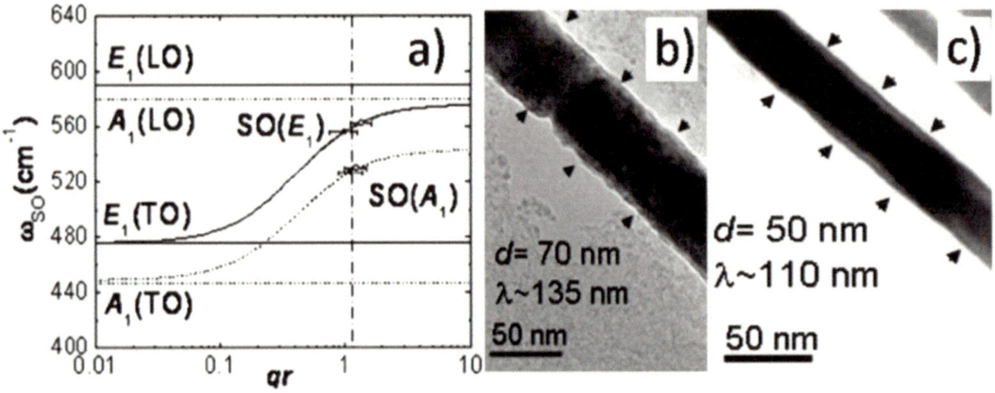

Fig. 11. a) Calculated SO phonon frequencies as a function of qr, full curve: SO (E_1), dashed curve: SO(A_1). Horizontal full and dashed lines are the TO and LO frequencies of E_1 and A_1 modes, respectively. Symbols are measured SO frequencies, open symbol: nanowires, filled symbol: nanobelts. Vertical line at qr=1.18 corresponds to d=100 nm. b) Surface roughness in typical nanowires and c) nanobelts. Arrows show the modulation in the surface structure. Diameters and the calculated wavelengths of SO potential are also inscribed. (Sahoo et al., 2008 Copyright © American Institute of Physics.)

2.1.5 AlN nanowires

In one of our recent studies (Sahoo et al., 2010) we have studied typical Raman spectra for optical phonons in AlN nanotips (Fig. 12a). Apart from the allowed optical phonons in AlN, an additional phonon mode at 850 cm^{-1} has been assigned to SO phonons, in view of its strong response to change in dielectric medium and good agreement with calculated SO phonon frequency. A prominent red shift of SO mode (23 cm^{-1} in CCl_4 medium) has been reported (inset of Fig. 12a).

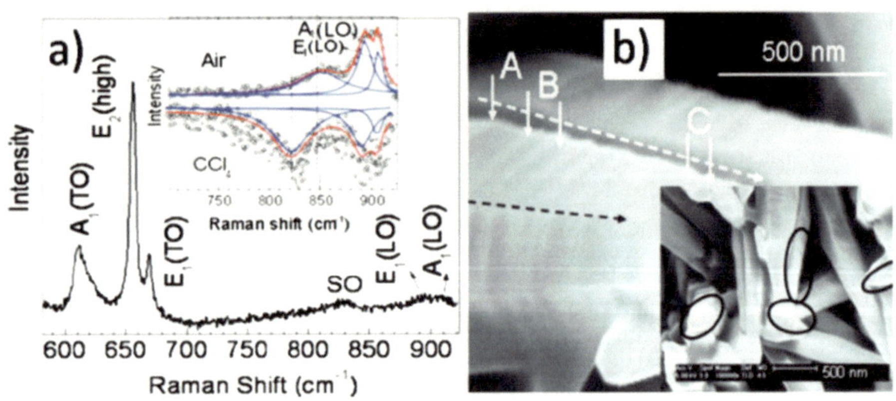

Fig. 12. a) Raman spectra of AlN nanotips recorded in CCl_4. Inset shows comparison of the position of SO phonon peak in air and CCl_4 medium, b) Morphological study at high resolution FESEM image of AlN nanotips with arrows showing the modulation of the surface. Inset shows corresponding marked region with distinct surface modulation. (Sahoo et al., 2010 Copyright © American Institute of Physics.)

The typical analysis of the nanotip surface (as directed by two doted arrows) shows that the steps (labeled as A, B, and C in Fig. 12b) are systematically decreasing along the growth direction. The width (d=2r) of the steps are 120, 90, and 50 nm with a corresponding length of 350, 260, and 140 nm, respectively. The wavelengths ($\lambda=2\pi/q$) of perturbation have been calculated taking into account of the corresponding r and x=1.2 value of each step. The values of wavelength corresponding to A, B, and C are reported to be 314, 235, and 130 nm, respectively. Thus these perturbation values, which are in close agreement with the length of the steps (Fig. 12b), absorb the required phonon momentum for the observation of the SO mode.

2.1.6 InGaN/GaN multi-quantum-well nanopillars

In another study (Wu et al,. 2009) we have reported the fabrication and the observation of SO phonon in aligned GaN nanopillars with spatial control of embedded InGaN/GaN multi-quantum-well (MQWs) using focused ion beam milling. Figure 13 (a) shows the SEM image of resulting InGaN/GaN MQWs structure milled with a serpentine-scanning beam and subsequently wet-etched with KOH solution. The pillar base diameter (Fig. 13a) is around 95 nm with aspect ratio of 7: 1. The pillar size is further reduced to 30 nm with enhanced high aspect ratio of 16: 1, with an extended KOH treatment time. The presence of surface roughness or even surface modulations of ~ 25 nm at two different regions of these 1-D MQW nanopillars have been observed (Fig. 13b).

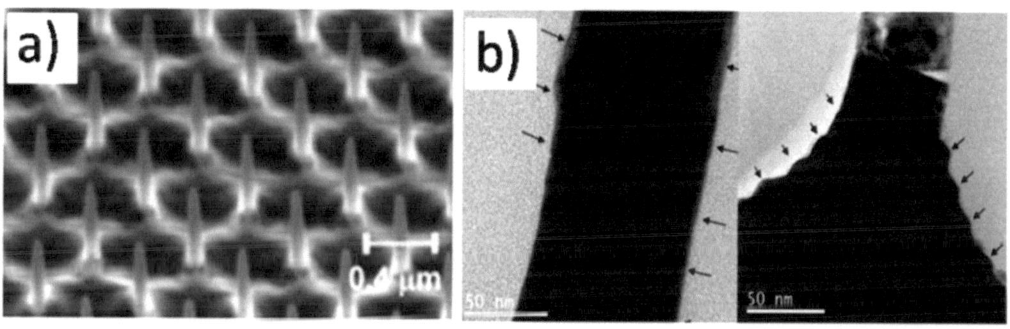

Fig. 13. a) SEM tilted-view images showing FIB milled nanopillars after KOH treatment. b) Surface roughness in typical MQW nanopillars. Arrows show the modulation in the surface structure at two different regions. (Wu et al., 2009 Copyright © John Wiley & Sons, Ltd.)

Figure 14 displays distinct Raman features of KOH etched nanopillars with 632.8 nm (Fig. 14a) and 514 nm (Fig. 14b) laser excitations. Along with all allowed Raman modes, the fitted broad peak in the range 665–699 cm^{-1} (inset in Fig. 14b) have been examined carefully to elucidate the possibility of SO phonon being responsible. The Raman spectra for nanopillars in a higher dielectric medium (ε_m = 2.56 for aniline) have been recorded and compared with those in air for the same excitation of 532 nm laser line (Fig.15a). It is reported that the peak positions of the A_1(LO)$_{GaN}$ and A_1(LO)$_{InGaN}$ modes did not change, whereas an apparent red shift of ~15 cm^{-1} for SO modes for GaN has been reported for the spectrum collected from the sample surrounded with a higher value of dielectric constant than that of air (ε_m = 1). These unidentified peaks had been assigned to SO phonon modes originating from surface roughness or even surface modulation in 1-D MQW a nanopillars (Fig. 13b).

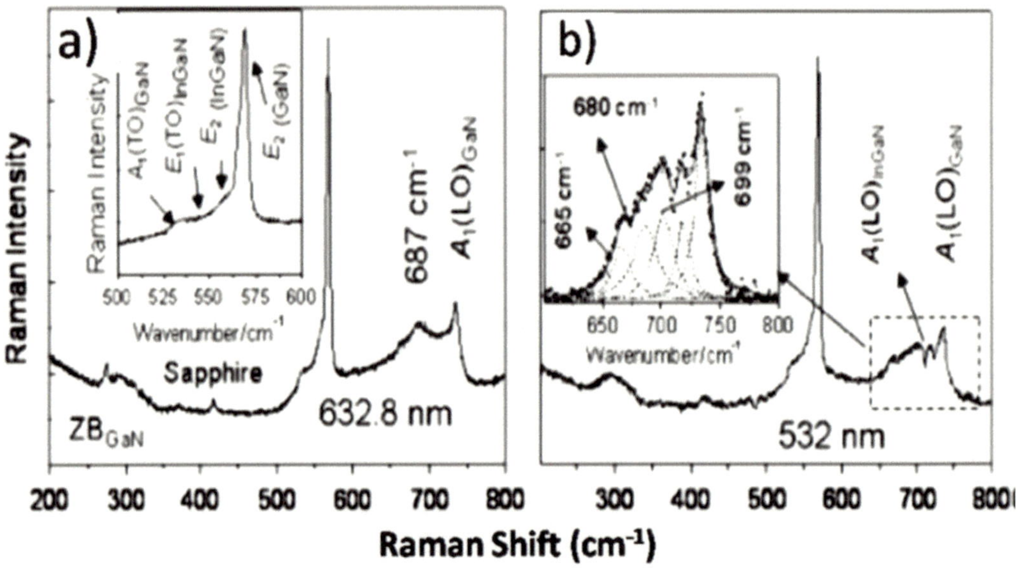

Fig. 14. Spectra for wet chemical (KOH) etched nanopillars with the excitations of (a) 632.8 nm, with inset showing the detailed spectrum in the 500–600 cm^{-1} range and (b) 532 nm laser lines with inset showing the Lorentzian fits for multiple peaks in the 625–775 cm^{-1} range. (Wu et al., 2009 Copyright © John Wiley & Sons, Ltd.)

The plots of ω_{SO} for GaN as a function of qr dispersion relation of SO modes for these 1-D MQW nanopillars, taking into consideration of cylindrical geometry, is shown in figure 15b.

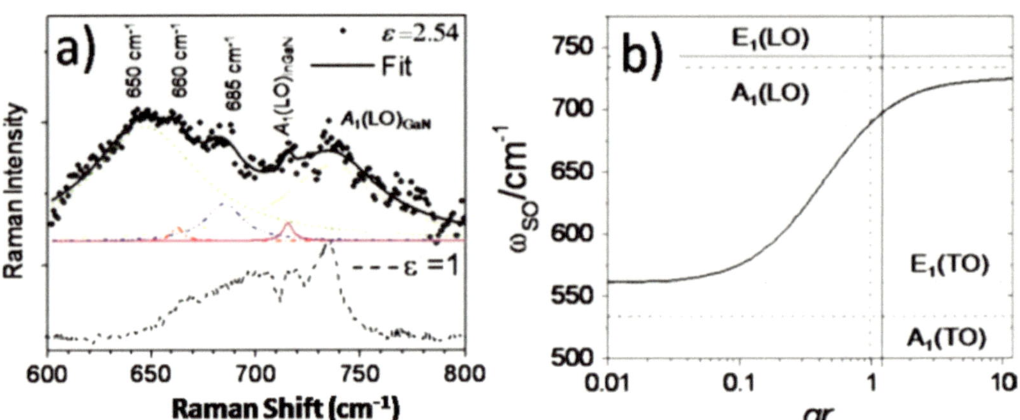

Fig. 15. a) Raman spectra for wet-etched nanopillars immersed in aniline ($\varepsilon = 2.54$) and in air ($\varepsilon = 1$) with the excitation of the 532 nm laser line. Lorentzian fits for multiple peaks are also shown for the spectrum recorded with aniline. b) Calculated SO phonon modes of GaN as a function of qr: full curve, SO(E_1); dashed curve, SO(A_1); horizontal full and dashed lines are the TO and LO wavenumbers of E_1 and A_1 modes, respectively. Vertical lines are marked for $qr = 0.64$ (dotted), $qr = 0.99$ (dash-dotted), and $qr = 1.18$. (Wu et al., 2008 Copyright © John Wiley & Sons, Ltd.)

The wavelength $\lambda = 2\pi/q$, corresponding to the SO potential (perturbation), has been estimated as ~250–300 nm for the nanopillars (using the dispersion relations in Fig. 15b with various excitations). Surface roughness with a modulation of~25 nm is observed (Fig. 14b) in these nanostructures. Modulation length with any integral multiple equivalent to the λ (10×25 nm=250 nm or 12×25 nm=300 nm) will be sufficient to initiate breakdown of translational symmetry for the contribution of the SO potential towards the SO phonons and to make the intensity of the SO modes comparable to that of other phonons. The origin of the broad new peak in the range of 665–699 cm^{-1} can be assigned to the SO modes corresponding to 1-D GaN and $In_{0.15}Ga_{0.85}N$.

2.2 SO phonons in rectangular geometry

The theoretical prediction for the SO modes varies from circular cross section wires to rectangular. However, there is no analytical expression for SO modes in rectangular cross section wires. Nevertheless, the dielectric continuum (DC) model approach that has been used for cylindrical nanowires remains the most elegant method and provides analytical expressions for $\omega_{SO}(q)$ vs ε_m for rectangular wires with further approximation,. An approximate DC model for rectangular wires has been introduced by Stroscio and co-workers (Stroscio et al,. 1990; Stroscio et al., 2001) that neglects the exponentially decaying electrostatic fields emanating from the corner regions. This assumption makes the problem separable in the plane perpendicular to the wire axis. The following dispersion relations is obtained for the SO phonons upon imposition of the usual electrostatic boundary conditions;

$$\epsilon_w(\omega)\tanh(q_i L_i/2) + \epsilon_m = 0, \tag{8a}$$

$$\epsilon_w(\omega)\coth(q_i L_i/2) + \epsilon_m = 0, \tag{8b}$$

Equation 8a is the symmetric (S) mode and equation 8b is the asymmetric (AS) mode, $\varepsilon_w(\varepsilon_m)$ is the dielectric function inside (outside) the wire, $L_i(i = x,y)$ is the edge width of the rectangular wire whose growth direction is along z, and $q_i(i = x,y)$ is the phonon wavevector. We must have,

$$q_x^2 + q_y^2 = q^2 \tag{9a}$$

$$q_x L_x = q_y L_y \tag{9b}$$

where equation 9b is the requirement that potentials of optic phonons in the x and y directions should have the same parity. Neglecting the damping and crystal anisotropy, the dielectric function $\varepsilon_w(\omega)$ can be expressed as,

$$\epsilon(\omega) = \epsilon_\infty \frac{\omega_{LO}^2-\omega^2}{\omega_{TO}^2-\omega^2} \tag{10}$$

and the Lyddane-Sachs-Teller (Yu and Cardona, 1999) relation gives

$$\frac{\epsilon_0}{\epsilon_\infty} = \frac{\omega_{LO}^2}{\omega_{TO}^2} \tag{11}$$

where ε_0 and ε_∞ are low and high-frequency values of $\varepsilon(\omega)$, respectively. We can solve for the S and AS mode SO phonon dispersions from equations 1-4:

$$\omega_{SO}^2(q)_S = \omega_{LO}^2 \left[\frac{\epsilon_\infty[\epsilon_0 \tanh(q_i L_i/2) + \epsilon_m]}{\epsilon_0[\epsilon_\infty \tanh(q_i L_i/2) + \epsilon_m]}\right]$$

$$= \omega_{TO}^2 \left[\frac{\epsilon_0 \tanh(q_i L_i/2) + \epsilon_m}{\epsilon_\infty \tanh(q_i L_i/2) + \epsilon_m}\right] \quad (12)$$

$$\omega_{SO}^2(q)_{AS} = \omega_{LO}^2 \left[\frac{\epsilon_\infty[\epsilon_0 \coth(q_i L_i/2) + \epsilon_m]}{\epsilon_0[\epsilon_\infty \coth(q_i L_i/2) + \epsilon_m]}\right]$$

$$= \omega_{TO}^2 \left[\frac{\epsilon_0 \coth(q_i L_i/2) + \epsilon_m}{\epsilon_\infty \coth(q_i L_i/2) + \epsilon_m}\right] \quad (13)$$

for nanowires with infinite edge length, i.e. $L\to\infty$, tanh $(qL/2)\to 1$, and coth $(qL/2)\to 1$; both symmetric (12) and asymmetric modes (13) recover again the SO phonon frequency of an infinite flat semiconductor surface expressed by (5).

2.2.1 ZnS nanowire

High crystalline, rectangular cross-sectional uniaxial wurtzite ZnS nanowires is grown along either *a*-axis ([100]) or the *c* axis ([001]) and details phonon spectra have been discussed by Xiong et al., (2004). The crystal anisotropy and the two growth orientations make the calculations even more complicated. Isotropic equation has been employed to capture the essential physical ideas without too much complication. An exact treatment for the dielectric function and the Lyddane-Sachs-Teller relation for uniaxial wurtzite crystals can be found. For wurtzite ZnS, $\varepsilon_{11}(\infty)$ and $\varepsilon_{33}(\infty)$ vary from 8.25 to 8.76. As an isotropic approximation of ε_0 = 8.3 and ε_∞ = 5.11 from and ω_{LO} = 346.5 cm^{-1} are made.

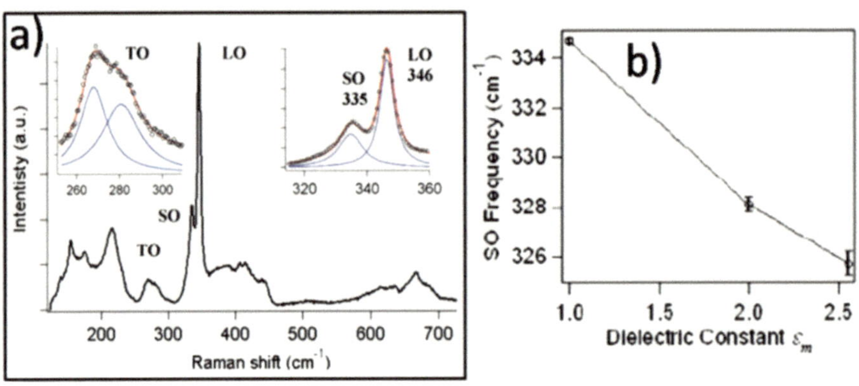

Fig. 16. (a) Raman spectrum of ZnS nanowires collected in air (ε_m=1). Two insets show Lorentzian line shape analysis of LO (346 cm^{-1}) and SO (335 cm^{-1}) modes and TO doublet (269, 282 cm^{-1}), (b) Variation of SO band peak positions as a function of the dielectric constants of the overlaying media. (Xiong et al., 2004 Copyright © American Chemical Society; Applied for permission).

Raman spectrum of rectangular cross section ZnS nanowires collected at room temperature in air is show in figure 16. Strong first-order LO scattering has been observed at 346 cm^{-1} and a TO doublet has been reported at 269 and 282 cm^{-1}. Some broad continuum scattering intensity has been observed in the range 350–450 cm^{-1} along with other features due to combination and overtone modes. The insets in figure 16 show the Lorentzian line shape analysis of the TO doublet and the LO–SO bands. Magnified spectrum is shown (Fig. 16) for the line shape analysis of the LO-SO region. As shown in figure 16 an additional Raman band have been reported at 335 cm^{-1} (in air) for square cross section ZnS nanowires. It has also been identified with SO modes, as it red shifts in dielectric media (dichloromethane and aniline). In air, the SO band found at 335 cm^{-1} red shifts to 328 cm^{-1} in dichloromethane and 326 cm^{-1} in aniline. The variation of experimental SO frequencies ZnS nanowires with the dielectric constants for the surrounding liquids are plotted in figure 16b. A linear decrease of the SO frequency has been reported as the dielectric constant of the overlaying medium is increased.

Figure 17(a) shows the SO dispersions for both S and AS asymmetric modes of ZnS for three overlaying dielectric media (ε_m). The S and AS modes exhibit different dispersion at small qL, but they converge to the same frequency as qL approaches very large values. Three horizontal dashed lines represent the actual experimental SO phonon frequencies in the three dielectric media. The three experimental frequencies connect to approximately the same wave vector q, showing an important self-consistency proof of the validity of the dielectric continuum model (Givargizov et al., 1973; Johansson et al., 2006). The surface modulation in TEM images (inset in Fig. 17a) along the wire axis, however, appears to be not as clearly periodic as in cylindrical nanowires. Nevertheless, still it confirms that the theoretical calculations based on the dielectric continuum model are consistent with TEM images; i.e., the wavelength of the perturbation that breaks the symmetry and activates SO mode is λ~70 nm.

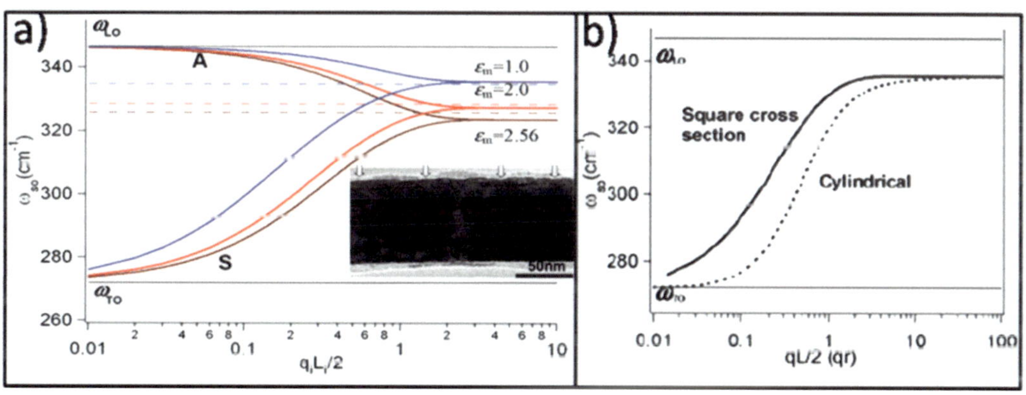

Fig. 17. (a) SO phonon dispersion for rectangular cross section wires for three dielectric media calculated from eqs 12 and 13, corresponding to S and AS modes, respectively. Three dashed lines represent the measured SO frequencies in air (ε_m = 1), dichloromethane (ε_m=2), and aniline (ε_m =2.56). Two horizontal solid lines represent LO and TO frequencies. Inset is the TEM images of ZnS nanowires. The surface modulation along the wire axis is clearly identified. The hollow arrows indicate the position where the minimum takes place; the average distance between the arrows is about 70 nm. (b) Calculated SO dispersions for rectangular wires with square cross sections (solid curve) and a supposed cylindrical ZnS nanowire (dashed curve) with a radius r in air. Two horizontal lines represent the LO and TO phonon branches, assuming they are dispersion less. (Xiong et al., 2004 Copyright © American Chemical Society; Applied for permission).

The effect of the nanowire morphology, looking into the cross-sectional shape, on the SO phonon dispersions has been made by computing the SO mode dispersion for ZnS cylindrical nanowires with diameter (d) against that of square cross section nanowires with sides of length $L = d$. The dashed curve shows the calculated SO dispersion for cylindrical ZnS nanowires in air (Fig. 17b). From the measured SO frequency in air, e.g., $\omega_{SO} = 334.7$ cm^{-1} and using the SO dispersion (Fig. 17a), the important wave vector q responsible for activating the SO modes is estimated. It has been reported that $qd = 25$ predicts the experimental SO frequency in air. The wavelength for the perturbation is therefore $\lambda=2\pi/q$ =13 nm. This λ differs by a factor of about 5 from the λ obtained from the rectangular cross section SO analysis. So it is obvious that the correct nanowire shape must be used in the analysis. The importance of the effect of the nanowire shape on the SO analysis is apparent from figure 17b.

2.2.2 GaN nanoribbons

The higher wave number fitted Raman spectrum (Fig. 18a) has been obtained from the GaN nanoribon grown inside the nanochannels of Na-4 in our study (Bhattacharya et al., 2011). The dispersion relation $\omega_{SO}(q)$ for a GaN nanoribbon encapsulated in Na-4 mica (Fig. 18b) shows a plot of both the S and AS modes of SO phonon dispersion curves for A_1 and E_1 symmetries. Horizontal lines corresponding to the experimentally observed wavenumbers at 633 and 678 cm^{-1} intersect the dispersion curves of AS$_{SO}$ modes at $q_iL_i/2$ values (marked with vertical lines). The wavevectors q of the surface perturbation that is responsible for activating the SO modes are estimated with the experimental 633 and 678 cm^{-1} ω_{SO} values (Fig. 18b). We find that $q_iL_i/2$ = 0.16 and 0.63 correspond to A_1 and E_1 symmetries, respectively. In the case of nanostructures, or for that matter on the surface, both even and odd modes of polarization eigen vectors are allowed for the breakdown in translational symmetry. Thus, both AS and S modes can respond to SO phonon modes. In the present study, we find the intersection only with AS mode, which is only incidental.

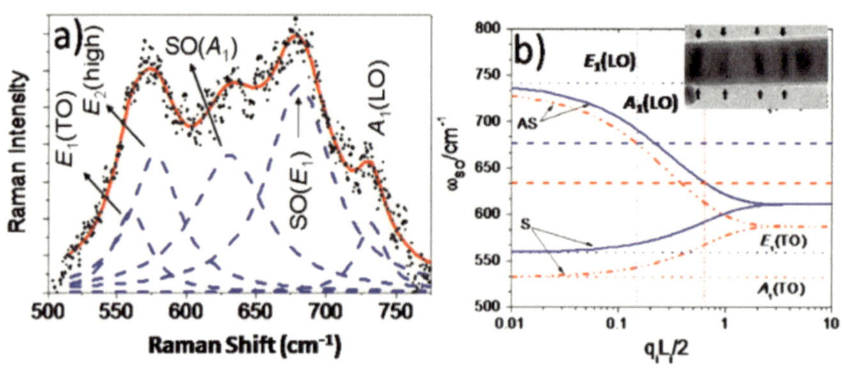

Fig. 18. (a) Higher wave numbers fitted Raman spectra of Na-4 mica–GaN with 325 nm excitation with Lorentzian lineshapes corresponding to various known modes of GaN along with SO modes.(b) SO phonon dispersion curves calculated for Na-4 mica encapsulated GaN nanoribbons (calculated for rectangular cross-section). The continuous and dashed-dot-dot curves correspond to E_1 and A_1 symmetries, respectively. AS and S modes are indicated in the plot. Two dashed horizontal lines represent the experimentally measured SO wavenumbers at 633 and 678 cm^{-1}. Four other horizontal dotted lines represent the LO and TO wavenumbers of bulk GaN corresponding to the E_1 and A_1 symmetries. Inset in figure 18b shows a typical GaN straight nanoribbon (width ~50 nm) with observed surface modulations. (Bhattacharya et al., 2011 Copyright © John Wiley & Sons, Ltd.)

The observed SO phonons intensities are comparable to those of the other phonons. The wavelengths for the SO potential perturbation (with the average nanoribbon cross-section dimensions as 50 × 1 ($L_x \times L_y$) nm²), that breaks the symmetry and activates the SO mode is $\lambda = 2\pi/q_i$ ($i = x, y$) 20–1000 nm (A_1 symmetry) and 5–250 nm (E_1 symmetry). Surface roughness with a modulation of ~50 nm has been reported (indicated by arrows in the inset of Fig. 18b), and any integral multiple equivalent to the λ (5 × 50 nm = 250 nm or 20 × 50 nm = 1000 nm) will be sufficient to initiate breakdown of translational symmetry for the contribution of SO potential to the SO modes.
The interaction potentials for the SO modes scales as $(q_zL_x)^{-1}$ and $(q_zL_y)^{-1}$ resulting in the electron–SO phonon scattering in nanoribbons dominating over electron–LO confined phonon scattering rate for $q_z \ll L_x^{-1}(L_y^{-1})$ (Stroscio et al., 1990). The observed $q_z \approx 2\times10^5$–6×10^4 cm^{-1} (λ = 250–1000 nm) $\ll L_y^{-1} \approx 1 \times 10^7$ cm^{-1} ($L_y \approx 1$ nm) makes the intensity of the SO mode comparable to that of the other optical modes as a result of relatively narrow distribution of surface perturbation in these samples about 50 nm. However, other rigid modes corresponding to even L (quantum number corresponding to axial modes) in the continuum model of the elongated nanostructures contribute to the width of the SO peak (Nobile et al., 2007).

2.3 SO phonons in spheroidal geometry

It is convenient to use cylindrical coordinates $(\rho\theta, \emptyset, z)$ for a spheroidal nanorod with semi-axes a and c and the dispersions of the anisotropic dielectric function $\epsilon_\perp(\omega)$ and $\epsilon_z(\omega)$ in the following denotations (Fonoberov & Baladin, 2004, 2005)

$$g_1 = \frac{\epsilon_z(\omega)}{\epsilon_\perp(\omega)};\quad g_2 = 1;\quad f_n = c^2 - g_n a^2$$

$$p_n = f_n + z^2 + g_n\rho^2; \qquad t_n = \sqrt{p_n^2 - 4f_n z^2}$$

$$\xi_n = |z|\sqrt{\frac{2}{p_n - t_n}};\quad \eta_n = \mathrm{sign}(z)\sqrt{\frac{p_n \quad t_n}{2f_n}}$$

the phonon potential inside (in) and outside (out) of the nanorod as

$$V^{(in)}_{L,M,\omega}(\rho,\emptyset,z) = \frac{P_L^M(\xi_1)}{P_L^M(c/\sqrt{f_1})}\,P_L^M(\eta_1)e^{iM\emptyset} \tag{14}$$

$$V^{(out)}_{L,M,\omega}(\rho,\emptyset,z) = \frac{Q_L^M(\xi_2)}{Q_L^M(c/\sqrt{f_2})}\,P_L^M(\eta_2)e^{iM\emptyset} \tag{15}$$

Here P_L^M and Q_L^M are associated Legendre functions of the first and second kind, respectively. The integers L (L≥ 1) and M ($|M| \le L$) are quantum numbers of the phonon mode, and the eigen frequencies ω of the phonon modes (Fonoberov & Baladin; 2005).

2.3.1 CdSe nanorod

The optical phonon excitations of randomly oriented and laterally aligned nanorod arrays (Fig. 19a) have been studied using resonant Raman scattering by Nobile and co-authors (Nobile et al., 2007). Colloidal wurtzite CdSe nanorods are fabricated by chemical synthesis in a hot mixture of surfactants. Nanorods are aligned by drop-caste on to an interdigitated electrode device and applying an AC electric field of 10^5 V.cm^{-1} at a frequency of 10 MHz during solvent evaporation (Fig. 19b).

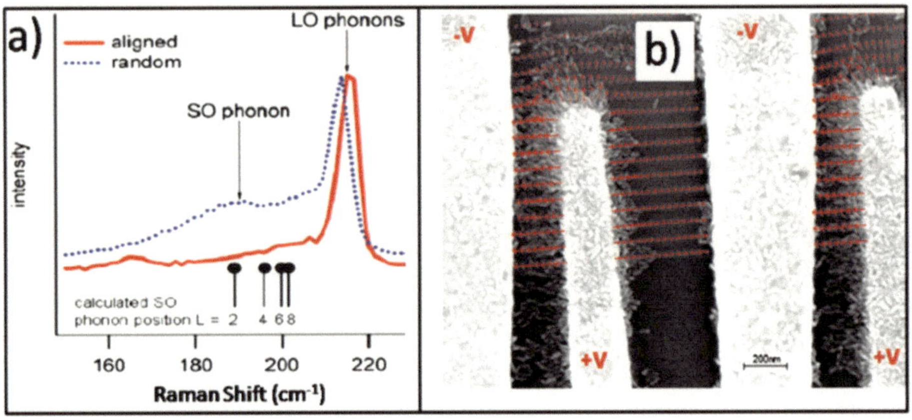

Fig. 19. (a) Resonant Raman spectra of randomly oriented (blue dotted line) and laterally aligned (solid red line) nanorods at T=15 K recorded in parallel polarization with respect to incident and detected light. The black bars indicate the calculated energies for the first four Raman active SO phonon modes. (b) Scanning electron microscope image of laterally aligned nanorods in between interdigitated electrodes defined by electron beam lithography. The red arrows show the calculated electric field for bias $\pm$V applied to neighboring electrodes for alignment. In the lower part of the image, the electric field lines are not shown in order to enhance the visibility of the nanorods. The rods accumulate in the regions of strongest electric field and observe excellent alignment with the electric field. (Nobile et al., 2007 Copyright © American Chemical Society; Applied for permission).

Raman spectra of randomly oriented (dotted) and closely packed laterally aligned (solid) nanorods (diameter 10 nm, and length 50 nm), which are taken on regions outside and within the interdigitated electrodes, respectively, on the same sample are shown in figure 19a. In the spectrum of the randomly oriented rods, the LO phonon at 214 cm^{-1}, and at the low-energy side, a broad peak centered around 185 cm^{-1} originating from the SO phonon modes are observed. The spectrum of the laterally aligned nanorods (with c-axis perpendicular to the incident light polarization) shows a very small blue-shift of the LO phonon peak and a significant decrease in the SO phonon intensity. In this contest, the experimental data is compared with first principle calculations of nanorods with corresponding aspect ratio (c/a of CdSe nanorods taken around 5). The theory predicts that only phonon modes with even L and with M =0 can be active in a resonant Raman spectrum.

According to the formalism reported above, figure 20 shows the calculated (phenomenological) potential of the SO phonon modes with M = 0 and even angular quantum numbers L. The energy of these modes is depicted in figure 19a by the black bars with full heads (energy increasing with higher angular momentum number). The calculated Raman active energies with the data obtained from the randomly oriented rods have shown excellent agreement. The phonon potentials of the SO modes with low L extend significantly outside of the rod surface as shown in figure 20.

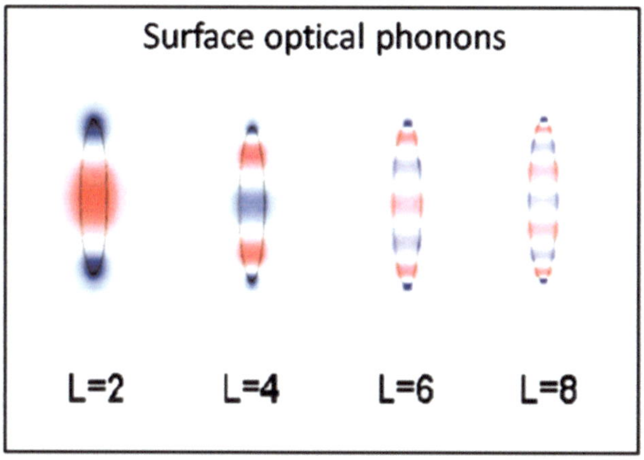

Fig. 20. Cross sections of the calculated phonon potential for the SO modes with M =0 and different angular momentum. Only SO phonon modes with even quantum number L are Raman active. Note that the L = 2 mode has significant phonon potential outside the nanorod surface. (Nobile et al., 2007 Copyright © American Chemical Society; Applied for permission).

The SO or interface phonon modes are stimulated by the change in dielectric medium at the interface. In a close-packed 2D array of nanorods, the space adjacent to the rods is occupied by other nanorods. Thus the dielectric medium surrounding the individual rods is severely modified, leading to the observed suppression of the SO phonon modes. The fact that SO phonon modes with higher angular momentum extend less outside the rod region accounts for the low-energy shoulder of the LO phonon peak around 200 cm^{-1}. The observation agrees well with the spatial distribution of potential of the SO phonon modes with long axis orientation of nanowires

3. Conclusion

The essential for understanding SO phonons is presented in this report for a wide range of compound semiconductor nanowires. The experimental analysis, as well as, theoretical calculations concluded that the SO mode in semiconductor nanowires is primarily associated with the surface modulation and geometry. The SO mode is clearly identified by its shift in the Raman spectrum as the overlaying medium dielectric constant is varied. The characteristic wave length of perturbation of the SO potential can be readily measured by the position of SO mode in Raman spectra. We believe that growth instabilities leading to the appearance of SO modes in the phonon spectra should be a general indicator of the presence of a corrugated surface. The instability of surface modulation mostly arises during vapour-liquid-solid growth of most nanowires system, and strongly depends on growth conditions, such as temperature, impurities, and supersaturation. In general, Raman scattering from SO phonons not only enlighten us on the atomic dynamics on surface but allows also help exploring the interaction of SO phonons with electronic surface states, and hence, can be exploited as a primary tool for surface analysis of one- (nanowire) and two-dimensional systems.

4. Acknowledgement

We acknowledge the contributions from S. Sahoo and A. K. Arora of MSG, IGCAR; M. S. Hu, K. H. Chen and L. C. Chen of Center for Condensed Matter Sciences, NTU, Taiwan; S. E. Wu, and C. P. Liu of National Cheng Kung University, Taiwan for their active involvement in the study of surface optical modes. We also thank S. Dash of SND, IGCAR for his general support and encouragement.

5. Reference

Adu, K.W.; Xiong Q.; Gutierrez, H.R.; Chen G. & Eklund, P.C. (2006). *Raman Scattering as a Probe of Phonon Confinement and Surface Optical Modes in Semiconducting Nanowires,* Appl. Phys. A 85, 287, ISSN: 0947-8396

Algra, R.E.; Verheijen, M.A.; Borgstrom, M.T.; Feiner, L.F.; Immink, G.; van Enckevort, W.J.P.; Vlieg, E. & Bakkers, E.P.A.M. (2008). *Twinning Superlattice in Indium Phosphide Nanowires,* Nature, 456, 369-372, ISSN: 0028-0836

Arguello, C. A.; Rousseau, D. L.; & Porto, S. P. S. (1969). *First-Order Raman Effect in Wurtzite-Type Crystals,* Phys. Rev. 181, 1351.

Benedek, G. & Toennies, J. P. (1994). *Helium Atom Scattering Spectroscopy of Surface Optical Phonons: Genesis and Achievements,* Surf. Sci., 299/300, 587

Bhattacharya, S.; Datta, Anindya.; Dhara, S. & Chakravorty, D. (2011). *Surface Optical Raman Modes in GaN Nanoribbons,* J. Raman Spectrosc., 42, 429-433 ISSN:1097-4555

Cao, L.; White J. S.; Park J. S.; Schuller J. A.; Clemens B. M. & Brongersma M. L. (2009). *Engineering Light Absorption in Semiconductor Nanowire Devices,* Nat. Mat., 8, 643-647, ISSN: 1476-1122

Cao, L.Y.; Laim, L.; Valenzuela, P. D.; Nabet, B. & Spanier, J.E. (2007). *On the Raman Scattering from Semiconducting Nanowires,* J. Raman. Spectrosc. 38, 697-703, ISSN: 0377-0486

Cao, L.Y.; Nabet, B. & Spanier, J.E. (2006). *Enhanced Raman Scattering from Individual Semiconductor Nanocones and Nanowires,* Phys. Rev. Lett., 96, 157402-5, ISSN: 0031-9007

Falge, H. J.; Borstel, G.; Otto, A. (1974). *Dispersion of Generalized Surface Optical Phonon Polaritons on α-Quartz Observed by Attenuated Total Reflection,* Phys. Stat. Solid B 65, 123

Fonoberov, V. A. & Balandin, A. A. (2004). *Interface and Confined Optical Phonons in Wurtzite Nanocrystals,* Phys. Rev. B 70, 233205.

Fonoberov, V. A. & Balandin, A. A. (2005). *Polar Optical Phonons in Wurtzite Spheroidal Quantum Dots: Theory and Application to ZnO and ZnO/MgZnO Nanostructures,* J. Phys.: Condens. Matter, 17, 1085-1097.

Frechette, J. & Carraro, C. (2006). *Diameter-Dependent Modulation and Polarization Anisotropy in Raman Scattering from Individual Nanowires,* Phys. Rev. B, 74, 2161404, ISSN: 1098-0121

Fuchs, R.; & Kliewer, K. L. (1965). *Optical Modes of Vibration in an Ionic Crystal Slab,* Phys. Rev. 140, A2076

Givargizov, E. I. (1973). *Periodic Instability in Whisker Growth,* J. Cryst. Growth, 20, 217.

Gupta, R.; Xiong, Q.; Mahan, G.D. & Eklund, P.C., (2003). *Surface Optical Phonons in Gallium Phosphide Nanowires,* Nano Lett., 3, 1745, ISSN: 1530-6984

Hartschuh, A.; Pedrosa, H.N.; Novotny, L. & Krauss, T.D. (2003). *Simultaneous fluorescence and Raman scattering from single carbon nanotubes,* Science, 301, 1354-6, ISSN:0036-8075

Hayden, O.; Agarwal, R. & Lu, W. (2008). *Semiconductor Nanowire Devices,* Nanotoday, 3, 12, ISSN:1748 01321

Hsiao, C. L.; Tu, L.W.; Chi, T. W.; Chen, M.; Young, T. F.; Chia, C.T.; Chang, Y. M. (2007). *Micro-Raman Spectroscopy of a Single Freestanding GaN Nanorod Grown by Molecular Beam Epitaxy,* Appl. Phys. Lett. 90, 043102

Ibach, H. & Mills, D. L. (1982). *Electron Energy Loss Spectroscopy and Surface Optical Vibrations,* Academic, New York

Johansson, J.; Karlsson, L.S.; Svensson, C.P.T.; Martensson, T.; Wacaser, B.A.; Deppert, K.; Samuelson, L. & Seifert, W. (2006). *Structural Properties of (111)B -Priented III–V Nanowires,* Nat. Mater. 5, 574

Lin, H.M.; Chen, Y.L.; Yiang, J.; Liu, Y.C.; Yin, K.M.; Kai, J.J.; Chen, L.C.; Chen, Y.F. & Chen, C.C. (2003). *Synthesis and Characterization of Core-Shell GaP@GaN and GaN@GaP Nanowires,* Nano Lett., 3, 4, 537-541, ISSN: 1530-6984

Livneh, T.; Zhang, J.P.; Cheng, G.S. & Moskovits, (2006). *Polarized Raman Scattering from Single GaN Nanowires,* Phys. Rev. B, 74, 035320, ISSN: 1098-0121

Lu, W.; & Lieber, C. M. (2007). *Nanoelectronics from the Bottom Up,* Nat. Mater., 6, 841 - 850 ISSN: 1476-1122

Mahan, G.D.; Gupta, R.; Xiong, Q.; Adu, C.K. & Eklund, P.C., (2003). *Optical Phonons in Polar Semiconductor Nanowires,* Phys. Rev. B, 68, 073402, ISSN: 0163-1829

Nobile, C.; Fonoberov, V. A.; Kudera, S.; Torre, A. D.; Ruffino, A.; Chilla, G.; Kipp, T.; Heitmann, D.; Manna, L.; Cingolani, R.; Balandin, A. A. & Krahne, R. (2007) , *Confined Optical Phonon Modes in Aligned Nanorod Arrays Detected by Resonant Inelastic Light Scattering,* Nano Lett. 7, 476.

Patolsky, F.; Timko, B. P.; Zheng, G.; & Lieber, C. M. (2007). *Nanowire Based Nanoelectronic Devices in the Life Sciences,* MRS Bulletin, 32, 142.

Pauzauskie, P.J., Talaga, D., Seo K., Yang, P.D. & Lagugne-Labarthet F. (2005). *Polarized Raman Confocal Microscopy of Single Gallium Nitride Nanowires,* J.A.C.S., 127, 49, 17146, ISSN: 0002-7863

Pauzauskie, P. J.; & Yang, P. (2006). *Nanowire Photonics,* Materials Today, 9, 36, ISSN:1369 7021

Richter, W.; Hinrichs, K. & Esser, N. (2000). *Raman Scattering from Surface Phonons on Semiconductors,* Proc. Int. Symp. on Progress in Surface Raman Spectroscopy, Editors: Tian, Z.Q.; Ren, B.; Xiamen University Press, S.133-136, ISBN 7-56151629

Sahoo, P.; Dhara, S.; Dash, S.; Tyagi, A.K.; Raj, B.; Das, C.R.; Chandramohan, P.; & Srinivasan, M.P. (2010). *Surface Optical Modes in GaN Nanowires,* Int. J. Nanotechnol., 7, 823-832.

Sahoo , S.; Dhara, S.; Arora, A. K.; Krishnan, R.; Chandramohan, P. & Srinivasan, M. P. (2010). *Raman Scattering from Surface Optical Phonon in Diameter Modulated AlN Nanotips,* Appl. Phys. Lett. 96, 103113

Sahoo, S.; Hu, M. S.; Hsu, C. W.; Wu, C. T.; Chen, K. H.; Chen, L. C.; Arora, A. K. & Dhara, S. (2008). *Surface Optical Raman Modes in InN Nanostructures,* Appl. Phys. Lett. 93, 233116

Sernelius, B.E. (2001). *Surface Optical Modes in Physics,* 1st edn. Wiley-VCH, New York, p. 350

Shan, C.X.; Liu, Z.; Zhang, X.T.; Wong, C.C. & Hark S.K. (2006). *Wurtzite ZnSe Nanowires: Growth, Photoluminescence, and Single-Wire Raman Properties,* Nanotechnology, 17, 5561-4, ISSN: 0957-4484

Spirkoska, D.; Abstreiter, G. & Fontcuberta i Morral A. (2008). *Size and Environment Dependence of Surface Optical Phonon Modes of Gallium Arsenide Nanowires as*

Measured by Raman Spectroscopy, Nanotechnology, 19, 435704, ISSN: 0957-4484. http://iopscience.iop.org/0957-4484/19/43/435704/

Stroscio, M.A.; Dutta, M.; ebrary Inc. (2001). *Phonons in Nanostructures*, Cambridge University Press: Cambridge; New York

Stroscio, M. A.; Kim, K. W.; Littlejohn, M. A.; Chuang, H. H., (1990). *Polarization Eigenvectors of Surface-Optical Phonon Modes in a Rectangular Quantum Wire*, Phys. Rev. B, 42, 1488.

Wallis, R.F. (1994). *Surface Optical Phonons: Theoretical Developments*, Surf. Sci. 299/300, 612

Wang, J.; Gudiksen, M. S.; Duan, X.; Cui, Y. & Lieber, C. M. (2001). *Highly Polarized Photoluminescence and Photodetection from Single Indium Phosphide Nanowires*, Science, 293, 1455-7, ISSN: 0036-8075

Wu, S.E.; Dhara, S.; Hsueh, T.H.; Lai, Y.F.; Wang, C.Y.; & Liu, C.P. (2009). *Surface Optical Phonon Modes in Ternary Aligned Crystalline InGaN/GaN Multiquantum-Well Nanopillar Arrays*, J. Raman Spectrosc., 40, 2044–2049

Xiong, Q.; Chen, G.; Gutierrex, H.R. & Eklund, P.C., (2006). *Raman Scattering Studies of Individual Polar Semiconducting Nanowires: Phonon Splitting and Antenna Effects*, Appl. Phys. A, 85, 299-305, ISSN: 0947-8396

Xiong, Q.; Wang, J.; Reese, O.; Yoon, L. C. & Eklund, P.C. (2004). *Raman Scattering from Surface Optical Phonons in Rectangular Cross-Sectional w-ZnS Nanowires*, Nano Lett. , 4, 10, 1991-6, ISSN: 1530-6984

Yang, P.; Yan, R. & Fardy, M. (2010). *Semiconductor Nanowire: What's Next?*, Nano letter, 10, 1529-36

Yu, P. Y.; Cardona, M. (1999). *Fundamentals of Semiconductors: Physics and Materials Properties*, Springer, Berlin; New York.

Zardo, I.; Conesa-Boj S.; Peiro F.; Morante, J.R.; Arbiol, J.; Abstreiter, G. & Fontcuberta i Morral A. (2009). *Raman Spectroscopy of Wurtzite and Zinc-Blende GaAs Nanowires: Polarization Dependence, Selection Rules and Strain Effects*, Phys. Rev. B, 80, 245324

Zeng, H.; Cai, W.; Cao, B.; Hu, J.; Li, Y. & Liu P. (2006). *Surface Optical Phonon Raman Scattering in Zn/ZnO Core-Shell Structured Nanoparticles*, Appl. Phys. Lett., 88, 181905, ISSN: 0003-6951.

2

Large-Scale Synthesis of Semiconductor Nanowires by Thermal Plasma

Peng Hu, Fangli Yuan and Yajun Tian
State Key Laboratory of Multi-phase Complex Systems,
Institute of Process Engineering,
Chinese Academy of Sciences, Beijing,
China

1. Introduction

One-dimensional (1D) semiconductor nanostructures, classified as nanowires, nanobelts and nanotubes, have shown superior properties in electric, chemical, mechanical and optical fields (Li et al., 2004; Pan et al., 2001; Pavesi et al., 2000; Q. Wang et al., 2008), and can be served as the fundamental building blocks for constructing advanced inter-connectors and nanoscale electronic, optoelectronic and sensing nanodevices (Duan et al., 2001; Huang et al., 2001; Lao et al., 2003; X.D. Wang et al., 2007). In general, the design of 1D nanostructure emphasizes to strength the growth of crystal along one direction and restrict other two dimensions to nanosize. In recent years, considerable achievements have been made on preparation of 1D nanostructures and different synthetic strategies have been developed to fabricate various 1D nanostructures, including element (Nikoobakht & El-Sayed, 2003; Lu et al., 2005; Vivekchand et al., 2004; Wirtz & Martin, 2003), oxide (Dai et al., 2003; Ohgi et al., 2005; Wu et al., 2005; Zheng et al., 2002), nitride (Kim et al., 2002; Toury et al., 2003), sulfide (Gao et al., 2005; Kar & Chaudhuri, 2005; Zhu et al., 2003), and others (Hornbostel et al., 1995; Jun et al., 2006; X.Y. Wang et al., 2002), which makes it possible to further investigate their properties and applications in nanodevices. In this chapter, we will review the recent development achieved primarily from the author's studies addressing the key issues of semiconductor 1D nanostructures by thermal plasma, including the growth, characterization and applications of obtained products based on their unique properties related to this special synthetic route.

2. General background of thermal plasma

Plasma typically refers to the fourth state of matter similar to gas in which a certain portion of the species is ionized, containing ions, electrons, atoms and neutral species (Boulos, 1991, 2004). Based on the thermodynamic equilibrium between the gas species, plasmas can be classified as equilibrium, local equilibrium and non-equilibrium plasma. In equilibrium plasma, the gas species are nearly fully ionized with same temperature and the plasma temperature is extremely high, as we called astrophysical plasmas (such as sun). In local equilibrium plasma, only a small fraction of the gas molecules are ionized while the electrons and heavy particles approximately have the same temperature and show local

electrical neutrality, in which the plasma temperature is still typically several thousand degrees and generally is called thermal plasma. While in non-equilibrium plasma, the electron temperature is still typical several thousand degrees but the temperature of heavy neutrals is rather low, and the plasma exhibits ambient temperature so we call them cold plasma.
Thermal plasmas can be generated at atmospheric pressure or under soft vacuum conditions for a wide range of gases, providing an inert, oxidizing or reducing atmosphere for the needs of materials processing. Typical examples of thermal plasmas include various forms of direct current (DC) arcs and radio frequency induction (RF) plasma discharges. Compare with other technology routes, the typical features of thermal plasma are high temperature up to 2000～20000K and charged particle density in the range of 10^{19}～10^{21} per m^3. The high temperature, enthalpies and heat fluxes in the plasma torch make it suitable to apply in many chemical synthesis and metallurgical processes, and the presence of active ions species in plasma can also promote the chemical activity of reaction species. The flexibility of starting materials with solid, liquid, gas or in combination and react environment with inert or reactive gas in the plasma reactor make it versatile in producing a variety of materials including metals, carbides, nitrides, oxides or composites of materials with well dispersed and definite size. It's fast quenching rate at flame tail can 'freeze' the particles in an intermediate state and gives the materials with novel structures and properties.

3. Growth of 1D semiconductor nanostructures by RF thermal plasma

The RF thermal plasma generally includes a refractory and a quartz tube in which working gas flows, and the plasma is generated through electromagnetic coupling of the RF power supply through the induction coil surrounding the tube by partially ionizing the working gas within the tube. The plasma gas does not come in contact with electrodes and eliminates possible sources of contamination and allows for operation with a wide range of gases, including inert, reducing, oxidizing and other corrosive atmospheres. Accordingly, the plasma so generated is called inductively coupled plasma or induction plasma, and Figure 1 shows the experimental plasma torch by argon discharge.

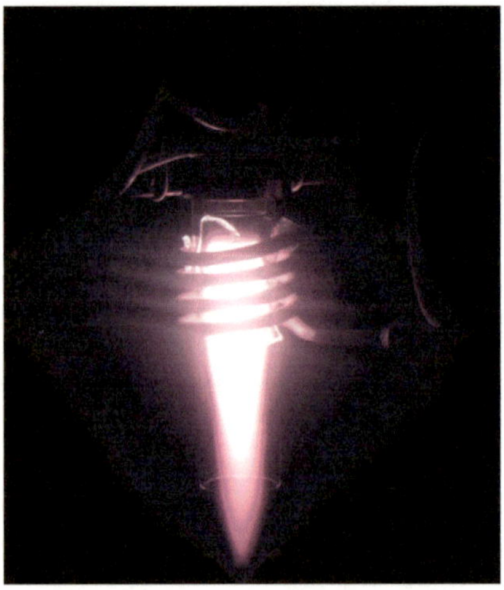

Fig. 1. Photograph of induction plasma generated by argon discharge

In plasma synthesis process, the starting materials are feed into the plasma by carrier gas and subsequently heated to melt ever vaporize. The axial feeding way and long particle residence time within the discharge zone provide the enough heat transfer from the plasma to the particles. During the following process, reaction and growth processes are occurred in flowing gas in the reactor, and the final products are collected at the bottom of the collector. The schematic illustration of plasma system is shown in Figure 2 (Hu et al., 2007).

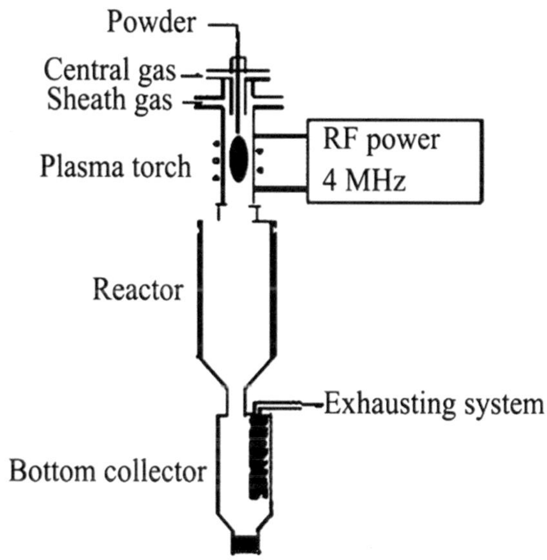

Fig. 2. Schematic illustration of the plasma system (Hu et al., 2007)

3.1 ZnO nanowires

3.1.1 Synthesis and structure control

For ZnO nanowires synthesis, zinc was used as starting materials and supplied into the plasma flame by carrying gas with a feed rate of 20~50 g/min and then underwent vaporization, oxidation, crystallization and growth processes following the flowing gas, finally obtained products. Argon (1.0 m^3/h) and nitrogen gas (5.0 m^3/h) were injected as the plasma-forming gas and sheath gas, respectively. The chamber pressure was maintained at atmospheric pressure. Specially, oxygen gas was injected into the system together with the sheath gas to oxidize the zinc powder.

After the experiment, white woollike ZnO products were obtained in the collector, and the morphology and structural investigation of synthesized products are shown in Figure 3. Figure 3a gives the typical SEM image of obtained products, which reveals the formation of large scaled ZnO with uniform morphology in the products. Magnified SEM image shown in Figure 3b indicates that as-synthesized products are composed of rodlike nanostructures with uniform diameter of about 50 nm along the stem. All the peaks in the XRD pattern of the sample can be well indexed to the hexagonal ZnO wurtzite structure as shown in Figure 3c. From the TEM images shown in Figure 3d and e, we can see that the nanowires with typical length about 2 μm and diameter about 50 nm are obtained, and the tips of two ends exhibit round shape. HRTEM image in Figure 3f reveals the perfect monocrystalline structure of obtained products.

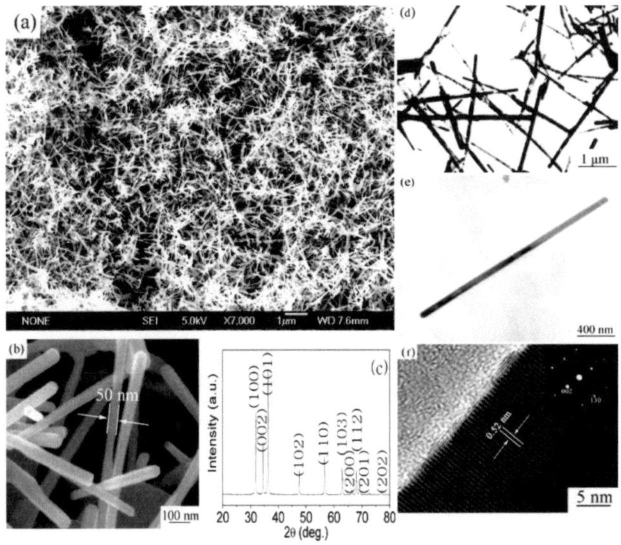

Fig. 3. (a) SEM image, (b) Magnified SEM image, displaying the uniform diameter, (c) XRD pattern, (d) low-magnification TEM image, (e) TEM image of single crystal and (f) HRTEM image of synthesized ZnO products (Hu et al., 2007)

In plasma synthesis process, starting materials, oxygen partial pressure and supersaturation of zinc vapor in the plasma play key roles in the growth of the nanowires. By controlling the species and feedstock rate of the starting material, and the flow rate of oxygen gas, the 1D structure with different morphology could be well adjusted, such as nanowires with different ratio of length to diameter, tetrapods and multipods.

3.1.2 Growth rate of nanowires in plasma

In our synthesis process, the resident time of particles in reactor is very short and as a result the growth rate of products should be very fast to grow into rodlike nanostructure with length of several micrometers. According to the reactor length and diameter, and the gas flow rate in it, the resident time can be calculated as 0.15 seconds under the typical experimental conditions. In another word, the axial growth rate is about 13 μm/s for the nanowire with length about 2 μm, which is rather fast than the time needed for nanowire synthesized by the predominant vapor deposition process and wet chemist routes.

The high growth rate of nanowire in plasma synthesis process can be explained by the crystal growth dynamics theory similar to that occurred in solvent, in which supercooling degree promotes the nucleation and growth of the crystal, and high level degree of undercooling provides intensive growth driver for reactive species to nucleate and grow into crystal. However, too much degree of undercooling decreases the diffusion of nucleating substances in solvent, and lead to the formation of crystal with small size. In plasma synthesis process, vapor species with several thousands degrees are first formed due to the high processing temperature in the flame zone and then cooled to form ultrahigh-level supercooling vapor in the plasma tail, which provides intensive growth driver for ZnO to nucleate and grow. In addition, part of zinc and oxygen was ionized in the plasma zone, which accelerates the transmittability of electrons from zinc to oxygen. All of these provide intensive growth driver for ZnO to nucleate and growth with high growth rate. However, the nanowires with small length are achieved when too much degree of undercooling was realized by introducing excess cooling gas to the reactor, and the results are well in accordance with the crystallization process in solvent.

3.1.3 Growth mechanism of nanowires in plasma

In our experiment, no substrate or catalysts are used to confine the growth direction of nanorods, and the basic phenomena involved in the crystal growth are an in-flight way. That is to say, all the growth processes of nanowires are occurred exclusively in flow gas. Since there is no involvement of catalysts or templates to serve as energetic favorable sites for the absorption of reactant molecules or to confine the growth direction of nanowires, it would be reasonable to imagine that the morphology of obtained products is determinate by the growth habit of crystals. ZnO belongs to hexagonal phase crystal which has the anisotropic growth nature. When ZnO crystal grows, anisotropic growth habit prefers the crystal to grow along the C axis and forms columnar structure. As a result, the crystal structure of the products may be the underlying factor that determines the formation of 1D nanostructure in plasma.

The growth process of nanowire via gas phase reaction typically involves the vapor-liquid-solid (VLS) process and vapor-solid (VS) process (Wagner & Ellis, 2004; Yang & Lieber, 1996; Zhang et al., 2001). In our experiment, no metal catalysts were used and the typically round tips in the end of the synthesized nanorods confirm that growth mechanism of nanorods is governed by the VS mechanism, but this process is exactly not the same as the references reported previously, in which the nanorod grows along one direction. In our products, the round tips on the two ends of the nanorods suggested that the nanorods were grown in two directions along the [0001] axis. The schematic of the growth mechanism is shown in Figure 4a. To investigate the growth mechanism more clearly, a contrasting experiment was carried out with increased temperature gradient. Figure 4b shows the TEM image of as-synthesized products, and a double-directional needlelike nanostructures were obtained, which reveals that the nanorods are governed by a wall-free and two direction growth mechanism.

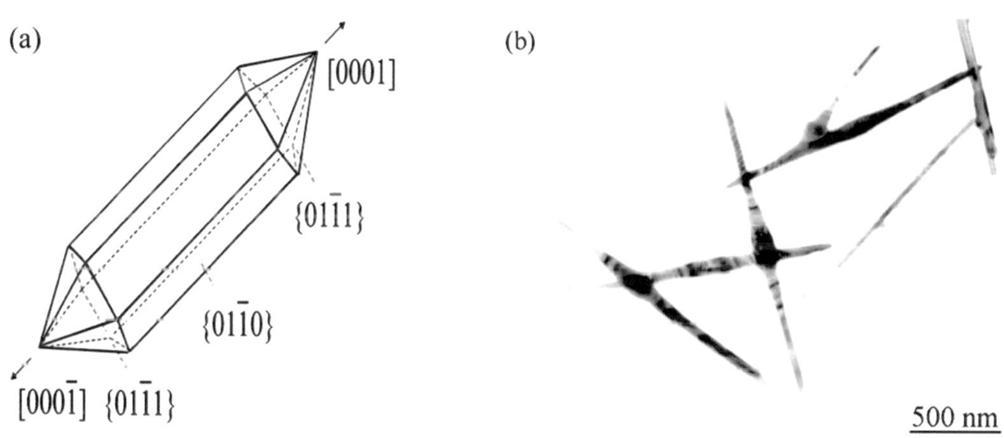

Fig. 4. (a) the schematic of the growth mechanism and (b) TEM image of synthesized double-direction needlelike nanostructures (Hu et al., 2007)

3.1.4 Properties and applications

3.1.4.1 Photocatalytic activities under visible light

Some interesting properties are found for our synthesized nanowires and Figure 5a gives the photocatalytic activities of synthesized ZnO nanocrystals under visible light irradiation. From the patterns we can see that as-synthesized products exhibit excellent photodegradation effect for methyl red solution and photocatalytic activity can be significantly improved by increasing the length of nanowires. After visible light irradiation for 12 h, the concentration for nanowires with length of 2 μm is about 45%, which is about 2-fold lower than that of nanowires with length of 200 nm. Figure 5b demonstrates the color change of methyl red solution using 2 μm nanowires as photodegradation agent and a translucence solution can be observed after 20 h visible light irradiation, which indicates the obvious photocatalytic effect of obtained ZnO nanowires.

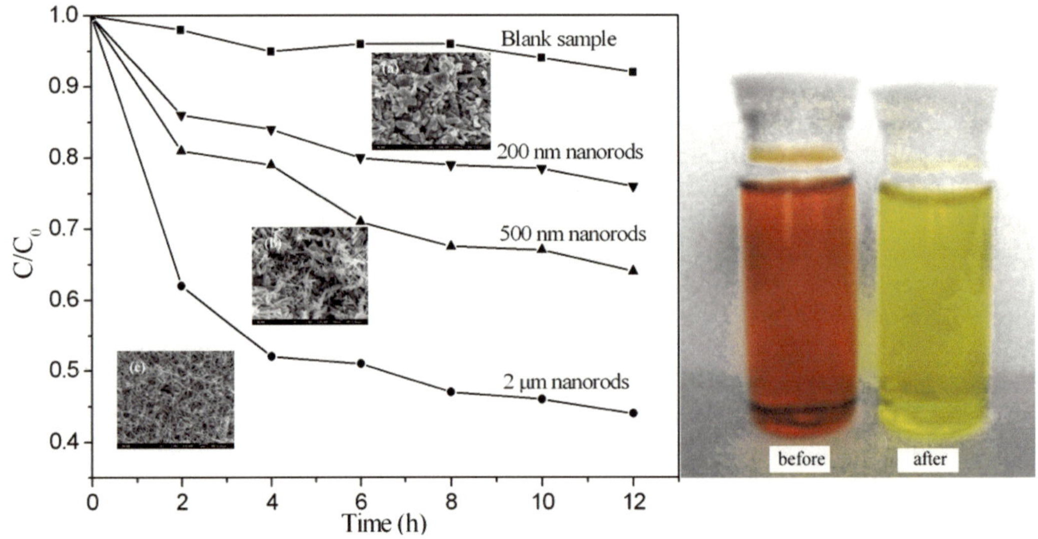

Fig. 5. (a)Photocatalytic activities of synthesized ZnO nanocrystals under visible light irradiation and (b) photodegradation effect of methyl red solution under 20 h visible light irradiation

It is known that ZnO is an n-typed semiconductor with band-gap energy of 3.2 eV and thus only absorbs UV light with the wavelength less than 385 nm. However, as far as photocatalytic efficiency and practical applications are concerned, it is desirable that ZnO absorbs not only UV but also visible light because visible light accounts for about 45% energy of solar radiation. Because of the semiconducting property of ZnO crystal, the visible-light absorption could be obtained by n-typed doping (such as transitional metal ions) but hard to achieved by p-typed doping (such as N). In our experiment, implanting nitrogen into ZnO crystals was achieved by introducing nitrogen gas into the plasma sheath gas and carrier gas, and visible light absorbing was obtained by narrowing or splitting the band gap of ZnO into several sub-gaps. The doping content of nitrogen element is measured to be several hundreds ppm by chemical composition analysis of obtained nanowires.

3.1.4.2 Nanowire based gas sensor

The gas sensing property of obtained nonowires (shown in Figure 6) was tested using formaldehyde gas as working gas in a home-made instrument as we reported earlier (Han et al., 2009), which demonstrates the high gas sensing property of plasma synthesized ZnO nanowires comparing to the literature (Bie et al., 2007; Patil et al., 2007). From the patterns

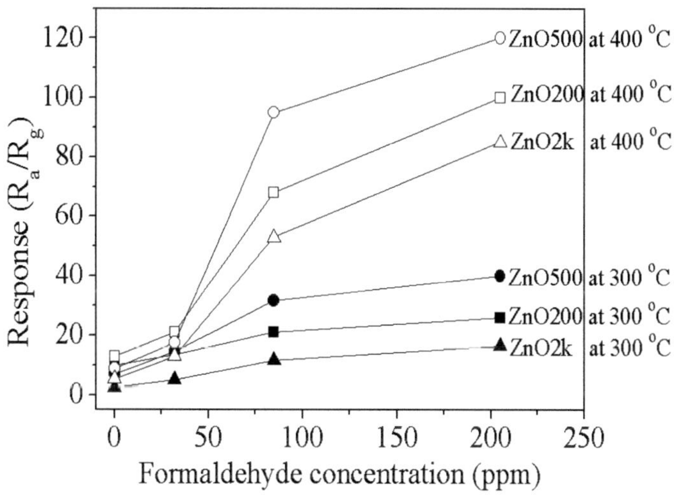

Fig. 6. Formaldehyde responses of the ZnO nanorods (tested at 300 °C and 400 °C at RH 70%) (Han et al., 2010)

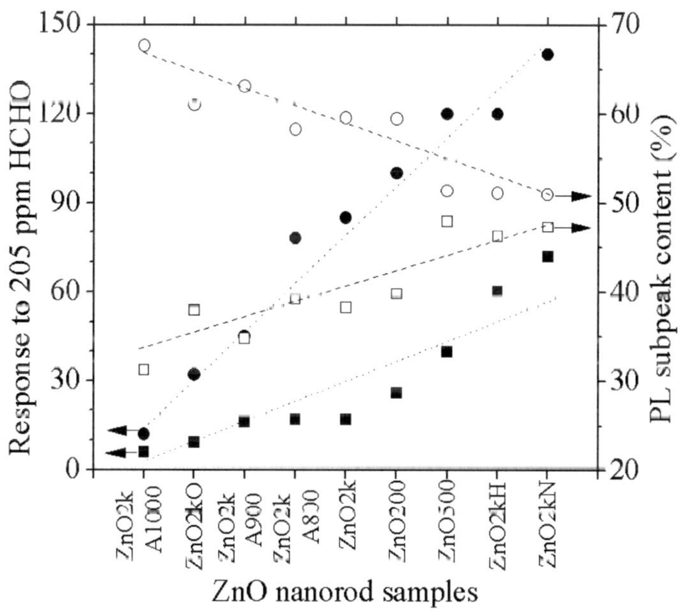

Fig. 7. Relationship between formaldehyde response of the ZnO nanowires at 300 °C (■) & 400 °C (●) and the DL (□) & AL (○) subpeaks contents (Han et al., 2010)

we can see that ZnO with length about 500 nm performed the highest gas response at both 300 °C and 400 °C, followed by ZnO nanowires with length about 200 nm and 2 μm, respectively. The difference in their gas response of obtained nanowires could be explained by the intrinsic defects in obtained products and the highest gas response of the nanowires with length about 500 nm could be contribute to the high donor-related and low acceptor-related crystal defects. Further, this conclusion is verified by annealing the ZnO nanowires in different atmosphere to adjust different defect density and related gas sensing property shown in Figure 7.

3.2 ZnS 1D nanostructures

ZnS is an important II–VI group semiconductor compound with hexagonal crystal structure that exhibits excellent properties in optical and optoelectronic fields, and its 1D nanostructures can also be obtained in plasma synthesis process. In a typical synthesis process, the staring materials of Zn and S powder (-200 mesh) were firstly mixed together (according to the mole ratio of 1:1) and then fed into the plasma by carrier gas (nitrogen gas, at a flow rate of 150 L/h) in a continuous way at different experimental parameters, such as feed rate and cooling gas rate. Figure 8 represents the ZnS 1D nanostructures synthesized at

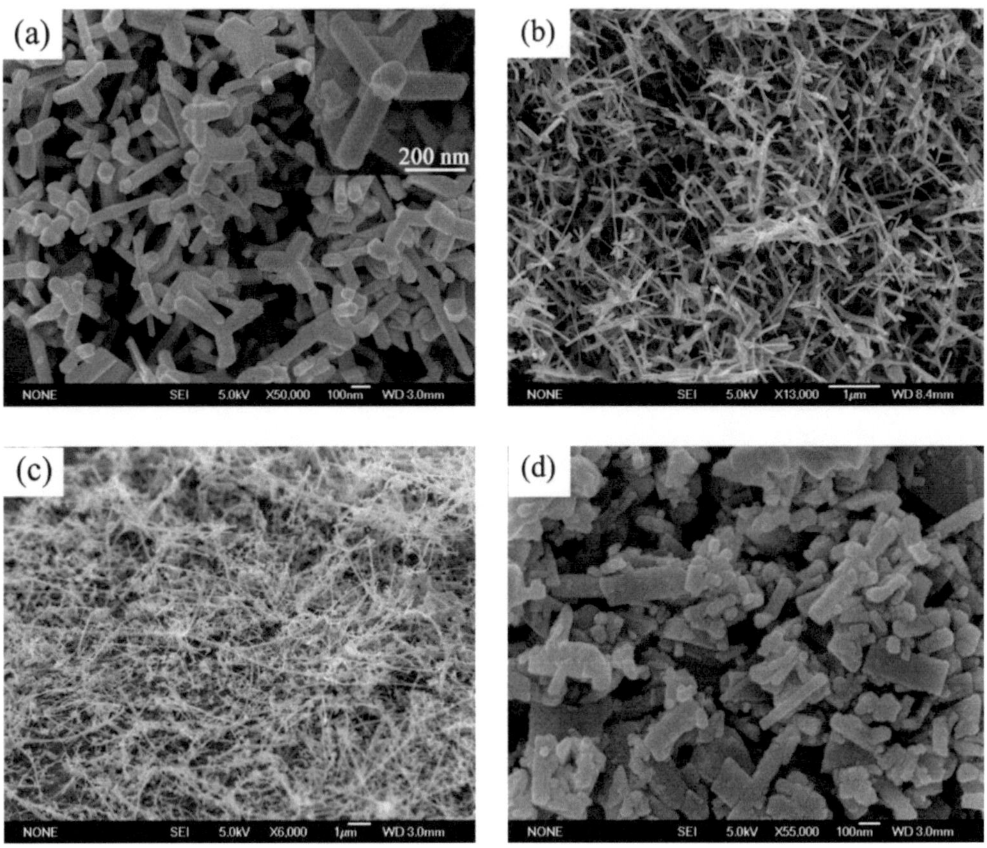

Fig. 8. SEM images of ZnS nanostructures produced with different sets of feeding rate and cooling gas flow rate (a) 12 g/min and without cooling gas, (b) 24 g/min and without cooling gas, (c) 21 g/min and with cooling gas 3 m^3/h and (d) 21 g/min and with cooling gas 6 m^3/h (Hu et al., 2009)

different conditions, and from the images we can conclude that different 1D nanostructures, including tetrapods, nanorods, nanobelts and nanoslices could be selectively synthesized by controlling the experimental parameters. Further investigation on their structures by TEM and HRTEM reveals that the 1D nanostructures are all single crystalline and hexagonal elongation along the c-axis according to the theoretical and crystal habit of the ZnS.

One of the important features of plasma synthesis route is its rather rapid growth rate compared with the conventional vapor deposition process. Meanwhile, rapid growth rate also causes the formation of intrinsic defects in crystal and Figure 9 demonstrates the existence of two different crystal defects in synthesized products. Figure 9a illustrates the stacking fault existed in a single nanobelt, which is parallel to the axis and runs through out the nanobelt. Figure 9b shows two types of structure zone existed in a leg of tetrapod ZnS. One consists of the wurtzite structure (hcp, hexagonal close-packed) in zone a and the other is sphalerite structure (fcc, face centered cubic) in zone b, and the two different structures were formed by changing the stacking sequence of the closed-packed planes of the ZnS crystal and resulted from the crystallogenesis of ZnS. It should be noted that some properties, such as the photoluminescence, gas sensing property of obtained products are closely related to the intrinsic defects in crystals (Jiang et al., 2003).

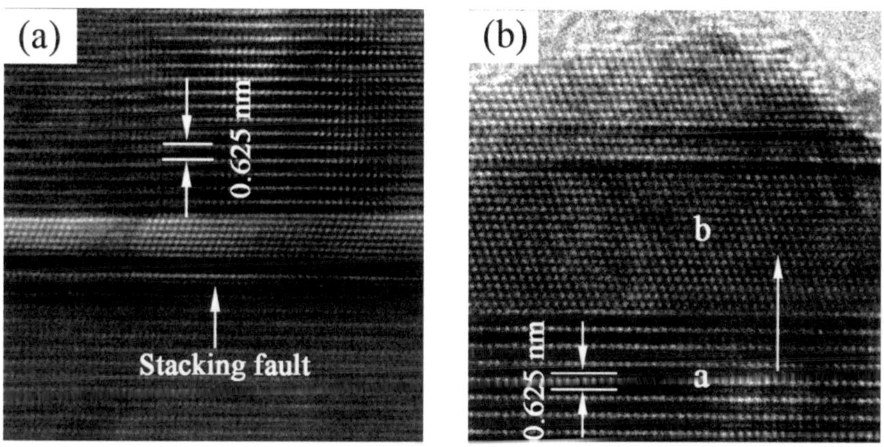

Fig. 9. The crystal defects exist in the ZnS nanocrystals (a) stacking fault in a nanobelt and (b) the sphalerite structure and the wurtzite structure polytypes in a leg of tetrapod (Hu et al., 2009)

It is worth noting that hollowing effect of 1D nanostructures was observed from products collected inner the reactor wall closed to plasma torch and the TEM images are shown in Figure 10. Structure analysis confirms that the hollow structure is resulted from the oxidation process of the ZnS nanocrystals. In our experiment, low concentration of oxygen gas was introduced into the plasma due to the impurity of the nitrogen gas and argon gas, and the oxygen gas was then heated to an ultra-high temperature and partially atomized even ionized in the plasma flame due to the effect of plasma electromagnetic field, which makes oxygen with high activity. The oxidation reaction was occurred on the surface of the crystals when the atomized oxygen contacts with the products attached on the reactor wall, and the core evacuation was underwent with the diffusion of S from inner core to outward. The hollow structures were final formed when the core was consumed completely, which is kind of similar to the formation of hollow structures by Ostwald ripening process observed in solution synthesis (Ball & Ruben 2004; Hu et al., 2008; Liu & Zeng, 2005; Zott, 2003).

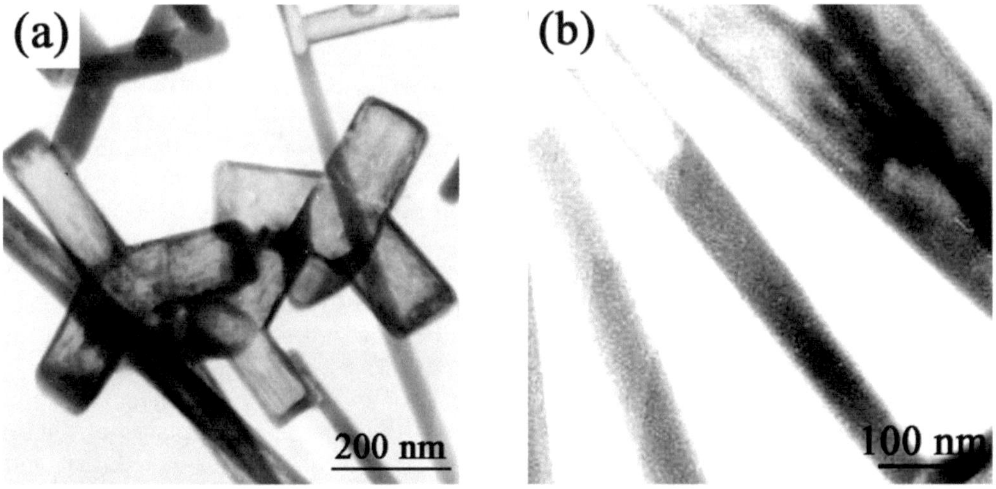

Fig. 10. TEM images of hollowing process in (a) tetrapod and (b) nanowire (Hu et al., 2009)

3.3 Transition metal doped 1D nanostructures

In addition, transition metal doped 1D nanostructures could also be obtained by adding the transition metal corresponding halides to starting materials. The synthesis process is same to that for pure binary compounds and Figure 11 demonstrates an example of the Ni doped ZnS 1D nanostructures, which reveals the versatility of plasma route for 1D nanostructures synthesis. The hysteresis loop of obtaind 1D structures shows a clear ferromagnetic behavior caused by Ni doping (shown in Figure 12) and can be adjusted by controlling the morphology of as-synthesized products.

Fig. 11. SEM images of Ni doped ZnS 1D nanostructures: (a)nanorods, (b)nanobelts and (c) nanowires

3.4 Other 1D nanostructures synthesized by RF plasma

Since the morphology of products in plasma synthesis is governed by the growth habit of materials, it is reasonably believed that other materials with hexagonal crystal structure could grow into 1D nanostructures by plasma process and here we give some examples to prove this view. Figure 13 represents some examples of the 1D materials with hexagonal crystal structure grown by plasma synthesis route, and the results well confirm the growth mechanism we proposed previously. Accordingly, the plasma synthesis method provides a facile and large scale way to synthesize 1D nanostructures by a wall-free growth process.

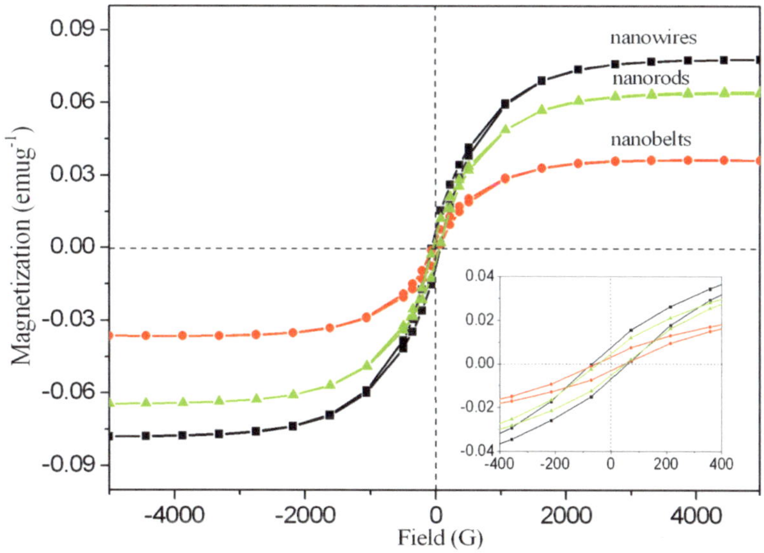

Fig. 12. Hystersis loops of Ni doped ZnS nanostructures with different morphology (a) nanowires (b) nanobelts and (c) nanorods

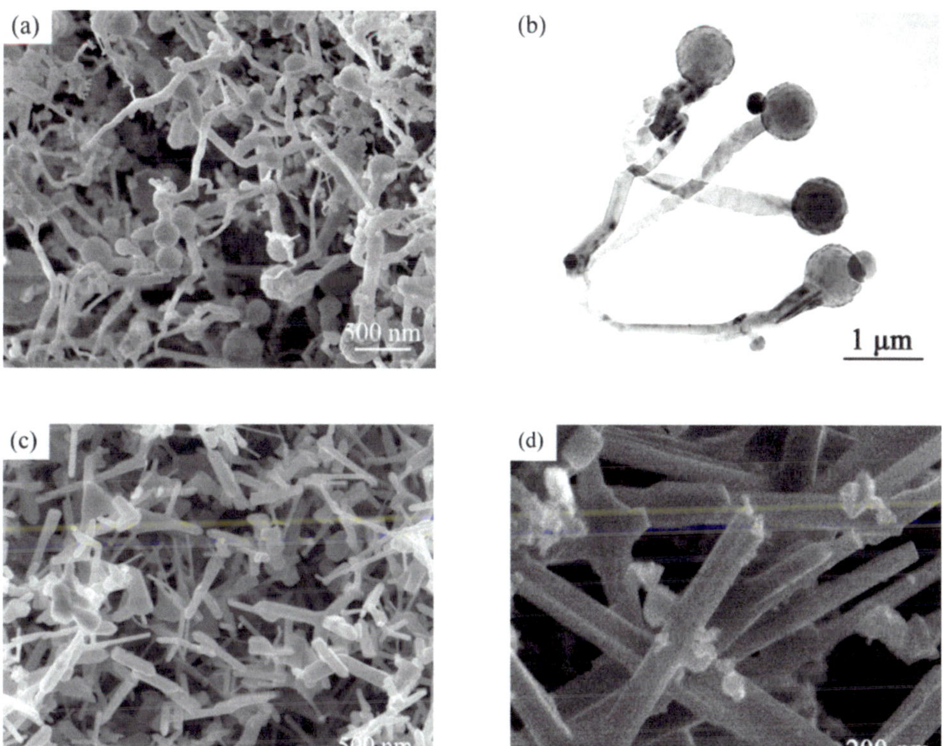

Fig. 13. Different 1D nanostructures synthesized by RF plasma (a) SEM and (b) TEM of AlN, SEM of (c) Zn and (d) WO_3

4. Conclusions

Semiconductor nanowires possess attractive electronic and photonic properties for potential applications in various nanodevices. In order to realize their device applications in real life, several problems must be solved, including controlled synthesis of semiconductor nanowires with commercial cost and achievement of nanodevices with high efficiency. This needs to develop the facile synthetic strategies and well understand the properties of obtained semiconductor nanowires. In this chapter, an overview of the recent progress in research on the large scale synthesis of semiconductor nanowires by plasma was presented, and the growth mechanism, fundamental properties and their applications are also involved. In general, the plasma synthesis process involves the vaporization, reaction, crystallization and growth processes of reactive species by an in-flight way with yield of about 10-50 g/min., which also gives the products some novel properties comparing to other synthetic strategies. With the promising approaches reviewed in the chapter and more have be achieved and to be developed, practical applications of semiconductor nanowires are imminent.

5. Acknowledgements

This work was supported financially by the National Natural Science Foundation of China (NSFC 50974111, 10905068, 20976182)

6. References

Ball, P. & Ruben, M. (2004). Color Theory in Science and Art: Ostwald and the Bauhaus. *Angewandte Chemie International Edition*, Vol.43, No.37, (September 2004), pp.4842-4847, ISSN 0570-0833

Bie, L. J.; Yan, X. N.; Yin, J.; Duan, Y. Q. & Yuan, Z. H. (2007). Nanopillar ZnO Gas Sensor for Hydrogen and Ethanol. *Sensors and Actuators B: Chemical*, Vol.126, No.2, (October 2007), pp.604-608, ISSN 0925-4005

Boulos, M. (1991). Thermal Plasma Processing. *IEEE Transactions on Plasma Science*, Vol.19, No.6, (December 1991), pp.1078-1089, ISSN 0093-3813

Boulos, M. (2004). Plasma Power Can Make Better Powders. *Metal Powder Report*, Vol.59, No.5, (May 2004), pp.16-21, ISSN 0026-0657

Dai. Z. R.; Pan. Z. W. & Wang. Z. L. (2003). Novel Nanostructures of Functional Oxides Synthesized by Thermal Evaporation. *Advanced Functional Materials*, Vol.13, No.1, (January 2003), pp.9-24, ISSN 1616-301X

Duan, X. F.; Huang, Y.; Cui, Y.; Wang, J. F & Lieber, C. M. (2001). Indium Phosphide Nanowires as Building Blocks for Nanoscale Electronic and Optoelectronic Devices. *Nature*, Vol.409, No.6816, (January 2001), pp. 66-69, ISSN 0028-0836

Gao, P.; Xie, Y.; Ye, L. N.; Chen, Y. & Guo, Q. X. (2006). From 2D Nanoflats to 2D Nanowire Networks: A Novel Hyposulfite Self-Decomposition Route to Semiconductor FeS_2 Nanowebs. *Crystal Growth & Design*. Vol.6, No.2, (Februry 2005), pp.583-587, ISSN 1528-7483

Han, N.; Tian, Y. J.; Wu, X. F. & Chen, Y. F. (2009). Improving Humidity Selectivity in Formaldehyde Gas Sensing by a Two-Sensor Array Made of Ga-Doped ZnO. *Sensors and Actuators B: Chemical*, Vol.138, No.1, (April 2009), pp.228-235, ISSN 0925-4005

Han, N.; Hu, P.; Zuo, A. H.; Zhang, D. W.; Tian, Y. J. & Chen, Y. F. (2010). Photoluminescence Investigation on the Gas Sensing Property of ZnO Nanorods Prepared by Plasma-Enhanced CVD Method. *Sensors and Actuators B: Chemical*, Vol.145, No.1, (March 2010), pp.114-119, ISSN 0925-4005

Hornbostel, M. D.; Hillyard, S.; Silcox, J. & DiSalvo, F. (1995). Nanometer width molybdenum selenide fibers. *Nanotechnology*, Vol.6, No.3, (July 1995), pp87-92, ISSN 0957-4484

Hu, P.; Yuan, F. L.; Bai, L. Y.; Li, J. L. & Chen, Y. F. (2006). Plasma Synthesis of Large Quantities of Zinc Oxide Nanorods. *The Journal of Physical Chemistry C*, Vol.111, No.1, (January 2007), pp.194-200, ISSN 19327447

Hu, P.; Yu, L. J.; Zuo, A. H.; Guo, C. Y. & Yuan, F. L. (2008). Fabrication of Monodisperse Magnetite Hollow Spheres. *The Journal of Physical Chemistry C*, Vol.113, No.3, (January 2009), pp.900-906, ISSN 1542-3050

Hu, P.; Bai, L. Y.; Yu, L. J.; Li, J. L.; Yuan, F. L. & Chen, Y. F. (2009). Shape-Controlled Synthesis of ZnS Nanostructures: A Simple and Rapid Method for One-Dimensional Materials by Plasma. *Nanoscale Research Letters*, Vol.4, No.9, (June 2009), pp.1047-1053, ISSN 1556-276X

Huang, M. H.; Mao, S.; feick, H.; Yan, H. Q.; Wu, Y. Y.; Kind, H.; Weber, E.; Russo, R. & Yang, P. D. (2001). Room-Temperature Ultraviolet Nanowire Nanolasers. *Science*, Vol.292, No.5523, (June 2001), pp.1897-1899, ISSN 0036-8075

Jiang, Y.; Meng, X. M.; Liu, J.; Xie, Z. Y.; Lee, C. S. & Lee, S. T. (2003). Hydrogen-Assisted Thermal Evaporation Synthesis of ZnS Nanoribbons on a Large Scale. *Advanced Materials*, Vol.15, No.4, (February 2003), pp.323-327, ISSN 1521-4095

Liu, B. & Zeng, H. C. (2005). Symmetric and Asymmetric Ostwald Ripening in the Fabrication of Homogeneous Core-Shell Semiconductor. *Small*, Vol.1, No.5, (May 2005), pp.566-571, ISSN 1613-6829

Jun, Y. W.; Choi, J. S. & Cheon, J. (2006). Shape Control of Semiconductor and Metal Oxide Nanocrystals through Nonhydrolytic Colloidal Routes. *Angewandte Chemie International Edition*, Vol.45, No.21, (May 2006), pp.3414-3439, ISSN 0570-0833

Kar, S. & Chaudhuri, S. (2005). Controlled Synthesis and Photoluminescence Properties of ZnS Nanowires and Nanoribbons. *The Journal of Physical Chemistry B*, Vol.109, No.8, (March 2005), pp.3298-3302, ISSN 0022-3654

Kim, H. M.; Kim, D. S.; Kim, D. Y.; Kang, T. W.; Cho, Y. H. & Chung. K. S. (2002). Growth and characterization of single-crystal GaN nanorods by hydride vapor phase epitaxy. *Applied physics Letters*, Vol.81, No.12, (September 2002), pp. 2193-2195, ISSN 0003-6951

Lao, J. Y.; Huang, J. Y.; Wang, D. Z. & Ren, Z. F. (2002). ZnO Nanobridges and Nanonails. *Nano Letters*, Vol.3, No.2, (February 2003), pp.235-238, ISSN 1530-6984

Li, Q. H.; Wan, Q.; Chen, Y. J.; Wang, T. H.; Jia, H. B. & Yu, D. P. (2004). Stable Field Emission from Tetrapod-like ZnO Nanostructures. *Applied physics Letters*, Vol.85, No.4, (July 2004), pp. 636-638, ISSN 0003-6951

Lu, X. M.; Fanfair, D. D.; Johnston, K. P. & Korgel, B. A. (2005). High Yield Solution-Liquid-Solid synthesis of Germanium Nanowires. *Journal of American Chemical Socierty*, Vol. 127, No.15, (November 2005), pp.15718-15719, ISSN 0002-7863

Nikoobakht, B. & El-Sayed, M. A. (2003). Preparation and Growth Mechanism of Gold Nanorods (NRs) Using Seed-Mediated Growth Method. *Chemistry of Material*, Vol.15, No.10, (April 2003), pp.1957-1962, ISSN 0897-4756

Ohgi, H.; Maeda, T.; Hosono, E.; Fujihara, F. & Imai, H. (2005). Evolution of Nanoscale SnO_2 Grains, Flakes, and Plates into Versatile Particles and Films through Crystal

Growth in Aqueous Solutions. *Crystal Growth & Design.* Vol.5, No.3, (May 2005), pp.1079-1083, ISSN 1528-7483

Pan, Z. W.; Dai, Z. R. & Wang, Z. L. (2001). Nanobelts of Semiconducting Oxides. *Science,* Vol.291, No.5510, (March 2001), pp.1947-1949, ISSN 0036-8075

Pavesi, L.; Dai Negro, L.; Mazzoleni, C.; Franzo, G. & Priolo, F. (). Optical Gain in Silicon Nanocrystal. *Nature,* Vol.408, No.6811, (November 2000), pp. 440-444, ISSN 0028-0836

Patil, D. R.; Patil, L. A. & Patil, P. P. (2007). Cr_2O_3-Activated ZnO Thick Film Resistors for Ammonia Gas Sensing Operable at Room Temperature. *Sensors and Actuators B: Chemical,* Vol.126, No.2, (October 2007), pp.368-374, ISSN 0925-4005

Toury, B.; Bernard, S.; Cornu, D.; Chassagneux, F.; Letoffe, J. M. & Miele, P. (2002). High-Performance Boron Nitride Fibers Obtained from Asymmetric Alkylaminoborazine. *Journal of Materials Chemistry.* Vol.13, No.2, (February 2003), pp274-279, ISSN 0959-9428

Vivekchand, S. R. C.; Gundiah, G.; Govindaraj, A. & Rao, C. N. R. (2004). A New Method for the Preparation of Metal Nanowires by the Nebulized Spray Pyrolysis of Precursors. *Advanced Materials,* Vol.16, No.20, (October 2004), pp.1842-1845, ISSN 1521-4095

Wagner, R. S. & Ellis, W. C. (1964). Vapor-Liquid-Solid Mechnism of Single Crystal Growth. *Applied physics Letters,* Vol.4, No.5, (December 2004), pp. 89-90, ISSN 0003-6951

Wang, Q.; Huang, J.; Xie, Z.; Wang, T. H.; Dattoli, E. N. & Lu, W. (2008). Branched SnO_2 Nanowires on Metallic Manowire Backbones for Ethanol Sensors Application. *Applied physics Letters,* Vol.92, No.10, (September 2008), pp. 102101-102103, ISSN 0003-6951

Wang, X. D.; Song, J. H.; Liu, J. & Wang, Z. L. (2007). Direct-Current nanogenerator Driven by Ultrasonic Waves. *Science,* Vol.316, No.5812, (April 2007), pp.102-105, ISSN 0036-8075

Wang, X. Y.; Zhang, J. Y.; Nazzal, A.; Darragh, M. & Xiao, M. (2002). Electronic structure transformation from a quantum-dot to a quantum-wire system: Photoluminescence decay and polarization of colloidal CdSe quantum rods. *Applied physics Letters,* Vol.81, No.25, (December 2002), pp. 4829-4831, ISSN 0003-6951

Wirtz, M. & Martin, C. R. (2003). Template-Fabricated Gold Nanowires and Nanotubes. *Advanced Materials,* Vol.15, No.5, (March 2003), pp. 455-458, ISSN 1521-4095

Wu, J. M.; Shih, H.C.; Wu, W. T.; Tseng, Y. K. & Chen, I. C. (2005). Thermal evaporation growth and the luminescence property of TiO_2 nanowires. *Journal of Crystal Growth.* Vol.281, No.2, (August 2005), pp. 384-390, ISSN 0022-0248

Yang, P. D. & Lieber, C. M. (1996). Nanorod-Superconductor Composites: A Pathway to High Critical Current Density Materials. *Science,* Vol.273, No.5283, (September 1996), pp.1836-1840, ISSN 0036-8075

Zhang, X. Y.; Zhang, L. D.; Meng, G. W.; Li, G. H.; Jin-Phillipp, N. Y. & Phillipp, F. (2001). Synthesis of Ordered Single Crystal Silicon Nanowire Arrays. *Advanced Materials.* Vol.13, No.16, (August 2001), pp.1238-1241, ISSN 1521-4095

Zheng, B.; Wu, Y. Y.; Yang, P. D. & Liu, J. (2002). Synthesis of Ultra-Long and Highly Oriented Silicon Oxide Nanowires from Liquid Alloys. *Advanced Materials.* Vol.14, No.2, (January 2002), pp.122-124, ISSN 1521-4095

Zhu, Y. C.; Bando, Y.; Xue, D. F. & Golberg, D. (2003). Nanocable-Aligned ZnS Tetrapod Nanocrystals. *Journal of American Chemical Socierty,* Vol. 125, No.52, (September 2003), pp.16196-16197, ISSN 0002-7863

Zott, R. (2003). Friedrich Wilhelm Ostwald (1853–1932), Now 150 Years Young..., *Angewandte Chemie International Edition,* Vol.42, No.34, (September 2003), pp.3990-3995, ISSN 0570-0833

3

Template-Assisted Electrochemical Synthesis of Semiconductor Nanowires

İlkay Şişman
Department of Chemistry, Faculty of Arts and Sciences,
Sakarya University, Sakarya,
Turkey

1. Introduction

Semiconductor nanostructures occupy the center of scientific interest because of their unique electronic nature. Among them, nanowires (NWs), also known as "nanorods" or "nanowhiskers," are especially attractive for nanoscience studies. While their diameters are in the nanometer range, they can have lengths in the tens of micrometers. Because of their unique density of electronic states, NWs in the limit of small diameters are expected to exhibit significantly different optical, electrical, and magnetic properties from their corresponding bulk materials. Consequently, many physical properties of semiconductors are significantly altered within the confines of the NW surfaces. On the other hand, their large surface-to-volume ratio allows for distinct structural and chemical behavior. For example, they can emit laser light, act like optical fibers, and change conductance when bound to different molecules [1].

The synthesis of NWs has been studied intensively worldwide for a wide spectrum of materials. Several chemical and physical methods are commonly used to produce NWs. Among them, electrochemical synthesis or electrodeposition to produce NWs has a number of advantages. Since electrochemical synthesis is usually controllable in the direction normal to the substrate surface, this technique can be readily extended to fabricate NWs. The length of the deposited NWs can be controlled by varying the duration of the electrodeposition process. In addition, electrochemical methods are inexpensive; operate at ambient temperatures and pressure. This chapter describes the some methods to semiconductor NWs production via electrochemical synthesis. Several electrochemical approaches have been reported to synthesize NWs including anodic aluminum oxide (AAO) template-assisted electrodeposition, lithographically patterned nanowire electrodeposition, surfactant-assisted electrodeposition, and template-free electrodeposition, etc. A great number of nanomaterials with diverse morphologies such as nanodots, nanoparticles, NWs, nanorods, nanobelts, nanotubes, nanospheres, nanoring, and nanoarrays, etc. have been synthesized based on template synthesis. All of these nanostructures were synthesized by deposition or growth of materials either inside the pores of the template or on the surface of the template. Among them the AAO template can stand high temperatures, is insoluble in organic solvents and geometrical parameters can be easily tuned by changing the synthesis conditions. This chapter is by no means comprehensive in covering all the relevant literatures. In the first part of this chapter, the AAO template-assisted electrochemical synthesis to make semiconductor NWs is presented and discussed. In the later parts, other template-assisted electrochemical synthesis methods are presented. For quick reference, examples of typical NWs that have been electrodeposited in the templates are listed in Table 1.

2. AAO template-assisted electrodeposition of semiconductor nanowires

In fact, electrodeposition techniques have been used to grow semiconductor thin films on conductive surfaces. Among the electrochemical approaches, semiconductor thin films can be formed by electrochemical surface-limited reactions. Electrochemical surface-limited reactions are generally referred to as underpotential deposition (UPD) [2]. As a result of the UPD process, which takes place at more positive potentials than the equilibrium (Nernst) potential, the electrode surface is partially or completely (up to an atomic layer) covered by a deposit. However, the overpotential deposition (OPD or bulk deposition) process is determined by electrode potential, deposit growth kinetics and mechanism, electroactive species concentration, and deposit-substrate and deposit-deposit interactions. OPD takes place at more negative potentials than the Nernst potential. Briefly, UPD may involve deposition onto substrate while OPD would involve deposition onto a substrate surface modified by an atomic layer, which was formed during the UPD process. Generally, deposits reach more than one atomic layer in the OPD regions [3].

Semiconductor	Band gap (eV)	Electrodeposition method	Reference
CdS	2.42	AAO, ESED	[12, 27–32, 34, 39]
CdSe	1.7	AAO, ESED, LPNE	[45–47, 49, 73]
CdTe	1.5	AAO	[52, 53]
ZnO	3.37	AAO	[57, 58, 60]
Cu_2O	2.1	AAO	[62–64]
In_2O_3	3.7	AAO	[65]
Bi_2Te_3	0.15	AAO, ESED, "track-etch"	[68, 72, 76]
PbTe	0.33	LPNE	[78]
PbSe	0.29	LPNE	[79]

Table 1. Selected electrodeposition methods for the synthesis of semiconductor NWs.

Traditionally, semiconductor thin films have been electrodeposited by the so-called induced codeposition method, where both elements are deposited at the same time from the single solution [4]. Stoichiometry is maintained by having the more noble element as the limiting reagent, and poising the potential where the less noble element will underpotentially deposit only on the more noble element. Codeposition holds great promise if greater control can be achieved. At present, the main points of control are solution composition and the deposition potential. Recently, we have reported electrodeposition of PbS, CdS, $Bi_2Te_{3-y}Se_y$, $Bi_{1-x}Sb_x$, $Bi_{2-x}Sb_xTe_3$, $Bi_{2-x}Sb_xSe_3$, SnSe, and CdTe thin films by codeposition method [5-11].
For semiconductor NWs, the electrodeposition technique was used in 1996 for fabricating arrays of CdS NWs in nanoporous AAO membranes [12]. Template-assisted electrodeposition is one of the important methods for synthesis of semiconductor NWs with controlled shape and size. The template contains very small cylindrical pores within the host material, and the empty spaces are filled with the chosen material to form nanowires. Such

host materials include nanoporous (AAO) membranes [13], "track-etch" membranes [14], polymer membranes [15], diblock copolymers [16], nanochannel array glasses [17] and SiO_2 nanocapillary arrays [18].

The aim of this chapter is not to give a complete overview of all the host materials; the focus is on AAO template-assisted electrodeposition of semiconductor NWs. Hand made or commercial AAO membrane is one of the most popular templates in material science studies. Although AAO membranes can be brittle, they are used as templates for NWs. AAO membranes have many properties that make them especially valuable as templates. These features including [19]: (a) the pore sizes can be controlled; (b) highly transparent in the visible region of the optical spectrum that allows the investigation of the optical properties of the nanomaterials deposited within the nanopores; (c) alumina is amphoteric in nature and can be easily dissolved in an acidic or basic solution, by which the pure arrays of NWs can be obtained. Hand made porous AAO templates are produced by mild anodizing pure aluminum films in various acids [20, 21]. Aluminum anodization has been shown to be an easy and inexpensive way of template fabrication. The resulting oxide film possesses a regular hexagonal array of parallel and cylindrical channels under anodization conditions. The pores form with uniform diameters because of a delicate balance between electric-field-enhanced diffusion, which determines the growth rate of the alumina, and dissolution of the alumina into the acidic electrolyte. Depending on the anodization conditions, the pore diameter can be systematically varied from < 10 nm up to 200 nm [20, 22]. However, commercial porous AAO templates are produced by hard anodizing pure aluminum films. This method is faster but it produces films with disordered pore structures. In addition, commercially available anodic alumina offers only a limited number of pore diameters and therefore for experimental purposes requiring very thin pore sizes one has to synthesize his own template. Fig. 1 shows a representative SEM image of the top surface of porous AAO membrane (Whatman International Ltd.) with an average pore diameter of 100 nm. On the other hand, hand made template synthesis process involves handling of very corrosive chemicals and cryogenic liquids, making it a very dangerous procedure.

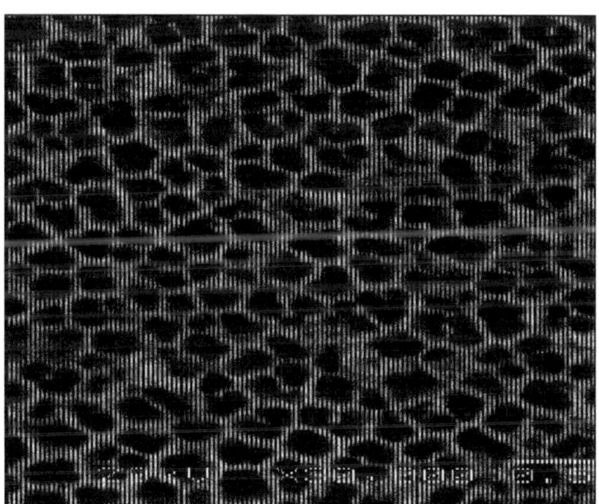

Fig. 1. A typical SEM image of the top surface of commercial AAO membrane. Marker is 0.5 µm.

In the electrodeposition processes, a thin conducting metal film, such as gold, is first evaporated on one side of the porous AAO membrane to serve as the cathode. The template is attached to the cathode, which is subsequently brought into contact with the deposition solution. When a current or potential is applied, cations and anions from the deposition solution diffuse towards the pore mouth, then inside towards the bottom of pores where electrochemical reactions of the species occur on gold surface. Then NWs grow along the pores to the top. The length of the NWs can be tuned by the time of electrodeposition process. After pore filling, the pure arrays of NWs can be obtained by dissolution of the template membrane in a basic solution. NW arrays are produced by the AAO template-assisted electrodeposition method as schematically shown in Fig. 2. In this fashion, semiconductor NWs can be deposit into the pores of AAO templates.

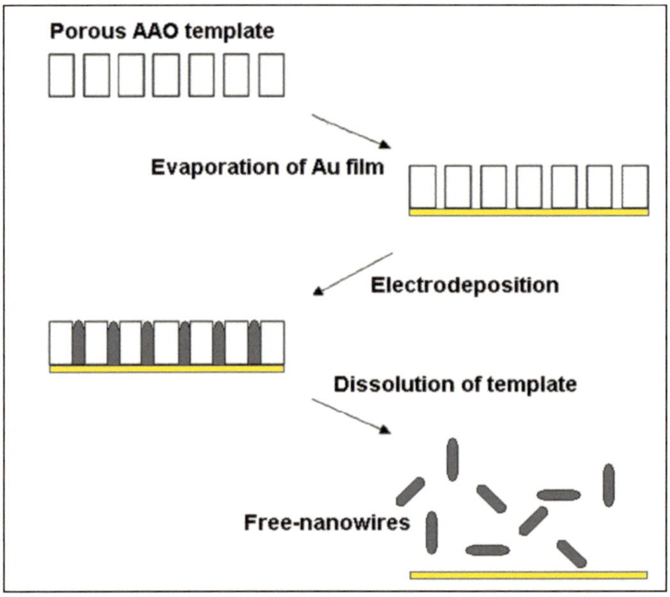

Fig. 2. Schematic diagram illustrating the growth of NWs by the AAO template-assisted electrodeposition method.

2.1 Cadmium chalcogenide NWs

Among the semiconductors explored in NW growth, cadmium chalcogenides (CdS, CdSe, and CdTe), being important direct band gap II-VI semiconductors with quantum-confined size-tunable properties, have found ample utilization opportunities in nanophotonics [23-25]. The AAO template-assisted electrodeposition technique is used for synthesizing cadmium chalcogenide (CdE = CdS, CdSe, and CdTe) NWs after the pioneering work by Routkevitch and co-workers [12] on ac electrodeposition of CdS from nonaqueous solution containing $CdCl_2$ and elemental S. But high temperature and expensive organic reagent (dimethylsulfoxide, DMSO) are necessary in order to dissolve elemental S in this method. Furthermore, a high density of defects such as stacking faults and twinned segments have been observed in those nanowires and the structure appears to be mainly the hexagonal form interleaved with domains of cubic structure [12, 26]. In a similar way, many researchers [27-31] deposited CdS NWs onto AAO membranes from nonaqueous solution containing $CdCl_2$, elemental S, and DMSO at high temperatures (100-120 °C). The formation of CdS NWs in the AAO template in DMSO solution containing $CdCl_2$ and elemental S may be written in the following equations [12]:

$$\text{At the cathode } Cd^{2+} + 2e^- \rightarrow Cd, \quad (1)$$

$$Cd + S \rightarrow CdS. \quad (2)$$

When a current or potential is applied between the cathode and anode, Cd^{2+} ions enter inside the cylindrical pores. On the other hand, elemental S enters into the pores by the diffusion process and forms CdS with the reaction of Cd atoms and they nucleate at the pore walls. As the reaction occurs inside the cylindrical pores, the wire diameter is controlled by the pore size, which grows from the surface of the electrode to the top of the template. However, Xu et al. [32] used an aqueous solution containing $CdCl_2$ and thioacetamide to produce CdS NWs onto AAO membranes at 70 °C. The overall reaction for the CdS NWs growth is written as eq. (3) [33].

$$Cd^{2+} + CH_3CSNH_2 + 2H_2O \rightarrow CdS + CH_3COOH + NH_4^+ + 1/2H_2 \quad (3)$$

Recently, we have reported the synthesis of CdS NWs onto AAO membranes by an electrodeposition method from aqueous solution containing $CdSO_4$, Na_2S, and ethylenediamine tetraacetic acid (EDTA) at room temperature (25 °C) [34]. EDTA is used to form a $CdEDTA^{2-}$ complex in order to prevent the chemical precipitation of Cd^{2+} and S^{2-} as CdS. The method starts with the investigation of electrochemical behaviors of each element in the presence of EDTA. A representative cyclic voltammogram for S^{2-} on the Au(111) working electrode in 0.1 M EDTA and 2 mM Na_2S solutions is shown in Fig. 3a.

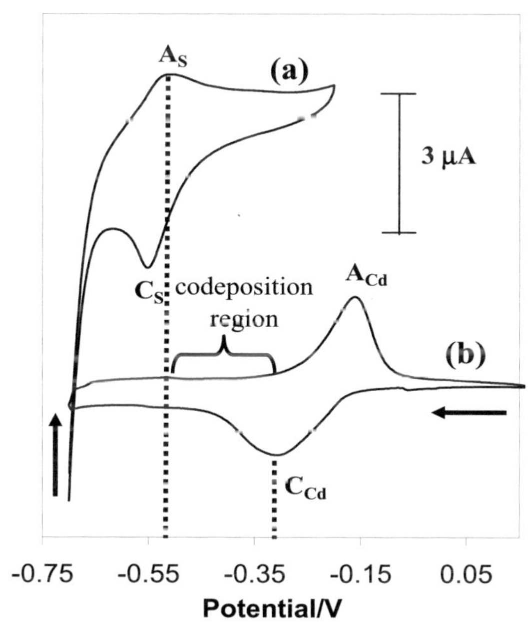

Fig. 3. Overlapped cyclic voltammograms of Au(111) electrode in 0.1 M EDTA solution containing: (a) 2 mM Na_2S and (b) 10 mM $CdSO_4$ at 100 mV/s.

Based on previous reports [6, 35, 36], the A_S and C_S peaks correspond to the S UPD and dissolution of S UPD, respectively. Fig. 3b shows a cyclic voltammetric curve of Cd on Au(111) in aqueous solution containing $CdEDTA^{2-}$ ions. The broad cathodic peak (C_{Cd}) corresponds to Cd UPD. On the reverse scan, the stripping peak (A_{Cd}) corresponds to dissolution of Cd UPD. The electrochemical behaviors of Cd in this work are similar to that reported by previous authors [6, 37, 38].

Based on the above results, for electrodeposition potential of Cd and S in UPD range can be determined from overlapped cyclic voltammograms of Cd and S. If the potential of the working electrode was kept constant at a potential between the reductive UPD wave of $CdEDTA^{2-}$ and the oxidative UPD wave of S^{2-} (codeposition region), Cd and S would deposit underpotentially at the electrode surface. These underpotentially deposited Cd and S atoms react to form the CdS. Since the potential values of the co-deposition region are not enough for the bulk deposition (OPD) of Cd and S, deposition of $CdEDTA^{2-}$ on Cd or S^{2-} on S will not occur. As a consequence, it should promote the electrochemical atom-by-atom growth of CdS compound. This concept was introduced and analyzed by Öznülüer et al. [5] and is a key concept to explain the success of the deposition of PbS. The electrodeposition starts with the preparation of AAO membrane as a template, with one side of the membrane coated with Au film as an electrode. After inserting the coated membrane into a solution of $CdSO_4$ + Na_2S + EDTA, CdS NWs can be obtained by applying a suitable potential value. The electrochemical quantification of the deposits was investigated by stripping voltammetry. The coulometric stripping of the resulting CdS indicated that the coverages of S and Cd were nearly 1:1 under most of the deposition conditions studied. This confirms the CdS nature of the deposited film. In addition, the deposition current during CdS nanowire arrays electrodeposition into the AAO template at −0.50 V was investigated. When the electrodeposition was performed for 6 h, it was observed that the current values dropped from 60 to 15 μA slowly. This decrease suggests the formation of a poorly conducting structure in the template. Fig. 4 shows TEM image of CdS NWs obtained −0.50 V for 2 h after etching the template. The diameters of the NWs within this array are about 100 nm and the lengths are up to 1 μm.

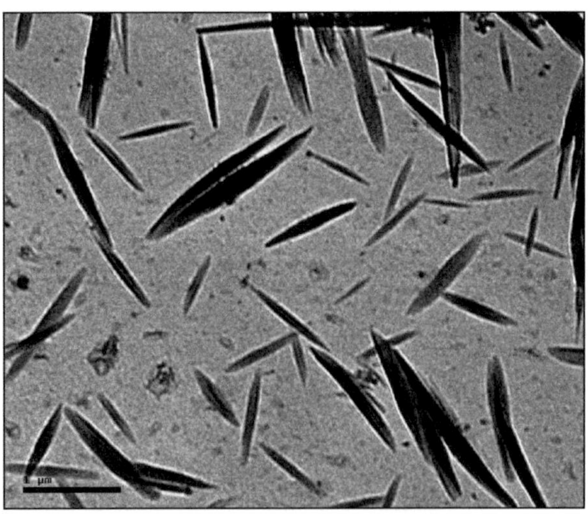

Fig. 4. TEM image of CdS NWs with a diameter of about 100 nm. Marker is 1 μm.

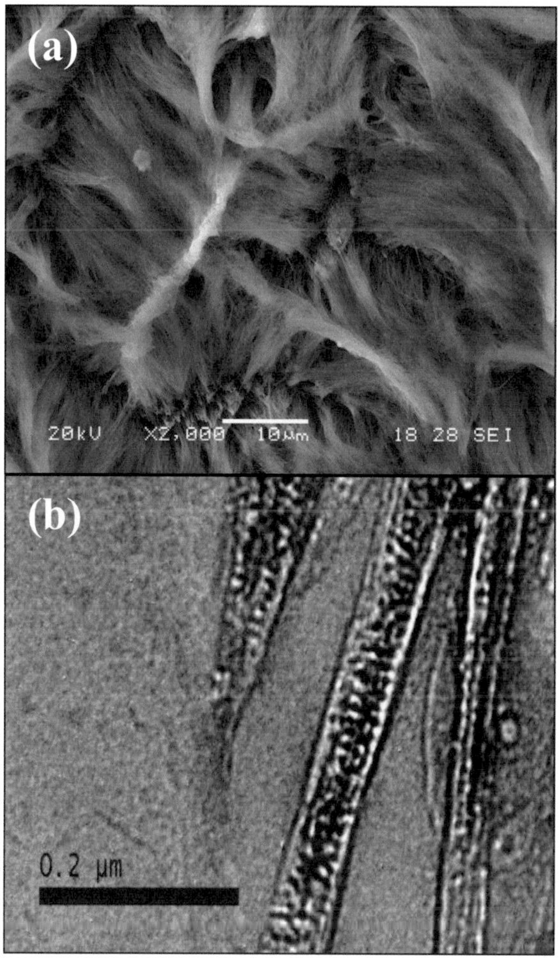

Fig. 5. (a) A typical SEM image of CdS NW arrays on the AAO template after etching times of 10 min. (b) TEM image of the a few CdS NWs electrodeposited in the template shown after the complete dissolution of AAO in 1 M NaOH.

SEM and TEM images of the CdS NWs electrodeposited on AAO template at −0.50 V for 6 h are shown in Fig. 5. The diameters of the NWs within this array are about 100 nm and the lengths are up to 10 µm. It is clear that the length of the CdS NWs increases with an increase in the deposition time.

Consequently, formation of CdS NWs in the AAO template in EDTA solution containing Cd^{2+} [39, 40] and S^{2-} at an acidic medium may be written in the following equations:

$$Cd^{2+} + (EDTA)^{4-} \rightarrow Cd(EDTA)^{2-} \quad (4)$$

$$Cd(EDTA)^{2-} + 2e^{-} \rightarrow Cd + EDTA^{4-} \quad (5)$$

$$H_2S \rightarrow S + 2e^{-} + 2H^{+} \quad (6)$$

$$Cd + S \rightarrow CdS \quad (7)$$

According to this sequence, CdS formation inside of AAO involves firstly diffusion of $Cd(EDTA)^{2-}$ and H_2S species from solution towards the pore mouth, then inside towards the bottom of pores where electrochemical reactions of the species occur on gold surface. Then CdS nanowires grow along the pores to the top.

XRD pattern of the CdS NWs electrodeposited on amorphous AAO is shown in Fig. 6. XRD analysis reveals that the electrosynthesized NWs have a preferential orientation along (110) direction for hexagonal crystal (JCPDS, 41-1049). The broad XRD peak is ascribed to the small grain size.

Fig. 7 shows the UV–vis spectra of CdS NWs inside the deionized water after complete dissolution of template. The blue shifted absorption at 489 nm as compared to the ~515 nm absorption for bulk CdS crystal may be attributed to the quantum confinement effect of CdS NWs [41].

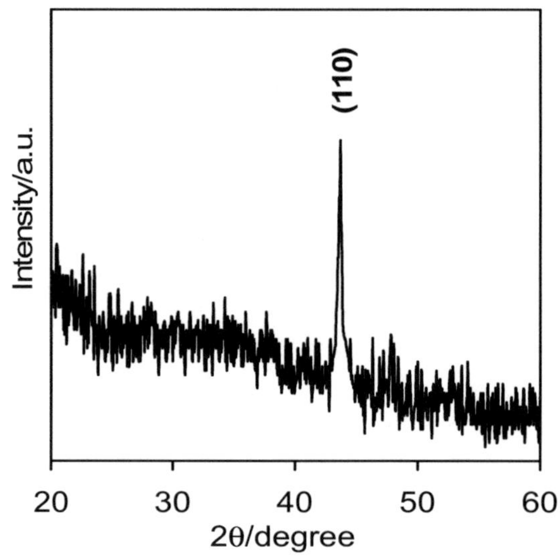

Fig. 6. A representative XRD pattern of the CdS NWs embedded in the AAO template.

These results indicate that highly crystalline CdS NWs can be grown by this method. By fixing the deposition potential range, stoichiometric CdS NWs can be produced in this method. Furthermore, the length of the NWs can be tuned by the controlling the deposition or etching time. As a result, this method is different than other AAO template-assisted electrodeposition approaches.

AAO membranes are frequently employed as templates for fabricating other cadmium chalcogenides (CdSe and CdTe). CdSe has been investigated over the past decade for applications in optoelectronics [42], luminescent materials [43], lasing materials [44], and biomedical imaging. CdSe NWs have been produced by direct current electrodeposition from both aqueous and nonaqueous electrolytes [45-47]. On the other hand, metal–CdSe–metal NWs have been synthesized by the electrochemical replication of porous AAO and polycarbonate membranes with pore diameters of 350 and 700 nm respectively [48]. The overall reaction for the CdSe NW growth is written as eq. (8) [49].

$$Cd^{2+} + H_2SeO_3 + 4H^+ + 6e^- \rightarrow CdSe + 3H_2O \quad (8)$$

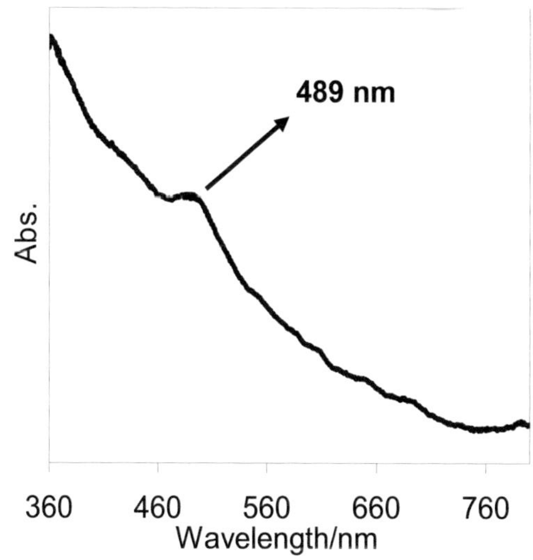

Fig. 7. Typical UV-vis absorption spectrum of CdS NWs.

CdTe NWs can also be produced by this method. Cadmium telluride (CdTe) is one of the most important II-VI group compound semiconductors due to its interesting photoelectric properties. It has a direct energy band gap [50], a relatively high optical absorption coefficient, and advanced photovoltaic properties, which make it suitable for fabricating photoelectron devices and solar cells [51]. It is expected that CdTe NWs would be of interest in technological application for nanodevices. However, few studies of the electrosynthesis of CdTe NWs on AAO membranes are reported [52, 53]. In that case the deposition proceeds directly by the global reaction [54]:

$$HTeO_2^+ + Cd^{2+} + 3H^+ + 6e^- \rightarrow CdTe + 2H_2O \quad (9)$$

2.2 Oxide NWs

Most of the oxides are semiconductors. Among these, ZnO NWs are especially attractive due to their tunable electronic and optoelectronic properties, and the potential applications in the nanoscale electronic and micro-optoelectronic devices [1, 55, 56]. Numerous reports on the preparations of ZnO films and nanoparticles have been reported. In contract, very little work has been focused on ZnO NWs. Ordered ZnO NWs embedded in AAO templates have already been fabricated by electrodepositing Zn into nanopores of AAO to form metallic nanowire arrays and then to oxidize them [57]. But the long time thermal oxidization at 300 °C needed in this method limits its application. ZnO NWs have also been synthesized by one-step electrodeposition [58] based on AAO templates conducted in aqueous Zn^{2+} solutions. The mechanism of electrodeposition of ZnO NWs from oxygenated aqueous Zn^{2+} solution could be summarized as followings [59]:

$$O_2 + 2H_2O + 4e^- \rightarrow 4OH^- \quad (10)$$

$$Zn^{2+} + 2OH^- \leftrightarrow Zn(OH)_2 \quad (11)$$

$$Zn(OH)_2 = ZnO + H_2O \quad (12)$$

The total reaction may be written as:

$$Zn^{2+} + 0.5O_2 + 2e^- \rightarrow ZnO \quad (13)$$

It can be seen that the aqueous solution induces the deposition of $Zn(OH)_2$ competing with the formation of ZnO. Wang et al. [60] reported a different approach to prepare the ZnO NWs into AAO templates involving nonaqueous DMSO. It is important that the formation of $Zn(OH)_2$ is prevented in a nonaqueous media.

AAO templates can be employed for electrosynthesis of Cu_2O NWs. Cu_2O is a relatively non-toxic p-type semiconductor with a band gap of 2.1 eV [61]. It can be used in solar energy converting devices. Cu_2O NWs have been produced by electrodeposition on AAO templates from alkaline aqueous solutions [62-64]. In alkaline solution, the reaction leading to the deposition of copper oxide is [64]

$$2Cu^{2+} + 2e^- + 2OH^- \rightarrow Cu_2O + H_2O \quad (14)$$

In addition, In_2O_3 NWs can also be produced by this method. However, to the best of our knowledge, there is only one report found on the electrodeposition of In_2O_3 NWs into AAO templates [65].

2.3 Bismuth-based NWs

The VA-VIA group semiconductors are commonly used for thermoelectric devices such as thermoelectric generators, coolers, radiation detectors, and so forth [66]. Among them, Bi_2Te_3 and its derivative compounds are considered to be the best materials used in thermoelectric refrigeration at room temperature [67]. High-quality Bi_2Te_3, $Bi_2Te_{3-y}Se_y$, $Bi_{1-x}Sb_x$, and $Bi_{2-x}Sb_xTe_3$ NWs have been prepared by electrodeposition in AAO membranes from both aqueous and nonaqueous electrolytes [68-71]. The obtained Bi_2Te_3 arrays are dense with a narrow distribution of NW diameters. On the other hand, by tuning the solution concentrations and controlling the growth rate by controlling the potential, the composition, crystallinity, and morphology of the $Bi_{1-x}Sb_x$ NWs can be tuned. Other bismuth-based NWs ($Bi_2Te_{3-y}Se_y$ and $Bi_{2-x}Sb_xTe_3$) are single phase and crystalline.

3. Electrochemical Step Edge Decoration (ESED)

The method, developed by Penner et al. [39], is claimed to have the capability to synthesize CdS NWs which are long, uniform in diameter, and removable from the surface on which they are synthesized. The method starts with selective electrodeposition of metal from an aqueous solution as a precursor, along step edges present on the stepped surface, such as graphite, to form into NWs in the first step (Fig. 8), according to the reaction

$$Cd(EDTA)^{2-} + 2e^- \rightarrow Cd + EDTA^{4-} \quad (15)$$

In the second step, Cd NWs were converted to CdS hemicylindrical shell NWs by exposure to flowing H_2S at 280–300 °C according to the reaction

$$Cd + H_2S \rightarrow CdS + H_2 \quad (16)$$

This method has been successfully applied to obtain NWs of Bi_2Te_3 [72] and CdSe [73].

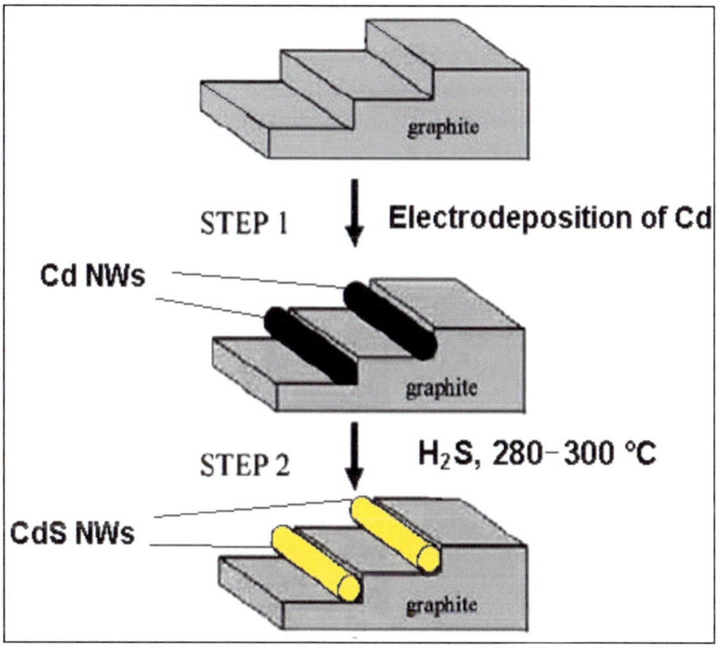

Fig. 8. A schematic diagram for the synthesis of CdS NWs through ESED method.

4. "Track-etch" membranes

Another type of porous template that is commonly used for NW synthesis is the template type fabricated by chemically etching particle tracks originating from ion bombardment, such as track-etched polymeric membranes [14]. These membranes contain cylindrical pores of uniform diameter. The most commonly used material to prepare membranes of this type is polycarbonate; however, a number of other materials are amenable to the track-etch process [14]. A number of companies sell nanoporous polymeric filtration membranes that have been prepared via the "track-etch" method. The method involves the irradiation of polycarbonate foils (PCF) using a gold-ion beam at the heavy-ion accelerator to damage the polymer structure. Then an ultraviolet treatment is applied to each face of the PCF. The pore density corresponds to the ion fluency, whereas the length of the pores is equal to the thickness of the membrane. The pores are randomly distributed across the membrane surface. The pore diameter is adjusted by a partial chemical dissolution. Membranes with a wide range of pore diameters (down to 10 nm) are available commercially. High-quality $Bi_2Te_3/(Bi_{0.3}Sb_{0.7})_2Te_3$, Si_xGe_{1-x}, and Bi_2Te_3 NWs have been prepared by electrodeposition in "track-etch" membranes from both aqueous and ionic liquid mediums [74-76]. These organic membranes may be preferred over anodized alumina membranes since they exhibit greater flexibility.

5. Lithographical Patterned Nanowire Electrodeposition (LPNE)

Lithographically patterned nanowire electrodeposition (LPNE), developed Menke et al. [77], is a new method for synthesizing semiconductor NWs on glass or oxidized silicon surfaces. The method involves the fabrication of a sacrificial, horizontal template using conventional microfabrication methods. LPNE uses photolithography to prepare a three-sided "nanoform" into which a nanowire can be electrodeposited using the horizontal nickel edge that defines the back surface of this nanoform. The LPNE synthesis of PbTe NW arrays was carried out according to the twelve-step process flow: (Step 1) A 1-2 μm thick photoresist (PR) layer is spin-coated onto a glass or oxidized silicon surface, (Step 2) a nickel layer 20 - 100 nm in thickness is thermally deposited, (Step 3) a positive-tone PR is coated onto this nickel layer and, (Step 4) photopatterned using 365 nm illumination and a contact mask, (Step 5) the exposed PR is removed, (Step 6) now the remaining PR on the same is flood-exposed. A crucial point is that after this flood-exposure, no developing is carried out; (Step 7) the exposed nickel was etched using HNO_3 to produce a 500 nm deep undercut at the edges of the exposed regions. The horizontal trench formed by this undercut is the "nano-form" into which the PbTe NWs will be electrodeposited. The height of this nano-form equaled the thickness of the nickel layer which defines the vertical back of the trench. This nano-form follows the contour of the photopatterned region. (Step 8) The PbTe NW is electrodeposited into the trench, (Step 9) the previously exposed top-PR layer is removed using developer, (Step 10) the nickel layer is removed by etching in HNO_3 leaving free-standing PbTe NWs on top of an intact and unexposed layer of bottom-PR, (Step 11) this bottom PR layer is exposed as before, (Step 12) the exposed PR is removed leaving suspended PbTe NW segments up to 25 μm in length [78]. This method could also be used to grow NWs of CdSe [49] and PbSe [79].

6. Conclusion

In conclusion, we have presented a brief overview of the selected template-assisted electrodeposition methods for the synthesis of semiconductor NWs. Compared to conventional methods, the high degree of reproducibility makes template-assisted methods the most attractive way. In this chapter, four groups of template-assisted electrodeposition methods have been investigated. Among these templates the nanoporous AAO is the most promising. For example, the AAO template can stand high temperatures, is insoluble in organic solvents and geometrical parameters can be easily tuned by changing the synthesis conditions. The suggested electrodeposition mechanisms for some semiconductor NWs onto templates were presented and discussed. As an example of semiconductor NWs, we were focused on AAO template-assisted electrodeposition of CdS NWs under the conditions of UPD. Using this method, many material parameters such as dimensions and compositions can be tuned. The method may serve as a model for synthesis of other semiconductor NWs.

7. Acknowledgment

Sakarya University is gratefully acknowledged for the financial support of this work.

8. References

[1] Huang, M.H.; Mao, S.; Feick, H.; Yan, H.Q.; Wu, Y.Y.; Kind, H.; Weber, E.; Russo, R. & Yang, P.D. (2001). Room-temperature ultraviolet nanowire nanolasers. *Science* 292, 1897-1899, ISSN: 0036-8075.

[2] Kolb, D.M. (1978). Physical and electrochemical properties of metal monolayers on metallic substrates, In: *Advances in Electrochemistry and Electrochemical Engineering*, H. Gerischer and C.W. Tobias (Ed.), pp. 125–271, Wiley, New York.

[3] Jeffrey, C.A.; Harrington, D.A. & Morin, S. (2002). In situ scanning tunneling microscopy of bismuth electrodeposition on Au(111) surfaces. *Surface Science* 512, L367-L372, ISSN: 0039-6028.

[4] Panicker, M.P.R.; Knaster, M. & Kröger, F.A. (1978). Cathodic deposition of CdTe from aqueous electrolytes. *Journal of the Electrochemical Society* 125, 566-572, ISSN: 0013-4651.

[5] Öznülüer, T.; Erdoğan, İ.; Şişman, İ. & Demir, Ü. (2005). Electrochemical atom-by-atom growth of PbS by modified ECALE method. *Chemistry of Materials* 17, 935–937, ISSN: 0897–4756.

[6] Şişman, İ.; Alanyalıoğlu, M. & Demir, Ü. (2007). Atom-by-atom growth of CdS thin films by an electrochemical co-deposition method: Effects of pH on the growth mechanism and structure. *Journal Physical Chemistry C* 111, 2670–2674, ISSN: 1932–7447.

[7] Köse, H.; Biçer, M.; Tütünoğlu, Ç.; Aydın, A.O. & Şişman, İ. (2009). The underpotential deposition of $Bi_2Te_{3-y}Se_y$ thin films by an electrochemical co-deposition method. *Electrochimica Acta* 54, 1680-1686, ISSN: 0013-4686.

[8] Biçer, M.; Köse, H. & Şişman, İ. (2010). Selective electrodeposition and growth mechanism of thermoelectric bismuth-based binary and ternary thin films. *Journal Physical Chemistry C* 114, 8256–8263, ISSN: 1932–7447.

[9] Şişman, İ. & Biçer, M. (2011). Structural, morphological and optical properties of $Bi_{2-x}Sb_xSe_3$ thin films grown by electrodeposition. *Journal of Alloys and Compounds* 509, 1538–1543. ISSN: 0925–8388.

[10] Biçer, M. & Şişman, İ. (2011). Electrodeposition and growth mechanism of SnSe thin films. *Applied Surface Science* 257, 2944–2949, ISSN: 0169–4332.

[11] Şişman, İ. & Demir, Ü. (2011). Electrochemical growth and characterization of size-quantized CdTe thin films grown by underpotential deposition. *Journal of Electroanalytical Chemistry* 651, 222–227, ISSN: 1572–6657.

[12] Routkevitch, D.; Bigioni, T., Moskovits, M.; Xu & J.M. (1996). Electrochemical fabrication of CdS nanowire arrays in porous anodic aluminum oxide templates. *Journal Physical Chemistry* 100, 14037–14047, ISSN: 0022–3654.

[13] Masuda, H. & Fukuda, K. (1995). Ordered metal nanohole arrays made by a 2-step replication of honeycomb structures of anodic alumina. *Science* 268, 1466-1468, ISSN: 0036-8075.

[14] Fleischer, R.L.; Price, P.B. & Walker, R.M. (1975). Nuclear tracks in solids: Principles and Applications. *University of California Press*. Berkeley.

[15] Kautek, W.; Reetz, S. & Pentzien, S. (1995). Template electrodeposition of nanowire arrays on gold foils fabricated by pulsed-laser deposition. *Electrochimica Acta* 40, 1461-1468, ISSN: 0013-4686.

[16] Thurn-Albrecht, T.; Schotter, J.; Kastle, G.A.; Emley, N.; Shibauchi, T.; Krusin-Elbaum, L.; Guarini, K.; Black, C.T.; Tuominen, M.T. & Russell, T.P. (2000). Ultrahigh-density nanowire arrays grown in self-assembled diblock copolymer templates. *Science* 290, 2126-2129, ISSN: 0036-8075.

[17] Tonucci, R.J.; Justus, B.L.; Campillo, A.J. & Ford, C.E. (1992). Nanochannel array glass. *Science* 258, 783-785, ISSN: 0036-8075.

[18] Rajendra Kumar, R.T.; Badel, X.; Vikor, G.; Linnros, J. & Schuch, R. (2005). Fabrication of silicon dioxide nanocapillary arrays for guiding highly charged ions. *Nanotechnology* 16, 1697-1700, ISSN: 0957-4484.

[19] Martin, C.R. & Mitchell, D.T. (1999). Template-synthesized nanomaterials in electrochemistry. In: *Electroanalytical Chemistry*. A. J. Bard & I. Rubinstein (Ed.), pp. 1-74, Marcel Dekker Inc.

[20] Diggle, J.W.; Downie, T.C. & Goulding, C.W. (1969). Anodic oxide films on aluminum. *Chemical Review* 69, 365-405, ISSN: 0009-2665.

[21] Li, A.P.; Müller, F.; Birner, A.; Neilsch, K. & Gösele, U. (1998). Hexagonal pore arrays with a 50-420 nm interpore distance formed by self-organization in anodic alumina. *Journal Applied Physics* 84, 6023-6026, ISSN: 0021-8979.

[22] Zhang, Z.; Gekhtman, D.; Dresselhaus, M. S. & Ying, J.Y. (1999). Processing and characterization of single-crystalline ultrafine bismuth nanowires. *Chemistry of Materials* 11, 1659-1665, ISSN: 0897–4756.

[23] Huynh, W.U.; Dittmer, J.J. & Alivisatos, A.P. (2002). Hybrid nanorod-polymer solar cells. *Science* 295, 2425-2427, ISSN: 0036-8075.

[24] Duan, X.F.; Huang, Y.; Agarwal, R. & Lieber, C.M. (2003). Single-nanowire electrically driven lasers. *Nature* 421, 241-245, ISSN: 0028-0836.

[25] Barrelet, C.J.; Greytak, A.B. & Lieber, C.M. (2004). Nanowire photonic circuit elements. *Nano Letters* 4, 1981-1985, ISSN: 1530-6984.

[26] Hutchison, J.L.; Routkevitch, D.; Moskovits, M. & Nayak, R.R. (1997). TEM and HREM structural studies of non-lithographically-produced CdS nanowires. *Institute of Physics Conference Series* 157, 389-392, ISSN: 0951-3248.

[27] Suh, J.S. & Lee, J.S. (1997). Surface enhanced Raman scattering for CdS nanowires deposited in anodic aluminum oxide nanotemplate. *Chemical Physics Letters* 28, 384-388, ISSN: 0009-2614.

[28] Xu, D.S.; Xu, Y.J.; Chen, D.P.; Guo, G.L.; Gui, L.L. & Tang, Y.Q. (2000). Preparation and characterization of CdS nanowire arrays by dc electrodeposit in porous anodic aluminum oxide templates. *Chemical Physics Letters* 325, 340-344, ISSN: 0009-2614.

[29] Xu, D.S.; Chen, D.P.; Xu, Y.J.; Shi, X.; Guo, G.L.; Gui, L.L. & Tang, Y.Q. (2000). Preparation of II-VI group semiconductor nanowire arrays by dc electrochemical deposition in porous aluminum oxide templates. Pure and Applied Chemistry 72, 127-135, ISSN: 0033-4545.

[30] Yang, Y.; Chen, H.L.; Mei, Y.F.; Chen, J.B.; Wu, X.L. & Bao, X.M. (2002). Anodic alumina template on Au/Si substrate and preparation of CdS nanowires. *Solid State Communications* 123, 279-282, ISSN: 0038-1098.

[31] Mondal, S.P.; Dhar, A. & Ray, S.K. (2007). Optical properties of CdS nanowires prepared by dc electrochemical deposition in porous alumina template. *Materials Science in Semiconductor Processing* 10, 185-193, ISSN: 1369-8001.

[32] Xu, D.S.; Xu, Y.J.; Chen, D.P.; Guo, G.L.; Gui, L.L. & Tang, Y.Q. (2000). Preparation of CdS single-crystal nanowires by electrochemically induced deposition. *Advanced Materials* 12, 520-522, ISSN: 0935-9648.

[33] Yamaguchi, K.; Yoshida, T.; Sugiura, T. & Minoura, H. (1998). A novel approach for CdS thin-film deposition: Electrochemically induced atom-by-atom growth of CdS thin

films from acidic chemical bath. *Journal of Physical Chemistry B* 102, 9677-9686, ISSN: 1089-5647.

[34] Biçer, M.; Aydın, A.O. & Şişman, İ. (2010). Electrochemical synthesis of CdS nanowires by underpotential deposition in anodic alumina membrane templates. *Electrochimica Acta* 55, 3749–3755, ISSN: 0013–4686.

[35] Colletti, L.P.; Teklay, D. & Stickney, J.L. (1994). Thin-layer electrochemical studies of the oxidative underpotential deposition of sulfur and its application to the electrochemical atomic layer epitaxy deposition of CdS. *Journal of Electroanalytical Chemistry* 369, 145–152, ISSN: 0022–0728.

[36] Alanyalıoğlu, M.; Çakal, H.; Öztürk, A.E. & Demir, Ü. (2001). Electrochemical studies of the effects of pH and the surface structure of gold substrates on the underpotential deposition of sulfur. *Journal of Physical Chemistry B* 105, 10588–10593, ISSN: 1089–5647.

[37] Inzelt, G. & Horanyi, G. (2000). On the alloy formation in the course of upd of Cd on gold. *Journal of Electroanalytical Chemistry* 491, 111–116, ISSN: 0022–0728.

[38] Lay, M.D.; Varazo, K.; Srisook, N. & Stickney, J.L. (2003). Cd underpotential deposition (upd) from a sulfate electrolyte on Au(111): studies by in situ STM and UHV-EC. *Journal of Electroanalytical Chemistry* 554, 221–231, ISSN: 0022–0728.

[39] Li, Q. & Penner, R.M. (2005). Photoconductive cadmium sulfide hemicylindrical shell nanowire ensembles. *Nano Letters* 5, 1720–1725, ISSN: 1530–6984.

[40] Yang, W.B.; Wu, Z.; Lu, Z.Y.; Yang, X.P. & Song, L.X. (2006). *Microelectronic Engineering* 83, 1971–1974, ISSN: 0167–9317.

[41] Mo, D.; Liu, J.; Yao, H.J.; Duan, J.L.; Hou, M.D.; Sun, Y.M.; Chen, Y.F.; Xue, Z.H. & Zhang, L. (2008). Preparation and characterization of CdS nanotubes and nanowires by electrochemical synthesis in ion-track templates. *Journal of Crystal Growth* 310, 612–616, ISSN: 0022–0248.

[42] Hodes, G.; Albuyaron, A.; Decker, F. & Motisuke, P. (1987). 3-dimensional quantum-size effect in chemically deposited cadmium selenide films. *Physical Review B* 36, 4215–4221, ISSN: 0163–1829.

[43] Bawendi, M.G.; Kortan, A.R.; Steigerwald, M.L. & Brus, L.E. (1989). X-ray structural characterization of larger CdSe semiconductor clusters. *Journal Chemical Physics* 91, 7282–7290, ISSN: 0021–9606.

[44] Ma, C.; Ding, Y.; Moore, D.F.; Wang, X. & Wang, Z.L. (2004). Single-crystal CdSe nanosaws. *Journal the American Chemical Society* 126, 708-709, ISSN: 0002-7863.

[45] Xu, D.S.; Shi, X.S.; Guo, G.L.; Gui, L.L. & Tang, Y.Q. (2000). Electrochemical preparation of CdSe nanowire arrays. *Journal of Physical Chemistry B* 104, 5061–5063, ISSN: 1089–5647.

[46] Peng, X.S.; Zhang, J.; Wang, X.F.; Wang, Y.W.; Zhao, L.X.; Meng, G.W. & Zhang, L.D. (2001). Synthesis of highly ordered CdSe nanowire arrays embedded in anodic alumina membrane by electrodeposition in ammonia alkaline solution. *Chemical Physics Letters* 343, 470–474, ISSN: 0009–2614.

[47] Sankar, P.R.; Tiwari, P.; Kumar, R.; Ganguli, T.; Mukherjee, C.; Srivastava, A.K.; Oak, S.M. & Pathak, R.K. (2010). Synthesis and characterization of cadmium selenide nanostructures on porous aluminum oxide templates by high frequency alternating current electrolysis. *Applied Surface Science* 256, 2097-2103, ISSN: 0169-4332.

[48] Pena, D.J.; Mbindyo, J.K.N.; Carado, A.J.; Mallouk, T.E.; Keating, C.D.; Razavi, B. & Mayer, T.S. (2002). Template growth of photoconductive metal-CdSe-metal nanowires. *Journal of Physical Chemistry* B 106, 7458–7462, ISSN: 1520–6106.

[49] Kung, S.C.; van der Veer, W.E.; Yang, F.; Donavan, K.C. & Penner, R.M. (2010). 20 μs Photocurrent Response from Lithographically Patterned Nanocrystalline Cadmium Selenide Nanowires. *Nano Letters* 10, 1481–1485, ISSN: 1530–6984.

[50] Rakhshani, A.E. (1997). Electrodeposited CdTe - Optical properties. *Journal of Applied Physics* 81, 7988–7993, ISSN: 0021–8979.

[51] Basol, B.M. (1988). Electrodeposited CdTe and HgCdTe solar-cells. *Solar Cells* 23, 69–88, ISSN: 0379–6787.

[52] Zhao, A.W.; Meng, G.W.; Zhang, L.D.; Gao, T.; Sun, S.H. & Pang, Y.T. (2003). Electrochemical synthesis of ordered CdTe nanowire arrays. *Applied Physic A-Materials Science & Processing* 76, 537–539, ISSN: 0947–8396.

[53] Ohgai, T.; Gravier, L.; Hoffer, X. & Ansermet, J.P. (2005). CdTe semiconductor nanowires and NiFe ferro-magnetic metal nanowires electrodeposited into cylindrical nano-pores on the surface of anodized aluminum. *Journal of Applied Electrochemistry* 35, 479–485, ISSN: 0021-891X.

[54] Lincot, D. (2005). Electrodeposition of semiconductors. *Thin Solid Films* 487, 40–48, ISSN: 0040–6090.

[55] Kong, Y.C.; Yu, D.P.; Zhang, B.; Fang, W. & Feng, S.Q. (2001). Ultraviolet-emitting ZnO nanowires synthesized by a physical vapor deposition approach. *Applied Physics Letters* 78, 407–409, ISSN: 0003–6951.

[56] Liu, C.H.; Zapien, J.A.; Yao, Y.; Meng, X.M.; Lee, C.S.; Fan, S.S.; Lifshitz, Y. & Lee, S.T. (2003). High-density, ordered ultraviolet light-emitting ZnO nanowire arrays. *Advanced Materials* 15, 838–841, ISSN: 0935–9648.

[57] Li, Y.; Meng, G.W.; Zhang, L.D. & Phillip, F. (2000). Ordered semiconductor ZnO nanowire arrays and their photoluminescence properties. *Applied Physics Letters* 76, 2011–2013, ISSN: 0003–6951.

[58] Zheng, M.J.; Zhang, L.D.; Li, G.H. & Shen, W.Z. (2002). Fabrication and optical properties of large-scale uniform zinc oxide nanowire arrays by one-step electrochemical deposition technique. *Chemical Physics Letters* 363, 123–128, ISSN: 0009–2614.

[59] Peulon, S. & Lincot, D. (1998). Mechanistic study of cathodic electrodeposition of zinc oxide and zinc hydroxychloride films from oxygenated aqueous zinc chloride solutions. *Journal of the Electrochemical Society* 145, 864-874, ISSN: 0013-4651.

[60] Wang, Q.T.; Wang, G.Z.; Xu, B.; Jie, J.S.; Han, X.H.; Li, G.P.; Li, Q.S. & Hou, J.G. (2005). Non-aqueous cathodic electrodeposition of large-scale uniform ZnO nanowire arrays embedded in anodic alumina membrane. *Materials Letters* 59, 1378–1382, ISSN: 0167-577X.

[61] Luo, J.; Zhang, L.; Zhang, Y.J. & Zhu, J. (2002). Controlled growth of one-dimensional metal-semiconductor and metal-carbon nanotube heterojunctions. *Advanced Materials* 14, 1413–1414, ISSN: 0935–9648.

[62] Liu, X.M. & Zhou, Y.C. (2005). Electrochemical deposition and characterization of Cu_2O nanowires. *Applied Physics A-Materials Science & Processing* 81, 685–689, ISSN: 0947–8396.

[63] Ko, E.; Choi, J.; Okamoto, K.; Tak, Y. & Lee, J. (2006). Cu_2O nanowires in an alumina template: Electrochemical conditions for the synthesis and photoluminescence characteristics. *ChemPhysChem* 7, 1505–1509, ISSN: 1439–4235.

[64] Inguanta, R.; Piazza, S. & Sunseri, C. (2008). Template electrosynthesis of aligned Cu_2O nanowires - Part I. Fabrication and characterization. *Electrochimica Acta* 53, 6504–6512, ISSN: 0013–4686.

[65] Zheng, M.J.; Zhang, L.D.; Zhang, X.Y.; Zhang, J. & Li, G.H. (2001). Fabrication and optical absorption of ordered indium oxide nanowire arrays embedded in anodic alumina membranes. *Chemical Physics Letters* 334, 298–302, ISSN: 0009-2614.

[66] Venkatasubramanian, R.; Siivola, E.; Colpitts, T. & O'Quinn, B. (2001). Thin-film thermoelectric devices with high room-temperature figures of merit. *Nature* 13, 597–602, ISSN: 0028–0836.

[67] Ioffe, A.F. (1957). Semiconductors Thermoelements and Thermoelectric Cooling. p. 39, Infosearch Limited, London.

[68] Sander, M.S.; Gronsky, R.; Sands, T.& Stacy, A.M. (2003). Structure of bismuth telluride nanowire arrays fabricated by electrodeposition into porous anodic alumina templates. *Chemistry of Materials* 15, 335–339, ISSN: 0897–4756.

[69] Martin-Gonzalez, M.; Snyder, G.J.; Prieto, A.L.; Gronsky, R.; Sands, T. & Stacy, A.M. (2003). Direct electrodeposition of highly dense 50 nm $Bi_2Te_{3-y}Se_y$ nanowire arrays. *Nano Letters* 3, 973–977, ISSN: 1530–6984.

[70] Prieto, A.L.; Martin-Gonzalez, M.; Keyani, J.; Gronsky, R.; Sands, T. & Stacy, A.M. (2003). The electrodeposition of high-density, ordered arrays of $Bi_{1-x}Sb_x$ nanowires. *Journal of the American Chemical Society* 125, 2388–2389, ISSN: 0002–7863.

[71] Martin-Gonzalez, M.; Prieto, A.L.; Gronsky, R.; Sands, T. & Stacy, A.M. (2003). High-density 40 nm diameter Sb-rich $Bi_{2-x}Sb_xTe_3$ nanowire arrays. *Advanced Materials* 15, 1003–1006, ISSN: 0935–9648.

[72] Menke, E.J.; Li, Q. & Penner, R.M. (2004). Bismuth telluride (Bi_2Te_3) nanowires synthesized by cyclic electrodeposition/stripping coupled with step edge decoration. *Nano Letters* 4, 2009–2014, ISSN: 1530–6984.

[73] Li, Q.; Brown, M.A.; Hemminger, J.C. & Penner, R.M. (2006). Luminescent polycrystalline cadmium selenide nanowires synthesized by cyclic electrodeposition/stripping coupled with step edge decoration. *Chemistry of Materials* 18, 3432–3441, ISSN: 0897–4756.

[74] Yoo, B.; Xiao, F.; Bozhilov, K.N.; Herman, J.; Ryan, M.A. & Myung, N.V. (2007). Electrodeposition of thermoelectric superlattice nanowires. *Advanced Materials* 19, 296–299, ISSN: 0935–9648.

[75] Al-Salman, R. & Endres, F. (2009). Template-assisted electrodeposition of Si_xGe_{1-x} nanowires with varying length and composition from two different ionic liquids. *Journal of Materials Chemistry* 19, 7228–7231, ISSN. 0959–9428.

[76] Frantz, C.; Stein, N.; Gravier, L.; Granville, S. & Boulanger, C. (2010). Electrodeposition and Characterization of Bismuth Telluride Nanowires. *Journal of Electronic Materials* 39, 2043–2048, ISSN: 0361–5235.

[77] Menke, E.J.; Thompson, M.A.; Xiang, C.; Yang, L.C. & Penner, R.M. (2006). Lithographically patterned nanowire electrodeposition. *Nature Materials* 5, 914–919, ISSN: 1476–1122.

[78] Yang, Y.; Kung, S.C.; Taggart, D.K.; Xiang, C.; Yang, F.; Brown, M.A.; Guell, A.G.; Kruse, T.J.; Hemminger, J.C. & Penner, R.M. (2008). Synthesis of PbTe nanowire arrays using lithographically patterned nanowire electrodeposition. *Nano Letters* 8, 2447-2451, ISSN: 1530-6984.

[79] Hujdic, J.E.; Taggart, D.K.; Kung, S.C. & Menke, E.J. (2010). Lead Selenide Nanowires Prepared by Lithographically Patterned Nanowire Electrodeposition. *Journal of Physical Chemistry Letters* 1, 1055-1059, ISSN: 1948-7185.

4

Semiconducting Oxide Nanowires: Growth, Doping and Device Applications

Qing Wan[1,2], Jia Sun[2] and Huixuan Liu[2]
[1]*Ningbo Institute of Materials Technology and Engineering, Chinese Academy of Sciences, Ningbo,*
[2]*Key Laboratory for Micro-Nano Optoelectronic Devices of Ministry of Education, Hunan University, Changsha,*
[1,2]*People's Republic of China*

1. Introduction

During the past decade, semiconducting oxide nanowires have been extensively investigated due to their unique features, such as the ultrahigh surface-to-volume ratios, quantum confinement effect, less scattering of the carrier, and higher mobilities compared to their bulk counterparts. These extraordinary properties of semiconducting oxide nanowires have led many researchers to pursue synthesis, doping, and novel device applications, due to their potential applications in nanoscale electronic and optoelectronic devices [1-23]. Such applications include high-performance nanowire-based field-effect transistors (FETs) [1-11], sensors [12-21] and vacuum electron field emitters [22-29], etc.

This chapter provides a comprehensive perspective on research efforts made on synthesis, doping, and device applications of semiconducting oxide nanowires. The contents have been divided into three sections. The first part briefly introduces the synthesis of semiconducting oxide nanowires. The second part introduces the doping schemes of semiconducting oxide nanowires, and then studied the electrical and optical properties of the doped nanowires. The last part mainly describes the applications of semiconducting oxide nanowires, such as FETs, sensors and field emission.

2. Semiconducting oxide nanowires growth

Up to now, many techniques have been developed in the synthesis of semiconducting oxide nanowires. Basically, they can be described as two different types: the "top-down" approaches and the "bottom-up" approaches. The growth techniques of nanowires discussed in this section are based primarily on bottom- up approaches. As compared with top-down techniques, the bottom-up approaches would go far beyond the limits of top-down technology in terms of future physical and economic limits [30]. Therefore, they are capable of creating nanoscale features. In general, a self-assembly growth could be a self-limited process. For example, it could be a growth spatially defined by the template or a self-ordering process established on a dynamic balance of two opposite physical or chemical interactions such as attractive and repulsive forces, diffusion and dissolution, and so on.

2.1 Growth mechanisms

The growth mechanisms for the most of semiconducting oxide nanowires include vapor-liquid-solid (VLS), vapor-solid (VS) mechanisms and so on.

2.1.1 VLS mechanism

Among the various mechanisms for the growth of semiconducting oxide nanowires, VLS mechanism is the well established and the most widely accepted model. As mentioned, the VLS mechanism was reported by Wagner and Ellis in the mid-1960s to explain the growth of Si whiskers using Au as a metal catalyst [31]. The name of VLS mechanism refers to the fact that the source material from the vapor passes through a liquid droplet and finally ends up as a solid. So, the VLS process can be divided into three steps. The first step is the formation of the liquid alloy droplet which contains Au catalyst and source materials. The second step is the crystal nucleation upon gas adsorption and supersaturation. The last step is the axial growth from the crystalline seeds to form nanowires [32]. Figure.1 illustrates the growth of nanowires by a VLS process [6]. According to the VLS mechanism, we know that the liquid phase is formed initially due to the presence of a low-melting-point phase in an alloy system which consists of the substrate and gas phase constituent. The liquid surface adsorbs the reactant gaseous species much more efficiently than does the solid surface. On supersaturation of the liquid alloy, a nucleation center forms and serves as a preferred site for the axial growth of the nanowires. The adsorbed gas reactants then diffuse through the liquid phase to the solid-liquid interface, resulting in the growth of the solid phase. The growth at the solid-liquid interface is much faster than at the solid-vapor interface due to the much larger accommodation coefficient of the reactants in the liquid [33-34]. During this process, a vapor phase reactant is solubilized by a liquid catalyst particle to form solid wire-like structures and the catalyst is envisioned as growths site that defines the diameter of nanowires. Therefore, the characteristic of VLS growth is the existence of gold nanoparticles on the top of the obtained nanowires [35], typically observed by transmission electron microscopy as shown in Fig.2. At the same time, the diameter of the nanowires is well controlled by the size of the gold nanoparticle as a result of the fact that the diameter of a nanowire via VLS growth is primarily determined by the liquid alloy droplet.

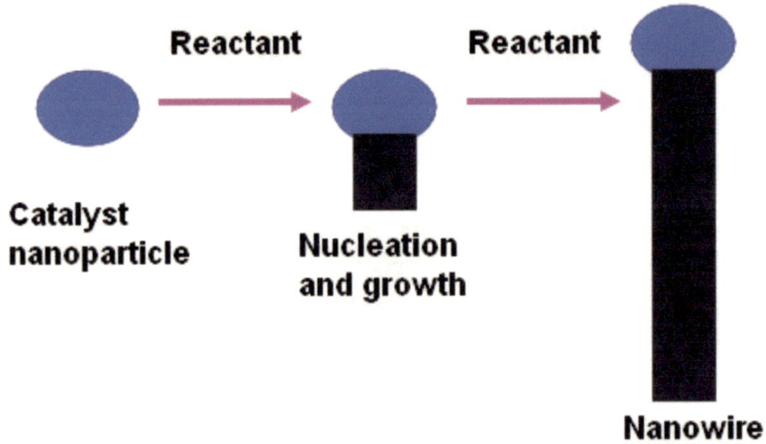

Fig. 1. Schematic illustrations of the growth process for a VLS process.

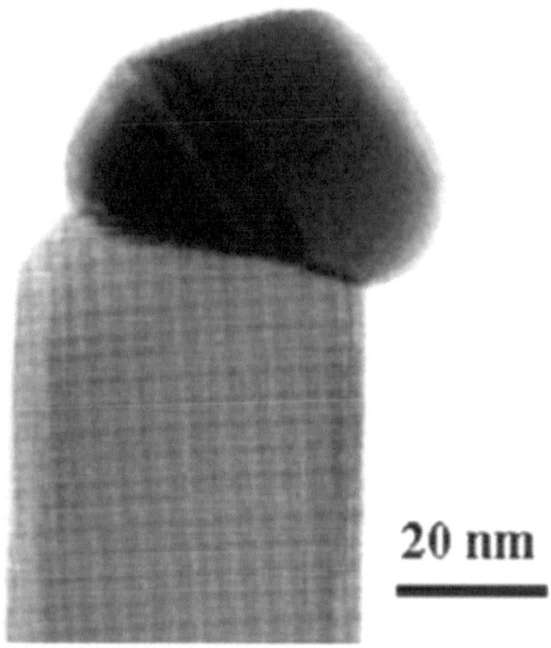

Fig. 2. TEM image of a tin-doped indium oxide nanowire grown along the (001) direction. (From ref. [35])

2.1.2 VS mechanism

VS mechanism is a catalyst-free process that forms nanowires. It involves the direct vaporization of the solid at a higher temperature followed by the deposition of the nanowires at a lower temperature. During VS growth, no catalyst is used and the nanowires are directly grown on the solid particles. This simple method has been widely used to synthesize a lost of semiconducting oxide nanowires [12, 36-37].

2.2 Growth techniques

Semiconducting oxide nanowires with various compositions have been synthesized using a wide variety of methods, from the simple thermal evaporation to the complicated epitaxial growth techniques. Table 1 gives a summary of the methods and mechanisms for semiconducting oxide nanowires growth. The common technique reported is vapor phase transport, including thermal evaporation, pulse laser deposition and chemical vapor deposition (CVD) [38-39]. Other techniques include template-assisted growth [40] and solution synthesis [41]. We devote our most attention to vapor phase transport and thermal evaporation methods.

Materials (nanowire)	Growth method	Growth mechanism	Ref
ZnO	Directly evaporation	VS	36,37,12,43, 45,46
Cd-ZnO	Thermal evaporation	VS	42,44
SnO_2	Laser ablation	VLS	49
SnO_2	Thermal evaporation	VLS	47
Sb-SnO_2	Thermal evaporation	VLS	48,52,54,55,18
Sb-SnO_2	Vapor-transport	VLS	10, 11, 56
Ta-SnO_2	Vapor-transport	VLS	53
In_2O_3	Laser ablation	VLS	50
Sn-In_2O_3	Vapor-transport	VLS	35
Sn/Mo-In_2O_3	Vapor-transport	VLS	51
Vertically aligned Sn-In_2O_3	Thermal evaporation	VLS	23
β-Ga_2O_3	Chemical vapor deposition	VLS	38
$ZnGa_2O_4$	Chemical vapor deposition	VLS	39

Table 1. A summary of the methods and mechanisms for semiconducting oxide nanowires growth

2.2.1 Vapor Phase Transport

Vapor phase transport (VPT) is an atomic deposition process. During this process, source species physically or chemically vaporized from a solid source are transported onto a substrate where they condenses and deposits. According to the approaches for the source vaporization, vapor phase transport can be in the form of thermal evaporation [42-48] and laser ablation [49-50]. When the source species are gas, evaporating liquids or chemically gasified solids, the deposition process is categorized as chemical vapor deposition [38-39]. So far, a majority of the vapor deposition techniques have been developed for the growth of single-crystal semiconducting oxide nanowires. The growth mechanism of VPT has been attributed to either VLS or VS, depending on whether the metal catalysts are presented or not.

2.2.2 Thermal evaporation

Thermal evaporation is one of the simplest methods available to synthesize semiconducting oxide nanowires. Deposition of nanowires occurs on the substrates placed downstream of the carrier gas (such as nitrogen and argon).

2.3 Summary

Among various growth mechanisms, the VLS mechanism has been combined with a wide variety of growth techniques, including catalyst-assisted thermal evaporation, solution and VPT method. Through carefully controlled growth conditions, the vertically aligned tin-doped indium oxide nanowire arrays have been demonstrated using the thermal evaporation. The solution growth method is of low cost, however it is complicated in terms of the control of the growth process. So far, VPT is the widely adopted method, probably due to the simplicity of the required facilities. At the same time, the VLS growth also leads to the incorporation of doping impurities in the as-grown semiconducting oxide nanowires.

3. Doping of semiconducting oxide nanowires

3.1 Doping

Doping in semiconductor has been a very well established process. The doping means the introduction of impurity atoms in a semiconductor, which is fundamental to the control of the bulk properties of a semiconductor. Impurity atoms incorporated into a semiconductor can only exist in two ways. For the first case, the impurity atoms replace the semiconductor atoms at their lattice positions. While for the second case, the impurity atoms exit the semiconductor atoms at their lattice of gap positions, which is a gap type of doping. Generally, the size of the replace atoms is similar to the size of substituted semiconductor atoms, while the size of gap atoms is much smaller. Figure.3 shows the schematics for a replace impurity atom and a gap impurity atom.

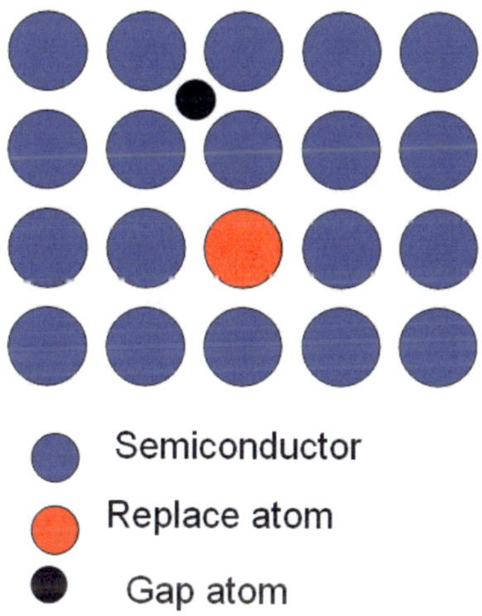

Fig. 3. The schematics for a replace impurity atoms and gap impurity atoms.

With the Incorporation of donors, a semiconductor becomes an n-type conductor. While with the incorporation of acceptors, the semiconductor becomes a p-type conductor. Whether the dopants successfully change the density of carriers or not depends on not only the energy of the electrons in the dopant atom but also the temperature. Incorporation of both donors and acceptors in a semiconductor does not lead to the increased concentrations of both electrons and holes since the density of the electrons and the density of holes are equal to the intrinsic density in the doped semiconductors. Typical doping concentrations are in the range of $10^{15}/cm^3$ to $10^{19}/cm^3$. This is relatively small as compared with the density of lattice atoms, which is about $10^{23}/cm^3$. The introduction of intentional impurities or dopants to control the behaviors of materials lies at the heart of many technologies. The impurity doping is an effective method for manipulating the physical properties of semiconductors, including the electronic, optical, and magnetic properties. A substitutional impurity with one more valence electron than the host atom it replaces can be ionized by thermal energy and donate its extra electron to the semiconductor (n-type doping). Similarly, an impurity with one less valence electron can provide an extra hole (p-type doping). These electrons and holes are then available as carriers of a electrical current. Since a thin conducting film is required for the applications of the nanocrystals, the ability to introduce these carriers is becoming important. So, doping process is critical for semiconductors. For these reasons, researchers have begun to explore the effects of dopants on the semiconductor nanocrystals with unusual and size-specific electronic, optical, and magnetic behaviors. However, extending doping process from films to nanoscale counterparts still remains a big challenge because of both fundamental synthetic issues and "self-purification" mechanisms. Moreover, dopants in nanocrystals are also expected to exhibit phenomena not found in the bulks since their electronic states are confined to a small volume.

3.2 The electrical properties of the doped semiconducting oxide nanowires

3.2.1 Tin-doped indium oxide

For unintentionally doped semiconducting oxide nanowires, the carriers are normally generated by structural defects such as oxygen deficiencies. As a result, the devices behave as wide band gap semiconductors whose performance is influenced by the surrounding environment. On the other hand, intentional doping can greatly modify the device properties and yield new device applications. One such example is tin-doped indium oxide (ITO), in which metal-like behavior is achieved when In_2O_3 is degenerately doped by Sn. Due to its high conductivity and high transmittance in the visible spectral region, ITO has become by far the most important transparent conducting oxide material, and ITO films have found applications in various optoelectronic devices such as flatpanel displays, solar cells, and light-emitting diodes. The ability to obtain highly transparent and highly conducting ITO nanowires may potentially further enhance the performance of such devices due to the increased effective device area using nanowire electrodes. Furthermore, the highly conducting ITO nanowires may also be used as interconnects in integrated nanocsale devices.

The growth of ITO nanowires/nanorods has been reported by several groups since the first study on semiconducting oxide nanobelts in 2001.[57] However, detailed electrical characterization was rarely reported, and it is not clear whether these ITO nanowire/nanorods have the desired electrical properties. In 2004, Q. Wan et.al. reported the resistivity value of ITO nanowires to be ~0.4 $\Omega\cdot$cm [58]. Subsequently, they optimized the doping growth and electrical characterization, the lowest resistivity value of the individual ITO nanowire was estimated to be 6.29×10^{-5} $\Omega\cdot$cm, which is comparable to the best values achieved in high-quality ITO films.[23, 35,51]

The electrical transport properties of the ITO nanowires were studied on devices consisting of individual nanowires. The nanowires were first removed from the growth substrate by sonication in isopropyl alcohol and deposited onto a degenerately doped n-type silicon substrate capped with a 50 nm silicon dioxide layer. Photolithography or electron-beam (e-beam) lithography processes were used to define pairs of metal electrodes on the SiO_2/Si substrate, followed by metal deposition of Ti/Au (10 nm/100 nm) by electron beam evaporation to complete the device structure. Prior to metal evaporation, the samples were cleaned with O_2 plasma (50 W, 30−60 s) to remove possible resist residue. Postannealing processes were not performed in the devices.

More than 40 ITO nanowire devices were fabricated and the electrical properties of individual nanowires were investigated both in ambient air and in vacuum (5 × 10-5 Torr). A linear current (Ids) versus voltage (Vds) curve (Figure 4a) was observed in all measurements and on samples defined by photolithography as well as e-beam lithography methods. This indicates that good Ohmic contacts can be readily achieved between the ITO nanowire and Ti/Au electrodes. To further characterize this observation, four-probe measurements were carried out to study the effects of the contacts. As shown in Figure 4a, the $I_{ds}-V_{ds}$ curve obtained from the four-probe method is almost identical to that obtained from the two-probe method using the inner pair of electrodes.

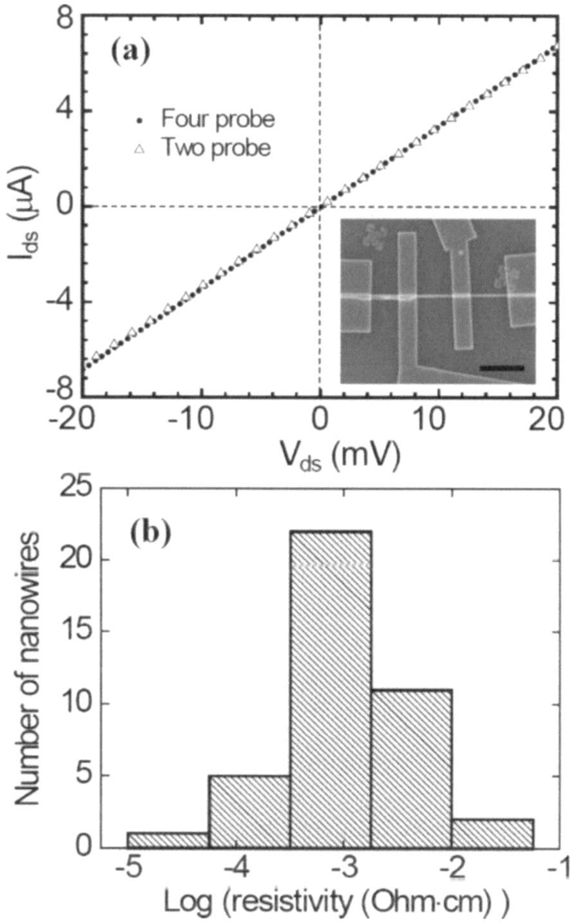

Fig. 4. (a) Comparison of the current–voltage characteristics of an individual ITO nanowire using four-probe and two-probe configurations. Inset, SEM image of the ITO nanowire device. Scale bar: 2 μm. (b) Histogram of the resistivities measured over 40 ITO nanowire.

The small difference between the measured resistances 2936 Ω (four-probe) and 2984 Ω (two-probe) can be mostly attributed to the resistance of the metal leads (~55 Ω) connecting the nanowire devices to the contact pads, and verifies that contacts to the ITO nanowires are indeed Ohmic with negligible values of resistance. The ability to produce reliable Ohmic contacts is very desirable in nanoscale device applications and also affords us to focus our studies on the simpler two-probe device structure in the following discussions. From the measured resistance value of the device in Figure 4a and the cross-section size (75 nm) and the length (1.8 μm) of the ITO nanowire (measured by SEM imaging), we can calculate the resistivity of the ITO nanowire to be 9.18×10^{-4} Ω·cm. The resistivities of about 40 ITO nanowires were obtained and plotted in Figure 4b. Significantly, the median resistivity value 7.15×10^{-4} Ω·cm, and lowest resistivity value 6.29×10^{-5} Ω·cm, are comparable to the best values achieved in high-quality ITO films.

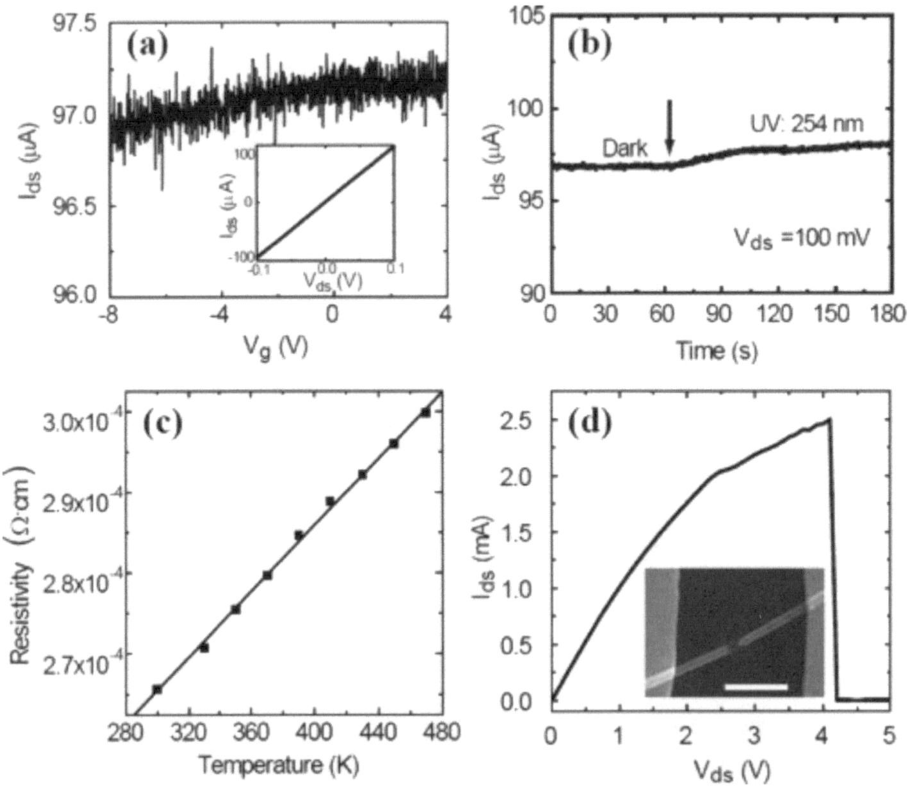

Fig. 5. Electrical characterization of a single ITO nanowire device. (a) Transfer curve (I_{ds}−V_g) of the ITO nanowire device at V_{ds}=100 mV. Inset, Ids−Vds curve of the device. The two-probe resistance of the device is ~1.05 kΩ. (b) UV response of the same device. (c) Temperature-dependent resistivity values of the ITO nanowire (300−470 K). The solid line is a linear fit. (d) I_{ds}−V_{ds} curve for the same ITO nanowire at large biases. The nanowire can carry a current of more than 2.5 mA before breaking down. Inset, SEM image of the failed ITO nanowire, showing that failure occurs in the middle of the nanowire. Scale bar: 500 nm.

Such high Sn doping levels will in fact render the ITO nanowires degenerately doped. Indeed, metal-like behaviors were observed in these ITO nanowire devices, as shown in Figure 5. Figure 5a shows the gate response curve of an ITO nanowire device with 90 nm lateral size and 3.2 μm channel length. The current Ids shows very little relative change as the gate voltage Vg is changed from −8 to 4 V. As a first-order estimation, the carrier density of the ITO nanowire can be calculated to be ~4.3×10^{20} cm^{-3} from the slope of the Ids−Vg curve. Such high carrier densities easily put the ITO nanowires in the degenerately doped regime. Assuming all Sn atoms are activated at room temperature, the Sn:In ratio can then be estimated to be 1.4:100, on the same order of the EDS estimations. Considering the simplicity of the model being used to calculate the carrier density and the small Sn signal used in the EDS estimations, such level of agreement is remarkable and verifies that a Sn-doping level of a few percent has been achieved in the ITO nanowires. The high doping level also explains the lack of UV response of these ITO nanowire devices (Figure 5b) and the insensitivity to the ambient environment. Contrary to undoped In_2O_3 nanowires which are very sensitive to UV light due to photogeneration of carriers, the ITO nanowire device shows only ~1.0% change in conductance when illuminated by UV light, even though the photon energy of the applied UV source (254 nm, 4 W) is clearly above the band gap of ITO. This lack of UV response is a direct result of the high starting carrier concentration. The metallic behavior was further verified by temperature-dependence measurements (Figure 5c), in which the resistivity of the ITO nanowire increases linearly from 2.66×10^{-4} to 2.99×10^{-4} $\Omega\cdot cm$ as the temperature is increased from 300 to 470 K. This behavior agrees well with the expected linear resistivity−temperature relationship at high temperatures for a metal when scattering is dominated by electron−acoustic phonon scattering. Detailed temperature dependence studies on another device with room-temperature resistivity of 6.69×10^{-5} $\Omega\cdot cm$ show that the resistivity of the ITO nanowires over a wide temperature range (50−300 K) can be well described by the Block−Grüneisen formula, as expected for a metal or a degenerately doped semiconductor. Finally, these "metallic" ITO nanowires can also carry a very high current density due to the single-crystalline nature. As shown in Figure 5d, the ITO nanowire device can endure a current of 2.5 mA before failure, corresponding to a current density of 3.1×10^7 A/cm^2.

3.2.2 Antimony -doped tin dioxide

Tin dioxide (SnO_2) represents an important metal oxide semiconductor that can be suitable for a range of applications through the incorporation of dopants, the electrical properties of SnO_2 nanowires as a function of antimony (Sb) doping were studied by Q. Wan [10]. Undoped, lightly (<0.5% Atomic mole ratio) and heavily (2–4%) Sb-doped SnO_2 nanowires were synthesized by a vapor-transport method via VLS mechanism under the same conditions. Figure.6 (a) shows a low-magnification transmission electron microscopy (TEM) image of the as-synthesized heavily Sb-doped SnO_2 nanowires. The catalyst nanoparticles are clearly visible on the tips of the SnO_2 nanowires, which confirm the VLS growth mechanisms. High-resolution TEM (HRTEM, Figure.6 (b)) confirms that the SnO_2 nanowires are single crystals with a tetragonal rutile structure.

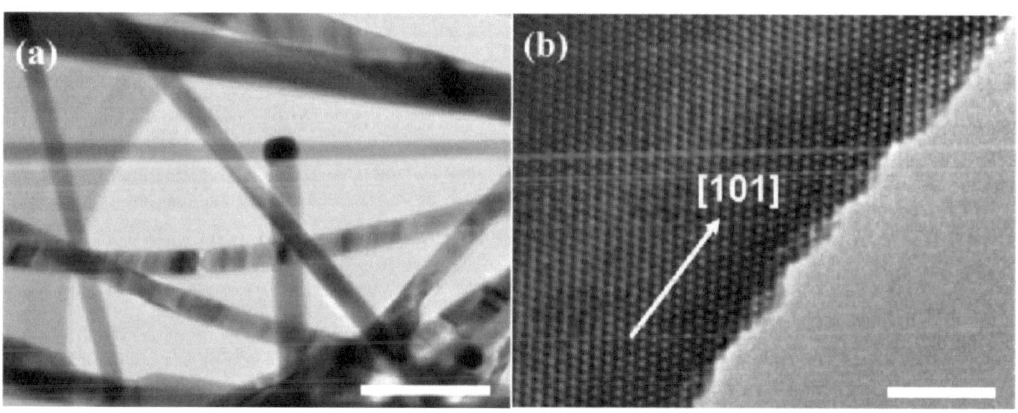

Fig. 6. (a) Low-magnification TEM image of degenerately Sb-doped SnO_2 nanowires. Scale bar: 100 nm. (b) HRTEM image of a single SnO_2 nanowire. Scale bar: 5 nm. (From ref. [10]).

Figure.7 (a) shows the current–voltage characteristics of an undoped SnO_2 nanowire device. A rectifying behavior (curve a) is normally observed in the air without illuminations, which indicates the formation of Schottky barriers at the metal/semiconductor interface. Upon illumination by UV light (254 nm), a linear current–voltage (ohmic-like) response (curve c) was observed along with a much higher current level. The ohmic-like behavior can be explained by enhanced electron tunneling at the metal/semiconductor interface due to an increase of the carrier density and the reduction of the depletion width upon illumination with above-bandgap photons, as reported for ZnO nanowire Schottky diodes [59-60]. Figure 7(b) shows the response of the device as a function of time (t) as the UV source was switched on and off. The current rapidly increases from 2 pA to 1mA upon UV illumination. The device can be reversibly switched between the low and high-conductance states with the response and recovery times estimated to be 1 s. The photocurrent sensitivity was estimated to be 5×10^5, a value much higher than that reported for undoped SnO_2 nanowire devices with ohmic contacts [49], due partly to the small dark current.

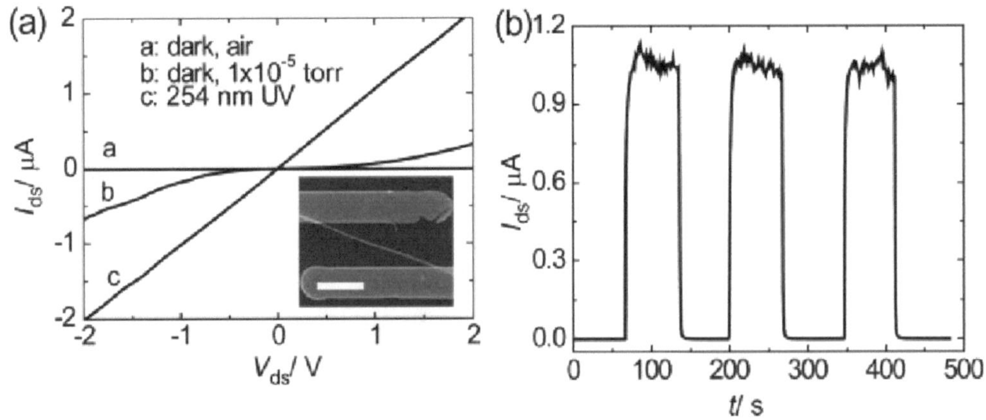

Fig. 7. Electrical properties of an undoped SnO_2 nanowire. (a) I_{ds}-V_{ds} curves of a nanowire device measured in the dark (curve a, in air and curve b, in vacuum) and exposed to 254-nm UV (curve c, in air). Inset: scanning electron microscopy (SEM) image of the SnO_2 nanowire device. Scale bar: 3mm. (b) Reversible switching of the nanowire device between low- and high-resistivity states as the UV light was turned on and off at different times t. The voltage bias on the nanowire was 1V (From Ref. [10]).

Form the above results, it could be concluded that the undoped SnO_2 nanowire devices show very poor transistor behavior in the air due to their high resistivity and the formation of Schottky contacts. The electrical performance of SnO_2 nanowires can be significantly improved by Sb doping. Figure 8 (a) shows typical I_{ds} -V_{ds} curves at different gate–source (Vgs) for a lightly Sb-doped SnO_2 nanowire configured as a field-effect transistor (FET) device. Linear behaviours in I_{ds} -V_{ds} curves are always observed for Sb-doped SnO_2 nanowire devices at low biases, which indicate low-resistance ohmic contacts between the nanowire channel and the Ti/Au electrodes. I_{ds}-V_{gs} curves obtained for the Sb-doped SnO_2 nanowire FET show that the device operates in the depletion mode as a result of the effective n-type doping with a transconductance (g) of 236 nS and an electron mobility of 550 $cm^2V^{-1}s^{-1}$ at V_{ds} = 0.1 V. The on/off ratio and sub-threshold slope for the Sb-doped SnO_2 FET device are 1×10^5 and 0.17 V decade^{-1} at Vds = 0.1 V, respectively. These excellent electrical properties confirm that the lightly doped SnO_2 nanowires are well-suited for transistor applications [54-55].

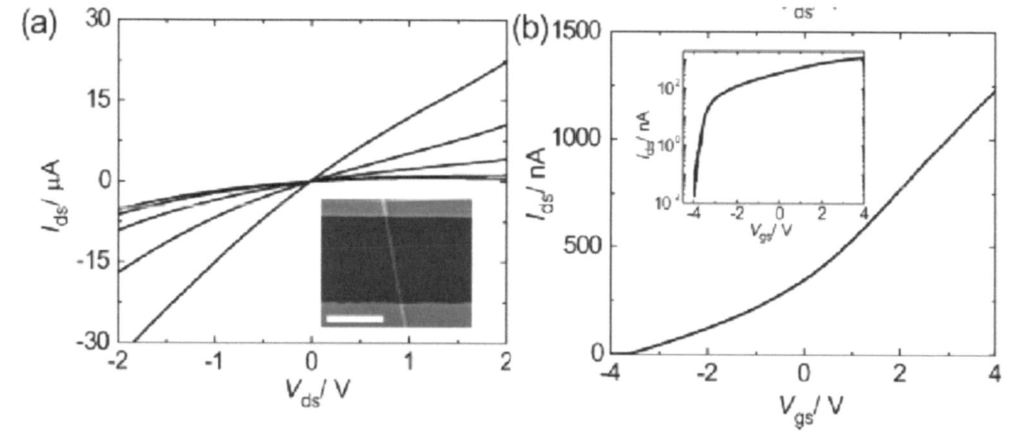

Fig. 8. Electrical properties of a lightly Sb-doped SnO_2 nanowire configured as a FET device. (a) Family of I_{ds} -V_{ds} curves at different gate–source Vgs. Vgs was varied from +4 to -4 V in steps of -2V and the device was measured in air at room temperature. Inset: SEM image of the device. Scale bar: 1.5mm. (b) Transfer (Ids -Vgs) curve of the same nanowire device at Vds =0.1 V. Inset: semilog plot of the Ids -Vgs curve (From Ref. [10]).

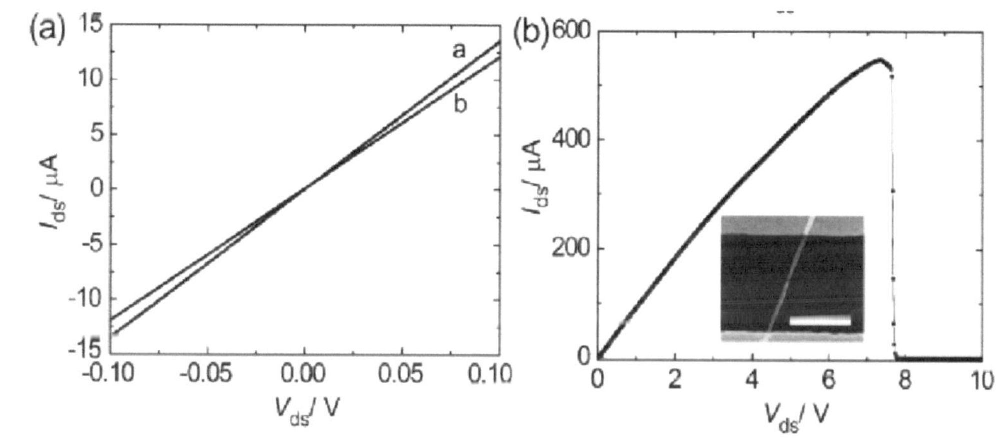

Fig. 9. (a) I_{ds} -V_{ds} characteristics of a SnO_2:Sb nanowire device measured in air without (curve b) and with (curve a) illumination by 254-nm UV light. (b) I_{ds} -V_{ds} curve recorded for the same SnO_2:Sb nanowire device, which shows the breakdown at very high current densities. The SnO_2:Sb nanowire can carry a current of more than 0.5 mA before breaking down. Inset: SEM image of the failed nanowire, which shows that failure occurs in the middle of the nanowire. Scale bar: 1 mm (From Ref.[10]).

The SnO_2 nanowires can be further doped to show metallic behavior by increasing the Sb:Sn ratio in the source during nanowire growth. The electrical properties of a degenerately (2–4 at%) Sb-doped SnO_2 (SnO_2:Sb) nanowire are shown in Figure 9 (a) with an estimated resistivity of $5.8 \times 10^{-4}\ \Omega$ cm in the dark. Due to the much higher carrier concentration, the SnO_2:Sb nanowires show a much smaller response (about 11.0%) to the same UV light compared to that for the undoped nanowires. As reported earlier [56], the SnO_2:Sb nanowires can also endure a high current density before failure. The failure current of 0.572 mA in Figure 9 (b) corresponds to a current density of $1.95 \times 10^{-7} A/cm^2$. The low resistivity and high failure density in turn make the SnO_2:Sb nanowires desirable for applications in nanoscale interconnects and electron field emitters.

These results show that Sb doping has significant influences on the electrical properties of SnO_2 nanowires. Undoped SnO_2 nanowires are hardly conducting and form Schottky contacts with metal electrodes. The nanowire Schottky devices are suitable in UV photodetector applications. On the other hand, lightly Sb-doped SnO_2 nanowires are well-suited to high-mobility transistor applications. A further increase in the doping level will result in the degenerately Sb-doped SnO_2 nanowires that show metallic behaviors with resistivity as low as $5.8 \times 10^{-4}\,\Omega$ cm and failure-current density as high as 1.95×10^{7}A/cm^2.

3.3 The optical properties of doped semiconducting oxide nanowires

Due to the wide gap of 3.37 eV and the large exciton binding energy of 60 meV, ZnO has been recognized as a promising photonic material in the blue-UV region [61]. ZnO has been widely used for its optoelectronic properties [62]. At the same time, the dopants can also strongly influence optical properties of the nanowires. Kouklin's group studied the optical properties of Cu-doped ZnO Nanowires [63]. The Cu dopants have been found to be both electrically and optically active within the nanowires, and more significantly are also able to act as visible-light photoconduction activators. Energy dispersive spectroscopy (EDS) measurements of the nanowires are shown in Figure.10 and point to the presence of Zn, O, and Cu (the Si signal arises from the substrate) in the nanowires. The EDS spectrum of nanowires grown by an analogous mechanism but outside the boat is presented in the inset of Figure.10. These results suggest that the incorporation of Cu is distance sensitive and stronger for samples placed in close proximity to the target (for the conditions used in this work). Owing to the very high temperatures (800°C) during the growth of the nanowires, doping has likely been achieved by two distinct routes: diffusion through sidewalls and/or direct deposition via solid–liquid interfaces present during VLS growth.

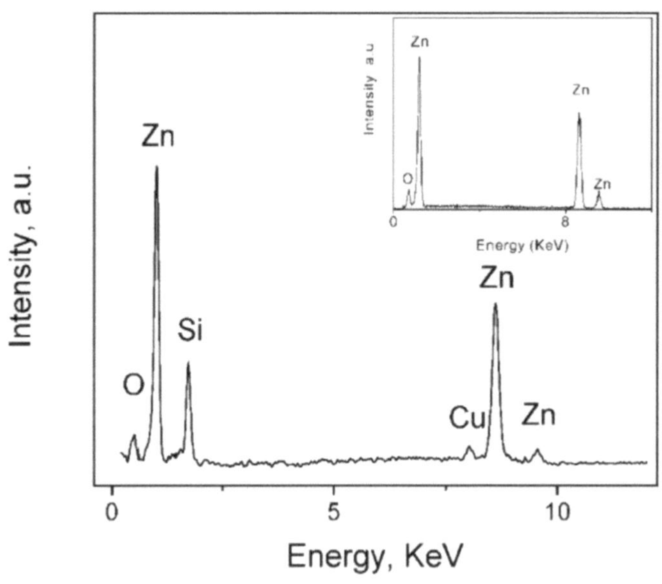

Fig. 10. EDS spectrum obtained for Cu:ZnO nanowires grown by VLS on top of a (111) Si substrate; the inset shows an EDS spectrum of ZnO nanowires grown on substrates located outside the boat (From Ref. [63]).

A typical PL spectrum is presented in Figure.11 and is clearly dominated by two peaks centered at 382 nm and 493 nm, referred to as free exciton and blue-green (BG) emission bands, respectively. A low-intensity shoulder peak at 455 nm has also been observed. The two main bands exhibit small red-shifts of 7 nm and 3 nm in their peak positions as compared to undoped samples (data not shown). Additionally, a two orders of magnitude increase in the ratio of BG to excitonic intensities of Cu:ZnO samples has been observed upon Cu doping. The appearance of a BG band and strong quenching of the exciton peak have been previously reported for thin films and nanowires of ZnO containing Cu impurities [64-65]. Therefore, the observed spectral changes upon doping can be primarily attributed to a drastic increase in the BG PL emission originating from the introduction of Cu centers.

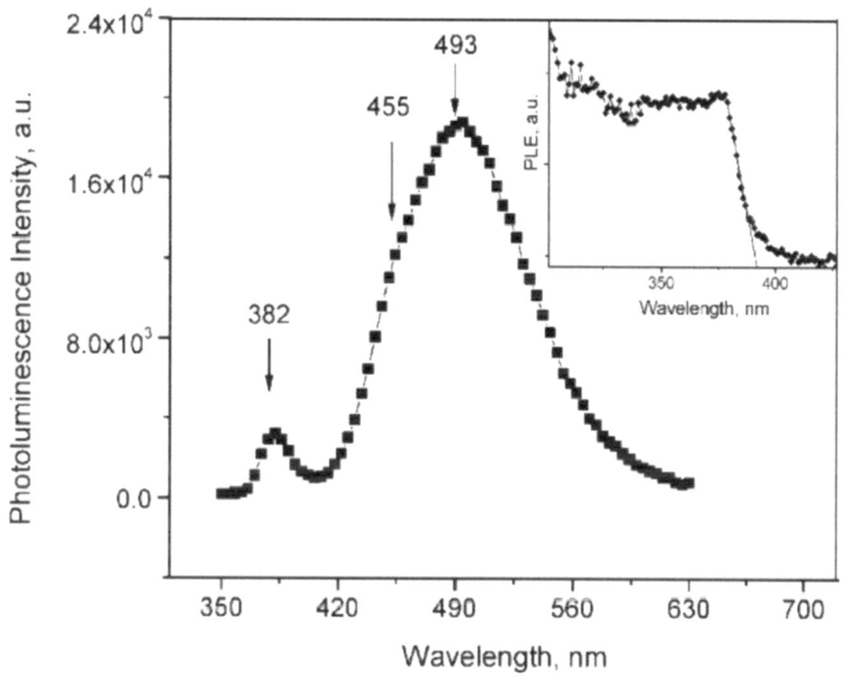

Fig. 11. RT micro PL spectrum of Cu:ZnO nanowires excited with a 325 nm light source. The inset displays the corresponding PLE spectrum with the emission monitored at 493 nm (From Ref. [63]).

PLE spectra have been obtained for Cu:ZnO nanowires at RT by monitoring their emission at 493 nm. The inset of Figure 11 shows the measured PLE data, clearly demonstrating the presence of a well-defined step in the emission intensity corresponding to an excitation wavelength close to 390 nm. The extrapolated intercept on the $\hbar\omega$-axis yields an effective bandgap of 3.17 eV, which is slightly red-shifted (by ca. 50meV) compared to that of pure ZnO nanowires. No light emission has been obtained for excitation energies less than E_{gap}, which is consistent with the previous results for Cu:ZnO materials. The studies presented here suggest the possibilities of engineering low-power nano-photodetectors and optical switches spanning multiple spectral ranges with improved gain characteristics by employing individual Cu-doped ZnO nanowires.

The effects of different doping level on the optical properties of doped ZnO nanowire need to be studied further. For example, the optical properties of S-doped ZnO nanowires were reported by Geng *et al* [66]. Figure.12 shows the results of the EDX measurements which were made on a different individual nanowire. The EDX pattern indicates that the nanowire is mainly composed of Zn and O with a small amount of S. S-doped ZnO nanowires with different content of sulfur have been obtained by changing the time of oxidation. Figures.12 (a), 12 (b), and 12 (c) correspond to 60, 90, and 100 min of oxidation time, respectively. When the time of oxygen stream lasted for 120 min, the pure ZnO nanowires have been obtained as shown in Figure.12 (d).

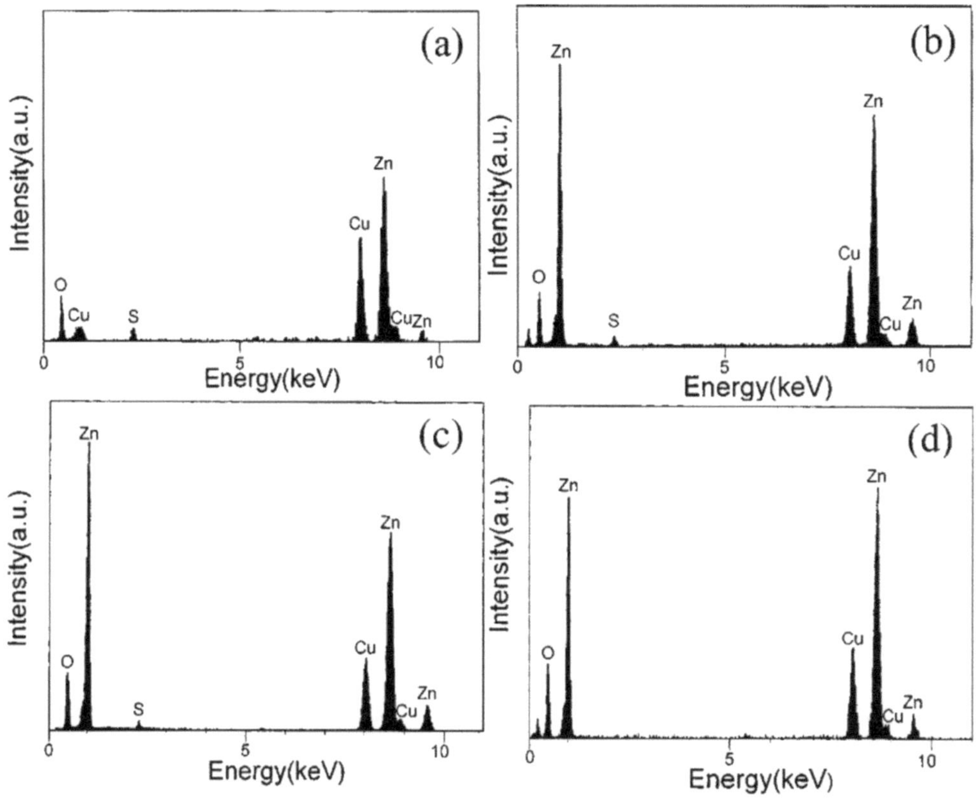

Fig. 12. EDX spectra of individual nanowire. Fig. 12 (a), 12 (b), 12 (c), and 12 (d) correspond to 60, 90, 100, and 120 min of oxidation time, and the corresponding content (atomic percent) of sulfur is 5.40%, 1.47%, 1.41%, and 0%, respectively (From Ref. [66]).

The room temperature photoluminescence spectra from ZnO nanowires and S-doped ZnO nanowires under the same experiment conditions are shown in Figure.13. Two luminescence bands have been observed for ZnO nanowire (curve d). One is an UV emission peak at 380 nm, which corresponds to the near band edge emission. The other luminescence band, a green light emission peak, was observed at 520 nm, which was attributed to the single ionized oxygen vacancy in ZnO. The PL spectra of S-doped ZnO nanowires with a different carrier concentration appear at an obviously blue shifted position. When ZnO is doped with S, the excess carriers supplied by the impurities to the conduction band contribute to increase the electrical conductivities of ZnO. Due to a small density of states of ZnO near the conduction-band minimum, the conduction band edge is filled by excessive carriers provided by the impurities, leading to a blue shift of optical band-to-band transitions, known as the Burstein-Moss effect. It is worth noting that the blue shift increases with increasing carrier concentration, and the changes of PL intensity are observed at the same time. The UV emission peak weakens and the green light emission is enhanced. When S enters the ZnO crystal lattice, it introduces lattice distortion. Since S has a larger Bohr radius than O, it influences the energy structure of ZnO. The band gap broadens for the reasons mentioned above leads to the blue shift of the PL spectra and changes the intensity of near band edge emission. The probabilities of forming a new band structure deformation increases with the carrier concentration, which gives rise to some new defects, such as oxygen vacancies, therefore should result in the changes of the PL intensities of green light emission.

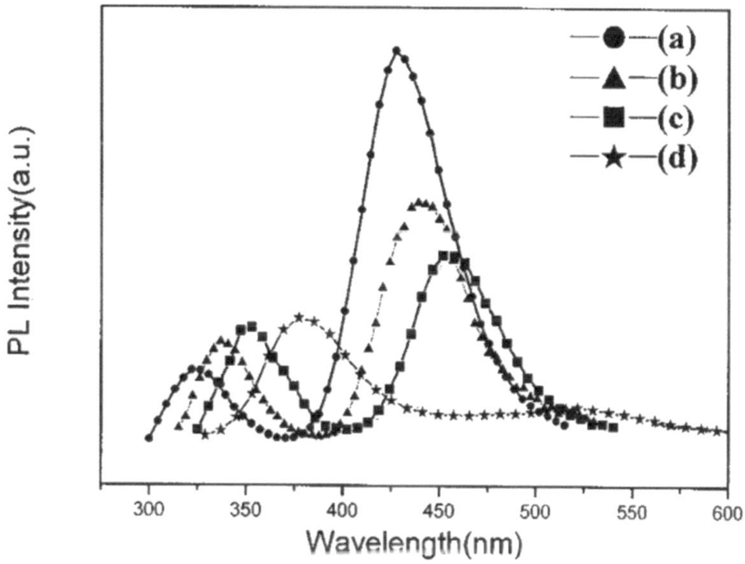

Fig. 13. The PL spectrum on a bulk quantity of nanowires measured at room temperature under 325 nm excitation. Curves (a), (b), (c), and (d) correspond to 60, 90, 100, and 120 min of oxidation time, and the corresponding content (atomic percent) of sulfur is 5.40%, 1.47%, 1.41%, and 0%, respectively (From Ref.[66]).

4. Applications of semiconducting oxide nanowires

4.1 Field-effect transistors based on semiconducting oxide nanowires

As one of the most functional components, field-effect transistors (FETs) using semiconducting oxide nanowires as active channels have been extensively investigated in the past few years [1-11]. A typical semiconducting oxide nanowire-based FET structure is shown in Fig. 14. In its simplest form, a semiconducting oxide nanowire is transferred onto a gate dielectric/gate electrode/substrate and is contacted with metallic source and drain electrodes. The nanowire and the gate electrode are capacitively coupled such that the application of a bias on the gate electrode induces charges in the nanowire. The voltage applied between the source and drain is referred to as the source-drain voltage, V_{DS}. For a given V_{DS}, the amount of current that flows through the nanowire from source to drain is a strong function of the gate voltage V_G. Since most of the oxide nanowires are n-type semiconductor materials, the gate electric field can induce electrons to be mobile in the nanowire channel of the FETs when applying positive voltage to the gate electrode. Thus, the device is "ON". While, a decrease in nanowire channel conductance is experienced when applying negative voltage to the gate electrode. Thus, the device is "OFF". Up to now, various semiconducting oxide nanowires have been studied for the fabrication of FETs, including ZnO [1-6], In_2O_3 [7-9], SnO_2 [10-11], and so on. In this section, some typical works on the FETs built on semiconducting oxide nanowires are summarized.

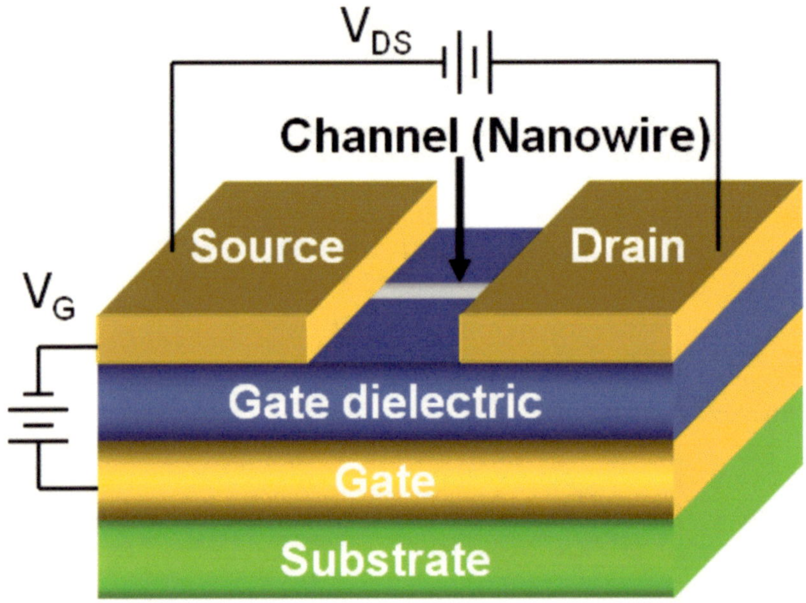

Fig. 14. Schematic image of the nanowire-based field-effect transistors

4.1.1 Device architecture of semiconducting oxide nanowire-based FETs

1. Back-gate FETs

Due to its fabrication simplicities, most of semiconducting oxide nanowire-based FETs are back-gated. The typical device fabrication processes are described as follows. Semiconducting oxide nanowires were synthesized by a variety of methods. Then, nanowires were transferred onto the degeneratively doped Si wafers coated with a layer of gate dielectric [2]. The source/drain electrodes were patterned by optical lithography and the metal contacts were deposited via electron beam evaporation. Figure 15 (a) shows a scanning electron microscope (SEM) image of a typical semiconducting oxide nanowire FETs. According to the cylinder-on-plate capacitance model, the back gate capacitance can be calculated using the relationship: $C_i = \dfrac{2\pi\varepsilon\varepsilon_0 L}{\ln(4h/d)}$.

Here L is the length of the nanowire device channel, ε_0 is the vacuum dielectric constant, h (nm) is the thickness of the gate dielectric layer, and d (nm) is the lateral size of the nanowire, ε is the relative dielectrics constant. The typical field-effect mobility of back-gate FETs is 10-100 cm^2/Vs. And these devices shows a reasonable current ON/OFF ratio of 10^3~10^6. Figure.15 (b) shows the output characteristic plot of the device. In general, the I_{DS} versus V_{DS} curves are linear at low bias, suggesting the ohmic contact at nanowire-electrode interface.

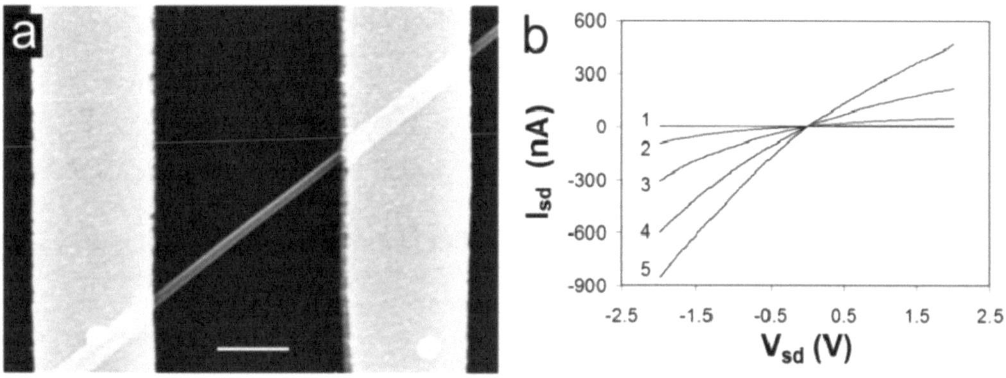

Fig. 15. (a) SEM image of a back-gated ZnO nanowire FET (scale bar: 1μm). (b) I_{DS} versus V_{DS} curves recorded at different gate voltages (From Ref. [2]).

2. *Top-gate FETs*

It is impossible to individually control each FET for the back-gated semiconducting oxide nanowire-based FETs, because heavily doped substrates are usually worked as the gate electrodes. Thus, top-gate nanowire FETs were developed. S. J. Pearton *et al* have reported single top-gate ZnO nanowire FET [3]. Figure 16 shows the SEM image of the top-gate nanowire FETs. The depletion-mode ZnO nanowire FET exhibits excellent pinch-off and saturation characteristics and a strong UV photo-response effect. A threshold voltage of ~-3 V and a maximum transconductance of 0.3 mS/mm were obtained. These devices look promising for the novel nanodevice applications requiring low leakage current.

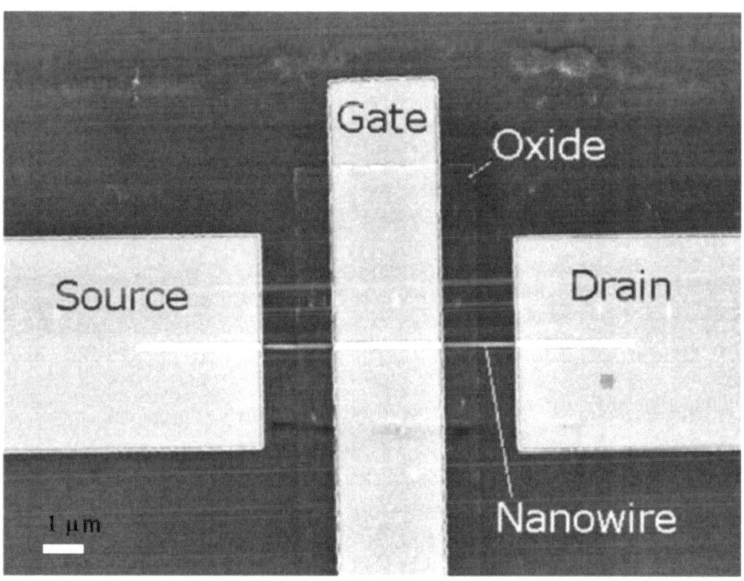

Fig. 16. SEM image of the top-gate nanowire FETs (From Ref. [3]).

3. Surround-gate FETs

As device dimensions continue to shrink into nanometer scale regime, fundamental physical limits and economic are likely to hinder further scaling according to Moore's law. New strategies including the usage of innovative device architectures are needed to extend the current capabilities. It is known that the fabrication of surround-gate FETs with Si nanowire channel was firstly achieved by the Lieber's group. In addition to the traditional back and top-gate FET, vertical surround-gate FETs was also achieved in semiconductor oxide nanowires by Hou T. Ng *et al.* The realization of a vertical surround-gate nanowire FETs takes the advantages of the vertical dimension unlike planar nanowire-based FETs and traditional metal-oxide-semiconductor (MOS) FETs [4]. Importantly, the device performance can be improved by applying this device architecture. Figure17 shows the fabrication processes of the vertical surround-gate FET. This fabrication method has allowed a lithography-free means of defining the vertical channel length and reduced the footprint of the devices since the source, drain and channel are stacked on top of each other vertically. The fabricated devices show a subthreshold swing of ~130 mV/decade and a large current ON/OFF ratio of larger than 10^5.

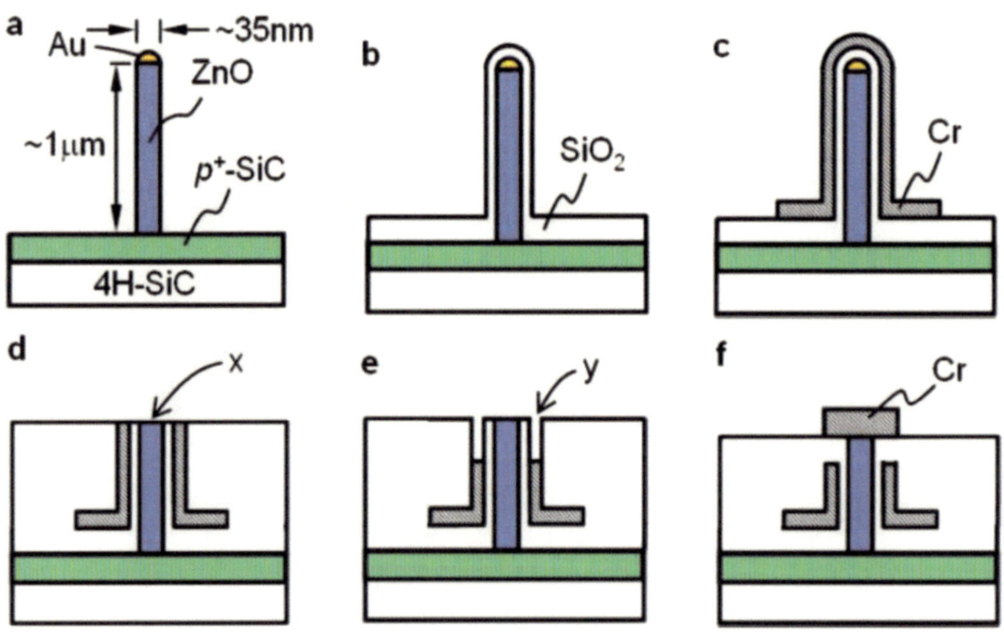

Fig. 17. The fabrication processes of the vertical surround-gate FET (From Ref. [4]).

4. Omega-shaped-gate FETs

Although surround-gate nanowire FETs are interesting and promising for novel nanodevice fabrications, there are only limited reports on the surround-gate nanowire FETs, due to the complexity associated with their fabrication processes. Theoretical simulations have predicted that omega-shaped-gate FETs should exhibit high device performances than surround-gate FETs, despite of their relatively simpler fabrication procedures. Thus, omega-shaped-gate device architectures were realized in semiconductor oxide nanowire-based FETs by Kihyun Keem *et al* [5]. Figure 18 shows the Schematic and top view SEM image of omega-shaped-gate FET fabricated by the photolithographic process. About 80% of the surfaces of the nanowires coated with gate dielectric are covered with the metal gate to form omega-shaped-gate FETs. This device showed high performance with a mobility of ~ 30.2cm^2/Vs, a peak transconductance of 0.4 μS, and a high current ON/OFF ratio of 10^7. The observed improvements of electrical performances are mostly attributed to the omega-shaped-gate geometry and the passivation of the surface of the nanowires.

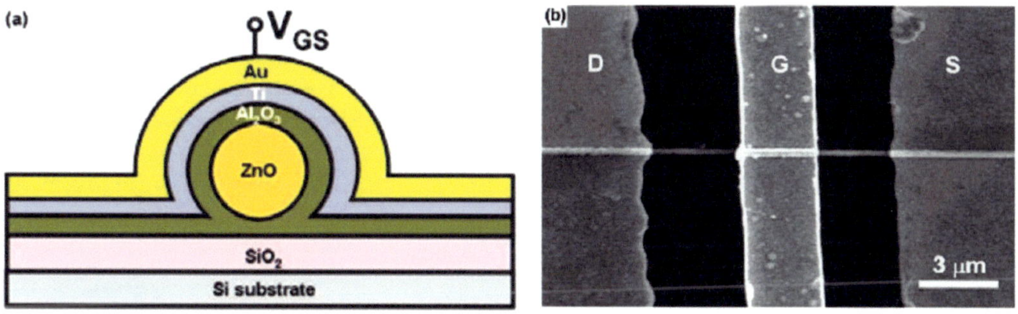

Fig. 18. (a) Schematic and (b) top view SEM image of omega-shaped-gate FET fabricated by the photolithographic process (From Ref. [5]).

4.1.2 Low-voltage semiconducting oxide nanowire-based FETs

Semiconducting oxides nanowire-based FETs are good building blocks for the functional nanodevices and are expected to lead to novel electronic and optoelectronic device applications [6]. To apply nanowires to various portable sensors and portable electronic device applications, there are several key challenges must be addressed [1]. In addition to the issues related to the integration of the nanowires into device structures with high yields and well-controlled contacts, development of compatible gate dielectrics with high specific capacitance and high gating efficiency is a significant issue in realizing device with low power consumption for portable applications. Some of the current strategies for increasing the gate specific capacitance have been reported, including using high *k* materials, reducing dielectric thickness and using solid electrolytes or ion gels as gate dielectrics.

1. High-k dielectrics gated nanowire-based FETs

The use of high-k gate dielectrics could result in the large capacitive coupling between the gate electrode and channel, therefore reducing the operating voltage of semiconductor oxide nanowire-based FETs. S. T. Lee group have demonstrated the fabrication of low-voltage Zn-doped In_2O_3 nanowire FETs with a SiN_x as a high-k dielectric [9]. The SiN_x dielectric was deposited by magnetron sputtering. The operating voltage of the FETs was reduced to as low as 2.0 V. The devices showed a high current ON/OFF ratio of ~10^6, a high ON-state current of 10^{-5} A, a small subthreshold swing of 120 mV/decade, and a near zero threshold voltage. Importantly, these devices show high reproducibility and stable performance. The high-performance nanowire-based FETs would enable the potential opportunities in the future multifunctional nanodevice applications.

2. Self-assembled organic nanodielectrics gated nanowire-based FETs

Most of the reported semiconductor oxide nanowires-based FETs have utilized gate dielectrics including thermally grown oxide layer and sputtering or chemical vapor deposited dielectric films. However, these gate dielectrics have limitations associated with high temperature processing. It is not suitable for low-power flexible electronic applications. Sanghyun Ju *et al* have reported that self-assembled organic nanodielectric has been used as the gate dielectric for individual ZnO nanowire FETs fabrications [1]. Figure 19 (a) shows the cross-sectional view of the self-assembled organic nanodielectric gated ZnO nanowire FET device structure. The self-assembled organic nanodielectric films used in their studies consist of layer-by-layer organic monolayers. Such dielectrics exhibit excellent insulating properties with a large specific capacitance of 0.18 μF/cm^2 and a low leakage current density of 10^{-8} A/cm^2. Figure 19 (b) and (c) show the output and transfer characteristic curves of the ZnO nanowire FETs with a self-assembled organic nanodielectrics. The fabricated device display excellent I_{DS} saturation at V_{DS} of 0.5 V, a threshold voltage of -0.4 V, a carrier mobility of 196 cm^2/Vs, a current ON/OFF ratio of 10^4, and a subthreshold slope of 400 mV/decade. The operating voltage of these devices is as low as 2.0 V. These results indicate that self-assembled organic nanodielectric gated ZnO nanowire FETs are promising for future flexible display and logic technologies.

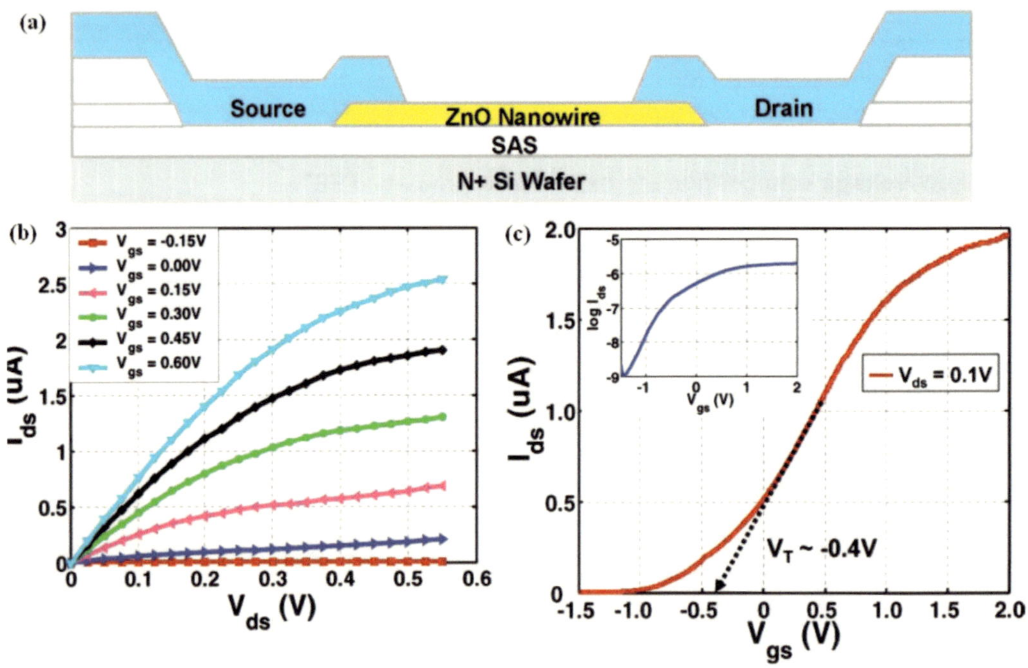

Fig. 19. (a) Cross-sectional view of the self-assembled organic nanodielectric gated ZnO nanowire FET device structure. (b) Output and (c) transfer characteristic curves of the ZnO nanowire FET with a self-assembled organic nanodielectric (From Ref. [1]).

3. *Solid-electrolyte gated nanowire-based FETs*

Another simple alternative for achieving low operating voltage is to exploit solid-electrolyte or ion gels as gate dielectrics associated with a large electric-double-layer (EDL) capacitance resulting from the short Debye lengths of a few nanometers. More importantly, these dielectrics gating eliminates hysteresis and provides a simple yet versatile method of studying fundamental electron transport in semiconductor oxide nanowire-based devices. Jia Sun *et al* have reported the fabrication of microporous SiO_2 solid-electrolyte gated low-voltage SnO_2 nanowire FETs by nickel mask method [11]. The operating voltage is found to be as low as 1.5 V due to the extremely large specific capacitance (~ 2 μF/cm²) related to the mobile ion-induced electric-double-layer effect.

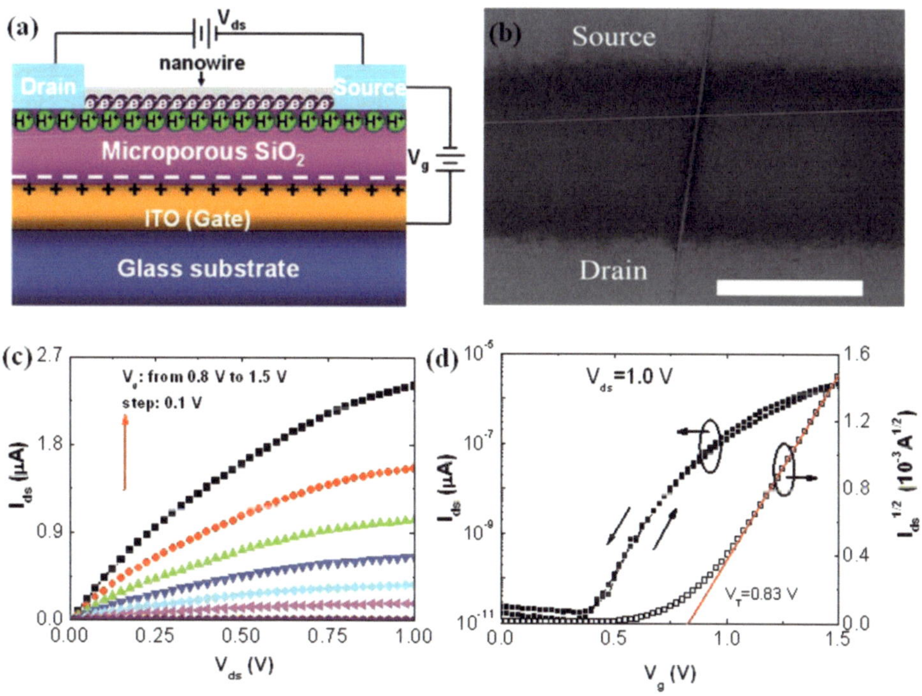

Fig. 20. (a) A schematic picture of the low-voltage operation of microporous SiO2 solid-electrolyte gated SnO_2 nanowire FETs. (b) A low-magnification SEM micrograph of the fabricated device. (c) Output and (d) transfer characteristic curves of the SnO_2 nanowire FET with a microporous SiO_2 solid-electrolyte gate dielectric (From Ref. [11]).

Figure 20 (a) shows the schematic diagram of the low-voltage operation mechanism of the SnO_2 nanowire FETs. When a positive gate voltage is applied, protons are drifted toward the SiO_2/SnO_2 nanowire channel interface where they accumulate over a Debye length L_D, and the positive charges induces an image charge and opposite sign in the SnO_2 nanowire channel layer, which is similar to the case of the EDL formation in organic transistors gated by ion gel or polyelectrolyte. Figure 20 (b) shows a SEM image of SnO_2 nanowire FET. Figure 20 (c) and (d) shows the output and transfer characteristic curve of the microporous SiO_2 solid-electrolyte gated SnO_2 nanowire FET. The field-effect mobility, current on/off ratio and sub-threshold swing are estimated to be 175 cm²/Vs, 10^5, and 116 mV/decade, respectively. We believe that such high-performance low-voltage SnO_2 nanowire transistors hold promise for novel device applications, such as portable ion-sensitive sensors.

4.1.3 Transparent semiconducting oxide nanowire-based FETs

Transparent electronics is a technology-in-a-hurry and its aim is to realize invisible electronic circuits in which transparent transistors are the most fundamental functional component. To achieve this objective and satisfy multifunctional applications in our daily life, the ideal transparent electronic devices must achieve high optical transparency and be compatible with diverse low-cost substrates. Traditional poly-silicon and amorphous-silicon devices as well as the circuits are widely used to fabricate the commercial display. However, their lack of transparency limits their utilities in transparent electronics. Transparent oxide nanowires with wide band gap could meet the above-mentioned requirements. Recently, some researchers have reported transparent transistors fabricated with ZnO, SnO_2 and In_2O_3 nanowires. David B. Janes group has reported the fabrication of fully transparent In_2O_3 and ZnO nanowire FETs on both glass and flexible plastic substrates [7]. Figure 21 shows the optical and electrical characteristics of fully transparent nanowires FETs. These devices exhibit high-performance n-type transistor characteristics with ~ 82% optical transparency. Fully transparent nanowire FETs are also attractive as pixel-switching and driving transistors in active-matrix organic light-emitting diode displays. The same group has firstly reported the demonstration of active-matrix light-emitting diode displays driven by the nanowire FETs and shows that such displays can be optically transparent. These results indicate that nanowire FETs devices strategies are promising for windshield displays, head-mounted displays, transparent screen monitors, mobile phones, and handheld personal computer.

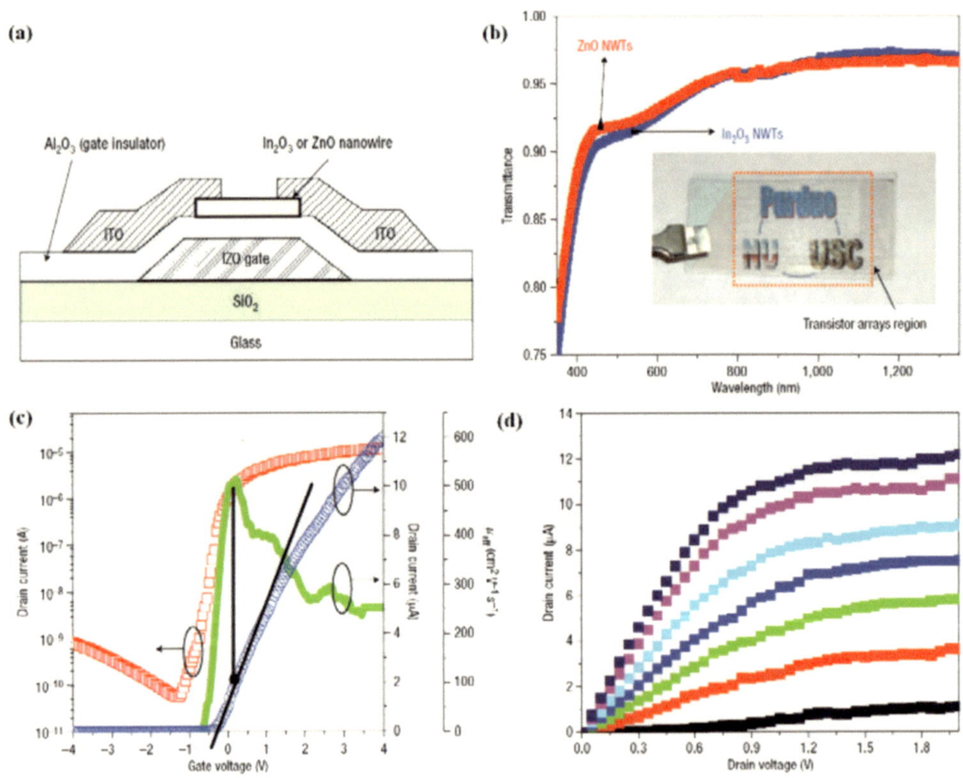

Fig. 21. (a) A schematic picture of fully transparent nanowire FETs. (b) Optical transmission spectra through entire nanowire FET sturctures. (c) Transfer and (d) output characteristic curves of the fully transparent nanowire FET (From Ref. [7]).

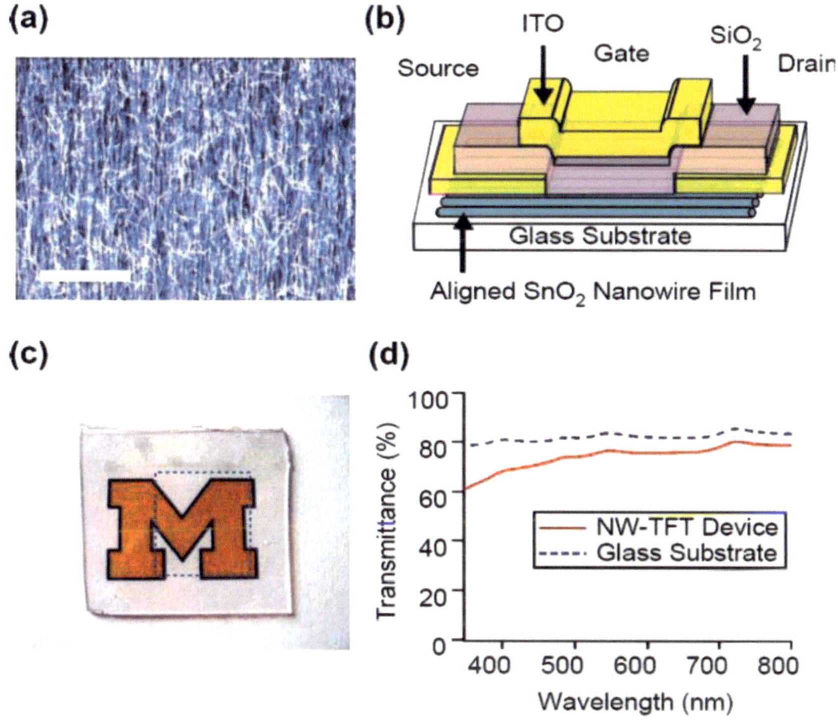

Fig. 22. (a) Dark-field optical microscope image of a SnO_2 nanowires films obtained through the physical transfer method. Scale bar: 20 μm. (b) Schematic of the transparent SnO_2 nanowire thin-film transistor device. (c) Digital photograph of SnO_2 nanowire thin-film transistor on glass. (d) Optical transmittance spectrum of the SnO_2 nanowire thin-film transistor device array (From Ref. [10]).

Semiconducting oxide nanowires have been used as interesting materials for the fabrication of high-performance transistors with electronic performance comparable to and in some cases exceeding that of single-crystal materials. A new concept of nanowire thin-film transistors has been proposed using oriented nanowire thin films as semiconductor channels. One of the key advantages of the nanowire thin-film transistors approaches compared to conventional thin-film transistor techniques is the clear separation of the device fabrication stage from the material growth stage. Therefore, it is no longer needed to concern with compatibilities with device substrates during growth. Wei Lu *et al* have fabricated fully transparent nanowire thin-film transistors based on lightly Ta-doped SnO_2 nanowires [10]. The nanowires-based devices exhibit uniform characteristics with average field-effect mobilities exceeding 100 cm^2/Vs. As shown in Figure 22, prototype nanowires thin-film transistors devices on glass substrates show excellent optical transparency and transistor performance. They found that fully transparent nanowires thin-film transistors exhibit similar high-mobilities even with low nanowire coverages.

At last, Table 2 summarizes the performance parameters of different doped and undoped semiconducting oxide nanowires based transparent FETs. We believe that these high-performance transparent nanowires FETs are promising for a broad range of existing applications from flat-panel displays to image sensor arrays and enable a whole new range of flexible, wearable, or disposable electronics.

Channel materials	Gate dielectric	Operation voltage (V)	Subthreshold swing (V/dec)	Mobility (cm^2/Vs)	On/off ratio	Ref
ZnO nanowires	18 nm/Al_2O_3	4	0.3	70~96	10^6	30
In_2O_3 nanowires	18 nm/Al_2O_3	4	0.16	300~514	10^6	30
In_2O_3 nanowires	22 nm/self-assembled nanodielectric	4	0.25	258	10^5	7
Ta-doped SnO_2 nanowires	75 nm/SiO_2	10	0.312	179	10^5	39
Zn-doped In_2O_3 nanowires	150 nm/SiN_x	2	0.12	631	10^6	35
As-doped In_2O_3 nanowires	50 nm/Al_2O_3	5	0.088	1490	5.7×10^6	32
Sb-doped SnO_2 NW	4.0 μm/microporous SiO_2 solid-electrolyte	1.5	0.116	175	10^5	41

Table 2. The performance parameters summary of different doped and undoped semiconductor oxide nanowire based transparent FETs.

4.2 Sensors based on semiconducting oxide nanowires

Solid-state sensors play an important role in medical diagnosis, environmental monitoring, chemical process controlling, personal safety and national security. There is a strong interest in the developments of high-performance sensors with high sensitivities, high selectivities and low detection limits. Semiconducting oxides have been known as good sensing materials due to their low cost and high compatibilities with microelectronic processing. In recent years, semiconducting oxide naowires have been attracted much attention due to their great potential applications. The novel properties of semiconducting oxide nanowires, associated with a large surface-to-volume ratio, finite or quantum size effects and a Debye length comparable to their dimensions, offer the basis for the exploration of new and interesting phenomena. The extremely large surface-to-volume ratio makes their electrical properties more sensitive to species. Superior stabilities of semiconducting oxide nanowires based sensors can be obtained duo to the high crystallinity. Up to now, various sensors have been already reported based on semiconducting oxide nanowires, including gas sensors, biosensors, and so on. Table 3 summarizes some typical sensors based on semiconducting oxide nanowires. These sensors have good response and recover characteristics. As one can see, the most widely studied substances are ZnO, SnO_2 and In_2O_3, probably due to their relatively simple synthesis and good sensitivities to species. In this section, we present a comprehensive perspective on research efforts made on these kinds of sensors based on semiconducting oxide nanowires. The fabrication and testing as well as the device performances of these sensors are outlined.

Types of sensor	Materials	Target	Detection limit	Response time	Ref.
Gas sensors	In_2O_3 nanowires	NO_2	5 ppb	1000 s	15
	ZnO nanowires	Ethanol	1 ppm (300°C)	10 s	12
	ZnO nanowires	NO_2	0.5 ppm (225°C)	24 s	14
	SnO_2 nanowires	Acetone	20 ppm (290°C)	7 s	67
	SnO_2 nanowires	Ethanol	0.5 ppm (300°C)	4 s	18
	In_2O_3 nanowires	NO_2	1 ppm (250°C)	20 s	73
	CuO nanowires	CO	5 ppm (room temperature)	10 s	68
	TeO_2 nanowires	NO_2 NH_3 H_2S	10 ppm (26°C) 100 ppm (26°C) 50 ppm (26°C)	10 min	69
	Single Zn-doped In_2O_3 nanowire	CO	2 ppm (room temperature)	20 s	70
	Single ZnO nanowire	H_2	500 ppm (room temperature)	55 s	72
Biosensnors	Single ZnO nanowire	streptavidin	2.5 nM	Not available	71
	Single ZnO nanowire	hemoglobin	2 fg/mL (negatively charged)	< 1s	20
	ZnO nanowires	Glucose	1 μM	10 s	16
Other sensors	Single SnO_2 nanowire	Humidity	30 % relative humidity (30°C)	120-170 s	19
	Single ZnO nanowire	UV	1.5 mW/cm^2	Less than 1 s	21
	Single ZnO nanowire	Strain	0.2 % strain	Less than 1 s	17

Table 3. Typical sensors based on semiconducting oxide nanowires.

4.2.1 Gas sensors based on semiconducting oxide nanowires

Up to now, semiconducting oxide nanowires based gas sensors are usually fabricated through resistors or FETs. These sensors are based on resistance changes caused by the exposure of the sensor surface to a target species. Chemisorbed gas molecules on a semiconducting oxide nanowires surface withdraw or donate electrons to the conduction channel, giving rise to a conductance, which is the fundamental working principle of these sensors. For the type of sensors based on resistors, the semiconducting oxide nanowire film is contacted with pairs of metal electrodes on a substrate of a ceramic tube. For the type of sensors based on FETs, a single or multiple nanowires bridge source and drain electrode on a substrate covered with gate insulator/gate electrode. The major advantages of these types of sensors are easy fabrication process, low cost and easy integration with heat transducers.

1. Semiconducting oxide nanowires gas sensors based on resistors

Wan *and co-workers* have demonstrated the fabrication of ethanol sensors based on ZnO nanowires by microelectromechanial system technologies. ZnO nanowires were ultrasonic dispersed in ethanol for 30 min and dried by an infrared light on a silicon-based membrane embedded with Pt interdigitated electrodes and a heater [12]. Figure 23 (a) shows the top-view SEM image of the fabricated sensors. Figure 23 (b) shows the sensing properties of ZnO nanowires upon exposure to 1-200 ppm ethanol in dry air at 300 °C working temperatures. The sensitivities at 200 ppm ethanol exposure is estimated to be 47, and at 1 ppm ethanol exposure to be 1.9. The response and recovery is relatively quick, which is desirable for practical application.

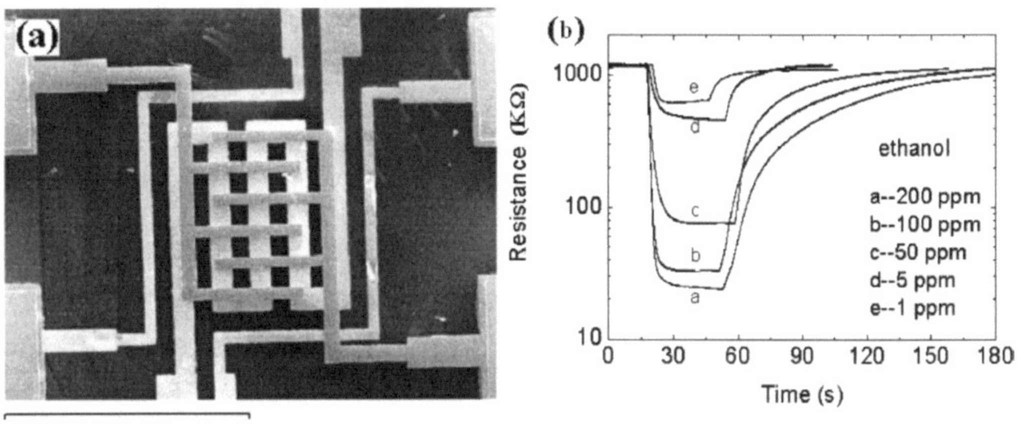

Fig. 23. (a) Top-view SEM image of the fabricated substrate embedded with Pt interdigitating electrodes and Pt heater (scale bar: 1mm). (b) Response and recovery characteristics of ZnO nanowires upon exposure to ethanol with a concentration of 1-200 ppm at 300°C (From Ref. [12]).

Recently, Wan and co-workers proposed an excellent performance of ethanol sensors based on the branched SnO_2 nanowire films [18]. The metallic backbones of Sb-doped SnO_2 nanowires dramatically reduced the resistance and the branched structure offered more pathways for electrical conductions in comparison with pure SnO_2 nanowire films, thus significantly reducing the detectable limits of ethanol to sub-ppm at 300 °C. These results indicate that semiconducting oxide nanowires can be used as the sensing materials for fabricating highly sensitive gas sensors.

Ahn *and co-workers* have fabricated novel ZnO nanowire gas sensors by a simple and efficient fabrication process and the gas-sensing capability is demonstrated for NO_2 gas [14]. Figure 24 (a) and (b) show a schematic illustration and SEM images of the fabricated sensors. The ZnO nanowires were synthesized and bridged the gap between two prepatterned Au catalysts. From the SEM image, we can see that the network-structured ZnO nanowires are floated on the SiO_2/Si substrate. The device structure proposed is very simple and efficient, because the electrical contacts to the nanowires are self-assembled during the synthesis of them. A good Ohmic behavior was obtained, which is important to the sensing properties because the sensitivities of a gas sensor can be maximized when the metal-semiconductor junction is Ohmic or has a negligible junction resistance. As shown in Figure 24 (c), the sensor displayed the fast responses and recovery behaviors with a maximum sensitivity to NO_2 gas at 225 °C. The response and recovery times are 24 s and 12 s for 0.5 ppm NO_2 gas.

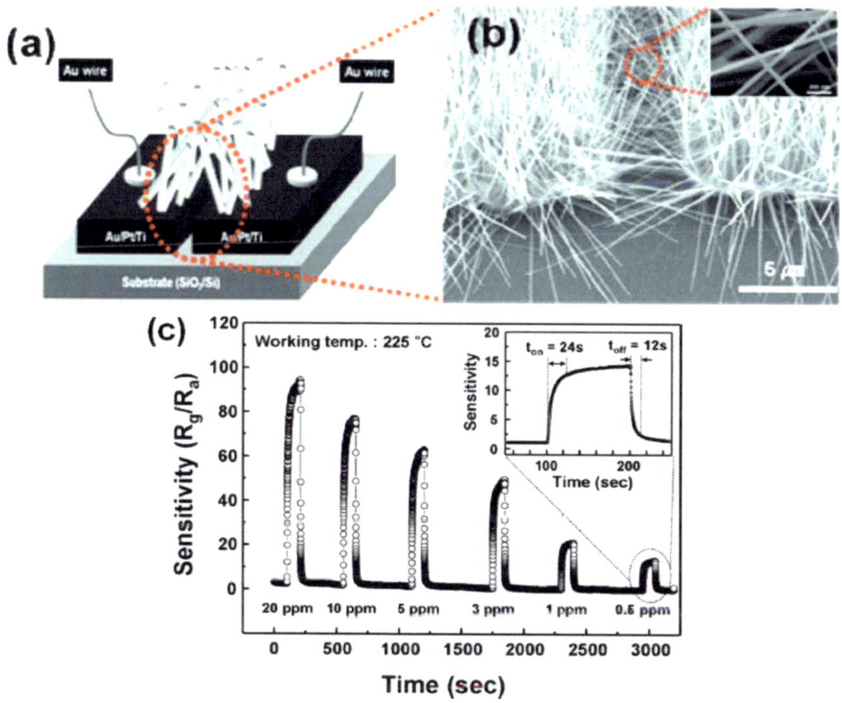

Fig. 24. (a) A schematic illustration and (b) SEM images for the network-structured ZnO nanowire floated SiO_2/Si substrate. (c) Gas sensitivity curves of the network-structured ZnO -nanowire gas sensor under exposure to 20, 10, 5, 3, 1, and 0.5 ppm NO2 gas at the measurement temperature of 225 °C (From Ref. [14]).

2. Semiconducting oxide nanowires gas sensors based on FET devices

The FET device configurations used for sensors can not only facilitate the monitoring of conductance change, but also utilize a transverse electric field to modulate the chemical sensoring behaviors of the nanowires. Fan *and co-workers* have reported that n-type single ZnO nanowire FETs are implemented as chemical gas sensors for the detection of NO_2, NH_3, CO and O_2 [13]. Single-crystalline ZnO nanowires were synthesized via a vapor trapping chemical vapor deposition method. The as-synthesized samples were suspended in alcohol and deposited onto p^{++}-Si substrates capped with a SiO_2 insulator. Photolithography or electron beam lithography was used to define source and drain electrode patterns. Figure 25 illustrates an atomic force microscopy image of a ZnO nanowire FET combined with the schematic of the measurement circuit.

Zhou *and co-workers* have firstly demonstrated room-temperature detection of NO_2 down to ppb levels using single and multiple In_2O_3 nanowires FET. In_2O_3 nanowires with diameters of ~ 10 nm and lengths of ~ 5 µm were synthesized by the laser ablation method [15]. The single or multiple In_2O_3 nanowires FET followed the traditional micro-fabrication technology. The conductance of In_2O_3 nanowires is directly related to the carrier concentration which can be altered by the adsorbed NO_2 molecules due to their strong electron withdrawing capabilities, therefore the NO_2 sensing is expected. Single In_2O_3 nanowire FETs based sensors displayed high NO_2 sensitivity down to 20 ppb level. As shown in Figure 26, the multi-nanowire sensors showed an even lower detection limit of 5 ppb, as compared to the 20 ppb limit of single nanowire sensors. They attribute this improved sensitivity to the formation of nanowire/nanowire junctions between the metal electrodes. Such junctions, when exposed to NO_2, should form a depleted layer around the intersection and thus block the electron flow in a way more prominent than the surface depletion of single nanowires with metal contacts. These high-performance sensors are promising for a new generation of NO_2 sensors in terms of their low detection limits to ppb levels and great simplicities in the device fabrications.

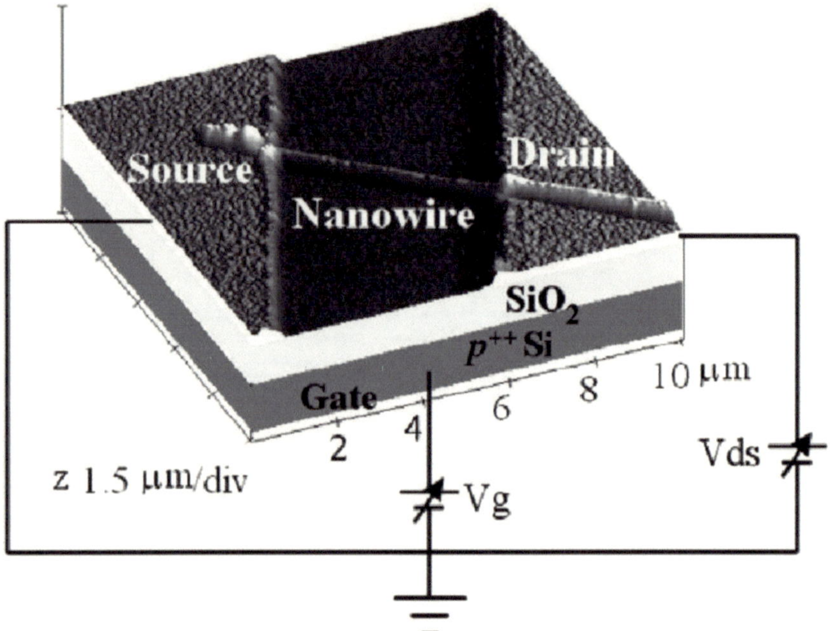

Fig. 25. Schematic of ZnO nanowire FET combined with the electrical transport measurement circuit (From Ref. [13]).

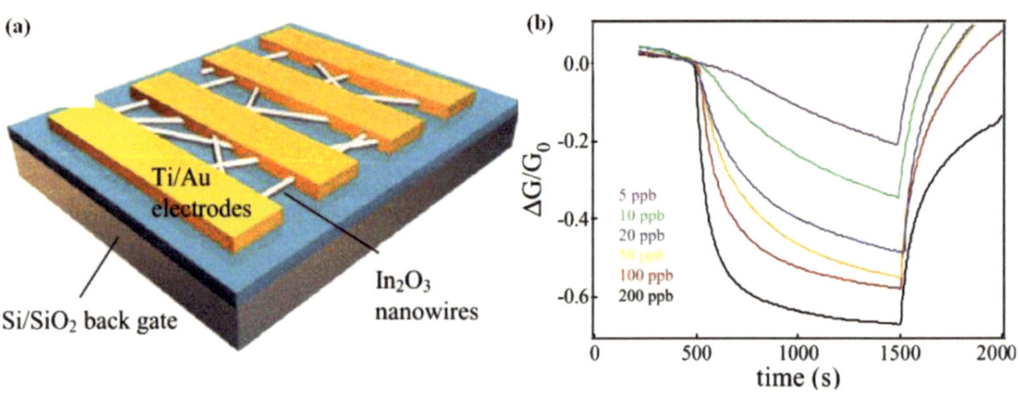

Fig. 26. (a) Schematic of a multi-nanowire FET. (b) Six sensing cycles of the multiwire device, corresponding to NO_2 concentrations of 5, 10, 20, 50, 100, and 200 ppb (From Ref. [15]).

4.2.2 Biosensors based on semiconducting oxide nanowires

Recording electrical signals from cells and tissues is a promising direction of investigations in the field of the fundamental biophysical studies and analytical biology. The combination of knowledges in bio- and electrochemistry, solid-state and surface physics, bioengineering, microelectronic technology and data processing offers the possibilities of a new generation of highly specific, sensitive, selective and reliable micro/nano-biochemical sensors and sensor arrays. Semiconductor oxide nanowires have been demonstrated as outstanding sensing-materials for the fabrications of ultra-sensitive nano-sensors due to their large surface-to-volume ratio. ZnO nanowire was tailored both physically and chemically to immobilize the enzyme glucose oxidase for construction of a glucose sensor, which was reported in the Ref. [16]. The sensor reported in this work demonstrated superior performance in sensitivities, dynamic linear range, response time, and detection limit. The high performance of the sensors was ascribed to its high specific surface area and high isoelectric point value for efficient immobilization of high concentration of acidic enzymes and the mediating effect by the redox reaction of ZnO nanowires.

FET configuration is the most typical configurations for the sensor applications. The contacts at the two ends of nanowires are usually chosen to be Ohmic contact in order to enhance the surface effect of the nanowires on the measured changes in the conductance. Wang *and co-workers* have deliberately introduced a nonsymmetric Schottky contact at one end of a ZnO nanowire device for the detection of biomolecules. A home-built microfluidic cell was placed over all biosensor devices to control the liquid environment [17]. For comparison, the Ohmic contact nanowire devices showed little changes in the electrical signals when the positively or negatively charged molecules are introduced at a concentration of 800 μg/mL. Figure 27 (a) shows a typical I-V curve of Schottky-contacted device. As shown in Figure 27 (b), a fast response and distinct current variations of Schottky-contacted nanowire biosensors can be observed when the device is exposed to a series of concentrations of negatively charged molecules. The detection limit of negatively charged biomolecules and positively charged biomolecules are as low as 2fg/mL and 20 mg/mL, respectively. They attributed the electrical response to the variation of the Schottky barrier height and barrier width as a result of biomolecule adsorption at the Schottky contact. Such nanosensors are probe-free, and it is not needed to introduce antibody-antigen interaction to enhance its sensitivity. This approach demonstrated here can serve as a guideline for designing more practical chemical and biochemical sensors.

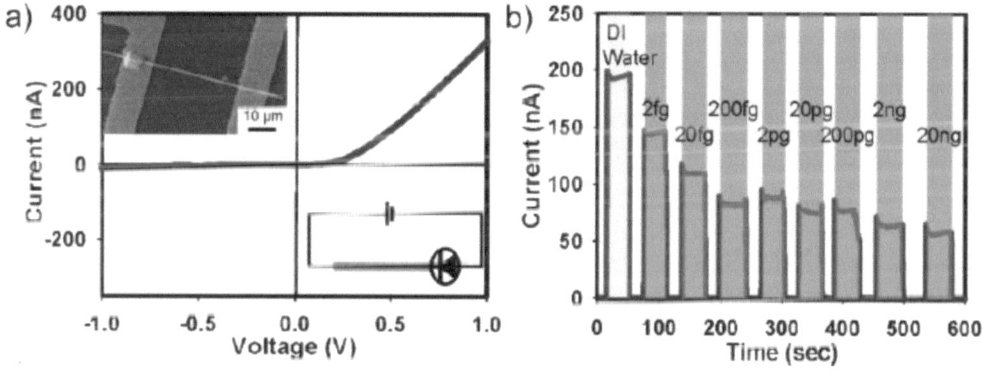

Fig. 27. (a) The I–V curve of a Schottky contact biosensor. The inset shown in upper-left is a SEM image of a Schottky contact biosensor device. (b) Current variations *vs* time curve of the sensor exposed to a series of concentrations of negatively charged molecules (From Ref. [17]).

4.3 Field Emission based on semiconducting oxide nanostructures

Field emission is based on the physical phenomenon of quantum tunneling, in which electrons are injected from the surface of materials into vacuum under the influence of an applied electric field. It is of great commercial interest in vacuum microelectronic devices such as field emission displays, x-ray sources, microwave devices, etc. One-dimensional nanosctructures such as nanotubes and nanowires are idea candidates for achieving high field emission current density at a low filed because of their high aspect ratio. In the last decade, field emission cold cathodes based on various oxide nanowires have been widely investigated.

It can be stated that within last few years, tremendous progress has already been made in field-emission applications of oxide nanowires. Here, we give some representative results on field-emission properties of ZnO nanowires reported so far. In 2002, Lee and co-workers reported the vacuum electron field emission properties of the vertically well-aligned ZnO nanowires.[74] Vertically aligned nanowires with a high density were grown on the Si substrate, as shown in Figure 28 (a). A magnified top view of nanowire array reveals that a high-purity nanowire has a sharp tip, as shown in Figure 28 (b). The synthesized ZnO nanowire indicates the average length of 13 μm and the typical diameter of 50 nm, revealing that the average aspect ratio is estimated to be larger than 250. As shown in Figure 29, the turn-on voltage of the ZnO nanowires was estimated to be about 6.0 V/ μm at current density of 0.1 μA/cm^2. The emission current density reached 1 mA/cm^2 at a bias field of 11.0 V/mm. Therefore, the well-aligned ZnO nanowires grown at such low temperature can promise the application of a glass-sealed flat panel display in a near future.

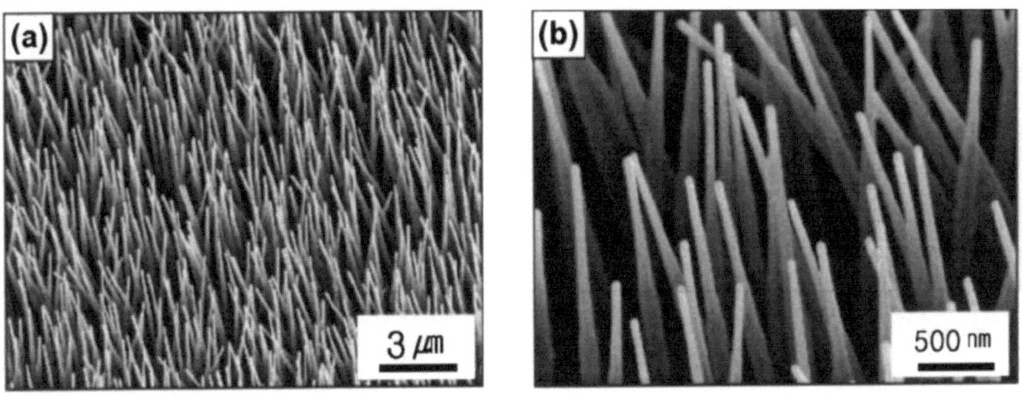

Fig. 28. SEM image of well-aligned nanowires grown on n-type silicon substrate (a) and a magnified SEM image of well-aligned nanowires (b).

2003, Yu and co-workers reported that ZnO nanoneedle arrays showed an emission current density of 2.4 mA/cm^2 under the field of 7 V/μm, and a very low turn-on field of 2.4 V/μm[75]. Such a high emission current density was attributed to the high aspect ratio of the ZnO nanoneedles. The high emission current density, high stability, and low turn-on field make the ZnO nanoneedle arrays one of the promising candidates for field-emission displays. In 2003, Wan and co-workers reported the synthesis and field emission properties of tetrapod-like ZnO nanostructures by thermal evaporation method via VS mechanism.[22] The field emission current versus applied electric field are shown in Figure 30 (a). The turn-on field is found to be as low as 1.6 V/μm at the current density of 1 μA/cm^2. The emission current density reaches 1mA/cm2 at a bias electrical field of 4.5 V/μm. The Fowler–Nordheim (F–N) plots corresponding to the data are shown in Figure 30 (b), from which the field enhancement factor β is estimated to be about 6285. ZnO nanostructures synthesized by our method are technologically useful for vacuum electron devices because they can be easily and economically synthesized and deposited on large substrates.

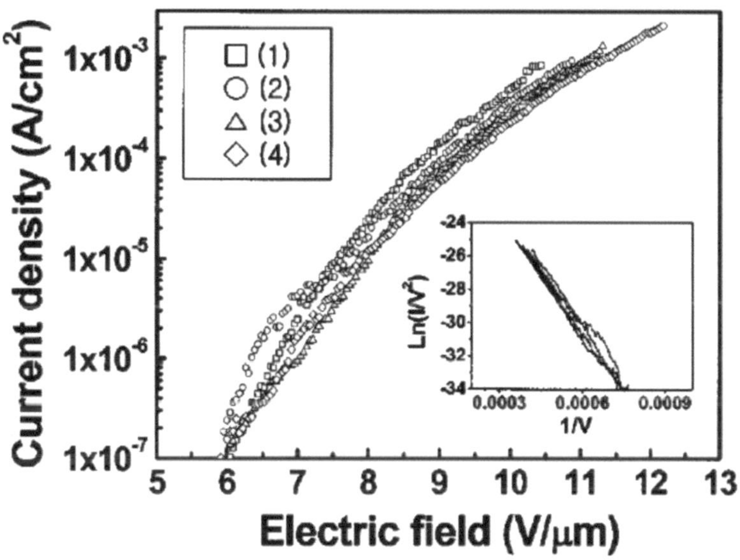

Fig. 29. Emission current density from ZnO nanowires grown on silicon substrate at 550 °C. The inset reveals that the field emission follows FN behavior.

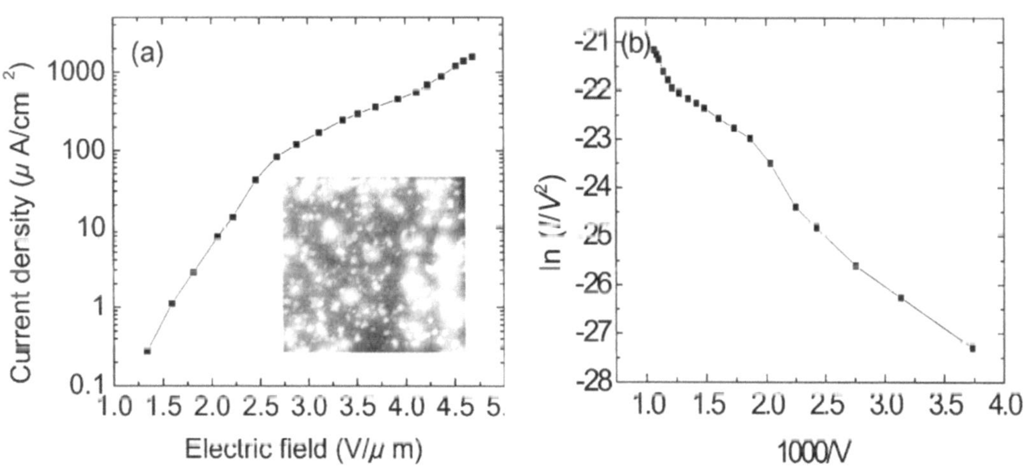

Fig. 30. The electron field emission characteristics of the tetrapod-like ZnO nanostructures. (a) The current-voltage characteristics at an anode-sample distance of 200 μm (the transparent anode image of the electron emission is shown in the inset picture) and (b) Fowler–Nordheim plot of the tetrapod-like ZnO nanostructures (From Ref. [22]).

Thermal vaporization and condensation was used to grow the ZnO nanowires on carbon cloth from a mixture source of ZnO and graphite powders in a tube furnace by Z. F. Ren and co-workers.[76] In this process, the zinc and zinc sub-oxide vapor were produced by carbothermal reduction and subsequently condensed in the lower temperature zone as the catalyst on which ZnO nanowires were grown. As shown in Figure 31, ZnO nanowire field emitters grown on carbon cloth show an extremely low operating electric field of 0.7 V/μm at an emission current density of 1 mA/cm^2. Such low field results from an extremely high field enhancement factor of about 4x0^4 due to a combined effect of the high intrinsic aspect ratio of ZnO nanowires and the woven geometry of carbon cloth.

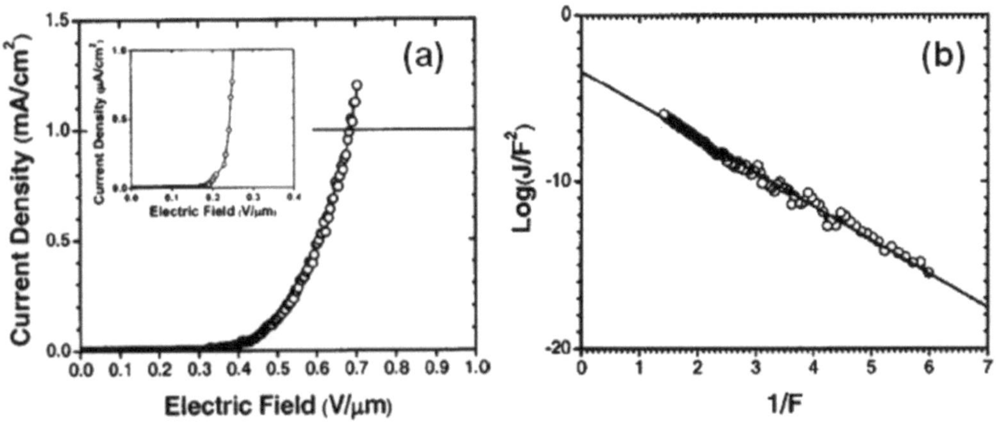

Fig. 31. The measured field emission current density of ZnO nanowires grown on carbon cloth as a function of the macroscopic electric field. The electric field required to obtain 1 mA/cm2 is 0.7 V/mm. (b) The Fowler–Nordheim plot of the field emission current density of ZnO nanowires grown on carbon cloth.

Tin-doped indium oxide (ITO) nanowires are also promising for field emitters because of their inherent properties of very low resistivities, thermally stable, and oxidation resistant. The epitaxial growth and vacuum electron field emission properties of vertically aligned ITO single-crystalline nanowire arrays on ITO/YSZ substrates have been studied by vapor transport method [23]. Figure 32 (a) shows the schematic view of our growth processes for vertically aligned ITO nanowire arrays. First, an ITO (In:Sn=95:5) buffer layer of 200 nm in thickness was deposited on (100) YSZ single-crystalline substrates (1×1 cm^2) by pulsed laser deposition at 600 °C in the oxygen ambient of 2×10^{-2} Torr. A KrF excimer laser (λ =248 nm, pulse duration=20 ns, repetition frequency=5 Hz, and photon energy density=2 J/cm^2 pulse) was used as the laser source. ITO films grown by pulsed laser deposition method were single-crystalline with a cube-on-cube epitaxial relationship with the YSZ substrate. Subsequently, a gold film with the thickness of about 10 nm was deposited on ITO/YSZ substrates by sputtering. At last, vertically aligned ITO nanowires were grown by a vapor transport method based on the vapor-liquid-solid (VLS) growth mode, as shown by the SEM image in figure 32 (b).

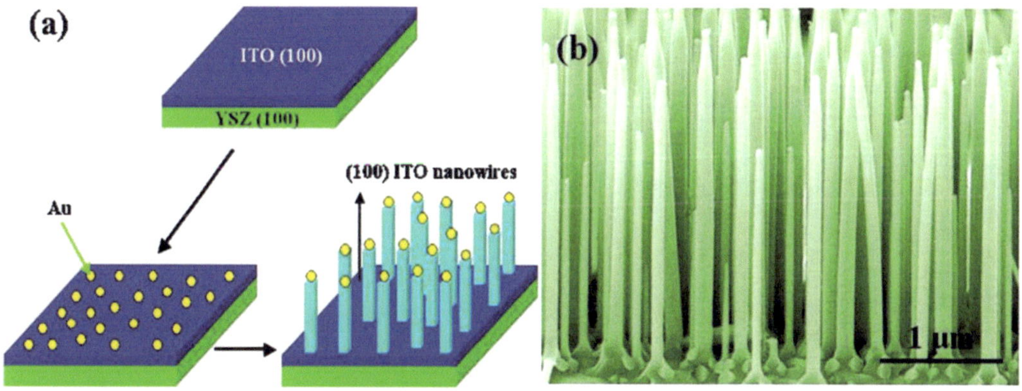

Fig. 32. (a) Schematic view of processes for vertically aligned ITO nanowire arrays grown on ITO buffer layer/(100) YSZ substrate. (b) A side-view SEM image of the vertically aligned ITO nanowires arrays grown on ITO (100) YSZ substrate (From Ref. [23]).

Field emission measurements were carried out in a vacuum chamber with the cathode-anode distance of 200 μm at room temperature. β is the field enhancement factor, E is the applied field (E=V/d), and Φ is the work function of the emission tip. Vertically aligned ITO nanowires arrays showed an enhanced field emission with a turn-on field of about 2.0 V/μm at a current density of 1 μA/cm². The β of vertically aligned ITO nanowires arrays was estimated to be bout 3850. Two curves in Fig. 35 (a) illustrate the field emission current density as a function of the applied field in semilogarithmic scale for two vertically aligned ITO nanowires arrays samples, respectively. Based on the data in Fig. 33 (a), a turn-on electrical field of about 2.0 V/μm was obtained for vertically aligned ITO nanowires arrays at a current density of 1μA/cm². Vertically aligned ITO nanowires have a stronger local electric field due to their orientation parallel to the electric-field direction. At the same time, vertically aligned ITO nanowires possess more emitter tips and the tips have better field emission performance than their bodies. Thus, vertically aligned ITO nanowire arrays show better field emission properties than randomly distributed ITO nanowires. The exponential dependence between the emission current density and the applied field indicates that the field emission from both ITO nanowire arrays follows the FN relationship, as shown in the In(J/E²)-1/E relationships in Fig.33 (b). From the slope of the FN curve, we can evaluate the value of β ultimately when the work function of the emission tip is known.

The field emission current dependence on vacuum pressure of the vertically aligned ITO nanowires arrays has been studied since environmental stabilities are also one of the main requirements for the practical applications. As shown in Figure 34 (a), a steady emission current density of about 2 mA/cm² was measured at the applied field of 6.0 V/μm for the vacuum level higher than 2×10^{-4}Pa. As shown in Figure 34 (b), no obvious degradations of current density are found and the current fluctuation is as low as ±5% during 2 hours continuous operation at 1×10^{-5} Pa. However, the current fluctuation becomes bigger when the vacuum level is degraded. We found that the current fluctuation was ±28% at 1×10^{-1} Pa. The emission current drops slowly in the vacuum range from 5×10^{-4} to 10^{-1}Pa. The current density at 10^{-1} Pa is about one-third of the current density at 10^{-5}Pa. A field emission recovery was observed when ITO nanowires arrays were operated under the initial pressure of 10^{-5}Pa.

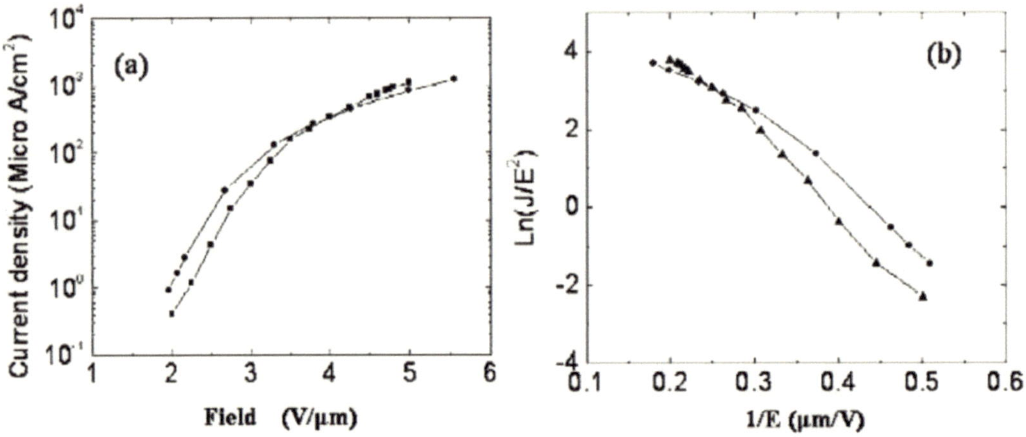

Fig. 33. (a) Field emission current density–field curves of the ITO nanowire arrays measured in the vacuum of 10^{-5}Pa. (b) The corresponding FN plots of (a) (From Ref. [23]).

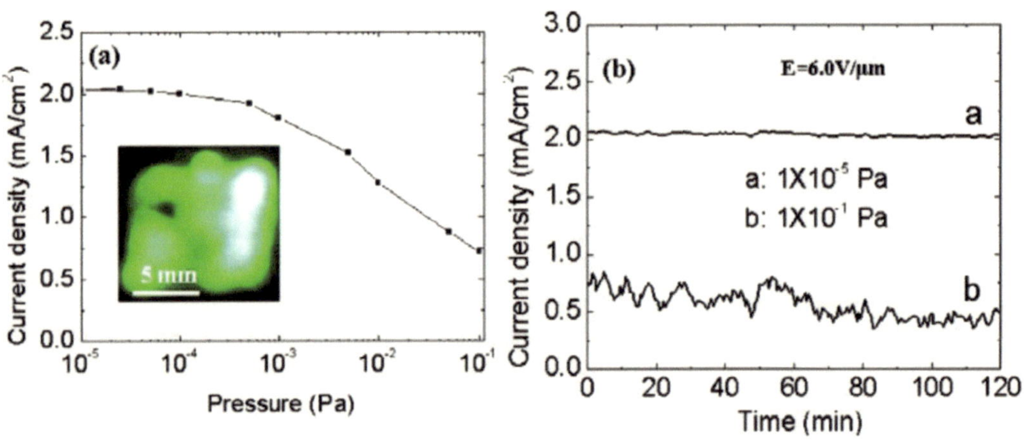

Fig. 34. (a) The dependence of emission current density of vertically aligned ITO nanowire arrays with the vacuum pressure in the range from 10^{-5} to 10^{-1}Pa. The inset shows the spatial distribution of emission of ITO nanowires at the 10^{-5}Pa. (b) Field emission current stability of ITO nanowire arrays at 10^{-5} and 10^{-1}Pa, respectively (From Ref. [23]).

These results indicate that vertically aligned ITO nanowires show relatively high environmental stabilities. Moreover, vertically aligned ITO nanowire arrays showed an enhanced field emission with a turn-on field of about 2.0 V/μm at a current density of 1 μA/cm^2. These results demonstrate that vertically aligned ITO nanowires arrays are very promising for vacuum electron field emission applications.

5. Summary

Semiconducting oxide nanowires, whose lateral dimensions fall anywhere in the range of 1 to 100 nm are not only interesting for fundamental research due to their unique structural and physical properties relative to their bulk counterparts, but also offer fascinating potential for future technological applications. Deeper understanding the growth, doping and applications of the oxide nanowires are central to the current research interest. This chapter has provided a comprehensive review of current research activities that concentrate on semiconducting oxide nanowires. We first gave an introduction to the synthesis and doping of semiconducting oxide nanowires. The influence of doping to the electrical and optical properties of the oxide nanowires was also described. Most attention was then devoted to the transistors, sensors and vacuum electron field emitters applications of oxide nanowires.

6. References

[1] Ju, S.; Lee, K. & Janes, D. B. "Low operating voltage single ZnO nanowire field-effect transistors enabled by self-assembled organic gate nanodielectrics," *Nano Lett.*, Vol. 5, No. 11, 2005, pp. 2281-2286.

[2] Goldberger, J.; Sirbuly, D. J.; Law, M. & Yang, P. "ZnO nanowire Transistors," J. *Phys. Chem. B.* Vol. 109, No. 1, 2005, pp.9-14.

[3] Heo, W.; Tien, L. C.; Kwon, Y.; Norton, D. P.; Pearton, S. J.; Kang, B. S. & Ren, F. "Depletion-mode ZnO nanowire field-effect transistor," *Appl. Phys. Lett.* Vol. 85, No.12, 2004, pp. 2274-2276.

[4] Ng, H. T.; Han, J.; Yamada, T.; Nguyen, P.; Chen, Y. P. & Meyyappan, M. "Single crystal nanowire vertical surround-gate field-effect transistor," *Nano Lett.* Vol. 4, No. 7, 2004, pp.1274-1252.

[5] Keem, K.; Jeong, D. Y. & Kim, S. "Fabrication and device characterization of omega-shaped-gate ZnO nanowire field-effect transistors," *Nano Lett.* Vol. 6, No. 7, 2006, pp.1454-1458.

[6] Shen, G. Z.; Chen, P. C.; Ryu, K. & Zhou, C.W. "Devices and chemical sensing applications of metal oxide nanowires" *J. Mater. Chem.*, 19, 2009, pp. 828-839.

[7] Ju, S.; Facchetti, A.; Xuan, Y. Liu, J.; Ishikawa, F.; Ye, P.; Zhou, C.; Marks, T. J. & Janes, D. B. "Fabrication of fully transparent nanowire transistors for transparent and flexible electronics," *Nat. Nanotech.* Vol. 2, 2007, pp. 378-384.

[8] Chen, P. C.; Shen, G.; Chen, H.; Ha, Y.; Wu, C.; Sukcharoenchoke, S.; Fu, Y.; Liu, J.; Facchetti, A.; Marks, T. J.; Thompson, M. E. & Zhou, C. "High-performance single-crystalline arsenic-doped indium oxide nanowires for transparent thin-film transistors and active matrix organic light-emitting diode display," *ACS Nano,* Vol. 3, No. 11, 2009, pp. 3383-3390.

[9] Zhang, F.; He, Z. B.; Yuan, G. D.; Jie, J. S.; Luo, L. B.; Zhang, X. J.; Chen, Z. H.; Lee, C. S.; Zhang, W. J.& Lee, S. T. "High-performance, fully transparent, and flexible zinc-doped indium oxide nanowire transistors," *Appl. Phys. Lett.* Vol. 94, No. 12, 2009, pp.123103.

[10] Wan, Q.; Dattoli, E. & Lu, W. "Doping-Dependent Electrical Characteristics of SnO_2 Nanowires," *small.*, Vol. 4, No. 4, 2008, pp. 451–454.

[11] Sun, J.; Liu, H. X.; Jiang, J.; Lu, A. & Wan, Q. "Low-voltage transparent SnO_2 nanowire transistors gated by microporous SiO_2 solid-electrolyte with improved polarization response." *J. Mater. Chem.*, Vol. 20, No. 37, 2010, pp. 8010-8015.

[12] Q. Wan, Q. H. Li, Y. J. Chen, and T. H. Wang, X. L. He and J. P. Li, C. L. Lin, "Fabrication and ethanol sensing characteristics of ZnO nanowire gas sensors," *Appl. Phys. Lett.*, Vol. 84, No. 18, 2004, pp. 3654-3656.

[13] Fan, Z. & Lu, J. G. "Chemical sensing with ZnO nanowire field-effect transistor," *IEEE Trans. Nanotech.* Vol. 5, No. 4, 2006, pp.393-396.

[14] Son, J. Y.; Lim, S. J.; Cho, J. H.; Seong, W. K. & Kim, H. "Synthesis of horizontally aligned ZnO nanowires localized at terrace edges and application for high sensitivity gas sensor," *Appl. Phys. Lett.* Vol. 93, No. 5, 2008, pp. 053109.

[15] Zhang, D.; Liu, Z.; Li, C.; Tang, T.; Liu, X.; Han, S.; Lei, B. & Zhou, C. "Detection of NO_2 down to ppb levels using individual and multiple In_2O_3 nanowire devices," *Nano Lett.*, Vol. 4, No. 10, 2004, pp.1919-1924.

[16] Zang, Z.; Li, C. M.; Cui, X.; Wang, J.; Sun, X.; Dong, H. & Sun, C. Q. "Tailoring Zinc oxide nanowires for high performance amperometric glucose glucose sensor," *Electroanalisis.* Vol. 19, No.9, 2007, pp.1008-1014.

[17] Zhou, J.; Gu, Y.; Fei, P.; Mai, W.; Gao, Y.; Yang, R.; Bao, G. & Wang, Z. L. "Flexible piezotronic strain sensor," *Nano Lett.* Vol. 8, No. 9, 2008, pp.3035-3040.

[18] Wan, Q.; Huang, J.; Xie, Z.; Wang, T.; Dattoli,E. N. & Lu, W. "Branched SnO_2 nanowires on metallic nanowire backbones for ethanol sensors application," *Appl. Phys. Lett.*, Vol. 92, No. 10, 2008, pp. 102101.

[19] Kuang, Q.; Lao, C.; Wang, Z. L.; Xie, Z. & Zheng, L. "High-sensitivity humidity sensor based on a single SnO_2 nanowire," *J. Am. Chem. Soc.* Vol. 129, No.19, 2007, pp. 6070-6071.

[20] Yeh, P. H.; Li, Z. & Wang, Z. L. "Schottky-gated probe-free ZnO nanowire biosensor," *Adv. Mater.* Vol. 21, No. 48, 2009, pp. 4975-4978.

[21] Das, S. N.; Moon, K. J.; Kar, J. P.; Choi, J. H.; Xiong, J.; Lee, T. I. & Myoung, J. M. "ZnO single nanowire-based UV detectors," *Appl. Phys. Lett.* Vol. 97, No. 2, 2010, pp. 022103.

[22] Wan, Q.; Yu, K.; Wang, T. H. & Lin, C. L. "Low-field electron emission from tetrapod-like ZnO nanostructures synthesized by rapid evaporation," *Appl. Phys. Lett.*, Vol. 83, No. 11, 2003, pp.2653-2655.

[23] Wan, Q.; Feng, P. & Wang, T. H. "Vertically aligned tin-doped indium oxide nanowire arrays: Epitaxial growth and electron field emission properties," Appl. Phys. Lett., Vol. 89, No. 12, 2006, pp. 123102.

[24] Lee, C. Y.; Tseng, T. Y.; Li, S. Y. & Lin, P. "Effect of phosphorus dopant on photoluminescence and field-emission characteristics of $Mg_{0.1}Zn_{0.9}O$ nanowires," *J. Appl. Phys.*, Vol. 90, No. 2, 2006, pp. 024303.

[25] Lee, C. J.; Lee, T. J.; Lyu, S. C.; Zhang, Y.; Ruh H. & Lee, H. J. "Field emission from well-aligned zinc oxide nanowires grown at low temperature," *Appl. Phys. Lett.*, Vol. 81, No. 19, 2002, pp. 3648-3650.

[26] Tseng, Y. K.; Huang, C. J.; Cheng, H. M.; Lin, I. N.; Liu, K. S. & Chen, I. C. "Characterization and Field-Emission Properties of Needle-like Zinc Oxide

Nanowires Grown Vertically on Conductive Zinc Oxide Films," *Adv. Funct. Mater.* Vol. 13, No. 10, 2003, pp.811-814.
[27] Jo, S. H.; Lao, J. Y.; Ren, Z. F.; Farrer, R. A.; Baldacchini, T. & Fourkas, J. T. "Field-emission studies on thin films of zinc oxide nanowires," *Appl. Phys. Lett.*, Vol. 83, No. 23, 2003, pp. 4821-4823.
[28] Li, S. Y.; Lin, P.; Lee, C. Y. & Tseng, T. Y. "Field emission and photofluorescent characteristics of zinc oxide nanowires synthesized by a metal catalyzed vapor-liquid-solid process," *J. Appl. Phys.*, Vol. 95, No. 7, 2004, pp. 3711-3713.
[29] Jo, S. H.; Banerjee, D. & Ren, Z. F. "Field emission of zinc oxide nanowires grown on carbon cloth," *Appl. Phys. Lett.*, Vol. 85, No. 8, 2004, pp. 1407-1409.
[30] Xia, Y. N.; Yang, P. D.; Sun, Y. G.; Wu, Y. Y.; Mayers, B.; Gates, B.; Yin, Y. D.; Kim, F. & Yan, H. Q. "One-dimensional nanostructures: Synthesis, Characterization, and Applications." vol. 15, No. 5, 2003, pp. 353-389.
[31] Wagner, R. S. & Ellis, W. C. "Vapor-liquid-solid mechanism of single crystal growth." *Appl. Phys. Lett.*, Vol. 4, No. 5, 1964, pp. 89-90.
[32] Eric, A.; Stach, P. J.; Pauzauskie,T.; Kuykendall,J.; Goldberger, R. H. & Peidong, Y. "Watching GaN Nanowires Grow" *Nano. Lett.*, Vol. 3, No. 6, 2003, pp.867- 869.
[33] Wang, Y., et al., "Catalytic Growth and Photoluminescence Properties of Semiconducting Single-Crystal ZnS Nanowires," *Chem. Phys. Lett.*, Vol. 357, No. 3-4, 2002, pp.314-318.
[34] Duan, X, & C. M. Lieber, "General Synthesis of Compound Semiconducting Nanowires," Adv. Matterials, Vol. 12, No. 4, 2000, pp. 298-302.
[35] Wan, Q.; Wei, M.; Zhi, D.; MacManus-Driscoll, J. L. & Blamire, M. G. "Epitaxial Growth of Vertically Aligned and Branched Single-Crystalline Tin-Doped Indium Oxide Nanowire Arrays," *Adv. Mater.*, Vol. 18, No. 2, 2006, pp. 234–238.
[36] Li, Q. H.; Wan, Q.; Liang, Y. X. & Wang, T. H. "Electronic transport through individual ZnO nanowires," *Appl. Phys. Lett.*, Vol. 84, No. 22, 2004, pp. 4556-4558.
[37] Chen, Y. J.; Cao, M. S.; Wang, T. H. & Wan, Q. "Microwave absorption properties of the ZnO nanowire-polyester composites," *Appl. Phys. Lett.*, Vol. 84, No. 17, 2004, pp 3367-3369.
[38] Feng, P.; Xue, X. Y.; Liu,Y. G.; Wan, Q. & Wang, T. H. "Achieving fast oxygen response in individual β-Ga_2O_3 nanowires by ultraviolet illumination," *Appl. Phys. Lett.* Vol. 89, No. 11, 2006, pp. 112114.
[39] Feng, P.; Zhang, J. Y.; Wan, Q. & Wang, T. H." Photocurrent characteristics of individual $ZnGa_2O_4$ nanowires." *J. Appl. Phys.* Vol. 102, No.7, 2007, pp. 074309.
[40] Li, Y.; Meng, G. W.; Zhang, L. D. & Phillipp, F. "Ordered semiconducting ZnO nanowire arrays and their photoluminescence properties," *Appl. Phys. Lett.*, Vol. 76, No. 15, 2000, pp.2011-2013.
[41] Greene, L. E.; Yuhas, B. D.; Law, M.; Zitoun, D. & Yang, P. D. "Solution-Grown Zinc Oxide Nanowires," *Inorganic Chemistry*, Vol. 45, No. 19, 2006, pp.7535-7543.
[42] Li, Q. H.; Wan, Q.; Wang, Y. G. & Wang, T. H. "Abnormal temperature dependence of conductance of single Cd-doped ZnO nanowires," *Appl. Phys. Lett.* Vol. 86, No. 26, 2005, pp. 263101.

[43] Li, Q. H.; Liang, Y. X.; Wan, Q. & Wang, T. H. "Oxygen sensing characteristics of individual ZnO nanowire transistors," *Appl. Phys. Lett.*, Vol. 85, No. 26, 2004, pp. 6389-6391.

[44] Wan, Q.; Li, Q. H.; Chen, Y. J.; Wang, T. H.; He, X. L.; Gao, X. G. & Li, J. P. "Positive temperature coefficient resistance and humidity sensing properties of Cd-doped ZnO nanowires," *Appl. Phys. Lett.*, Vol. 84, No. 16, 2004, pp. 3685-3687.

[45] Wan, Q.; Lin, C. L.; Yu, X. B. & Wang, T. H. "Room-temperature hydrogen storage characteristics of ZnO nanowires," *Appl. Phys. Lett.*, Vol. 84, No. 1, 2004, pp.124-126.

[46] Feng, P.; Wan, Q.; Fu, X. Q.; Tian, Y. & Wang, T. H. "Anomalous electrorheological behavior of ZnO nanowires," *Appl. Phys. Lett.* Vol. 87, No. 3, 2005, pp. 033114.

[47] Ying, Z.; Wan, Q.; Cao, H.; Song, Z. T. & Feng, S. L. "Characterization of SnO_2 nanowires as an anode material for Li-ion batteries," *Appl. Phys. Lett.* Vol. 87, No.11, 2005, pp. 113108.

[48] Wan, Q. & Wang, T. H. "Single-crystalline Sb-doped SnO_2 nanowires: synthesis and gas sensor application," *Chem. Commun.*, No. 30, 2005, pp. 3841-3843.

[49] Liu, Z.; Zhang, D.; Han, S.; Li, C.; Tang, T.; Jin, W.; Liu, X.; Lie, B. & Zhou, C. W. "Laser Ablation Synthesis and Electron Transport Studies of Tin Oxide Nanowires," *Adv. Mater.*, Vol. 15, No. 20, 2003, pp. 1754–1757.

[50] Li, C.; Zhang, D. H.; Han, S.; Liu, X. L.; Tang, T. & Zhou, C. W. "Diameter-Controlled Growth of Single- Crystalline In_2O_3 Nanowires and Their Electronic Properties," *Adv. Mater.*, Vol. 15, No. 2, 2003, pp. 143–146.

[51] Wan, Q.; Dattoli, E. N.; Fung, W.Y.; Guo, W.; Chen, Y. B.; Pan, X. Q. & Lu, W. "High-Performance Transparent Conducting Oxide Nanowires," *Nano Lett.*, Vol. 6, No. 12, 2006, pp. 2909-2915.

[52] Huang, J.; Lu, A.; Zhao, B. & Wan, Q. "Branched growth of degenerately Sb-doped SnO_2 nanowires," *Appl. Phys. Lett.*, Vol. 91, No.7, 2007, pp. 073102.

[53] Dattoli, E. N.; Wan, Q.; Guo, W.; Chen, Y.; Pan, X. & Lu, W. "Fully Transparent Thin-Film Transistor Devices Based on SnO_2 Nanowires," *Nano Lett.*, Vol. 7, No. 8, 2007, pp. 2463-2469.

[54] Sun, J.; Tang, Q.; Lu, A.; Jiang, X. & Wan, Q. "Individual SnO_2 nanowire transistors fabricated by the gold microwire mask method," *Nanotechnology.*, Vol. 20, No. 25, 2009, pp. 255202.

[55] Liu, H. X.; Sun, J.; Tang, Q. & Wan, Q. "Ultralow-Voltage Electric Double-Layer SnO_2 Nanowire Transistors Gated by Microporous SiO_2 -Based Solid Electrolyte," *J. Phys. Chem. C*, Vol. 114, No. 28, 2010, pp. 12316–12319.

[56] Wan, Q.; Dattoli, E. N. & Lu, W. "Transparent metallic Sb-doped SnO_2 nanowires," Appl. Phys. Lett., Vol. 90, No.22, 2007, pp.22210.

[57] Pan, Z.W.; Dai, Z.R.; & Wang, Z.L."Nanobelts of semiconducting oxides", Science, Vol 291 No.5510 2001, pp.1947-1949.

[58] Wan, Q.; Song, Z. T.; Feng, S. L.; Wang, T. H.. "Single-crystalline tin-doped indium oxide whiskers: Synthesis and characterization," *Appl. Phys. Lett.*, Vol. 85, No. 20, 2004, pp. 4759-4761.

[59] Keem, K.; Kim, H.; Kim, G. T.; Lee, J. S.; Min, B.; Cho, K.; Sung, M. Y.& Kim, S. "Photocurrent in ZnO nanowires grown from Au electrodes," *Appl. Phys. Lett.*, Vol. 84, No. 22, 2004, pp. 4376-4378.

[60] Heo, Y. W.; Tien, L. C.; Norton, D. P.; Kang, B. S.; Ren, F.; LaRoche, J. R. & Pearton, S. J. "Pt /ZnO nanowire Schottky diodes," *Appl. Phys. Lett.*, Vol. 85, No. 15, 2004, pp. 3107-3109.

[61] Huang, M. H.; Wu, Y. Y.; Feick, H.; Tran, N.; Weber, E. & Yang, P. "Catalytic Growth of Zinc Oxide Nanowires by Vapor Transport," *Adv. Mater.*, Vol. 13, No. 2, 2001, pp.113-116.

[62] Kind, H.; Yan, H. Q.; Messer, B. J.; Law, M. & Yang, P. "Nanowire Ultraviolet Photodetectors and Optical Switches," *Adv. Mater.*, Vol. 14, No. 2, 2002, pp. 158-160.

[63] Kouklin, N. "Cu-Doped ZnO Nanowires for Efficient and Multispectral Photodetection Applications," *Adv. Mater.* Vol. 20, No. 11, 2008, pp. 2190–2194.

[64] Garces, N. Y.; Wang, L.; Bai, L.; Giles, N. C. & Halliburton, L. E. "Role of copper in the green luminescence from ZnO crystals," *Appl. Phys. Lett.*, Vol. 81, No. 4, 2002, pp. 622-624.

[65] Xu, C.; Sun, X. W.; Zhang, X. H.; Ke, L. & Chua, S. J. "Photoluminescent properties of copper-doped zinc oxide nanowires," *Nanotechnology*, Vol. 15, No. 7, 2004, pp. 856-861.

[66] Geng, B. Y.; Wang, G. Z.; Jiang, Z.; Xie, T.; Sun, S. H.; Meng, G. W. & Zhang, L. D. "Synthesis and optical properties of S-doped ZnO nanowires," *Appl. Phys. Lett.*, Vol. 82, No. 26, 2003, pp. 4791-4793.

[67] Qin, L.; Xu, J.; Dong, X.; Pan, Q.; Cheng, Z.; Xiang, Q. & Li, F. "The template-free synthesis of square-shaped SnO_2 nanowires: the temperature effect and acetone gas sensors," Nanotechnology, Vol. 19, No. 18, 2008, pp. 185705.

[68] Liao, L.; Zhang, Z.; Yan, B.; Bao, Q. L.; Wu, T.; Li, C. M.; Shen, Z. X.; Zhang, J. X.; Gong, H.; Li, J. C. & T. Yu, "Multifunctional CuO nanowire devices: p-type field effect transistors and CO gas sensors" Nanotechnology, Vol. 20, No. 8, 2009, pp.085203.

[69] Liu, Z.; Yamazaki, T.; Shen, Y.; Kikuta, T.; Nakatani, N. & Kawabata, T. "Room temperature gas sensing of p-type TeO_2 nanowires," Appl. Phys. Lett. Vol. 90, No. 17, 2007, pp.173119.

[70] Singh, N.; Yan, C. & Lee, P. S. "Room temperature CO gas sensing using Zn-doped In_2O_3 nanowire field effect transistors," Sens. Actuators B, Vol. 150, No.1, 2010, pp.19-24.

[71] Choi, A.; Kim, K.; Jung, H. I. & Lee, S. Y. "ZnO nanowire biosensors for detection of biomolecular interactions in enhancement mode," Sens. Actuators B, Vol. 148, No.2, 2010, pp.577 582.

[72] Das, S. N.; Kar, J. P.; Choi, J. –H.; Lee, T. I.; Moon, K. –J. & Myoung, J. –M. "Fabrication and characterization of ZnO single nanowire-based hydrogen sensor," J. Phys. Chem. C. Vol. 114, No. 3, 2010, pp.1689-1693.

[73] P. Xu, Z. Cheng, Q. Pan, J. Xu, Q. Xiang, W. Yu, and Y. chu, "High aspect In_2O_3 nanowires: synthesis, mechanism and NO_2 gas-sensing properties," Sens. Actuators B, Vol. 130, No.2, 2008, pp.802-808.

[74] Lee, C. J.; Lee, T. J.; Lyu, S. C.; Zhang, Y.; Ruh, H.; & Lee, H. J." Field emission from well-aligned zinc oxide nanowires grown at low temperature", *Appl. Phys. Lett.*, Vol. 81, No. 19, 2002, pp.3648-3650.

[75] Zhu, Y. W.; Zhang, H. Z.; Sun, X. C.; Feng, S. Q.; Xu, J.; Zhao, Q.; Xiang, B.; Wang, R. M. & Yu, D. P. "Efficient field emission from ZnO nanoneedle arrays," *Appl. Phys. Lett.*, Vol. 83, No. 1, 2003, pp.144-146.

[76] Jo, S. H.; Banerjee, D.; & Ten, Z. F. "Field emission of zinc oxide nanowires grown on carbon cloth", *Appl. Phys. Lett.*, Vol. 85, No. 8, 2004, pp.1407-1409.

Part 2

Oxide Nanowires

5

Application of the Kirkendall Effect to Morphology Control of Nanowires: Morphology Change from Metal Nanowires to Oxide Nanotubes

Ryusuke Nakamura and Hideo Nakajima
The Institute of Scientific and Industrial Research, Osaka University
Japan

1. Introduction

In recent years, considerable effort has been put into the design and fabrication of one-dimensional nanomaterials such as nanorods and nanowires with functional properties. As reviewed in other Chapters, a wide variety of nanowires including inorganic and organic materials have been fabricated and their functional properties have been examined.

Additionally, it is important to obtain nanostructures with a specific size and morphology for expanding the applications of one dimensional nanomaterials. Therefore, the control of the shape of nanowires is one of the most important topics in the current research on nanomaterials. In particular, there is an increasing interest in methods to fabricate hollow nanostructures because their unique shape makes them applicable as delivery vehicles, fillers as well as for catalysis, and is expected to bring about changes in chemical, physical, and catalytic properties. The recent progress in synthesis and applications of hollow nanomaterials including nantoubes has been already reviewed (Fan at al., 2007; Lou et al., 2008; An & Hyeon, 2009).

In general, the most popular concept of methods to fabricate hollow nanostructures can be recognized as the combination of templating, coating and chemical etching. By coating the surface of the template nanowires and removing the template by chemical etching, a wide variety of materials with hollow structures have been successfully synthesized. For example, GaN nanotubes were fabricated through the preparation of ZnO/GaN core/shell structures and the subsequent removal of ZnO nanowires, as shown in Fig. 1 (Goldberger et al., 2003). The arrays of ZnO nanowires were grown on (110) sapphire wafers using a vapour deposition process and then GaN chemical vapor was deposited epitaxially on the ZnO nanowire arrays placed inside a reaction tube. The core part of ZnO can be removed from ZnO/GaN wires by chemical etching with ammonia at high temperature or thermal reduction at high temperatures (for example, 873 K in H_2). The synthesis of Zn_3P_2, Cd_3P_2 (Shen et al., 2006) and Al_2O_3 (Ras et al., 2008) nanotubes based on templating and etching method were also reported.

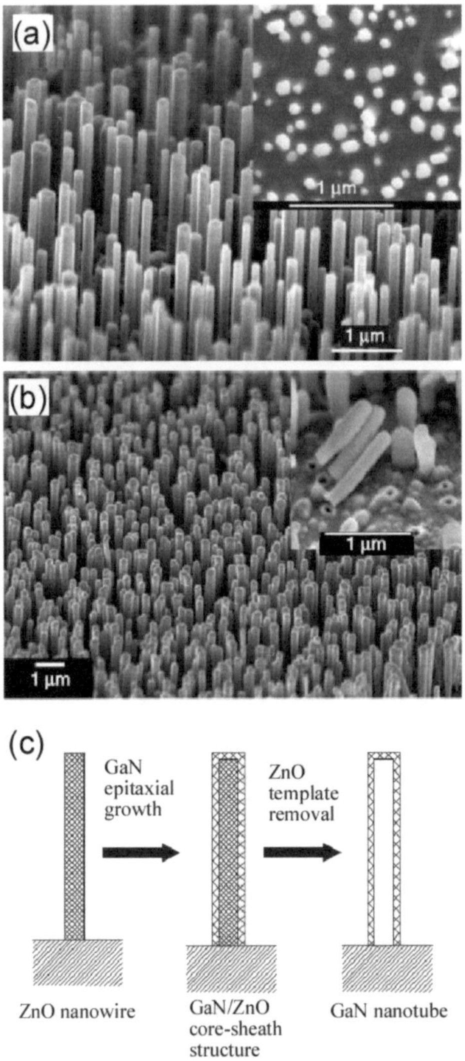

Fig. 1. SEM images of the ZnO nanowire template arrays (a), and the resulting GaN nanotube array (b). Inset in (a) shows cross-sections of the ZnO nanowires. Inset in (b) shows the fractured interface between the GaN nanotubes and the substrate. (c) Schematic illustration of the 'epitaxial casting' process for making single-crystal GaN nanotubes. (Reproduced with permission from the reference (Goldberger et al., 2003))

In addition, different ideas have been applied to the fabrication of hollow nanomaterials. For example, the Kirkendall-effect (Smigelskas & Kirkendall, 1947) related processes have been recognized as one of the useful methods to fabricate oxides, sulfides and phosphides. The Kirkendall effect as a classical phenomenon in metallurgy was established in 1947. It basically refers to a mutual diffusion process through an interface of two metals so that vacancy diffusion occurs to compensate for the unequality of the mass flow and that the initial interface moves. Smigelskas and Kirkendall observed a movement of the initial interface in a Cu/brass diffusion couple, as a result of a faster diffusion of zinc into the copper than that of copper into the brass (the intrinsic diffusion coefficient of Zn is 2.5 times larger than that of Cu at 785°C.) The Kirkendall experiment demonstrated two important facts: (i) atomic diffusion occurs via vacancies and (ii) each metal diffuses at a different mobility. In some cases, condensation of excess vacancies can give rise to void formation, called 'Kirkendall voids', near original interface and within the faster diffusion side.

Formation of the Kirkendall voids is basically unfavorable from the viewpoint of technological application. The Kirkendall voids should be avoided in the case that interdiffusion occurs at the bonded-interface because they deteriorate the bonding strength of the interface or may cause wire bond failure in integrated circuits. On the other hand, chemists applied the destructive effect constructively for synthesizing hollow nanostructures in a way that the Kirkendall voids coalesce into a single hollow core. In 2004, Yin et al. (2004) demonstrated that initially solid Co nanoparticles transformed into hollow nanoparticles through the reaction with sulfur, oxygen and selenium. The sulfidation of cobalt nanoparticles resulted in the formation of hollow cobalt sulfide of either Co_3S_4 or Co_8S_9, depending on the molar ration of sulfur and cobalt. During the reaction, the surfaces of cobalt nanoparticles are covered with the sulfide layers and the diffusion of cobalt and sulfur atoms in opposite direction occurs through the sulfide layers. As the reaction proceeds, voids are formed in the cobalt side of the interface because the outward diffusion of cobalt ions is much faster than the inward diffusion of sulfur.

Since Yin et al. (2004) published the results of the synthesis of hollow nanoparticles using the Kirkendall effect, which is a diffusional phenomenon at the interface between different solids, the Kirkendall effect has been applied to the fabrication of a variety of hollow nanoparticles and nanotubes. The first example of the fabrication of nanotubes by the Kirkendall effect is the formation of $ZnAl_2O_4$ spinel nanotubes from ZnO/Al_2O_3 core/shell nanowires (Fan et al, 2006). Figure 2 shows TEM images for $ZnAl_2O_4$ nanotubes together with the schematic illustrations of the formation process. As shown in the illustration, the starting material is a ZnO nanowire (10 30 nm thick, up to 20 μm long) coated by Al_2O_3(10 nm thick). By annealing the core/shell nanowires in air at 973 K for 3 h, a cylindrical interior nanopore is introduced during the solid state diffusion between ZnO and Al_2O_3. ZnO/Al_2O_3 reaction is considered as a one-way transfer of ZnO into Al_2O_3 in a pseudo binary system.

The Kirkendall effect is related closely to the formation processes of oxide layers on the several metals such as Cu, Ni and Fe; the rapid outward diffusion of metal ions rather than inward oxygen ions through the oxide layers can induce the formation of vacancy clusters, i.e., the Kirkendall voids at the metal sides. Therefore, our focus has been on the fabrication of oxide nanotubes via the oxidation of metal nanowires such as Cu, Ni and Fe, taking vacancy clustering inside metal nanowires due to the Kirkendall effect into consideration.

In this chapter, we will overview our recent results on the morphology changes of metal nanowires via oxidation reactions (Nakamura et al., 2009a, 2009b). The formation mechanisms of oxide nanotubes and porous nanowires will be discussed in terms of diffusion in metals and oxides. Furthermore, the results on the structural stability of nanotubes at high temperatures will be introduced; the morphology change associated with the shrinkage of interior nanopores will be reviewed.

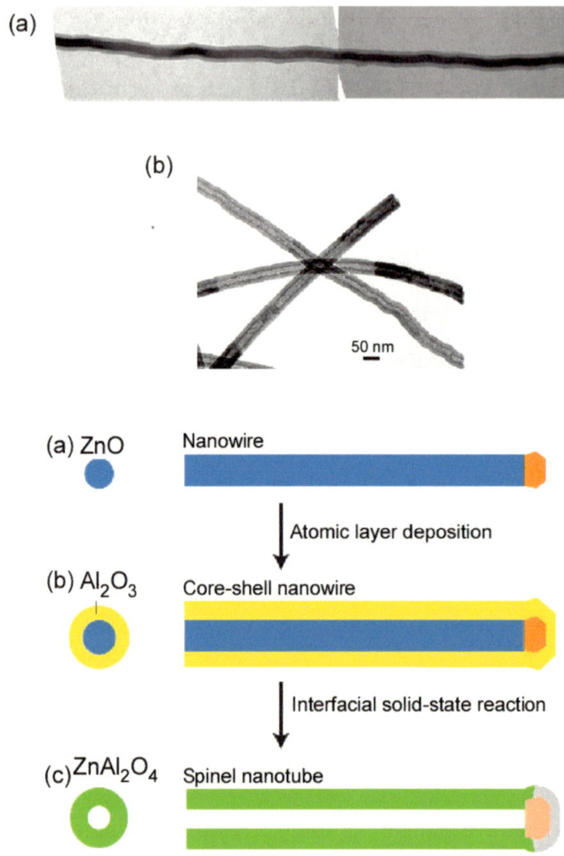

Fig. 2. Transformation of core-shell nanowires to nanotubes by means of the Kirkendall effect (upper). TEM image of an example (a) ZnO-Al_2O_3 core-shell nanowire and (b) $ZnAl_2O_4$ spinel nanotubes. Schematic diagram of the formation process of $ZnAl_2O_4$ spinel nanotubes (lower). (a) Single-crystal ZnO nanowires are grown by the vapour-liquid-solid mechanism using Au nanoparticles as a catalyst. (b) The nanowires are coated with a uniform layer of Al_2O_3 by ALD, forming core-shell ZnO-Al_2O_3 nanowires. (c) Annealing the core-shell nanowires leads to the formation of $ZnAl_2O_4$ nanotubes by a spinel-forming interfacial solid-state reaction involving the Kirkendall effect. (Reproduced with permission from the reference (Fan et al, 2006))

The purpose of this chapter is to show how to fabricate oxide nanotubes through the Kirkendall effect by utilizing metal nanowires as a precursor.

2. Shape evolution of metal nanowires due to the Kirkendall effect

2.1 Morphology change from metal nanowires during oxidation

In this section, the formation behaviour of interior nanovoids during the oxidation of metal nanowires such as Fe, Cu and Ni will be introduced in 2.1.1 and 2.1.2. The mechanism of interior nanovoids and the difference in void formation behaviour will be discussed in 2.1.3 in terms of diffusion in oxides and metals.

2.1.1 Formation of oxide nanotubes via the oxidation of Fe and Cu nanowires

A typical example of the changes in the morphology after the oxidation of iron nanowires from 473 to 573 K is shown in Fig. 3. In our study, metallic nanowires were prepared by electrodeposition through a polycarbonate membrane with a nominal cylindrical pore size of 15 nm and the thickness of 6 μm (Nakamura et al., 2009b). Figure 3 shows bright field images (BFIs) of Fe nanowires (a) before and after oxidation at (b) 473 K for 7.2 ks, (c) 523 K for 3.6 ks, and (d) 573 K for 3.6 ks and their corresponding selected area electron diffraction (SAED) patterns (a')-(d'). Fe nanowires before oxidation had a diameter of 55 nm and a body-centred-cubic structure, as shown in Figs. 3(a) and (a'). Voids were clearly observed along the interface between the Fe nanowire and the outer oxide layer at 473 K for 7.2 ks (Fig. 3(b)) and these were larger at 523 K for 3.6 ks (Fig. 3(c)). Finally, a Fe nanowire turned into a nanotube with a cylindrical nanopore of 40 nm in diameter. The crystal structure of the nanotube was identified to be Fe_3O_4 (magnetite) by analyzing the SAED of Fig. 3 (d').

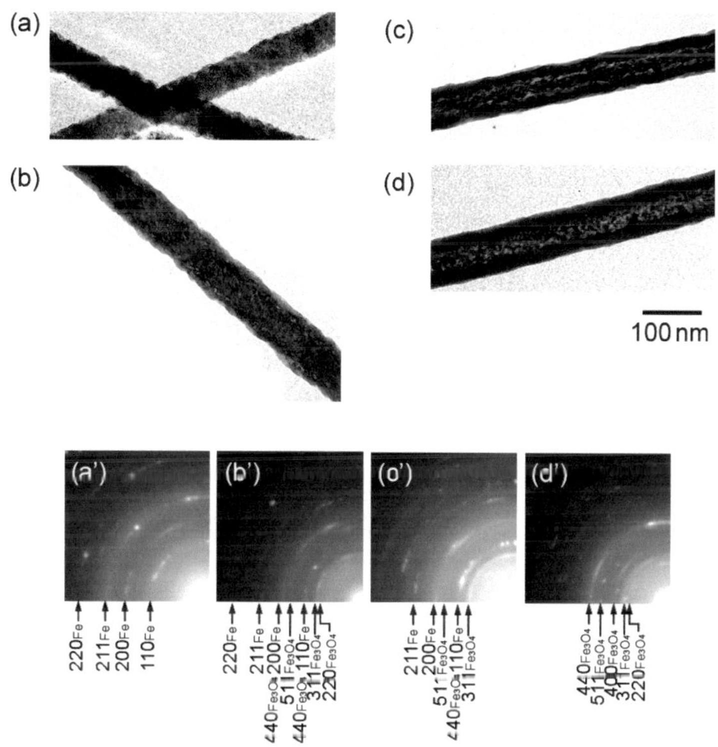

Fig. 3. (a) A bright field image (BFI) of as-deposited Fe nanowires on an amorphous alumina film. (b-d) BFIs of Fe nanowires after oxidation: (b) at 473 K for 7.2 ks, (c) at 523 K for 3.6 ks, and (d) at 573 K for 3.6 ks. (a'-d') are the corresponding selected area electron-diffraction patterns of (a-d). (Reproduced with permission from the reference (Nakamura et al, 2009b))

Figure 4 shows the BFIs of Cu nanowires before (a) and after oxidation at (b,c) 423 and (d) 573 K and their corresponding SAED patterns (a')-(d'). Interior voids with different sizes are fragmentarily generated inside the Cu nanowires after oxidation at 423 K for 1.2 ks, as shown in Fig. 4(b). The oxidized Cu nanowires become nanotubes with a uniform inner and outer diameter at 423 K for 5.4 ks (Fig. 4(c)), as is the case with the oxidation of Fe nanowires. The crystal structure of the nanotubes obtained via the oxidation of the Cu nanowires at 423 K for 5.4 ks was identified to be Cu_2O. Cu_2O transformed into CuO at 573 K while maintaining the nanotube structure, as can be seen in Figs. 4 (d) and (d'). The similarity of the changes in morphology and crystal structure can be also seen for the oxidation of Cu nanoparticles (Nakamura et al., 2008); Cu nanoparticles turn into hollow Cu_2O at around 423 K, followed by the phase transformation from Cu_2O to CuO at 573 K with the hollow structure being maintained.

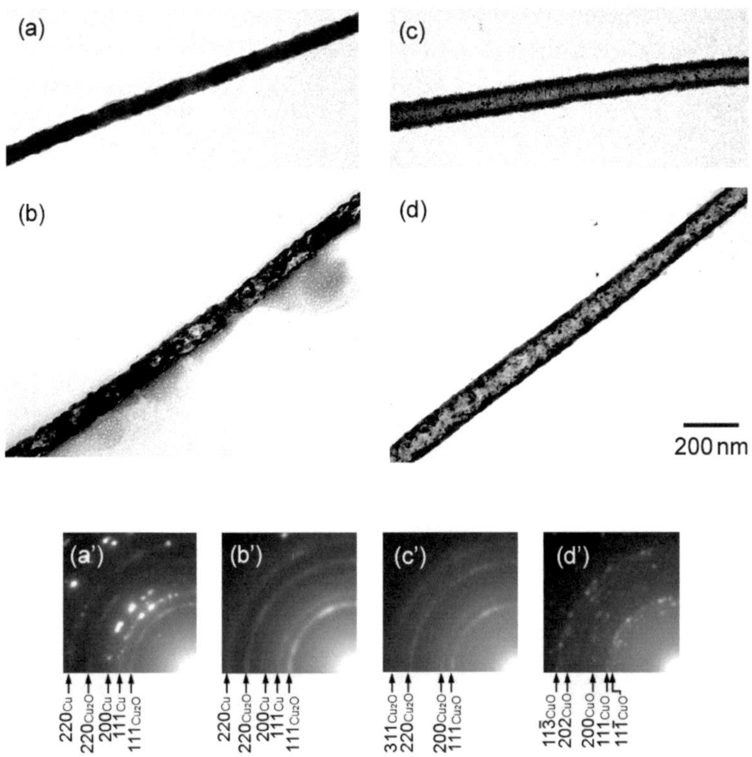

Fig. 4. BFIs of Cu nanowires (a) before oxidation and after oxidation (b) at 423 K for 1.2 ks, (c) at 423 K for 5.4 ks and (d) at 573 K for 3.6 ks and their corresponding SAED patterns (a'-d'). (Reproduced with permission from the reference (Nakamura et al, 2009b))

2.1.2 Formation of bamboo-like porous NiO nanowires from Ni nanowires

Figure 5 shows representative electron micrographs of Ni nanowire after oxidation at 673 and 773 K. After oxidation at 673 K for 1.8 ks, large voids are formed at certain places at the interface between an inner remaining Ni wire and the outer oxide layer, as indicated by arrows in Fig. 5(b). The formation behaviour of voids at the interface between Ni and NiO is different from that between Fe and Fe_3O_4, where voids are uniformly formed along the interface as apparent from Figs. 3(b) and (c). Ni nanowires became irregularly-shaped bamboo-like NiO with separate interior nanovoids of irregular diameters at 773 K for 3.6 ks, as shown in Fig. 5(c). After the original work carried out by us, Ren at al. (2010) reported similar results on morphology change during the oxidation of Ni nanowires in the temperature range between 623 and 923 K.

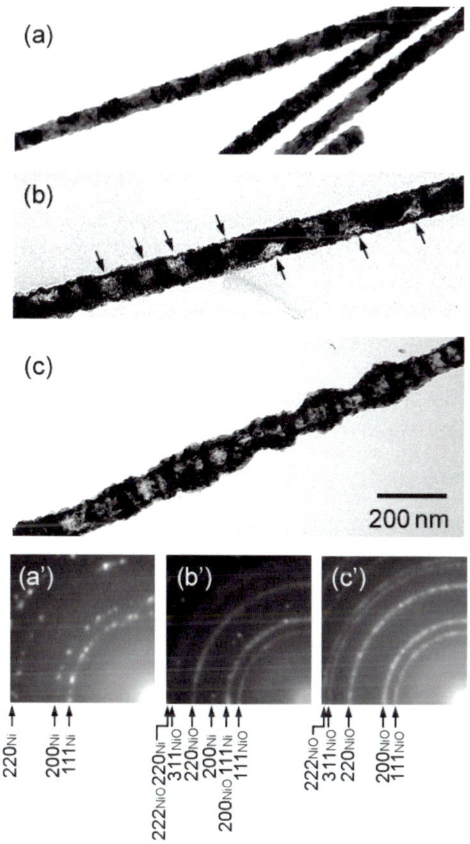

Fig. 5. A typical example of change in morphology of Ni nanowires via oxidation: (a) before oxidation and after oxidation (b) at 673 K for 1.8 ks and (c) at 773 K for 3.6 ks. (Reproduced with permission from the reference (Nakamura et al, 2009b))

2.1.3 Formation mechanism of interior nanopores

The formation mechanism of Fe_3O_4 and Cu_2O nanotubes is illustrated in Fig. 6 (a); (i) an oxide layer is formed due to the oxidation reaction dominated by the outward diffusion of metal ions through the oxide layer, (ii) voids are generated along the interface due to the inward diffusion of vacancies, which results from the Kirkendall effect, as the oxide layer grows, and (iii) an oxide nanotube with a cylindrical pore is formed after all the metal atoms are consumed by reacting with oxygen.

On the other hand, peculiar behaviour in void formation can be seen in the oxidation of Ni nanowires; the voids grow larger at certain places of the interface between inner Ni and the outer oxide layer during oxidation. Therefore, in the case of Ni, the final morphology is a irregularly-shaped bamboo-like structure with separate interior nanovoids. It can be expected from the TEM image of Fig. 5 (b) that vacancies, which flow into the Ni side as a counter to the outward diffusion of Ni, migrate over a long-range distance, resulting in the growth of voids at certain places. A schematic illustration describing the void formation behaviour during the oxidation of a Ni nanowire is shown in Fig. 6(b). Vacancies, which are formed due to the outward diffusion of Ni at the interface between Ni and NiO, migrate toward a void (i). The growth of a localized void causes inhomogeneous oxide growth; the growth of the NiO layer is suppressed at the region with a void compared with the region without void in Ni (ii). As a result, a bamboo-like structure with separate interior voids and an irregular shape is formed (iii). In our previous study (Nakamura et al., 2008a), the localization of void formation, which was observed in the oxidation of Ni nanoparticles, was discussed in terms of the mobility of vacancies: vacancies in Ni have sufficient mobility to diffuse over a long-range distance considering the slow growth rate of the NiO layer, resulting in the growth of a single large void at a certain site on the Ni/NiO interface and the formation of an interior nanopore at an off-centred position of hollow NiO. Both the localization of an interior void during the oxidation of Ni nanoparticles and nanowires seem to originate in the high mobility of vacancies in Ni during its oxidation.

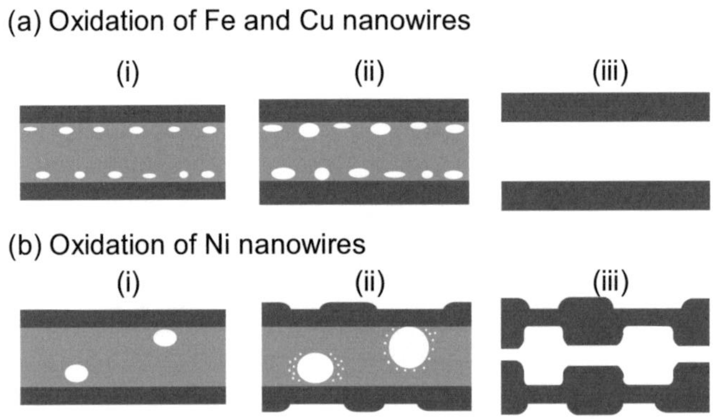

Fig. 6. Schematic illustrations of morphology change during oxidation of (a) Fe and Cu and (b) Ni nanowires: (i) the formation of oxide layer due to outward diffusion of metal ions through the initial oxide layer and the formation of voids at the interface between inner metal and outer oxide layer, (ii) the growth of the voids at the interface and (iii) the formation of oxide nanotubes and nanoporous wires with interior nanopore. (Reproduced with permission from the reference (Nakamura et al, 2009b))

2.2 Shape evolution via annealing at higher temperatures

According to Gusak et al.(2005), hollow nanostructures are energetically unstable because they include the extra inner surface. They predicted theoretically that hollow nanostructures tend to shrink and collapse as a result of annealing at high temperatures. The authors obtained the experimental results that hollow oxide nanoparticles shrunk and collapsed at higher temperatures (Nakamura et al., 2008b). In order to understand the structural stability of nanotubes of Fe_3O_4 and CuO and the bamboo-like structure of NiO, the morphology change after annealing in air at higher temperatures was also investigated (Nakamura et al., 2009b). The results will be reviewed.

2.2.1 Shrinkage of interior nanopores

Figure 7 shows the bright field images of (a) CuO nanotubes and (b) the NiO bamboo-like structures after annealing in air at 773 K for 3.6 ks and at 923 K for 3.6 ks, respectively. It is evident that the uniform cylindrical pore in a CuO nanotube, as shown in Fig. 4(d), is almost annihilated, with separate interior voids partly remaining, suggesting that the shrinkage and collapse of CuO nanotubes occur during annealing. In the case of NiO bamboo-like structures as well as CuO nanotubes, the size of interior separate voids in Fig. 7(b) clearly decreases, compared with those in Fig. 5(d).

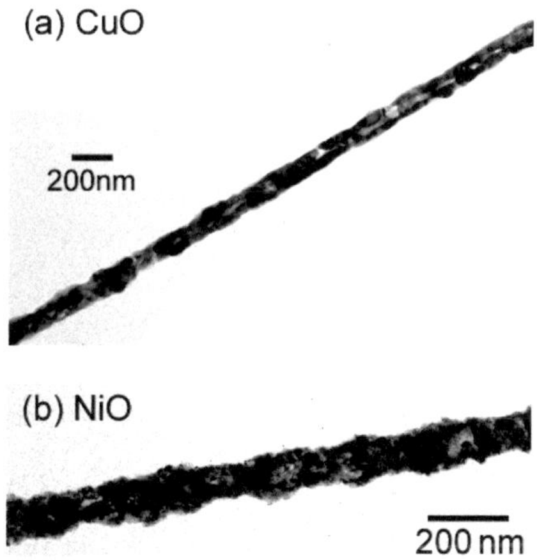

Fig. 7. BFIs of (a) CuO and (b) NiO nanotubes after annealing in air at 773 K for 3.6 ks and at 923 K for 3.6 ks, respectively. (Reproduced with permission from the reference (Nakamura et al, 2009b))

It was demonstrated theoretically (Tu & Gösele, 2005; Gusak et al., 2005; Evteev et al., 2007; Fischer & Svoboda, 2008) and experimentally (Nakamura et al., 2008b) that hollow nanoparticles tend to shrink and collapse at high temperatures because atomic movement occurs to annihilate the inner surface, which originates in the instability of hollow nanoparticles. The shrinkage of interior nanopore(s) was also observed for the oxide nanowires of Fe, Cu and Ni with interior nanopore(s) in our study. It can be concluded, therefore, that both the nanotubes and the hollow nanoparticles tend to shrink and collapse because they are energetically unstable. In our previous study on the structural stability of hollow CuO and NiO nanoparticles (Nakamura et al., 2008b), it was found that hollow CuO and NiO shrank and collapsed into solid oxides in air at 673 and 923 K, respectively, where the diffusion coefficient of slower component reaches about 10^{-22} m^2s^{-1}. The temperatures at which CuO nanotubes and NiO bamboo-like structures shrank are almost consistent with these temperatures, supporting the idea that shrinking process is controlled by slower diffusion species in the case of two-component system (Gusak et al., 2005). It is not easy to estimate the diffusion coefficient of slower components of iron oxides in the shrinking process at the temperatures where shrinkage occurs because the phase transformation from Fe_3O_4 to γ-Fe_2O_3 is accompanied by shrinkage. Assuming that the shrinkage mechanism of Fe_3O_4 nanotubes can be simply described by a slower diffusion coefficient in Fe_3O_4, the diffusion coefficient of slower diffusion components, oxygen ions, can be calculated from diffusion data by Castle and Surman (1967) to be around 10^{-23} m^2s^{-1} at 673-773 K where the shrinkage starts. The estimated value seems to be not so different from the 10^{-22} m^2s^{-1} obtained for the slower diffusion coefficients in CuO and NiO.

2.2.2 Transition from iron oxide nanotube to nanoporous wires

Although Fe_3O_4 nanotubes shrink and collapse by annealing at higher temperatures in air as is the case with Cu- and Ni-oxides, peculiar changes in morphology occur in the process from hollow structures to collapsed structures. In this section, the unique behaviour is reviewed. Figure 8 shows a typical example of the morphology transition of Fe_3O_4 nanotubes during annealing in air for 3.6 ks at (a) 673, (b) 773, and (c) 873 K. Fe_3O_4 nanotubes also show a tendency to shrink with increasing annealing temperature, with the nanotube structure collapsing at 873 K. It should be noted, however, that spherical nanovoids of several nanometers in diameter are also generated along the inner wall of the nanotubes at 673 K (Fig. 8(a)) and they become larger at 773 K (Fig. 8(b)). In our previous paper, we found out the experimental result supporting that the formation of the additional spherical nanovoids is closely correlated with the shrinkage of the cylindrical nanopore; the additional nanovoids appeared at 673 K, where the diameter of nanotubes started to decrease (Nakamura et al., 2009a). Furthermore, from the line profile of Fig. 8 (d) converted from the SAED patterns (a')-(c'), it is clear that Fe_3O_4 transforms into γ-Fe_2O_3 above 673 K.

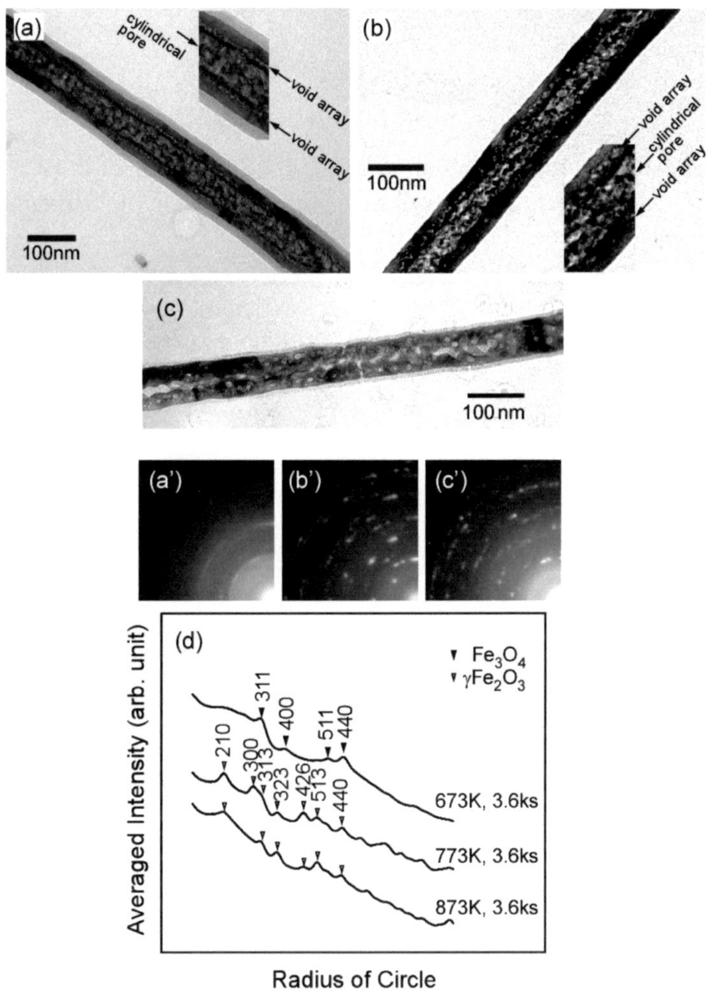

Fig. 8. A typical example of change in morphology of Fe_3O_4 nanotubes through annealing in air: (a) 673 K, (b) 773 K and (c) 873 K for 3.6 ks. (a')-(c') The corresponding SAED patterns and (d) the line profile of (a')-(c'). (Reproduced with permission from the references (Nakamura et al, 2009a,b))

There is a definite difference in morphology change during shirking between CuO and NiO, and Fe_3O_4; additional nanovoids are introduced along the inner wall of the nanotubes only in the shrinking process of Fe_3O_4 nanotubes. It was reported in our previous papers (Nakamura et al., 2008b) that a similar tendency can be seen in the shrinking process of hollow oxide nanoparticles; hollow CuO and NiO nanoparticles become solid particles without the formation of additional voids, whereas hollow Fe_3O_4 nanoparticles turn into solid γ-Fe_2O_3 via a porous structure (Nakamura et al., 2009a).

It should be pointed out, as mentioned above, that the phase transformation from Fe_3O_4 to γ-Fe_2O_3 proceeds when the formation of additional voids and the shrinkage of a cylindrical nanopore occur simultaneously while the crystal structure of CuO and NiO nanotubes remains unchanged in the process of shrinking. The significant difference in the shrinking process between Fe_3O_4 nanotubes and CuO and NiO nanotubes is attributed to the phase transformation of iron oxides.

According to Gusak et al. (2005), the shrinkage of hollow nanoparticles can be described as the outward diffusion of vacancies from the inner pores (vacancy source) to the surface (vacancy sink) of hollow nanoparticles. It is possible, therefore, to speculate that the vacancies, which generate from an inner pore, diffuse outward and then contribute to the formation of additional nano-voids. The formation of many voids of several nanometers in length along the inner wall of the Fe_3O_4 nanotube is regarded as evidence of the outward diffusion of vacancies from the inner cylindrical pore.

The morphology change from a Fe_3O_4 nanotube to porous structures is illustrated in Fig. 9; (i) vacancies diffuse outward from an interior pore as the pore shrinks, (ii) additionally spherical voids are introduced due to vacancy clustering, (iii) the shrinkage of the inner pore proceeds and the additional spherical voids grow larger, and (iv) γ-Fe_2O_3 nanoparticles and nanowires without nanopores were formed. This is the process by which the phase transformation from Fe_3O_4 to γ-Fe_2O_3 gradually takes place.

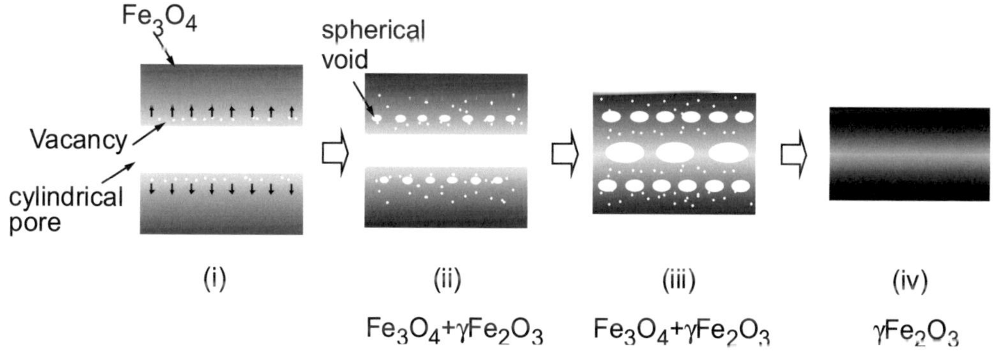

Fig. 9. Schematic illustrations of transition in the morphology of a Fe_3O_4 nantube: (i) dissociation and outward diffusion of vacancies at the inner surface of a nanotube of Fe_3O_4, (ii) formation of additional spherical voids associated with shrinkage of an interior pore and phase transformation from Fe_3O_4 to γ-Fe_2O_3, (iii) growth of voids and shrinkage of an interior pore, and (iv) the formation of nonporous γ-Fe_2O_3.Vacancies are generated from inner surface and diffuse outward. The outward vacancy diffusion accompanies the phase transformation from Fe_3O_4 to γ-Fe_2O_3. The vacancies combine to form voids during the diffusion process. (Reproduced with permission from the reference (Nakamura et al, 2009a))

As mentioned earlier, the phase transformation from Fe_3O_4 to γ-Fe_2O_3 coincides with the shrinkage of the cylindrical pore, in other words, the outward diffusion of vacancies. Sato et al. (1991) suggested that stress is induced around the boundary between Fe_3O_4 and γ-Fe_2O_3 in the oxidation process of Fe_3O_4 thin film into γ-Fe_2O_3. In the shrinking process of hollow Fe_3O_4, the formation of the γ-Fe_2O_3 phase seems to proceed simultaneously from the surface side to the inner side, inducing a strain field around the interface between Fe_3O_4 and γ-Fe_2O_3. If the strain field at the phase boundary is compressive, the vacancies that dissociate from the inner pore of the hollow particles assemble and then form a cluster around the boundary due to the tensile field of the vacancy, resulting in the relaxation of the strain around the phase boundary. Thus, the formation of additional nanovoids during the annealing of Fe_3O_4 nanotubes in air at high temperatures is closely related to the phase transformation from Fe_3O_4 and γ-Fe_2O_3. The possible formation mechanism of duplex porous iron-oxides can be described both by the outward vacancy diffusion from the nanopore and the recombination of the vacancies in the vicinity of the interface between Fe_3O_4 and γ-Fe_2O_3.

2.3 Other examples of nanotube formation via the Kirknedall effect

Another example of formation of nanotubes through the Kirkendall effect is the formation of Ag_2Se nanotubes from the reaction Ag and Se ions (Bernard et al., 2006). TEM images of obtained Ag_2Se nanotubes are shown in Fig. 10, together with the schematic illustrations of

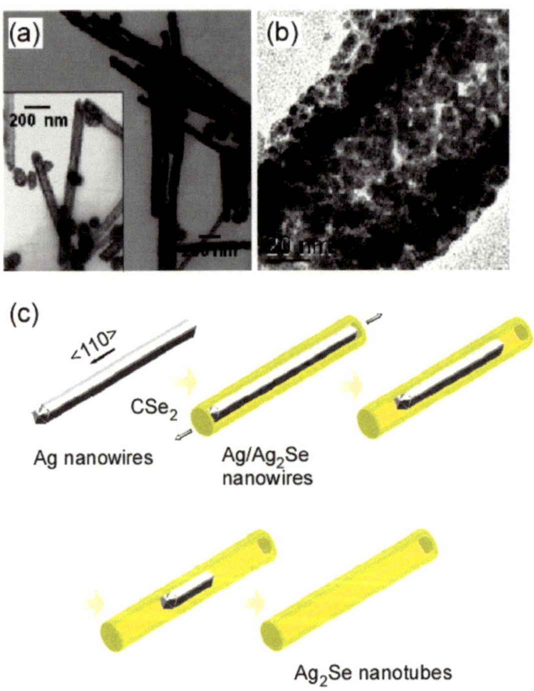

Fig. 10. Ag_2Se nanotubes formed based on the Kirkendall effect. (a) TEM image of the tubes. (b) Higher-magnification view of the tube showing that the tube wall is composed of multiple grains. (c) Schematic images of the diffusion process. Due to the higher concentration of CSe_2 adsorbed at the end (111) faces than on the side (100) faces of the Ag nanowires, the void grows along the longitudinal <110> direction from the ends. (Reproduced with permission from the reference (Bernard et al., 2006)).

the formation process. CSe_2 was first adsorbed onto the surfaces of the Ag nanowire templates, and subsequent photolysis causes the dissociation of C=Se bonds, giving rise to a high surface concentration of Se atoms. Ag_2Se layers are formed on the surface of Ag nanowires at 433 K and then growth of Ag_2Se layers is dominated by the outward diffusion of Ag^+ ions through the layer. The authors mentioned that the voids were observed to grow horizontally along the wire axis rather than isotropically because the adhesion of CSe_2 is preferred on the ends of the Ag nanowire rather than on its side faces. Hence, the high concentration of Se atoms at the nanowire ends promotes the diffusion of Ag^+ along the longitudinal axis, leading to the observed anisotropic growth of vacancies along <110>. This represents a different strategy of applying the Kirkendall effect to nanotube formation from the one presented in Fig. 10.
Raidongia & Rao (2008) reported the synthesis of nanotubes of SiO_2, Co_3O_4, ZnS, CdS and CdSe by the reaction of the corresponding elemental nanowires with oxygen, sulfer and selenium due to the mechanisms of the Kirkendall effect.

3. Conclusion

In the past few years, there have been considerable advancements concerning the synthesis of nanowires and nanotubes. The synthesis routes based on chemical reactions such as chemical etching and galvanic replacement are broadly used for a wide variety of materials. More information on the chemical synthesis routes can be obtained in the references (Lou et al., 2008; An & Hyeon, 2009). In addition, different ideas have been applied to the fabrication of hollow nanomaterials. For example, the Kirkendall-effect related processes have been recognized as one of the useful methods for the fabrication of nanotubes of oxides, sulfides and phosphides. Since the Kirkendall effect is related to the mass transport at the interface between different solids, there is much possibility to control hollow nanostructures of a variety of alloys, intermetallic compounds and mixed oxides. The result by Fan et al. (2006), who successfully fabricated $ZnAl_2O_4$ spinel nanotubes from ZnO/Al_2O_3 core/shell nanowires, strongly suggests the formation of A-B alloys, compounds and mixed oxides if A/B core/shell type nanostructures are easily obtained. We can find the attempts to apply the interdiffusion processes between different oxides, which are accompanied by formation of the Kirkendall void, to the introduction of nanopipes or nanochannel inside nanowires (Marcu et al., 2010). Such basic studies are expected to increase the interests in the application nanopipes or nanochannel for DNA molecules' manipulation and investigation.
Our recent studies on the shrinkage behaviour of nanotubes are also introduced. Nanotubes of metal oxides tend to shrink and collapse at higher temperatures in oxidation atmosphere because hollow nanostructures with an inner surface are energetically unstable. Shrinking of hollow oxides occurs at temperatures where the diffusion coefficients of slower diffusing ions in the oxides are of the order of 10^{-22} m^2s^{-1}. These results suggest that it is possible to control the size of interior nanopores and particles using annealing in high temperatures. In the case of hollow iron oxide, however, a peculiar morphology change is induced in the process of shrinkage; hollow oxides with an interior spherical and cylindrical nanopore transform into porous structures with additional multiple nanovoids. Transition in porous structure seems to be related to the outward diffusion of vacancies from interior pore and

the phase transition from magnetite to maghemite. It was found by our experiments that the outward diffusion of vacancies that takes place during shrinkage contributes to the formation of a unique nanoporous structure. It should be pointed out that annealing of hollow nanostructures at high temperatures may induce further change in morphology in some cases. Further research works should be carried out to investigate the morphology change of different nanostructures at high temperature.

4. Acknowledgement

The authors would like to thank Prof. H. Mori and Mr. E. Taguchi of Research Center for Ultra-High Voltage Electron Microscopy, Osaka University, for TEM observation and Dr. H. Tsuchiya and Prof. S. Fujimoto of Graduate School of Engineering, Osaka University, for their kind support of the preparation of metal nanowires via electrochemical method. This work was supported by Grant-in-Aid for Scientific Research (S) (No. 17106009) and also by Priority Assistance for the Formation of World Wide Renowned Centers of Research- The Global COE Program (Project: Center of Excellence for Advanced Structural and Functional Materials Design), from the Ministry of Education, Culture, Sports, Science and Technology, Japan.

5. References

An, K. & Hyeon, T. (2009). Synthesis and biomedical applications of hollow nanostructures. *Nano Today*, Vol. 4, No. 4, pp. 359-373, ISSN 1748-0132.

Bernard Ng, C. H. Tan, H. & Fan, W. Y. (2006). Formation of Ag_2Se aanotubes and Dendrite-like structures from UV irradiation of a CSe_2/Ag colloidal solution. *Langmuir*, Vol. 22, No. 23, pp. 9712-9717, ISSN 0743-7463

Castle, J. E. & Surman, P. L. (1967). The Self-Diffusion of Oxygen in Magnetite. *The Journal of Physical Chemistry*, Vol. 71, No. 13, pp. 4255-4259, ISSN 0022-3654

Evteev, A. V. Levchenko, E. V. Belova, I. V. & Murch, G. E. (2007). Shrinking kinetics by vacancy diffusion of a pure element hollow nanosphere. *Philosophical Magazine*, Vol. 87, No. 25, pp. 3787-3796, ISSN 1478-6435

Fan, H. J. Gösele, U. & Zacharias, M. (2007). Formation of Nanotubes and Hollow aanoparticles based on Kirkendall and diffusion processes: A review. *Small*, Vol. 3, No. 10, pp. 1660-1671, ISSN 1613-6829

Fan, H. J. Knez, M. Scholz, R. Nielsch, K. Pippel, E. Hesse, D. Zacharias, M. & Gösele, U. (2006). Monocrystalline spinel nanotube fabrication based on the Kirkendall effect. *Nature Materials*, Vol. 5, No. 8, pp. 627-631, ISSN 1476-1122

Fischer, F. D. & Svoboda, J. (2008). High temperature instability of hollow nanoparticles. *Journal of Nanoparticle Research*, Vol. 10, No. 2, pp. 255-261

Goldberger, J. He, R. Zhang, Y. Lee, S. Yan, H. Choi, H.-J. & Yang, P. (2003). Single-crystal gallium nitride nanotubes. *Nature*, Vol. 422, No. 6932, pp. 599-602, ISSN 0028-0836

Gusak, A. M. Zaporozhets, T. V. Tu, K. N. & Gösele, U. (2005). Kinetic analysis of the instability of hollow nanoparticles. *Philosophical Magazine*, Vol. 85, No. 36, pp. 4445-4464, ISSN 1478-6435

Lou, X. W. Archer, L. A. & Yang, Z. (2008). Hollow micro-/nanostructures: synthesis and applications. *Advanced Materials*, Vol. 20, No. 21, pp. 3987-4019, ISSN 1521-4095

Marcu, A. Yanagida, T. & Kawai, T. (2010). Nanochannels' fabrication using Kirkendall effect. *Solid State Sciences*, Vol. 12, No. 6, pp. 978-981, ISSN 1293-2558

Nakamura, R. Lee, J. G. Mori, H. & Nakajima, H. (2008a). Oxidation behaviour of Ni nanoparticles and formation process of hollow NiO. *Philosophical Magazine*, Vol. 88, No. 2, pp. 257 - 264, ISSN 1478-6435

Nakamura, R. Tokozakura, D. Lee, J. G. Mori, H. & Nakajima, H. (2008b). Shrinking of hollow Cu_2O and NiO nanoparticles at high temperatures. *Acta Materialia*, Vol. 56, No. 18, pp. 5276-5284, ISSN 1359-6454

Nakamura, R. Matsubayashi, G. Tsuchiya, H. Fujimoto, S. & Nakajima, H. (2009a). Transition in the nanoporous structure of iron oxides during the oxidation of iron nanoparticles and nanowires. *Acta Materialia*, Vol. 57, No. 14, pp. 4261-4266, ISSN 1359-6454

Nakamura, R. Matsubayashi, G. Tsuchiya, H. Fujimoto, S. & Nakajima, H. (2009b). Formation of oxide nanotubes via oxidation of Fe, Cu and Ni nanowires and their structural stability: Difference in formation and shrinkage behavior of interior pores. *Acta Materialia*, Vol. 57, No. 17, pp. 5046-5052, ISSN 1359-6454

Raidongia, K. & Rao, C. N. R. (2008). Study of the transformations of elemental nanowires to nanotubes of metal oxides and chalcogenides through the Kirkendall effect. *The Journal of Physical Chemistry C*, Vol. 112, No. 35, pp. 13366-13371, ISSN 1932-7447

Ras, R. H. A. Kemell, M. de Wit, J. Ritala, M. ten Brinke, G. Leskelä, M. & Ikkala, O. (2007). Hollow inorganic nanospheres and nanotubes with tunable wall thicknesses by atomic layer deposition on self-assembled polymeric templates. *Advanced Materials*, Vol. 19, No. 1, pp. 102-106, ISSN 1521-4095

Ren, Y. Chim, W. K. Chiam, S. Y. Huang, J. Q. Pi, C. & Pan, J. S. (2010). Formation of nickel oxide nanotubes with uniform wall thickness by low-temperature thermal oxidation through understanding the limiting effect of vacancy diffusion and the Kirkendall phenomenon. *Advanced Functional Materials*, Vol. 20, No. 19, pp. 3336-3342, ISSN 1616-3028

Sato, M. Namikawa, T. & Yamazaki, Y. (1991). Cross-sectional observation of Fe_3O_4-gamma-Fe_2O_3 intermediate thin-films. *Japanese Journal of Applied Physics*, Vol. 30, No. 3B, pp.L489-491, ISSN 0021-4922

Shen, G. Bando, Y. Ye, C. Yuan, X. Sekiguchi, T. & Golberg, D. (2006). Single-Crystal Nanotubes of II3-V2 Semiconductors. *Angewandte Chemie International Edition*, Vol. 45, No. 45, pp. 7568-7572, ISSN 1521-3773

Smigelskas, A. D. & Kirkendall, E. O. (1947). Zinc diffusion in alpha brass. Trans. AIME, Vol. 171, No., pp. 130-142

Tu, K. N. & Gösele, U. (2005). Hollow nanostructures based on the Kirkendall effect: Design and stability considerations. *Applied Physics Letters*, Vol. 86, No. 9, pp., ISSN 0003-6951

Yin, Y. Rioux, R. M. Erdonmez, C. K. Hughes, S. Somorjai, G. A. & Alivisatos, A. P. (2004). Formation of hollow nanocrystals through the nanoscale Kirkendall effect. *Science*, Vol. 304, No. 5671, pp. 711-714

6

Formation of Oxide Nanowires by Thermal Evaporation and Their Application to Gas Sensors

Toshinari Yamazaki
Graduate School of Science and Engineering,
University of Toyama, Toyama
Japan

1. Introduction

Oxide semiconductor gas sensors are based on their change in resistance upon exposure to oxidizing or reducing gases. While gas sensors made of conventional particles and thin films of oxide semiconductors have been studied for these several tens of years, gas sensors made of nanowires have been studied for these ten years because they have possibility of indicating excellent gas sensing properties, owing to the discriminating structure of the nanowires. The high surface-to-volume ratio of the nanowires is expected to provide high sensitivity. The one dimensionally long shape of the nanowires may lead to favourable sensing properties. There are two types of nanowire sensor. One is made of nanowires dispersed on a substrate equipped with a pair of electrodes while the other is made of a single nanowire. In our study, the former type was investigated.

We have first developed three types of evaporation method for forming oxide nanowires and then investigated their structure. The first one is a conventional vapour transport method. The second one is simple evaporation in ambient atmosphere. The last one is a method using a conventional vacuum evaporator with its chamber filled with prescribed pressure of oxidizing gas such as ambient air; we call this method a modified gas evaporation method.

To clarify the feature of nanowire sensors, we have investigated the properties of the gas sensors made of nanowires. The effect of Pd doping was examined in the sensors made of SnO_2 nanowires formed by the vapour transport method, and their performance was compared with sputtered film sensor. It is of great interest to investigate the effect of the thickness and length of the nanowires. The effect of the thickness was investigated for WO_3 nanowires formed by the vapour transport method and SnO_2 nanowires formed by the modified gas evaporation method, while the effect of the length was investigated for the SnO_2 nanowires formed by the modified gas evaporation method in comparison with the gap lengths of electrodes. Gas sensors made of Ga_2O_3 and TeO_2 nanowires, which have been rarely studied, were also investigated for exploring new nanowire sensor materials. In this article, the methods for forming nanowires are first explained, and then the structure of the nanowires and gas sensing properties are introduced.

2. Methods for forming nanowires

2.1 Vapor transport method

In the conventional vapor transport method, which is shown in Fig. 1(a), an alumina boat loaded with a raw material was placed in a quartz tube inserted into a horizontal electric furnace, and Ar gas was introduced through the quartz tube. Although oxygen gas was not introduced, oxide nanowires were obtained probably because of unintentional introduction of a minute amount of oxygen gas. This method was used for the formation of SnO_2 (Shen et al.,2009), WO_3, and Ga_2O_3 (Liu et al., 2008) nanowires, where the position of the substrate, the flow rate and pressure of Ar, and the furnace temperature was adjusted, depending on the materials.

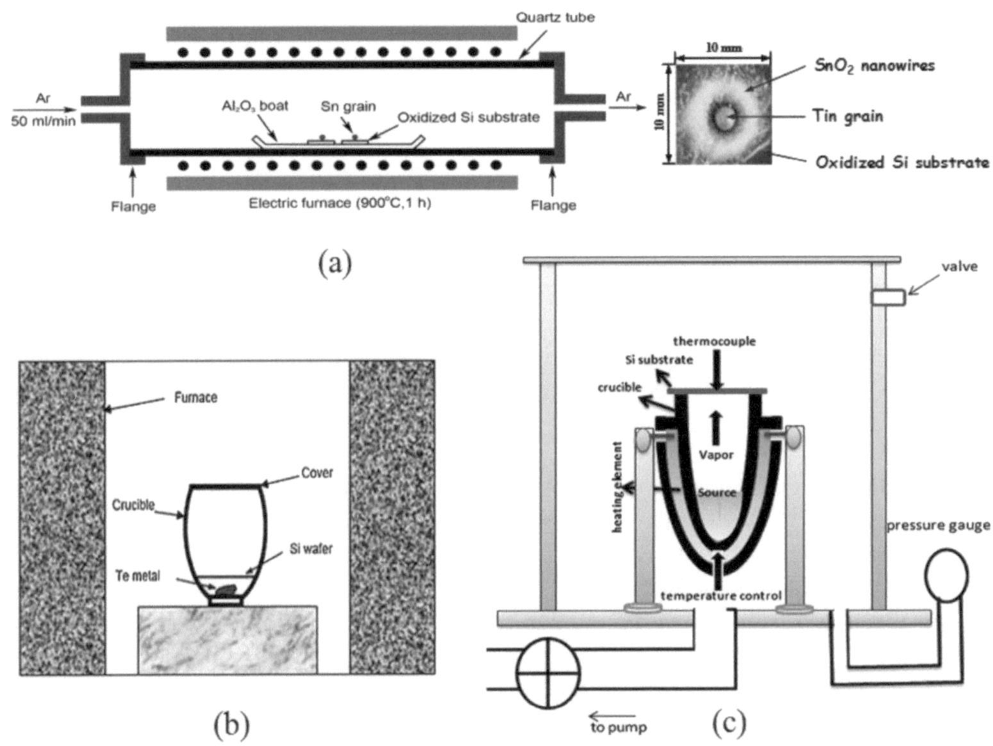

Fig. 1. Formation of oxide semiconductor nanowires. (a) Vapor transport method. (b) Evaporation in ambient air. (c) Modified gas evaporation with a conventional vacuum evaporator.

2.2 Evaporation in ambient air

This method, in which nanowires are formed in an ambient air, is very simple and was applied to the formation of TeO_2 nanowires (Liu et al., 2007; Liu et al., 2008). As shown in Fig.1(b), a desirable amount of tellurium grains at a high purity of 99.999% was placed in an alumina crucible 30mm in height and 20mm in diameter. A piece of oxidized Si substrate was placed about 2mm above the tellurium grains to collect the products. The crucible was covered and heated in a box type furnace at a temperature of 300–500°C for 2 hours. A layer of white product was obtained on the under surface of the substrate.

2.3 Modified gas evaporation

A method, shown in Fig. 1(c), uses a conventional vacuum evaporator. This method, which we call modified gas evaporation method, was applied to the formation of SnO_2 nanowires. The gas evaporation method has been usually used for the formation of metal nanoparticles, which is attained by evaporating metals in vacuum chamber filled with a low pressure (10–1000 Pa) of an inert gas such as helium or argon (Kimoto et al., 1963). It can also be applied to the formation of oxide nanoparticles (Kaito et al., 1977). In the method used here, Sn grains, SnO powder, or SnO_2 powder as a raw material was placed at the bottom of an alumina crucible, and the crucible was covered with an oxidized Si substrate. The crucible was then heated at a high temperature under a low pressure of ambient air. In this configuration, discriminating structures such as nanowires were formed because the substrates on the crucible were also heated at a high temperature.

3. Structure and gas sensing properties of nanowire sensors

3.1 SnO_2 nanowires formed by vapor transport

SnO_2 is the gas sensing material that has been most extensively studied (Windischmann & Mark, 1979; Heiland, 1982; Tamaki et al., 1989). SnO_2 nanowires have also recently been studied vigorously as a gas sensing material (Ramirez et al., 2007; Kumar et al., 2009). In addition to undoped SnO_2 nanowires, we investigated Pd-doped and Pt-doped nanowires, for noble metal doping such as Pd or Pt often improves sensitivity by activating the surface reaction of the gas sensing materials (Yamazoe et al., 1983; Tsang et al., 2001). Because the effect of Pd doping and that of Pt-doping on sensor response were similar to each other, we introduce here only the undoped and Pd-doped nanowire sensors, leaving out Pt-doped nanowire sensors (Shen et al., 2009).

In order to obtain Pd-doped nanowires, 40 µl of ethanol solution of $PdCl_2$ at a concentration of 1×10^{-3} mol/ml or 2.5×10^{-3} mol/ml were dropped on nanowires dispersed on an oxidized Si substrate. The nanowires thus obtained were annealed at 350°C in air for half an hour. The concentration of Pd was estimated to be 0.8 and 2.0 wt%. According to the measurement of X-ray diffraction (XRD) patterns, the SnO_2 nanowires were identified to be tetragonal SnO_2 (JCPDS card No. 41-1445). Figure 2(a) shows a typical scanning electron microscopy (SEM) image of undoped SnO_2 nanowires. The inset shows a high magnification image. SnO_2 nanowires of 30–200 nm in diameter and several tens of micrometers in length are observed. A transmission electron microscopy (TEM) image of an undoped SnO_2 nanowire is shown in Fig. 2(b). A high-resolution TEM (HRTEM) image of this nanowire and its selected area electron diffraction (SAED) pattern represented in Fig. 2(c) indicate that a nanowire is a single crystal. The interplanar spacing of 0.34 nm indicated in Fig. 2(c) corresponds to the (110) plane of a tetragonal SnO_2. The growth direction of SnO_2 nanowire is [301], which is consistent with previous reports (Chen et al., 2003; Wang et al., 2004). The TEM image of a 2 wt% Pd-doped SnO_2 nanowire in Fig. 3(a) shows that dark spots distribute randomly on the surface. An HRTEM image (Fig. 3(b)) shows that they are 3–7nm in diameter. The microbeam diffraction pattern measured for a spot shown in the inset of Fig. 3(b) indicates that the dark spots are PdO particles. According to EDX analysis (Fig. 3(c)), SnO_2 nanowires are surely doped with Pd. No Cl-related signal was observed in the spectrum.

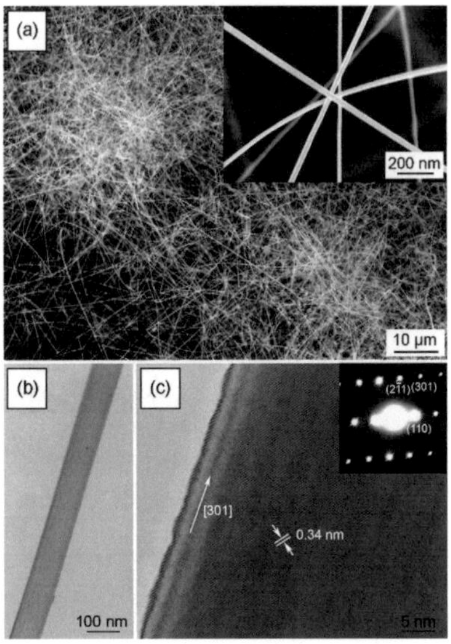

Fig. 2. SEM (a), TEM (b), and HRTEM (c) observation of undoped nanowires.

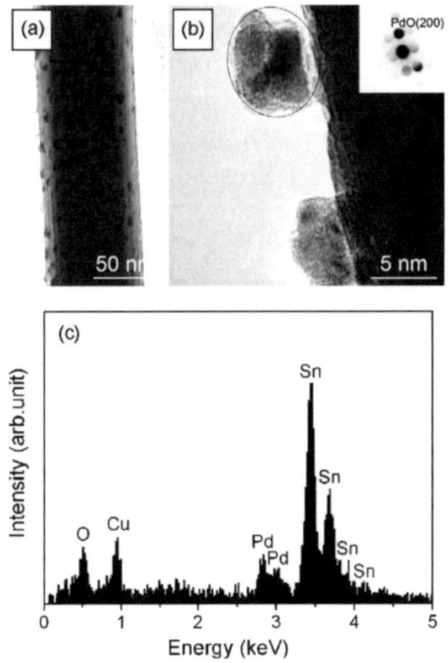

Fig. 3. TEM (a), HRTEM (b), and EDX (c) observation of Pd-doped nanowires.

Sensor devices, illustrated in Fig. 4, were fabricated by dropping nanowire-suspended ethanol on an oxidized Si substrates equipped with a pair of interdigitated Pt electrodes. The gap length of the electrodes was 120 μm. The weight of SnO_2 nanowires dispersed on a sensing element with sizes of 7 mm × 10 mm was 500 μg. The gas sensing properties were measured for the sensors placed in a quartz tube inserted into a tubular electric furnace. Dry synthetic air, mixed with a desired concentration of H_2, flowed at a rate of 200 ml/min.

The changes in the resistance of the sensors made of undoped, 0.8 wt%Pd-doped, and 2 wt%Pd-doped SnO_2 nanowires upon exposure to H_2 at 150°C are shown in Figs. 5(a). The resistance decreases upon exposure to H_2. The decrease in the resistance is expanded as the H_2 concentration increases. The resistance recovers to its initial value after H_2 removal, indicating a good reversibility of these gas sensors. The decrease in the resistance is greatly enhanced by introducing Pd. In Fig. 5(b), the sensitivities defined as Ra/Rg, where Ra and Rg are the resistance before and after exposure to a gas (H_2), respectively, are shown as functions of operating temperature for the three sensors described above. Pd doping improves the sensitivity and lowers the operating temperature at which the sensitivity is maximized.

Now, we compare nanowire sensor with sputtered film sensor. Figure 6 shows the temporal change in the resistance upon exposure to H_2 at a temperature of 100°C. The sensors compared are the one made of nanowires doped with 2 wt% Pd and the one made of porous sputtered film doped with 8 wt% Pd. The operating temperature is 100°C. The latter sensor was made of a Pd-doped porous sputtered film. It showed a high sensitivity to H_2. Although the sensitivity of the Pd-doped nanowire sensor is lower than that of the porous Pd-doped sputtered film, the response time and recovery time of the former are shorter than those of the latter. The short response time and recovery time of the nanowire sensor may be related to the single crystalline character of the nanowires. Because a sputtered film is composed of columnar grains made of fine crystallites, it has narrow voids between columnar grains as well as grain boundaries between the crystallites. When the sputtered film sensor is exposed to H_2, the reconstruction of surface atoms and adsorbed species spreads into the voids and boundaries, being accompanied by atomic diffusion through voids and boundaries. Such a diffusion process may delay the completion of the reconstruction process. On the other hand, the nanowire sensor, which is not accompanied by such a process, may show a quick response and recovery.

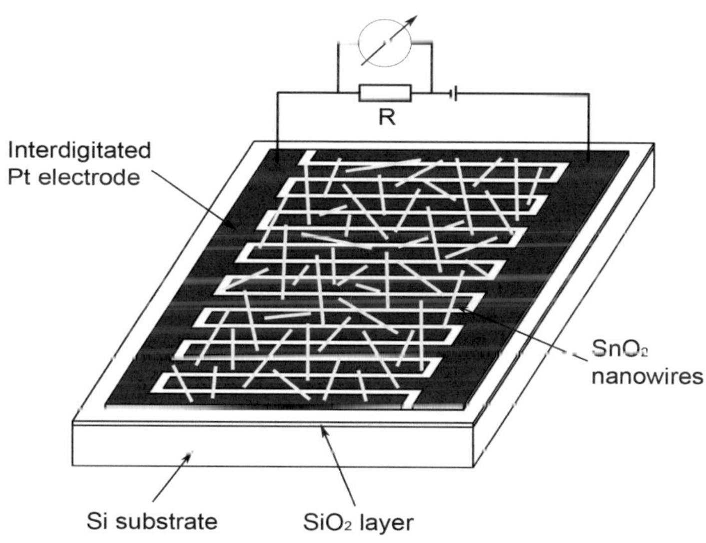

Fig. 4. Schematic diagram of a gas sensor made of nanowires dispersed on a pair of interdigitated electrodes.

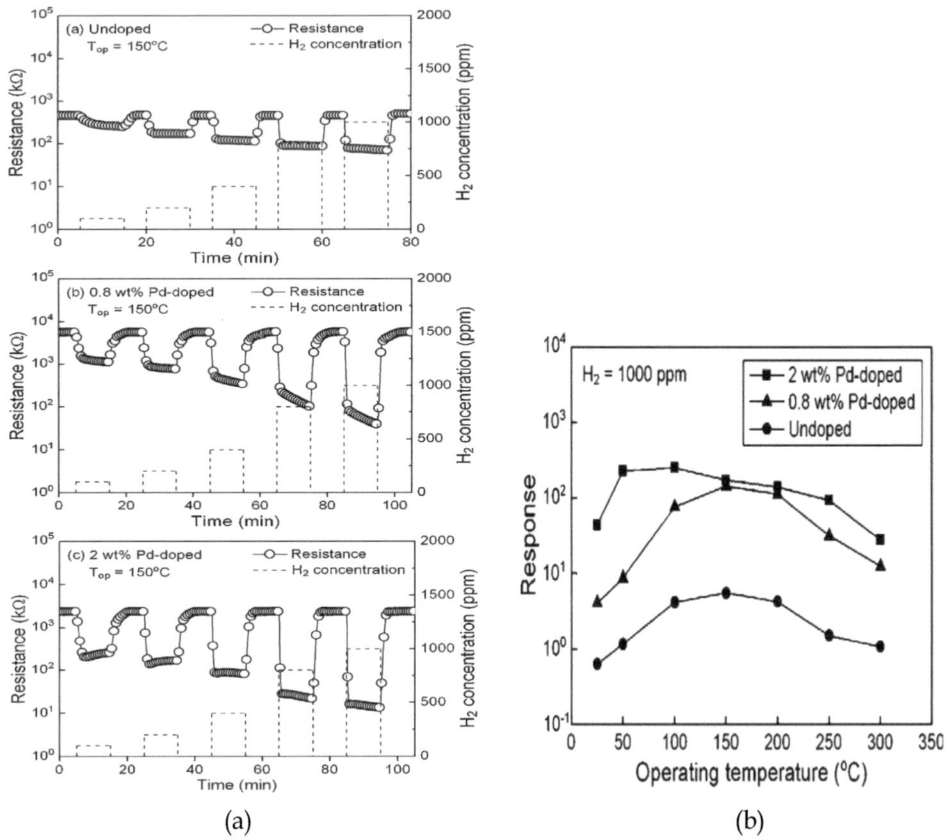

Fig. 5. (a) Changes in the resistance upon exposure to H_2 measured at 150°C. (b) Sensitivity as a function of temperature.

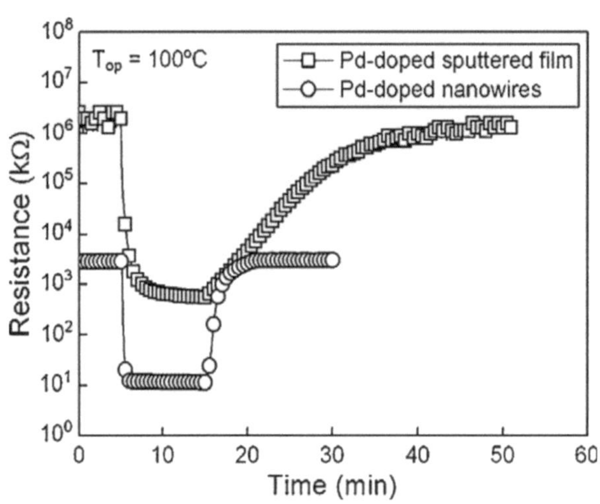

Fig. 6. Changes in the resistance of the 2 wt% Pd-doped nanowire sensor and a porous Pd-doped sputtered film sensor upon exposure to 1000ppm H_2 gas at 100°C .

To clarify the dependence of the response on the operating temperature, the resistances of the undoped and 2 wt% Pd-doped sensors are shown as a function of temperature in Fig. 7, where the values before and after exposure to H_2 are indicated by closed and open symbols, respectively. We see that both the resistance and its temperature dependence are relatively small for the undoped sensor whereas both are large for the 2wt% Pd-doped sensor. In the undoped sensor, oxygen atoms adsorbed on the surface of SnO_2 nanowires in air obtain electrons from the surface region of SnO_2. Thus, a substantial region of each nanowire is depleted of electrons, which results in the high resistance of the sensor. With increasing temperature, the electrons trapped by adsorbed oxygen are thermally excited to the conduction band, causing a decrease in the resistance. When SnO_2 nanowires are exposed to hydrogen, the density of adsorbed oxygen decreases because of the reaction between hydrogen and adsorbed oxygen, which results in a decrease in the resistance. When SnO_2 nanowires are decorated with PdO particles, electrons transfer from SnO_2 to PdO as well as to adsorbed oxygen atoms because the work function of PdO is larger than that of SnO_2 (Jang. et al., 2006; Shimizu et al., 2007). Thus, each nanowire is depleted of electron in its whole region. Therefore, the resistance of the sensor is extremely high in air. When hydrogen is introduced, PdO is reduced to metallic palladium, returning electrons to SnO_2. Hydrogen molecules adsorbed on palladium simultaneously spill over the surface of SnO_2, activating the reaction between hydrogen and the adsorbed oxygen (Yamazoe et al., 1983; Tsang et al., 2001). These phenomena result in an effective shrinkage of the depletion region and lower the resistance of the Pd-doped sensor compared with that of the undoped sensor, leading to a higher response. Because the effect of Pd is active at a relatively low temperature, the sensitivity of Pd-doped sensor is rather high even at room temperature.

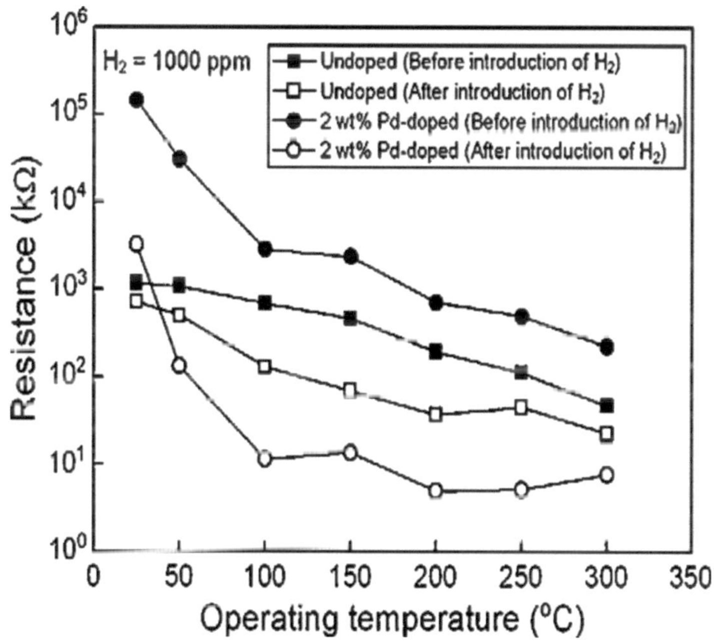

Fig. 7. Temperature dependence of the resistances measured for the undoped and 2 wt% Pd-doped SnO_2 nanowire sensors before and after introduction of H_2.

3.2 WO_3 nanowires formed by vapor transport

WO_3 has been extensively studied as a material suitable for NO_2 sensing (Chung et al., 1999; Jin et al., 2005) since it was shown that it indicated a higher sensitivity to NO_2 than to CO, H_2, CH_4, and i-C_4H_{10} (Akiyama et al., 1991). NO_2 sensors made of WO_3 nanowires have also been studied; the sensors made of nanowires formed by using thermal evaporation (Ponzoni et al., 2006) and solvothermal method (Rout et al., 2006) indicated high sensitivities to NO_2 in the range of parts per billion at high temperatures above 200°C. Sensors made of WO_3 nanowires formed by using the vapor transport method have been investigated (Cao et al., 2009). They could detect NO_2 in the range of parts per billion at 180°C. Sensors made of WO_3 nanowires formed by using soft chemistry indicated high sensitivities to NO_2 in the range of parts per billion at 150°C (Polleux et al., 2006).

In our study, WO_3 nanowires were formed by the vapor transport method using WO_3 powder as a raw material. An alumina boat loaded with WO_3 powder (100 mg, 99.9% purity) was placed at the center of a quartz tube. A substrate on another alumina boat was placed 16, 17, or 17.5 cm away from the WO_3 powder. After the quartz tube was evacuated from the end of the quartz tube at the side of substrate to a pressure of approximately 30 Pa, Ar gas was introduced at a flow rate of 900 ml/min into the quartz tube, which resulted in a pressure of 270 Pa. The temperature of the furnace or the temperature of WO_3 powder was then increased to 1050, 1100, or 1150°C. After maintaining the furnace at these temperatures for 2 hours, the furnace was cooled naturally to room temperature. It should be noted that the substrate temperature was lower than the furnace temperature. Table 1 shows the temperatures of the substrates placed 16, 17, or 17.5 cm away from the WO_3 powder at various furnace temperatures. Our attempt to deposit a product on an oxidized Si substrate failed. Thus, an oxidized Si substrate coated with Au film 5 nm thick was used to deposit the product.

SEM images of the products are shown in Fig. 8. The morphology of the products depends on the furnace temperature and substrate position. We refer to all of the products as nanowires for convenience, although the products deposited at 1050°C are better described as nanostrips. It was difficult to peel the nanowires deposited at 1050°C from the substrate. The nanowires formed at 1100 and 1150°C were in a powdered form, i.e., they were easily peeled from the substrates. The diameter of the nanowires decreased as the distance of the substrate from the WO_3 powder as a raw material increased. Thin nanowires were formed at a furnace temperature of 1100°C. In this condition, the thinnest nanowires with a diameter of 50 nm were formed 17.5 cm away from the WO_3 powder. According to XRD, all products were hexagonal WO_3 listed in JCPDS card No. 85-2460. Note that this structure is different from those reported for the nanowires formed using the vapor transport method by other research groups. WO_3 nanowires with a monoclinic structure (Cao et al., 2009) and $W_{18}O_{49}$ nanowires with a monoclinic structure (Hong et al., 2006) have been reported. The hexagonal structure that we obtained agrees with that of the nanowires formed by a solvothermal method (Choi et al., 2005). According to TEM observation, a nanowire was a single crystal and its growth direction was [001].

Sensors were fabricated by poring a few drops of nanowire-suspended ethanol onto oxidized Si substrates equipped with a pair of Pt interdigitated electrodes with a gap length of 35 μm and an element area of 9 × 10 mm^2. The weight of the dispersed nanowires was 200 μg. The measurement of the sensing properties was performed in the same manner as that for the SnO_2 nanowire sensors.

Furnace temperature (°C)	Substrate temperature (°C)		
	Position of 16 cm	Position of 17 cm	Position of 17.5 cm
1050	490	300	230
1100	520	330	250
1150	550	350	270

Table 1. Substrate temperature at different positions for different furnace temperatures.

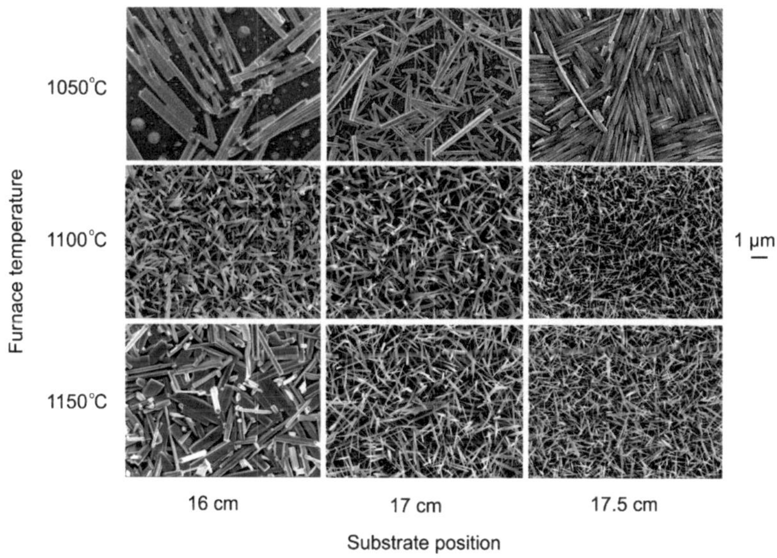

Fig. 8. SEM images of the products deposited at different positions and different furnace temperatures.

The temporal responses to NO_2 at 3 ppm measured for a sensor made of nanowires as thin as 50 nm, deposited on a substrate 17.5 cm away from the WO_3 powder at a furnace temperature of 1100°C, are represented in Fig. 9(a). The results are shown for the operating temperatures of 50, 150, and 250°C. The resistance decreases as the temperature increases. When the sensor is exposed to NO_2, the resistance quickly increases. After the removal of NO_2, the resistance recovers to the initial value in approximately 10 min at a high temperature of 250°C. However, it recovers more slowly at a lower temperature. Figure 9(b) shows the temperature dependence of the sensitivity defined as (Rg-Ra)/Ra. The results for the sensors made of the nanowires deposited at different positions at 1100°C (upper figure) and those for the sensors made of the nanowires deposited on a substrate at 17.5 cm at different temperatures(lower figure) are represented. All sensors, made of nanowires with different diameters, have a peak operating temperature at which their sensitivities are maximized. While NO_2 sensing due to electron transfer from oxide to adsorbed NO_2, which forms NO_2^- (Ruhland et al., 1998), is thermally activated with increasing temperature, the decrease in sensitivity at temperatures above the peak temperature suggests the change in surface reaction. Recent spectroscopic studies (Sergent et al., 2007; Lebranc et al., 2000) demonstrated that at higher temperatures adsorbed species for NO_2 gas is NO_3^- rather than NO_2^-, which suggests that NO_2 gas does not react directly with oxide but react with a high density of adsorbed O^-, forming NO_3^- (Francioso et al., 2006) and not contributing to increase in the resistance. The peak temperature decreases as the diameter of the nanowires decreases. The highest sensitivity obtained in this study was 37, which was observed at 100°C for the thinnest nanowires with a diameter of 50 nm.

The sensitivities at 100°C are summarized as a function of the average diameter of the nanowires in Fig. 10. The relationship between the sensitivity and diameter is expressed by a unique curve, indicating that the sensitivity increases with decreasing diameter.

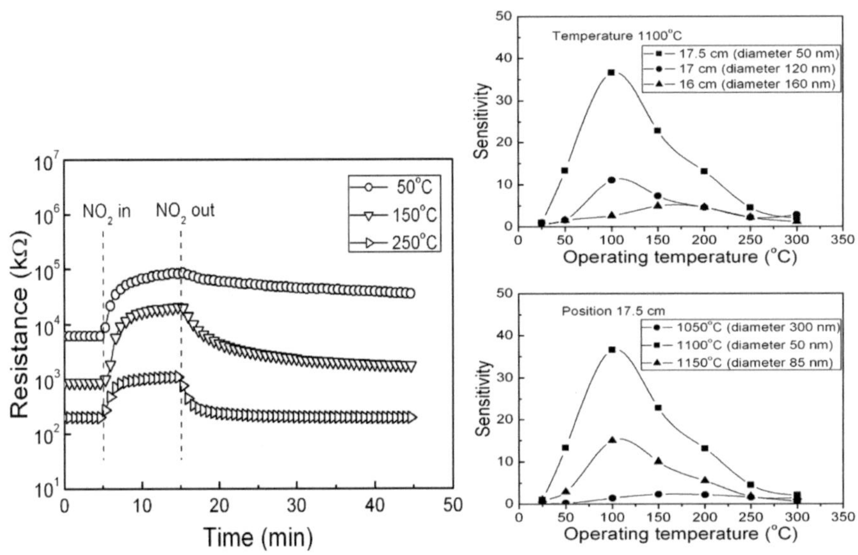

(a) (b)

Fig. 9. (a) Change in resistance of WO_3 nanowire sensor upon exposure to NO_2 at 3 ppm at different operating temperatures. (b) Operating temperature dependence of sensitivity to NO_2 measured for the nanowires wih different diameters.

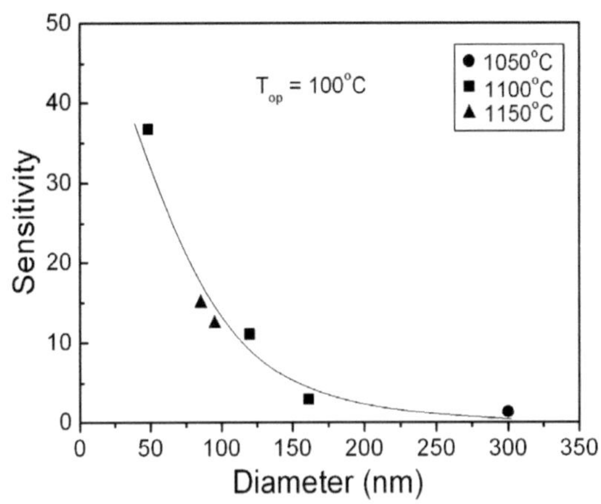

Fig. 10. Sensitivity at 100°C as a function of the diameter of nanowires. The sensitivities of the sensors made of nanowires deposited at different furnace temperatures are plotted.

3.3 Ga_2O_3 nanowires formed by vapor transport

Ga_2O_3 thin films have been studied as a gas sensor operating at a high temperature because of their stability at a high temperature. For example, they have been studied as oxygen sensors operating at 600–1000°C (Babana et al, 2005; Schwebel et al., 2000). They have also been studied as sensors for detecting reducing gases such as H_2, CO, CH_4 (Fleisher et al., 1992; Schwebel et al., 1998). Ga_2O_3 nanowires have also been studied as gas sensing materials (Liu et al., 2008; Mazeina et al., 2010).

In our study, a gallium grain (99.999%, 0.5 g) was placed on an alumina substrate in an alumina boat (Liu et al., 2008). The grain was then heated in a tube furnace at 900°C for 1 hour in an Ar flow of 100 ml/min at ambient pressure. After cooling to room temperature, a layer of white product was obtained on the alumina substrate around the gallium grain.

An SEM image in Fig. 11(a) shows that the obtained nanowires are about 50–150 nm in diameter and tens of micrometers in length. The XRD pattern in Fig. 11(b) agrees with the pattern of monoclinic β-Ga_2O_3 (JCPDS card No. 43-1012). A TEM image of the nanowires with a diffraction pattern, shown in Fig. 11(c), confirms the formation of single-crystal monoclinic Ga_2O_3. In our study, the Ga_2O_3 nanowires were formed in an Ar flow, which provided a lean oxygen environment. Therefore, the Ga_2O_3 nanowires obtained contain a high concentration of oxygen vacancies. This is confirmed by photoluminescence (PL). As shown in Fig. 11(d), the PL spectrum of the nanowires shows a broad strong PL peak at about 440 nm. According to literature (Binet & Gourier, 1998), the broad blue luminescence at 400–500 nm is ascribed to the electron transition mediated by oxygen vacancies in the band gap. These oxygen vacancies contribute to the n-type electrical conduction of the Ga_2O_3 nanowires.

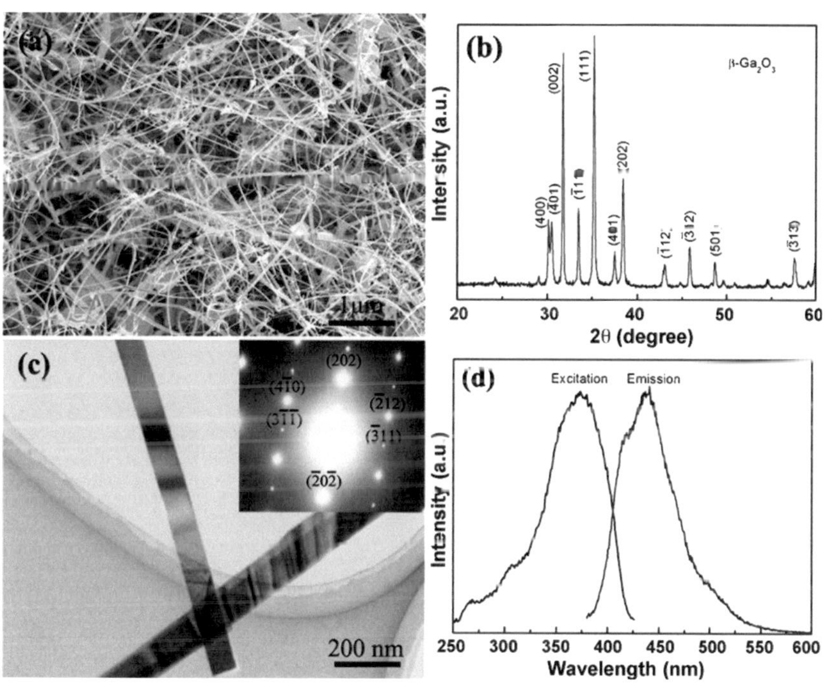

Fig. 11. (a) SEM image, (b) XRD pattern, (c) TEM image with diffraction pattern, and (d) photoluminescence spectrum of Ga_2O_3 nanowires.

Gas sensors were fabricated by dispersing nanowires on an oxidized Si substrate equipped with Pt electrodes with a gap length of 40 μm. Their gas sensing property was evaluated for O_2 in nitrogen and CO in dry synthetic air. The sensors showed obvious responses to O_2 and CO at temperatures much lower than that reported for Ga_2O_3 film sensors. Reversible changes in resistance upon introduction and removal of O_2 at 300°C is shown in Fig.12.
Figure 13 shows the sensitivity, defined by (Rg-Ra)/Ra for O_2 at 1% and (Ra-Rg)/Rg for CO at 200 ppm, measured at 100–500°C. It shows that the temperature dependence indicates a peak for both O_2 and CO. The highest sensitivity to 1% O_2 is 4.8 at 300°C, while the value to CO is 4.0 at 200°C. It is important to note that the operating temperature of our Ga_2O_3 nanowire sensor is much lower than 600–1000°C reported for Ga_2O_3 film sensors. This fact expands the application of Ga_2O_3 sensors to low operating temperatures.

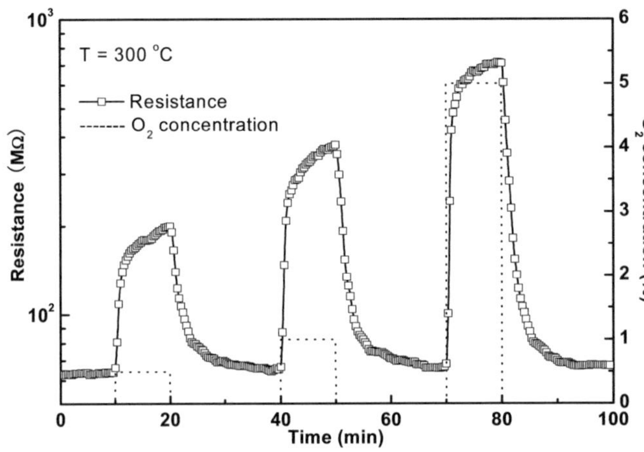

Fig. 12. Temporal changen in the resistance of a Ga_2O_3 nanowire sensor upon exposure to O_2 at 300°C.

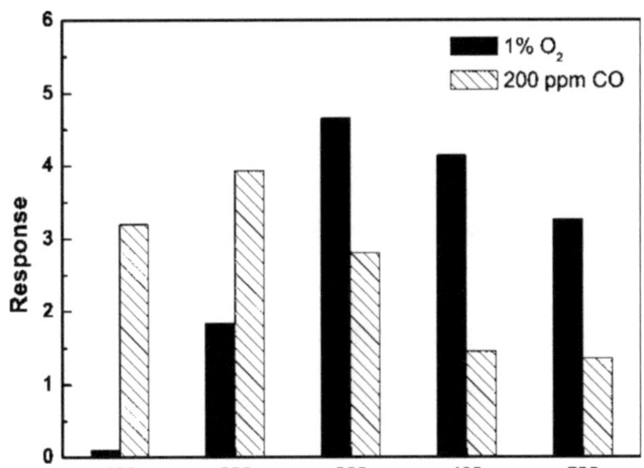

Fig. 13. Temperature dependence of the sensitivity of the Ga_2O_3 nanowire sensor to O_2 and CO gases.

3.4 TeO_2 nanowires formed by evaporation in ambient air

There have been no reports on the formation of TeO_2 nanowires. Figure 14 shows the typical SEM image of the nanowires that we formed by thermal evaporation of Te grains in ambient air (Liu et al., 2007; Liu et al., 2008). The nanowires are several tens of micrometers in length and 30–200 nm in diameter. According to XRD, the nanowires obtained were a tetragonal TeO_2 (JCPDS No. 11-0693). Figure 15(a) shows a typical TEM image of the TeO_2 nanowires. The HRTEM image and diffraction pattern in Figs. 15(b) and 15(c) show that a nanowire is a single crystal. The lattice spacings 0.47 and 0.29 nm observed in the HRTEM image correspond to those of (010) and (102) planes of a tetragonal TeO_2, respectively.

It was found that heating temperature and time were two crucial parameters in the determination of the morphology of the products. According to XRD, all products obtained were tetragonal TeO_2. Micro-sized wires were obtained at a temperature higher than 450°C, while only a small amount of particles were obtained at 300°C. The heating temperature of 400°C was optimal for the growth of TeO_2 nanowires. However, it is noticeable that a thin layer of TeO_2 particles was formed before the growth of TeO_2 wires in all cases. These particles may have played an important role in the growth of TeO_2 nanowires.

Gas sensors were fabricated using electrodes with a gap length of 40 μm. Test gases were added to a dry synthetic air with a total gas flow rate fixed at 200 ml/min. The temperature of the sensors was kept at room temperature (26°C). Figures 16(a)–16(c) show the change in the resistance upon introduction and exhaust of NO_2, NH_3, and H_2S at different concentrations. The resistance decreases quickly upon introduction of NO_2. The change in resistance can reach 90% of its final change about 2 min after the introduction of NO_2. The response to NH_3 is relatively slow. The resistance increases even 10 min after exposure to NH_3. It does not return to the initial value even 30 min after exhaust of NH_3, implying a slow surface reaction on the TeO_2 surface. The resistance changes reversibly upon introduction and exhaust of H_2S. However, the change in the resistance upon exposure to H_2S is much reduced at a concentration lower than 50 ppm, and the sensitivity is negligible at a concentration of 10 ppm. Note that the resistance increases upon exposure to oxidizing NO_2 while it decreases upon exposure to reducing NH_3 and H_2S. This implies that the TeO_2 nanowires are a p-type semiconductor.

Fig. 14. SEM image of TeO_2 nanowire.

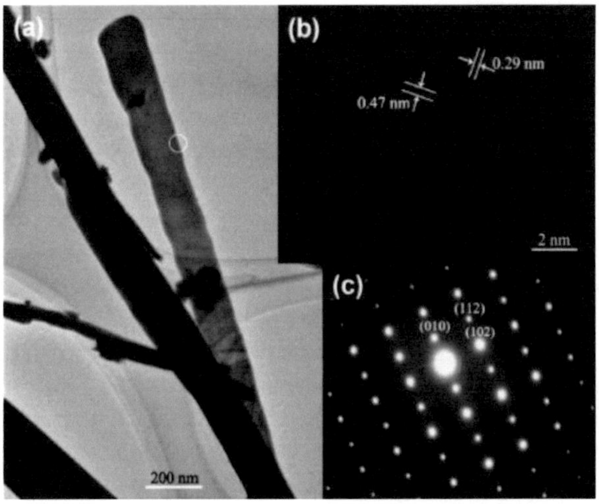

Fig. 15. TEM image of TeO_2 nanowires (a), HRTEM image of a TeO_2 nanowire (b), and its diffraction pattern (c).

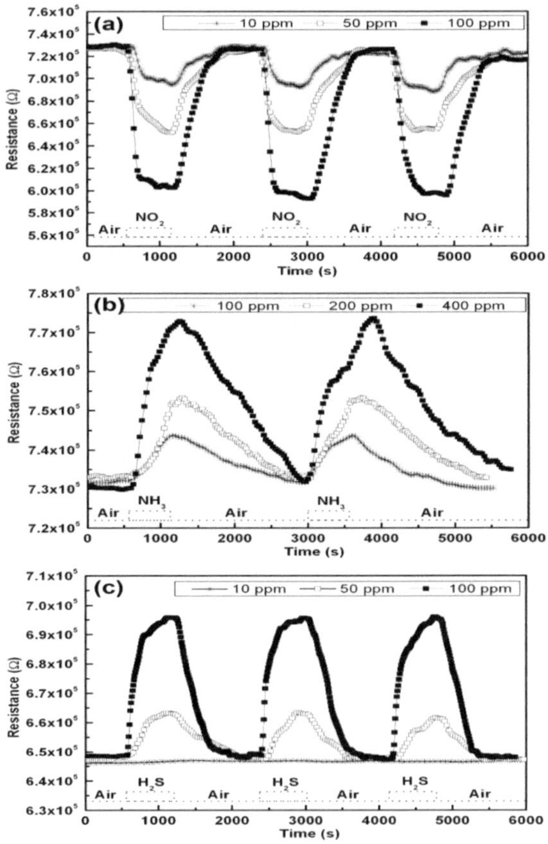

Fig. 16. Change in resistance upon exposure to NO_2 (a), NH_3 (b), and H_2S (c) at different concentrations.

The observation of n-type conduction of TeO_2 at high temperatures above 400°C or low oxygen partial pressure was reported, and it was explained in terms of oxygen vacancies (Jain & Nowick, 1981). However, p-type TeO_2 has also been reported. That is, p-type character has been observed for a single crystal (Hartmann & Kovacs, 1982), thin film (Jain & Garg, 1979), and polycrystal (Doi et al., 1975) by the measurement of Seebeck coefficient or photoconductivity at normal condition. The p-type conduction was ascribed to interstitial oxygen atoms (Doi et al., 1975), which trap electrons, generating holes as majority carriers.

3.5 SnO_2 nanowires formed by modified gas evaporation

A crucible containing Sn grains, SnO powder, or SnO_2 powder, each with a purity higher than 99.99%, were heated at 900°C for 90 min under a low pressure of ambient air. The pressures of the air during deposition were 140, 86, and 86 Pa for Sn grains, SnO powder, and SnO_2 powder, respectively.

SEM images in Fig. 17 show that the morphology of the product depends on the raw materials. Microwires (MWs), formed using Sn grains, are about 2 μm in diameter and 20 μm in length. They have rough surface and rectangular cross-section. Nanowires (NWs), formed using SnO powder, are 20 μm in average length and have smooth surface. Their circular cross-sections are 50-100nm in diameter. The rice-shaped nanoparticles (NPs), formed using SnO_2 powder, are about 100 nm in average diameter. XRD patterns showed that all of these materials were SnO_2 with a rutile structure (JCPDS card No.41-1445).

The changes in the resistances of the sensors made of SnO_2 microwires, nanowires, and nanoparticles upon exposure to 2ppm NO_2 at 200°C are shown in Fig. 18(a). The resistances increase upon exposure to NO_2 and recover to their initial values after exhaust of NO_2, indicating a good reversibility. The temperature dependencies of the sensitivities Rg/Ra of these sensors are shown in Fig. 18(b). The sensitivity of nanoparticle sensor is relatively low. The highest sensitivity 5.5 of this sensor was observed at 200°C. The sensitivity of the microwire sensor is slightly higher than that of nanoparticle sensor despite the smaller surface-to-volume ratio of microwires than nanoparticles. The sensitivity of the nanowire sensor is much higher than that of the microwire sensor. The peak temperatures of the sensitivity observed for the microwire and nanowire sensors, namely 100 and 150°C, are lower than 200°C observed for the nanoparticle sensor. The change in the diameter of nanostructure leads to a change in both the sensitivity and peak temperature (Korotcenkov, 2002). The highest sensitivity obtained was 88 at 150°C measured for the nanowire sensor. Thus, we can believe that a sensor made of thin nanowires is suitable for monitoring a low concentration of NO_2.

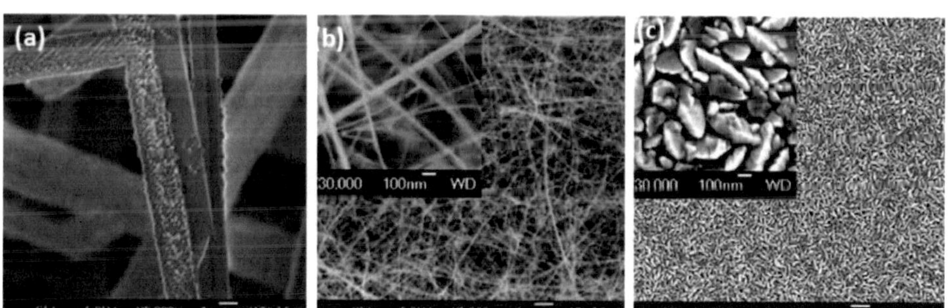

Fig. 17. SEM images of microwires, nanowires, and rice-shaped nanoparticles.

The response and recovery times, defined as the time necessary for the change in resistance upon exposure and exhaust of NO_2 to reach 90% of the whole change, are shown as functions of temperature in Fig. 19. The response and recovery times of microwires and nanowires are shorter than those of nanoparticles. The behavior of the sensitivity as well as response and recovery time shown in Figs. 18 and 19 suggests that the surface reaction depends on temperature. The electron transfer from SnO_2 to adsorbed NO_2, which forms NO_2^-, occurs at relatively low temperatures of 100 and 150°C in microwires and naowires, respectively. This results in the high sensitivities of the microwire sensor and nanowier sensor at these temperatures. At higher temperatures the sensitivity is lowered similarly to WO_3 sensors.

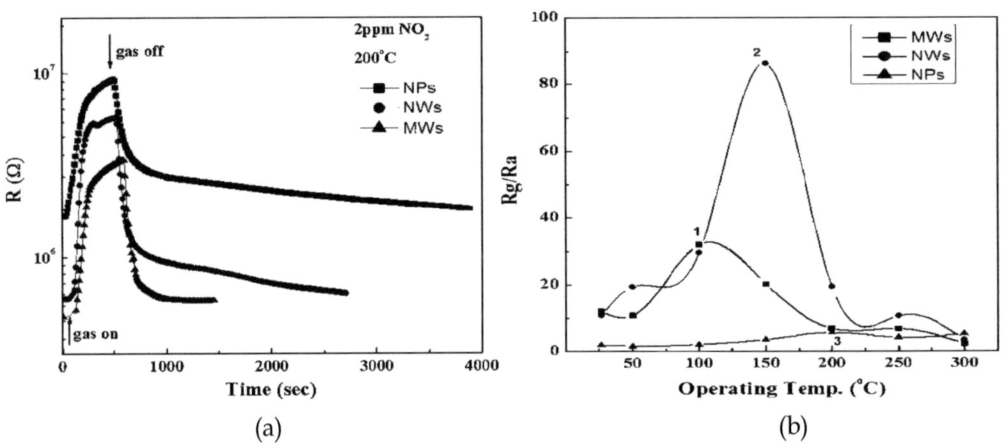

Fig. 18. (a) Change in the resistances of the sensors made of microwires, nanowires, and rice-shaped nanoparticles upon exposure to 2 ppm NO_2 at 200°C. (b) Sensitivities of the sensors to 2ppm NO_2 as functions of temperature.

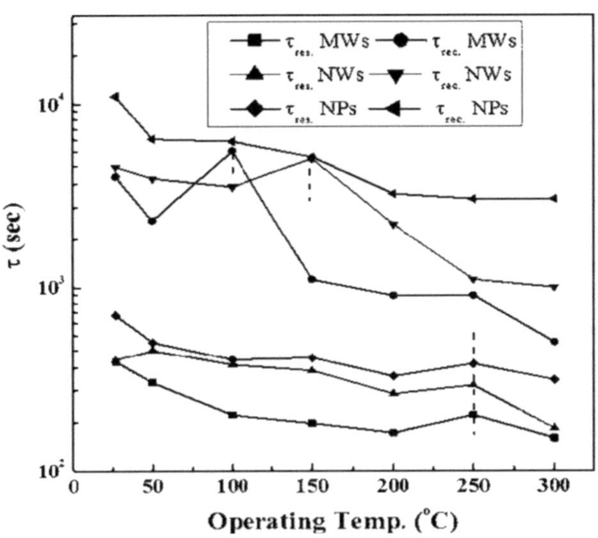

Fig. 19. Recovery and response time of SnO_2 nanowire sensors exposed to 2 ppm NO_2 at a function of temperatures.

Now, we consider the effect of grain size, namely the effect of the diameter of particles or the length of wires. Spherical or wire grains arrayed linearly in the gap between a pair of electrodes are illustrated in Fig. 20, where the grains are assumed to be depleted of electrons only at the surface region of the thickness of L_D. The number of grains in the gap is much smaller in the microwire and nanowire sensors than in the nanoparticle sensor. Therefore, in the microwire and nanowire sensors, the contribution to the sensor resistance of the bulk and the interface between metal electrode and oxide is large while the contribution of grain boundaries is small. As shown in Fig. 20, if the diameter of the grains is small, the highly conductive bulk region is narrow. Thus, the conductance of the bulk region is more affected by its own shrinkage upon exposure to NO_2, leading to nanowires more sensitive to NO_2 than microwires. Although the surface-to-volume ratio of microwires is smaller than that of nanoparticles, the sensitivity of the microwire sensor is larger than that of the nanoparticle sensor. In the sensor of microwires or nanowires, the influence of the resistance of metal-oxide interface on the sensitivity may be larger than that of the resistance of grain boundaries. This seems to lead to the higher sensitivities observed for microwires and nanowires.

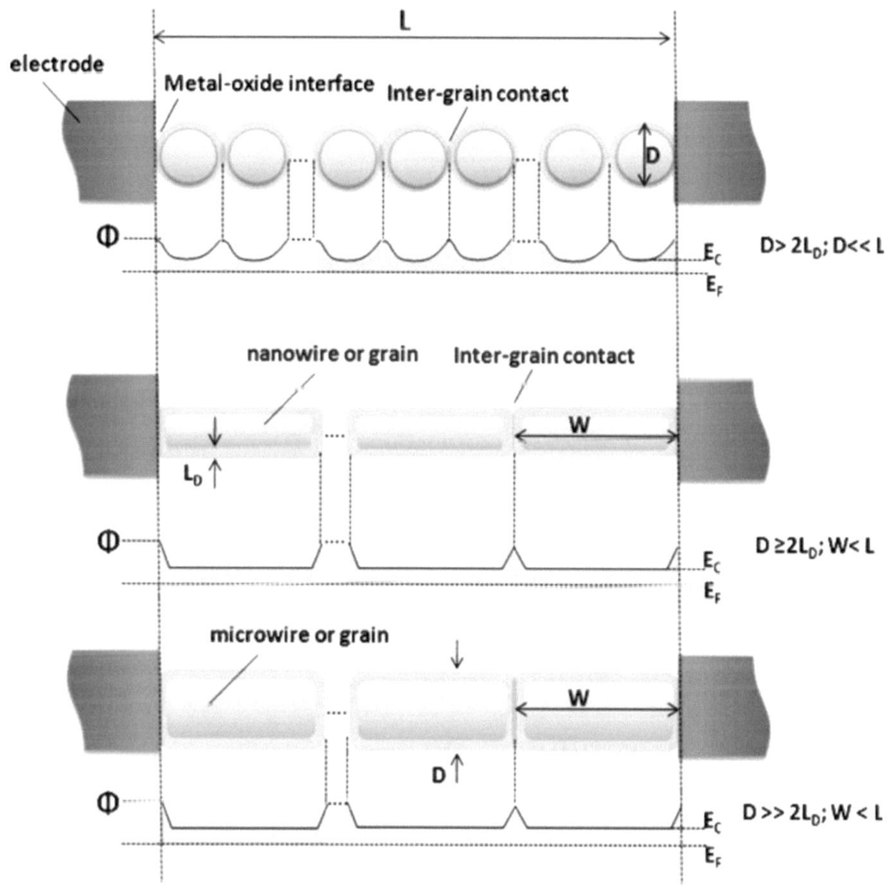

Fig. 20. Model for the distributions of SnO_2 grains. The length and diameter of SnO_2 grains affect the density of grain boundaries and potential distribution.

3.6 Effect of gap length for SnO_2 nanowires

If the interface between the metal electrode and oxide is intrinsically important in the determination of the sensitivity of nanowire sensors, as described in the previous section, the sensitivity must be affected by the gap length of a pair of electrodes. Thus, the influence of the gap length was investigated using nanowires formed by heating an Sn grain in a crucible at 800°C at a pressure of 40 Pa of ambient air. The fabrication of a pair electrode, made of an Au sputtered film, is illustrated in Fig. 21. Au-film strip with a width of 2 mm, equipped with a gap formed using field ion beam technology (FIB), is shown at the bottom. Four sensors thus fabricated had a gap length of 1, 5, 10, and 30 μm, respectively.
An SEM image represented in Fig. 22(a) shows that the nanowires obtained are straight with rather uniform length (~5 μm) and diameter (~100 nm). An example of the SEM image of a gap formed by FIB is shown in Fig. 22(b).

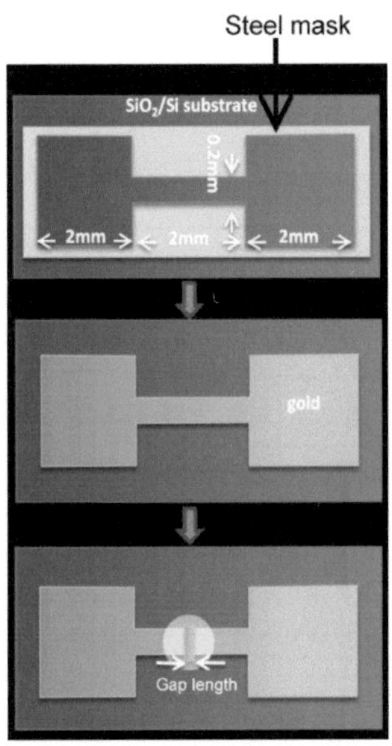

Fig. 21. Fabrication of a pair of gold electrodes. After an Au-film pattern was formed using a steel mask, a short gap was formed using FIB.

Fig. 22. (a) SEM images of SnO_2 nanowires. (b) A micro-gap with a length of 1μm formed by using the method illustrated in Fig. 21.

The change in resistance upon exposure to NO_2 at different concentrations was measured for the four sensors with various gap lengths. The results at 250°C are shown in Fig. 23. The resistance increases upon exposure to NO_2 and decreases upon exhaust of NO_2. The response and recovery speeds increase with decreasing gap length. We notice that the resistance of the sensors with longer gaps does not return to the initial value precisely and that the signal of these sensors is noisy for a higher concentration of NO_2.

Figure 24 shows the sensitivity, Rg/Ra, to NO_2 at 2 ppm as a function of the gap length. The results at various operating temperatures are shown. The sensitivity of the sensor with a gap length of 1 μm, which is shorter than the length of the nanowires, is highest at every temperature. Temperature dependencies of the sensitivity are shown in the inset of this figure. Similarly to the sensors made of WO_3 nanowires, the temperature dependence shows a peak. As the temperature increases, the sensitivities of all sensors increase drastically at a temperature up to 100°C and then decrease at a temperature above 100°C. Thus, the sensitivity at 100°C measured for the sensor with the shortest gap length of 1 μm is as high as 600.

Although the role of the metal-oxide interface is still not clear, it may be due to some kind of chemical sensitization. The so called chemical sensitization in oxide semiconductor gas sensor is mediated by a "spillover effect", and it was experimentally studied for hydrogen and oxygen (Yamazoe et al., 1983; Boudart & Mol, 1999; Bowker et al., 2000; Kolmakov et al., 2005). It is probable that an Au electrode can also act as a catalyst mediating some chemical sensitization. References (Tamaki et al., 2005; Tamaki et al., 2008) discussing NO_2 sensors made of WO_3 nanoparticles equipped with Au nano-gap electrodes suggested that NO_2 gas might be adsorbed in a high density at the interface due to catalytic effect of Au. It was also suggested that the adsorption of NO_2 on an Au electrode might increase its work function. These reactions cause a higher Schottky barrier upon exposure of the sensor to NO_2 gas, leading to a higher resistance or the higher sensitivity, as observed in the sensitivity of the sensor with a short gap length of 1μm.

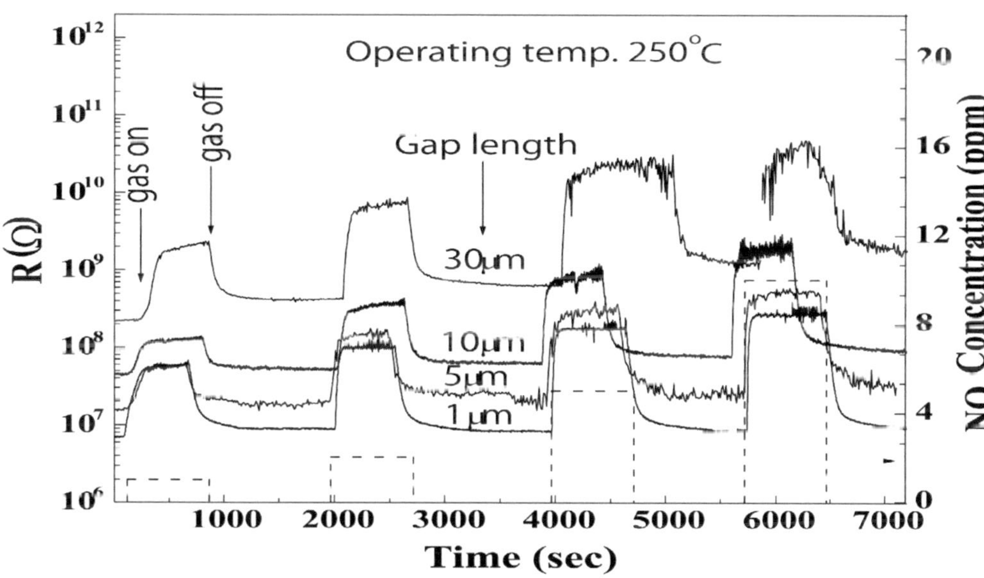

Fig. 23. Change in the resistances of SnO_2 sensors with different gap lengths upon exposure to different concentrations of NO_2 gas at 250°C.

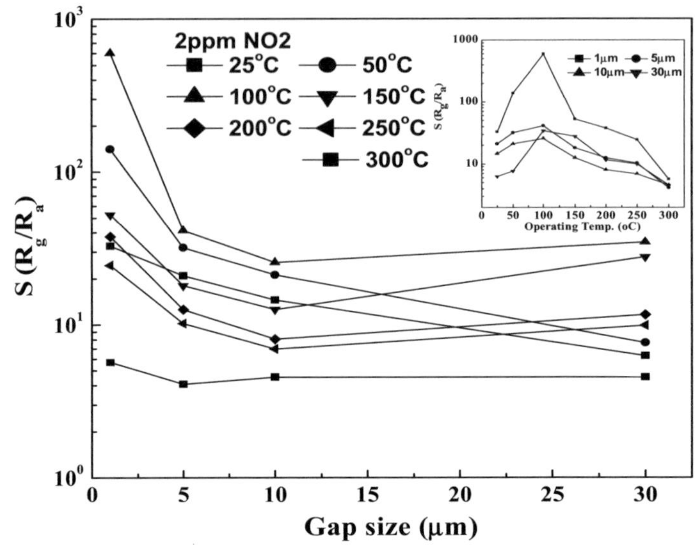

Fig. 24. Sensitivity to NO_2 at 2 ppm as a function of the gap length. The results at various temperatures are shown. The inset shows the same data but versus the operating temperature.

4. Conclusions

These ten years, various kinds of oxide semiconductor nanowires have been studied as gas sensing materials because of their large surface-to-volume ratio leading to a high sensitivity and also because of their discriminating morphology which may lead to unexpectedly excellent properties. We attempted to develop various types of thermal evaporation method for making various kinds of oxide nanowires and also attempted to clarify the feature of the gas sensing properties of nanowire sensors. Conventional vapour transport method was used for the formation of SnO_2, WO_3, and Ga_2O_3 nanowires. A very simple thermal evaporation in ambient air was used for forming TeO_2 nanowires. A modified gas evaporation method using vacuum evaporator, which rather easily provides a high raw material temperature and well-controlled atmosphere, was used for forming SnO_2 nanowires. In these methods, thin nanowires could be formed by adjusting both the atmosphere and heating temperature of raw materials.

A sensor made of SnO_2 nanowires formed by a vapour transport method showed a high sensitivity at a low temperature when doped with Pd, similarly to conventional SnO_2 sensors. It was found that a nanowire sensor indicated quicker response and recovery to H_2 than sputtered-film sensor, although the sensitivity was lower than sputtered film. Sensors made of SnO_2 nanowires and microwires formed by modified gas evaporation were superior to a sensor made of nanoparticles in sensitivity to NO_2, despite the smaller surface-to-volume ratio than nanparticles in the case of microwires. These sensors were also superior to the nanoparticle sensor in response and recovery speed to NO_2. In addition, a sensor with a gap length shorter than the length of nanowires was confirmed to be superior to sensors with longer gap length in sensitivity as well as response and recovery speed, indicating the importance of the interface between the electrode and nanowires. As for WO_3 nanowire sensors, the sensitivity to NO_2 was confirmed to increase as the diameter of the nanowires decreased.

Ga_2O_3 nanowire sensor showed a sufficient sensitivity and reversible behaviour at relatively low temperature of 100-300°C. TeO_2 nanowires were first formed and evaluated as a gas sensing material. The sensor indicated p-type semiconductor character of TeO_2 nanowires, showing increase and decrease in resistance upon exposure to reducing gases (NH_3 and H_2S) and oxidizing gas (NO_2), respectively.

5. Acknowledgements

The author is grateful to Dr. Zhifu Liu, Dr. Yanbai Shen, and Dr. Meng Dan for their cooperation in this study. He is also grateful to Mr. Yukichi Takayasu and Ms. Akiko Hirata for their cooperation in experiments. He is also grateful to Venture Business Laboratory, University of Toyama, for its promoting this study.

6. References

Akiyama, M.; Tamaki, J.; Miura, N. & Yamazoe, N. (1991). Tungsten oxide-based semiconductor sensor highly sensitive to NO and NO_2. *Chem. Lett.* 20, 9, (1991) 1611-1614, ISSN: 1348-0715 (print) , 0366-7022 (online).

Babana, C.; Toyodac, Y. & Ogita, M. (2005). Oxygen sensing at high temperatures using Ga_2O_3 films. *Thin Solid Films,* 484, (July 2005) 369-373, ISSN: 0040-6090.

Binet, L. & Gourier, D. (1998). Origin of the blue luminescence of β-Ga_2O_3. *J. Phys. Chem. Solids*, 59, (August 1998) 1241-1249, ISSN: 0022-3697.

Boudart, M. (1999). On the nature of spilt-over hydrogen. *J. Mol. Catal. A-Chem.*, 138, 2-3, (February 1999) 319-321, ISSN: 1381-1169.

Bowker, M.; Bowker, L.J.; Bennett, R.A.; Stone, P. & Ramirez-Cuesta, A. (2000). In consideration of precursor states, spillover and Boudart's 'collection zone' and of their role in catalytic processes. *J. Mol. Catal. A-Chem.*, 163, 1-2, (December 2000) 221-232, ISSN: 1381-1169.

Cao, B.; Chen, J.; Tang, X.J. & Zhou, W.L. (2009). Growth of monoclinic WO_3 nanowire array for highly sensitive NO_2 detection . *J. Mater. Chem.*, 19, 16, (2009) 2323-2327, ISSN: 0959-9428.

Chen, Y.Q.; Cui, X.F.; Zhang, K.; Pan, D.Y.; Zhang, S.Y.; Wang, B.; Hou, J.G. (2003). Bulk-quantity synthesis and self-catalytic VLS growth of SnO_2 nanowires by lower-temperature evaporation. Bulkquantity, *Chem. Phys. Lett.*, 369, (February 2003) 16-20, ISSN: 0009-2614.

Choi, H.G.; Jung, Y.H. & Kim, D.K. (2005). Solvothermal Synthesis of Tungsten Oxide Nanorod/Nanowire/Nanosheet. *J. Am. Ceram. Soc.*, 88, 6, (June 2005) 1684-1686, ISSN: 0002-7820.

Chung, Y.K.; Kim, M.H.; Um, W.S.; Lee, H.S.; Song, J.K.; Choi, S.C.; Yi, K.M.; Lee, M.J. & Chung, K.W. (1999). Gas sensing properties of WO_3 thick film for NO_2 gas dependent on process condition. *Sens. Actuator B-Chem.*, 60, (November 1999) 49-56, ISSN: 0925-4005.

Doi, K.; Sasaki, T. & Hijikata, K. (1975). Electrical properties of TeO_2. *Bull. Chem. Soc. Jap.*, 48, 1, (1975) 144-146.

Fleischer, M.; Giber, J. & Meixner, H. (1992). H_2-induced changes in electrical conductance of β-Ga_2O_3 thin-film systems. *Appl. Phys. A*, 54, 6, (1992) 560-566, ISSN: 0947-8396 (print), 1432-0630 (online).

Francioso, L.; Forleo, A.; Capone, S.; Epifani, M.; Taurino, A.M. & Siciliano, P. (2006). Nanostructured In_2O_3–SnO_2 sol–gel thin film as material for NO_2 detection. *Sens. Actuator B-Chem.*, 114, (April 2006) 646-655, ISSN: 0925-4005.

Hartmann, E. & Kovacs, L. (1982). Electrical conductivity of paratellurite (TeO_2) crystals. *Phys. Status Solidi. A*, 74, (Novembre 1982) 59-64, ISSN: 1862-6319 (online).

Heiland, G. (1982). Homogeneous semiconducting gas sensors. *Sens. Actuator.*, 2, (1982) 343-361.

Hernandez-Ramirez, F.; Tarancon, A.; Casals, O.; Arbiol, J.; Romano-Rodriguez, A. & Morante, J.R. (2007). High response and stability in CO and humidity measures using a single SnO_2 nanowire. *Sens. Actuator B-Chem.*, 121, (January 2007) 3-17, ISSN: 0925-4005.

Hong, K.Q.; Xie, M.H. & Wu, H.S. (2006). Tungsten oxide nanowires synthesized by a catalyst-free method at low temperature. *Nanotechnology*, 17, 19, (October 2006) 4830-4833, ISSN: 0957-4484.

Jain, H. & Nowick, A.S. (1981). Electrical conduction in paratellurite (TeO_2) crystals. *Phys. Status Solidi. A*, 67, (October 1981) 701-707, ISSN : 1862-6319 (online).

Jain, D.K. & Garg, J.C. (1979). Field-assisted conduction mechanism in tellurium dioxide films. *Indian J. Pure Appl. Phys.*, 17, 10, (1979) 642-645, ISSN: 0019-5596.

Jang, H.S.; Kang, S.O. & Kim, Y.I. (2006). Enhancement of field emission of SnO_2 nanowires film by exposure of hydrogen gas. *Solid State Commun.*, 140, (December 2006) 495-499, ISSN: 0925-4005.

Jin, C.J.; Yamazaki, T.; Shirai, Y. Yoshizawa, T.; Kikuta, T.; Nakatani, N. & Takeda, H. (2005). Dependence of NO_2 gas sensitivity of WO_3 sputtered films on film density. *Thin Solid Films*, 474, (March 2005) 255-260, ISSN: 0040-6090.

Kaito, C.; Fujita, K.; Shibahara, H. & Shiojiri, M. (1977). Electron microscopic study of metal oxide smoke particles prepared by burning metals in Ar– O_2 gas. *Jpn. J. Appl. Phys.*, 16, (May 1977) 697-704, ISSN: 0021-4922 (print), 1347-4067 (online) .

Kimoto, K.; Kamiya, Y. & Nonoyama, R. (1963). An electron microscope study on fine metal particles prepared by evaporation in argon gas at low Pressure. *Jpn. J. Appl. Phys.*, 2, (November 1963) 702-713, ISSN: 0021-4922 (print), 1347-4067 (online).

Kolmakov, A.; Klenov, D.O.; Lilach, Y.; Stemmer, S. & Moskovits, M. (2005). Enhanced gas sensing by individual SnO_2 nanowires and nanobelts functionalized with Pd catalyst particles. *Nano Lett.*, 5, 4, (April 2005) 667-773, ISSN: 1530-6984.

Korotcenkov, G.; Brinzari, V.; Schwank, J. & Cerneavschi, A. (2002). Possibilities of aerosol technology for deposition of SnO_2-based films with improved gas sensing characteristics. *Mater. Sci. Eng.* C, 19, (January 2002) 73-77, ISSN: 0928-4931.

Kumar, V.; Sen, S.; Muthe, K.P.; Gaur, N.K.; Gupta, S.K. & Yakhmi, J.V. (2009). Copper doped SnO_2 nanowires as highly sensitive H_2S gas sensor. *Sens. Actuator B-Chem.*, 138, (May 2009) 587-590, ISSN: 0925-4005.

Leblanc, E.; Perier-Camby, L ; Thomas, G ; Gibert, R ; Primet, M. & Gelin, P. (2000). NO_x adsorption onto dehydroxylated or hydroxylated tin dioxide surface. Application to SnO_2-based sensors, *Sens. Actuator B-Chem.*, 62, (January 2000) 67–72, ISSN: 0925-4005.

Liu, Z.; Yamazaki, T.; Shen, Y.; Kikuta, T.; Nakatani, N. & Kawabata, T. (2007). Room temperature gas sensing of p-type TeO_2 nanowires. *Appl. Phys. Lett.*, 90, 17, (April 2007) 173119, ISSN: 0003-6951.

Liu, Z.; Yamazaki, T.; Shen, Y.; Kikuta, T. & Nakatani, N. (2008). Synthesis and characterization of TeO_2 Nanowires. *Jpn. J. Appl. Phys.*, 47, 1, (January 2008) 771-774, ISSN: 0021-4922 (print), 1347-4067 (online).

Liu, Z.; Yamazaki, T.; Shen, Y.; Kikuta, T.; Nakatani, N. & Li, Y. (2008). O_2 and CO sensing of Ga_2O_3 multiple nanowire gas sensors.*Sens. Actuator B-Chem.*, 129, (February 2008) 666-670, ISSN: 0925-4005.

Mazeina, L.; Bermudez, V.M.; Perkins, F.K.; Arnold, S.P. & Prokes, S.M. (2010). Interaction of functionalized Ga_2O_3 NW-based room temperature gas sensors with different hydrocarbons. *Sens. Actuator B-Chem.*, 151, (November 2010) 114-120, ISSN: 0925-4005.

Polleux, J.; Gurlo, A.; Barsan, N.; Weimar, U.; Antonietti, M. & Niederberger, M. (2006). Template-free synthesis and assembly of single-crystalline tungsten oxide nanowires and their gas-sensing properties. *Angew. Chem. Int. Ed.*, 45, 2, (2006) 261-265, ISSN: 1433-7851.

Ponzoni, A.; Comini, E.; Sberveglieri, G.; Zhou, J.; Deng, S.Z.; Xu, N.S.; Ding, Y. & Wang, Z.L. (2006). Ultrasensitive and highly selective gas sensors using three-dimensional tungsten oxide nanowire networks. *Appl. Phys. Lett.*, 88, 20, (May 2006) 203101, ISSN: 0003-6951.

Rout, C.S.; Ganesh, K.; Govindaraj, A. & Rao, C.N.R. (2006). Sensors for the nitrogen oxides, NO_2, NO and N_2O, based on In_2O_3 and WO_3 nanowires. *Appl. Phys. A*, 85, (Novembre 2006) 241-246, ISSN: 0947-8396 (print), 1432-0630 (online).

Ruhland, B.; Becker, T. & Müller, G. (1998). Gas-kinetic interactions of nitrous oxides with SnO_2 surfaces. *Sens. Actuator B-Chem.*, 50, (July 1998) 85-94, ISSN: 0925-4005.

Schwebel, T.; Fleischer, M.; Meixner, H. & Kohl, C.D. (1998). CO-sensor for domestic use based on high temperature stable Ga_2O_3 thin films. *Sens. Actuator B-Chem.*, 49, 1-2, (June 1998) 46-51 ISSN: 0925-4005.

Schwebel, T.; Fleischer, M. & Meixner, H. (2000). A selective, temperature compensated O_2 sensor based on Ga_2O_3 thin films. *Sens. Actuator B Chem.*, 65, (June 2000) 176-180, ISSN: 0925-4005.

Sergent, N.; Epifani, M ; Comini, E ; Faglia, G & Pagnier, T. (2007). Interactions of nanocrystalline tin oxide powder with NO_2: A Raman spectroscopic study, *Sens. Actuator B-Chem.*, 126, (Septembre 2007) 1–5, ISSN: 0925-4005.

Shen, Y.; Yamazaki, T.; Liu, Z.; Meng, D.; Kikuta, T.; Nakatani, N.; Saito, M. & Mori, M. (2009). Microstructure and H_2 gas sensing properties of undoped and Pd-doped SnO_2 nanowires. *Sens. Actuator B-Chem.*, 135, (January 2009) 524-529, ISSN : 0925-4005.

Shen, Y.; Yamazaki, T.; Liu, Z.; Meng, D.; Kikuta, T. (2009). Hydrogen sensors made of undoped and Pt-doped SnO_2 nanowires. *J. All. Comp.*, 488, (Novembre 2009) L21-L25, ISSN: 0925-8388

Shimizu, Y.; Hyodo, T. & Egashira, M. (2007). H_2 sensing performance of anodically oxidized TiO_2 thin films equipped with Pd electrode. *Sens. Actuator B-Chem.*, 121, (January 2007) 219-230, ISSN: 0925-4005.

Tamaki, J.; Nagaishi, M.; Teraoka, Y.; Miura, N. & Yamazoe, N. (1989). Adsorption behavior of CO and interfering gases on SnO_2. *Surf. Sci.*, 221, (Octobre1989) 183-196, ISSN: 00396028.

Tamaki, J.; Miyaji, A.; Makinodan, J.; Ogura, S.; & Konishi S. (2005). Effect of micro-gap electrode on detection of dilute NO_2 using WO_3 thin film microsensors. *Sens. Actuator B-Chem.*, 108, (July 2005) 202-206, ISSN: 0925-4005.

Tamaki, J.; Hashishin, T.; Uno, Y.; Dao,D.V.; & Sugiyama, S. (2008). Ultrahigh-sensitive WO3 nanosensor with interdigitated Au nano-electrode for NO_2 detection. *Sens. Actuator B-Chem.*, 132, (May 2008) 234-238, ISSN: 0925-4005.

Tsang, S.C.; Bulpitt, C.D.A.; Mitvhell, P.C.H. & Ramirez-Cuesta, A.J. (2001). Some new insights into the sensing mechanism of palladium promoted tin (IV) oxide sensor. *J. Phys. Chem. B.*, 105, (June 2001) 5737-5742, ISSN: 1520-6106.

Yamazoe, N.; Kurokawa, K. & Seiyama, T. (1983). Effects of additives on semiconductor gas sensors. *Sens. Actuator* 4, (1983) 283-289.

Wang, J.X.; Liu, D.F.; Yan, X.Q.; Yuan, H.J.; Ci, L.J.; Zhou, Z.P.; Gao, Y.; Song, L.; Liu, L.F.; Zhou, W.Y.; Wang, G. & Xie, S.S. (2004). Growth of SnO_2 nanowires with uniform branched structures. *Solid State Commun.*, 130, (April 2004) 89-94, ISSN: 0038-1098.

Windischmann, H. & Mark, P. (1979). A model for the operation of a thin-film SnOx conductance-modulation carbon monoxide sensor. *J. Electrochem. Soc.*, 126, (April 1979) 627-633, ISSN: 0013-4651 (print), 1945-7111 (online) .

7

Electrodeposited Copper Oxide and Zinc Oxide Core-Shell Nanowire Photovoltaic Cells

Dante DeMeo, Samuel MacNaughton,
Sameer Sonkusale and Thomas E. Vandervelde
Tufts University
USA

1. Introduction

Uncertainty in energy capacity, limited fossil fuel resources, and changes in climate predicate a need for increased research and development into alternative and sustainable energy solutions. Solar energy is one solution to this problem and many variations of it exist; however, the majority of them are prohibitively expensive. We propose a low-cost solar energy generation method which is cost-effective both in materials and production. Our solution will utilize cheap, abundant materials as well as lower-cost fabrication methods to produce photovoltaic (PV) cells. Although it is unlikely that the efficiency of such cells will be record-breaking, its low cost should make its price-per-watt-produced competitive, which is one of the most important metrics for the commercialization of any solar technology.

Our design consists of a radial heterojunction comprised of p-type copper oxide and n-type zinc oxide nanowires, which are oxides of earth-abundant materials. The nanowires have a core-shell design to minimize carrier travel distance and maximize junction area. Furthermore, we utilize a wet chemistry fabrication process, making the production of such cells inexpensive, easily scalable and non-demanding in terms of fabrication energy. The process involves growing copper nanowires, oxidizing, plating zinc oxide, and depositing a top contact.

2. The case for solar

Solar energy is an attractive solution to demanding energy problems because of the amount of energy it can produce, its low maintenance and operating costs, and its clean nature. The sun has the ability to provide more than enough energy to satiate the world's demands, even accounting for future growth. Once in place, solar panels will continue to provide electricity for decades with minimal human interaction; systems can still operate after 40 years (King, Quintana et al. 2000) and have the potential to last even longer. There are no fuel costs or harmful pollutants during operation, and setup needs only little training.

The amount of energy coming from the sun is greater than any other form of renewable energy. Solar radiation accounts for 173,000 TW of energy, dwarfing all other renewables such as wind and waves at 3600 TW, geothermal energy at over 32 TW, and tidal energy at 3

TW (Da Rosa 2005). Although these numbers do not reflect the amount of electricity able to be produced from their respective sources of energy, it does demonstrate a clear dominance of solar over all other forms of energy.

Solar energy has the potential to supply all of the worlds power needs as demonstrated by Figure 1. If the six black circles were covered in solar panels of only 8% efficiency, 18 TW of energy could be produced (Loster 2010), which is more than the total energy consumption in 2007 of 16.6 TW (U.S. Energy Information Administration 2010). The total land area needed is approximately 910,000 sq. km. (Loster 2010) or about 0.6% of the total land area on earth. It should be noted that many solar cell designs which exceed 8% conversion efficiency have been produced. (Green, Emery et al. 2010)

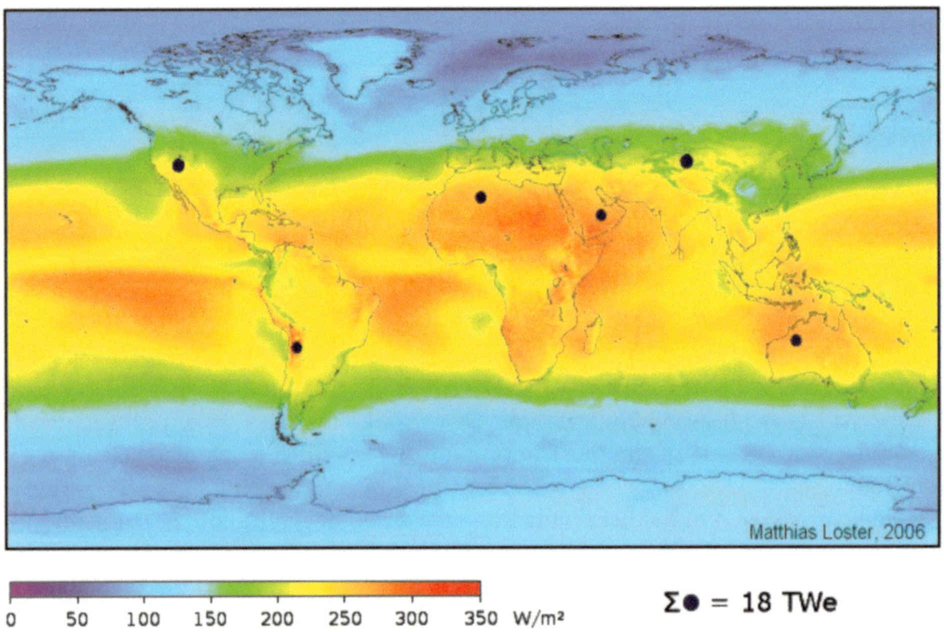

Fig. 1. "Sunlight hitting the dark disks could power the whole world if installed in areas marked by the six discs in the map, solar cells with a conversion efficiency of only 8% would produce, on average, 18TW electrical power. That is more than the total power currently available from all our primary energy sources, including coal, oil, gas, nuclear, and hydro. The colors show a three-year average of solar irradiance, including nights and cloud coverage." (Loster 2010)

Although solar photovoltaics have a substantial up-front cost, the operation and maintenance costs are minimal. Since solar cells are solid-state devices and have no moving parts (except in the case of tracking systems), they are highly reliable.

Photovoltaics generate no harmful emissions during use, nor do they produce any noise or unwanted by-products. (Luque and Hegedus 2003) This clean quality as well as the highly modular and scalable nature of the technology makes it an excellent choice for a myriad of power systems. Solar panels can be used by households to provide energy for one's home, for off grid remote power generation, or for large-scale utility plants. Outside of the manufacturing process, solar panels are very environmentally friendly and do not contribute

CO_2 to the atmosphere. Utilizing end-of-life recycling and more benign industrial processing will make the technology even more environmentally friendly in the future.

2.1 Solar cell basics

Solar photovoltaic cells are devices which turn radiant energy from the sun directly into electricity. Factors which affect the amount of energy created include the intensity of radiation, the spectral distribution of the radiation, and the specific materials, design, and quality of the solar cell at hand. Most inorganic solar cells, however, work in the same manner and will be described presently.

Electromagnetic radiation (primarily in the visible and near-infrared regions of the spectrum) is emitted from the sun and absorbed by the solar cell. A photon will then excite a negatively charged electron from the valence band (low energy state) to the conduction band (a higher energy state) leaving behind a positively charged vacancy, called a hole. For this energy transfer to create any usable energy, the photon must have an energy greater than the band-gap of the material, or else the electron will immediately relax down and recombine with the hole and the energy will be lost as heat. Upon excitation above the bandgap the photon creates an electron and a hole which are now free to move throughout the semiconductor crystal. These act as charge carriers which transport the energy to the electrical contacts, which results in a measurable external current. These processes are shown in Figure 2.

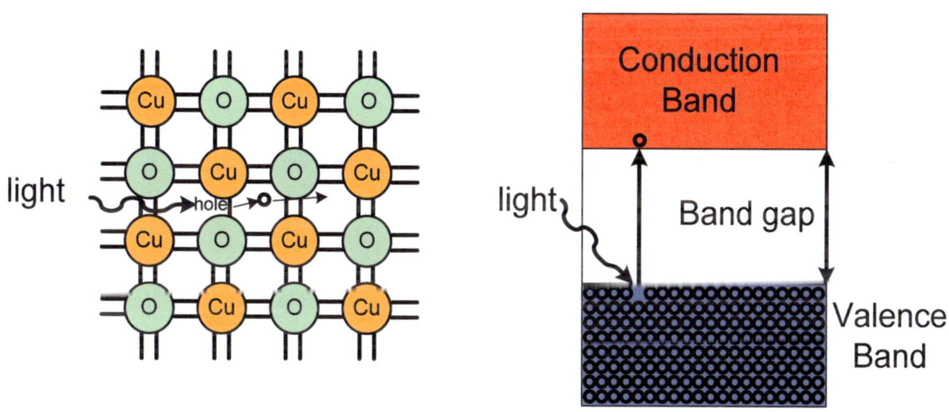

Fig. 2. (Right) A photon excites an electron (black circle) from the valence band to the conduction band. (Left) Each black line represents a valence electron in a covalent bond. When a photon breaks one of these bonds, a free electron and a hole are created.

The materials and structure of the solar cell are very important in this process. A solar cell is made out of semiconductor material which facilitates the creation and motion of charge carriers. The specific material determines the band-gap and thus which wavelengths of light the cell can absorb, as well as many other optoelectronic properties[1]. Once the electron-hole

---

[1] For more information on optoelectronic properties of materials, one may refer to virtually any relevant textbook on the subject. Several recommended texts include: Kasap, S. O. (2001). Optoelectronics and Photonics Principles and Practices. Upper Saddle River, New Jersey, Prentice-Hall Inc., Kasap, S. O. (2006). Principles of Electronic Materials and Devices. New York, McGraw-Hill., and Saleh, B. E. A. and M. C. Teich (2007). Fundamentals of Photonics. Hoboken, New Jersey, John Wiley & Sons.

pair is created from incident sunlight, the solar cell must separate these charges or else they will quickly recombine and lose their energy. To do this, two layers of semiconductor are used: an n-type, which has an excess of fixed negative charges, and a p-type, which has an excess of fixed positive charges. When these two semiconductor layers are placed next to each other, opposing charge carriers self-annihilate in the region of the junction. The remaining fixed charges cannot move to recombine and thus create a built-in potential to the junction. This potential causes the charges to separate and move toward the contacts. If the electrons and holes survive long enough, then the solar cell can generate energy. This process can be seen in Figure 3, where light creates electron-hole pairs which are then carried towards the contacts due to the voltage difference between the n and p layers. Due to the opposite polarity of the different charge carriers, even though they travel in opposite directions, their currents add in the same direction.

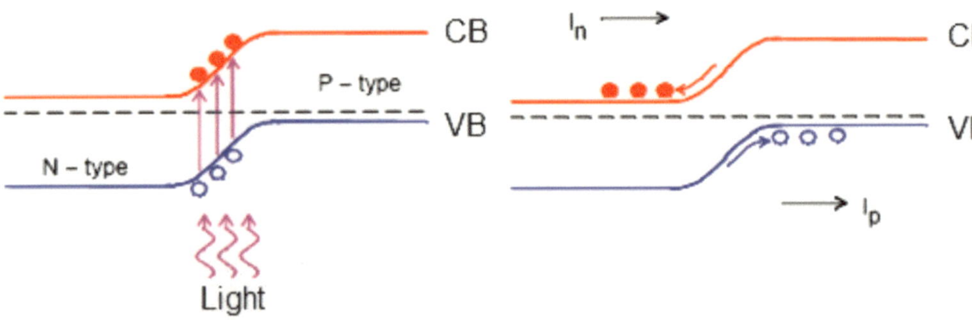

Fig. 3. Charge separation in a solar cell. The red circles are electrons, and the blue circles are holes. CB stands for conduction band, and VB stands for valence band. Note that although the electrons and holes travel in opposite directions, since they have opposite polarities, their currents, denoted as In and Ip, add in the same direction

The band-gap presents a distinct trade-off in the performance of a material in a solar cell device. A wide band-gap material does not absorb the energy of lower-energy photons. A small band-gap material may absorb more solar energy; however, the photo-generated charge carriers lose much of this energy due to thermalization down to the band-gap energy as the electrons travel out of the material. This is illustrated in Figure 4. Thus, regardless of the bandgap, a significant portion of solar energy is lost. As discussed in the following section, a nanowire array has the potential to resolve this issue, vastly increasing the maximum efficiency of the photovoltaic (Kempa, Naughton et al. 2009).

The ultimate figure of merit for solar cells is cost per watt; this decides whether a particular solar energy technology will become commercially viable. Our use of almost exclusive wet chemistry and abundant materials has the potential to be incredibly cheap and created on a massive scale. Solar energy has the potential to accommodate much of the world's energy needs, but for it to actually become a major source of energy the materials used must be in sufficient supply. The materials we present are abundant enough to provide electricity for the entire world (Wadia, Alivisatos et al. 2009). This abundance, when combined with the inexpensive mass manufacturing possibilities of the wet chemistry process, indicates that this device could help achieve energy independence while safeguarding our environment.

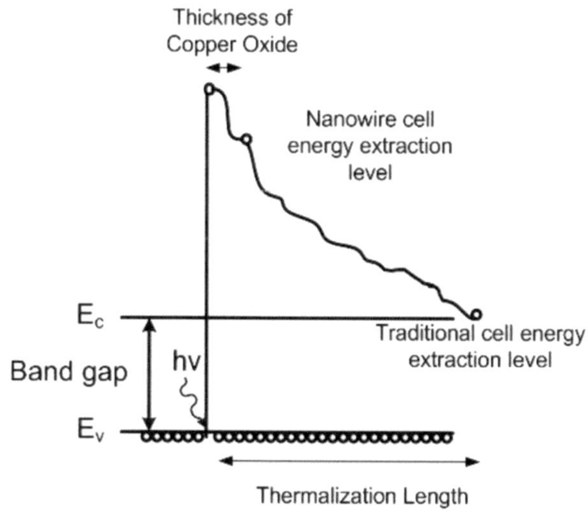

Fig. 4. Thermalization loss in a solar cell. Ec stands for conduction band energy level, and Ev stands for valence band energy level.

Our design focuses on the use of cheap, abundant materials, as well as inexpensive fabrication methods in order to drive down the cost of producing solar cells. The use of a nanowire geometry should help to increase absorption, as well as decrease the carrier travel distance. Details of our design and the benefits therein will be discussed in the following sections.

2.2 How nanowires benefit solar cells

Nanostructures are constantly being found to improve the performance of objects and devices in many different fields, including photovoltaics. Specifically, nanowires are being used to enhance solar cells and have several key benefits including: decreased net reflectance (increased absorbance), increased junction area, and decreased carrier travel distance.

The increased surface area of a nanowire array and the inherent surface features result in decreased reflection and consequently higher absorption of incident light. (Hu and Chen 2007) This is largely due to light trapping and scattering effects where light gets reflected in between the nanowires, thus largely reducing the likelihood that it will be reflected outwards. This effect is shown in Figure 5 and is explained in more detail in reference (Muskens, Rivas et al. 2008). Reference (Cao, Fan et al. 2010) demonstrates resonant effects based on nanowire diameter and the weak angle dependence of incident light inherent in nanowires which contribute to increased absorption.

The nanowire core-shell geometry causes carriers to travel radially across the wire, thus reducing the distance the carrier must travel before it can reach the contact. The junction is also close to the contacts due to the nanowires, thus the separated carriers have a shorter distance to travel within the semiconductor and consequently a lower likelihood of recombining. Additionally this also means that the material quality can be lower than a traditional device (Tian, Zheng et al. 2007), an important factor with our material and growth procedures. We are examining the carrier transport mechanisms in these photovoltaic cells to see if the decreased carrier travel distance caused by the lateral movement reduces carrier extraction times. This may lead to increased energy extracted per photon, as hot carriers could be extracted before they lose their above-bandgap energy to thermalization.

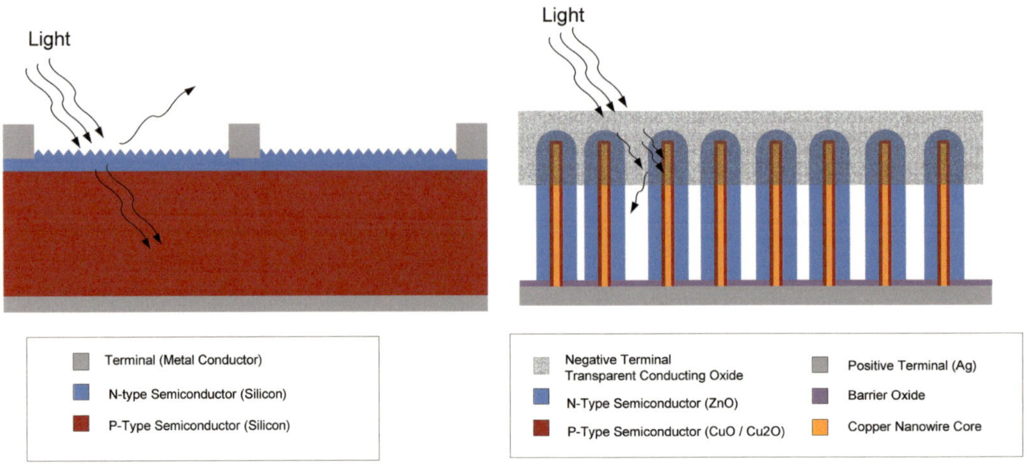

Fig. 5. Bulk solar cell and nanowire solar cell, notice the light trapping effect of the nanowire cell.

The core-shell nanowire array allows for the use of ultrathin semiconductor layers, which reduce the amount of material used, while not sacrificing photon absorption. Although the materials used here are good absorbers already, Figure 6 demonstrates how the nanowires greatly improve light absorption. Since absorption is high while using ultrathin layers, the added benefit of reducing the amount of material used will lower overall costs of the design.

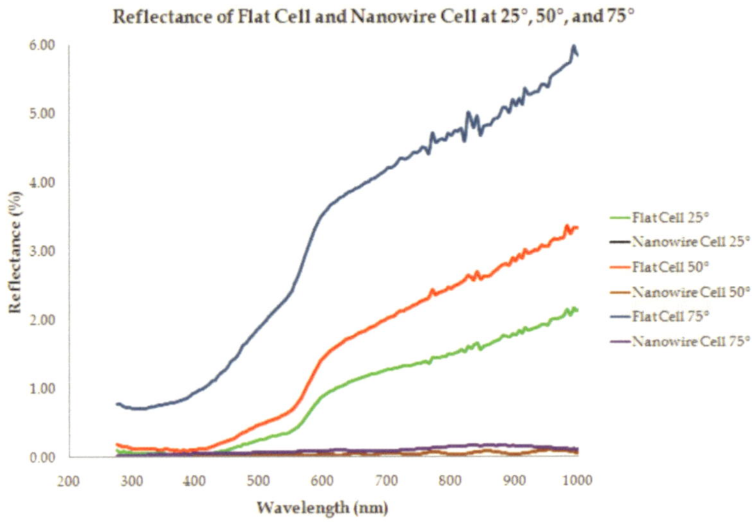

Fig. 6. Reflectance measurements for nanowire and flat cell structure at multiple angles with 45° polarized light. Measurements were taken on a JA Woollam VASE ellipsometer. The black indicating the nanowire cell at 25° is mostly obscured by the axis and other lines. The nanowire cell at 25° resulted in the least amount of reflection.

Yet another benefit to the core-shell geometry is that no matter where the photon is absorbed, it will be close to the junction. So if a photon travels halfway down a wire before being absorbed, or is absorbed at the tip of the wire, it has a similar likelihood of creating an electron-hole pair. In a bulk cell this is not the case as the further into a cell the photon travels, the further it is from the junction and the less likely it is that it will contribute to carrier generation.

3. Nanowires and their fabrication

Due to their geometry and scale, nanowires have generated massive research interest in the recent decades. Nanowires demonstrate exceptional surface to volume ratios, which increase inversely with the diameter of the nanowire. Carbon nanotubes have surface to volume ratios exceeding 10^6 cm^2/cm^3. As such, nanowires have attracted much attention in the realms of catalytic science, energy storage, and electrochemistry; however, nanowire applications for energy generation are less forthcoming.

Many processes have been developed to fabricate nanowires. In general, 'bottom-up' methodologies are employed because 'top-down' techniques are limited in the achievable critical dimension and aspect ratios.

The technique of electrospinning forms nanofibers by taking advantage of the repulsive electrostatic forces present in a charged droplet of fluid. In the right conditions, a charged droplet can effuse a nanoscale stream that forms into a nanofiber. Composites may be used such that further processing can yield nanowires composed of metal or other materials (Shui and Li 2009).

Exceptionally high aspect ratio nanowires can also be grown along the step edges found in natural or manufactured structures. Zach et al. have demonstrated the ability to selectively electroplate at the atomic step edges of highly ordered pyrolytic graphite (Zach, Ng et al. 2000). Similarly, Menke et al. pioneered nanowire fabrication by electroplating along the undercut region of photoresist after wet etching an electrode masked by a lithographically-patterned photoresist film (Menke, Thompson et al. 2006).

Single-crystal semiconducting nanowires can be grown by the vapor-liquid-solid method (VLS growth). This popular method involves depositing a thin (1-10nm), often gold, seed layer on a substrate of the desired nanowire material. The substrate is then put into a vacuum chamber into which appropriate vapor-phase precursors for the material to be grown are introduced. The seed layer forms nucleation points that, at certain temperatures, facilitate the transfer of vapor species through the liquid phase to condensation on the substrate below the gold (Givargizov 1975).

Nanowires can also be fabricated by solution based processing. This is analogous to VLS growth except occurring in a solution rather than a vacuum chamber. In general, a catalyzing agent at the tip of the growing nanowire aids the transfer from ionic species in solution to solid species in the nanowire. Other components in solution inhibit growth on the substrate and sidewalls of the nanowire (Wang, Dong et al. 2006). Also of note is the similar polyol synthesis method (Sun, Mayers et al. 2003). This method involves the passivation of all but one crystal plane, allowing for the single dimensional growth along one crystal plane. High aspect ratio silver, gold, and palladium nanowires have been synthesized in this manner.

Standard lithography may also be used to create nanowires. However, the critical dimension (diameter) and aspect ratio are limited by the technology. The aspect ratios and dimensions achieved by the aforementioned techniques far outmatch those used in standard lithographic processes. However, the use of lithography does present two distinct advantages when dealing with electronic devices. For one, the nanowires can be easily integrated into other devices and structures that are fabricated by lithographic methods. For example, interconnects, transistors, and MEMS structures can be made together with nanowires. Secondly, for standard top-down fabrication methods, there exist many well-characterized means to make electrical connections with the outside world (wirebonding, solder bumping, etc). This is often an issue when working with bottom-up methodologies.

Finally, templated growth is a popular and versatile nanowire production method. This process is used in this work to fabricate core-shell nanowires. In this method, nanowires are masked by a template such as an anodized alumina membrane or a track etched polycarbonate membrane. Both of these membranes have extremely high aspect ratio pores. Anodized aluminum oxide (AAO) membranes have pores that range from nanometers to microns in diameter and up to hundreds of microns in length. To make nanowires, the desired material is filled into the template. A variety of methods have been used to achieve this including atomic layer deposition, centrifugation, electrophoretic deposition, and, most commonly, electrochemical deposition (ECD) (Cao and Liu 2008). In electrochemical growth a conductive seed layer is deposited on one side of the template, and the other side is introduced to a plating solution for the desired material. The final step in fabrication is the removal of the template, which is accomplished by dissolving the membrane in an appropriate solution (halogenated solvents for polycarbonate membranes or strong bases for AAOs).

3.1 Copper oxide - zinc oxide heterojunction nanowires:

The materials investigated here are copper (II) oxide and zinc oxide. These have bandgaps of 1.2 eV (Jiang, Herricks et al. 2002) and 3.3 eV (Ozgur, Alivov et al. 2005), respectively. There is a fair amount of copper (I) oxide present as well which has a band-gap of 2.0 eV (Rakhshani 1986), but this material should decrease with optimization of our fabrication process. These materials were chosen for their ease of use in electrodeposition and their relative abundance. They have the added benefit of being natural n and p type materials. Undoped zinc oxide has a residual n-type conductivity (Look, Hemsky et al. 1999; Ozgur, Alivov et al. 2005), and CuO is naturally p-type. (Jiang, Herricks et al. 2002) Currently an indium tin oxide (ITO) thin film is used for the top contact, but future work will include investigating more economical alternatives.

The structure of the nanowire consists of co-axially stacked layers creating a core-shell arrangement as seen in Figure 7. The benefits of the structure have been discussed in section 2.2. The base and core of the nanowires consist of a copper core, which acts as the bottom contact, next is a shell of copper oxide around the copper core, and then a second shell of zinc oxide. Finally, a layer of indium tin oxide is deposited on the top. Due to the sputtering deposition, this top layer is not very conformal and covers mostly the top of the wires. This process can be replaced by an ALD deposition to achieve a conformal coating.

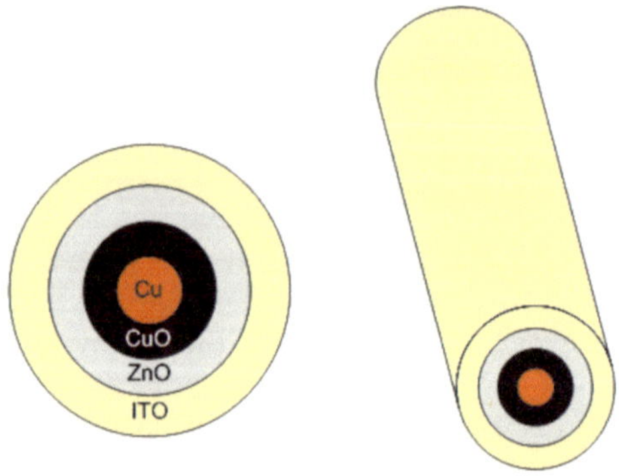

Fig. 7. Core-shell diagram

The band diagram of the cell can be seen in Figure 8. This shows the band-gaps of the materials and the voltage drop between them. This voltage drop is what pushes the electrons and holes away from each other to prevent them from immediately recombining. The band-gap of ZnO is 3.3eV and as such can absorb only UV light out to 376nm. Cuprous oxide (Cu_2O) can absorb light out to 620nm, and cupric oxide, CuO, can absorb radiation out to 1033nm. This encompasses the vast majority of the solar spectrum. The ZnO layer is very thin and has a wide band-gap. Therefore its purpose is to act as the n-type material to create the built in voltage of the diode, which serves to extract charge carriers. The copper oxides serve as the absorber and p-type material in the diode.

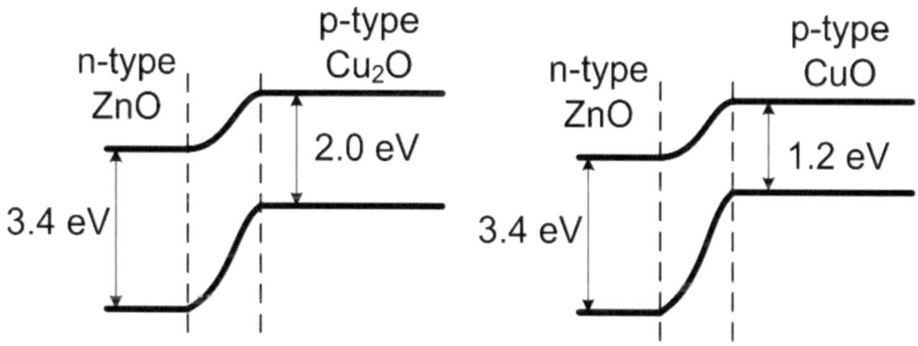

Fig. 8. Band-structure for both copper oxides present.

The process of creating our nanowires will be discussed in section 3.3, however the latest iteration (before top contact deposition) can be seen in Figure 9.

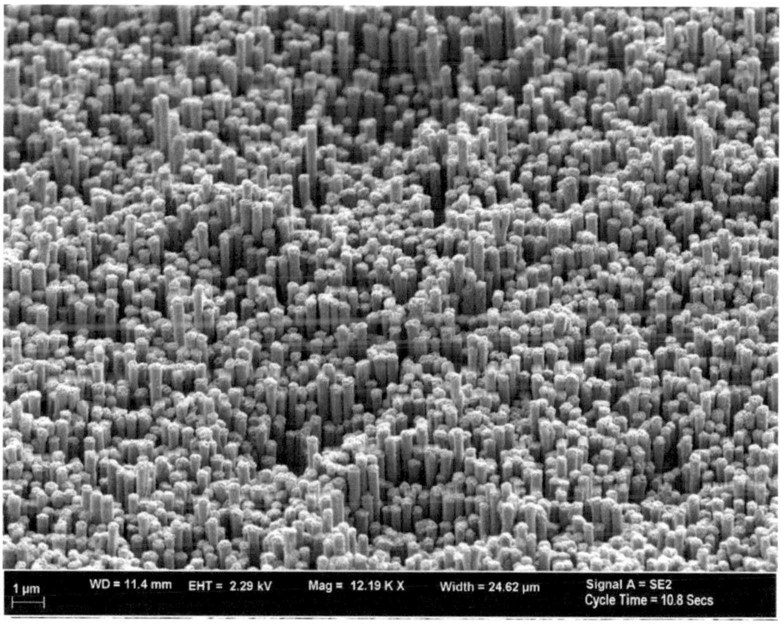

Fig. 9. SEM image of our latest iteration of core-shell nanowires

3.2 Electrochemical and other fabrication methods for nanowire arrays

The geometric constraints when fabricating nanowires (especially core-shell structures) make the use of many common top-down techniques impractical. The dimensions of nanowires place constraints on fabrication with which many conventional deposition methods cannot comply. For example, physical vapor deposition (PVD) processes are line of sight methods, and, as such, these methods never yield impressive step coverage in high aspect ratio (length to width ratio) features such as nanowires (Madou 2002). Chemical vapor deposition performs better and is routinely used to coat or fill features with aspect ratios of five or ten (Gordon, Hausmann et al. 2003). However, this is still orders of magnitude removed from the typical nanowire aspect ratios. The following section details fabrication methods well-suited to the creation and alteration of nanowire arrays with special attention to copper and zinc oxides and core-shell structures.

Electrochemical deposition, shown in Figure 10, is the utilization of electrically driven redox reactions to solidify ions out of solution. Most often, metallic cations are reduced at a cathode, while oxidation at an anode of the same metal replaces the reduced ions. It is possible to plate alloys and compounds through careful manipulation of the plating bath. For example, many metal oxides, including copper and zinc oxides, can be deposited by careful control of the solution and electrical conditions (Golden, Shumsky et al. 1996)

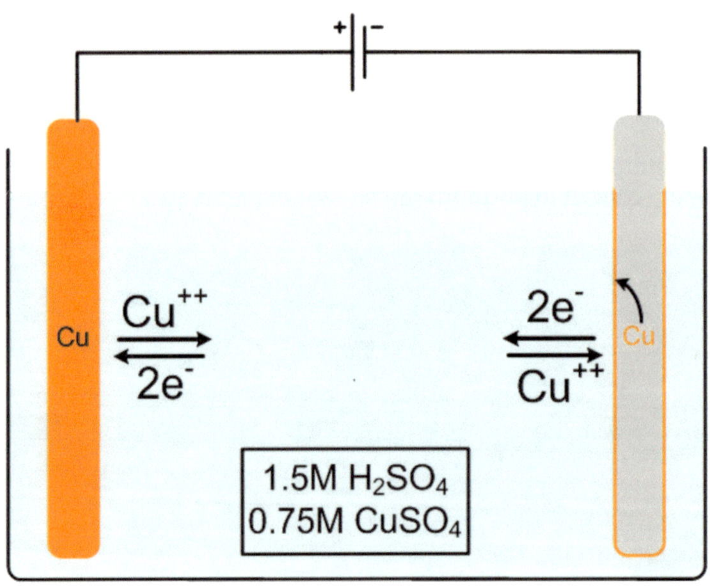

Fig. 10. Mechanism of electrochemical deposition

One drawback of electro-deposition is its tendency towards non-conformal growth on non-planar surfaces. This is not an issue when growing nanowires through a template, which serves to constrain the growth. However, on rough or textured surfaces (such as a nanowire array), electrons gather in the asperities on the surface. More deposition occurs at these points, which only exacerbates the issue as this make the surface rougher. Additives into the plating bath can alleviate this issue by gathering at the points of highest current density and inhibiting deposition, but this introduces impurities into the deposited material. The power of additives in creating conformal deposits is evidenced by the advent of copper interconnects in the integrated circuit industry where high-aspect ratio vias are filled by ECD (Andricacos 1999). However, aspect ratios are even more pronounced for the extreme geometries observed in nanowire arrays. There are also mass transfer concerns as the diffusion length for ions in solution exceeds the inter-nanowire spacing. Therefore, it is difficult (or improbable) for ions to travel the length of the nanowire to its base without being reduced and deposited prematurely. While ECD is difficult to implement for conformal coatings on nanowire arrays, it remains the method of choice to make high-quality nanowires through templated growth.

Electroless and immersion deposition are chemical methods to deposit metal films without applying an electric potential and have traditionally been used to conformally coat difficult geometries. Electroless plating (also known as autocatalytic deposition) utilizes reducing agents in solution to drive the reaction at the surface, which acts as a catalyst. Electroless deposition does not suffer from the current crowding effects of electrolytic deposition, but mass transfer effects can influence the deposition rates at different points in a nanowire geometry (Paunovic, Schlesinger et al. 1998).

Immersion plating is the displacement reaction of a more noble metal replacing a more active metal on a surface. For example, gold ions in solution would reduce and plate onto an iron bar. The iron surface would oxidize and dissolve into solution to maintain charge conservation. The reaction is self-limiting with deposits being only a few monolayers thick. Unlike autocatalytic and electrolytic plating, immersion deposits are exceptionally uniform regardless of surface topology. As such, immersion deposits are an attractive option for creating the 'shell' in nanowire core-shell applications. The main drawback is the material limitations: the depositing material must be higher in the galvanic series than the substrate (Langdon 1988).

The material constraints of immersion plating can be solved through a process known as contact plating. In this process, the substrate to be coated is put into electrical contact with a more electropositive (less noble) metal. Oxidation occurs at the electropositive metal, driven by the constituents of the bath. This oxidation yields electrons, which travel through the electrical contact and allow for reduction of ions out of solution. If the less noble metal is also the coating material, this reaction is self-limiting in that the reaction ceases when the work is coated. Thus, conformal coatings of less-noble metals may be electrochemically deposited on difficult geometries (Durney 1984).

While not electrochemical in nature, atomic layer deposition (ALD) should be mentioned as the ultimate method of conformal coating. Unlike the previous processes, ALD occurs from reactants in gas phase, generally at low pressures. In ALD a gas precursor is introduced and allowed to form a monolayer on the surface of a sample before being pumped out. Then, a second gas is introduced, which reacts with the monolayer to yield the desired film. This process is repeated to build up a film atomically, monolayer by monolayer. Because the monolayers are exceptionally uniform, conformal coatings on difficult topologies are easily achieved. Aspect ratios of almost 50 are easily coated uniformly (Ritala and Leskela 2001). The main drawback to ALD is the extremely low rate of deposition, which typically reaches a maximum at a few Angstroms per minute.

Finally, oxidation reactions are perhaps the most facile and effective way of creating exceptionally uniform and conformal layers on complex structures. Oxidation can be performed in either aqueous or gaseous environments. The latter case, generally referred to simply as thermal oxidation, has been well-characterized over a wide-range of temperatures

(Cabrera and Mott 1949). Thin, uniform oxides can be formed on metals simply by applying heat in an oxygen atmosphere (Rusu, Gırtan et al. 2007; Njeh, Wieder et al. 2002). However, the high temperatures needed in thermal oxidation can have some undesired affects with regards to annealing, coefficients of thermal expansion, and diffusion. For example, a copper-zinc structure cannot be converted into a copper oxide – zinc oxide heterojunction by thermal oxidation because the materials will diffuse into one another. This forms brass, not a diode.

In the case of wet oxidation, the reaction can occur spontaneously (Tam and Robinson 1986) or can be driven electrically in a process known as anodization (Yamaguchi, Yamazaki et al. 1998). These processes have the advantage of taking place at low temperatures. Anodization can be viewed as the reverse process as electroplating. In anodization, a positive potential applied to the piece will cause oxygen ions in solution to oxidize (in the electrochemical sense) atoms at the surface. Whereas, in ECD, the oxidized ions at the anode dissolve into solution, in anodization, the oxides remain at the anode and build up a conformal film. Incidentally, it is by this process that the AAO templates are made for the initial nanowire growth in the process presented in this paper.

3.3 Fabrication

Our fabrication takes place at the Tufts Micro and Nano Fabrication Facility (TMNF) at Tufts University, with some work being done at the Center for Nanoscale Systems (CNS) at Harvard University. TMNF is a class 1000 cleanroon where most of our wet chemical and sputtering processes are performed. CNS is a large 10,000 sq. ft. class 1000 cleanroom where most of our imaging is done. We also perform our atomic layer deposition at CNS as well.

Our fabrication flow process can be seen in Figure 11. This diagram shows the process from start to finish and includes the creation of the copper nanowires, oxidation of the copper oxide layer, deposition of the ZnO layer, and finally the deposition of the top contact layer. The process starts by obtaining an anodized aluminum oxide template which can either be purchased commercially or created by anodizing an aluminum film. Copper is then sputtered on one side of the template become the electrode and act as a seed layer for the nanowires. The AAO then acts as a mask for the through-plating of copper. Five to ten microns of copper is electroplated into the pores. The AAO is then dissolved in a 10% solution of potassium hydroxide (KOH) for 20 minutes, revealing a densely packed copper nanowire array (as shown in Figure 9 above).

The electrochemical deposition (ECD) of copper is performed using a solution of 0.75 M copper sulfate ($CuSO_4$) and 1.5 M sulfuric acid (H_2SO_4) at room temperature with air agitation. This reaction is current limited with the supplied current at less than or equal to $5mA/cm^2$. The zinc oxide ECD is performed using a solution of 0.1 M zinc nitrate ($ZnNO_3$) at 70° - 80° C at a potential of 0.75 volts.

4. Future research

Nanowires and especially nanowire solar cells are still in an early stage of development and are found almost exclusively in the laboratory, although there are several companies planning products which utilize nanowires, i.e. Bandgap Engineering, QDSoleil, Illuminex. As with so many new technologies, there is much work yet to be done and there are still many improvements, optimizations, and options to be explored in the research presented here. Several areas which require further research and investigation are with conformal coatings of the nanowires and general process optimization, difference material choices, and long-term stability.

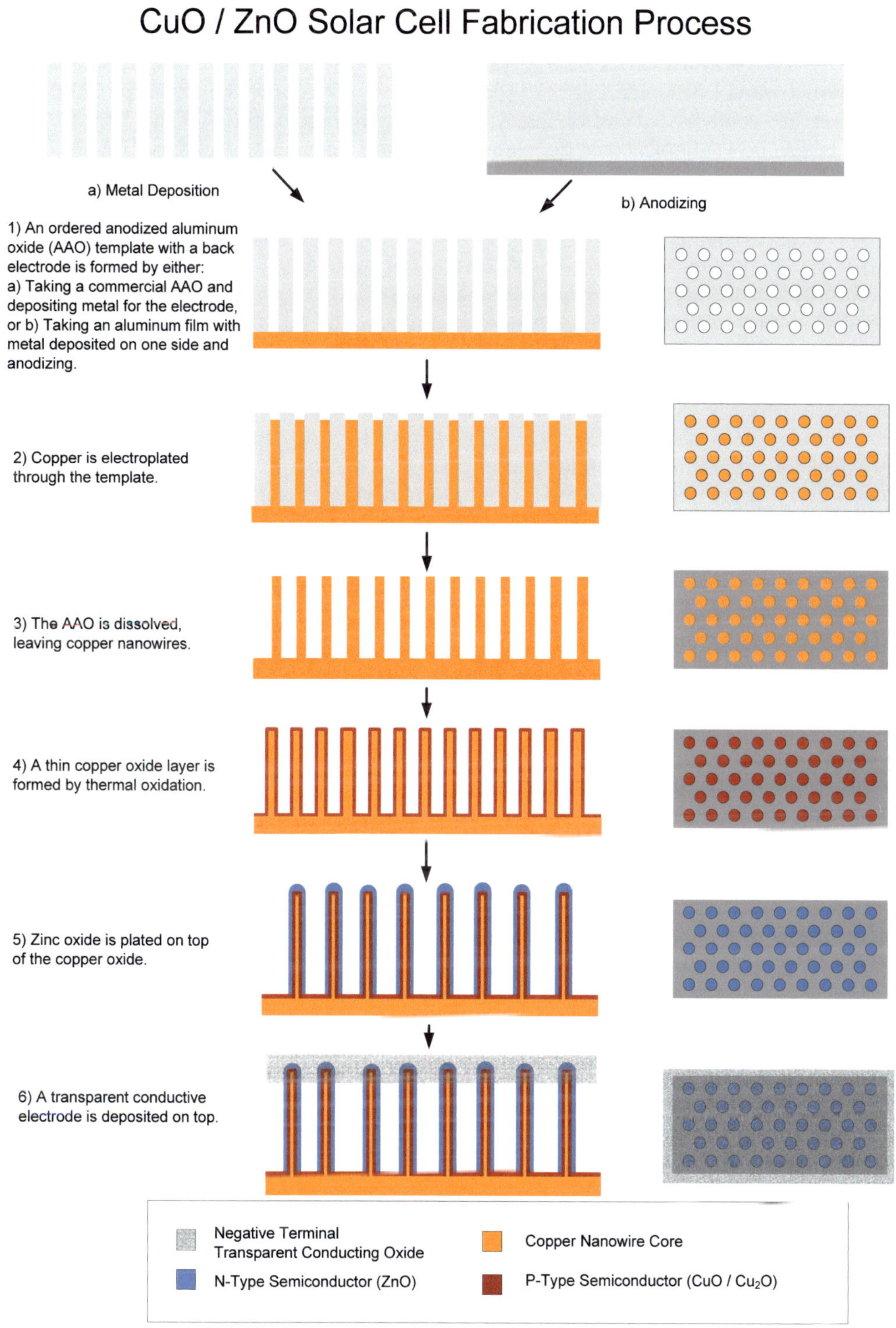

Fig. 11. Fabrication process for core-shell nanowire array photovoltaics

One of the biggest difficulties in growing these core-shell nanowires is achieving conformal coatings to create the core-shell structure. Electrochemical methods, besides being cheap, have proven to be much better than physical vapor deposition techniques, however due to the extremely small distances between the nanowires, often only the upper portions of the nanowires are coated. Several of the ways we are trying to mitigate this problem is by testing to see how atomic layer deposition compares to the ECD process. Also, contact plating shows promise in conformal, electrochemical coating of nanowire geometries. Much of the fabrication process already developed needs further optimization. Several factors which require tweaking are the specific parameters to obtain optimal widths and lengths of the nanowires, specific temperature and duration of thermal oxidation, specific voltage and concentration of the electrochemical baths, and especially our top contact deposition. We currently sputter coat an indium tin oxide transparent conductor, however this is an expensive material and a very non-conformal deposition. Sol-gel deposition techniques and alternate transparent conduction oxides such as doped ZnO should be investigated.

The oxides currently used were chosen specifically for their ease of use with ECD as well as their abundance in the earth's crust. This allowed us to make quick initial progress inexpensively, as well as to follow our goal of creating a low-cost solar alternative. However the copper and zinc oxides are not particularly well suited to the solar spectrum as they have relatively wide band-gaps and it would behoove us to look into other material choices. The fact that copper forms a Schottky barrier with copper oxide precludes the creation of an optimal ohmic contact reducing the efficiencies we may obtain, and as such a different bottom contact material or absorber material should be investigated. Another option is to optimize this process for a Schottky barrier PV cell, where the junction is formed not from a difference in doping but from the depletion layer induced by the metal at the metal-semiconductor junction.

One potential problem to be wary of is that the nanowires are extremely pressure sensitive. Locations on the cell where they have been handled are visibly damaged under a microscope, and in some cases visible to the eye. Although unavoidable in the early stages of research, this is an important problem which must be addressed down the line. Work must be done to provide adequate handling techniques during the fabrication process, as well as sufficient encapsulation to ensure long-term operation when deployed.

5. Conclusion

In this chapter, we have presented a novel fabrication process for the creation of oxide based core-shell nanowire photovoltaic cells. The major focus of this design process has been to keep costs low. To do this, earth-abundant materials have been investigated as well as wet chemical fabrication methods. As it is early in the design, our samples are plagued by issues which should be rectified by optimization of the fabrication process. Although we have seen a diode response from the fabricated samples, we are unsure of whether this is due to the possible schottky diode formed between the copper and copper oxide layers, or from the pn junction. These devices have the potential to become incredibly cheap solar cells but to achieve this goal further research and development is necessary and warranted.

6. Acknowledgements

The authors would like to sincerely thank the Tufts Micro and Nanofabrication facility as well as Harvard CNS for use of their clean room facilities. We would like to thank the

Wittich Energy Sustainability Research Initiation Fund for their funding as well as the National Science Foundation as this material is based upon work supported under a National Science Foundation Graduate Research Fellowship.

7. References

Andricacos, P. C. (1999). "Copper On-Chip Interconnections." The Electrochemical Society Interface 8(1): 6.

Cabrera, N. and N. Mott (1949). "Theory of the oxidation of metals." Reports on Progress in Physics 12: 163.

Cao, G. and D. Liu (2008). "Template-based synthesis of nanorod, nanowire, and nanotube arrays." Advances in Colloid and Interface Science 136(1-2): 45-64.

Cao, L. Y., P. Y. Fan, et al. (2010). "Semiconductor Nanowire Optical Antenna Solar Absorbers." Nano Letters 10(2): 439-445.

Da Rosa, A. V. (2005). Fundamentals of Renewable Energy Processes, Elsevier Acedemic Press.

Durney, L. J. (1984). Electroplating engineering handbook. New York, Van Nostrand Reinhold.

Givargizov, E. I. (1975). "Fundamental aspects of VLS growth." Journal of Crystal Growth 31: 20-30.

Golden, T. D., M. G. Shumsky, et al. (1996). "Electrochemical deposition of copper(I) oxide films." Chemistry of Materials 8(10): 2499-2504.

Gordon, R., D. Hausmann, et al. (2003). "A kinetic model for step coverage by atomic layer deposition in narrow holes or trenches." Chemical Vapor Deposition 9(2): 73-78.

Green, M. A., K. Emery, et al. (2010). "Solar cell efficiency tables (version 35)." Progress in Photovoltaics: Research and Applications 18(2): 144-150.

Hu, L. and G. Chen (2007). "Analysis of Optical Absorption in Silicon Nanowire Arrays for Photovoltaic Applications." Nano Letters 7(11): 3249-3252.

Jiang, X., T. Herricks, et al. (2002). "CuO Nanowires Can Be Synthesized by Heating Copper Substrates in Air." Nano Letters 2(12): 1333-1338.

Kasap, S. O. (2001). Optoelectronics and Photonics Principles and Practices. Upper Saddle River, New Jersey, Prentice-Hall Inc.

Kasap, S. O. (2006). Principles of Electronic Materials and Devices. New York, McGraw-Hill.

Kempa, K., M. J. Naughton, et al. (2009). "Hot electron effect in nanoscopically thin photovoltaic junctions." Applied Physics Letters 95(23): 233121.

King, D. L., M. A. Quintana, et al. (2000). "Photovoltaic module performance and durability following long-term field exposure." Progress in Photovoltaics: Research and Applications 8(2): 241-256.

Langdon, P. H., Ed. (1988). Metal finishing. Guidebook & directory issue. Hackensack, N.J., Metals and Plastics Publications.

Look, D. C., J. W. Hemsky, et al. (1999). "Residual Native Shallow Donor in ZnO." Physical Review Letters 82(12): 2552.

Loster, M. (2010). "Total Primary Energy Supply - From Sunlight." Retrieved January 18, 2011, from http://www.ez2c.de/ml/solar_land_area/.

Luque, A. and S. Hegedus (2003). Handbook of Photovoltaic Science and Engineering, John Wiley & Sons.

Madou, M. J. (2002). Fundamentals of microfabrication: the science of miniaturization. Boca Raton, CRC Press.

Menke, E. J., M. A. Thompson, et al. (2006). "Lithographically patterned nanowire electrodeposition." Nat Mater 5(11): 914-919.

Muskens, O. L., J. G. m. Rivas, et al. (2008). "Design of Light Scattering in Nanowire Materials for Photovoltaic Applications." Nano Letters 8(9): 2638-2642.

Njeh, A., T. Wieder, et al. (2002) "Reflectometry studies of the oxidation kinetics of thin copper films." Surface and Interface Analysis, 626-628 DOI: 10.1002/sia.1421.

Ozgur, U., Y. I. Alivov, et al. (2005). "A comprehensive review of ZnO materials and devices." Journal of Applied Physics 98(4): 041301-103.

Paunovic, M., M. Schlesinger, et al. (1998). Fundamentals of electrochemical deposition. New York, Wiley.

Rakhshani, A. E. (1986). "Preparation, characteristics and photovoltaic properties of cuprous oxide--a review." Solid-State Electronics 29(1): 7-17.

Ritala, M. and M. Leskela (2001). "Atomic layer deposition." Handbook of thin film materials 1: 103-159.

Rusu, G. G., M. Gırtan, et al. "Preparation and characterization of ZnO thin films prepared by thermal oxidation of evaporated Zn thin films." Superlattices and Microstructures 42(1-6): 116-122.

Saleh, B. E. A. and M. C. Teich (2007). Fundamentals of Photonics. Hoboken, New Jersey, John Wiley & Sons.

Shui, J. and J. C. M. Li (2009). "Platinum Nanowires Produced by Electrospinning." Nano Letters 9(4): 1307-1314.

Sun, Y., B. Mayers, et al. (2003). "Polyol Synthesis of Uniform Silver Nanowires: A Plausible Growth Mechanism and the Supporting Evidence." Nano Letters 3(7): 955-960.

Tam, T. M. and R. D. Robinson (1986). "ANALYSIS OF FACTORS CONTROLLING PEEL STRENGTH OF BLACK OXIDE CONVERSION COATINGS." Plating and Surface Finishing 73(4): 74-77.

Tian, B., X. Zheng, et al. (2007). "Coaxial silicon nanowires as solar cells and nanoelectronic power sources." Nature 449(7164): 885-889.

U.S._Energy_Information_Administration (2010). International Energy Outlook 2010. Washington DC, US Government Printing Office.

Wadia, C., A. P. Alivisatos, et al. (2009). "Materials Availability Expands the Opportunity for Large-Scale Photovoltaics Deployment." Environmental Science & Technology 43(6): 2072-2077.

Wang, F., A. Dong, et al. (2006). "Solution-Liquid-Solid Growth of Semiconductor Nanowires." Inorganic Chemistry 45(19): 7511-7521.

Yamaguchi, Y., M. Yamazaki, et al. (1998). "Photocatalytic ZnO films prepared by anodizing." Journal of Electroanalytical Chemistry 442(1-2): 1-3.

Zach, M. P., K. H. Ng, et al. (2000). "Molybdenum Nanowires by Electrodeposition." Science 290(5499): 2120-2123.

8

ZnO Nanowires and Their Application for Solar Cells

Qiang Peng and Yuancheng Qin
School of Evironmental & Chemical Engineering
Nanchang Hangkong University
Nanchang, 330063
China

1. Introduction

Nanowires (NW) are defined here as metallic or semiconducting particles having a high aspect ratio, with cross-sectional diameters « 1 μm, and lengths as long as tens of microns. Well-aligned one-dimensional nanowire arrays have been widely investigated as photoelectrodes for solar energy conversion because they provide direct electrical pathways ensuring the rapid collection of carriers generated throughout the device (Tang et al., 2008), as well as affording large junction areas and low reflectance owing to light scattering and trapping (Muskens et al., 2008).

Solar energy conversion is a highly attractive process for clean and renewable power for the future. Excitonic solar cells (SCs), including organic and dye-sensitized solar cells (DSSC), appear to have significant potential as a low cost alternative to conventional inorganic photovoltaic (PV) devices. The synthesis and application of nanostructures in solar cells have attracted much attention. Metal oxide nanowire (NW) arrays with large surface area and short diffusion length for minority carriers represent a new class of photoelectrode materials that hold great promise for photoelectrochemical (PEC) hydrogen generation applications. Up to now, various metal oxide nanostructures such as TiO_2, ZnO, Fe_2O_3, ZrO_2, Nb_2O_5, Al_2O_3, and CeO_2 have been successfully employed as photoelectrodes in SCs. Among the above-mentioned metal oxide nanostructures, the study of TiO_2 and ZnO is of particular interest due to the fact that they are the best candidates as photoelectrode used in SCs. However, the advantage offered by the increased surface area of the nanoparticle film is compromised by the effectiveness of charge collection by the electrode. For DSSCs, the traditional nanoparticle film was replaced by a dense array of oriented, crystalline nanostructures to obtain faster electron transport for improving solar cell efficiency. A typical high-efficiency DSSC (Grätzel, 2009) consists of a TiO_2 nanocrystal thin film that has a large surface area covered by a monolayer of dye molecules to harvest sunlight. Comparedwith TiO_2, ZnO shows higher electron mobility with similar bandgap and conduction band energies. ZnO is a direct wide bandgap semiconductor (Eg = 3.4 eV) with large exciton binding energy (~60 meV), suggesting that it is a promising candidate for stable room temperature luminescent and lasing devices. Therefore, ZnO nanowires is an alternative candidate for high efficient SCs.

So far, various ZnO nanostructures have been extensively investigated for SCs. In the early reports on ZnO-based DSSCs, ZnO nanoparticles were often used as the photoanode prepared by a conventional doctor blade technique (Keis et al., 2002; Keis et al., 2002). Lévy-Clément et al. (2005) reported experimental results on a new ETA solar cell fabricated from an electron-accepting layer of free-standing ZnO nanowires. Law et al. (2005) presented first the ordered nanowire DSSC. The nanowire DSSC is an exciting variant of the most successful of the excitonic photovoltaic devices. As an ordered topology that increases the rate of electron transport, a nanowire electrode may provide a means to improve the quantum efficiency of DSSCs in the red region of the spectrum, where their performance is currently limited. Raising the efficiency of the nanowire cell to a competitive level depends on achieving higher dye loadings through an increase in surface area. Law et al. (2006) described the construction and performance of DSSCs based on arrays of ZnO nanowires coated with thin shells of amorphous Al_2O_3 or anatase TiO_2 by atomic layer deposition. Wu et al. (2007) employed mercurochrome and N3 dyes to be the sensitizers in the ZnO-nanowire DSSCs. A lower fill factor is obtained in the N3-sensitized cell which results in comparable efficiencies in both ZnO-nanowire DSSCs although the N3 molecules possess a wider absorptive range for light harvesting. Hsueh et al. (2007) deposited p-Cu_2O onto vertical n-ZnO nanowires prepared on ZnO:Ga/glass templates. With the sputtered Cu_2O, the nanowires became clublike (i.e. nanowire with a head). Leschkies et al. (2007) combine CdSe semiconductor nanocrystals (or quantum dots) and single-crystal ZnO nanowires to demonstrate a new type of quantum dot-sensitized solar cell. An array of ZnO nanowires was grown vertically from a fluorine-doped tin oxide conducting substrate. A significant improvement of the efficiency of the ZnO nanowire DSSC has been achieved by the chemical bath deposition of the dense nanoparticles within the interstices of the vertical ZnO-NW anode (Ku et al., 2007). Greene et al. (2007) evaluated an ordered organic-inorganic solar cell architecture based on ZnO-TiO_2 core-shell nanorod arrays encased in the hole-conducting polymer P3HT. Chen et al. (2009) studied a two-stage growth of a ZnO hierarchical nanostructure consisting of ZnO nanorod at bottom and ZnO nanowire atop in the low pH solutions. The mechanism was explained by hydrothermal reactions and dissolution dynamics. Chen et al. (2009) reported vertically aligned zinc oxide (ZnO) nanorod arrays coated with gold nanoparticles have been used in Schottky barrier solar cells. The nanoparticles enhance the optical absorption in the range of visible light due to the surface plasmon resonance. Recently, Briseno et al. (2010) demonstrated the basic operation of an organic/inorganic hybrid single nanowire solar cell. End-functionalized oligo and polythiophenes were grafted onto ZnO nanowires to produce p-n heterojunction nanowires. Gan et al. (2010) prepared hybrid ZnO/TiO_2 photoanodes for dye-sensitized solar cells by combining ZnO nanowire arrays and TiO_2 nanoparticles with the assistance of the ultrasonic irradiation assisted dipcoating method. Myung et al. (2010) synthesized high-density ZnO-CdS core-shell nanocable arrays by depositing CdS outerlayers on pregrown vertically aligned ZnO (wurtzite) nanowire arrays using the chemical vapor deposition method. Seol et al. (2010) a novel CdSe/CdS/ZnO nanowire array fabricated by a 3-step solution-based method was used as a photoanode of a quantum dot sensitized solar cell, which generated a maximum power conversion efficiency of 4.15%. Wu et al. (2010) reported a 74% enrichment of the efficiency of ZnO nanowire DSSCs is achieved by the addition of a novel light-scattering nanocrystalline film.

In this Chapter, after the brief review of the research progress on ZnO nanowires, some important results obtained on ZnO nanowires are summarized. In the second section, the latest progress in the growth of ZnO nanowires will be described. Their solar cell application will be discussed in detail in the third section. This Chapter ends with a brief summary, which also includes our personal remarks on future research of ZnO nanowires.

2. The growth of ZnO nanowires

A number of methods have been employed to achieve ZnO nanostructured arrays, including chemical and physical vapor deposition, hydrothermal process, metallorganic vapor-phase epitaxial growth, templated growth method and electrochemical deposition technique. Vapor-liquid-solid growth and vapor-solid-solid growth has been conventionally used to synthesize ZnO arrays as reviewed recently by Wang et al. (2009) and Haller et al. (2010).

Chao et al. (2010) reported the ZnO nanowires were grown in a furnace by chemical vapor deposition with gold as catalyst. Figure 1 shows the tilt-view SEM image of ZnO nanowire arrays on the sapphire substrate. This image revealed that the ZnO wires are vertically aligned, the length of nanowire is around 1-2 mm and the diameter is in the range of 70-100 nm. the synthesis and characterization of three-dimensional hetergeneous graphene nanostructures comprising continuous large-area graphene layers and ZnO nanostructures, fabricated via chemical vapor deposition, are reported by Lin et al. (2010). Electron microscopy investigation of the three-dimensional heterostructures shows that the morphology of ZnO nanostructures is highly dependent on the growth temperature. The morphology of the large-area graphene layers was identified via SEM as shown in Figure 2a, indicating regions of monolayers and few-layers. boundary structures of graphene layers enhanced the growth of dense array of ZnO nanostructures, observed as bright regions shown in Figure 2b. The nanowires obtained by chemical and physical vapor deposition have generally a good crystalline quality and an important length (more often they are in nanobelt morphology).

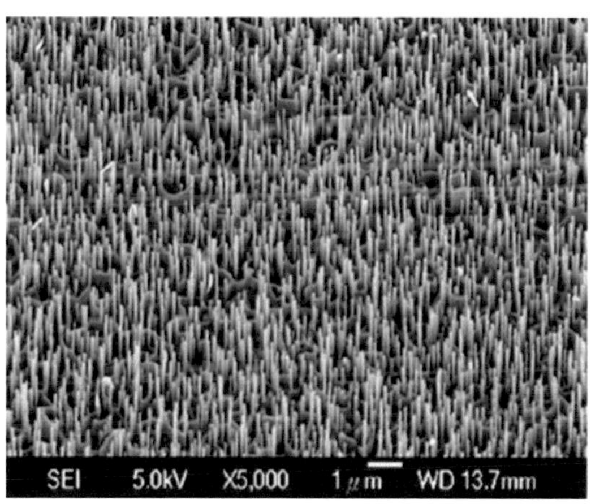

Fig. 1. Tilt-view SEM images of aligned ZnO nanowire arrays. From Ref. (Chao et al., 2010).

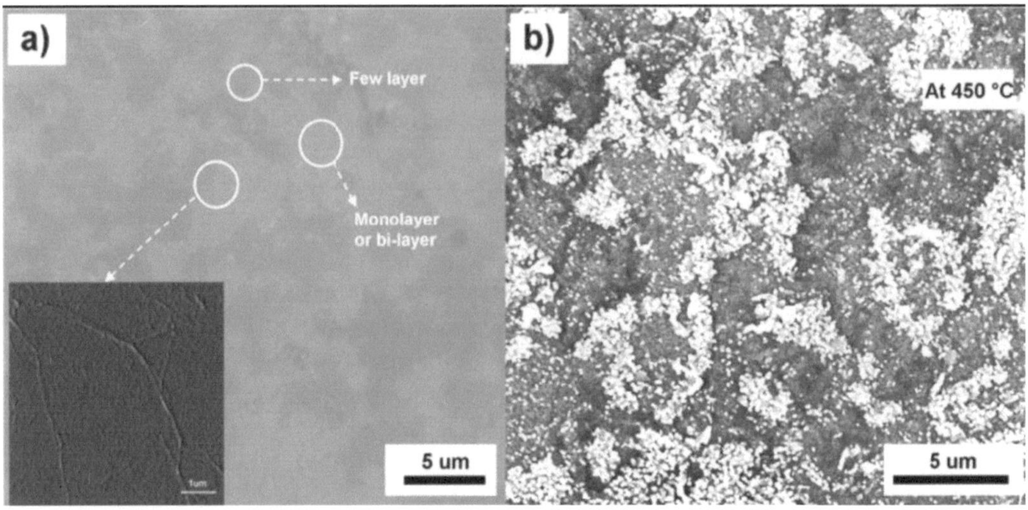

Fig. 2. a) SEM image of a chemical vapor deposition-grown graphene layer on a SiO_2/Si substrate. b) SEM image of ZnO nanostructures grown on the same chemical vapor deposition-graphene film at 450 °C. From Ref. (Lin et al., 2010).

Fang et al. (2006) have successfully synthesized aligned ZnO nanofibers in a dense array from and on a Zn substrate by hydrothermal treatment of Zn foil in an ammonia/alcohol/water mixed solution. Notably, the ZnO nanofibers are ultrathin (3-10 nm) with a length of ≈500 nm. This is the first time that uniform, aligned, and ultrathin ZnO nanofibers have been obtained via a hydrothermal method in the absence of catalysts and at a relatively low temperature. The photoluminescence measurements at room temperature revealed a significantly blue-shifted near-band-edge emission at 373 nm (3.32 eV), which was ascribed to quantum confinement arising from the reduced size of the ultrathin ZnO nanofibers. Then, the hydrothermal synthesis of large-scale, ultralong ZnO nanowire and nanobelt arrays with honeycomb-like micropatterns has been realized by simple surface oxidation of zinc foil in aqueous solutions of NaOH and $(NH_4)_2S_2O_8$ at 150 °C (Lu et al., 2006). This solution approach to fabricate 1D ZnO nanostructures with controlled morphologies and micropatterns can be easily scaled up and potentially extended to the fabrication and assembly of 1D nanostructures of other oxide systems. As shown in Figure 3a, a large scale thin film of long ZnO nanowire (20–50 mm) arrays formed unique micropatterns of honeycomb-like structures typically ranging from 10 to 30 mm in size on the Zn substrate after the Zn foil was immersed in the reaction solution containing 0.48 M NaOH and 0.095 M $(NH_4)_2S_2O_8$ and hydrothermally treated at 150 °C for 2 days. An enlarged image of the honeycomb-like structure is shown in Figure 3b, which indicates that these structures were formed when the collapsing ZnO nanowires from opposite directions met to bundle together between two neighboring areas. The diameter of the ZnO nanowires is measured to range from 60 to 200 nm (Figure 3c) and the electron diffraction (ED) pattern of a single nanowire suggests that each ZnO nanowire is a single crystal oriented along the c-axis (Figure 3d), similar to the growth direction of the ZnO nanorods obtained at room temperature.

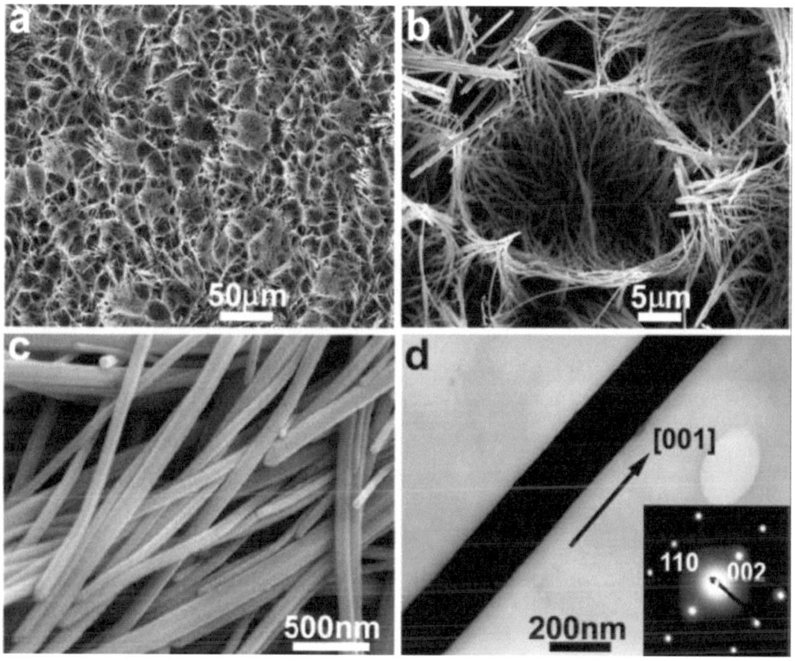

Fig. 3. SEM (a–c) and TEM (d) images of ZnO nanowire arrays hydrothermally grown on Zn foil at 150 °C for 2 days. From Ref. (Lu et al., 2006).

A new method of epitaxial ZnO column deposition that exhibits uniformity and reproducibility over a large surface area were demonstrated by Peterson et al. (2004). This method employs an aqueous solution containing NaOH and $Zn(NO_3)_2$ and substrates coated with sputtered ZnO and does not require the use of complexing agents and can produce micrometer-thick films in less than 1 h. The resulting quasi epitaxial films of highly ordered columns have the reproducibility and uniformity over large areas to be employed in the development of solar cells and other devices. Liu et al. (2003) reported the fabrication of ordered and homogeneous arrays of ZnO nanowires with a narrow diameter distribution, a high aspect ratio, high wire density, and large-area homogeneity by using anodic aluminum oxide templates and vapor deposition. The nanowire arrays were dense (~10^{10}-10^{11} cm^{-2}) with a high aspect ratio up to ~5×10^6, and homogeneous over a large aera (~20 mm^2). Ultraviolet lasing action of the arrays was observed by photoluminescence at room temperature above an exciting laser (λ = 335 nm) threshold of ~100 kW cm^{-2}.

Because chemical and physical vapor deposition need to work in vacuum and/or at high temperature, these techniques require sophisticated and expensive equipments. The electrochemical deposition technique is becoming an important means for the fabrication of ZnO nanowires due to the low cost, mild conditions, accurate process control and widely used in industry. Electrodeposition of different oxides has been an increasingly active research area in recent years and has been widely used by many research teams because the preparation from aqueous solutions has several advantages over the above techniques. Leprince-Wang et al. (2006) reported on the structure study of ZnO nanowires grown via electrochemical deposition, a simple and low temperature approach. Figure 4 presents typical TEM images showing the general morphology of the electrodeposited ZnO nanowires. It is found that the nanowires formed from the M90 type membrane are about 130-150 nm in diameter and 2-3 mm in length. Yang et al. (2007) investigated systematically the evolution of electrochemical deposition produced ZnO nanostructures by varying electrochemical conditions, identify important factors for the growth of the ZnO nanorod/nanobelt arrays, deduce possible electrode reactions and pathways, and discuss possible growth mechanisms under the alkaline conditions (Figure 5). Pradhan et al. (2010) have demonstrated a simple electrochemical deposition technique for growing ZnO nanostructures on ITO-glass substrates at 70 °C in an aqueous $Zn(NO_3)_2 \cdot 6H_2O$ (mixed with KCl) solution. By judiciously manipulating the deposition conditions, the mean ledge thickness of the nanowalls and the diameter of the nanowires can be controlled over the ranges of 50-100 and 50-120 nm, respectively. The KCl supporting electrolyte concentration can be used to control the morphology of ZnO nanostructures growth. Elias et al. (2010) have developed a novel low-cost method to produce large area of single crystal and perfectly-ordered hollow urchin-like ZnO nanowire arrays by a combined colloidal patterning and electrochemical approach at temperature as low as 80 °C. The process enables a versatile control of dimensions and morphologies of ZnO nanowires as well as control of the core diameter and spatial arrangement (by changing the size of PS spheres) for the first time. The key mechanism for the formation of these architectures is the treatment of PS with $ZnCl_2$ at high concentration which renders them electrically conductive, enabling the deposition of ZnO on their surface. Figure 6 shows the schematic view of the fabrication processes of hollow urchin-like ZnO nanowires. This mechanism opens up new opportunities for processing novel metal oxide or hydroxide materials based on a similar growth mechanisms to that of ZnO.

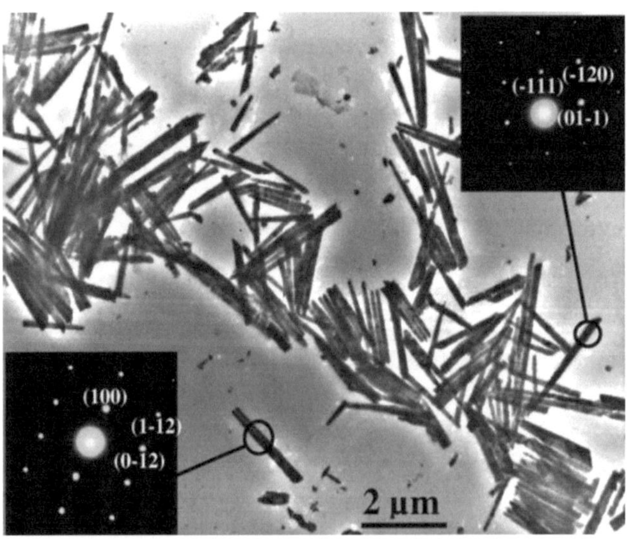

Fig. 4. TEM images showing a general morphology of the electrodeposited ZnO nanowires from M90 type membranes. From Ref. (Leprince-Wang et al., 2006).

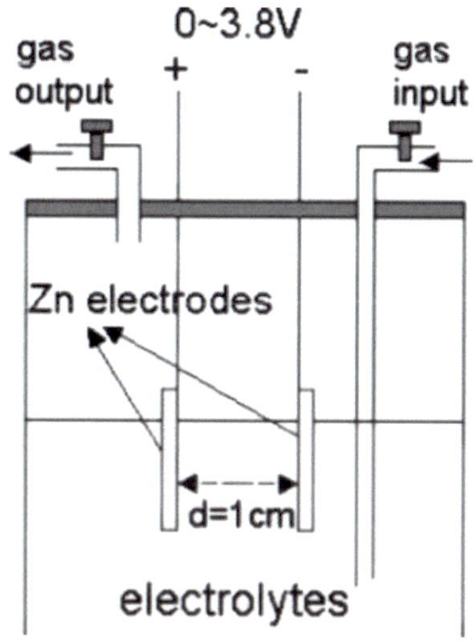

Fig. 5. Schematic diagram of the glass cell setup for the electrochemical synthesis of ZnO nanorod/nanobelt arrays. From Ref. (Yang et al., 2007).

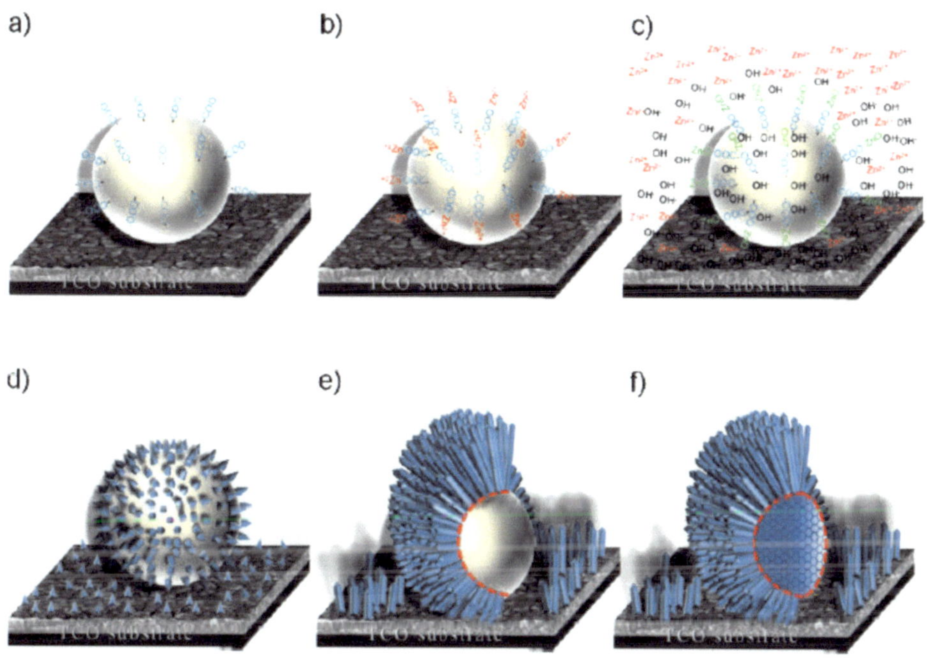

Fig. 6. Schematic view of the fabrication processes of hollow urchin-like ZnO nanowires. From Ref. (Elias et al., 2010).

3. ZnO nanowires for solar cell application

Nanostructured metal oxide materials is an intensive research area today with many potential applications. Dye-sensitized nanoporous materials are especially of interest for solar cells. ZnO/CdSe/CuSCN extremely thin absorber (*eta*)-solar cells based on ZnO nanowires have been successfully realized using easily accessible electrochemical and solution deposition techniques (Tena-Zaera et al., 2006). The ZnO/CdSe nanowire layer exhibited a high light-trapping effect, with an effective absorbance of ~89% and effective reflectance of ~8% in the 400–800 nm region of the solar spectrum (AM1.5). High density aligned ZnO nanotube arrays were synthesized using a facile chemical etching of electrochemically deposited ZnO nanorods (Luo et al., 2010). Cadmium selenide nanoparticles as sensitizers were assembled onto the ZnO nanotube and nanorods arrays for solar cell application. A conversion efficiency of 0.44% was achieved for CdSe/ZnO nanotube-based solar cell under the white light illumination intensity of 85 mW/cm^2. An 8% enhancement in conversion efficiency was observed between the CdSe/ZnO nanotube-based and nanorod-based solar cell due to the enhancement of the photocurrent density. This approach to design photovoltaic electrode would give a direction in the field of multi-junction solar cell materials.

Due to its low cost and high efficiency, DSSC is a promising candidate to be a new renewable energy device. Several attempts have been made to use ZnO in DSSC. Baxter et al. (2006) used ZnO nanowires as the photo-electrode in DSSCs. A schematic diagram of the nanowire DSSC is shown in Figure 7. Typical solar cell photocurrent, photovoltage, fill factor and overall efficiencies were 1.3 $mAcm^{-2}$, 0.7 V, 0.35–0.40 and 0.3%, respectively. Electron transport through the nanowires is not constrained by nanowire dimensions.

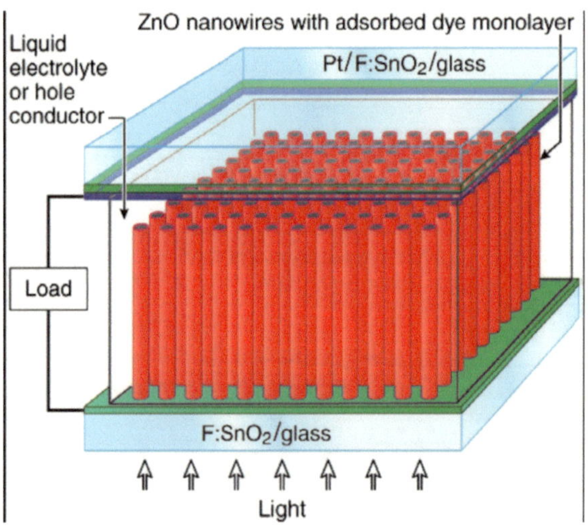

Fig. 7. Schematic diagram of the nanowire-based dye-sensitized solar cell. From Ref. (Baxter et al., 2006).

Pradhan et al. (2007) have fabricated DSSCs by using vertical nanowires of ZnO. Vertical growth of the nanowires was achieved via a simpler chemical route. In addition, they have introduced a new organic dye, namely Rose Bengal in xanthene class, as a photosensitizer. The new class of dye, whose energy matches the ZnO and usual KI-I_2 redox couple for DSSC applications, provides an alternative over conventional ruthenium complex-based DSSCs. Figure 8 shows the schematic band diagram showing the working principle of the DSSC. The maximum IPCE of the system is about 5.3% at 570 nm. The short circuit current of the cells increased linearly with illumination intensity.

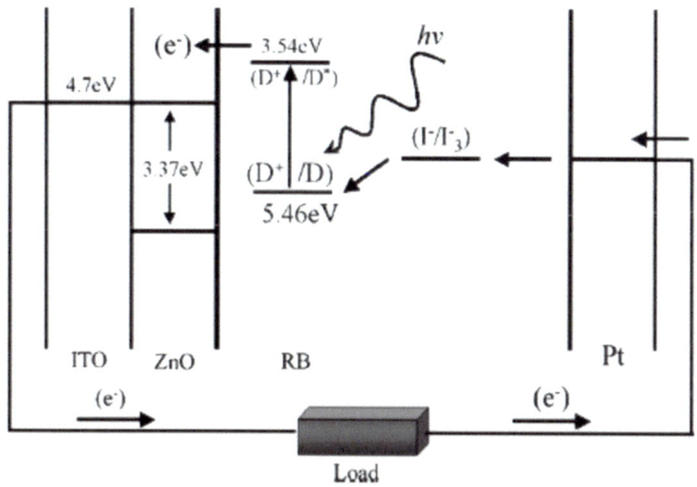

Fig. 8. Schematic band diagram showing the working principle of the DSSC. From Ref. (Pradhan et al., 2007).

Wang et al. (2008) reported success in synthesizing ZnO/ZnSe core/shell nanowires on a large-area, transparent, conducting substrate, using a relatively simple and low-cost approach. They have characterized their structural and optical properties by applying a comprehensive set of techniques. Absorption and photoconductivity studies show an extension of photoresponse into the region well below the ZnO bandgap.

Recently, Qiu et al. (2010) used synthesized ZnO nanowires to assemble DSSC, and to see if the further increase of nanowire length can improve the device performance. Figure 9a shows the photocurrent–voltage (*J*–*V*) characteristics of ZnO nanowire DSSCs dye-sensitized solar cells with various lengths. The Nyquist plots of both the measured data points and fitted curves are shown in Figure 9b. The performance of DSSCs increased with increasing length of the ZnO nanowire arrays, indicating that the ultralong ZnO nanowire arrays have great potential in improving the performance of DSSCs.

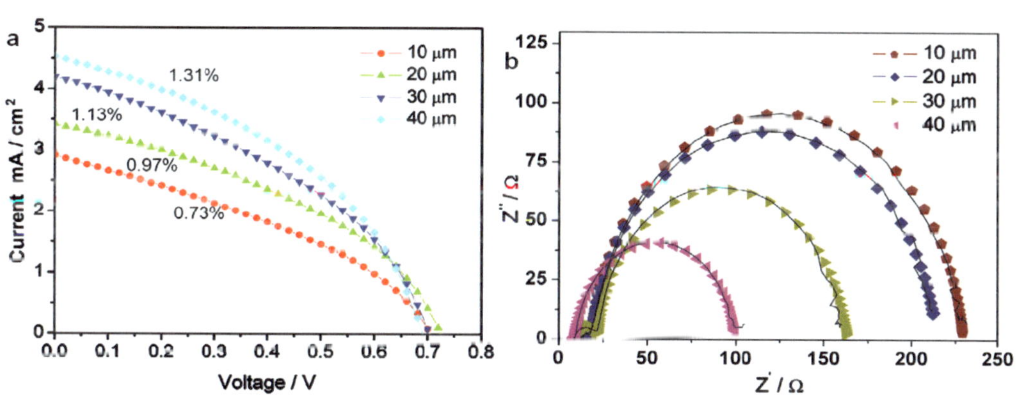

Fig. 9. a) *J* – *V* curves of the ZnO nanowire DSSCs fabricated from different lengths; and b) Nyquist plots of the impedance data of the ZnO nanowire DSSCs constructed from different lengths. From Ref. (Qiu et al., 2010).

Umar et al. (2009) reported a very rapid and large-scale synthesis and DSSCs application of well-crystallized hexagonal-shaped ZnO nanorods. The as-grown nanorods based DSSCs exhibited an overall light to electricity conversion efficiency (ECE) of 1.86% with a fill factor of 74.4%, shortcircuit current of 3.41 mA/cm^2 and open-circuit voltage of 0.73 V. Qurashi et al. (2010) demonstrated DSSCs performance of dumb bell-shaped hexagonal nanorods and well-aligned ZnO nanorod arrays. ZnO nanostructures were used as the wide band gap semiconducting photoelectrode in DSSCs. Well-aligned ZnO nanorod arrays were greatly enhances dye adsorption, leading to improved light harvesting and overall efficiencies. Solar cells made from aligned ZnO nanorod arrays showed photocurrents of 2.08 mA/cm^2, internal quantum efficiencies of 34.5%, and overall efficiencies of 0.32%. However, DSSC made from the randomly formed dumbbell-shaped hexagonal ZnO nanorods showed efficiency about 0.26%, with internal quantum efficiency of 31.5% respectively. A 74% enrichment of the efficiency of ZnO nanowire (NW) dye-sensitized solar cells (DSSCs) is achieved by the addition of a novel light-scattering nanocrystalline film (nanofilm) (Wu et al., 2010). *J*sc, *V*oc, and FF of the nanofilm/ZnO NW DSSCs are all enhanced compared to those of the ZnO NW DSSCs. The significant enhancement of the efficiency of the ZnO NW DSSC by the lightscattering layer of nanofilm is mainly attributed to the successful reflection of unabsorbed photons back into the NW anode.

Xu et al. (2010) have demonstrated that the DSSC based on hierarchical ZnO nanowire-nanosheet architectures with better dye loading and light harvesting showed a power conversion efficiency of 4.8%, which is nearly twice as high as that of the DSSCs constructed using the primary ZnO nanosheet arrays. The the schematic diagram of the DSSC based on the hierarchical ZnO nanoarchitectures is illustrated in Figure 10. The improvement in the photovoltaic performance can benefit from the enlargement of the internal surface area within the photoanode without sacrificing a direct conduction pathway for the rapid collection of photogenerated electrons. Further, the concept of the hierarchical nanowire-nanosheet architectures is anticipated to be applicable to other semiconductor

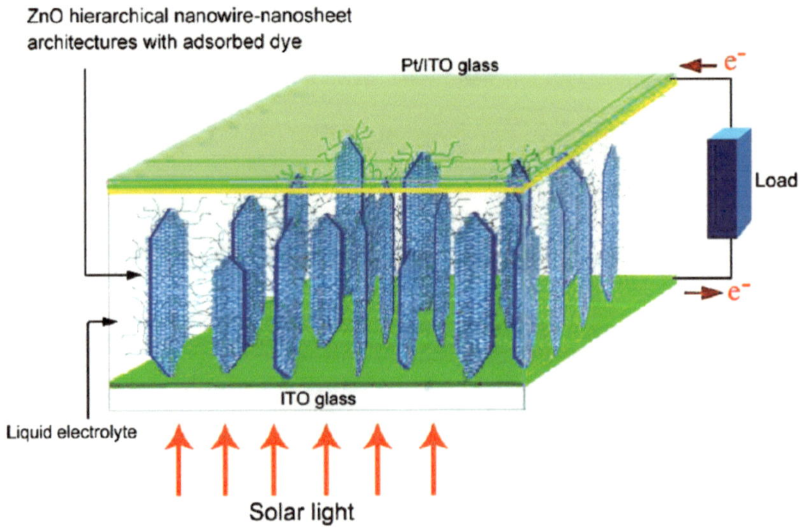

Fig. 10. Schematic diagram of the DSSCs based on the hierarchical ZnO nanowire-nanosheet architectures. From Ref. (Xu et al., 2010).

photoelectrodes in organic-inorganic nanocomposite solar cells and hierarchical heterojunction nanostructures for future potential applications. Hierarchical ZnO nanostructures with high surface to volume ratio are built in direct contact with conductive FTO substrate (Fu et al., 2010). The formation of the hierarchical ZnO structure follows an organic amine assisted growth mechanism. The bifunctional diamine, EDA, causes slight etching of the primary ZnO crystal, thus initiating the site-specific heterogeneous nucleation of hierarchical structure. The morphology and the branch density of hierarchical structure can be tuned easily by changing the molar ratio of [EDA]/[Zn^{2+}], and the growth temperature can be efficiently lowered by addition of HMT. Current density-voltage characterizations on DSSCs suggest that the conversion efficiencies were improved by increasing the surface area with hierarchical ZnO nanowires.

Recently, ZnO nanorod-based DSSCs have been investigated. The strategy to increase the surface area was investigated because ZnO nanorods have a surface area that is one-fifth of TiO_2 particles. DSSCs were fabricated using vertically aligned ZnO nanorod arrays on FTO glasses. The DSSCs with an annealed ZnO seed layer exhibited greater cell performance than those that were not annealed. It is noted that annealing of the seed layer improved adhesion between the FTO and the seed layer, and ZnO nanorods were grown into effective electrodes. The DSSCs with annealed ZnO nanorods produced greater cell efficiency than those that did not receive annealing. Figure 11 illustrates the ZnO nanorod-based DSSC fabrication process. Zinc oxide nanorods annealed in N_2/H_2 or O_2 exhibited greater dye loading due to a higher OH concentration and a hydrophilic surface property. In addition, annealing of ZnO nanorods slightly increased the grain size of the ZnO crystal and greatly reduced the defectd ensity in the ZnO crystal. Therefore, the improved cell efficiency of the DSSC in which ZnO nanorods were annealed resulted from the increases in J_{SC} and the fill factor.

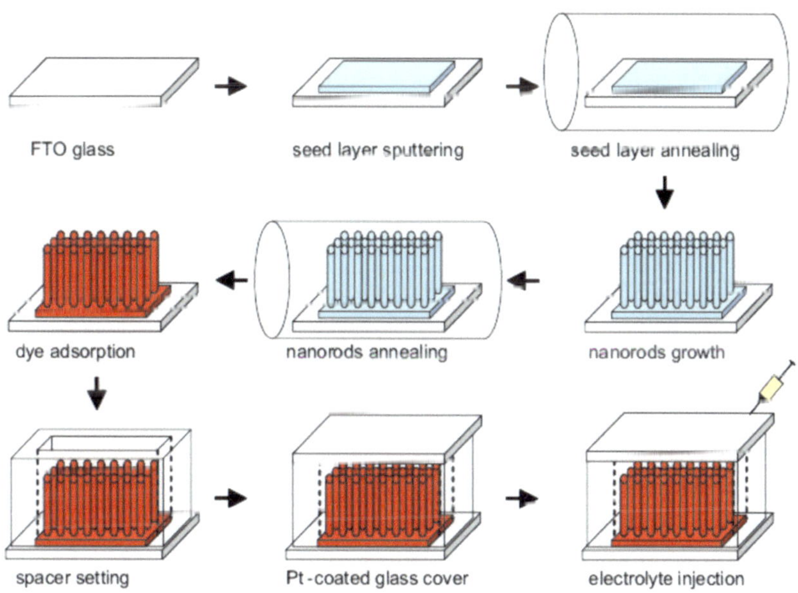

Fig. 11. Schematic diagrams of the fabrication process of ZnO nanorod-based DSSCs. From Ref. (Chung et al., 2010).

Organic solar cells are the another attracting enormous attentions to date, as low cost, light weight, solution process ability. Conjugated low band gap polymer is blended with fullerene to achieve the bulk heterojunction devices. Low band gap polymer materials and their application in organic photovoltaics have been reviewed by Bundgaard et al. (2007). Krebs et al. (2009) presented a complete polymer solar cell module prepared in the ambient atmosphere using all-solution processing with no vacuum steps and full roll-to-roll (R2R) processing. The completed modules (Figure 12) were flexible and quite robust. They have detailed the transfer of the P3CT/ZnO technology to methods giving full R2R compatibility in the ambient atmosphere with no vacuum coating steps being involved during the processing of the five layers of the modules.

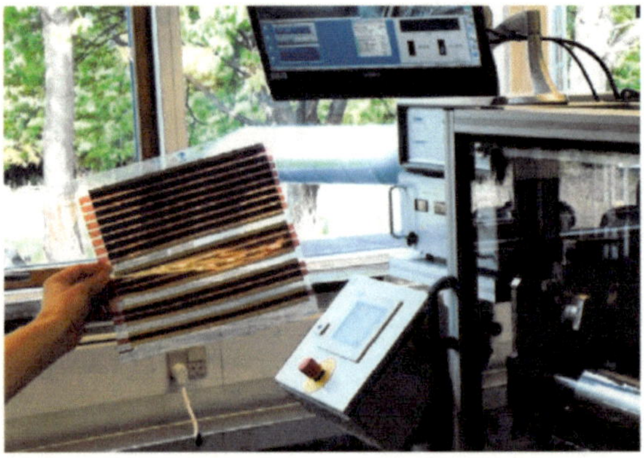

Fig. 12. Photograph of one of the final modules in front of the R2R coater. From Ref. (Krebs et al., 2009).

Organic solar cell devices were fabricated using P3HT and PCBM, which play the role of an electron donor and acceptor, respectively (Park et al., 2009). Organic solar cells based on Al-doped ZnO as an alternative to ITO. Organic solar cells with intrinsic ZnO inserted between the P3HT/PCBM layer and AZO were also fabricated (Figure 13). The intrinsic ZnO layer prevented the shunt path in the device. The performance of the cells with a layer of intrinsic ZnO was superior to that without the intrinsic ZnO layer.

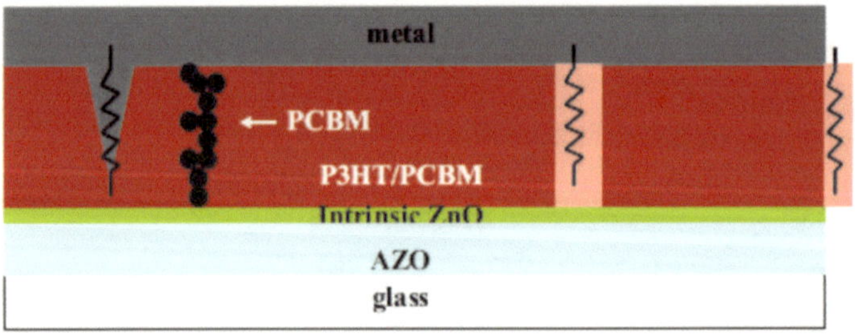

Fig. 13. Schematic diagram of the shunt path in organic solar cells with the intrinsic ZnO buffer layer. From Ref. (Park et al., 2009).

An inverted polymer solar cell geometry comprising a total of five layers was optimized using laboratory scale cells and the operational stability was studied under model atmospheres (Krebs et al., 2009). The device geometry was substrate-ITO-ZnO-(active layer)-

PEDOT:PSS-silver with P3HT-PCBM as the active layer. The inverted model device was then used to develop a new process giving access to fully R2R processed polymer solar cells entirely by solution processing starting from a PET substrate with a layer of ITO. All processing was performed in air without vacuum coating steps and modules comprising eight serially connected cells gave power conversion efficiencies as high as 2.1% for the full module with 120 cm^2 active area (AM1.5G, 393 W m^{-2}) and up to 2.3% for modules with 4.8 cm^2 active area (AM1.5G, 1000 W m^{-2}). An inverted-type organic bulk-heterojunction solar cell inserting ZnO as an electron collection electrode, fluorine-doped tin oxide (FTO)/ZnO/(PCBM:P3HT)/(PEDOT:PSS)/Au, was fabricated in air and characterized by an alternating current impedance spectroscopy (Kuwabara et al., 2009). The photo *I-V* measurement gave a PCE of 2.49%, and the impedance spectroscopy measurements in the dark and under light irradiation gave Nyquist plots consisting of two components. According to this result, it was proven that the depletion layer functioning to take out the photocurrent to the external circuit was formed in both of the ZnO and PCBM:P3HT layers at the ZnO/PCBM:P3HT interface. The inverted polymer solar cell based on a sol-gel derived ZnO thin film as an electron selective layer is investigated by Liu et al. (2010). Figure 14 shows the schematic diagram of inverted polymer solar cell based on ZnO thin film. the average grain size of the P3HT/PCBM layer and oxidation of the Ag electrode have a direct relationship with the evolution of device performance. The improvement of Jsc and FF is attributed to the grain growth of the P3HT/PCBM layer and the enhancement of Voc is due to the increase in work function of the Ag electrode. The highest PCE of 3.8% is thus achieved for the device placed in air for six days without the use of PEDOT:PSS and encapsulation.

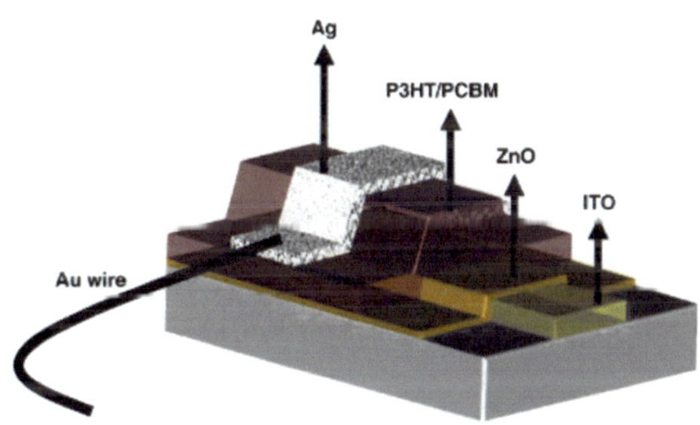

Fig. 14. Schematic diagram of inverted polymer solar cell based on ZnO thin film. From Ref. (Liu et al., 2010).

Hybrid solar cell is an alternative type of the promising device, which combined the organic semiconductor as donor material together with inorganic semiconductor as acceptor material. Huynh et al. (2002) demonstrated that semiconductor nanorods can be used to fabricate readily processed and efficient hybrid solar cells together with polymers. Hybrid P3HT/nanostructured oxide devices were fabricated using solution-based methods with efficiencies greater than 0.5% (Olson et al., 2006). The P3HT/ZnO device was limited in photocurrent due to the large spacing between the ZnO fibers. This was overcome by blending PCBM into the P3HT film. As shown in Figure 15, the polymer can be effectively intercalated into the ZnO fiber film thus making a hybrid nanostructured oxide/conjugated polymer composite device.

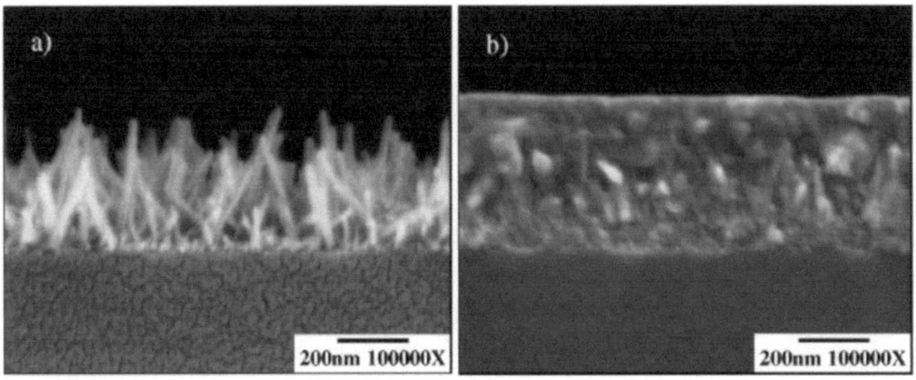

Fig. 15. (a) SEM image of a glass/ZnO nucleation layer/ZnO nanocarpet structure. (b) SEM image of P3HT intercalated into the nanocarpet structure. From Ref. (Olson et al., 2006).

Different layered ZnO/MEH:PPV composite solar cells have been fabricated by Plank et al. (2008) to assess the role of the ZnO backing layer on the open circuit voltage of nanowire composite solar cells. The thickness dependence of the blocking sputtered blocking layer is investigated. A 130 nm ZnO layer gives in a cell configuration with MEH-PPV an open circuit voltage of 0.41 V, which decreases with thicker ZnO layers to 0.28 V at 650 nm. The sputter ZnO/MEH:PPV devices have been shown to have reproducible I-V characteristics over many pixels indicating the high quality of the sputtered ZnO films. It has been clearly observed that the quality and thickness of the ZnO backing layer influence the device performance of simple geometry flat junction semiconductor and polymer composite solar cells. Uniform, pinhole-free oxide films are essential for the fabrication of working solar cells and to ensure reproducibility of results. 3D-ordered mesoporous ZnO films were fabricated by electrodeposition in DMSO solution (Wang et al., 2008). The porous electrode hybrid solar cells were made by infiltrating P3HT or P3HT:ZnO composite into the ordered porous ZnO films. The photocurrent of the ITO/ZnO(IO)/P3HT/Al device was limited because of the large diffusion distance for exciton to reach ZnO frameworks. This was obviously improved by using the P3HT:ZnO composite. A significant higher photocurrent was observed owing to the enhanced exciton dissociation and electron transfer efficiency. Solar decay analyses showed lifetime of ITO/ZnO(IO)/P3HT:ZnO/Al device could be improved by the application of a UV filter. a soluble perylene-derivative dye, N, N′-didodecyl-3,4,9,10-perylene tetracarboxylic diimide (PDI) was used in this work to enhance the photoelectric efficiency of the hybrid P3HT/ZnO bulk-heterojunction solar cells (Wang et al., 2008). PDI can absorb the sunlight in a broad wavelength range. By blending with PDI, the light absorption and exciton separation of the P3HT/ZnO solar cells can be significantly improved.

Thitima et al. (2009) demonstrated the charge injection efficiency of hybrid solar cell consisting of P3HT and PCBM/ZnO with and without N719 dye molecule. After the modification of ZnO nanorod arrays with N719, short-circuit current density (J_{sc}) of 8.89 mA/cm^2 was obtained, and it was 1.5 times higher than that of without the N719. The power conversion efficiency was enhanced from 1.16% to 2.0% through the additional surface modification of the ZnO nanorod array with N719 dye. Lin et al. (2009) have demonstrated an improvement of photovoltaic performance based on the nanostructured ZnO/P3HT hybrid through interface molecular modification on ZnO nanorod surface (Figure 16). By probing the carrier dynamics at ZnO/P3HT interfaces, they have found that the interfacial molecules can play the role of assisting charge separation and suppression of back recombination at interfaces, which accounts for the observed enhanced short circuit current (*Jsc*) and open circuit voltage (*Voc*) in photovoltaic performance. Ji et al. (2010) demonstrated a hybrid solar cell which was made by blending nanocrystalline ZnO and conjugated polymer regioregular P3HT as the active layer of the solar cell. It can be seen that the efficiency of this new type of solar cells obviously varied as the size and morphology of ZnO nanostructures. The short-circuit photocurrent, fill factor, and power conversion efficiency were enhanced while the smaller nc-ZnO was utilized in such a device.

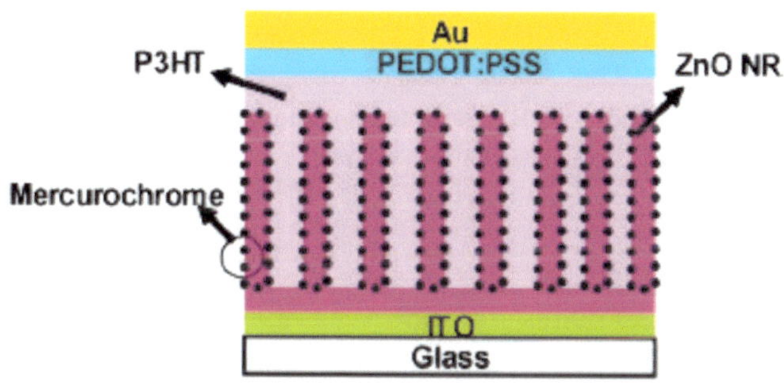

Fig. 16. The schematic structure of the nanostructured ZnO/P3HT hybrid photovoltaic device. From Ref. (Lin et al., 2009).

Wu et al. (2010) investigated hybrid photovoltaic devices based on P3HT and an ordered electrospun ZnO nanofibrous network (Figure 17). The performance of the P3HT/ZnO hybrid solar cell is dependent on fabrication conditions, especially the thickness of the nanofibrous film. It has been found that the lifetime of carriers is lower in the device consisting of thicker ZnO nanofibrous films due to the higher density of surface traps in the ZnO nanofibers. The device with optimum fabrication conditions exhibits a power conversion efficiency of 0.51%.

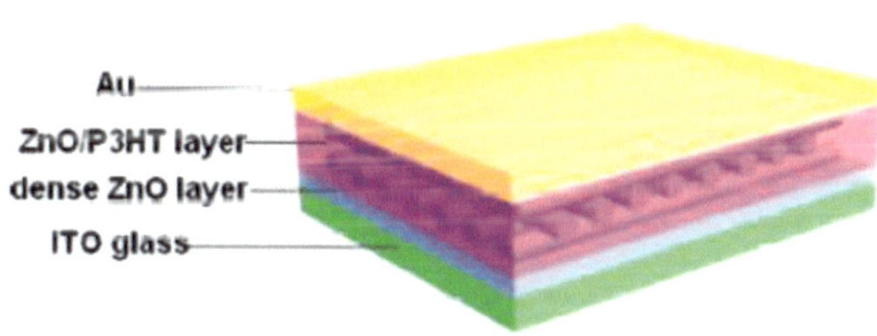

Fig. 17. Stucture of the hybrid solar cell based on electrospun ZnO nanofibers and P3HT. From Ref. (Wu et al., 2010).

A hybrid solar cell is designed and proposed as a feasible and reasonable alternative, according to acquired efficiency with the employment of ZnO nanorods and ZnO thin films at the same time (Hames et al., 2010). Both of these ZnO structures were grown electrochemically and P3HT:PCBM was used as an active polymer blend, which was found to be compatible to prepared ITO substrate base. This ITO base was introduced with mentioned ZnO structure in such a way that, the most efficient configuration was optimized to be ITO/ZnO film/ZnO nanorod/P3HT:PCBM/Ag. Efficiency of this optimized device is found to be 2.44%. All ZnO works were carried out electrochemically, that is indeed for the first time and at relatively lower temperatures. Das et al. (2010) fabricated hybrid polymer-metal oxide bulk heterojunction solar cell by blending of ZnO and regioregular P3HT through solution process and flow coating on the flexible substrate. The decrease in the photoluminescence emission intensity more than 79% for ZnO:P3HT composites film indicates high charge generation efficiency. The cell shows the *Voc* and *Isc* of 0.33 V and 6.5 mA/cm^2, respectively. Cheng et al. (2010) fabricated the P3HT/ZnO NWs hybrid prototype device. The I-V and time-depend photocurrent wasmeasured for both the pristine ZnO NWs array and the hybrid structure. An ultraviolet (UV) light of λ = 350nm was used to investigate the photo-electric properties of the pristine ZnO NWs array and the hybrid structure in measurement. The P3HT coating process resulted in a higher and faster photoelectric response for the hybrid structure, which is benefit from the charge transfer process and the eliminating of adsorbed oxygen. The present work provides profound understandings on the electron transport of ZnO NWs array in a working hybrid polymer solar cell.

Nadarajah et al. (2008) reported first results on a new solar cell structure which incorporates n-type ZnO nanowires, an undoped CdSe layer, obtained from quantum dot precursors, and a p-type polymer layer as the main components (Figure 18). In the fabrication process the quantum dot layer is converted to a conformal ~30 nm thick polycrystalline film. MEH-PPV as well as P3HT have been explored for this contact, best results were obtained with P3HT. The fabrication of the cell occurs in lab air at temperatures below 100 °C. Several intermittent annealing steps raise the energy conversion efficiency to approximately 1%.

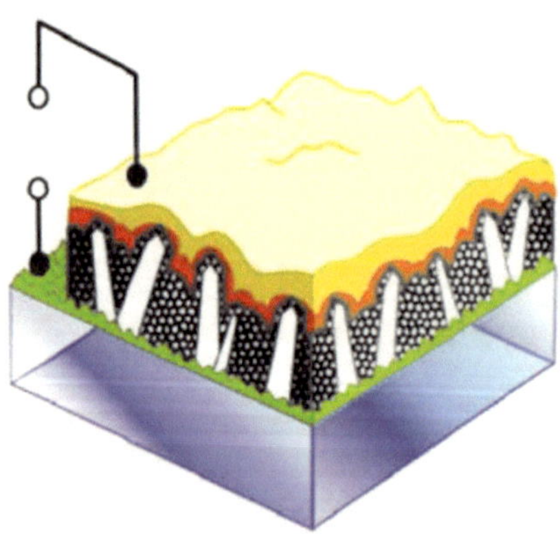

Fig. 18. Schematic diagram of the solar cell structure. From Ref. (Nadarajah et al., 2008).

4. Summary

In summary, we have reviewed the research progress on ZnO nanowires and some important results obtained on ZnO nanowires. And we have reviewed the latest progress in the growth of ZnO nanowires and their solar cell application. A number of methods have been employed to achieve ZnO nanostructured arrays. Several attempts have been made to use ZnO in solar cells. In our opinion, more work is still needed to make further progress on this topic. First, the quality and stability of the ZnO nanowires need to be further improved. This will require even better control the background conductivity, development of new growth methods and search for new acceptor dopants. Second, The techniques used in fabricating these solar cells have still to be optimized. Once these milestones are achieved, the ZnO nanowire arrays have great potential in improving the performance of SCs.

5. Acknowledgments

The authors thank the Natural Science Foundation of China (NSFC, 20802033), the Ministry of Education of China (MEC, NCET-10-0170) and the Science and Technology Department of Jiangxi Province (STDJP, 2008DQ00700) for continuous financial support.

6. References

Baxter, J. B.; Walker, A. M.; van Ommering, K. & Aydil, E. S. (2006). Synthesis and characterization of ZnO nanowires and their integration into dye sensitized solar cells. *Nanotechnology*, Vol. 17, No. 11, (June 2006), pp. S304–S312, ISSN: 0957-4484.

Briseno, A. L.; Holcombe, T. W.; Boukai, A. I.; Garnett, E. C.; Shelton, S. W.; Fréchet, J. J. M. & Yang, P. (2010). Oligo- and Polythiophene/ZnO Hybrid Nanowire Solar Cells. *Nano Lett.*, Vol. 91, No. 11, (July 2007), pp. 954-985, ISSN: 1530-6984.

Bundgaard, E. & Krebs, F. C. (2007). Low band gap polymers for organic photovoltaics. *Solar Energy Materials & Solar Cells*, Vol. 73, No. 1, (May 2002), pp. 51–58, ISSN: 0927-0248.

Chao, H. Y.; Cheng, J. H.; Lu, J. Y.; Chang, Y. H.; Cheng, C. L. & Chen, Y. F. (2010). Growth and characterization of type-II ZnO/ZnTe core-shell nanowire arrays for solar cell applications. *Superlattices and Microstructures*, Vol. 47, No. 1, (January 2010), pp. 160-164, ISSN: 0749-6036.

Chen, J.; Lei, W.; Song, J. L.; Sun, X. W.; Zhang, X. B. & Deng, W. Q. (2009). Two-step growth of a hierarchical ZnO nanostructure by aqueous thermal decomposition in a neutral solution and its photovoltaic property. *Physica E: Low-dimensional Systems and Nanostructures*, Vol. 41, No. 5, (March 2009), pp. 822–827, ISSN: 1386-9477.

Chen, Z. H.; Tang, Y. B.; Liu, C. P.; Leung, Y. H.; Yuan, G. D.; Chen, L. M.; Wang, Y. Q.; Bello, I.; Zapien, J. A.; Zhang, W. J.; Lee, C. S. & Lee, S. T. (2009). Vertically Aligned ZnO Nanorod Arrays Sentisized with Gold Nanoparticles for Schottky Barrier Photovoltaic Cells. *Journal of Physical Chemistry C*, Vol. 113, No. 30, (May 2009), pp. 13433–13437, ISSN: 1932-7447.

Cheng, K.; Cheng, G.; Wang, S.; Fu, D. Zou, B. & Du, Z. (2010). Electron transport properties in ZnO nanowires/poly(3-hexylthiophene) hybrid nanostructure. Materials Chemistry and Physics, Vol. 124, No. 2-3, (August 2010), pp. 1239-1242, ISSN: 0254-0584.

Chung, J.; Lee, J. & Lim, S. (2010). Annealing effects of ZnO nanorods on dye-sensitized solar cell efficiency. *Physica B: Condensed Matter*, Vol. 405, No. 11, (June 2010), pp. 2593-2598, ISSN: 0921-4526.

Das, N. Ch. & Sokol, P. E. (2010). Hybrid photovoltaic devices from regioregular polythiophene and ZnO nanoparticles composites. *Renewable Energy*, Vol. 35, No. 12, (May 2010), pp. 2683-2688, ISSN: 0960-1481.

Elias, J.; Lévy-Clément, C.; Bechelany, M.; Michler, J.; Wang, G.; Wang, Z. & Philippe, L. (2010). Hollow Urchin-like ZnO thin Films by Electrochemical Deposition. *Advanced Materials*, Vol. 22, No. 14, (May 2010), pp. 1607-1612, ISSN: 0935-9648.

Fang, Y.; Pang, Q.; Wen, X.; Wang, J. & Yang, S. (2006). Synthesis of Ultrathin ZnO Nanofibers Aligned on a Zinc Substrate. *Small*, Vol. 2, No. 5, (March 2006), pp. 612-615, ISSN: 1613-6810.

Fu, Y.; Sun, J.; Xie, Y.; Liu, J.; Wang, H. & Du, X. (2010). ZnO hierarchical nanostructures and application on high-efficiency dye-sensitized solar cells. *Materials Science and Engineering B*, Vol. 166, No. 3, (February 2010), pp. 196-202, ISSN: 0921-5107.

Gan, X.; Li, X.; Gao, X.; Zhuge, F. & Yu, W. (2010). ZnO nanowire/TiO_2 nanoparticle photoanodes prepared by the ultrasonic irradiation assisted dip-coating method. *Thin Solid Films*, Vol. 518, No. 17, (June 2010), pp. 4809-4812, ISSN: 0040-6090.

Grätzel, M. (2009). Recent Advances in Sensitized Mesoscopic Solar Cells. *Accounts of Chemical Research*, Vol. 42, No. 11, (November 2009), pp. 1788-1798, ISSN: 0001-4842.

Greene, L. E.; Law, M.; Yuhas, B. D. & Yang, P. (2007). ZnO-TiO_2 Core-Shell Nanorod/P3HT Solar Cells. *Journal of Physical Chemistry C*, Vol. 111, No. 50, (November 2006), pp. 22652-22663, ISSN: 1932-7447.

Haller, S.; Suguira, T.; Lincot, D. & Yoshida, T. (2010). Design of a hierarchical structure of ZnO by electrochemistry for ZnO-based dye-sensitized solar cells. *Physica Status Solidi A-Applications and Materials Science*, Vol. 207, No. 10, (July 2010), pp. 2252–2257, ISSN: 1862-6300.

Hames, Y.; Alpaslan, Z.; Kösemen, A.; San, S. E. & Yerli, Y. (2010). Electrochemically grown ZnO nanorods for hybrid solar cell applications. *Solar Energy*, Vol. 84, No. 3, (February 2010), pp. 426-431, ISSN: 0038-092X.

Hsueh, T.; Hsu, C.; Chang, S.; Guo, P.; Hsieh, J. & Chen, I. (2007). Cu_2O/n-ZnO nanowire solar cells on ZnO:Ga/glass templates. *Scripta Materialia*, Vol. 57, No. 1, (July 2007), pp. 53-56, ISSN: 1359-6462.

Huynh, W. U.; Dittmer, J. J. & Alivisatos, A. P. (2002). Hybrid Nanorod-Polymer Solar Cells. Science, Vol. 295, No. 5564, (March 2002), pp. 2425-2427, ISSN: 0036-8075.

Ji, L. W.; Shih, W. S.; Fang, T. H.; Wu, C. Z.; Peng, S. M. & Meen, T. H. (2010). Preparation and characteristics of hybrid ZnO-polymer solar cells. *Journal of Materials Science*, Vol. 45, No. 12, (March 2010), pp. 3266-3269, ISSN: 0022-2461.

Keis, K.; Bauer, C.; Boschloo, G.; Hagfeldt, A.; Westermark, K.; Rensmob, H. & Siegbahn, H. (2002). Nanostructured ZnO electrodes for dye-sensitized solar cell applications. *Journal of Photochemistry and Photobiology A: Chemistry*, Vol. 148, No. 1-3, (May 2002), pp. 57–64, ISSN: 1010-6030.

Keis, K.; Magnusson, E.; Lindström, H.; Lindquist, S. & Hagfeldt, A. (2002). A 5% efficient photoelectrochemical solar cell based on nanostructured ZnO electrodes. *Solar Energy Materials & Solar Cells*, Vol. 73, No. 1, (May 2002), pp. 51–58, ISSN: 0927-0248.

Krebs, F C. (2009). Polymer solar cell modules prepared using roll-to-roll methods: Knife-over-edge coating, slot-die coating and screen printing. *Solar Energy Materials & Solar Cells*, Vol. 93, No.4, (January 2009), pp. 465-475, ISSN: 0927-0248.

Krebs, F. C.; Gevorgyan, S. A. & Alstrup, J. (2009). A roll-to-roll process to flexible polymer solar cells: model studies, manufacture and operational stability studies. *Journal of Materials Chemistry*, Vol. 19, No. 30, (May 2009), pp. 5442-5451, ISSN: 0959-9428.

Ku, C. & Wu, J. (2007). Electron transport properties in ZnO nanowire array/nanoparticle composite dye-sensitized solar cells. *Applied Physics Letters*, Vol. 91, No. 9, (August 2007), pp. 0931171-3, ISSN: 0003-6951.

Kuwabara, T.; Kawahara, Y.; Yamaguchi, T. & Takahashi, K. (2009). Characterization of Inverted-Type Organic Solar Cells with a ZnO Layer as the Electron Collection Electrode by ac Impedance Spectroscopy. *ACS Applied Materials & Interfaces*, Vol. 1, No. 10, (September, 2009), pp. 2107-2110, ISSN: 1944-8244.

Law, M.; Greene, L. E.; Johnson, J. C.; Saykally, R. & Yang, P. (2005). Nanowire dye-sensitized solar cells. *Nature Materials*, Vol. 4, No. 6, (June 2005), pp. 455-459, ISSN: 1476-1122.

Law, M.; Greene, L. E.; Radenovic, A.; Kuykendall, T.; Liphardt, J. & Yang, P. (2006). ZnO-Al_2O_3 and ZnO-TiO_2 Core-Shell Nanowire Dye-Sensitized Solar Cells. *Journal of Physical Chemistry B*, Vol. 110, No. 45, (September 2006), pp. 22652-22663, ISSN: 1520-6106.

Leprince-Wang, Y.; Wang, G. Y.; Zhang, X. Z. & Yu, D. P. (2006). Study on the microstructure and growth mechanism of electrochemical deposited ZnO nanowires. *Journal of Crystal Growth*, Vol. 287, No. 1, (January 2006), pp. 89-93, ISSN: 0022-0248.

Leschkies, K. S.; Divakar, R.; Basu, J.; Enache-Pommer, E.; Boercker, J. E.; Carter, C. B.; Kortshagen, U. R.; Norris, D. J. & Aydil, E. S. (2007). Photosensitization of ZnO Nanowires with CdSe Quantum Dots for Photovoltaic Devices. *Nano Lett.*, Vol. 7, No. 6, (May 2007), pp. 1793-1798, ISSN: 1530-6984.

Lévy-Clément, C.; Tena-Zaera, R.; Ryan, M. A.; Katty, A. & Hodes, G. (2005). CdSe-Sensitized p-CuSCN/Nanowire n-ZnO Heterojunctions. Advanced Materials, Vol. 17, No. 12, (September 2005), pp. 1512-1515, ISSN: 0935-9648.

Lin, J.; Penchev, M.; Wang, G.; Paul, R. K.; Zhong, J.; Jing, X.; Ozkan, M. & Ozkan, C. S. (2010). Heterogeneous Graphene Nanostructures: ZnO Nanostructures Grown on Large-Area Graphene Layers. *Small*, Vol. 6, No. 21, (February 2010), pp. 2448–2452, ISSN: 1613-6810.

Lin, Y.; Lee, Y.; Chang, L.; Wu, J. & Chen, C. (2009). The influence of interface modifier on the performance of nanostructured ZnO/polymer hybrid solar cells. *Applied Physics Letters*, Vol. 94, No. 6, (February 2009), pp. 0633081-3, ISSN: 0003-6951.

Liu, C.; Zapien, J. A.; Yao, Y.; Meng, X.; Lee, C. S.; Fan, S.; Lifshitz, Y. & Lee, S. T. (2003). High-Density, Ordered Ultraviolet Light-Emitting ZnO Nanowire Arrays. Advanced Materials, Vol. 15, No. 10, (May 2003), pp. 838-841, ISSN: 0935-9648.

Liu, M.; Chang, C.; Chang, C.; Tsai, K.; Huang, J.; Chou, C.; Wang, I.; Wang, P.; Lee, C.; Chao, C.; Yeh, C.; Wu, C. & Lin, C. (2010). Morphological evolution of the poly(3-hexylthiophene)/[6,6]-phenyl-C_{61}-butyric acid methyl ester, oxidation of the silver electrode, and their influences on the performance of inverted polymer solar cells with a sol-gel derived zinc oxide electron selective layer. *Thin Solid Films*, Vol. 518, No. 17, (April 2010), pp. 4964-4969, ISSN: 0040-6090.

Lu, C.; Qi, L.; Yang, J.; Tang, L.; Zhang, D. & Ma, J. (2006). Hydrothermal growth of large-scale micropatterned arrays of ultralong ZnO nanowires and nanobelts on zinc substrate. *Chemical Communications*, Vol. 42, No. 33, (July 2006), pp. 3551-3553, ISSN: 1359-7345.

Luo, L.; Lv, G.; Li, B.; Hu X.; Jin, L.; Wang, J. & Tang, Y. (2010). Formation of aligned ZnO nanotube arrays by chemical etching and coupling with CdSe for photovoltaic application. *Thin Solid Films*, Vol. 518, No. 18, (July 2010), pp. 5146-5152, ISSN: 0040-6090.

Muskens, O. L.; Rivas, J. G.; Algra, R. E.; Bakkers, E. P. A. M. & Lagendijk, A. (2008). Design of Light Scattering in Nanowire Materials for Photovoltaic Applications. *Nano Lett.*, Vol. 8, No. 9, (August 2008), pp. 2638-2642, ISSN: 1530-6984.

Myung, Y.; Jang, D. M.; Sung, T. K.; Sohn, Y. J.; Jung, G. B.; Cho, Y. J.; Kim, H. S. & Park, J. (2010). Composition-Tuned ZnO-CdSSe Core-Shell Nanowire Arrays. *ACS NANO*, Vol. 4, No. 7. (June 2010), pp. 3789-3800, ISSN: 1936-0851.

Nadarajah, A.; Word, R. C.; VanSant, K. & Könenkamp R. (2008). Nanowire–quantum-dot–polymer solar cell. *Physica status solidi (b)*, Vol. 245, No. 9, (August 2008), pp. 1834-1837, ISSN: 0370-1972.

Olson, D. C.; Piris, J.; Collins, R. T.; Shaheen, S. E. & Ginley, D. S. (2006). Hybrid photovoltaic devices of polymer and ZnO nanofiber composites. *Thin Solid Films*, Vol. 496, No. 1, (September 2005), pp. 26-29, ISSN: 0040-6090.

Park, S.; Tark, S. J.; Lee, J. S.; Lim, H. & Kim, D. (2009). Effects of intrinsic ZnO buffer layer based on P3HT/PCBM organic solar cells with Al-doped ZnO electrode. *Solar Energy Materials* & *Solar Cells*, Vol. 93, No. 6-7, (January 2009), pp. 1020-1023, ISSN: 0927-0248.

Peterson, R. B. Fields, C. L. & Gregg, B. A. (2004). Epitaxial Chemical Deposition of ZnO Nanocolumns from NaOH Solutions. *Langmuir*, Vol. 20, No. 12, (May 2004), pp. 5114-5118, ISSN: 0743-7463.

Plank, N. O. V.; Welland, M. E.; MacManus-Driscoll, J. L. & Schmidt-Mende, L. (2008). The backing layer dependence of open circuit voltage in ZnO/polymer composite solar cells. *Thin Solid Films*, Vol. 516, No. 20, (December 2007), pp. 7218 - 7222, ISSN: 0040-6090.

Pradhan, B.; Batabyal, S. K. & Pal, A. J. (2007). Vertically aligned ZnO nanowire arrays in Rose Bengal-based dye-sensitized solar cells. *Solar Energy Materials* & *Solar Cells*, Vol. 91, No. 9, (May 2007), pp. 769-773, ISSN: 0927-0248.

Pradhan, D.; Sindhwani, S. & Leung, K. T. (2010). Parametric Study on Dimensional Control of ZnO Nanowalls and Nanowires by Electrochemical Deposition. *Nanoscale Research Letters*, Vol. 5, No. 11, (July 2010), pp. 1727-1736, ISSN: 1931-7573.

Qiu, J.; Li, X.; Zhuge, F.; Gan, X.; Gao, X.; He, W.; Park, S.; Kim, H. & Hwang, Y. (2010). Solution-derived 40 μm vertically aligned ZnO nanowire arrays as photoelectrodes in dye-sensitized solar cells. *Nanotechnology*, Vol. 21, No. 19, (April 2010), pp. 195602, ISSN: 0957-4484.

Qurashi, A.; Hossain, M.; Faiz, M.; Tabet, N.; Alam, M. W. & Reddy, N. K. (2010). Fabrication of well-aligned and dumbbell-shaped hexagonal ZnO nanorod arrays and their dye sensitized solar cell applications. *Journal of Alloys and Compounds*, Vol. 503, No. 2, (August 2010), pp. L40-L43, ISSN: 0925-8388.

Seol, M.; Kim, H.; Tak, Y. & Yong, K. (2010). Novel nanowire array based highly efficient quantum dot sensitized solar cell. *Chemical Communications*, Vol. 46, No. 30, (June 2010), pp. 5521-5523, ISSN: 1359-7345.

Tang, Y. B.; Chen, Z. H.; Song, H. S.; Lee, C. S.; Cong, H. T.; Cheng, H. M.; Zhang, W. J.; Bello, I. & Lee, S. T. (2008). Vertically Aligned p-Type Single-Crystalline GaN Nanorod Arrays on n-Type Si for Heterojunction Photovoltaic Cells. *Nano Lett.*, Vol. 8, No. 12, (October 2008), pp. 4191-4195, ISSN: 1530-6984.

Tena-Zaera, R.; Ryan, M. A.; Katty, A.; Hodes, G.; Bastide, S. & Lévy-Clément, C. (2006). Fabrication and characterization of ZnO nanowires/CdSe/CuSCN *eta*-solar cell. *Comptes Rendus Chimie*, Vol. 9, No. 5-6, (May-June 2006), pp. 717-729, ISSN: 1631-0748.

Thitima, R.; Patcharee, C.; Takashi, S. & Susumu, Y. (2009). Efficient electron transfers in ZnO nanorod arrays with N719 dye for hybrid solar cells. *Solid-State Electronics*, Vol. 53, No. 2, (December 2008), pp. 176-180, ISSN: 0038-1101.

Wang, K.; Chen, J.; Zhou, W.; Zhang, Y.; Yan, Y.; Pern, J. & Mascarenhas, A. (2008). Direct Growth of Highly Mismatched Type II ZnO/ZnSe Core/Shell Nanowire Arrays on Transparent Conducting Oxide Substrates for Solar Cell Applications. *Advanced Materials*, Vol. 20, No. 17, (September 2008), pp. 3248-3253, ISSN: 0935-9648.

Wang, L. Z. (2009). ZnO nanowire and nanobelt platform for nanotechnology. *Materials Science and Engineering R*, Vol. 64, No. 3-4, (April 2009), pp. 33-71, ISSN: 0927-796X.

Wang, M. & Wang, X. (2008). Electrodeposition zinc-oxide inverse opal and its application in hybrid photovoltaics. *Solar Energy Materials & Solar Cells*, Vol. 92, No. 3, (November 2007), pp. 357-362, ISSN: 0927-0248.

Wang, M. & Wang, X. (2008). P3HT/ZnO bulk-heterojunction solar cell sensitized by a perylene derivative. *Solar Energy Materials & Solar Cells*, Vol. 92, No. 7, (March 2008), pp. 766-771, ISSN: 0927-0248.

Wu, J.; Chen, G.; Yang, H.; Ku, C. & Lai, J. (2007). Effects of dye adsorption on the electron transport properties in ZnO-nanowire dye-sensitized solar cells. *Applied Physics Letters*, Vol. 90, No. 21, (May 2007), pp. 2131091-3, ISSN: 0003-6951.

Wu, J.; Chen,Y.; Liao, W.; Wu, C. & Chen, C. (2010). Construction of Nanocrystalline Film on Nanowire Array via Swelling Electrospun Polyvinylpyrrolidone-Hosted Nanofibers for Use in Dye-Sensitized Solar Cells. *ACS NANO,* Vol. 4, No. 10. (September 2010), pp. 5679-5684, ISSN: 1936-0851.

Wu, S.; Tai, Q. & Yan, F. (2010). Hybrid Photovoltaic Devices Based on Poly (3-hexylthiophene) and Ordered Electrospun ZnO Nanofibers. *Journal of Physical Chemistry C,* Vol. 114, No. 13, (March 2010), pp. 6197-6200, ISSN: 1932-7447.

Xu, F.; Dai, M.; Lu, Y. & Sun, L. (2010). Hierarchical ZnO Nanowire-Nanosheet Architectures for High Power Conversion Efficiency in Dye-Sensitized Solar Cells. *Journal of Physical Chemistry C,* Vol. 114, No. 6, (December 2009), pp. 2776-2782, ISSN: 1932-7447.

Yang, J.; Qiu, Y. & Yang, S. (2007). Studies of Electrochemical Synthesis of Ultrathin ZnO Nanorod/Nanobelt Arrays on Zn Substrates in Alkaline Solutions of Amine-Alcohol Mixtures. *Crystal Growth & Design,* Vol. 7, No. 12, (September 2007), pp. 2562-2567, ISSN: 1528-7483.

Part 3

Nanowire Devices

9

Ultrafast Photoluminescence in Nanowires

Zhao FL[1], Wang XF[1,4], Chen HJ[1,2] and Luo JY[1,3]
[1]*Sun Yat-sen University*
[2]*The Chinese University of Hongkong*
[3]*Wuyi University*
[4]*Guangdong University of Petrochemical Technology*
China

1. Introduction

As the boundary condition was dropped down to the nanometer scale, the photo electronic properties can be changed dramatically especially that the photon luminescence (PL) properties may be changed significantly. Because two of the dimensions is comparable to or even smaller than the wavelengths of the PL, the nanowires are expected to show a variety of quantum confinement effects, such as density of state singularities, macro molecular behaviour over the long distances, high coherence of the emission, much high luminescence efficiency and low lasing threshold which is essential to construct novel integrated photo electronic devices in nanoscale. Besides, the high surface-to-volume ratio of these nanostructures also allows studying the role of surface states (and their ambient-driven changes) in determining the nanostructure carrier transport and optical excitation/recombination properties. The surface state and quantum confinement are two fundamental effects responsible to the photo electronic characteristic of nanostructures. Naturally the photoelectronic properties determine the future application for nanostructures including nanoparticles, nanowires, nanorods and so on. For example photo electronic properties determine whether the nanowires fabricated are suitable for such electronic application as field emission and photo-detection etc. Meanwhile many kinds of nanowires whose bulk materials can be used as photo luminescent sources are taken granted as the nanolaser active media. Such an idea needs lots of experiment to examine and confirm by the data of both the steady-state spectroscopy and time-resolved spectroscopy dynamically.

Therefore we demonstrate a series studies on dynamic mechanism of nanowires (1 dimensional structure) and nanoparticles (0 dimensional structures) with respect to time, excitation density and photonic energy of PL. The sample focused are zinc oxide (ZnO)(Wang et al., 2009), carbon silicon (SiC)(Zhao et al., 2007), tungsten oxide (WO) covering both direct bandgap semiconductor and indirect bandgap semiconductors.

The following will present the surface state effect on nanoparticles, ultrafast light emission from nanowires and directional lasing behaviour from nanotree. The mechanism of the ultrafast light emission will be discussed qualitatively.

2. Experimental method

The ultrafast photoluminescence experiment demonstration was constructed with laser system, sample adjustment and detection system. The laser system was select based on the steady state PL spectroscopy of the nanowires. For example, we selected the femtosecond laser pulse at wavelength of 320nm to study the band-band transition of SiC nanowires and ZnO nanoparticles. But for the surface-state exploration we chose picosecond laser pulse at wavelength of 532nm for excitation to the ZnO nanosparticles and we selected picosecond laser pulse at wavelength of 300nm for the excitation to the nanodiamond particles(Zhao et al., 2004). The sample design includes the direction alignment of the sample, the consideration of the polarization of the excitation and the detection because the nanowires undergoes special orientation which is sensitive to the excitation polarization and gives light emission in certain direction and polarization. If necessary we would arrange for a set of microscope monitor system to infirm the correct excitation locally.The detection system was constructed by a set of selective streak cameras and the corresponding spectroscope. The streak camera can be changed with the different time-resolution from 20 picosecond to 700 femtosecond. The spectroscope can be selected whether the quartz prism spectroscope was used for ultraviolet detection for the high transmissivity or the optical grating spectroscope for high wavelength resolution was applied for the proper spectral limit.

2.1 Experiment set-up

The experimental set-up is shown in Figure 1. The laser system includes commercial available femtosecond laser system with optical parameter amplifier (OPA), picosecond

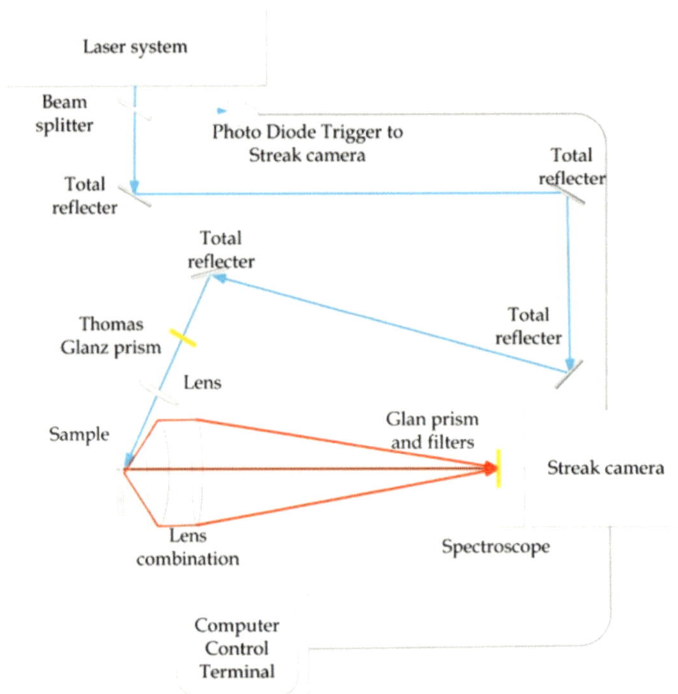

Fig. 1. Experimental set-up for ultrafast photoluminescence measurement

laser system with Cavity-damped dye laser and high energy picosecond laser system with OPA. Different laser sources are selected due to different motivations. The three total reflectors are adjusted for compensate the electrical delay of the photodiode triggering the streak camera in order to keep the synchronism between the optical induced signal and the triggering signal. But for the streak camera with high repetition rate such as 82 MHz, the external optical delay path is not necessary since the laser excitation keeps constantly both on the intensity and the profile in time domain. The excitation can be focused to the sample on face or from the behind according to the properties of the sample. Generally, the excitation incidence beam is injected on face to the sample fabricated on the substrate of silicon or some other opacities. Furthermore in order to avoid the excitation light into the detection system the excitation can be assigned as perpendicular to the emission collecting lens axis especially for the colloidal solution of nanostructures. The two Glan-Laser Calcite Polarizer prisms are adjusted for the polarization selected and the intensity manipulation for excitation. Additionally the cut-off filters for the excitation can not be neglected for preventing the detection system exposed under high irradiation of the laser.

2.1 Deconvolution of the time-resolved spectroscopy

The fluorescence detected by the streakcamera is the convolution of the instrumental response function and the fluorescence of the materials. Assume the instrumental response function is $I(t)$ and the fluorescence of the materials is $F_{theo}(t)$ the detected light emission is $F_{exp}(t)$, they can be described by the equation (1).

$$F_{\exp}(t) = I \otimes F_{theo} = \int I(t) F_{theo}(t - t')dt' \qquad (1)$$

Here the instrumental response function consists the excitation and the influence of the optical filters and lenses. The total influence tends to be in the distribution as Gaussian function profile in time-domain. Because the ideal transition rate at room temperature is proportional to the life-time reversely, we considered the theoretical fluorescence intensity $F_{theo}(t)$ as a sum of multi-exponential form:

Therefore we should find proper amplitude and corresponding life-time satisfying the following:

$$F_{exp}(\mathrm{t}) = \int I(t) \sum_{i=1,n} a_i exp(-t/\tau_i) \qquad (2)$$

We applied an iterative Monte-Carlo method to search the best result out of the fluorescence decays with consideration of a global optimization algorithm. It should be noticed that the parameter a_i are normalized percentage and the sign of a_i implying the energy out (positive) and in (negative). Energy out means that the photons are emitted from the materials and energy in means that energy has been transferred into the materials to bring them in excited states or leading to other active states alike.

2.2 ultrafast photoluminescence from ZnO nano particles

As the so-called nanowires is quasi 1-dimensional material. The photo electronic properties may undergo both bulk-like and nanodots-like. For example the laser cavity effect is originated from the bulk materials while quantum confinement effect and surface state effect are more dependent upon the quantum dots and the nanoparticles. In order to get the whole description of nanowires we should make it clear that the photo electronic properties of nanodots or nanoparticles.

2.2.1 Surface state induced ultrafast photoluminescence from ZnO nanoparticles

The quantum confinement effect (QCE) and surface states are two incompatible physical mechanisms in modifying the energy band structure of nanostructures. These two mechanisms compete with each other on the influence to PL spectra. For nanodots or nanostructures with diameters less than 10 nm, the QCE plays a dominant role while the large surface-to-volume ratio also brings much influence on the system's Hamiltonian when the material size is reduced to the nanometre scale. Therefore we demonstrated the size-dependent time-resolved PL spectra of ZnO nanoparticles with different sizes. The results show a key role of the surface states recombination in the PL of nanometre sized particles due to the large surface- to-volume ratio.

ZnO nanoparticles of different grain size were prepared by the homogenous precipitation method (HPM)(Wang et al., 2006). The average particle sizes for five groups of samples are 17 nm, 25 nm, 40 nm, 10 nm and 300 nm. The samples were excited by quasi-continuous picosecond (ps) laser pulses from Vanguard 2000-HM532 laser at 532 nm at room temperature. The laser pulse duration is 12 ps and the pulse repetition is 82 MHz. The maximal output laser power is 2 W. The excitation laser pulses were focused on the sample by a lens with a focal length of 75 mm. The emission was collected before being detected by a synchroscan streak camera (Hamamatsu M1587, resolution of 10 ps) for time-resolved photo luminescence (TRPL) detection. Fig. 2 gives the image of the streak camera of two kinds of nanoparticles representatively. Both images are obtained under the same spectral condition. The focused spot size of the laser beam is 100 micrometer in diameter, and the excitation power density is 2.59×10^5 W/cm^2. All the PL spectra exhibit broad bands covering the range from 1.2 eV to 2.3 eV. From Figure 2 we can see clearly that with the increased grain size of the ZnO nanoparticles, the PL band undergoes a red-shift significantly. By optimizing the peak position and half-width of the Gaussian peaks, it was possible to obtain a good fit for the four-peak combination coming from the surface state induced light emission.

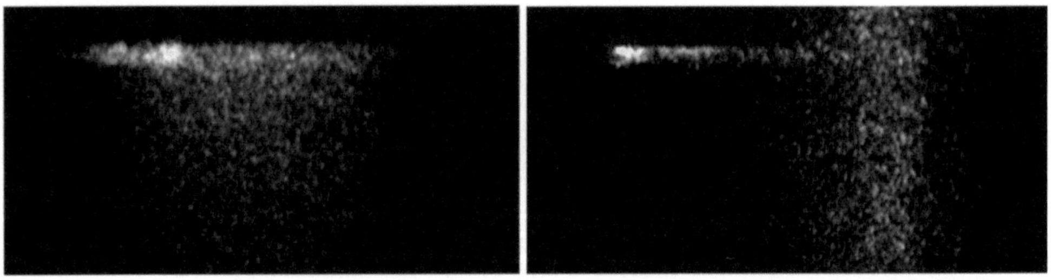

Fig. 2. The image of surface induced ultrafast light emission. (a) Particle grain size is 20nm, (b) Particle size is 300nm.

The band structure analysis are shown in table 1, Xc indicates the deconvoluted peaks according to Gaussian method with both the central position of the peak and the corresponding intensity. As labeled, the four Gaussian peaks are centered at Xc1, Xc2, Xc3, and Xc4 from the highest photonic energy to the lowest one. All the PL spectra are obtained under the same experimental condition.

Sample	Size (nm)	Xc1		Xc2		Xc3		Xc4	
		Peak (nm)	Inten. (a.u.)	Peak (nm)	Inten. (a.u.)	Peak (nm)	Inten. (a.u.)	Peak (nm)	Inten. (a.u.)
a	20	596	0.41	643	0.71	730	0.66	807	0.25
b	25	595	0.35	648	0.69	739	0.69	804	0.38
c	40	602	0.37	661	0.67	750	0.63	810	0.42
d	110	589	0.18	698	0.46	762	0.42	807	0.49
e	300	598	0.12	700	0.41	764	0.44	807	0.48

Table 1. The band structure analysis results of ultrafast light emission from nanoparticles with various grain sizes.

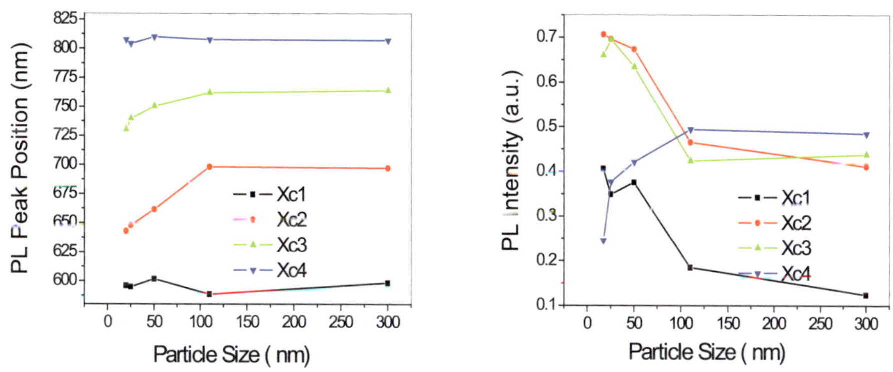

Fig. 3. The size threshold for surface-state light emission (left) relationship between the peak position and the grain sizes; (right) relationship between the PL intensity and the grain sizes.

Fig. 3 provides the luminescence peak position and the corresponding relative amplitude as a function of ZnO particle size. Upon increasing the grain size from samples ***a*** to ***e***, slight wavelength changes were observed for the Gaussian peaks denoted as Xc1 (~2.08 eV) and Xc4 (~1.54 eV), respectively. While the luminescence peaks Xc2 and Xc3 shift to lower energy as the grain size increases. The wavelength of Xc2 shifts from 1.93 eV (sample ***a***) to 1.77 eV (sample ***e***), and that of Xc3 shifts from 1.70 eV (sample ***a***) to 1.62 eV (sample ***e***). In addition, upon increasing the particle size, the relative amplitude of the Xc4 band is found to increase accordingly, while that of Xc3 band decreases. However, little change is observed for bands Xc1 and Xc2. Therefore, the Gaussian analysis indicates that the red-shift in the PL spectra is mainly the result of the increase in relative magnitude of the Gaussian combination in the lower energy region.

The energy band gap of ZnO is about 3.37 eV, corresponding to light emission at wavelength of 368 nm. ZnO luminescence mainly covers three bands: ultraviolet (UV), green and orange. Most scientists indicated that the green PL comes from intrinsic defects in ZnO samples, while the orange or yellow emission comes from lithium and other impurities (Djurisic et al., 2005). Under such a low photonic energy excitation of 2.33 eV (532nm), which is much lower than that for the band gap transition of ZnO, the PL spectra of all the nanoparticle samples clearly exhibit broad emissions in the range from 1.2 eV to 2.3 eV. The broad emission bands are not considered to come from the introduced impurities, because

no impurity was detected in our samples during identification. Such spectra indicate that certain states must lie in the fundamental energy band gap close to the Fermi level of ZnO nanoparticles. We attribute the broad bands obtained with the low photon energies to surface-state transitions, because the only difference in the preparation is the temperature corresponding to their particle sizes.
The existence of such surface-state bands has firmly been established both experimentally and theoretically. Moreover, the surface states located close to the Fermi level are due to the so-called dangling bond in elemental semiconductors. A dangling bond on a crystal surface likely gives rise to a localized state within the band gap. The unsaturated covalent bands can easily form states especially in the center of the gap. Although, it is suggested that no gap surface states exist on perfect ZnO surfaces, as no broken *sp*3 bulk bonds are present. In the case of nanoparticles, however, one of the key features is the large surface-to-volume ratio (S/V). That is, the large S/V greatly increases the possibilities of dangling bonds or broken bonds in ZnO nanoparticles, which is confirmed by our experimental results (Wang et al., 2007). Upon increasing particle size, however, the time-resolved optical emission process exhibits significant differences. For simplicity, Figure 4 shows the time decay curves of three typical different average sizes of samples ***a***, ***c*** and ***e***, with diameter of 17 nm, 40 nm and 300 nm, respectively. Upon increasing the particle diameter, the decay rate of the PL obviously slows down. With regards to the theoretical radiative recombination as a sum of exponential functions, we applied a deconvolution procedure based on the Monte-Carlo algorithm to search dynamic lifetimes. The bi-exponential decay process has been observed for samples ***a*** and ***c***. Figure 4a and Figure 4b show the corresponding fitted results, in which the scatters represent the experiment data and the solid line is the exponential fitting. The two lifetime constants of sample ***a*** are about 96 ps and 1.1 ns with relative amplitudes of 0.70 and 0.30, respectively, while those for sample ***c*** are about 190 ps and 2ns with relative amplitudes of 0.82 and 0.18, respectively. In the case of the largest particle size of sample ***e***, we cannot see the decay of the emission within the observed time scale as shown in Fig. 4c.

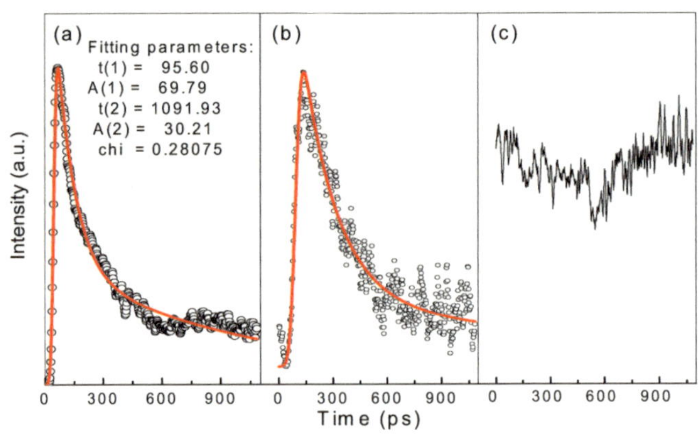

Fig. 4. The time decay curves of three typical different average sizes of samples a, c and e, with diameters 17 nm, 40 nm and 300 nm, respectively.

From our spectral results we can concluded that as the grain size is increased the surface state induce light emission turns to weak and the lifetime turns longer. It should be noted that as the grain size drops down to several nanometer the quantum confinement should be dominant to the ultrafast light emission because the bandgap enlargement will be much strong than the disturbance induce by surface state. As the size of the grain goes up to certain value that the bulk structure becomes dominant so that the surface-volume ratio is not much significant the surface state induced light emission should be diminished. Just at a proper size region the surface-state induced light emission cannot be neglected and this may bring important effect to the photo electronic integration on nanometer scale because of the size-peak-lifetime dependent light emission.

2.2.2 Lasing effect from ZnO nanoparticles

Besides Surface state effect in nanoparticles there is other interesting lasing effect from nanoparticles of ZnO. The experiment was demonstrated at room temperature. The excitation laser pulse was generated from commercial available OPA system pumped by Ti: sapphire system (Maitai, Spectral Physics) at wavelength of 320 nm with the pulse repetition of 1 kHz. The pulse duration is 160 fs. The excitation intensity was manipulated by a Glan-Laser Calcite Polarizer-Laser Calcite Polarizer prism. The light emission was collected by a lens composition before focused into the transient spectroscopy system (HAMAMATSU C 6860) connected with quartz prism spectroscope. The spectral resolution is less than 1 nm. In spite that time-resolution is 700 fs the laser pulse was indeed broadened by optical apparatus so that the overall time-resolution of the detection system is 1ps.

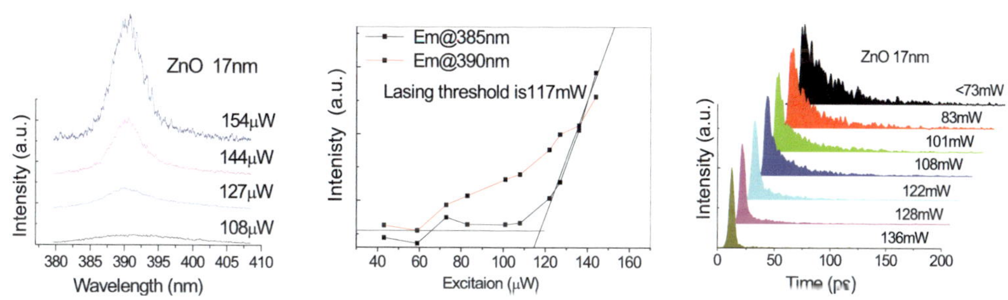

Fig. 5. The ultrafast light emission from ZnO nanoparticles.

Figure 5 shows that ultrafast lasing effect of ZnO nanoparticles with the size of about 17nm. The wavelength spectrum shows no sharp peak emerging at 385nm or at 390nm which is different from the lasing from ZnO bulk crystal materials. One reason is that electron hole plasma is generated in the nanoparticles and the nanometer boundary condition and the nanoparticles cannot provide any kinds of cavity to generate optical resonance. Furthermore the PL life time is shortened as the pump power increased until the pump power was increased to 136 micro Watt that the decay curve time-resolved PL turns to be symmetrical. Such a fast process implies that the EHP possesses an ultrafast behaviour in nanometer scale.

Therefore, surface state effects play important role in ultrafast light emission from the nanometer particles and the ultrafast light emission related to the transition from the conduction band to the valence band is rather an ultrafast lasing effect than nanolaser. Therefore nanoparticles are not proper candidates for special wavelength respected nanolaser; the nanowires may be a good choice. But the pumping intensity dependence of the PL intensity indicate that the PL at wavelength of 385nm corresponding to the band-band transition of the ideal ZnO crystal structure still under goes the linear dependence as the pumping power is above 117 micro. This is the important evidence for amplified stimulated light emission but not laser because the no cavity has been generated for any oscillation. That is nanoparticles can build amplified light emission easily but cannot construct the oscillating cavity.

2.3 Ultrafast photoluminescence from SiC nanowires and WO nanowires

Nanoparticles are quasi zero dimensional materials and the PL spectra imply that the nanolaser cannot be respected from just the nanoparticles. Nanowire is quasi 1D material and it undergoes the photo electronic properties of both nanoparticles and the bulk one. Therefore may the nanowire be the ideal candidate for more application in photo electronic.

We have studied the ultrafast light emission of SiC nanowires and WO nanowires and we have measured the ultrafast wide band blue light emission from SiC nanowires and super fluorescence from WO nanowires. The structural characterizations of SiC nanowires and WO nanowires have been shown in the Figure 6.

The experiment of Ultrafast photoluminescence spectroscopy for nanowires of SiC and WO are carried out under femtosecond laser pulse excitation at the wavelength of 320nm from an OPA of a femtosecond (fs) Ti: Sapphire laser with pulse duration of 160 fs (commercial OPA model 800CF, Amplifier model Hurricane, Seed model Maitai from Spectra-physics). By a fused silica lens (75.6-mm focal length), the excitation pulse was focused onto the sample surface at an angle of about 45º to the normal of the surface. The average power of the excitation on the sample was about up to 400 micro Watt (corresponding to a power density of about 10 GW/cm^2).

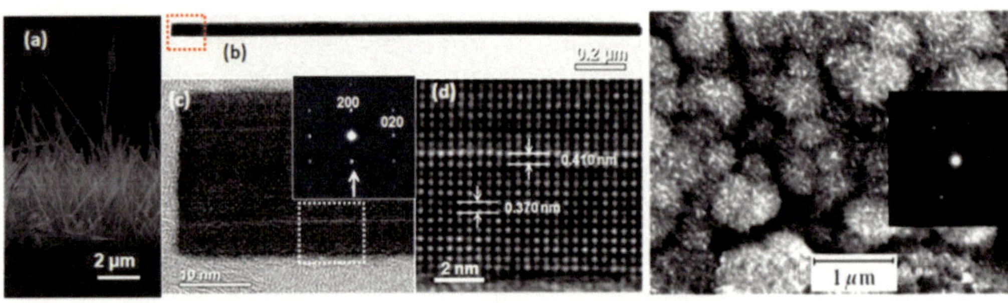

Fig. 6. Morphology of nanowires of WO and SiC, (a) is the SEM of WO nanowires, (b) (c) and (d) are the SEM, XRD and HRTEM for WO lattice characterization; (e, the right most) is the SEM and XRD characteristic of SiC nanowires.

The PL from SiC nanowires was collected at a direction normal to the sample surface and collected into the spectroscope connected with a streak camera (Hamamatsu C1587) with a time resolution of 2 ps. The temporal full-width at half- maximum (FWHM) of our instrument response function (IRF) was less than 10 ps.

For the PL from WO nanowires we can choose the spectral resolution from 0.5nm to 6.0nm and the time resolution from 10ps to less than 1ps.

2.3.1 Ultrafast light emission from SiC nanowires

The details of the spectroscopy of SiC nanowires and the powder of SiC are shown in Figure 7. The photoluminescence of SiC nanowires exhibits wide emission band and cannot be all included in the interested spectra range for the limitation of the spectrometer. From the figure, we can see that there is a large blue shift (~29 nm) between the emission centers of SiC powder and SiC nanowires. Obviously, the blue shift is caused by the nanowires. The photoluminescence of SiC powder can be well fitted by a single Gaussian line with its peak at 446.6nm. However, the emission of SiC nanowires is much broad and represents multiple peaks. We performed Gaussian multi-peak fitting process to the deconvolution and got the result of two peaks located at 418nm and 446.6nm respectively. Thus we can conclude that the peak at 446nm is due to the emission from the SiC powder, while the wide peak at 418nm is caused by the SiC nanowires. Because the diameter of the nanowires ranges from 20 to 50 nm and the curvature of the apex of the nanowires can be down to 2 nm, there should be great difference in the degree of quantum confinement and the surface-to-volume ratio of the nanowires, which accounts for the wide emission peak of the nanowires and the slight shift of the PL peak wavelength. It is well known that the band gap of the bulk 3C SiC is 2.3 eV (Bimberg et al., 1978), corresponding to the emission at the wavelength of 539 nm at room temperature, therefore, the PL peak energy of the nanowires is blue-shifted. But such large blue shift is not practical for quantum confinement. Another reason is that the SiC nanowire undergoes certain phase transition so that 6H SiC was present in the process of the irradiation under intense femtosecond laser pulse. Even for 6H SiC material the bandgap is 2.9 eV corresponding to the light emission at wavelength of 427nm our light emission at 418nm still present a blue shift significantly which due to 64 meV of the band gap widening. Similar results have been obtained in other low-dimensional SiC, the blue-shift emission peak can be explained by the energy gap widening due to the quantum confinement effect in these nanowires.

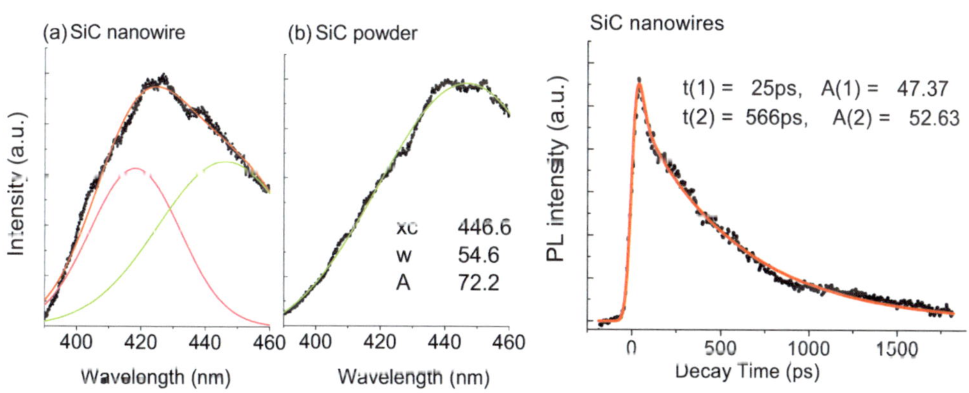

Fig. 7. Ultrafast light luminescence from nanowires and micro powder of SiC, (a) and (b) are the wavelength domain properties of nanowires and micro powder of SiC; (c) is the time domain curve of nanowires of SiC. All scatter dots are for the experimental results and the solid lines are the fitting results.

As the new peak at 418 nm is the dominant emission of the SiC nanowires, we will focus on its dynamic analysis. The red line in Fig. 7c shows the biexponential fitting result to the experimental data. The deconvolution result shows that the two decay constants of the SiC nanowires are 25 ps and 566 ps, respectively. It implies the coexistence of two different recombination paths. It should be noted that the decay time of 567 ps is on the same order of magnitude of that of the SiC nanocrystalline films reported and is nearly three orders of magnitude faster than that of the bound-exciton transitions of bulk SiC at low temperature (Hartman and Dean, 1970). The other decay time is 26 ps. Since quantum confinement effects is indicated occur in low-dimensional SiC, the geometrical restriction of the electron-hole pairs lead to a strong enhancement effect of the oscillator strength of the confined levels resulting in direct transitions in the nanowires and the faster decay process.

Similar explanation has been proposed in low-dimensional Si to account for the fast decay process. Similar results of Si nanowires at room temperature have comparable PL decay time (20 ps), which is also attributed to the luminescence due to the quantum-confinement effect. On the other hand, it's known that surface states play an important role in the carrier recombination process with large surface-to-volume ratio in low-dimensional Si. Theoretical calculation shows that the lifetime of self-trapped exciton localized at the surface state is close to 1 ns. Furthermore, similar lifetime (~1 ns) has been found in low-dimensional Si and SiC experimentally, which is also attributed to the recombination of the excitons localized at the surface states. Thus, the slower decay process in SiC nanowires could be subjected to the surface states recombination process. This can be further proved by the energy dispersive x-ray analysis, which shows that the nanowires are pure, containing only Si and C indicating that the slower process is not possible from impurity related emission centers.

In summary, the needle-shaped SiC nanowires give intense blue light emission band peaking at 418 nm under UV fs laser excitation at room temperature. TRPL analysis result shows a biexponential decay behaviour with a faster time constant of 26 ps and a slower one of 567 ps, respectively. The faster decay component is attributed to the direct recombination of the quantum-confined carriers, while the slower one corresponding to the recombination of excitons localized at surface states.

2.3.2 Super fluorescence from nanowires of WO

The ultrafast light emission from WO nanowires is analyzed by two steps. First we have detected the time-wavelength-intensity dependent spectroscopy. Under femtosecond laser excitation the WO nanowires undergo two significant effects. One is the PL intensity increases dramatically as the pumping intensity increases. The other is the life-time was shortened to almost to the same level of the instrumental response. To make it quantitatively, we have carried the pump intensity dependent PL measurement which is shown in Fig. 8 (b). We can find that under the low intensity of the pumping, the PL intensity is proportional to the pumping intensity which means that PL emission comes from the spontaneous emission. We applied the linear fitting curve of the function of $I_e=a_1I_p$ (the red scatter data, a_1 is a constant of 6.0) and find that under the pumping power of $100M/cm^2$, the linear fitting is quite accordant with the experiment. That is under such low excitation the PL comes from spontaneous emission dominantly. While for higher intensity of pumping, we can find that the PL-pumping relationship turns to be quadratic and the fitting curve of the function of $I_e=a_2I_p^2$ (the green dot line, a_2 is a constant of 0.28) matches the experimental results well with the only regret that not enough data points were obtained on the lack of intensity of the laser system. Such a quadratic relationship between the PL intensity and the excitation is one dominant factor for lasing effect. On the other hand we can get clear that the life-time is shorted dramatically from the spontaneous emission this is another factor for lasing effect of materials. But the bandwidth sharpening has not been detected under such a spectral resolution.

Another interesting result is that there is a band shift in Fig. 8 (a) when the pumping intensity increases the lasing like peak turns to shift to red. We applied the high spectral measurement to investigate the detail dynamics of the PL behaviour as shown in Figure 9. The left side in Figure is the integration of the whole image of the streak camera. As the spectral resolution improved the wavelength spectroscopy can be found that there be three bands from the wavelength from 385nm to 395nm corresponding to the transition of the oxygen vacancies. The bandgap of WO is 2.6~2.7 eV corresponding wavelength about 460nm. While the resonant states introduced by oxygen vacancies can generate an direct transition leading to photonic emission with much high efficiency than that of the intrinsic transition from the conduction band to the valence band because the WO is the indirect semiconductor (Karazhanov et al., 2003).

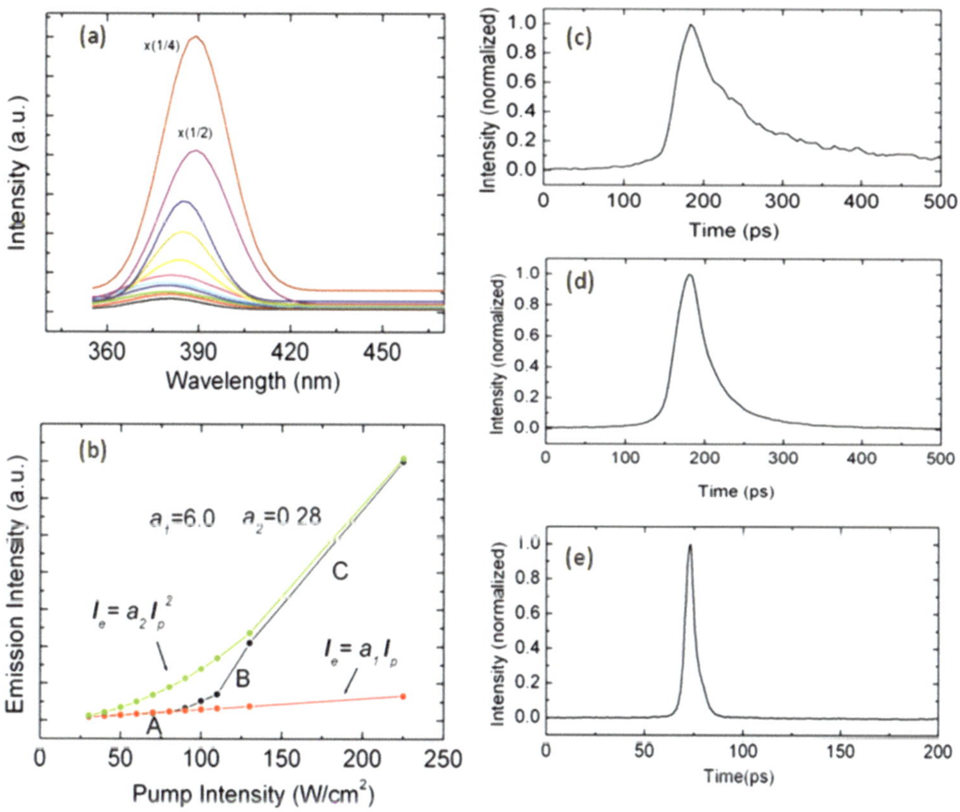

Fig. 8. (a) The emission intensity spectra of WO_{3-x} nanowires obtained under the pump intensities of femtosecond laser, correspondingly from low to high, of 30, 40, 50, 60, 70, 80, 90, 100, 110, 130, and 225 W/cm², respectively. (b) The intensity of emission peak versus the pump intensity (the black dot line), the fitting curve of the function of $I_e = a_1 Ip$ (the red dot line, a1 is a constant of 6.0), and the fitting curve of the function of $I_e = a_2 Ip^2$ (the green dot line, a2 is a constant of 0.28). (c), (d), and (e) are the emission decay curves of WO_{3-x} nanowires at low pump intensity of less than 90 W/cm²(section A), the midst pump density of 110 W/cm²((section B), and high pump density of more than 130 W/cm² (section C), respectively.

The most significant experimental finding of our study is the time-resolved spectroscopy measurement of SF which resolves in first time the dynamic process of a SF in picosecond-scale. As shown in Figs. 9(a) and 9(b), a red shift and a blue shift at the first and second halves of the SF pulse have been clearly detected, respectively, with use of high spectral resolution spectrometer (0.5 nm) and high time resolution of our streak camera (less than 1.0 ps). Three peaks with the centre wavelengths of 396nm (3.13 eV), 391nm (3.17 eV), and 386 nm (3.21 eV) (Fig. 9(c)) can be seen at the time of 4.3, 5.5 and 8.9 ps (Fig. 9(d)) after the initiation of excitation pulse, respectively. Fig. 9(d) reveals that for the majority of emitted photons, the first group is at the wavelength of 396 nm at time of 4.3ps, the second one at 391 nm at time of 5.5ps and the last one at 386 nm at time of 8.8ps. The peak at 391 nm may be attributed to the recombination of Frenkel excitons themselves, and the peaks at 396 and 386 nm are then a red shift and a blue shift at the first and second halves of the ultrafast photo luminescent pulse, respectively. This effect of frequency-shift during radiation is the fingerprint characteristic of cooperative emission in solid state, different from the case of atomic gas system.

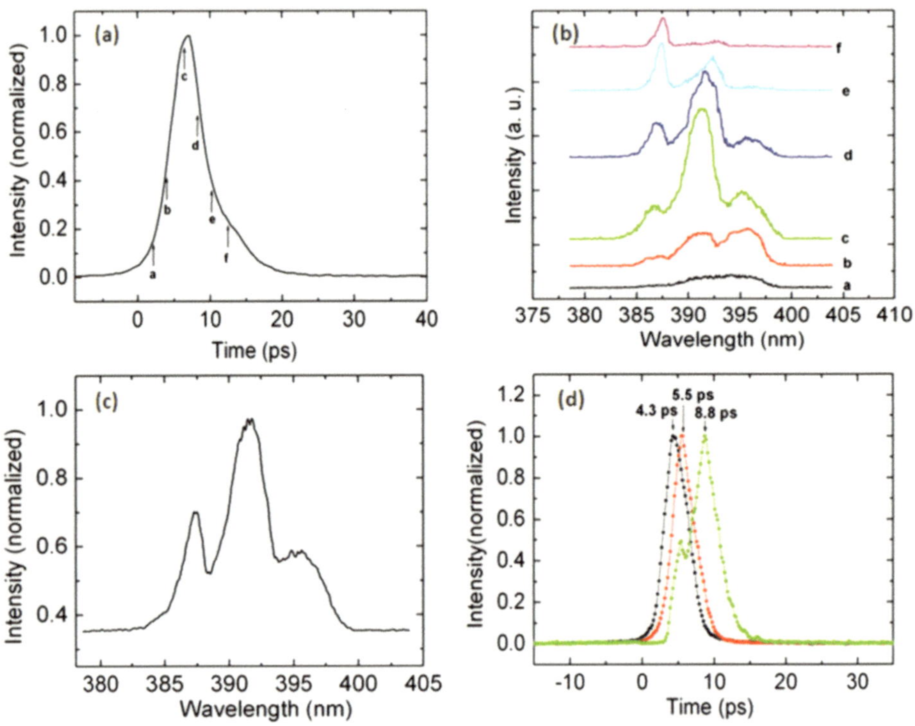

Fig. 9. (a) The ultraviolet SF pulse profile recorded under the pump intensity of 225 W/cm^2 (the peak of excitation pulse at the time of zero). (b) The time-resolved emission spectra of ultraviolet SF just recorded at the time of a, b, c, d, e, and f, which are indicated in (a). (c) The time-integrated spectrum of ultraviolet SF. (d) The emission decay curves of ultraviolet SF at the centre wavelength of 396 nm (black), 391 nm (red), and 386 nm (green); their maximal intensities occur at 4.3 ps, 5.5 ps and 8.8 ps after the peak of excitation pulse, respectively.

We have demonstrated a new new theoretical model for a two-dimensional system, and to apply it to explain the present findings (Luo et al. 2011).
As the theoretical model indicates that the first and last groups of emitted photons in SF process occur at $\hbar\omega_0 + 4\hbar\Omega$ and $\hbar\omega_0 - 4\hbar\Omega$ respectively which is independent on any N atoms system. It should be noted that the PL intensity is an ensemble emission from all nanowires with different large N atoms (oxygen vacancies) under illumination of the femtosecond laser beam. Thus, two strong peaks at $\hbar\omega_0 + 4\hbar\Omega$ and $\hbar\omega_0 - 4\hbar\Omega$ besides the strongest peak at $\hbar\omega_0$ should be observed in the time-integrated spectra as shown in Fig. 9(c). Furthermore, with assumption that the lattice constant of WO_3 cubic phase R_0 equals to 0.37 nm, the magnitude of the transition dipole of Frenkel exciton in WO nanowires is obtained as d = 4.25x10-30 m·C, which equals to the value of the dipole moment when the distance of an electron departure from the centre of positive charges is 0.265 Å (half of Bohr radius) which is quite a reasonable distance in practice.
Therefore ultrafast photo luminescent phenomena of various nanowires enrich the prospective application of nanowires in light sources and integrated photo electronics with both ideal structures of nanowires and the defects contained nanostructures. With the artificial nanostructures we can obtain various novel structures for further practical design of functional materials.

2.4 Directional lasing analysis form ZnO nanotrees

Compare with the uniform nanowires the composed structure of nanowires even with micro rods are a new kind of complex artificial materials with novel photo electronic features. Among various composed nano-micro structures ZnO nanotrees have attracted much attention in recent years. We have measured the ultrafast photo luminescent of ZnO nanotree with various spacial respects and found that there be different lasing behaviour inside the nanotree structures (Zhao et al., 2010).

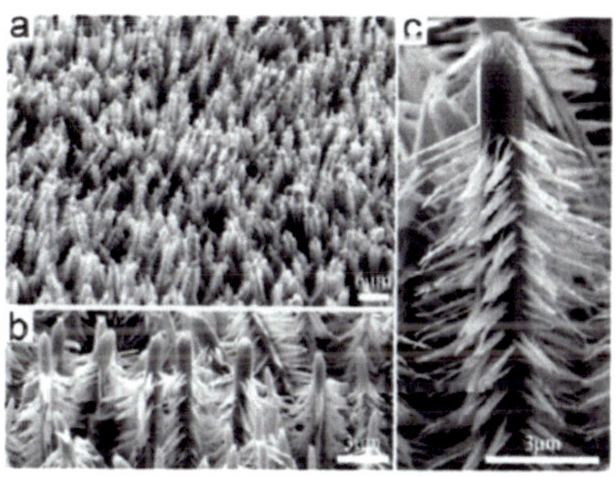

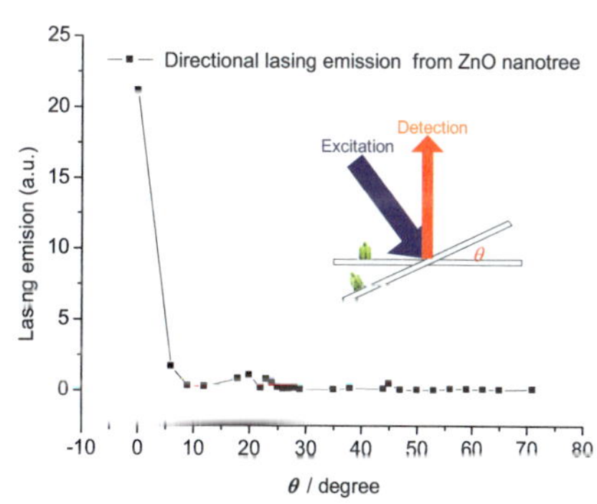

Fig. 10. SEM image of ZnO nanotree. a: title view; b: side view; c: magnification image of single nanotree. Fig. 11. Direction dependent lasing emission of ZnO nanotree structure.

From Figure 10 we can see clear that each ZnO nanotree is uniformly composed of a hexagonal prismatic microcrystalline trunk and radial-oriented nanobranches on the micro trunk's six columnar surfaces (Fig. 10 c), thus forming a nanostructured *pine tree forest* (Fig. 10, a-b). Time-resolved PL spectroscopy was demonstrated with a femtosecond optical parameter amplification system (Spectra-Physics 800CF) as the excitation and a femtosecond synchroscan streak camera FESCA (HAMAMATSU C6860) connected with a spectroscope C5094 as the spectrum detecting system. The time resolution of the whole system is less than 1.0 picosecond and the best spectral resolution is 0.3 nm. The excitation femtosecond laser pulse was focused on the sample by a lens with focal length of 75 mm with both a sharp angle and a glance angle between the incident light and the normal of the sample plane respectively, the light emission from the sample was collected with a large NA lens with the focal length of 50 mm with lens radius of 50 mm. The composition of lenses was applied to generate a conjugated point of sample into the slit of spectroscope C5094 which connected by the streak camera modeled as FESCA C6860 triggered by the femtosecond laser oscillator at 82MHz. The OPA output was set at the wavelength of 320nm, with the pulse energy of 100 nano Joule per pulse. In order to change the excitation intensity we applied Glan-Laser Calcite Polarizer in front of the focusing lens of the sample. A short wavelength cut-off filter at 340nm was adjusted before the input slit of the spectroscope.

The spacial distribution of the ultrafast photoluminescence of ZnO tree was measured with the set-up as the schematic inset in Figure 11. As the trunk is parallel to the normal line of the sample plane we adjusted the angle between the excitation laser beam and the normal line of the sample to obtain the relationship between the lasing emission and the geometry of excitation to the nanotrees. The spacial intensity distribution shows the maximum lasing emission at the direction nearly parallel to the trunks which increase dramatically as the angle bends to nearly parallel to the direction of the trunk of the tree. This means no matter which direction to get the excitation the end effective lasing comes from the trunk of the nanotrees. Such effective lasing emission implies that there may be energy transfer process from the branches to the trunk of the nanotrees.

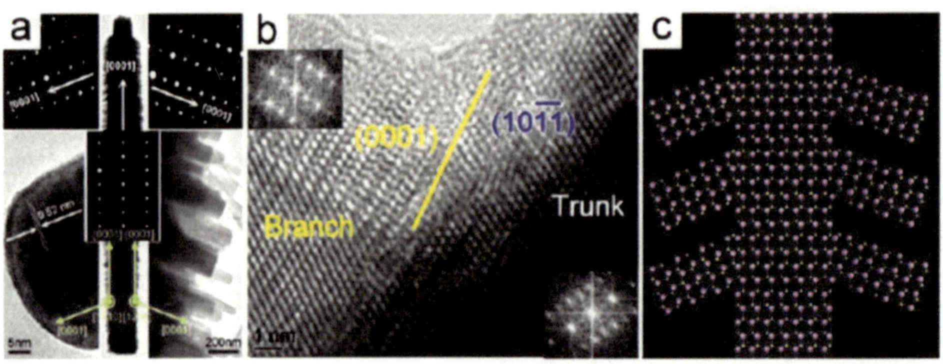

Fig. 12. XRD results of the ZnO nanotree, a: TEM image of nanotree, b: HRTEM of branch-trunk junction, c: schematic structure model of ZnO nanotree.

By checking the crystal structure of the ZnO nanotree we can find that both the microtrunks and the nanobranches have grown along the ***c*** axis (Fig. 12). All the microtrunks stand nearly vertical on the zinc substrate due probably to either the first few layers *in situ* formed on the Zn or the adsorbed *en's* interfacing. The smooth nanobranches, however, are not exactly perpendicular but oblique to the columnar surface of the microtrunk, with an angle of about 108° between growth direction of the branches and that of the microtrunk. Thus if the ***c*** axis is the proper lasing direction we should have detected the lasing effect at the angle along the nanobranches and trunks. There should be other behaviour dependent. We applied the excitation with different orientations and the corresponding detection for both the microtrunks and the nanobranches.

As the nanotree shows a specific spatial construction compared with other homogeneous nanowires and nanorods we should explore the detailed light emission on the basis of the novel structures. In order to explore the details of the excitation energy flow we demonstrated four extremely schematic setups for the detection of the dynamic lasing emission. According to the directions of the excitation and detection, we denote the E for excitation and D for detection which addressed letter V for the nearly parallel to the *c* axis of the trunk and H for the line forming a large angle of 78° with the *c* axis of the trunk, respectively. In order to practical adjustment of all lens and apparatus, we have assigned the setup EVDV and EHDH by narrowing to a sharp angle of 15° as shown in Figure 13.

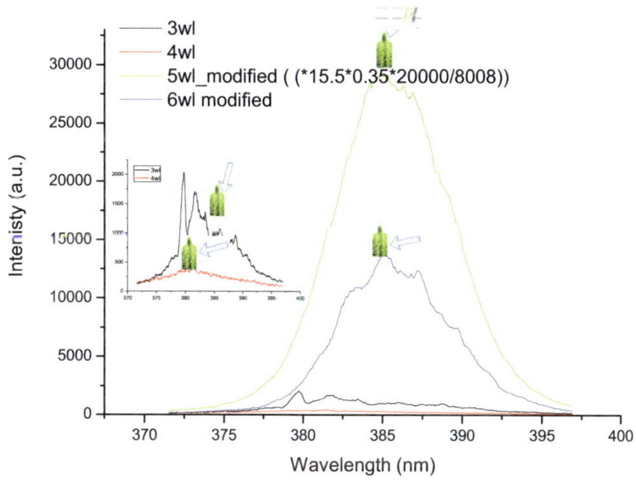

Fig. 13. Comparative lasing efficiency under different geometrical conditions. The blue arrow and the red arrow represent for the excitation and the emission correspondingly. The number is for modification in order to get proper ccomparisons among different conditions.

Figure 13 shows that when the excitation was set vertically, the lasing along the trunk presents in different separate modes. One of the most significant lasing comes from the mode at wavelength of 379 nm which was assigned as P band of ZnO nanostructures. The others separate modes should come from the cavity effect of the trunk. From the separation and the cavity analysis we can calculate the length trunk is about 15 micro meters long accordant with the morphology. But the horizontal light emission (EHDH) shows a wide band with very low intensity. Such ultrafast light emission is similar to amplified spontaneous emission. Furthermore, we can find that such light emission undergoes red-shift compared to that of trunk which is due to the nanostructures larger than 100nm scale. Therefore, we attribute such wide band random lasing to the light emission of the nanobranches growing on the trunk with angle of 108° to the *c* axis of the trunk.

Because the far field lasing emission are all demonstrated by the light with the polarization perpendicular to the *c* axis of the wurtzite structure of ZnO, while for the light emission from the direction of electric vector parallel to *c* axis has been vanished in far field detection. Therefore we turn to the component perpendicular to *c* axis of branches and trunk correspondingly. When the excitation beam was set at the sharp angle

(15°, EVDV), the projective factors of the electric components of the trunk and the branches are 0.966 and 0.052 respectively. While for the large angle excitation setup (78°, EHDV and EHDV), the excitation of the trunk has been decreased as the projective factor coming down to 0.208 and the branches excitation has been increased as the projective factor up to 0.866. Thus under semi-parallel setup the trunk got mainly excited and the lasing emission parallel to the trunk axis presents cavity modes significantly which is confirmed by our discussion above.

As for semi-perpendicular excitation (EH), we may expect that the branches get mainly excited and the trunk got excited much less than that of semi-parallel setup the light emission should be mainly from the branches of the nanotree. But the result has shown much different to such a deduction. Firstly, as the branches undergo the structure in nanometer scale, the light emission presents a wide band spectrum as detected under EHDH, but it demonstrated a very low intensity compared to the trunk emission of EVDV setup. Secondly, with the detection of EHDV separate modes are distinguished with the same wavelength separation as that of the EVDV as well as much higher intensity and a slight redshift than that of EVDV. Because the excitation keeps the same under these two conditions we should consider the energy transfer from the branches to the trunk of the nanotree. Since the PL only concerns about the electric vector perpendicular to the c axis the detected light emission far field is only the emission of the electric vector which is perpendicular to the corresponding *c* axis of the trunk and the branches. While for the connecting part of the trunk and branches there are structure defects there should be coupling between two directional electric vectors which dominating the energy transfer from the branches. As the interaction and energy dissipation of the energy processes there should be a redshift of the acceptor to the energy donor. Such a redshift matches the spectra from EVDV and EHDV.

As we known that photosynthesis is one of the most important energy transform process on earth and the initial energy transfer process is the key step to harvest the solar energy for the coming processes, the artificial ultrafast energy harvesting apparatus has attracted much more attention of both material experts and photonic scientists. Our nanotree of ZnO has the similar structure as the LHC of algae which has six antae linked onto the core structure. Since the core structure is arranged on the membrane of the thylakoid behaving as the acceptor of the excitation energy, this open a door for us to applied a certain structure to the trunk of the nano-pine-tree to link to the corresponding acceptor of such a coherent excitation leading to energy transferring apparatus with high efficiency and high speed.

The PL efficiency of trunk to branches has been calculated based on the structural statistics. Table 1 shows the detailed calculation of TB ratio (Data have been listed in Table 2.). The efficiency is the projection factor of the excitation beam. The structural T/B ration is the ratio of the absorption cross area based on the structural statistics. The experimental T/B ratio is calculated from the PL spectroscopy. Under the vertical excitation the structural T/B ratio is 0.355 while the experimental T/B ratio is 0.063 indicating that there is part of energy transferred from the trunk to the branches. But under parallel excitation, the structural T/B ratio is 0.496 while the experimental ratio is up to 20.6 showing that the majority of the excitation energy in the branches transferred into the trunk leading to a large PL ratio of T/B.

To study the dynamic process of the nanotree structures, we have carried out a excitation - dependent lasing detection, which under the parallel condition we adjusted the excitation laser power by Glan-Laser Calcite Polarizer prism from few micro Watt to about 120 micro Watt. The excitation dependent lasing emission is shown in Figure 14. Figure 14 (a) is the time-decay curves of both lower excited lasing emission and the higher excited lasing emission respectively. The slower delay curve belongs to the lower excitation at the mean optical power of 5.5 micro Watt which corresponding to the focal pumping density of 374MW/cm². Such a process was taken to the ultrafast lasing emission of high excitation but not relates to the correlated light emission. The fast decay curve belongs to the higher pumping at the mean focal density of 3.4GW/cm², which is closed to bring the correlated light emission such that we have obtained the almost Gaussian profile of the delay curve which means that the light emission response time is less than the excitation lasing pulse, about 4 picoseconds in our experiment.

Furthermore, a clear linear relationship has been given by the intensities of both excitation and lasing emission as shown in Figure 14 (b). The fitting result shows that the threshold of excitation for lasing emission is 20 micro Watt. The linear dependence of the light emission with the intensity matches the laser theory of a constant gain efficiency of the gain medium which is nanostructures ZnO in our experiment. Therefore the trunk of the nanotree behaves as both the gain medium and the cavity of the lasing emission. The branches play the role of light harvesting and transfer the harvested energy to the laser cavity to get the phase matched lasing emission. Such a geometrical restricting lasing structure is appropriated for photo electronic integration.

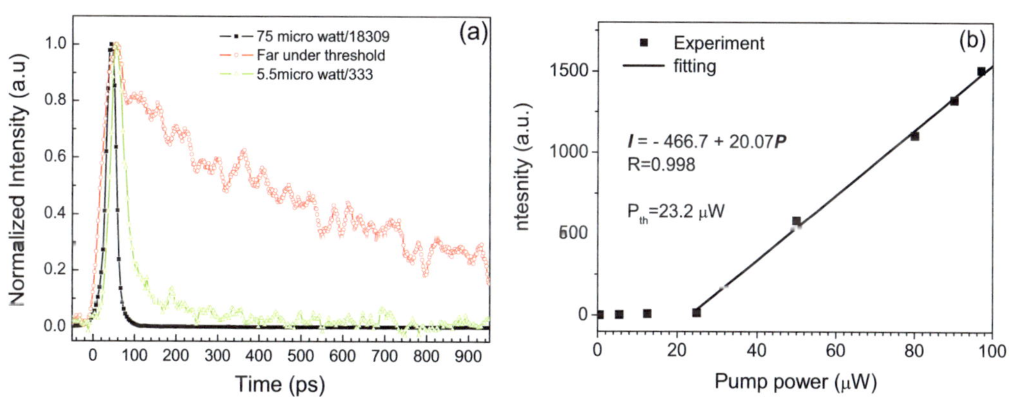

Fig. 14. Lasing effect of the nanotree under parallel excitation. a) Time resolved spectra of emission under different pumping intensity. b) Dependence of light emission on the excitation intensity.

One important comment is that on order to get the lasing emission evidently we have carried all the lasing detection under the pumping laser intensity at the power of 50 micro Watt corresponding to the focal light intensity of 3.4 GW/cm². Additionally, for the more detailed dynamic energy transfer study we have demonstrated the ultrafast ling detection under the fastest responding selection but have not obtained any significant improved results. It implies that the energy transfer between the branches and the trunk must be much faster than 1.0 picosecond, the time-resolution of our detection system, which may have a certain degree of coherent interaction between the trunk and branches of the ZnO nanotree.

From discussion above, we have demonstrated a directional lasing emission from structured ZnO nanotree with ultrafast time-resolved laser spectroscopy. The lasing shows a far field diffraction angle less than 0.04 rad (Figure 11) with separate cavity modes. The detailed direction excitation setup analysis gives an ultrafast energy transfer between the branches and the trunk of the nanotree evidently. Nano branches play the role of light harvesting and transferring the harvested energy to the trunk which behaves as both the gain medium and the laser cavity for phase matching of lasing emission. Such a geometrical restricting lasing structure is appropriated for photo electronic integration (Wang et al., 2009). The transferring time may be much faster than 1.0 picosecond. The results imply that with proper artificial techniques man can fabricate apparatus with featured high energy transfer function and ultrafast energy transfer time as well as high intensity lasing emission shedding light on the ultrahigh speed light communications and integrated photo electronic.

Ex	Em	Excitation efficiency	Trunk area	power	Branch area	power	Experimental detection	
			T	T	B	B	T	B
V	H	0.0955	0.335	0.021	0.665	0.064		270490
	V	0.9045	0.264	0.239	0.736	0.667	16958	
				0.260		0.732		
T/B ratio			0.355				0.063	
H	H	0.9568	0.335	0.320	0.665	0.636		5656
	V	0.0432	0.264	0.011	0.736	0.032	116700	
				0.331		0.668		
T/B ratio			0.496				20.6	

Table 2. The energy transfer calculation between the trunks and the branches of the ZnO nanotrees, Ex is for the excitation direction while Em for the detection direction.

3. Conclusion

In summary, nanowires as one-dimensional material have attracted much attention on the functional application research. Photo electronic properties determine its application on photo detection, integrated optical devices and novel kinds of light source.

We have constructed a series experiments for the ultrafast photo luminescence detection and completed the analysis of the mechanisms of the ultrafast light emission from various nanowires as well as beginning with the nanoparticles of ZnO. We have found the surface state induced ultrafast light emission from the transition within the bandgap in ZnO nanoparticles and measured that there is a size threshold for surface state induced light emission. Besides, we have detected the ultrafast blue light emission from the needle like SiC nanowires and the spectral results show the significant quantum confinement. Furthermore, we have accomplished the mechanism studies on the super fluorescence from WO nanowires and concluded that the SF is due to the oxygen defects resonance transition. To investigate the nanowires UPL more practically, we developed a complex set-up to study the ultrafast light emission of the nanotrees of ZnO and find the lasing effect from the trunks of the nanotrees and find the clue of the energy transfer from the branches to the trunks. From the ultrafast photo luminescence studies we have been clear for the following.

With proper artificial design man can find the wide-band ultrafast light emission due to the surface-state. Because the bandwidth of surface state induced light emission can be tuned through the size of the material in nanometre scale we may construct optional light sources by the size design. Certainly surface state can also bring trouble for the neat narrow band emission this should be pay attention in the material design.

The super fluorescence in the defect contained nanowires made of indirect semiconductors such as WO nanowires brings new hope for ultrafast integrated photo electronics. Several picoseconds response time means an ultrafast light transformation in practice.

Composition of the nanotree undergoes the ultrafast lasing effect and energy transfer phenomenon. Such structure may be useful for artificial solar energy harvesting apparatus development.

4. Acknowledgment

The authors gratefully acknowledge the financial support of the project from the National Natural Science Foundation of China (Grant No. 10574165), the project from the Natural Science Foundation of Guangdong (Grant No. 8151027501000017), the Science and Technology Department of Guangdong Province, the Department of Information Industry of Guangdong Province and the Science and Technology Department of Guangzhou City.

5. References

Bimberg, D., Skolnick, M.S., Choyke, W.J., 1978. Observation of an Electron-Hole Liquid in Cubic SiC. Physical Review Letters 40, 56.

Djurisic, A.B., Kwok, W.M., Leung, Y.H., Phillips, D.L., Chan, W.K., 2005. Stimulated emission in ZnO nanostructures: A time-resolved study. J. Phys. Chem. B 109, 19228-19233.

Hartman, R.L., Dean, P.J., 1970. Magneto-Optical Properties and Recombination Rate of the Green Luminescence in Cubic SiC. Phys. Rev. B 2, 951.

Luo, J.Y., Xu, N.S., Zhao, F.L., Deng, S.Z., Tao, Y.T., 2011. Ultraviolet superfluorescence from oxygen vacancies in WO3-x nanowires at room temperature. J. Appl. Phys 109, 024312.

Karazhanov, S.Z., Zhang, Y., Wang, L.W., Mascarenhas, A., Deb, S., 2003. Resonant defect states and strong lattice relaxation of oxygen vacancies in WO_{3}. Phys. Rev. B 68, 233204.

Wang, X.F., Xie, P.B., Zhao, F.L., Wang, H.Z., Wang, Y.M., 2009. Spectra of ZnO nanoparticles under low photon energy excitation. Particuology 7, 496-500.

Wang, X.F., Zhao, F.L., Xie, P.B., Deng, S.Z., Xu, N.S., Wang, H.Z., 2006. Surface emission characteristics of ZnO nanoparticles. Chem. Phys. Lett. 423, 361-365.

Wang, X.F., Zhao, F.L., Xie, P.B., Liang, S.D., Wang, X.F., Xu, N.S., 2007. Surface luminescence in ZnO nanoparticles. Optics Communications 276, 186-190.

Zhao, F.H., Zheng, J.G., Yang, X.F., Li, X.Y., Wang, J., Zhao, F.L., Wong, K.S., Liang, C.L., Wu, M.M., 2010. Complex ZnO nanotree arrays with tunable top, stem and branch structures. Nanoscale 2, 1674-1683.

Zhao, F.L., Chen, H.X., Deng, S.Z., Xu, N.S., Jia, T.Q., Xu, Z.Z., 2007. Ultrafast blue light emission from SiC nanowires. Chinese Optics Letters 5, 184-186.

Zhao, F.L., Gong, Z., Liang, S.D., Xu, N.S., Deng, S.Z., Chen, J., Wang, H.Z., 2004. Ultrafast optical emission of nanodiamond induced by laser excitation. Appl. Phys. Lett. 85, 914-916.

10

Numerical Simulation of Transient Response in 3-D Multi-Channel Nanowire MOSFETs Submitted to Heavy Ion Irradiation

Daniela Munteanu, Jean-Luc Autran and Sébastien Martinie
IM2NP Laboratory, CNRS (Centre National de la Recherche Scientifique), Aix-Marseille University France

1. Introduction

The bulk MOSFET scaling has recently encountered significant limitations, mainly related to the gate oxide (SiO_2) leakage currents (Gusev et al., 2006; Taur et al., 1997), the large increase of parasitic short channel effects and the dramatic mobility reduction (Fischetti & Laux, 2001) due to highly doped Silicon substrates precisely used to reduce these short channel effects. Technological solutions have been proposed in order to continue to use the "bulk solution" until the 32 nm ITRS node (ITRS, 2009). Most of these solutions envisage the introduction of high-permittivity gate dielectric stacks (to reduce the gate leakage, (Gusev et al., 2006; Houssa, 2004), midgap metal gate (to suppress the Silicon gate polydepletion-induced parasitic capacitances) and strained Silicon channel (to increase carrier mobility (Rim et al., 1998). However, in parallel to these efforts, alternative solutions to replace the conventional bulk MOSFET architecture have been proposed and studied in the recent literature. One solution is the radical change of the device architecture such as in Multiple-Gate devices introducing additional gate electrodes: 2 (double-gate), 3 (FinFET or trigate) or 4 (gate-all-around, completely surrounding the channel). Silicon nanowires MOSFETs with gate-all-around (GAA) provide an original and very promising architecture to further increase the integration density and performances of nano-devices (Park & Colinge, 2002). These structures exhibit a superior control of short channel effects resulting from an exceptional electrostatic coupling between the conduction channel and the surrounding gate electrode. As a result, intrinsic channels can be used leading to higher mobilities and drain currents.

3D Multi-Channel MOSFETs (MCFETs) have been recently proposed to achieve a higher current drivability and a significant enhancement of the on-state current over the off-state current ratio (I_{ON}/I_{OFF}) (Bernard et al., 2007, Ernst et al., 2006) compared to conventional single channel devices. MCFETs combine the advantages of excellent control of short-channel effects with a high on-state current due to a multiple-gate architecture and the 3-D integration of vertically stacked channels. GAA devices with ultra-thin and narrow channels (about 10 nm) are seen as the ideal architecture for off-state current control of sub-10 nm gate lengths (Ernst et al., 2006). Meanwhile, the current density per surface of such a device is limited by the lithography pitch, which dictates the distance between nanowires. The current density can be improved by the vertical integration of GAA devices. Thanks to vertical stacked channels, a 5× increase in current density per layout surface can be achieved compared to planar transistors with the same gate stack (Ernst et al., 2006).

As the MOSFET is scaling down, the sensitivity of integrated circuits to radiation, coming from the natural space or present in the terrestrial environment, has been found to seriously increase (Baumann, 2005; Dodd, 1996; Dodd & Massengill, 2003; Dodd, 2005). In particular, ultra-scaled memory ICs are more sensitive to single-event-upset (SEU) and digital devices are more subjected to digital single-event transient (DSETs). Single-event-effects (SEE) are the result of the interaction of highly energetic particles, such as protons, alpha particles, or heavy ions, with sensitive regions of a microelectronic device or circuit. These SEE may perturb the device/circuit operation (e.g., reverse or flip the data state of a memory cell, latch, flip-flop, etc.) or definitively damage the circuit (e.g. gate oxide rupture, destructive latch-up events).

The physical mechanisms related to the production of SEE in microelectronic devices consist in three main successive steps: (1) the charge deposition by the energetic particle striking the sensitive region, (2) the transport of the released charge into the device and (3) the charge collection in the sensitive region of the device. In the following we succinctly describe these different mechanisms, for a detailed presentation we invite the reader to consult (Baumann, 2005; Dodd, 1996; Dodd & Massengill, 2003; Dodd, 2005).

Charge deposition (or generation): When an energetic particle strikes the device, an electrical charge can be deposited by one of the following mechanisms: direct ionization by the interaction with the material or indirect ionization, by secondary particles issued from nuclear reactions with the atoms of the struck material. Direct ionization typically characterizes heavy ions ($Z \geq 2$) of the space environment. They interact with the target material mainly by inelastic interactions and transmit a large amount of energy to the electrons of the struck atoms. These electrons produce a cascade of secondary electrons which thermalize and create electron-hole pairs along the particle path. In a semiconductor or insulator, a large amount of the deposited energy is thus converted into electron-hole pairs, the remaining energy being converted into heat and a very small quantity in atoms displacement. It was experimentally shown that the energy necessary for the creation of an electron-hole pair depends on the material bandgap. In a Silicon substrate, one electron-hole pair is produced for every 3.6 eV of energy lost by the ion. Other particles, such as neutrons of the terrestrial environment, do not interact directly with target material since they do not ionize the matter on their passage. However, these particles should not be neglected, because they can produce SEE due to their probability of nuclear reaction with the atoms of materials which compose the microelectronic devices. This mechanism is called indirect ionization. The products resulting from a nuclear reaction can deposit energy along their traces, in the same manner as that of direct ionization. Since the creation of the column of electron-hole pairs of these secondary particles is similar to that of ions, the same models and concepts can be used.

Charge transport: When a charge column is created in the semiconductor by an ionizing particle, the released carriers are quickly transported and collected by elementary structures (e.g. p-n junctions). The transport of charge relies on two main mechanisms: the charge drift in regions with an electric field and the charge diffusion in neutral zones. The deposited charges can also recombine with other mobile carriers existing in the crystal lattice.

Charge collection: The charges transported in the device induce a parasitic current transient, which could induce disturbances in the device and associated circuits. The devices most sensitive to ionizing particle strikes generally contain reversely-biased p-n junctions, because the strong electric field existing in the depletion region of the p-n junction allows a very efficient collection of the deposited charge. The effects of ionizing radiation are different according to the intensity of the current transient, as well as the number of impacted circuit nodes. If the current is sufficiently important, it can induce a permanent damage on gate insulators (gate rupture, SEGR) or the latch-up (SEL) of the device. In usual low power circuits, the transient current may generally induce only an eventual change of the logical state (cell upset).

Modeling and simulating the effects of ionizing radiation has long been used for better understanding the radiation effects on the operation of devices and circuits. In the last two decades, due to substantial progress in simulation codes and computer performances which reduce computation times, simulation reached an increased interest. Due to its predictive capability, simulation offers the possibility to reduce radiation experiments and to test hypothetical devices or conditions, which are not feasible (or not easily measurable) by experiments. Physically-based numerical simulation at device-level presently becomes an indispensable tool for the analysis of new phenomena specific to short-channel devices (non-stationary effects, quantum confinement, quantum transport), and for the study of radiation effects in new device architectures (such as multiple-gate, Silicon nanowire MOSFET), for which experimental investigation is still limited. In these cases, numerical simulation is an ideal investigation tool for providing physical insights and predicting the operation of future devices expected for the end of the roadmap. A complete description of the modeling and simulation of SEE, including the history and the evolution of this research domain, have been presented in the survey papers by Dodd (Dodd, 1996; Dodd & Massengill, 2003; Dodd, 2005) and Baumann (Baumann, 2005). It is also important to note that phenomena related to an ionizing particle striking a microelectronic device are naturally three dimensional (3-D) mechanisms, due to both the tri-dimensional structure of the ion track and the 3-D structure of real devices. 3-D simulation is then necessary not only for actual short/narrow devices, but also for new device architectures for which 3 D electrostatic or quantum confinement effects cannot be taken into account in a 2-D simulation. 3-D simulation is also necessary when considering non-normal incidence of the ion strike on the device.

In this chapter, we investigate the transient response of MCFETs submitted to heavy ion irradiation using 3-D numerical simulation. The MCFET contains GAA and FinFET nanowire MOSFETs with ultra-thin, narrow channels (10 nm) and a 32 nm channel length. Recent simulation studies have shown that GAA MOSFET devices are less sensitive to single event transients (SET) than fully-depleted single gate SOI devices (Francis et al., 1995; Munteanu et al., 2006; Munteanu et al., 2007). This is due to the improved control of the channel potential which reduces floating body effects and improves the device immunity to single event transients. MCFET devices are then expected to show a very low sensitivity to ionizing radiation.

This chapter is organized as follows: after the presentation of the 3-D simulated structures and simulation models (sections II and III), we will analyze in section IV the effect of the ion strike on the main internal electrical parameters inside the structure and on the drain current transient. In this section we will compare the sensitivity to heavy ion irradiation of the MCFETs with that of other single and multiple-gate devices. Finally, in section V we will investigate the influence of the ion strike parameters (location, direction, and radius of the ion track) and of the lateral spacing between the nanowire stacks on the current transient and charge collection.

2. Description of simulated devices

The structure of the simulated MCFETs is a 3 x 3 nanowire matrix containing square cross-section nanowires. The description of the 3-D architecture considered here and the definition of the geometrical parameters are shown in Fig. 1. The MCFET matrix is composed of three parallel transistor stacks, each stack containing 3 vertically stacked nanowire devices (2 GAA and 1 FinFET). The MCFET is calibrated to fill the ITRS'2009 (ITRS, 2009) low-power (LP) requirements in terms of off-state current for the technology nodes corresponding to the year 2009 (I_{OFF} < 5×10^{-3} A/μm). The individual nanowire MOSFETs are designed with a 32 nm gate length, square cross-section with t_{Si}=W=10 nm, and a 3 nm-thick gate oxide. An intrinsic silicon film and a midgap gate are considered. Three lateral spacings, s, between the nanowire stacks are considered: 100, 75, and 50 nm.

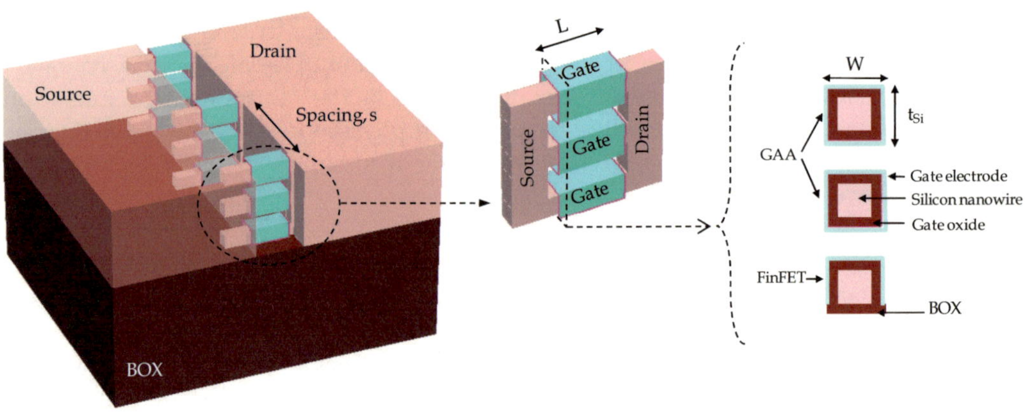

Fig. 1. Schematic description of the 3-D simulated MCFET structures and their main geometrical parameters. The MCFET matrix is composed of three parallel transistor stacks, each stack containing 3 vertically stacked nanowire devices (2 GAA and 1 FinFET). All nanowires have silicon film with square section (t_{Si}=W=10 nm). For a better view of the nanowires the gate material, spacers, isolation oxide and a part of the source and drain regions are not shown.

3. Description of the simulation approach

3-D numerical simulations have been performed with the DESSIS device simulator from the 3-D Sentaurus code (Sentaurus, 2009). The main models used in simulation are the doping-dependent Shockley-Read-Hall and Auger recombination models and the Fermi-Dirac carrier statistics. The model of the effective intrinsic density includes doping-dependent band-gap narrowing (Slotboom model (Sentaurus, 2009)) and a lattice temperature-dependent band gap. The hydrodynamic model was used for the carrier transport equations, including the energy balance equations for electrons, holes, and the lattice. The impact ionization and the carrier mobility models depend on carrier energy calculated with the hydrodynamic model. The mobility model includes dependencies on the lattice temperature, channel doping level and normal electric field through the Lombardi model (Sentaurus, 2009). In the following we succinctly describe the hydrodynamic transport model used in the present simulation approach.

3.1 Transport model

Historically, the first models used in carrier transport simulation described the physical phenomena taking place in the device as functions of the electric field, even if these phenomena depend on carrier energy (Selberrer, 1984). This is possible when considering that carrier energy is in permanent balance with the electric field. Carrier transport in MOSFET devices is mainly due to electrostatic potential gradients and/or gradients of carrier concentration (Selberrer, 1984). The current density in a biased device is then usually modeled by the sum of a conduction component (drift) and a diffusion component, as follows (for electrons):

$$J_n = q\mu_n nE + qD_n \nabla n \tag{1}$$

where q is the elementary charge, μ_n is the carrier mobility, D_n is the thermal diffusion coefficient, E is the electric field and n is the electron density. D_n and μ_n depend on material and electric field and are connected by the Einstein's equation:

$$D_n = \mu_n \frac{kT_L}{q} \tag{2}$$

with T_L the lattice temperature and k is the Boltzmann constant. Similar equations are considered for holes (see the paragraph "Drift-Diffusion" below).

This traditional description of electronic transport constitutes the "Drift Diffusion" (DD) model, the basic model used in CMOS devices simulation (Lundstrom, 2000; Selberrer, 1984). This modeling level is generally adapted for long devices, with either weak or strong electric fields (except for the modeling of impact ionization; see below in this paragraph). When the device feature size is reduced, the electronic transport becomes qualitatively different from the traditional transport model since the average carrier velocity does not depend on the local electric field. Average carrier velocity is a function of the carrier energy which depends on the variations in time and space of the electric field. In short devices, steep variations of electric field take place in the active area of the devices. Then, non-stationary phenomena (such as velocity overshoot (Baccarani & Wordeman, 1985; Jacoboni & Reggiani, 1983) occur following these rapid spatial or temporal changes of high electric fields. In small devices, non-stationary phenomena play an important role and may dominate the device operation. Since DD model neglects non-stationary effects, new advanced transport models become mandatory for accurate transport simulation in ultra-short devices (Apanovich et al., 1994; Blotekjaer, 1970; Stratton, 1962).

The DD model considers that carriers gain maximum energy instantaneously balanced with the electric field (Lundstrom, 2000). Consequently, non-stationary effects (velocity overshoot and carrier transport by thermal diffusion processes associated with electronic temperature gradients) specific to short devices are neglected in DD model, as well as the dependence of impact ionization on the carrier energy.

In reality, the carrier energy does not immediately respond to changes in electric field. Mobility and diffusion coefficients are tensor quantities that depend on several parameters besides electric field (Khanna, 2004). In ultra-short MOSFETs (deca-nanometre channel lengths), the high internal electric fields result in substantial electron heating. The hydrodynamic model, obtained by taking the first three moments of the Boltzmann Transport Equation (BTE), represents the carrier transport effects in short devices more accurately than the DD model. The hydrodynamic model is a macroscopic approximation to the BTE taking into account the relaxation effects of energy and momentum. In this model, the propagation of electrons in a semiconductor is treated as the flow of a charged, thermally conducting gas subjected to an electric field. This model removes several limiting assumptions of DD: the carrier energy can exceed the thermal energy and all physical parameters are energy-dependent. The current density and the energy flow are modelled in HD model by the following equations (given here for electrons (Sentaurus, 2009)):

$$\vec{J}_n = q\mu_n \left[-n\vec{\nabla}\phi + \frac{kT_n}{q}\vec{\nabla}n + \frac{k}{q}(1+\xi_n)n\vec{\nabla}T_n \right] \tag{3}$$

$$div\vec{S}_n = -\vec{J}_n\vec{\nabla}\phi - \frac{3k}{2}\frac{\partial}{\partial t}(nT_n) - W_n \tag{4}$$

$$\vec{S}_n = -K_n\vec{\nabla}T_n - \frac{k\Delta_n}{q}\vec{J}_n T_n \tag{5}$$

where T_n is the electron temperature, ξ_n is a model coefficient, $\vec{S}_n$ is the energy flow, W_n is the energy density loss rate, K_n is the thermal conductivity and:

$$\Delta_n = \frac{5}{2} + \xi_n \tag{6}$$

while the energy density loss W_n is given by:

$$W_n = \frac{3}{2}n\frac{k(T_n - T_L)}{\tau_{rel}} + \frac{3}{2}kT_n R_{SRH} + E_g(G_n - R_n{}^A) \tag{7}$$

where τ_{rel} is the energy relaxation time, R_{SRH} is the SRH recombination rate, G_n is the impact ionization rate, $R_n{}^A$ is the Auger recombination rate, E_g is the Silicon bandgap. Similar equations are used for holes. Usually, the mobility μ_n is modelled as a decreasing function of energy (because the scattering rate increases with the energy of the particle). Finally, the system of equations of the HD model is completed by the continuity equations:

$$div\vec{J}_n = qR + q\frac{\partial n}{\partial t} \tag{8}$$

$$div\vec{J}_p = -qR - q\frac{\partial p}{\partial t} \tag{9}$$

where R is the generation-recombination rate.

3.2 Modeling the effect of a particle strike

The radiation effects have been simulated using the HeavyIon module of the TCAD Sentaurus toolsuite (Sentaurus, 2009), considering an electron-hole pair column centred on the ion track axis to model the ion strike. The ion track structure to be used as input in simulation is presently a major issue for device simulation. The first representations included a simple cylindrical charge generation with a uniform charge distribution and a constant LET along the ion path. However, the real ion track structure is radial and varies as the particle passes through the matter. When the particle strikes a device, highly energetic primary electrons (called δ-rays) are released. They further generate a very large density of electron-hole pairs in a very short time and in a very small volume around the ion trajectory, referred as the ion track. These carriers are collected by both drift and diffusion mechanisms, and are also recombined by different mechanisms of direct recombination (radiative, Auger) in the very dense core track, which strongly reduces the peak carrier concentration. All these mechanisms modify the track distribution both in time and space. As the particle travel through the matter, it loses energy and then the δ-rays become less energetic and the electron-hole pairs are generated closer to the ion path. Then, the incident particle generates characteristic cone-shaped charge plasma in the device (Dodd, 2005).
The real ion track structure has been calculated using Monte-Carlo methods (Hamm et al., 1979; Martin et al., 1987; Oldiges et al., 2000). These simulations highlighted important differences between the track structure of low-energy and high-energy particles, even if the LET is the same (for details see (Dodd et al., 1998; Dodd, 2005)). High-energy particles are representative for ions existing in the real space environment, but they are not available in typical laboratory SEU measurements (Dodd, 1996). Then the investigation of the effects of high-energy particles by simulation represents an interesting opportunity, which may be difficult to achieve experimentally.
Analytical models for ion track structure have been also proposed in the literature and implemented in simulation codes. One of the most interesting models is the "non-uniform power law" track model, based on the Katz theory (Kobetich & Katz, 1968) and developed by Stapor (Stapor & McDonald, 1988). In this model, the ion track has a radial distribution of excess carriers expressed by a power law distribution and allows the charge density to vary along the track (Dussault et al., 1993). Other analytical models propose constant radius non-uniform track or Gaussian distribution non-uniform track.
In commercial simulation codes, the effect of a particle strike is taken into account as an external generation source of carriers. The electron-hole pair generation induced by the particle strike is included in the continuity equations via an additional generation rate. This radiation-induced generation rate can be connected to the parameters of irradiation, such as the particle Linear Energy Transfer (LET). The LET is the energy lost by unit of length (-dE/dl), which is expressed here in MeV cm²/mg (1pC/μm≈100 MeV cm²/mg). The particle LET can be converted into an equivalent number of electron-hole pairs by unit of length using the mean energy necessary to create an electron-hole pair (E_{ehp}) (Roche, 1999):

$$\frac{dN_{ehp}}{dl} = \frac{1}{E_{ehp}} \frac{dE}{dl} \tag{10}$$

where N_{ehp} is the number of electron-hole pairs created by the particle strike. By associating two functions describing the radial and temporal distributions of the created electron-hole pairs, the number of electron-hole pairs is included in the continuity equations (Munteanu & Autran, 2008) via the following radiation-induced generation rate:

$$G(w,l,t) = \frac{dN_{ehp}}{dl}(l) \cdot R(w) \cdot T(t) \tag{11}$$

where R(w) and T(t) are the functions of radial and temporal distributions of the radiation induced pairs, respectively. Equation (11) assumes the following hypothesis: the radial distribution function R(w) depends only on the distance traversed by the particle in the material and the generation of pairs along the ion path follows the same temporal distribution function in any point. Since function G must fill the condition:

$$\int_{w=0}^{\infty} \int_{\theta=0}^{2\pi} \int_{t=-\infty}^{\infty} G\, w\, dw\, d\theta\, dt = \frac{dN_{ehp}}{dl} \tag{12}$$

functions R(w) and T(t) are submitted to the following normalization conditions:

$$2\pi \int_{w=0}^{\infty} R(w)\, w\, dw = 1 \tag{13}$$

$$\int_{t=-\infty}^{\infty} T(t)\, dt = 1 \tag{14}$$

The ion track models available in commercial simulation codes usually propose a Gaussian function for the temporal distribution function T(t):

$$T(t) = \frac{e^{-(\frac{t}{t_C})^2}}{t_C \sqrt{\pi}} \tag{15}$$

where t_C is the characteristic time of the Gaussian function which allows one to adjust the pulse duration. The radial distribution function is usually modelled by an exponential function or by a Gaussian function:

$$R(w) = \frac{e^{-(\frac{w}{r_C})^2}}{\pi r_C^2} \tag{16}$$

where r_C is the characteristic radius of the Gaussian function used to adjust the ion track width. Previous works have demonstrated that the different charge generation distributions used for the radial ion track does affect the device transient response, but the variation is typically limited to ~5% for ion strikes on bulk p-n diodes (Dodd, 2005; Dussault et al., 1993). Considering a LET which is not constant with depth along the path has a more significant impact on the transient response in bulk devices. The key parameters of the single event transient (peak current, time to peak and collected charge) have up to 20% variation when LET is allowed to vary with depth compared to the case of a constant LET (Dussault et al., 1993). Nevertheless, the LET variation with depth has no influence on the transient response of actual SOI devices with thin Silicon film.

In the following simulations, two characteristic radii have been considered for the spatial Gaussian dependence of the ion track: 50 and 20 nm. The Gaussian time distribution (Eq. (15)) is centered on 10 ps and has a characteristic width of 2 ps. The linear energy transfer (LET) value is kept constant along the track. The ion strikes the middle of the channel between the source and the drain and perpendicular to the gate electrode, as shown in Fig. 2. The different ion strike locations considered in this work are schematically presented in Fig. 2.

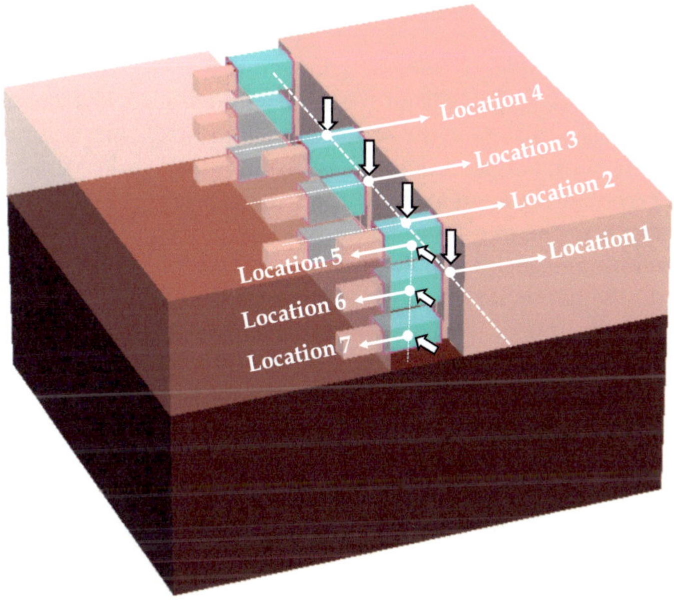

Fig. 2. Positions (arrows) of the ion strike considered in this work; the ion strikes in the middle of the channel (between the source and drain regions). For a better view of the nanowires, the gate material, spacers and isolation oxide are not shown.

Ion strike locations labelled "1", "2", "3", and "4" are parallel to the y-axis (perpendicular to the x-z plane) and the ion strike locations "5", "6", and "7" are parallel to the z-axis (perpendicular to the x-y plane). The lateral spacing between locations "1", "2", "3", and "4" are equal to s/2. The 3-D profile of heavy ion charge density generated in the structure is shown in Fig. 3 for the ion strike locations "1", "2", "3", and "4". The MCFET is biased in the off-state (V_G=0 V). The drain terminal is constantly biased at 0.8 V. The collected charge is derived by integrating the drain current over the transient duration.

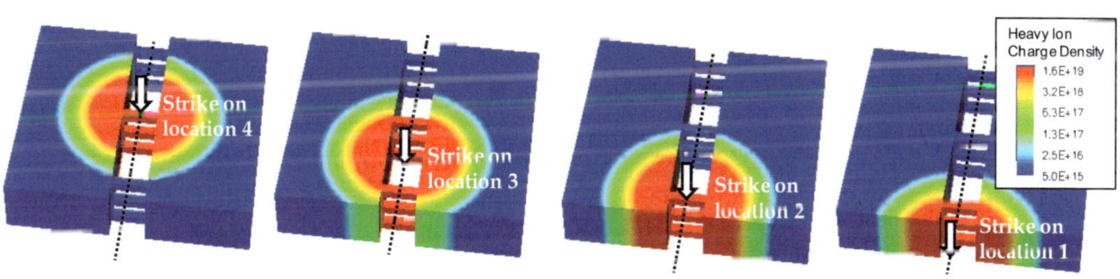

Fig. 3. 3-D profile of the heavy ion charge density for the ion strike locations "1", "2", "3", and "4". The positions (arrows) of the ion strike considered in this work are also shown; the ion strikes in the middle of the channel (between the source and drain regions). For a better view of the nanowires, the gate material, spacers and isolation oxide are not shown.

4. Transient simulation results

4.1 Potential and carrier density

Figure 4 shows the 3-D profiles of the electrostatic potential (Fig. 4(a)) and the electron density (Fig. 4(b)) in a 3-D MCFET with a lateral spacing s=100 nm for the ion strike location "2" at t=10 ps (maximum generated charge by the ion strike). The ion track radius considered here is equal to 50 nm. This figure shows that the electrostatic potential profile is perturbed by the ion strike, especially in the first nanowire stack that is struck directly. However, the impact is less visible in the second nanowire stack and is almost undetectable for the third nanowire stack (situated the farthest from the ion strike impact location). For a better view of the potential variation in the MCFET stacks, we plot in Fig. 5 the potential in a cutline along the x-axis in the middle of the channel at different times before and after the ion strike. Two devices are considered: the impacted transistor located in the first nanowire stack (Fig. 5(a)) and the nanowire symmetrically situated in the third nanowire stack (Fig. 5(b)). The variation with time of the potential in the impacted transistor indicates that the parasitic bipolar device is turning on. On the contrary, the potential in the third nanowire stack varies only very slightly, as shown in Fig. 5(b). This is due to the narrow ion track radius compared to the lateral spacing between nanowires.

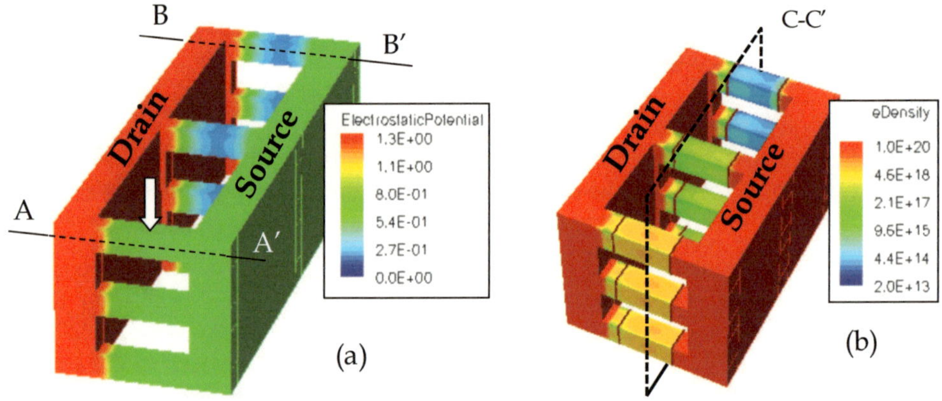

Fig. 4. 3-D profile of electrostatic potential (a) and of electron density (b) in the 3-D MCFET for the ion strike location "2" at t=10 ps. For a better view of the nanowires, the gate material, spacers, isolation oxide, and a part of the source and drain regions are not shown. The MCFET is biased in the off-state (V_G=0 V, V_D=0.8 V). The nanowire spacing is s=100 nm, the ion strike LET is 10 MeV/(mg/cm^2), and the ion track radius is 50 nm. The ion strike location is indicated by the arrow.

The 2-D profiles of the electron density in a cross-section in the MCFET (cut plane C-C′ indicated in Fig. 4(b)) are reported in Fig. 6 before and after the ion strike in location "2". These profiles give details concerning the distribution inside each nanowire. The electron density is centered in the middle of the film for the GAA devices, which is a typical feature of these devices where the gate is wrapped around the entire channel. In the off-state bias condition, carrier conduction in GAA is dominated by the volume inversion phenomenon (Munteanu & Autran, 2003): carriers flow from source to drain over the entire silicon film thickness. This is not the case for FinFET devices, where the electron density is not centered in the middle of the film. Figure 5(b) also shows that the electron density in the nanowires situated in the third stack is very slightly disturbed by the ion strike. Figure 6 also shows that the electron density is strongly enhanced in the first nanowire stack (the impacted stack) after the ion strike (for t=10 ps in Fig. 6) and decreases as long as the structure relaxes and the deposited charge is collected or recombined. In the same time, the electron density in the third nanowire stack is almost unchanged compared to the electron density before the ion strike. This confirms the above remarks concerning the variation of the electrostatic potential.

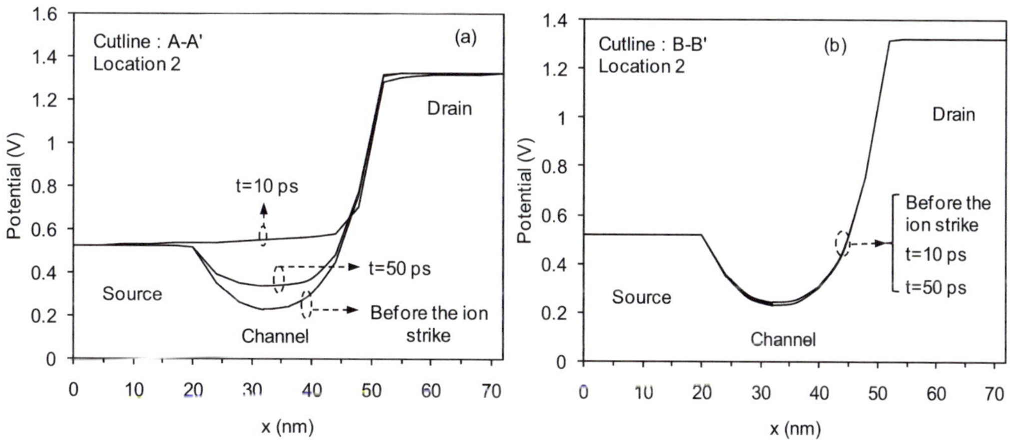

Fig. 5. 1-D potential profiles at different times in two cut-lines (indicated in Fig. 4(a)) along the x-axis in the middle of the nanowire: (a) cutline A-A′ in the impacted GAA nanowire and (b) cutline B-B′ in the GAA nanowire symmetrically situated in the third stack. The MCFET is biased in the off-state (V_G=0 V, V_D=0.8 V). The nanowire spacing is s=100 nm, the ion strike LET is 10 MeV/(mg/cm^2), and the ion track radius is 50 nm.

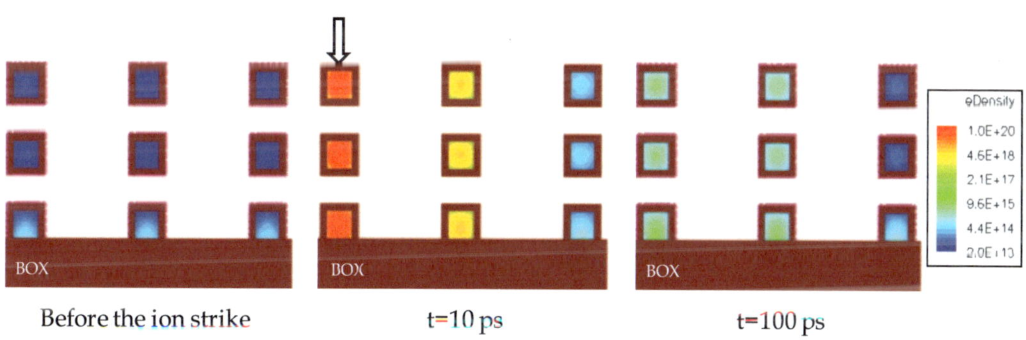

Fig. 6. 2-D profile of electron density along the cross-section C-C′ of Fig. 4(b) in the 3D MCFET before the ion strike, at t=10 ps and t=100 ps. The gate material is not shown. The ion strike location is indicated by the arrow. Other parameters as in Fig. 4.

4.2 Drain current density and charge collection

Figure 7 shows the drain current transient resulting from the ion strike in location "2". The variation of the collected charge with time is also reported. The charge collection is very fast in MCFETs, due to the GAA devices which have small active volumes that allow all the excess charge to be quickly evacuated. In addition, the GAA architecture allows the floating body-effects to be reduced, due to an excellent control of the body potential by the gate. It has been shown in (Munteanu et al., 2007) that the individual GAA device (with the same geometrical parameters as in the present work) has a total transient duration of 8 ps at 10% of the peak value for LET=10 MeV/(mg/cm^2). The single-event transient of the MCFET simulated here has a total transient duration of 10 ps at 10% of the peak value for a LET of 10 MeV/(mg/cm^2). This value is lower than that obtained in (Munteanu et al., 2006) for fully-depleted single-gate SOI devices with 50 nm gate length (where a total transient duration of 15 ps at 10% of the peak value is found). However, in (Munteanu et al., 2006) a LET of 30 MeV/(mg/cm^2) and a radius of 14 nm have been used. To facilitate the comparison, we considered here a fully-depleted single-gate SOI device having the same geometrical structure as the multiple-gate devices composing the MCFET. For this device, we have simulated the drain current transient for a LET of 10 MeV/(mg/cm^2) and a radius of 50 nm. A transient duration of 13.5 ps at 10% of the peak value has been found. The values obtained for MCFET are consistent with transient duration obtained in simulation in (Ferlet-Cavrois et al., 2005), but they are very low compared with those expected by extrapolation from simulations in (Dodd et al., 2004). This is probably due to the partially depleted structures used in (Dodd et al., 2004), whereas ultra-thin fully-depleted devices are considered here.

To resume, simulation results show that MCFET devices exhibit a quick charge collection, faster than that of fully-depleted/partially-depleted SOI devices with similar structure parameters. This short pulse width in MCFET devices could be interesting for single event transient hardening (Diehl et al., 1983; Dodd et al., 2004). From all these results, we could expect that MCFET devices being less sensitive to heavy ion irradiation than fully-depleted/partially-depleted SOI devices. This is mainly due to the excellent immunity to single-event effects of GAA and to the small active volume of individual nanowire devices.

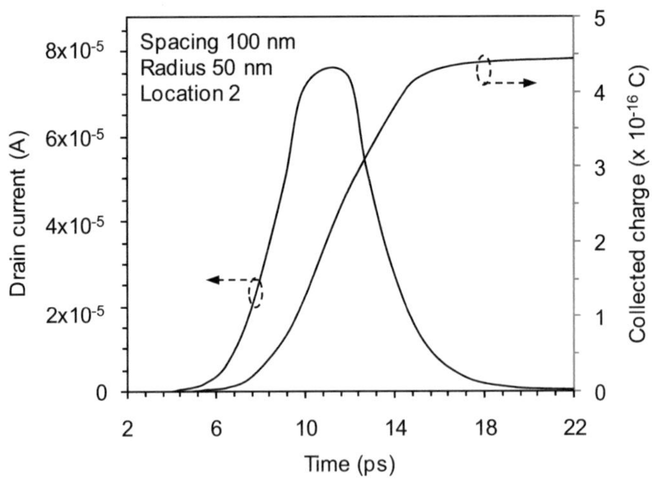

Fig. 7. Drain current transient and collected charge induced by an ion striking in the middle of the silicon film for the ion strike location "2". The MCFET is off-state biased (V_G=0 V, V_D=0.8 V). The nanowire spacing is s=100 nm, the LET value is 10 MeV/(mg/cm^2), and the ion track radius is 50 nm.

5. Impact of the ion strike parameters and lateral spacing between nanowires

5.1 Ion strike location and spacing between nanowire stacks

Drain current transients for a vertical ion strike and four ion strike locations are shown in Fig. 8(a) for a MCFET with a lateral spacing s=100 nm and an ion track radius of 50 nm. Figure 8(a) shows that for s=100 nm the ion striking in locations "2" and "4" produces identical drain current peaks and almost identical drain current transients. These locations correspond to a strike centered in the middle of a silicon nanowire, while locations "3" and "1" correspond to a strike between nanowires, on the isolation oxides. The drain current peak obtained for locations "2" and "4" is higher than that for location "3" which is higher than that for location "1". This is due to the higher deposited charge for locations "2" and "4" than for "3" and "1". These results are consistent with (Alles et al., 2005). It is interesting to note that for the considered lateral spacing (s=100 nm), which is large compared to the ion track radius (50 nm), the strikes centered on any nanowire (locations "2" or "4") produces almost the same current transient.

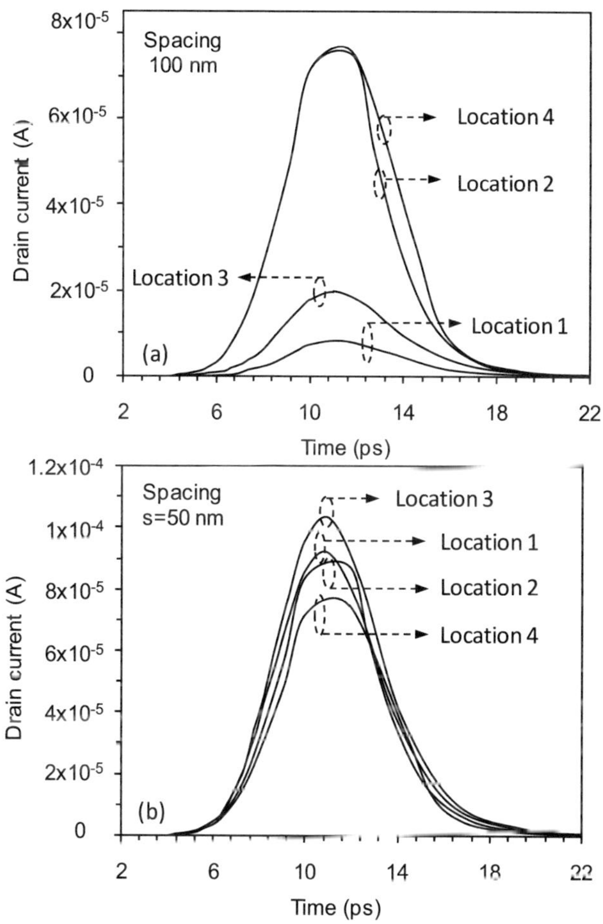

Fig. 8. Drain current transients induced by an ion striking vertically (parallel to y axis) in the middle of the structure. Four strike locations are considered ("1" to "4", as shown in Fig. 2). The MCFET is off-state biased (V_G=0 V, V_D=0.8 V). The lateral spacing is s=100 nm, the ion LET is 10 MeV/(mg/cm^2) and the ion track radius is 50 nm.

An interesting analysis concerns the influence on the drain current transient and collected charge of the lateral spacing between the nanowire stacks. Increasing the lateral spacing between nanowires (and keeping constant the ion track radius) will not change the results compared to those obtained for s=100 nm. The interesting case is when the lateral spacing is reduced and the ion track radius is kept constant. For a thorough investigation, we simulated two additional MCFET structures having lateral spacings of s=75 nm and s=50 nm, and we compare the results with those obtained for s=100 nm. The drain current transients produced by the ion strike are plotted in Fig. 8(b) for the ion strike locations "1" to "4" and for s=50 nm. In these figures the LET value is 10 MeV/(mg/cm²).

We can see that when reducing the lateral spacing from 100 nm to 50 nm, the drain current transients produced by the strikes centered in the middle of the nanowire are no longer identical (the drain transient peak for location "4" becomes lower than for location "2"). For this small spacing between nanowires, the ion strike between the nanowires (ion strike location "3") produces the highest current peak. These results are consistent with those of (Alles et al., 2005).

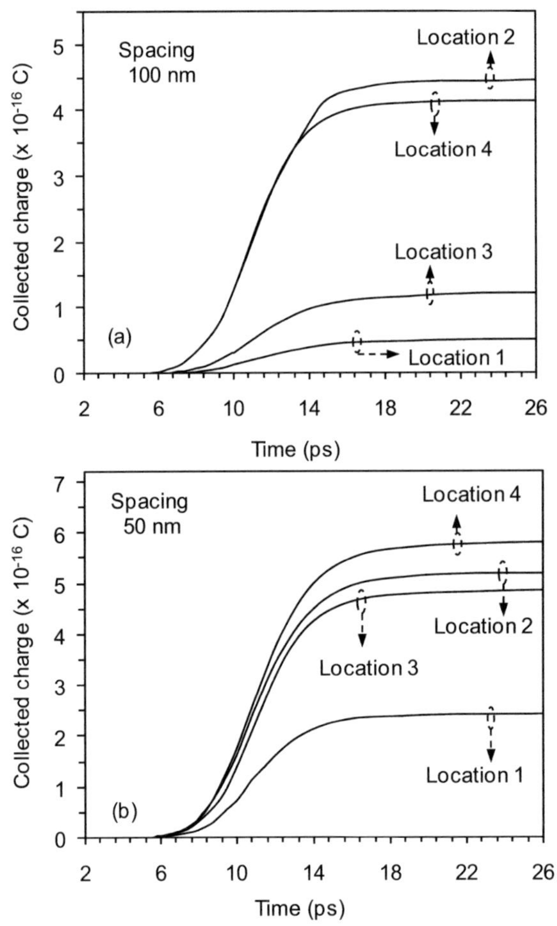

Fig. 9. Collected charge for the drain current transients shown in Fig. 8: (a) s=100 nm and (b) s=50 nm.

The collected charges extracted as function of time from the drain current transients presented above (Fig. 8) are shown in Fig. 9. For s=100 nm (Fig. 9(a)), the collected charge is slightly higher for location "2" than that for location "4". The strikes between nanowires give lower collected charges than the strikes on the silicon nanowires.

When the lateral spacing is reduced to s=50 nm (Fig. 9(b)), the collected charge of an ion striking at location "4" becomes higher than those for the other locations. Fig. 9(b) also shows that the collected charge for a ion striking between the silicon nanowires is enhanced when reducing the lateral spacing and becomes closer to that of strikes centered on the nanowire ("2" and "4") for the smallest spacing s=50 nm. For a better illustration of this point, we compare in Fig. 10 the drain current transients produced by the ion striking in locations "1" and "3" between the silicon nanowires, for the three lateral spacings (50, 75 and 100 nm). The collected charges corresponding to transients of Fig. 10 are plotted in Fig. 11. Figures 12 and 13 show the drain current transients and the corresponding collected charges for locations "2" and "4" for strikes centered on the nanowire.

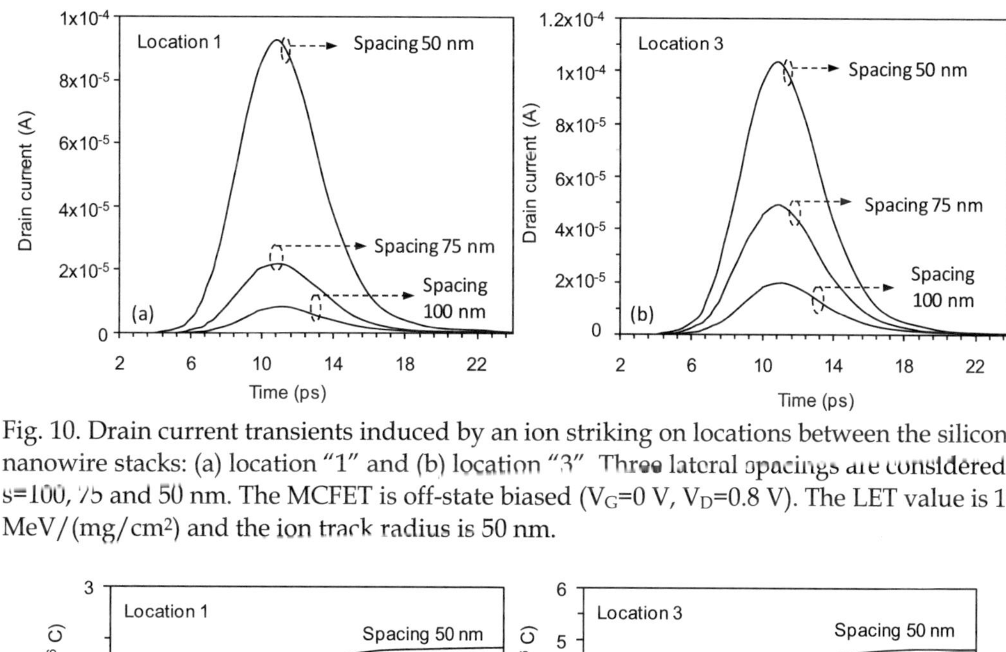

Fig. 10. Drain current transients induced by an ion striking on locations between the silicon nanowire stacks: (a) location "1" and (b) location "3". Three lateral spacings are considered s=100, 75 and 50 nm. The MCFET is off-state biased (V_G=0 V, V_D=0.8 V). The LET value is 10 MeV/(mg/cm^2) and the ion track radius is 50 nm.

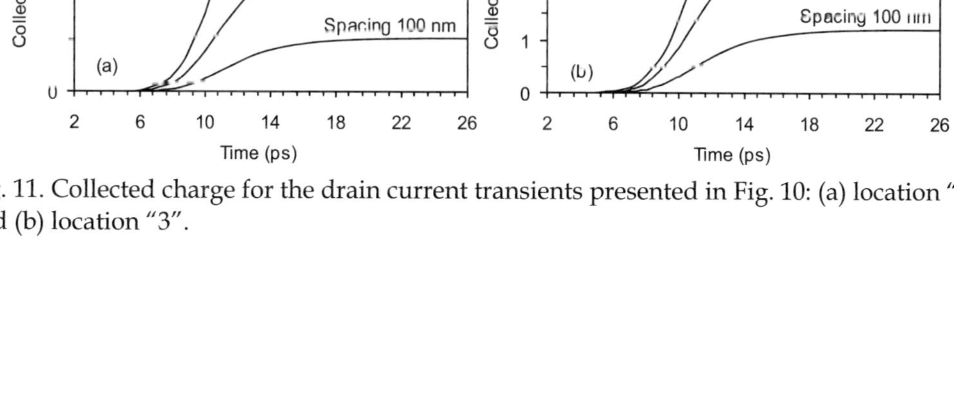

Fig. 11. Collected charge for the drain current transients presented in Fig. 10: (a) location "1" and (b) location "3".

Figures 10(a) and 10(b) indicate that for the locations "1" and "3" strike situated between the nanowires the drain current transients are different for the three lateral spacings. Both the current peak and the collected charge (Figs. 11(a) and 11(b)) increase when the lateral spacing decreases. Figure 12(a) shows that almost the same drain current transient is obtained for the location "2" for large spacing s=100 nm and s=75 nm. The transient peak is higher for s=50 nm than for the larger lateral spacings. For this location, the collected charge decreases when the lateral spacing is increased. For location "4" (Fig. 12(b)) the drain current transient is identical for the three locations; however the collected charge is higher for s=50 nm and is the same for s=75 nm and 100 nm.

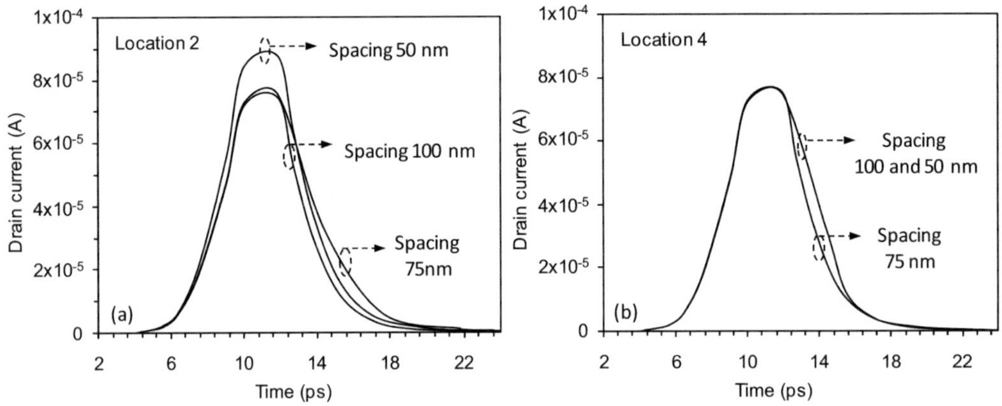

Fig. 12. Drain current transients induced by an ion striking on locations centered on the silicon nanowire: (a) location "2" and (b) location "4". Three lateral spacings are considered s=100, 75 and 50 nm. The MCFET is off-state biased (V_G=0 V, V_D=0.8 V). The LET value is 10 MeV/(mg/cm^2) and the ion track radius is 50 nm.

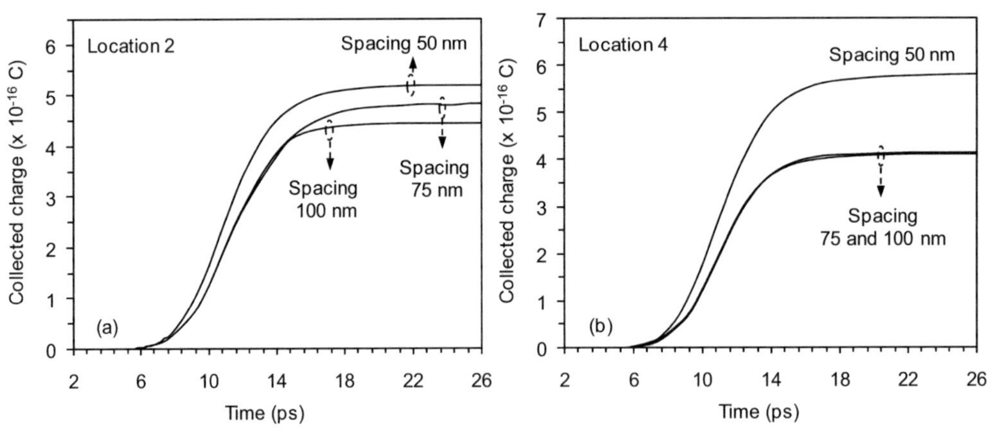

Fig. 13. Collected charge for the drain current transients presented in Fig. 12.

5.2 Ion strike direction and ion track radius

We analyze in this section the impact of the direction of the ion strike and the influence of the ion track radius. The results presented above have been obtained for ion striking vertically (parallel to y-axis). Then, we have simulated an ion horizontally striking (parallel to the z-axis) on the gate on locations "5", "6", and "7" shown in Fig. 2. Locations "5" and "6" represent a strike in the middle of the silicon GAA nanowire, perpendicular to the gate, while location "7" corresponds to an ion strike on the FinFET device. All these strikes are centered on the silicon nanowire. Figure 14 shows the drain current transients obtained for the three horizontal strikes described above. In this figure, the drain current transient for a vertical strike on location "2" is also reported for comparison. The results show that the transient peaks for horizontal strikes are higher than the peak obtained for a vertical strike on location "2". This is probably due to the higher deposited charge for a horizontal direction due to the large depth of the MCFET device in the z-direction.

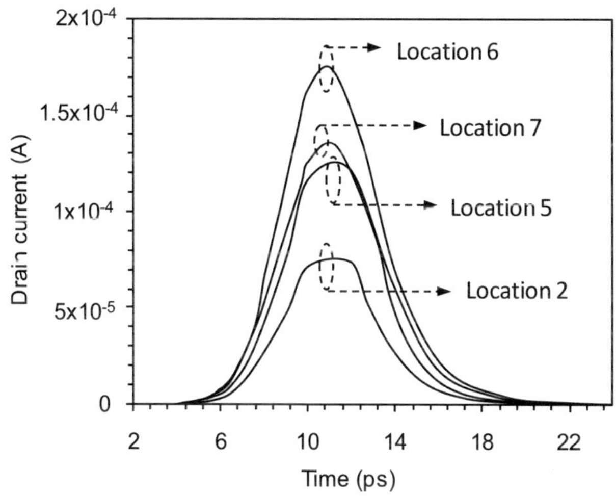

Fig. 14. Drain current transients induced by a horizontal ion strike (parallel to z-axis) in the middle of the silicon film. The drain current transient for a vertical strike on location "2" is also reported. The MCFET is off-state biased (V_G=0 V, V_D=0.8 V). The LET value is 10 MeV/(mg/cm^2), the lateral spacing is s=100 nm and the ion track radius is 50 nm.

For the horizontal direction of the ion strike previously considered, an interesting study concerns the impact of the ion track radius. We simulated the ion strike considering two ion track radii, 50 nm and 20 nm. Figure 15(a) shows the drain current transients obtained in these cases for an ion strike in location "6". Drain current transients obtained for ion strikes in locations "5", "6" and "7" for a radius of 20 nm are also shown in Fig. 15(b). The current peak is higher when considering a narrow radius, because more of the charge is localized in the nanowire with the smaller characteristic radius, instead of being generated in the wire-to-wire isolation. This is confirmed in Fig. 16, where the collected charge is plotted for different horizontal strike locations. The collected charge is higher for a 20 nm radius that for a 50 nm radius for the three locations of the ion strike. For a narrow radius (=20 nm) the lowest collected charge is obtained for an ion striking on the FinFET device, while for a large radius (=50 nm) the strike on the first GAA device of the stack gives the lowest collected charge. The highest charge is obtained for an ion striking the GAA device located in the middle of the vertical stack for both ion track radii considered here.

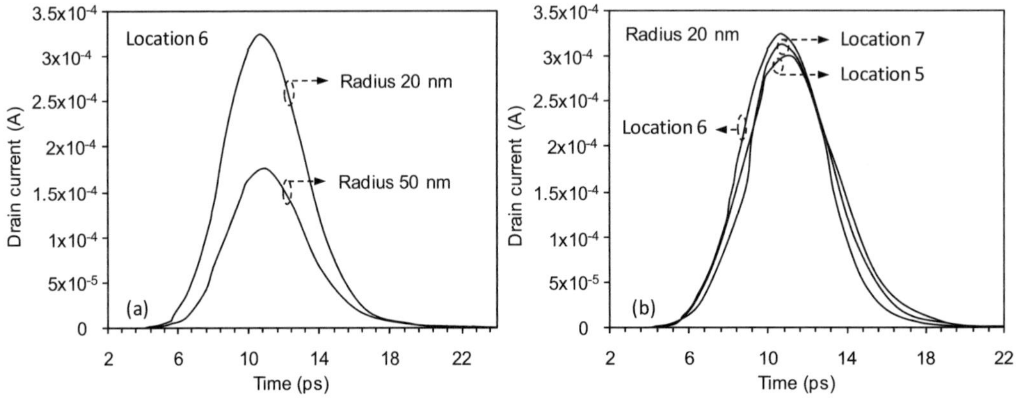

Fig. 15. Drain current transients induced for a horizontal ion strike (parallel to z axis) and two ion track radii. The MCFET is off-state biased (V_G=0 V, V_D=0.8 V). The LET value is 10 MeV/(mg/cm^2) and the lateral spacing between nanowires is s=100 nm.

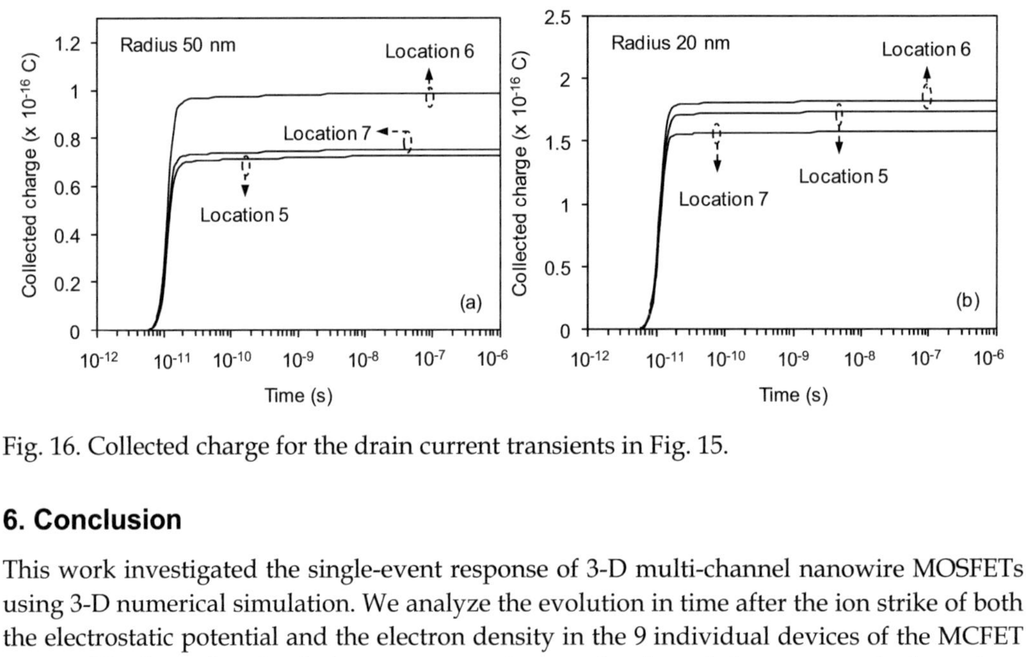

Fig. 16. Collected charge for the drain current transients in Fig. 15.

6. Conclusion

This work investigated the single-event response of 3-D multi-channel nanowire MOSFETs using 3-D numerical simulation. We analyze the evolution in time after the ion strike of both the electrostatic potential and the electron density in the 9 individual devices of the MCFET matrix. We show that the drain current transients and collected charge strongly depend on the ion strike location, direction and track radius. The lateral spacing between adjacent nanowire stacks is found to be a key-parameter in the analysis of the worst case location of the ion strike. We show that for a large lateral spacing between stacks compared with the ion track radius, the strikes centered on any nanowire produces almost the same current transients. In this case the transient peak is higher than that obtained for a strike between nanowires on the isolation oxide. On the contrary, for a small lateral spacing, comparable to the ion track radius, the highest current peak is obtained for a strike between the nanowires.

However, the highest collected charge is obtained for the strike on the nanowire situated on the center of the MCFET matrix. Finally, our results show that the charge collection is very fast in a MCFET for all ion strike parameters and configurations. This is due to the multiple-gate devices which have small active volumes that allow all the excess charge to be quickly evacuated. The MCFET simulated here has a total transient duration of 10 ps at 10% of the peak value for LET=10 MeV/(mg/cm^2), which is almost identical to that of a individual GAA device, but lower than that obtained in simulation of fully-depleted single-gate SOI devices. From these results, we could expect a better immunity to single-event phenomena of MCFETs compared to other conventional structures, such as fully-depleted/partially-depleted SOI devices. This will probably have a consequence on the behaviour under irradiation of circuits based on these devices. However, the single device behaviour is not enough to determine the circuit sensitivity to single-events because this also depends on the load capacitance. Since the single-event transients of MCFETs are high-bandwidth, they are very sensitive to inductive and reactive capacitance (i.e., node loading) in the circuit. More detailed study concerning this point is needed to exactly quantify the sensitivity to single-event of MCFETs-based circuits.

7. References

Alles, M.L.; Ball, D. R.; Massengill, L.W.; Schrimpf, R.D.; Warren, K.M. & Weller, R.A (2005). Considerations for Single Event Effects in Non-Planar Multi-Gate SOI FETs. *Proceedings of IEEE International SOI Conference*, pp. 191-193, Oct. 2005, IEEE.

Apanovich, Y.; Lyumkis, E.; Polski, B.; Shur, A. & Blakey, P. (1994). Steady-State and Transient Analysis of Submicron Devices Using Energy Balance and Simplified Hydrodynamic Models. *IEEE Transactions on Computer-Aided Integrated Circuits and Systems*, Vol. 13, No. 6, (Jun. 1994) pp. 702-711.

Baccarani, G & Wordeman M.R. (1985). An Investigation of Steady State Velocity Overshoot in Silicon. *Solid-State Electronics*, Vol. 28, no. 4, (Apr. 1985) pp. 407-416.

Baumann, R. C. (2005). Radiation Induced Soft Errors in Advanced Semiconductor Technologies. *IEEE Transactions on Device Material Reliability*, Vol. 5, no. 3, (Sept. 2005) pp. 305-316.

Bernard, E.; Ernst, T.; Guillaumot, B.; Vulliet, N.; Garros, X.; Maffini-Alvaro, V.; Andrieu, F.; Barral, V.; Allain, F.; Toffoli, A.; Vidal, V.; Delaye, V.; Vizioz, C., Campidelli, Y.; Kermarrec, O.; Hartmann, J.M.; Borel, S.; Faynot, O.; Souifi, A.; Coronel, P.; Skotnicki, T. & Deleonibus, S. (2007). Impact of the gate stack on the electrical performances of 3D Multi-Channel MOSFET (MCFET) on SOI. *Proceedings of ESSDERC Conference*, pp. 147-150, Sept. 2007, IEEE.

Blotekjaer, K. (1970). Transport Equations for Electron in Two-Valley Semiconductors. *IEEE Transactions on Electron Devices*, Vol. ED-17, No. 1, (Jan. 1970) pp. 38-47.

Diehl, S.E.; Vinson, J.E.; Shafer, B.D. and Mnich, T.M. (1983). Considerations for single event immune VLSI logic. *IEEE Transactions on Nuclear Science*, Vol. NS-30, No. 6, (Dec. 1983) pp. 4501-4507.

Dodd, P. E. (1996). Device Simulation of Charge Collection and Single-Event Upset. *IEEE Transactions on Nuclear Science*, Vol. 43, No. 2, (1996) pp. 561-575.

Dodd, P. E.; Musseau, O.; Shaneyfelt, M. R.; Sexton, F. W.; D'hose, C.; Hash, G.L.; Martinez, M.; Loemker, R. A.; Leray, J.-L. & Winokur, P. S. (1998). Impact of ion energy on single-event upset. *IEEE Transactions on Nuclear Science*, Vol. 45, No. 6, (Dec. 1998) pp. 2483–2491.

Dodd, P.E. & Massengill, L.W. (2003). Basic mechanisms and modeling of single-event upset in digital microelectronics. *IEEE Transactions on Nuclear Science*, Vol. 50, No. 3, (Jun. 2003) pp. 583–602.

Dodd, P. E.; Shaneyfelt, M. R.; Felix, J. A. & Schwank, J. R. (2004). Production and propagation of single-event transients in high-speed digital logic ICs. *IEEE Transactions on Nuclear Science*, Vol. 51, No. 6, (Dec. 2004) p. 3278-3284.

Dodd, P. E. (2005). Physics-Based Simulation of Single-Event Effects. *IEEE Transactions on Device Material Reliability*, Vol. 5, No. 3, (Sept. 2005) pp. 343-357.

Dussault, H.; Howard, Jr., J. W.; Block, R.C.; Pinto, M. R.; Stapor, W. J. & Knudson, A. R. (1993). Numerical simulation of heavy ion charge generation and collection dynamics. *IEEE Transactions on Nuclear Science*, Vol. 40, No. 6, (1993) pp. 1926–1934.

Ernst, T.; Dupré, C.; Isheden, C.; Bernard, E.; Ritzenthaler, R.; Maffini-Alvaro, V.; Barbé, J.-C.; De Crecy, F.; Toffoli, A.; Vizioz, C.; Borel, S.; Andrieu, F.; Delaye, V.; Lafond, D.; Rabillé, G.; Hartmann, J.-M.; Rivoire, M.; Guillaumot, B.; Suhm, A.; Rivallin, P.; Faynot, O.; Ghibaudo, G. & Deleonibus, S. (2006). Novel 3D integration process for highly scalable Nano-Beam stacked-channels GAA (NBG) FinFETs with HfO2/TiN gate stack. *Proceedings of IEDM Technical Digest*, pp. 997-1000, Washington, USA, Dec. 2006, IEEE.

Ferlet-Cavrois, V.; Paillet, P.; McMorrow, D.; Torres, A.; Gaillardin, M.; Melinger, J. S.; Knudson, A. R.; Campbell, A. B.; Schwank, J. R.; Vizkelethy, G.; Shaneyfelt, M. R.; Hirose, K.; Faynot, O.; Jahan, C. & Tosti, L. (2005). Direct Measurement of Transient Pulses Induced by Laser Irradiation in Deca-Nanometer SOI Devices. *IEEE Transactions on Nuclear Science*, Vol. 52, No. 6, (Dec. 2005) pp. 2104-2113.

Fischetti, M. V. & Laux, S. E. (2001). Long-Range Coulomb Interactions in Small Si Devices. Part I: Performance and Reliability. *Journal of Applied Physics*, Vol. 89, No. 2, (2001) pp. 1205–1231.

Francis, P.; Colinge, J.P. & Beger, G. (1995). Temporal Analysis of SEU in SOI/GAA SRAMs. *IEEE Transactions on Nuclear Science*, Vol. 42, No.6, (Dec. 1995) pp. 2127-2137.

Gusev, E. P.; Narayanan, V. & Frank, M. M. (2006). Advanced high-k dielectric stacks with polySi and metal gates: Recent progress and current challenges. *IBM Journal of Research and Development*, Vol. 50, No. 4/5, (2006) pp. 387–410.

Hamm, R. N.; Turner, J. E.; Wright, H. A. & Ritchie, R. H. (1979). Heavy ion track structure in Silicon. *IEEE Transactions on Nuclear Science*, Vol. 26, No. 6, (1979) pp. 4892–4895.

Houssa, M. (2004). *Fundamental and Technological Aspects of High-k Gate Dielectrics*, Institute of Physics, London.

ITRS 2009. International Technology Roadmap for Semiconductors. Available online: http://public.itrs.net.

Jacoboni, C. & Reggiani, L. (1983). The Monte Carlo method for the solution of charge transport in semiconductors with applications to covalent materials. *Review of Modern Physics*, Vol. 55, No. 3, (July 1983) pp. 645–705.

Khanna, V.K. (2004). Physics of carrier-transport mechanisms and ultra-small scale phenomena for theoretical modelling of nanometer MOS transistors from diffusive to ballistic regimes of operation. *Physics Reports*, Vol. 398, (2004) pp. 67–131.

Kobetich, E. J. & Katz, R. (1968). Energy Deposition by Electron Beams and δ Rays. *Physical Review*, Vol. 170, No. 2, (1968) pp. 391-396.

Lundstrom, M. (2000). *Fundamentals of Carrier Transport*. Cambridge University Press, USA.

Martin, R. C.; Ghoniem, N. M.; Song, Y. & Cable, J. S. (1987). "The size effect of ion charge tracks on single event multiple-bit upset", *IEEE Transactions on Nuclear Science*, Vol. 34, No. 6, (1987) pp. 1305–1309.

Munteanu, D. & Autran, J.L. (2003). Two-dimensional Modeling of Quantum Ballistic Transport in Ultimate Double-Gate SOI Devices. *Solid State Electronics*, Vol. 47, No. 7, (2003) pp. 1219-1225.

Munteanu, D.; Ferlet-Cavrois, V.; Autran, J.L.; Paillet, P.; Baggio, J.; Faynot, O.; Jahan, C. & Tosti, L. (2006). Investigation of Quantum Effects in Ultra-Thin Body Single- and Double-Gate Devices Submitted to Heavy Ion Irradiation. *IEEE Transactions on Nuclear Science*, Vol. 53, No. 6, (Dec. 2006) pp. 3363-3371.

Munteanu, D.; Autran, J.L.; Ferlet-Cavrois, V.; Paillet, P.; Baggio, J.; & Castellani, K. (2007). 3-D Quantum Numerical Simulation of Single-Event Transients in Multiple-Gate Nanowire MOSFETs. *IEEE Transactions on Nuclear Science*, Vol. 54, No. 4, (Aug. 2007) pp. 994-1001.

Munteanu, D. & Autran, J.L. (2008). Modeling of digital devices and ICs submitted to transient irradiations. *IEEE Transactions on Nuclear Science*, Vol. 55, no. 4, (Aug. 2008) pp. 1854-1878.

Oldiges, P.; Dennard, R.; Heidel, D.; Klaasen, B.; Assaderaghi, R. & Ieong, M. (2000). Theoretical determination of the temporal and spatial structure of α-particle induced electron-hole pair generation in Silicon. *IEEE Transactions on Nuclear Science*, Vol. 47, No. 6, (Dec. 2000) pp. 2575–2579.

Park, J.T. & Colinge, J.P. (2002). Multiple-gate SOI MOSFETs: device design guidelines. *IEEE Transactions on Electron Devices*, Vol. 49, No. 12, (Dec. 2002) pp. 2222-2229.

Rim, K.; Hoyt, J. L. & Gibbons, J. F. (1998). Transconductance Enhancement in Deep Submicron Strained-Si 12-MOSFETs. *Proceedings of IEDM Technical Digest*, pp. 707–710, Washington, USA, Dec. 1998, IEEE.

Roche, P. (1999). Etude du basculement induit par une particule ionisante dans une mémoire statique en technologie submicronique, Phd Thesis, 1999, in french.

Sentaurus (2009). Sentaurus TCAD Manuals, Synopsis, 2009.

Selberrer, S. (1984). *Analysis and Simulation of Semiconductor Devices*. Modular Series on Solid State Devices, USA.

Stapor, W. J. & McDonald, P. T. (1988). Practical approach to ion track energy distribution. *Journal of Applied Physics*, Vol. 64, No. 9, (1988) pp. 4430-4434.

Stratton, R. (1962). Diffusion of Hot and Cold Electrons in Semiconductor Barriers. *Physical Review*, Vol. 126, no. 6, (1962) pp. 2002-2013.

Taur, Y.; Buchanan, D.; Chen, W.; Frank, D.; Ismail, K.; Lo, S.-H.; Sai-Halasz, G.; Viswanathan, R.; Wann, H.-J. C.; Wind, S. & Wong, H.-S. (1997). CMOS scaling into the nanometer regime. *Proceedings of IEEE*, Vol. 85, (1997) pp. 486–504.

11

Co/Cu Nanowire Systems for GMR Sensing Applications

Daniele Pullini[1], David Busquets Mataix[2] and Alessio Tommasi[1,2]
[1]Centro Ricerche Fiat,
[2]Universidad Politécnica de Valencia,
[1]Italy
[2]Spain

1. Introduction

Since the discovery of giant magnetoresistance (GMR) in 1988 by Albert Fert and Peter Grünberg many companies have sought to develop practical applications for this intriguing effect of magnetic multilayer stacks. The applications of magnetic multilayer stacks are diverse, but in general they can serve as: magnetic field sensors (Simonds, 1995; Grünberg et al., 1986; Saurenbach et al., 1988; Baibich et al., 1988), non-volatile memories (Manalis et al., 1995; Chou et al., 1994), or variable resistors when small magnetic fields have to be measured. The first GMR sensor commercially available was introduced in 1995 by NVE Corporation. Since then, users can rely on a large variety of devices exploiting GMR: analogue and digital sensors, switches, gradient transducers for contactless positioning systems, e.g. gear tooth and encoder applications. Today, it is generally acknowledged that GMR sensors outperform the competing technologies such as AMR and Hall virtually in every application, and often at a significantly lower cost. The conventional GMR stacks are produced by vacuum deposition technology; more specifically, by this technique, a large variety of thin film materials are serially deposited, within a single vacuum chamber, one on top of another, to form multilayer stacks of tailored magnetic properties. The need to accurately control the nanometric thickness of each film, their extremely low roughness, and their purity do majorly contribute to the final cost of this type of systems. For this reason, the use of these devices, especially in low-margin markets such as the automotive, is still prohibitive for a number of potential field applications. In this context, alternative processes are envisaged to lower the final cost of GMR sensors. To this end, the pulsed electrodeposition into nanoporous templates (hereinafter template pulsed electrodepostion, TPED) is presented here as a competitive alternative to mass produce GMR sensors in the form of arrays of multilayer nanowires. As being a promising candidate for a low-cost production of sensors tailored for the aforementioned field uses (Pullini et al., 2007a), Co/Cu multilayer nanowire arrays fabricated by TPED (Fert & Piraux, 1999; Piraux et al., 1994) are discussed in this chapter. Specifically, the present chapter aims at presenting the criteria to model, design, develop Co/Cu-multilayer nanowire arrays, and discuss their potential exploitation perspectives to be used as magnetic field sensors.

In a first part of this chapter a quasi-analytical analysis (QAD) of cylindrical Co/Cu/Co tri-layer systems is detailed, and the equations drew to calculate the critical fields, the linearity,

and the eccentricity of their re-magnetization characteristics are reported. Systems differing by diameter, easy-axis, thickness of each layer employed are classified accordingly in order to provide to an end-user with the most significant choice parameters to fulfill a custom application. Micromagnetic simulation is then used to corroborate the QAD predicted properties for a large plurality of elementary trilayers and to simulate the magnetic behaviour of selected interacting trilayers both when piled up to form nanowires (up to 50 Co/Cu bilayers) and arranged in parallel, on the same plane, to mimic nanowire arrays (5 x 3 array).
The second part of the chapter is focused on TPED; in particular, this technique is here described and the most important parameters allowing the tailoring of the system characteristics are highlighted and their variability ranges detailed and linked to morphological, structural and magnetic properties. In particular, from a suitable electrolyte composition, the electrode potential, the current density, and the deposition temporization to be used in an pulsing regime to grow Co and Cu thin layers within nanoporous templates are reported. Always in this section, a description of the best practice to get large quantities of filled nanopores per area as well as large current efficiency is given.

2. Quasi-analytical modeling of Co/Cu/Co trilayer systems

The aim of the present section is to propose a rational for designing and classifying, as a function of selected parameters, cylindrical Co/Cu/Co trilayer systems of different diameters to have an insight into the properties of more complex nanowire based systems to be used for building magnetic field sensors (Ohgai et al., 2003). It is important to clarify here that the Co/Cu/Co trilayer has been considered instead of the Co/Cu nanowire's elementary-cell, that is the bilayer, in order to take into account the contribution of the interlayer exchange coupling energy to the system's total energy, which does not exist in an elementary bilayer.
To design such a system one can rely on different methods: phenomenological modeling from the experimental characterization and testing of physical samples, analytical modeling, and micromagnetic simulation, by which the rotation of the single dipoles can be described. Among these approaches, experimental is time consuming, theory is usually limited to verifying the correctness of the assumptions, while simulation is a helpful tool to be used when theory gives ambiguous responses (e.g. saddle points on the system's energy surface) and the experimental measures are not fully repeatable due to the size and shape dispersion of the nano-object samples fabricated very often in few units. According to the device addressed, whether being a sensor, a memory cell, or a magnetometer, the response to an applied magnetic field of an elementary Co/Cu/Co trilayer system has to be predicted with satisfactory confidence to justify the massive experimental workload typically needed for their development.
In principle, the responses of these systems have to be very different from one another according to the diverse applications; in fact, if a memory element is targeted a bistable hysteresis loop is desirable, while if a magnetic field sensor is wanted, one prefers to deal with a symmetric GMR and a linear analogue response within a defined dynamic range of operation. Generally speaking, elementary Co/Cu/Co trilayer systems show a symmetric GMR when they are in the parallel states of magnetization at the saturation field and their remanence is zero in the absence of a magnetic field applied. Differing in geometry parameters and material characteristics, the number of diverse systems exhibiting such a behaviour is huge; therefore, since the present chapter specifically refers to sensing

applications, to circumscribe the investigation, those systems whose first magnetization curve starts from zero in the absence of an applied field were searched for – these systems satisfy what is referred here as zero-field condition (hereinafter ZFC). The ZFC as well as the hysteresis characteristics of these systems depends on many factors such as the diameter (d), the thicknesses (t_{Cu}, t_{Co}), and crystal orientations of the separate layers. In the present work, in order to find the most useful systems, the energy problem of cylindrical Co/Cu/Co trilayers as a function of the magnetocrystalline-anisotropy easy-axis direction (hereinafter a), d, t_{Cu}, and t_{Co}, and their dynamic response to a changing applied field is studied. Specifically, to solve the energy problem in the ZFC, and to estimate the coercivity and saturation fields of cylindrical Co/Cu/Co trilayers a quasi-analytical design (QAD) approach was defined and adopted instead of running time consuming micromagnetic simulation. In the frame of the QAD, the demagnetization energy values of the elementary trilayer when in the fundamental configurations of magnetization (when the single layers' magnetizations are parallel/antiparallel to one another and to the trilayer plane respectively) were calculated numerically and used to interpolate the analytical expression of the demagnetization energy as a part of the analytical expression of the system's total energy. When the systems satisfying the ZFC were found, and their critical fields calculated in this way, the micromagnetic simulation was carried out for these selected systems to verify the QAD predictions and to know the systems' dynamic behaviour. As a result, a systemic data framework was completed to aid the design of custom Co/Cu/Co trilayer systems for the diverse sensing applications.

The prediction of the critical fields in the magnetization process results from the analysis of the total energy of the system. The energies involved in the Co/Cu/Co trilayer system are exchange, Zeeman, demagnetization, magnetocrystalline anisotropy, and RKKY interlayer exchange coupling energy. In order to find the systems which satisfy the ZFC, as well as to predict their critical fields, the system's total energy is defined by assuming that the-least-energy states occur when the magnetic moments belonging to each magnetic layer are lined up. This assumption leads to:

$$E_{ex} = 0. \tag{1}$$

The Zeeman energy is given by:

$$E_z = M_s H V_{Co}\left[(\mathbf{m_1}\cdot\mathbf{h}) + (\mathbf{m_2}\cdot\mathbf{h})\right], \tag{2}$$

where $M_s\mathbf{m_1}$ and $M_s\mathbf{m_2}$ are the magnetizatio n of the top and bottom layer respectively and $H\mathbf{h}$ is the applied magnetic field considered uniform over the system volume. The magnetocrystalline anisotropy of cobalt is uniaxial in most the fabricated systems reported in the literature (Sun et al., 2005) and it can be written as follows:

$$E_{anis} = K_1 V_{Co}\left(|\mathbf{m_1}\times\mathbf{a}|^2 + |\mathbf{m_2}\times\mathbf{a}|^2 \right), \tag{3}$$

where K_1 is the anisotropy constant and V_{Co} is the volume of the cobalt layer. Two magnetic layers spaced by a thin non-magnetic layer experience RKKY interlayer exchange coupling, which is described by the following energy term:

$$E_{IEC} = I_{12}(t_{Cu}) 2 A_{Co}(\mathbf{m_1}\cdot\mathbf{m_2}), \tag{4}$$

where A_{Co} is the area of the Co/Cu interface and I_{12} is the interlayer coupling constant (Bruno & Chappert, 1991) and is dependent on the thickness and orientation of the spacer layer (copper in the present work). The demagnetization energy is given by:

$$E_d = -\iiint_{V_{Co}} \mathbf{M} \cdot \iint_{A_{Co}} \mathbf{M} \cdot \mathbf{n} \frac{\mathbf{r} - \mathbf{r}'}{\left|\mathbf{r} - \mathbf{r}'\right|^3} d^2\mathbf{r}' d^3\mathbf{r}. \tag{5}$$

This expression can be approximated empirically by fitting with a linear function the values, calculated numerically, of the demagnetization energy of the system when in four fundamental states of magnetization. As a result, the following expression was found:

$$E_{demag} = \frac{1}{2}\left[\Delta E_{d_{||}} \mathbf{m}_{1_{||}} \cdot \mathbf{m}_{2_{||}} + \Delta E_{d_{\perp}} \mathbf{m}_{1_{\perp}} \cdot \mathbf{m}_{2_{\perp}} + \left(\Sigma E_{d_{\perp}} - \Sigma E_{d_{||}}\right)\left(\left|\mathbf{m}_{1_{\perp}}\right|^2 + \left|\mathbf{m}_{2_{\perp}}\right|^2\right)\right], \tag{6}$$

where $\mathbf{m}_{i||}$ and $\mathbf{m}_{i\perp}$ denote respectively the in-plane and perpendicular-to-plane components of $\mathbf{m}_i$ and $\Delta E_{d||} \equiv E_{d,P||} - E_{d,AP||}$, $\Delta E_{d\perp} \equiv E_{d,P\perp} - E_{d,AP\perp}$, $\sum E_{d\perp} \equiv E_{d,P\perp} + E_{d,AP\perp}$, and $\sum E_{d||} \equiv E_{d,P||} + E_{d,AP||}$ are coefficients linked to the demagnetization energy of a trilayer system when the Co layers are in the parallel/antiparallel configurations of magnetization, here indexed P/AP, and when their magnetizations are both parallel/orthogonal to the trilayer plane, here indexed ||/⊥. The coefficients aforementioned depend on the dimensions of the layers as they were interpolated from the calculated values of e_d, the energy per magnetic volume, as a function of t_{Co} / d and t_{Cu} / d. It was supposed that the system's layers are parallel to the *xy*-plane and the applied field is lined up along the *x* direction. The magnetocrystalline easy-axis, which is directly related to the crystal phase of the magnetic layers, can lie, in principle, along any direction depending on the lattice orientation of the deposited cobalt, and the system's hysteresis loop reflects a clear uniaxial-anisotropy behaviour. A simple test of correlation between the quasi-analytical predictions and the micromagnetic simulations was carried out at zero field applied; in particular, the evolution towards the equilibrium state of the systems which satisfies, and do not satisfy, the ZFC were simulated and verified. In this context, the system was assumed to be unmagnetized as initial condition and let evolve to the least energy state. To predict the equilibrium state of these systems at zero field we evaluated in which state of magnetization the system's total energy is the smallest. The simplest case is when the magnetocrystalline easy-axis is lined up along the same direction of the demagnetization one (here the *x* direction), therefore, only the P and AP in-plane magnetizations are candidates. In particular, the system is in the P state if:

$$4\, A_{Co} I_{12} > \Delta E_{d_{||}}, \tag{7}$$

meaning when the interlayer exchange coupling is sufficiently ferromagnetic to overcome the demagnetization, the latter tending to cause AP configuration. From equation (7), the parameters defining the geometry of the trilayer systems which satisfy the ZFC can be calculated. In a next step of our investigation the magnetization states of the systems which satisfy the ZFC were calculated in the presence of increasing applied magnetic field in the *x* direction. Then the equation $dE/d\mathbf{m}_1 = 0$ was solved in order to localize the systems' energy minima; as a result, it was found that the top layer flips either exclusively in the *xy*-plane (||) or exclusively in the *xz*-plane (⊥) when the applied field is sufficiently intense. The flip fields: $H_{||}$, $H_{\perp}$ (relative to both the flip directions) can be calculated by solving: $d^2E / d\mathbf{m}_1^2 = 0$; it should not astonish that the flip process which occurs correspond to an applied field equal to $H_c = \min(H_{||}, H_{\perp})$, where $H_{||}$ and $H_{\perp}$ are solutions of: $d^2E / d\mathbf{m}_{1||}^2 = 0$ and $d^2E / d\mathbf{m}_{1\perp}^2 = 0$, and they are respectively:

$$H_{||} = \frac{\Delta E_{d_{||}} + 4 A_{Co} I_{12} + 4 K_1 V_{Co}}{2 M_s V_{Co}}, \tag{8}$$

$$H_{\perp} = \frac{\left(E_{d,AP_{\perp}} - E_{d,AP_{||}}\right) + \left(E_{d,P_{\perp}} - E_{d,P_{||}}\right) + 4 A_{Co} I_{12} + 4 K_1 V_{Co}}{2 M_s V_{Co}}. \tag{9}$$

Analogously, when the easy-axis is oriented along the z direction we have the following expressions:

$$H_{||} = \frac{\Delta E_{d_{||}} + 4 A_{Co} I_{12}}{2 M_s V_{Co}}, \tag{10}$$

$$H_{\perp} = \frac{\left(E_{d,P_{||}} - E_{d,AP_{\perp}}\right) + \left(E_{d,P_{||}} - E_{d,P_{\perp}}\right) + 4 A_{Co} I_{12} + 4 K_1 V_{Co}}{2 M_s V_{Co}}. \tag{11}$$

3. Behaviour simulation

To simulate the dynamics of magnetization of Co/Cu/Co trilayer systems the Object Oriented Micro-Magnetic Framework (OOMMF) software package developed by NIST was used. In this framework, the cobalt layers can be represented by magnetic discs while the copper layer, being non-magnetic, is considered a part of the surroundings. The lattice orientation of copper is reflected in the function $I_{12}(t_{Cu})$ (Bruno & Chappert, 1991); however, this paper discusses only the (111) orientation of copper as this crystal order is more frequently reported in the literature experiments (Tan & Stadler, 2006; Jyoko et al., 1997). Concerning cobalt, the structure more frequently reported in the literature is the *hcp* although this material can also be grown as *fcc* by electrodeposition and other techniques. For this reason, the exchange stiffness and magnetocrystalline anisotropy of cobalt *hcp* have been used here and the respective values are given in Table 1.
The *c axis* of *hcp* cobalt is the easy axis of magnetization; therefore, when detailing the Co easy-axis orientation in the simulator framework, this point has to be carefully taken into account to avoid simulating systems which eventually cannot be fabricated for practical reasons. However, seeing that Piraux and co. demonstrated that the *c axis* direction of *hcp* cobalt can be controlled by process parameters when the material is grown into nanoporous templates by electrodeposition we considered here the easy-axis of Co oriented along the *x* and *z* directions (Piraux et al., 1994).

Magnetocrystalline anisotropy, K_1	4.5×10^5 J/m^3 [a]
Exchange stiffness, A	28×10^{-12} J/m [b]
Saturation magnetization, M_s	1.4459×10^6 A/m[a]

Table 1. Values of parameters given to the cobalt layers.

In this section only the systems which satisfy the ZFC are concerned, and for these systems, the dynamic behaviour is simulated by sweeping the applied field, from zero, towards the +/-*x* direction; therefore, if initially, in the absence of applied field, the system's bottom layer is magnetized along the +*x* direction its top layer must be magnetized along the -*x* one. From this condition, an external magnetic field is applied along the +*x* axis and, as a consequence, only the top layer can flip from its original orientation towards the +x direction over the course of the simulation. In particular, the applied field is increased of intensity in small steps (minimum step: 1 Oe), and for each field the system evolves to the equilibrium. The OOMMF can calculate, at each incremental step, the system evolution in two different ways: (i) by solving the Landau-Lifshitz (LL) equations (time driver method, hereinafter TD) or (ii) by searching for the nearest energy minimum by using a conjugate gradient minimizer (minimization driver method, hereinafter MD). The use of the TD is proper when knowing the remagnetization dynamics towards the equilibrium is needed and, as a result, the single magnetic moment precession in the time domain can be visualized (for these cases a damping constant equal to 0.5 was used). Unfortunately, this method is time consuming, and it is not worthwhile to use it when one is only interested in localizing the equilibrium final state; in this case instead, the MD is more convenient as it is considerably faster. A concern about MD, however, is that due to the nature of the algorithm implemented, the system tends to stay in the energy saddle point of the AP state, or similarly in the -P state at the top of the energy hill, while the TD algorithm lets the system precess out of the saddle. In other words, in the former case, the system tends to persist in a previous equilibrium state even when the applied field is significantly changed, a fact which results in a misleading prediction of the system's hysteresis loop (the prediction error of the flip field can also be as large as 10% for particular systems). In the present work, this problem was solved by adding a random noise field to the system in the MD case to break the symmetry explicitly.

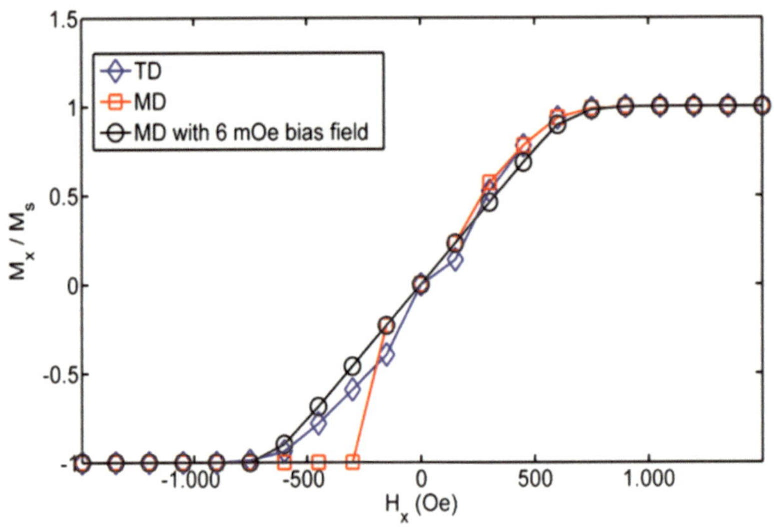

Fig. 1. Normalized magnetization curves of a Co/Cu/Co trilayer system of: d = 700 nm, t_{Co} = 1 nm, t_{Cu} = 2 nm and easy-axis oriented along the z direction.

The MD approach was used here with a quantified random noise which allowed us to estimate, saving more than 90% of computational time, the critical fields calculated by the TD. For a specific system, Figure 1 shows the effect of a 6 mOe field (random noise) on the predictions of the magnetization curve previously calculated by the TD. In terms of orders of magnitude, the intensity of the random noise used in this case is 10^{-6} smaller than the field to be applied to change the system magnetization (H_{sat}, about -800 Oe in the figure example), value corresponding to the maximum error committed when the flip field is calculated through the biased MD instead of the TD. In particular, in the Figure 1 the phenomenon of magnetization persistence when MD is used without random noise (square-dot curve) is represented.

3.1 Correlation domain between quasi-analytical prediction and micromagnetic simulation

To have a clue of the degree of confidence in using the quasi-analytical approach to predict the critical fields of a trilayer system, the coercivity and saturation fields of a large variety of system configurations of diameter 100 nm and differing in t_{Co} (from 1 to 40 nm) and t_{Cu} (from 3 to 40 Å) were calculated by QAD and compared with the relative output values of the OOMFF simulation. From the simulations performed it was noted that when the easy-axis of the magnetocrystalline anisotropy is oriented along the same direction of the applied field the flip from AP to P is immediate, therefore, the correlation analysis is reported here for easy-axis oriented along the *x* direction to show the accordance grade between modeling and simulation and to gauge their discrepancies; in fact, for this specific case, H_c and H_{sat} are coincident and give an unequivocal measurement of the switching field. In Figure 2 the switching fields calculated by QAD and the corresponding values calculated by OOMMF are reported for these systems.

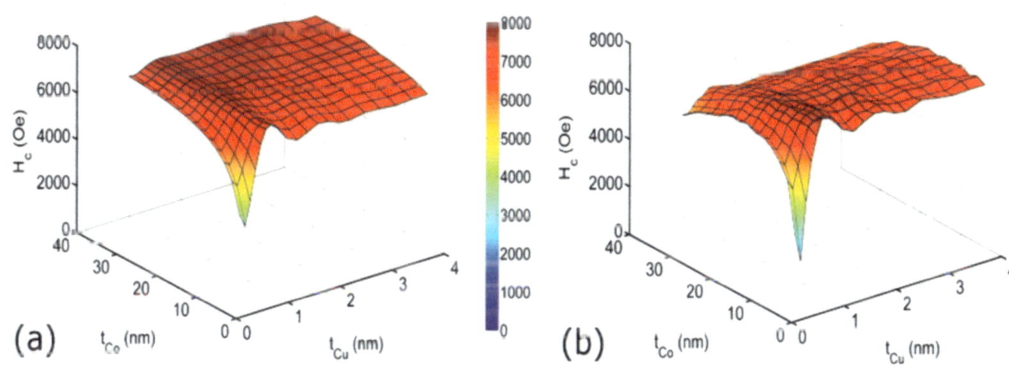

Fig. 2. Coercive field of trilayer systems of diameter 100 nm, and easy-axis oriented along the *x* direction, calculated against the thickness of the separated layers: a) values calculated by QAD, b) values simulated by OOMMF.

This argument suggests that for all the systems for which demagnetization dominates the exchange energy the QAD can be considered trustworthy independently of the easy-axis orientations of the magnets involved.

Figure 3 shows the percent error between the QAD values and the simulated data calculated in system of diameter 100 nm. At t_{Co} = 15 nm there is a cusp that marks the end of the QAD validity domain out of which the error rate increases proportionally to the aspect ratio. In systems characterized by Co layers thick a few nanometers the oscillatory nature of the exchange coupling energy affects the QAD model which still provides results with an error smaller than 5%. The rotation regime is clearly linked to the aspect ratio of the Co layer - it was observed that in systems of diameter less than 100 nm the coherency range is narrower and includes systems with Co layers thick a few nanometers; contrariwise in systems with a diameter larger than 100 nm the coherency range widens. As a conclusion, where this condition is validated, the QAD is a helpful tool to aid the choice of the proper system to be fabricated for a custom application.

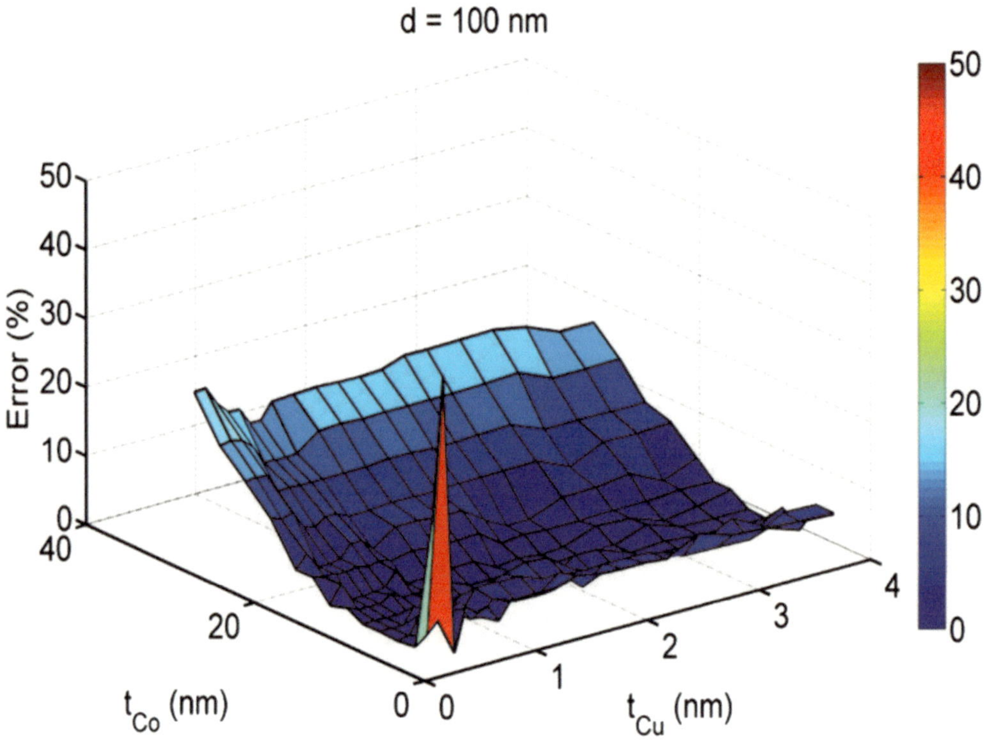

Fig. 3. Percentage error between the values calculated by QAD and simulated data for systems of diameter 100 nm and easy-axis oriented along the x direction.

3.2 The ZFC systems

In the present section the Co/Cu/Co trilayer systems which satisfy the ZFC are presented. In particular, the state of magnetization at zero field of cylindrical trilayer systems of diameters 50, 100 and 200 nm and easy-axis oriented along the z direction are reported, as a function of the layers thickness, in Figure 4.

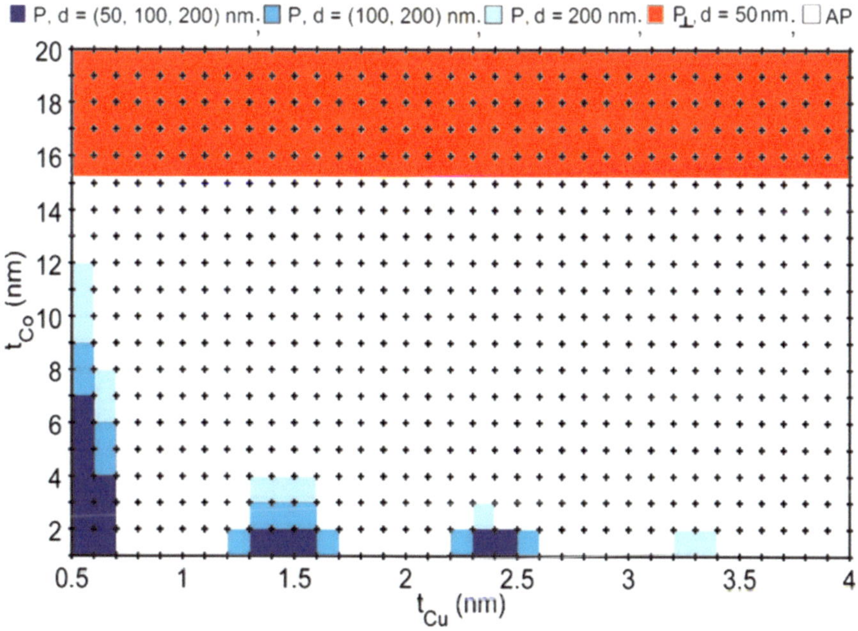

Fig. 4. State of magnetization at zero applied field of cylindrical Co/Cu/Co trilayer systems of diameters 50, 100 and 200 nm characterized by an easy-axis oriented along the *z* direction. The legend refers to different colored squares identifying the magnetization state of the cylindrical trilayer systems whose diameter is reported between brackets. The cross markers placed at the left bottom corners of the colored squares univocally intercept the thickness values of the Co and Cu layers of each system considered.

From this figure it is apparent that the trilayer systems which satisfy the ZFC are defined by selected combinations of *d*, t_{Cu} and t_{Co}; this fact results from the competition among the different elementary interactions taking place between two separate magnets which depend, in different fashions, on their size and shape.

The reason why only the systems characterized by the easy-axis oriented along the z-direction were considered here concerns the fact, widely acknowledged, that in such a system when the field is applied parallel to the easy-axis of magnetization the magnetization reversal is very abrupt, while when the field is applied parallel to the hard-axis, the magnetization generally rotates, gradually and in a reversible fashion, away from the easy direction (Du Trémolet de Lacheisserie et al., 2002). In this latter case, the magnetization processes involve the continuous rotation of magnetization from positive to negative saturation which is very useful for sensing purposes. Considering the current setup, to obtain systems with a analogue magnetic response is necessary to impose that the easy-axis be lined up along the *z* direction. The latter assertion is confirmed by micromagnetic simulation; the systems characterized by an easy-axis parallel to the applied-field's sweeping direction (*x* direction in our case), in most cases, can flip back and forth practically instantly from the AP to the P state of magnetization, and generally exhibit a square shaped hysteresis loop. For this reason, from here onwards, it is showed a value data-base of simulated quantities characterizing the response magnetization curve of a large number of system configurations (with easy-axis along *z* direction) to aid a prompt design of custom sensing devices.

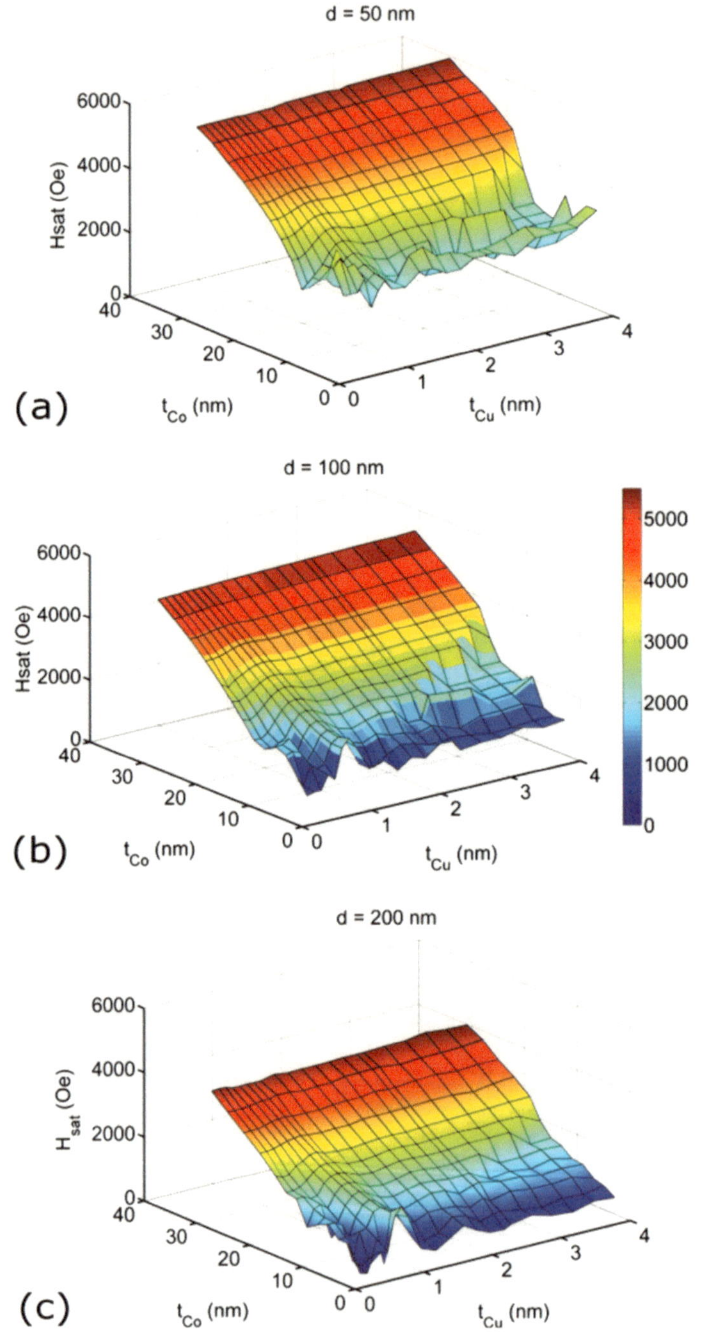

Fig. 5. Saturation field of cylindrical Co/Cu/Co trilayer systems of diameters 50, 100 and 200 nm characterized by an easy-axis oriented along the *z* direction. The mesh's intersection points of the 3D surface depicted correspond to the geometrical parameters of the simulated systems.

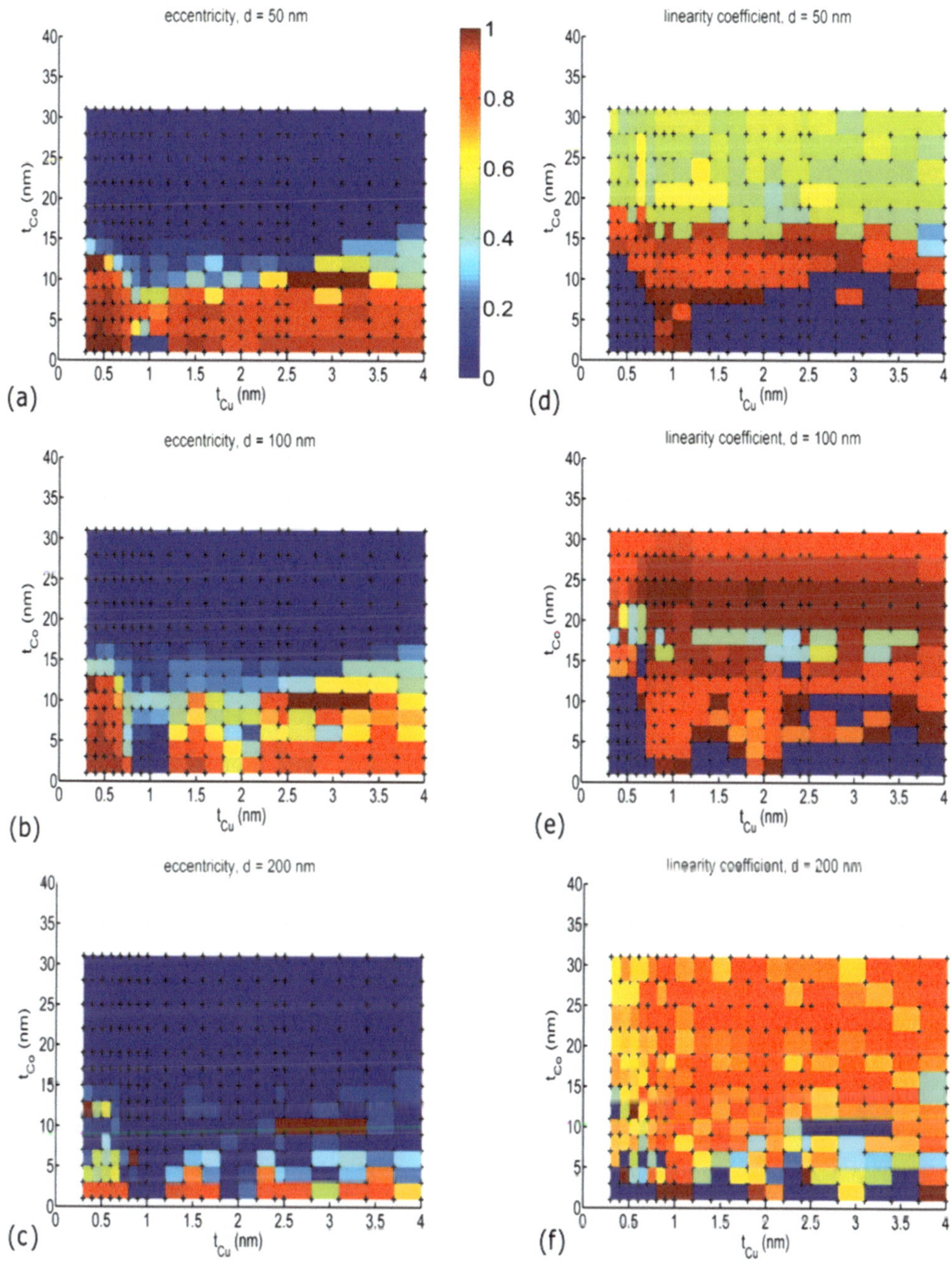

Fig. 6. Eccentricity and linearity coefficient of the magnetization curves of cylindrical Co/Cu/Co trilayer systems of diameters 50, 100 and 200 nm characterized by an easy-axis oriented along the *z* direction. Colors represent the value of eccentricity and linearity coefficient while the cross markers placed at the left bottom corners of the tone squares univocally intercept the thickness value of the system's Co and Cu layers.

3.3 Systems with easy-axis in the z direction

Respectively the figures 5 and 6 show, as a function of *d*, t_{Cu} and t_{Co}, the saturation field (H_{sat}), the eccentricity and the linearity coefficient of the magnetization curves of systems with diameters 50 nm, 100 nm and 200 nm and characterized by an easy-axis oriented along the *z* direction. Particularly, the eccentricity is defined here as the ratio between the intensity of the coercive and saturation fields, and the linearity coefficient is the Pearson product-moment correlation coefficient calculated from the simulated values of magnetization and the respective applied field ones within the dynamic range defined from - H_{sat} and H_{sat} (Rodgers & Nicewander, 1988). Figures 5 and 6 represent a database of parameters to choose from when a magnetic sensor of this type has to be designed for custom applications. It is apparent that, according to the envisaged application, the best design choice is a trade off among the magneto-resistive ratio, the H_{sat} (linked to the sensor dynamic range), the eccentricity and the linearity of the magnetization curve which have to be respectively minimized and maximized to avoid massive electronic post-corrections. In Figure 7 the magneto-resistive ratio of Co/Cu/Co trilayer systems, calculated from the Fert and Valet model is reported for the convenience of the sensor designer (Valet & Fert, 1993; Fert et al., 1994).

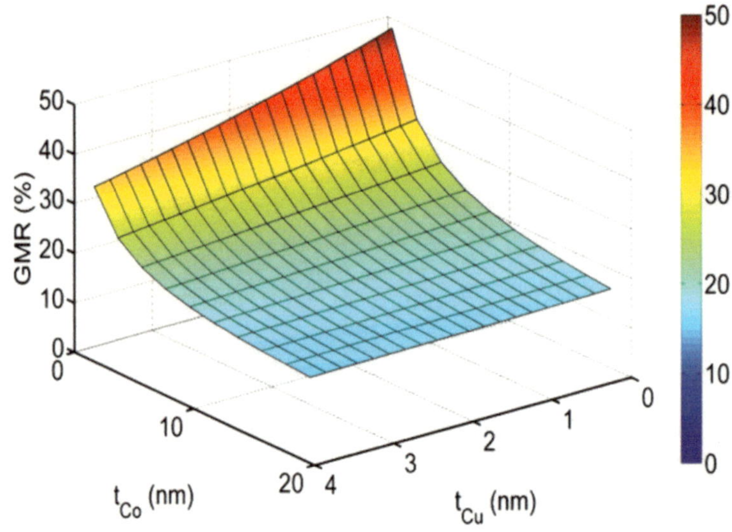

Fig. 7. Magneto-resistive ratio as a function of the Co and Cu layer thicknesses. The mesh's intersection points of the 3D surface depicted correspond to the geometrical parameters of the systems considered.

3.4 From the trilayer to the multilayers

The present chapter specifically refers to magnetic field transducers constituted by arrays of Co/Cu multilayer nanowires connected in parallel in a current perpendicular to plane (CPP) architecture. If the magneto-resistance ratio does not depend on the number of layers employed in the system, the system's critical-fields and the shape of its magnetoresistance curve depend both on the number of Co/Cu bilayers piled up to form each single nanowires and on the presence of identical systems located in the surroundings (as for arrays of nanowires). In particular, to gauge the influence of multiple bilayers (connected either serially or in parallel) on the dynamic range of the multilayer systems is of paramount importance for the practical use of the technology addressed. Therefore, for this section, micromagnetic simulation was used to corroborate the QAD predicted properties of a large plurality of elementary trilayers (section 3.3) and to simulate the magnetic behaviour of selected bilayers interacting when forming a nanowire (up to 50 Co/Cu bilayers - hereinafter simulated multilayer system, SMS), and arranged in parallel, on the same plane, to mimic a systems' arrays (5 x 3 array). The Figure 8 represents how the layers number affects the dynamic range of a single nanowire - the dashed line represents the logarithmic best fit interpolating the simulated data (rhombus). As apparent from the plot the nanowire's saturation field increases logarithmically as a function of the elementary bilayers. As apparent from this graph, if low dynamic ranges are needed, shortest possible nanowires have to be fabricated.

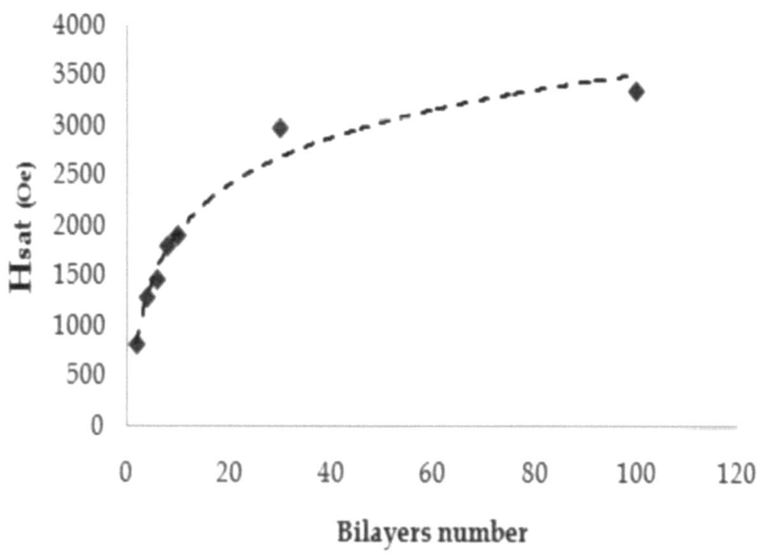

Fig. 8. Saturation magnetic field as a function of the bilayer number employed in the multilayer. The dashed line represents the logarithmic fit which best interpolates the simulated data (rhombus).

To take into account the mutual reciprocal effect of identical nanowires on the overall saturation, the magnetostatic interaction in the dipole approximation (Sun et al., 2005) was made explicit for the systems addressed. In this approximation a single nanowire is represented by a magnetic dipole characterized by a magnetization value obtained considering the contribution of each Co layer of the system; as a result, the potential energy of the interaction between two nanowires could be written as follow:

$$\mathbf{H} = -\frac{\mu_0}{4\pi r_{jk}^3}\left(3\left(\mathbf{m}_\mathrm{j} \cdot \mathbf{e}_\mathrm{jk}\right)\left(\mathbf{m}_\mathrm{k} \cdot \mathbf{e}_\mathrm{jk}\right) - \mathbf{m}_\mathrm{j} \cdot \mathbf{m}_\mathrm{k}\right), \tag{12}$$

where $\mathbf{m}_k$ and $\mathbf{m}_j$ are the magnetizations vector of two interacting nanowires, $\mathbf{r}_{jk}$ is their distance and $\mathbf{e}_{jk}$ is a unit vector parallel to the line joining their centers. Specifically, in this section, an array of nanowires arranged in a honeycomb configuration was considered to represent the most common array geometry encountered when using anodic porous alumina (APA, 10^{10} nanopores/cm^2) as a nanoporous template to have it fabricated. The Figure 9 shows a simplified array (hereinafter simulated array system, SAS) consisting of thirteen multilayer system of geometric and crystalline properties (d = 50 nm, t_{Co} = 2.5 nm, t_{Cu} = 2.5 nm and easy axis parallel to wire axis) used in the previous simulation.
To mimic the magnetostatic interaction of nanowires in parallel, simulations were carried out for the same SAS geometry but considering multilayers having respectively two, four, six and eight bilayers (the elementary GMR unit - one Co/Cu bilayer) each.

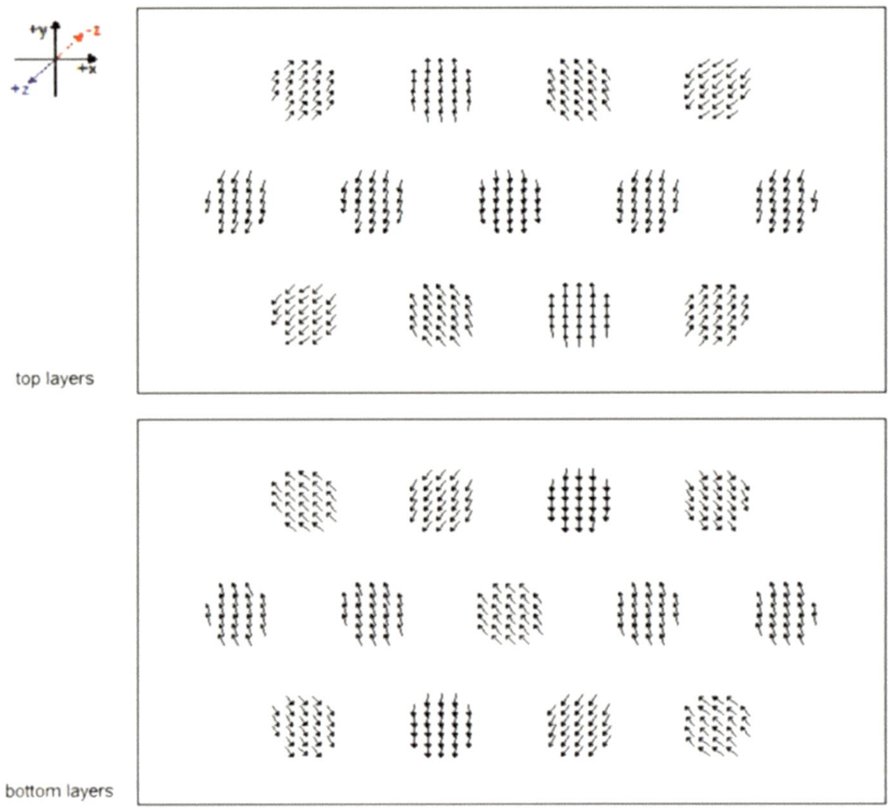

Fig. 9. An example of simulated micromagnetic configurations of the top and bottom layers of 13 trilayer systems in a honeycomb arrangement.

The Figure 10 reports the comparison of the saturation magnetic field between the SAS and the corresponding SMS as a function of the number of layers employed. As apparent in the figure the interaction between multilayers in parallel contributes to lower the saturation magnetic field. For the systems considered a maximum deviation of 10% is acknowledged.

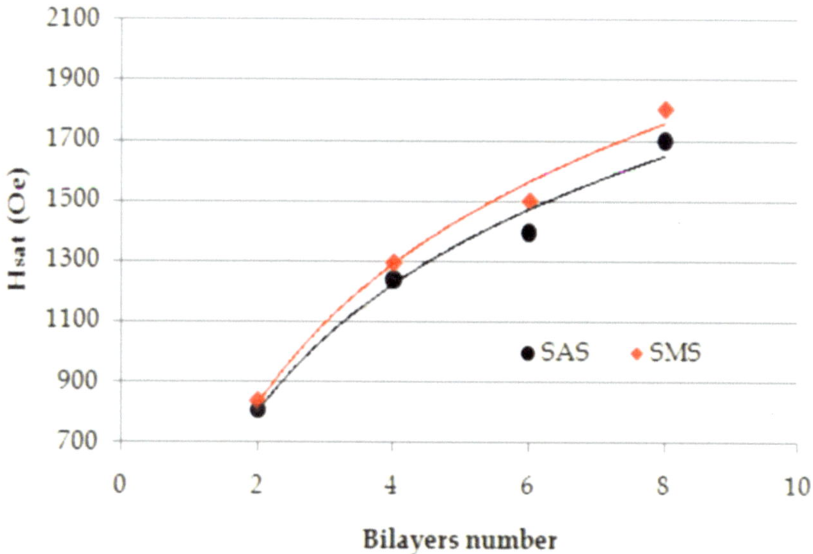

Fig. 10. Saturation magnetic field of SAS (rhombus) and SMS (circle) as a function of the number of bilayers employed each multilayer.

4. Template electrodeposition of Co/Cu-multilayer-nanowires arrays

To have Co/Cu-multilayer-nanowires arrays fabricated many approaches have being followed, among them, the chemical vapour deposition (CVD) is typically used to grow nanowires on nanocatalysts, top-down lithography and ion milling are exploited to prefabricate nanoporous membranes and lift-off masks to be filled, at a second stage, with a desired material to form nanowires and nanobelts respectively. In this context, electrodeposition has been widely utilized to fill nanoporous templates. In particular, this technique, elsewhere referred (Pullini & Busquets-Mataix, 2011) as template electrodeposition (TED) when used to fabricate single-element nanowires, or template pulsed electrodeposition (TPED) when used to fabricate multilayer nanowires, has met outstanding consideration; in fact, it is a simple method to manufacture nanomaterials on large areas both for fundamental studies and for the production of nanodevices in high volumes. Among the plethora of applications investigated, the aforementioned technique is acknowledged as being promising to mass produce magnetic-field sensors (Manalis et al., 1995; Chou et al., 1994; Simonds, 1995). The present section refers specifically to the use of the TPED for the fabrication of Co/Cu-multilayer-nanowire arrays exhibiting Giant Magnetoresistance (GMR) and in particular to provide a user with growth process parameters sized up to obtain system of desired characteristics. The conventional GMR systems are developed by vacuum techniques which require sophisticated and expensive equipments to be adopted; on the contrary, the TPED simply requires a galvanic cell controlled by a low cost potentiostat. The simplicity of this technology is considered a great advantage for industrialization; therefore, for this reason, the main objective of the present section is to detail the TPED process to pave the way for the mass production of multilayer-nanowire-array (for CPP architecture).

4.1 Process mastering for tailoring growth

By TPED each material building multilayer nanowires arrays can be deposited by applying the proper voltage to the cell (voltage cycling). The single bath technique consists in using a single electrolytic bath made with salts of both the metal ions to be deposited, where the ion-concentration ratio is highly unbalanced towards the less noble metal (Piraux et al., 1994; Blondel et al., 1994; Evans et al., 2000; Fert & Piraux, 1999; Ohgai et al., 2003; Liu et al., 1995). As Cu deposits at a lower negative potential than Co, the former is the nobler element, therefore, its concentration in the deposition solution has to be smaller. As a consequence of that, 100% purity copper can be deposited but traces of Cu are also deposited within the Co deposition step. According to the Co purity desired one can tailor the ratio of the species present in the galvanic bath. In this specific context, the Cu impurity contents in the Co layers badly affect the magnetic properties of Co, therefore, from the application standpoint the Cu precursor in the bath has to be lowered to the minimum. On the contrary, the smaller the Cu concentration is, the slower the growth of the Cu layers. As a matter of fact, the optimal Cu concentration was chosen from a trade off between a cost-effective fabrication throughput and the GMR performance of the multilayer stack. In this work, the electrolyte used was a sulphate bath prepared by dissolving 520 g/l (1.85 mol/l) of $CoSO_4.7H_2O$, 5.2 g/l (0.021 mol/l) of $CuSO_4.5H_2O$, and 52 g/l (0.84 mol/l) of H_3BO_3. The latter compound was added to buffer the pH of the electrolytic solution to an approximate constant value of 4.5 throughout the deposition duration. These values result in a ion concentration ratio $[Co^{2+}]$:$[Cu^{2+}]$ of about 90 to 1. Analytical grade chemicals and purified water with a resistivity higher than 16.8 MΩ cm were used throughout.

Two different types of templates are commonly used in the art, namely the ion-track etched (TE) polymers and the anodic porous alumina membranes. Although both membrane types can provide nanochannels tailored to a given diameter, their nature and morphology present some important differences influencing the filling process and, therefore, they must be considered with particular attention (Ohgai et al., 2003; Schönenberger et al., 1997; Ferain & Legras, 2001; Pullini et al., 2007a).

Even though it is commonly acknowledged from an industrialization standpoint that TPED is a cost-effective technique and easily scalable, mainly due to a wide plurality of factors affecting the nanowire growth, there are still contradicting process details in the published literature which make the technology not ready yet to be scaled up. In fact, there are great deal of process variables which influence the electrodeposition to a higher or lesser degree: concentration of species, pH, presence of complexants and/or, brighteners, temperature, voltage and current cycling, and the membrane used (material, pore diameter and length). Today a systemic analysis of TPED taking into account these variables is still missing.

Multilayer nanowires were then fabricated alternating square-pulse potentials by the computer controlled potentiostat AMEL 500 by using a three-electrode cell, properly designed to host flexible polycarbonate TE-templates, was made of PTFE. A Pt grid was employed as a counter electrode and a saturated calomel electrode (SCE) as the reference one. In Figure 11 is shown the architecture of the experiment (a, b and c) and the typical potential driving cycle used in the process (d) with the corresponding current-density cycle (e). The open end of nanopores was arranged to face upwards, i.e. anode over cathode configuration (Konishi et al., 2003) and the deposition area was of 2 cm^2. Prior to deposition, when the nanoporous membranes were arranged properly, the cell was placed in an ultrasonic agitation bath always for 5 minutes to allow the complete wetting of the membranes' nanopores. As reported elsewhere (Schönenberger et al., 1997), here we corroborate that this sonication time is sufficient to have the complete filling of the nanopores allowing a homogeneous growth of nanowires over the whole exposed area. The deposition was carried out at room temperature without stirring. As a result, the complete filling of the nanopores, resulting in the maximum fabrication rate of nanowires, could be achieved with success by TPED.

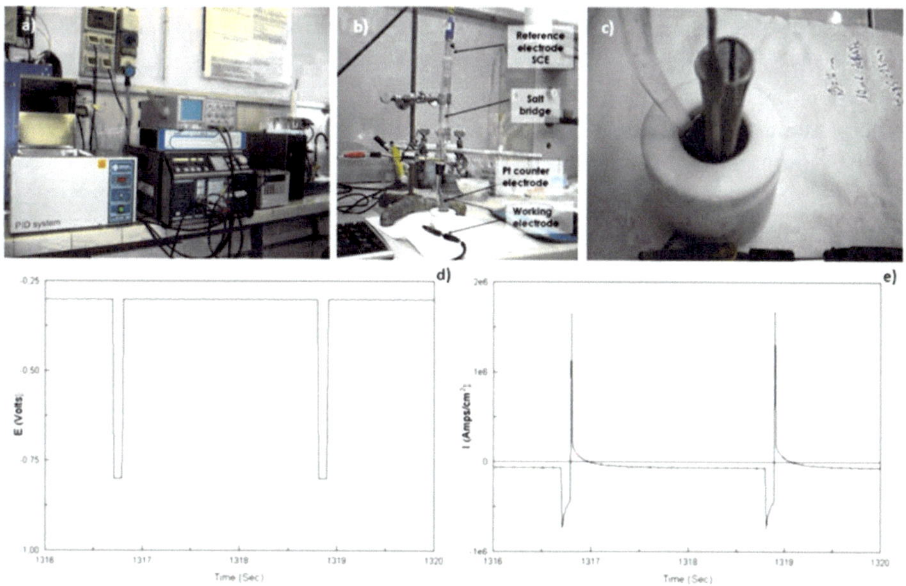

Fig. 11. (a) AMEL 500 potentiostat; (b) 3-electrodes electrodeposition cell; (c) PTFE cell detail; (d) a typical potential driving cycle (-0.3V for Cu deposition and -0.8V for Co deposition); (e) the corresponding current-density cycle.

For this section arrays of Co/Cu multilayer nanowires of 30, 50, and 100 nm diameters have been fabricated in order to provide a sensor developer with precise values for the most important growth process parameters to be used to tailor the geometry characteristics of the multilayer systems desired, particularly the layers' thickness. To this end, knowing the current efficiency of Co and Cu when deposited in alternate fashion by TPED is of fundamental importance to know the relationship between the deposition timing and the mass deposited to form each layer. In fact, by the Faraday's law the electrical charge employed within each deposition cycle is directly linked to mass deposited each layers if a 100% current efficiency is assumed. D. Pullini and D. Busquets-Mataix (Pullini & Busquets-Mataix, 2011) discusses the influence of the space confinement on the TPED and report the calculated values of current efficiencies used here to calculate the layer thicknesses to build nanowires of tailored characteristics. In this work, the current efficiencies of Co and Cu in nanotemplates were determined by the following method. The single layer volumes, which, at constant charge load, have to be theoretically invariant along the nanowire length, were determined by microscopy direct measurements of the layer diameters, tilt angle, and thicknesses. For each sample, a large number of layers belonging to central portions of nanowires were considered and the mean value of the volumes deposited each metal layer calculated. The average layer masses of Cu and Co were obtained and compared with those calculated theoretically by the Faraday's law after integrating the current recorded over the time interval of the corresponding deposition steps. The Co anodic peaks taking place at the beginning of the following Cu deposition cycles were not taken into account. In doing so, the theoretical to real deposition value could be directly compared. Finally, the current efficiencies of Cu and Co were calculated as the ratio between the measured values of layer mass and the theoretical ones. As the nanowires fabricated for this work present the narrowest distribution of layer diameters and thicknesses in correspondence of their central parts, therefore for the present analysis, only those ones were considered (Pullini et al., 2007b). Following the chemical dissolution of the polycarbonate membranes, the morphology analysis of the nanowires was performed by TEM and complemented in some cases with scanning electron microscopy. As explained elsewhere different methods to enhance the contrast between layers for a more accurate measurement were used (Pullini et al., 2007a). The time evolution of the current efficiencies of Co and Cu in the TPED is shown by the figure 4 of Pullini and Busquets-Mataix (Pullini & Busquets-Mataix, 2011) manuscript. Both for Co and Cu cases, one can observe a slight reduction of the current efficiency when longer deposition duration is experienced. More in depth, during short deposition cycles, cations are abundant in solution, the TPED is charge controlled and the growth rate is only limited by the rapidity which ions can be chemically reduced at the interface. For the shortest deposition cycles in fact the process only depends on the applied potential irrespectively of the diameter of pores to be filled. A possible explanation of this fact is the change of the deposition conditions due to the variation of the membrane polarization provoked by the potential alternation. In fact, the membrane polarization change affects the double-layer characteristics, and the migration and diffusion of ions, somehow resulting similar to those of a space-unconfined deposition. As a main results, it was shown that the current efficiency of Cu measured for all the potential range addressed here in the TPED mode is very close to 100%; instead, it is apparent for Co that a consistent reduction has been observed - the Co TPED efficiency ranges between 60% to nearly 70%. In a first instance, the significant reduction of Co current efficiency could be accounted as a re-dissolution of Co during the Cu deposition cycle (Valizadeh et al., 2002). This hypothesis is corroborated by the presence of a high anodic peak observed at the beginning of each Cu deposition cycle - the peak shows itself in the first instants after the potential switches. Co dissolution can drastically change the actual layer thicknesses with respect to the nominal values deduced from Faraday's law (V. Weihnacht et al. 2003; Bakonyi & Péter, 2010), and therefore, the direct measurement method presented in the manuscript is fairly suited to ascertain the real deposition efficiencies in multilayered nanowires. This peak can be reduced by the adequate selection of the Cu deposition potential as showed by Liu et al. (Liu et al., 2004) and Péter et al. (Péter et al., 2004) to higher absolute values (i.e. -0.6V). From the fact that the pore diameter does not play a significant role on the Co and Cu current efficiencies in the TPED mode and form the measured values of the current efficiencies reported in D. Pullini and D. Busquets-Mataix (D. Pullini & D. Busquets-Mataix, 2011) a growth parameter data base, to promptly allow a user to fabricated a Co/Cu multilayer nanowires array of tailored properties, the electric charge to be employed for deposit the material thickness desired to build the single layers of the array is given in this section as a function of the deposition potential.

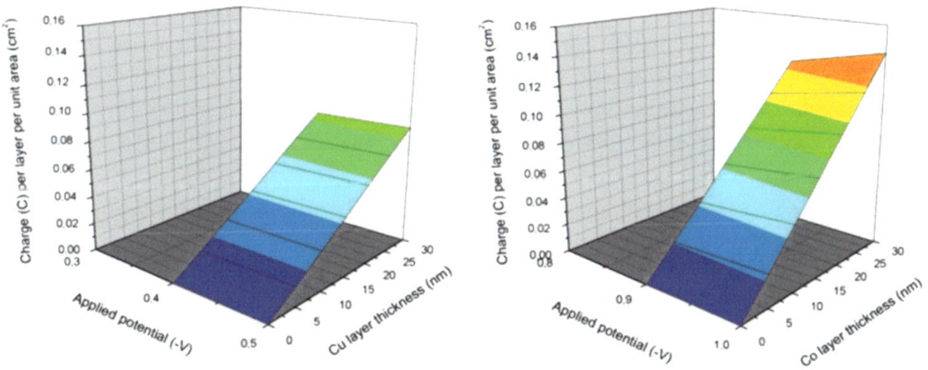

Fig. 12. Charge per cycle per unit area as a function of the electrode potentials to deposit Cu (left) and Co (right) in TPED.

In fact, using Figure 12 the electrode potential for Cu (left graph) and Co (right graph) in TPED is directly linked to the thickness desired of the single layer of either Co or Cu, the charge per cycle and per unit area is reported, and makes straightforward the use of TPED for the purpose. It is important to notice that the area to be considered for calculations is the effective area, i.e. not the area of the membrane wetted by the electrolyte but the product of the area of a single hole times the density of holes in the membrane. In these, current efficiency for Cu deposition was close to 100% whereas for Co deposition varied from 60% to 70%, being these differences mainly related to applied potential. A detailed discussion of the Co and Cu current efficiencies either when the elements are deposited in constant or in alternate mode is reported as a function of the respective electrode potentials and the deposition duration. [ref: Pullini, D.; Busquets, D.; (2011). Electrodeposition Efficiency of Co and Cu in the Fabrication of Multilayer Nanowires by Polymeric Track-Etched Templates. ACS Appl. Mater. Interfaces, 3 (3), pp 759–764- DOI: 10.1021/am1011222]

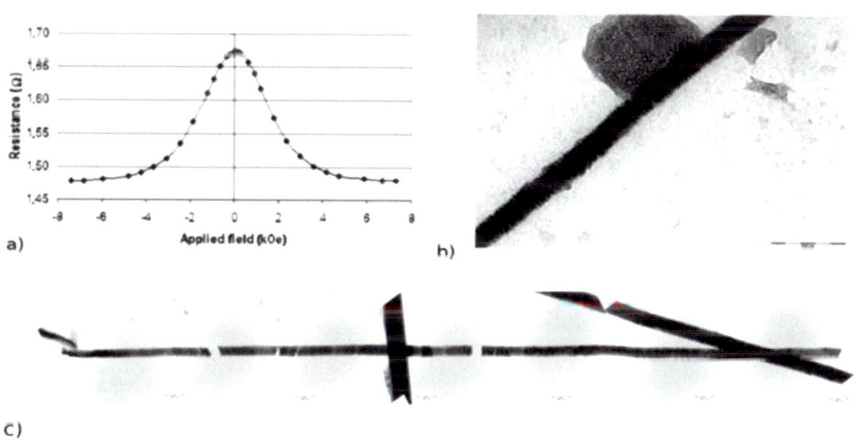

Fig. 13. (a) Magnetoresistive response of arrays of Co/Cu nanowire of 50 nm diameter; (b) TEM imagine of 60 nm diameter Co/Cu multilayer nanowire; (c) TEM image of an 100 nm diameter Co/Cu multilayer nanowire – in-full collage.

GMR measurements were carried out by a four-contact method on each sample fabricated for the study. A thin layer of gold was sputtered on the top side of the template membrane to ensure the electrical continuity. Magnetic field was applied orthogonally to the wire axes according to a conventional CPP configuration, whereas current flows along the wire axes. Saturation was always reached and GMR was calculated as $\Delta R/R_{max}$ (Fert & Piraux, 1999). A typical magnetoresistance to field curve is presented in Figure 13 where are also two TEM images of a Co/Cu multilayer nanowire of GMR properties.

5. Conclusions

In this chapter a quasi-analytical method to predict the magnetic properties of trilayer systems to aid the design of magnetic field sensors has been defined. In particular, the authors addressed here Co/Cu/Co trilayer systems of cylindrical shape to have an insight into the magnetic behaviour of more complex Co/Cu nanowire systems frequently reported in the literature for said purposes. The interest in Co/Cu multilayer nanowires is motivated by the fact that their properties are relatively easy to be tailored by controlling the process parameters if electrodeposition into nanoporous templates is used to have them grown. The QAD consists of defining and solving the total energy problem of these systems as a function of the trilayers diameter, the layers thickness and the crystal orientation of the materials used. The total energy of the system is built from the analytical expressions of all the involved energy terms except the demagnetization one whose expression is a function of the values, calculated numerically, which the demagnetization energy assumes when the system is in four fundamental states of magnetization. A large number of trilayer systems differing in d (50, 100 and 200 nm), **a** (oriented along the x and z axis), t_{Co} (ranging from 0 and 40 nm), and t_{Cu} (ranging from 0 to 4 nm) have been analyzed by this method. For all these systems, the micromagnetic simulation has been used to understand when the assumption of coherent rotation of the single magnetic moments belonging to the separate layers is valid. For the reasons explained, we can assume that in system with 100 nm of diameter when the Co layers are thinner than 15 nm the single dipoles rotate coherently and the QAD predictions fully match the simulation values. Instead, for larger t_{Co} the QAD is still trustworthy at predicting qualitatively the magnetization reversal process although discrepancies in the absolute values of H_c are acknowledged. At last, a data-base of values of simulated quantities which characterize the response magnetization curve is given for a large number of system configurations to aid a prompt design of custom sensing devices. Nevertheless, from the precise estimation for current efficiency of Co and Cu in TPED mode the growth-process parameters to tailor the morphology of custom systems have been detailed to provide a sensor developer with fundamental data to have GMR–nanowires array system fabricated.

6. References

Simonds, J. L. (1995). Magnetoelectronics: Today and Tomorrow. *Physics Today*, Vol.48, No.4, (April 1995), pp. 26-30, ISSN 0031-9228

Grünberg, P.; Schreiber, R.; Pang, Y.; Brodsky, M. B. & Sowers, H. (1986). Layered Magnetic Structures: Evidence for Antiferromagnetic Coupling of Fe Layers across Cr Interlayers. *Phys. Rev. Lett.*, Vol.57, No.19, (September 2009) pp. 2442-2445, ISSN 0031-9007

Saurenbach, F.; Walz, U.; Hinchey, L.; Grünberg, P. & Zinn, W. (1988). Static and dynamic magnetic properties of Fe-Cr-layered structures with antiferromagnetic interlayer exchange. *J. Appl. Phys.*, Vol.63, No.8, (May 1988), pp. 3473-3476, ISSN 0021-8979

Baibich, M. N.; Broto, J. M.; Fert, A.; Nguyen Van Dau, F. & Petroff, F. (1988). Giant Magnetoresistance of (001)Fe/(001)Cr Magnetic Superlattices, *Phys. Rev. Lett.*, Vol.61, No.21, (August 1988), pp. 2472-2475, ISSN 0031-9007

Manalis, S.; Babcock, K.; Massie, J.; Elings, V. & Dugas, M. (1995). Submicron studies of recording media using thin-film magnetic scanning probes. *Appl. Phys. Lett.*, Vol.66, No.19, (May 1995), pp. 2585-2587, ISSN 0003-6951

Chou, S. Y.; Wei, M. S.; Krauss, P. R. & Fisher, P. B. (1994). Single-domain magnetic pillar array of 35 nm diameter and 65 Gbits/in.^{2} density for ultrahigh density quantum magnetic storage. *J. Appl. Phys.*, Vol.76, No.10, (1994), pp. 6673-6675, ISSN 0021-8979

Pullini, D.; Innocenti, G.; Busquets, D. & Ruotolo, A. (2007a). Investigation of multilayer local tilt within long portion of single Co/Cu nanowires. *Appl. Phys. Lett.*, Vol.90, No.13 (October 2006), pp. 133106-133106-3 , ISSN 0003-6951

Fert, A. & Piraux, L. (1999). Magnetic nanowire. *J. Magn. Magn. Mat.*, Vol.200, No.1-3, (March 1999), pp. 338-358, ISSN 0304-8853

Piraux, L.; George, J.M.; Depres, J.F.; Leroy, C.; Ferain, E.; Legras, R.; Ounadjela, K. & Fert, A. (1994). Giant magnetoresistance in magnetic multilayered nanowires. *Appl. Phys. Lett.*, Vol.65, No.19, (November 1994), pp. 2484-2486, ISSN 0003-6951

Ohgai, T.; Hoffer, X.; Gravier, L.; Wegrowe, J. E. & Ansermet, J.-Ph. (2003). Bridging the gap between template synthesis and microelectronics: spin-valves and multilayers in self-organized anodized aluminum nanopores. *Nanotechnology*, Vol.14, Issue 9, (September 2003), pp. 978-982, ISSN 0957-4484

Sun, L.; Hao, Y.; Chien, C.L. & Searson, P.C. (2005). Tuning the properties of magnetic nanowires. *IBM Journal of Research and Development*, Vol.49, No.1, (January 2005), pp. 79-102, ISSN 0018-8646

Bruno, P. & Chappert, C. Oscillatory coupling between ferromagnetic layers separated by a nonmagnetic metal spacer. *Phys. Rev. Lett.*, Vol.67, No.12 (June 1991), pp. 1602-1605, ISSN 0031-9007

Du Trémolet de Lacheisserie, É.; Gignoux, D. & Schlenker, M. (2002). *Magnetism II: Materials and Applications*, Kluwer Academic Publishers, ISBN 1402072236, Norwell

Rodgers, J. L. & Nicewander, W. A. (1988). Thirteen ways to look at the correlation coefficient. *The American Statistician*, Vol.42, No.1, (February 1988), pp. 59-66, ISSN 0003-1305

Valet, T. & Fert, A. (1993). Theory of the perpendicular magnetoresistance in magnetic multilayers, *Phys. Rev. B*, Vol.48, No.10, (April 1993), pp. 7099-7113, ISSN ISSN 1550-235X

Fert, A.; Valet, T. & Barnas, J. (1994). Perpendicular magnetoresistance in magnetic multilayers: Theoretical model and discussion (invited). *J. Appl. Phys.*, Vol.75, No.10, (May 1994), pp. 6693-6698, ISSN 0021-8979

Pullini, D. & Busquets-Mataix, D. (2011). Electrodeposition Efficiency of Co and Cu in the Fabrication of Multilayer Nanowires by Polymeric Track-Etched Templates. *ACS Appl. Mater. Interfaces*, Article ASAP Publication Date (Web): February 22, 2011

Blondel, A.; Meier, J. P.; Doudin, B. & Ansermet, J. Ph. (1994). Giant magnetoresistance of nanowires of multilayers. *Appl. Phys. Lett.*, Vol.65, No.23, (July 1994), pp. 3019-3021, ISSN 0003-6951

Evans, P. R.; Yi, G. & Schwarzacher, W. (2000). Current perpendicular to plane giant magnetoresistance of multilayered nanowires electrodeposited in anodic aluminum oxide membranes. *Appl. Phys. Lett.*, Vol.76, No.4, (June 1999), pp. 481-483, ISSN 0003-6951

Liu, K.; Nagodawithana, K.; Searson, P.C. & Chien, C.L. (1995). Perpendicular giant magnetoresistance of multilayered Co/Cu nanowires. *Phys. Rev. B*, Vo.51, No.11, (November 1994), pp. 7381-7385, ISSN 1550-235X

Ferain, E. & Legras, R. (2001). Pore shape control in nanoporous particle track etched membrane. *Nucl. Instrum. Methods Phys. Res. B*, Vol.174, No.1-2, (September 2000), pp. 116-122, ISSN 0168-583X

Konishi, Y.; Motoyama, M.; Matsushima, H.; Fukunaka, Y.; Ishii, R. & Ito, Y. (2003). Electrodeposition of Cu nanowire arrays with a template. *J. Electroanal. Chem.*, Vol.559, (February 2003), pp. 149-153, ISSN 1572-6657

Schönenberger, C.; van der Zande, B. M. I.; Fokkink, L. G. J.; Henny, M.; Schmid, C.; Krüger, M.; Bachtold, A.; Huber, R.; Birk, H. & Staufer, U. (1997). Template Synthesis of Nanowires in Porous Polycarbonate Membranes: Electrochemistry and Morphology. *J. Phys. Chem. B*, Vol.101, No.28, (March 1997), pp. 5497-5505, ISSN 1089-5647

Pullini, D.; Busquets, D.; Ruotolo, A.; Innocenti, G. & Amigo, V. (2007b). Insights into pulsed electrodeposition of GMR multilayered nanowires. *J. Magn. Magn. Mater.*, Vol.316, No.2, (September 2007), pp. e242–e245, ISSN 0304-8853

Valizadeh, S.; George, J. M.; Leisner, P. & Hultman, L. (2002). Electrochemical synthesis of Ag/Co multilayered nanowires in porous polycarbonate membranes. *Thin Solid Films*, Vol.402, No.1-2, (September 2001), pp. 262–271, ISSN 0040-6090

V. Weihnacht, V.; Péter, L.; Tóth, J.; Pádár, J.; Kerner, Zs.; Schneider, C.M. & Bakonyi, I. (2003). Giant Magnetoresistance in Co-Cu/Cu Multilayers Prepared by Various Electrodeposition Control Modes. *J Electrochem. Soc.*, Vol.150, No.8, (February 2003), pp. C507-C515, ISSN 1945-7111

Bakonyi, I. & Péter, L. (2010). Electrodeposited multilayer films with giant magnetoresistance (*GMR*): Progress and problems. *Progr. Mater. Sci.*, Vol.55, No.3, (July 2009), pp. 107-245, ISSN 0079-6425

Liu, Q.X.; Péter, L.; Tóth, J.; Kiss, L.F.; Cziráki, Á. & Bakonyi, I. (2004). The role of nucleation in the evolution of giant magnetoresistance with layer thicknesses in electrodeposited Co–Cu/Cu multilayers. *J. Magn. Magn. Mater.*, Vol.280, No.1, (February 2004), pp. 60-74, ISSN 0304-8853

Péter, L.; Liu, Q.X.; Kerner, Zs. & Bakonyi, I. (2004). Relevance of the potentiodynamic method in parameter selection for pulse-plating of Co–Cu/Cu multilayers. *Electrochimica Acta*, Vol.49, No.9-10, (November 2003), pp. 1513-1526, ISSN 0013-4686

12

Surface-Directed Growth of Nanowires: A Scalable Platform for Nanodevice Fabrication

Babak Nikoobakht
*Surface and Microanalysis Science Division,
National Institute of Standards and Technology,
Gaithersburg, MD 20899,
USA*

1. Introduction

Optical lithography continues to be the popular technique for further miniaturization of electronic circuitry and its components. However, as further device miniaturization continues, the complexity of pattern generation and cost increase (ITRS 2006); therefore the use of such techniques becomes limited to high-end applications, such as microprocessor manufacturing. As an alternative, nanomaterials and non-conventional nanofabrication methods, such as "bottom-up" chemical approaches, offer great opportunities in producing useful nanostructure-based devices with potential advantages such as enhanced performance and/or lower cost. A majority of the produced nanomaterials for device applications require post processing (*e.g.* transfer from a source to a target substrate). For composite structures this transfer is not a concern, but for producing intricate nanodevices with a large scale hierarchical order, this step becomes a bottleneck. Generally, scaling up requires knowledge of surface registries of a large group of nanocrystals on a given surface that is needed for the subsequent fabrication steps. In addressing this issue, we combine a surface-directed nanocrystal growth with conventional lithography to control the registries of laterally grown semiconductor nanowires. In this process, a nanowire is formed using a surface-directed vapor-liquid-solid (SVLS) process inside a Au nanodroplet; the nanodroplet defines the starting point of the nanowire and a single crystal substrate defines its growth direction. We use substrates such as sapphire and gallium nitride and have shown that each crystal promotes its own specific growth directions. Using this platform, nanowires are grown where the nanodevices are expected to be fabricated. In the following, we intend to provide an account of the progress on directed deposition and growth of nanocrystals in the plane of a substrate and existing challenges to interface them with the outside world.

"Bottom-up" fabrication strategies and directed assembly methods have become increasingly attractive in arranging nanostructure in a hierarchical order. One of the key barriers to large-scale integration of functional nanowires and nanocrystals into devices and systems has been their difficulty in parallel interfacing with other device components. A basic prerequisite for reproducibly interfacing, on the order of millions of, nanocrystals is to be able to precisely control their growth sites. Shape of the nanocrystal significantly impacts the complexity of this process and thus nanocrystals with one elongated dimension fit in better with the available microfabrication protocols. In controlling the growth site of one-dimensional nanocrystals, also referred to as quantum wires, nanorods or nanowires, one popular approach is their growth in a free standing form using a metal–catalyst via a process called vapor-liquid-solid (VLS) (Wagner and Ellis 1964; Haraguchi, Katsuyama et al. 1991; Morales and Lieber 1998). The advantages of this growth technique include uniform composition and electronic structure of nanowires, and the ability to alter their composition by, for instance, introducing a quantum well (Bjork, Thelander et al. 2004) or a heterojunction (Bjork, Ohlsson et al. 2002; Wu, Fan et al. 2002). Furthermore, the VLS technique allows control over the nucleation site of a nanowire such that it is grown where the metal catalyst is deposited (Huang, Mao et al. 2001).

In applications requiring electron-hole recombination or separation, typically a n-p junction is needed that can be formed within a nanowire, either axially or radially (Putnam, Boettcher et al.; Garnett and Yang 2008). In another design, the n-p junction is formed at the contact point of nanowires with the underlying substrate (Dalui, Lin et al.; Fang, Zhao et al.; Park and Yi 2004; Chen, Chang et al. 2009). In both configurations, a top contact is necessary to complete the circuit. To have a functional device, prior to the top contact deposition, a non–conductive spacer is used to fill the void between nanowires (Park and Yi 2004; Sun, Huang et al. 2008; Kelzenberg, Boettcher et al. 2010). Different groups have used this technique for interfacing large ensembles of nanowires in photovoltaic applications; however, filling the void between nanowires can be a cumbersome step as it can force nanowires to collapse or lead to partial filling for closely packed nanowires. In other applications where nanowires need to be integrated in a planar format, the standing nanowires must be removed from the substrate and cast on a different surface. This step typically requires aligning nanowires to enhance the efficiency of interfacing them with other device components. Several strategies aiming at controlling hierarchical order of nanowires have been developed, such as electric field assisted orientation (Smith, Nordquist et al. 2000) and alignment with fluid flow in microchannels (Huang, Duan et al. 2001). Examples of other alignment techniques include the Langmuir-Blodgett technique(Kim, Kwan et al. 2001), assembly of densely packed nanowires using a pattern transfer process (Melosh, Boukai et al. 2003), shear force-alignment of nanowires in thin polymer films (Yu, Cao et al. 2007), and role-printing of nanowires (Yerushalmi, Jacobson et al. 2007). All of these techniques result in planar alignment of nanowires; however, they do not provide control over their surface registries for the successive patterning steps.

In this chapter, we discuss a fabrication technique that has the potential to address the limitations described above. This technique combines a "bottom-up" chemical approach with optical lithography and improves the scalability, precision, and fidelity of integrating nanowires to a platform. We discuss the surface-directed growth concept as well as the different steps necessary for scalable fabrication of some nanowire devices, such as field effect transistors (FETs) and light emitting diodes (LEDs). In the first section, we discuss the growth of ZnO nanowires on *a*-plane sapphire and their application in nanowire FETs. We provide some examples of fabricated nanowire FET devices with a single, two, or several nanowires. Furthermore, we analyze some of the electrical characteristics of individual nanowires and ensembles of nanowires. In the second part of this chapter, we discuss the growth of ZnO nanowires on gallium nitride (GaN), formation of planar arrays of n-p heterojunctions, and their LED behavior.

2. Surface-directed growth of ZnO nanowires on sapphire and their charge transport

Zinc oxide is a sensor, piezoelectric, UV light emitter, and transparent semiconductor in the visible spectrum, and as such is a technologically important material (Pan, Dai et al. 2002; Wang 2004; Wang, Song et al. 2007; Pearton, Lim et al. 2008). In its one-dimensional form, ZnO nanowires have been grown using a phase transport process at high temperature, or via a wet chemistry at low temperature (Wagner and Ellis 1964; Pan, Dai et al. 2002; Law, Greene et al. 2005). In the traditional VLS process, as described earlier, the nanocrystal growth is confined within a metal nanodroplet that dewets the substrate in early stages of growth, resulting in free-standing nanowires. In our technique, the metal nanodroplet stays in contact with the substrate during the ZnO growth, allowing the substrate to guide the horizontal growth of the nanocrystal; hence we call it a surface-directed VLS (SVLS) process. As the nanocrystal grows, the Au nanodroplet also moves on the surface while maintaining its interface with the substrate and nanocrystal. This method preserves the advantages of the VLS process while simultaneously allowing for careful control of the nanocrystal growth site and direction. Horizontal nanowires can be grown in a tube furnace at ambient pressure via a physical phase transport process. The anisotropic growth of a nanowire starts at the initial nanodroplet location and continues along the $[1\bar{1}00]_{sap}$ of an *a*-plane sapphire as the nanodroplet moves on the surface (Nikoobakht, Michaels et al. 2004). In this direction, the ZnO (*c*-plane) and sapphire (*a*-plane) have a better lattice match along their "**a**" and "*c*" axes, respectively, leading to anisotropic ZnO growth. Intentional anisotropic crystal growth has been previously observed, for instance, in the growth of one-dimensional rare-earth silicides or germanium on Si (Ogino, Hibino et al. 1999; Chen, Ohlberg et al. 2000; Ragan, Chen et al. 2003). These techniques, while resulting in planar growth of nanowires, do not offer the control over their growth site or orientation. Our nanowire-to-device integration strategy includes a two-step photolithography process which is schematically described in Fig. 1. First, small Au pads and global marks are placed on *a*-plane sapphire surface using conventional optical lithography (Fig. 1a). ZnO nanowires are then epitaxially grown on opposing sides of each Au pad in $[1100]_{sap}$ direction (Fig. 1b) inside a tube furnace. In the second photolithography step, patterns of metal electrodes are placed on the nanowires (Fig. 1c). To complete the fabrication of the field effect transistors, a third step of photolithography (not shown) is used to place the top gate electrode patterns on the nanowires. Prior to this patterning step, the nanowires are coated with a thin oxide (gate oxide) layer that protects nanowires from direct contact with the gate electrode.

Fig. 1. Fabrication flow for interfacing ZnO nanowires to metal contacts. a) Deposition of Au pads and global marks. b) Planar growth of ZnO nanowires from Au pads. c) Placing metal electrodes using conventional optical lithography.

For preparing microscale Au pads, following photolithography protocols, substrates are coated with a typical photoresist. Patterns of 1 µm x 5 µm pads along with global marks are created on the photoresist. Thin Au films, 1nm to 3 nm, are deposited on the photoresist using a thermal evaporator. Photoresist lift-off is carried out by sequential submersion of the substrates in two acetone containers (at 80 °C) for a period of about 25 minutes. The lift-off process is concluded by sequential washing the substrates in hot (75 °C) and room temperature ethanol for a total time of 10 minutes. To remove the organic residue, ozone cleaning is performed for about 3-4 minutes, followed by washing in deionized (DI) water, and nitrogen gas drying. At this point a Au-patterned substrate is transferred to the end of the small quartz tube and nanowires are grown according to the procedure described earlier. To grow horizontal ZnO nanowires, a modified approach (Nikoobakht, Michaels et al. 2004) of an earlier method for growing free standing nanowires (Wagner and Ellis 1964; Huang, Wu et al. 2001) was used. Briefly, a ZnO/graphite mixture (0.15 g, 1:1 mass ratio) is loaded on a Si substrate and positioned at the center of an inner tube (13 cm length, 1.9 cm inner diameter). The tube, containing a Au coated sapphire substrate, is inserted into a tube furnace such that the mixed powder is located at the center of the outer tube (80 cm length, 4.9 cm inner diameter). The furnace temperature is set at 890 °C (with a ramp rate of about 110 °C/min.) for 10 minutes under 0.6 standard liters per minute (SLPM) flow of 99.99% Ar or N_2 gas. 8 mm x 8 mm *a*-plane sapphire pieces are washed with small cotton swabs and DI water and then blown dry with nitrogen (99.99%). For Au nanodroplets less than 25 nm in size, we observe an in-plane and oriented growth of small diameter nanowires on *a*-plane sapphire. For larger size Au nanodroplets, the population of free-standing nanowires increases.

2.1 Directed growth of nanowires on sapphire

In the first photolithography step, the Au pads are deposited such that their short sides are perpendicular to the $[1\bar{1}00]_{sap}$ direction of the sapphire wafer (Fig. 2a-c). We have shown that this is a major growth direction of nanowires on the sapphire surface. The identified pad orientation results in a smaller number of nanowires per unit length (discussed later). A closer view of the positioning of the Au pads relative to the nanowire growth direction is shown in Fig. 2(a, b). An atomic force microscopy (AFM) image of a Au pad and its height profile are illustrated in Fig. 2(c-d), which shows an average height of 2.5 nm. Typically, a Au thickness ranging from 2 nm to 8 nm is suitable for horizontal nanowire growth. The thin Au film transforms to nanodroplets that, at higher temperatures, nucleate the nanowires.

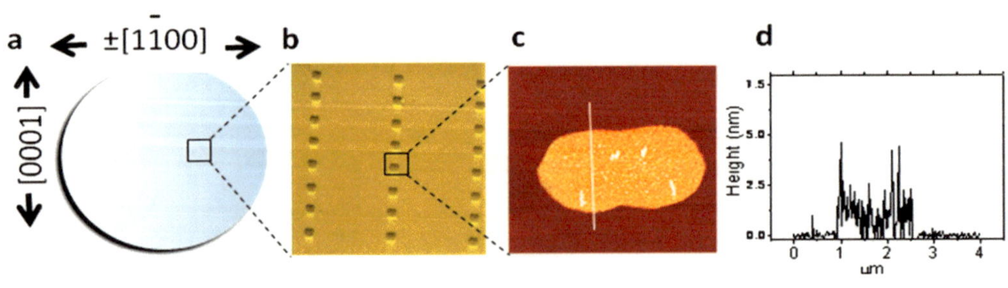

Fig. 2. Orientation of the deposited Au pads on the sapphire wafer. a) Sapphire wafer with its typical *c*-plane cut. b) SEM image of arrays of Au pads deposited on sapphire. c) AFM image of an individual Au pad after photoresist removal. d) Height profile of the Au pad shown in (c).

Only those nanodroplets residing at the two short sides of each Au pad produce horizontal ZnO nanowires (Fig. 3a-b). The group of nanowires shown in the scanning electron microscopy (SEM) image of Fig. 3b is also imaged by AFM (Fig. 3c) to characterize their dimensions and show the strength of this approach in locating a small group of nanowires on a large surface.

The AFM height profile of these nanowires shows that their diameter ranges from 8 nm to 13 nm. Typically, Au pads with 3 (±1) nm thicknesses result in nanowires with an average thickness of 11 (±3) nm and the nanowire density per pad width is found to be 10 nanowires/μm. The number of nanowires can be further reduced by decreasing the Au pad size or by increasing the resolution of the optical lithography. Based on SEM and AFM size measurements, the width-to-height ratio of the nanowires is typically found to be close to one; therefore, a semicircular profile or a faceted structure is assumed for horizontal nanowires. This is confirmed by examining the cross-sections of these nanowires, as demonstrated in Fig. 3(d, e). By increasing the number of deposited Au pads on the substrate, the scale of assembly of nanowires can be readily increased (Fig.4). The directionality of growth is mainly dictated by the underlying substrate, and so this unique growth direction is observed everywhere on the wafer as well as in sapphire wafers from different batches. As shown in Fig. 4, free-standing nanowires grow from the Au nanodroplets that are far from the periphery of the Au pads (Huang, Mao et al. 2001).

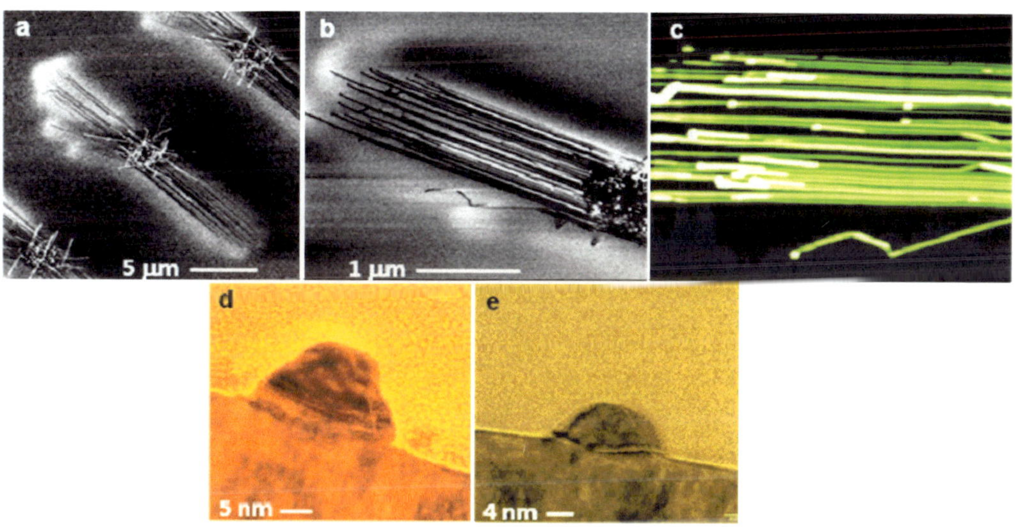

Fig. 3. a-b) SEM images of site-selective growth of horizontal nanowires. c) AFM image of the group of nanowires shown in part (b). d-e) Transmission electron microscope images of cross-sections of two individual nanowires.

2.2 Integration of horizontal nanowires with metal contacts

Since the relative positions of nanowires are known with respect to the global marks on the surface (from step 1), integration of the nanowires and metal electrodes can be carried out by aligning these marks with the complementary ones on the second layer of the photolithography mask. Metal electrode patterns are placed on nanowires across the whole 8 mm x 8 mm substrate with a precision that is limited to the resolution of the mask aligner.

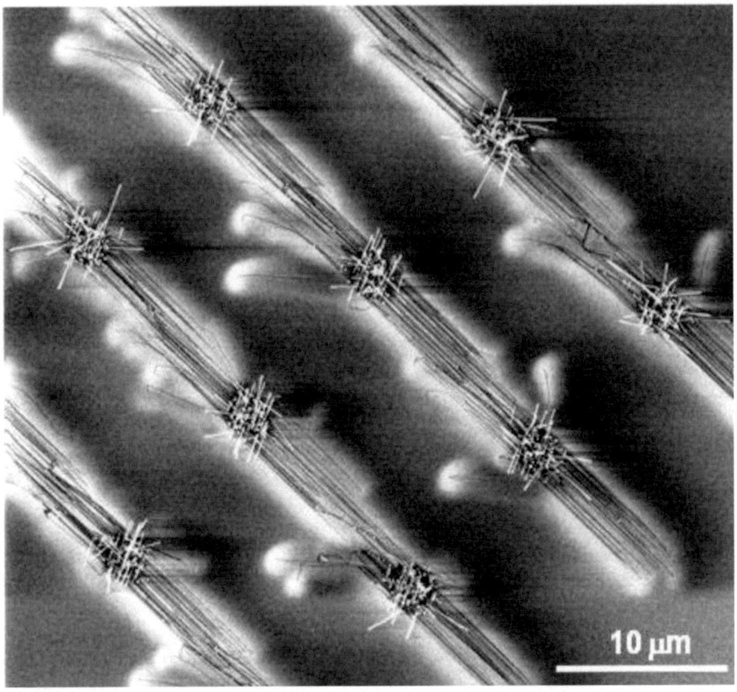

Fig. 4. Groups of horizontal nanowires made from planar arrays of Au pads.

Figure 5a shows a low magnification view of a sapphire wafer after deposition of metal electrodes and metal pads. The highlighted area in this figure is shown in the optical micrograph of Fig. 5b where the Au pads are seen as dark spots located underneath the metal electrodes. Nanowires are not resolved in this figure, but their direction is shown with the black arrow. In the used setup, alignment of nanowires and metal electrodes is such that the left metal contacts reside at one end of the nanowires. The highlighted box in Fig. 5b is further magnified in the SEM image of Fig. 5c in which the original Au pads are marked blue.

Depending on the number of nanowires grown from a given Au pad, a nanodevice can be comprised of single or multiple nanowires. Examples of devices with a few nanowires are shown in Figures 5(c) and 6(a-b). Devices containing double and single nanowires with multiple metal electrodes are shown in Figures 6(c- d). Using this technique, we are able to achieve parallel fabrication of *nanodevices* using a photolithography process with one micron feature resolution. This is in contrast to the current state-of-the-art nanowire-device fabrication in which registries of nanowires are not known and single nanowire devices are typically fabricated randomly in a serial fashion by electron-beam lithography. Single nanowire devices are of great interest, because of their enhanced characteristics that, most likely, smears out in devices containing multiple nanowires; however, single nanowire devices have not been well studied due to the limitation in their large scale formation. As mentioned earlier, by increasing the resolution of the optical lithography, better control over the number of nanowires per growth site is expected; considering the current advances in the semiconductor industry for feature miniaturization, if the deposited Au features can be reduced in size, the present technique has the potential to produce high density of such devices.

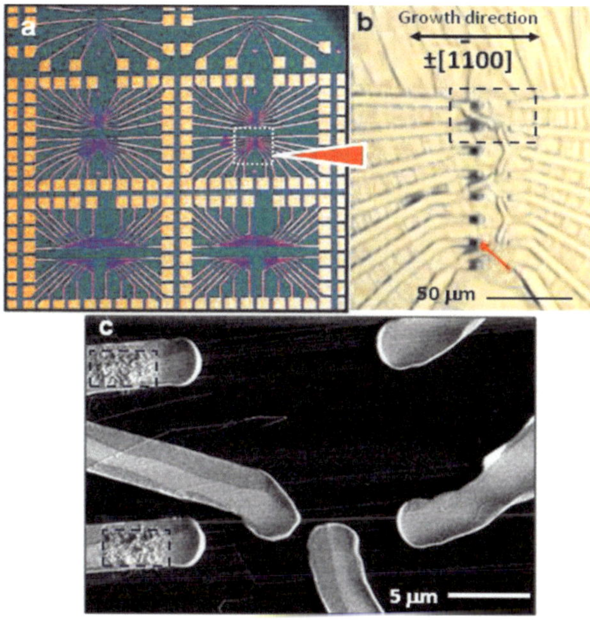

Fig. 5. Scale of placement of metal electrodes on nanowires. a) Low magnification optical image of electrodes and bonding pads. b) The highlighted box in part (a) is magnified here showing the overlap of the dark spots with the left hand side metal electrodes. c) the marked area in part (b) is shown in the SEM image depicting two groups of nanowires originiated from two Au pads (marked blue) and their orientation relative to the metal electrodes.

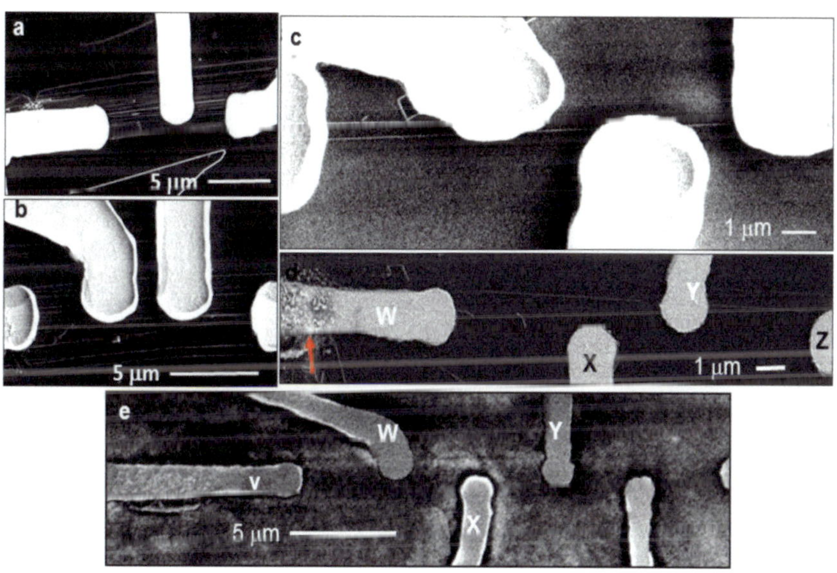

Fig. 6. Examples of fabricated multiple-, double, and single nanowire devices. The electrode patterns and their spacing are defined in the photomask. Aligning the global marks in the 1st and 2nd mask readily results in overlap of the nanowires with the metal electrodes. Letters V,W, X,Y, and Z in part (d-e) are explained in section 2.3.

2.3 Electrical measurements

In a single fabrication process, as described above, more than 600 nanodevices can be prepared composed of both single- and multi- nanowire devices. The results presented here are from three different fabrication processes; more than 20 nanodevices were tested repeatedly over the course of 3 months during which none showed any evidence of degradation or aging. Electrical measurements included two-terminal current vs. voltage (I-V) and transconductance (I_{ds}-V_G) were carried out using (Keithley 6430)[1] and (Hewlett-Packard 4140B)[1] source-measure units and a probe station. In the examined nanowires, the channel length ranged from 4 μm to 8 μm and nanowire diameters were between 7 (±1) nm to 22 (±1) nm. Depending on the length and diameter of the nanowires, the maximum current extracted at a 5V bias ranged from 2 nA to 180 nA. Among tested devices, different types of contacts are observed ranging from Ohmic to Schottky (Fig. 7). As an example, the I-V scan in Fig. 7a was obtained from the device in Fig. 6d (contacts WY) showing an Ohmic contact. In this device the nanowire length between the electrodes was 6.5 μm and its diameter was 12 (± 1) nm (measured by AFM). The I-V scan in Fig. 7b, corresponding to the device VW (Fig. 6e) with a length of about 4 μm and a width of 22 (±1) nm, displays a different behavior. In this case, the linear slopes of the I-V scans indicate the presence of an Ohmic contact, but with two different slopes for reverse and forward biases. The reasons behind the different slopes are not clear, but most likely are contact related. Although metal contact deposition is expected to be fairly homogenous for nanowires on a given substrate, the variation in electrical properties of the nanowires could be influenced by a number of other factors, such as contact contamination or inadequate thermal annealing. The results presented in this study are obtained without thermal annealing of the electrical contacts. Another device behavior is shown in Fig. 7c that has a rectifying behavior. Compared to previous nanowire devices, this nanowire has a diameter of about 7 nm, which is relatively small.

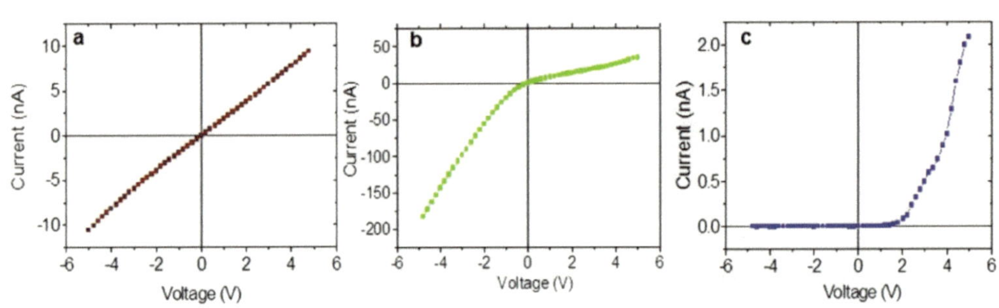

Fig. 7. Two-terminal current vs. source-drain voltage curves for three single NW devices.

The diode like behavior of this device could be due to its Schottky contacts or the smaller diameter of the nanowire. Nonetheless, information on charge transport at this size regime is very limited and more work is required to reach a comprehensive conclusion.

---

[1] Certain commercial equipment, instruments, or materials are identified in this paper to adequately specify the experimental procedure. In no case does such identification imply recommendation or endorsement by the National Institute of Standards and Technology nor does it imply that the materials or equipment identified are necessarily the best available for the purpose

2.4 Modulating the electrical conductivity of nanowires

For these measurements, a top-gated FET design (as illustrated in figure 8) was fabricated. In this design, a thin layer of silicon oxide (100 nm) is first deposited on nanowires using plasma-enhanced chemical vapor deposition (PECVD), followed by deposition of Au metal as gate electrode using optical lithography and thermal metal evaporation.

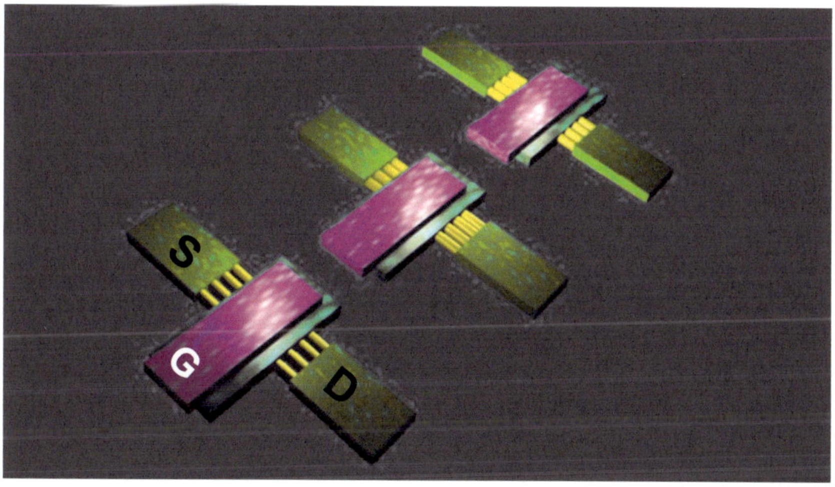

Fig. 8. Top-gated FET nanowires. Source, drain, and gate electrodes are labeled as S, D, and G, respectively.

Here the idea is to modulate the electrical current that passes through the nanowires by applying a voltage to a gate electrode that resides about 100 nm above the nanowires. As-grown ZnO nanowires typically have excess electrons, likely due to oxygen vacancies, and interstitials (Kohan, Ceder et al. 2000), and thus are n-type semiconductors. If that is the case, as gate electrode bias is increased to positive values, the conductivity of the nanowires increases. Typical current vs. source-drain voltage (I_{DS}-V_{DS}) scans are shown in Figures 9a and 9b for a multi-nanowire and a two-nanowire device, respectively. At a positive gate bias, the I_{DS} increases markedly; hence the channel is n-type (Sze 1969). A positive gate bias enables access to electrons (mostly in the valance band) of the nanowires, resulting in an increase in electron density in the conduction band and thus an increase in device conductivity. For a negative gate voltage, due to the repulsion of electrons by gate, the number of electrons in the conduction band decreases and thus the size of the conductive channel decreases. This is the trend that is also observed in nanowire devices at negative gate biases, as can be seen for the negative voltage region in Fig. 9c-d. The plots in this figure demonstrate the gate modulation efficiency (I_{DS}-V_G) for a variety of source drain voltages. For such devices the threshold voltage was found to be about -3V; this is the gate voltage at which nanowires stop charge transport. Since these nanowire devices have an intrinsic conductance at zero gate voltage, by definition they are classified as "depletion mode" FET devices. This behavior can be modeled using the following equation which portrays the idealized characteristic of a insulated-gate field effect transistor (Sze 1969):

$$\frac{dI}{dV_G} = \frac{\mu C}{L^2} V_{ds} \tag{1}$$

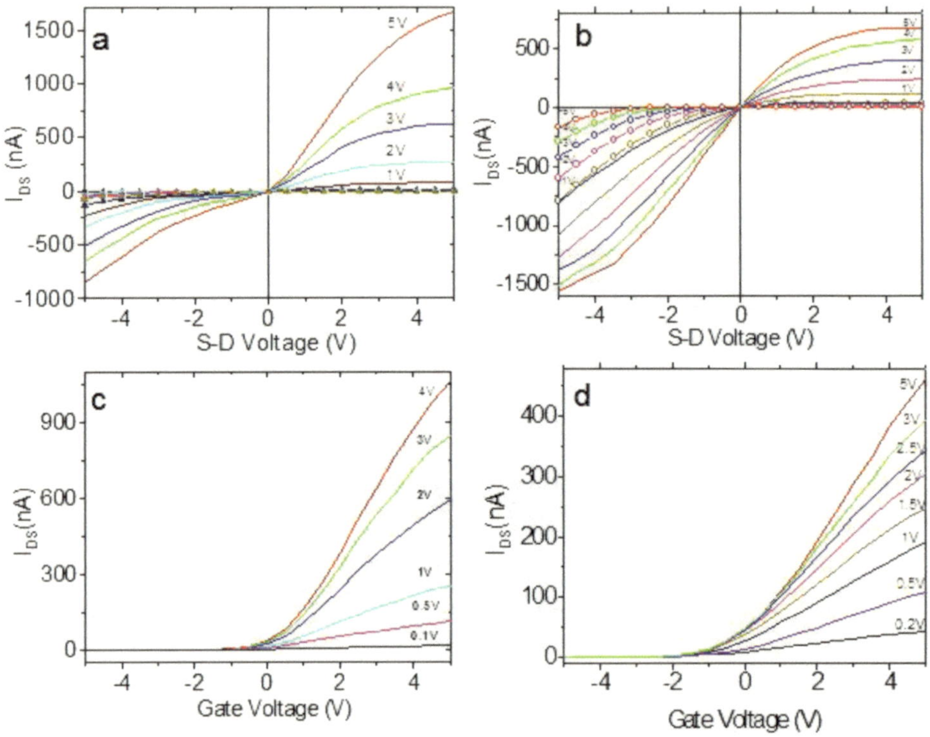

Fig. 9. Current-voltage and transconductance measurements of NW devices. I_{DS}-V_{DS} data for top-gated FET NW device containing: a) eight NWs, with diameter distribution of 13 (±5) nm and channel length of 8.5 μm. b) two NWs, 10 (±1) nm and 20 (±1) nm in diameter and a channel length of 6.3 μm. The scans were recorded at different gate voltages ranging from -5V to 5V. By increasing V_{DS}, I_{DS} linearly increased in both devices followed by device saturation due to a drastic drop-off in the charge carriers (electrons). When negative gate bias was applied (lines with symbols), channel conductivity gradually decreased to zero. For fixed V_{DS} values, I_{DS}-V_G scans were collected for: (c) an eight-NW device. d) a two-NW device. The absence of current saturation in this voltage range indicated low resistivity of the contacts.

In equation (1), C is the capacitance of the gate that is defined by $C = 2\pi\varepsilon_{SiO_2}\varepsilon_0 L / \ln(\frac{4h}{d})$; ε_0 is the permittivity of vacuum and ε_{SiO2} is relative dielectric of the gate oxide layer.
L is the device length, h is the oxide thickness, and d is the nanowire diameter. Using equation (1) and slopes of the graphs at V_{DS} of 1V in Fig. 9 (c-d), the field-effect mobilities, μ, for the multi-nanowire and two-nanowire devices are found to be ~ 15 cm^2/Vs and 20 cm^2/Vs, respectively. In calculating the mobility, the channel width was assumed to be the sum of the diameters of the nanowires. Although the profile of the nanowires is closer to a semicircular shape, the nanowire profile was considered circular when estimating the field effect mobilities. The on/off current ratios of the devices were found to be about 10^5 for a V_{DS} of 4V. Devices with multiple nanowires show characteristics comparable to those of thicker ZnO nanowires (*e.g.*, 100 nm diameter) (Fortunato, Barquinha et al. 2004). While having smaller individual

diameters (15 nm to 25 nm), they still net a larger total surface area. These could be suitable for sensing application where higher surface area is an advantage.
In examining single nanowire devices, I_{DS}-V_{DS} measurements were collected at different gate biases for a nanowire with 14(±1) nm diameter and a 7.1 μm channel length. As seen in figure 10, current intensity that goes through individual nanowires increases as the gate bias voltage increases (under forward bias). This device, compared to the two-nanowire device (Fig. 9b), shows a lower drain current, which could be due to Schottky contacts and the smaller nanowire diameter. Figure 10b shows that the device remains off at zero gate bias, indicating a very small conductive channel width. This defines this type of device as a FET in "enhancement mode". The threshold voltage was found to be about +1V.

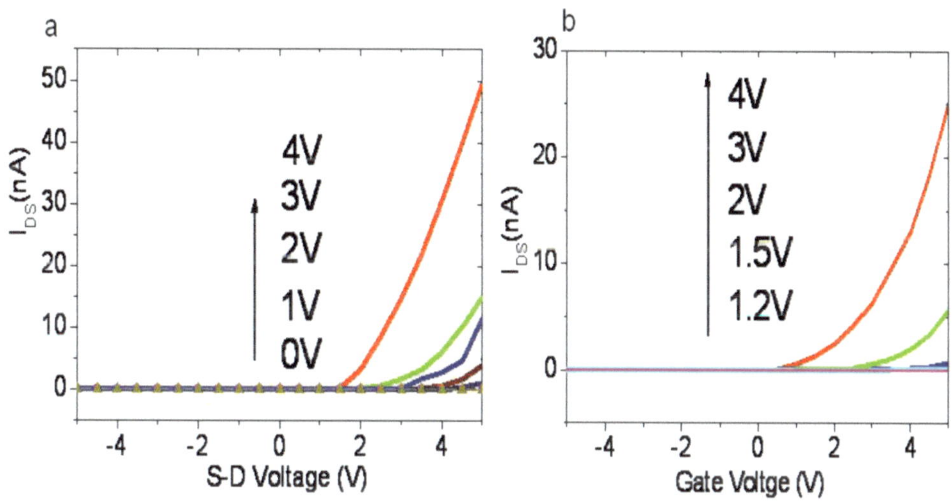

Fig. 10. a) I_{DS}-V_{DS} scans for a single NW FET. The maximum drain-current at V_{DS} of 5V increased from 1nA to about 50 nA within 0V to 4V of V_G. b) I_{DS}-V_G curves show that the single NW device is not conducting at zero gate bias, which is different from multi-NW devices (9c, d).

Among the tested devices it was noticed that single nanowires with diameters less than 15 nm tended to remain off under no gate bias. The observation of the two different modes in multiple and single nanowire devices is likely due to a decrease in the number of nanowires and also their smaller diameter. In low power applications, a device in the "enhancement mode" is more desirable over the "depletion mode" because there is no gate voltage necessary to turn the transistor off (Fortunato, Barquinha et al. 2004). This is a remarkable property that can be used to tune the characteristics of a nanowire FET. The field effect electron mobility measured for several single nanowire devices was found to be 4 (±2) cm^2/Vs with an on/off current ratio of at least ~5x10^4. It is notable that the values reported above were obtained for device lengths ranging from 4 μm to 8 μm. Nanowire surface engineering, e.g., overcoating nanowires with a material with suitable band gap and lattice constant, is expected to decrease the number of electron scattering sites in the nanowire and therefore improve the electron mobility. Compared to ZnO thin film transistors, ZnO nanowire devices show comparable electron carrier densities (10^{18} cm^{-3}) and field effect mobilities, but significantly lower threshold voltages (Fortunato, Barquinha et al. 2004). In all of our fabricated devices the threshold voltage (V_{th}) was found to be between 1V to -4V originating from a gate dielectric thickness of about 60 nm. ZnO nanowire devices are expected to have much lower V_{th} values and improved device behavior once a high-κ and structurally matched dielectric is used. More comprehensive transport studies on single and multi-nanowire are needed to further understand the electrical properties of this new class of nanomaterial.

A multitude of applications using this technique are possible. In the field of crystal growth, use of "mobile metal nanodroplets", such as Au, could be a new way for planar and localized growth of nanocrystals and their heterostructures. We have extended this approach to other (II-VI or III-V) semiconductor nanowires that have an anisotropic crystal mismatch with their underlying substrate, which is discussed in the following section.

3. Surface-directed growth of ZnO nanowires on GaN and their electro-optical properties

In addition to the growth of ZnO on sapphire, more recently SVLS was used for growth of GaAs nanowires on GaAs substrates (Fortuna, Wen et al. 2008). In the present work, we extend this process to formation of high quality heterojunctions of II-VI and III-V semiconductors; a characteristic that has been challenging to realize in homo- and heterojunction growth of these semiconductors (Kozodoy, Ibbetson et al. 1998; Vispute, Talyansky et al. 1998). To demonstrate this concept, we form addressable microarrays of p-n heterojunctions by lateral growth of n-type ZnO nanowires, a II-VI semiconductor with hexagonal structure, on a p-type GaN substrate with a similar crystal structure (Nikoobakht and Herzing 2010). ZnO, a wide band-gap semiconductor, is attracting renewed interest for UV light emission and lasing, piezoelectric devices (Wang and Song 2006), "invisible" circuitry (Nomura, Ohta et al. 2003), and photovoltaics (Law, Greene et al. 2005). Its band gap can be engineered by the addition of Mg or Cd dopants, which are inexpensive and abundant (Pearton, Lim et al. 2008). ZnO also exhibits one of the highest exciton binding energies (~ 60 meV), which makes it a good candidate for high efficiency UV-visible light emitters. To this end, we show a simple device design that allows charge injection to a large number of laterally grown heterojunctions. This approach could enable formation of a wide range of electrically addressable II-VI and III-V heterojunctions and their easy integration into photonic and lab-on-chip platforms with applications in energy generation and light detection.

The lateral growth of ZnO nanowires takes place on a Au-patterned *c*-plane GaN substrate. Au patterns were deposited either by dispersing Au nanoparticles (10 nm to 40 nm in size) or by using a thermal evaporator. The process is carried out inside a tube furnace that is held at 850 °C, with the substrate placed in a stream of ultra-dry N_2 carrier gas containing Zn and O precursors. The lateral growth of nanowires on GaN in the growth chamber is more sensitive relative to the growth of standing nanowires; for a reproducible growth, care must be taken to place Au coated substrates at the outlet of the small quartz tube (refer to section 2). At high temperature, the resulting Au nanodroplets promote the planar growth of ZnO nanocrystals via the SVLS process, which is influenced by the lattice match and crystal symmetry of the nanocrystals and the substrate. Crystal symmetry clearly demonstrates its effect in a system like ZnO (*a, b*= 0.32489 nm) on GaN (*a, b*=0.31894nm), which contains a lattice mismatch of ~ 1.8%. In the simplest case, we observed that using 10 nm to 40 nm Au nanoparticles dispersed on GaN, the substrate directs the growth in six equivalent directions of $<10\bar{1}0>$ with a hexagonal symmetry as shown in Figure 11(a,b).

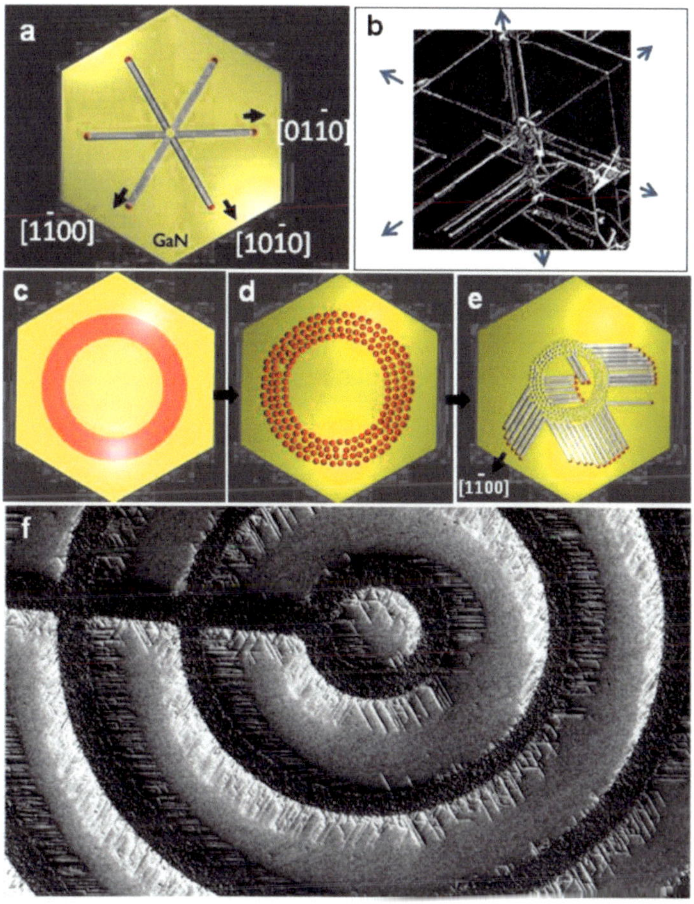

Fig. 11. a) Schematic showing the six-fold symmetry of the surface-directed VLS growth of ZnO nanocrystals using Au nanoparticles (red circles) on c-plane GaN and b) its corresponding electron micrograph illustrating these growth directions. c) The circular drawing in red represents the Au pattern deposited on GaN via standard photolithography, which in part (d) converts to nanodroplets at elevated temperatures. e) The Au nanodroplets at the boundaries of the circular pattern result in horizontal nanowires. f) Arrays of p-n heterojunctions made from n-type ZnO nanowires grown on p-type GaN.

The observed six-fold symmetry confirms the contribution of the underlying substrate in determining the growth direction. The site selectivity of the growth is shown by Au patterning a GaN surface and growth of ZnO nanostructures. In this process, a photolithographically-generated Au pattern with a thickness of 8 (±2) nm (Fig. 11c) at elevated temperatures (500 °C -700 °C) converts to packs of Au nanodroplets (Fig. 11d). As shown in Figure 11e, at a temperature range of 840 °C- 900 °C, while the ZnO phase transport is in progress only Au nanodroplets formed at the perimeter of a Au pattern have the opportunity to laterally move on the substrate and form nanowires. Nanowires preferentially grow in three directions and are all electrically connected via the ZnO backbone that can be seen as the dark-gray circular pattern.

The nanowire width stays constant during its lateral growth, indicating that diffusion of Au to the nanowire facets and reduction in the volume of the Au nanodroplets remains insignificant for nanodroplets smaller than 20 nm (Hannon, Kodambaka et al. 2006). When the diameter of the Au nanodroplets increases above 20 nm, the nanowires acquire a new dimension by growing in the $[0002]_{ZnO}$ direction (normal to the substrate plane). Utilizing this observation, by depositing Au patterns as shown in Figure 12a with thickness of 10 (±2) nm to 17(±2) nm (step 1), it is possible to laterally grow ZnO nanowalls. Steps 2 and 3 in this figure represent the formation of Au nanodroplets followed by growth of ZnO nanowalls, respectively. The schematic in step 3 illustrates tilted view of an assembly. Following these steps results in assemblies of ZnO nanowalls and standing nanowires as seen in Figure 12b. This assembly is grown from 5 mm long Au lines with 5 μm of width and 20 μm pitch. In this figure, patterned Au lines are deposited parallel to the $[10\bar{1}0]$ direction of GaN in order to allow the lateral growth of ZnO nanowalls in only two directions ($[01\bar{1}0]$ and $[1\bar{1}00]$).

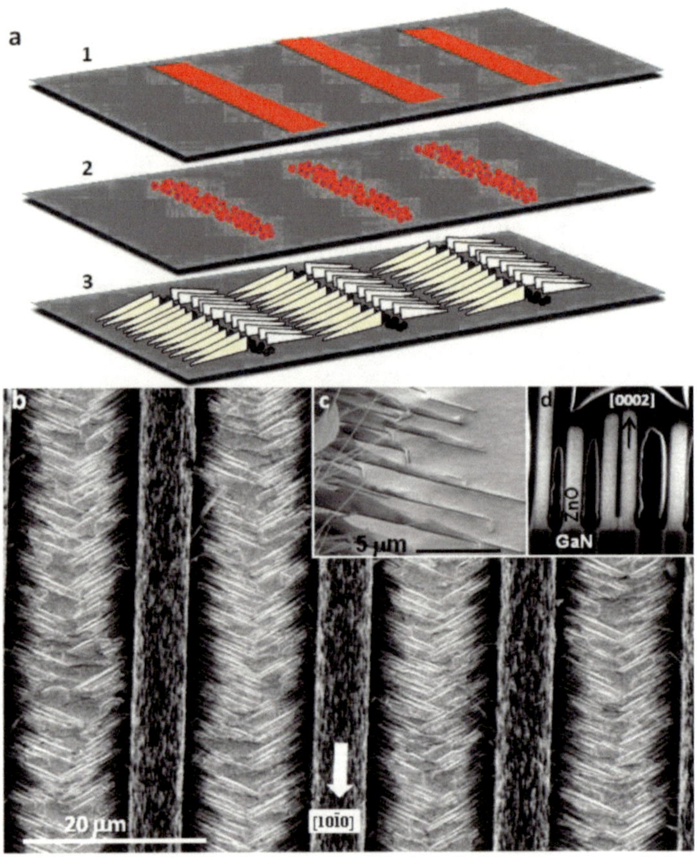

Fig. 12. a) Steps for growing assemblies of laterally grown ZnO nanowalls. b) An assembly of ZnO nanowalls grown from a periodic array of Au lines with a film thickness of 15 (±2) nm, width of 5 μm and length of 5 mm. Au lines are deposited along the $[10\bar{1}0]_{GaN}$. c) A tilted SEM view of ZnO nanowalls in a small section of the assembly from part (a). d) STEM image of cross-sections of a group of ZnO nanowalls showing their upright growth in *c*-direction and their four major facets.

Examining the cross-sections of closely packed ZnO nanowalls (Figure 12b) shows they are vertical ZnO slabs with four dominant facets that can grow a few microns in the direction normal to the interface. Due to the large number of Au nanodroplets formed at the perimeter of each Au line, a high density of nanowalls can be grown, which is advantageous in applications such as photovoltaics and photodetection where the surface area is important.

Figure 13 shows that the height of a nanowall gradually decreases toward its leading end, indicating that its vertical growth most likely is due to a slower self-catalytic process (Wang, Kong et al. 2003). Previously, in the case of ZnO nanocantilevers, self-catalytic growth was attributed to the formation of Zn clusters on the (0002) polar surface of ZnO nanowires (Wang, Kong et al. 2003). Since nanowalls are only observed in the case of larger sized Au nanodroplets, it is possible that a larger volume Au nanodroplet also results in formation of Zn cluster (at its eutectic point) that promotes the growth in the ZnO [0002] direction.

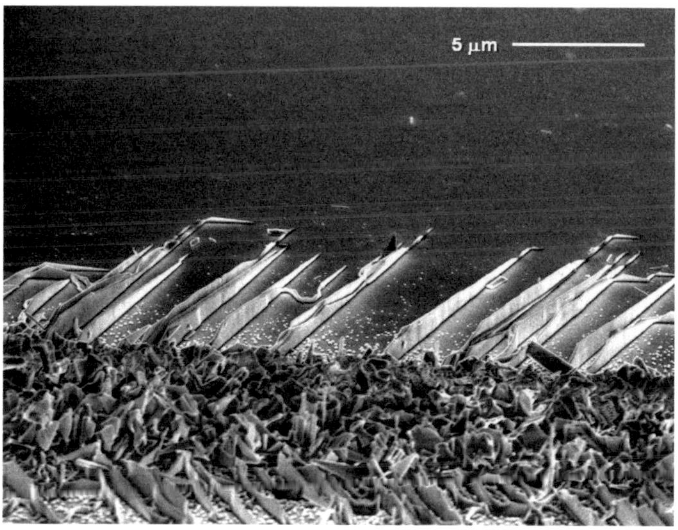

Fig. 13. Tilted view of grown ZnO nanowalls on GaN. The height decreases toward the leading end of nanowalls.

3.1 ZnO-GaN p-n heterojunctions and fabrication of nano-LEDs

The present technique is suitable for fabricating II-VI nanowire-based devices on III-V semiconductor substrates including GaN and its other combinations, such as InGaN and AlGaN, to realize platforms for tunable light emission and detection. As an example, in Figure 14a, a simple strategy is demonstrated for fabricating arrays of electrically-driven, light-emitting, p-n heterojunctions formed at the interface between ZnO nanowires or nanowalls and the GaN substrate.

The GaN substrate is *c*-plane and 5 µm thick, which is grown on *c*-plane sapphire. The GaN layer is doped with Mg at a concentration of 5 x 10^{17} cm^{-3} (purchased from TDI[1]). In this technique, since the relative location of nanowire arrays are known, metal contacts can be readily deposited via standard photolithography over the ZnO backbones. A representative image of a three-layer (metal/ ZnO nanowalls /GaN) structure is shown in Figure 14b, where the top metal electrode (Ti-Au) deposited on the ZnO backbone is highlighted in yellow. During the device operation, the metal contacts on the ZnO backbones allow electron injection into nanowires or nanowalls and hole injection to the p-GaN side. The location, size and shape of the light emitting devices can be controlled by depositing Au catalyst patterns ranging from a few microns to several millimeters with triangular, circular, and square shapes (Nikoobakht and Herzing 2010).

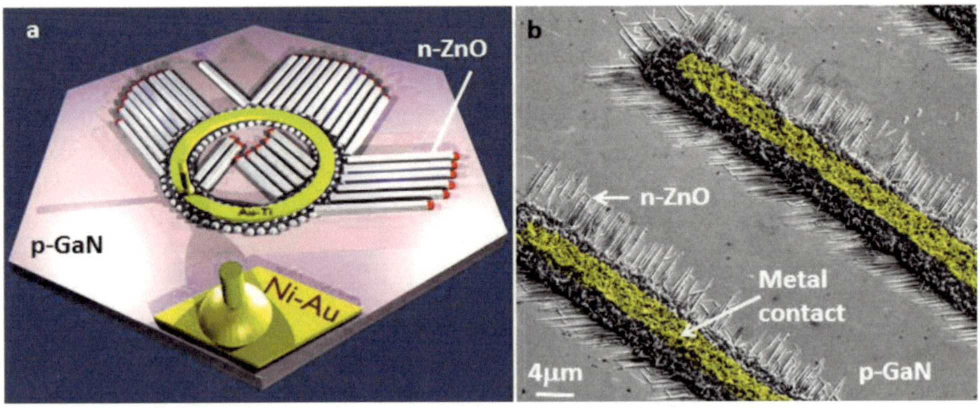

Fig. 14. a) Schematic showing a tilted view of an electrically-driven array of p-n heterojunctions made from n-type ZnO nanowalls on p-type GaN. In this cartoon, the GaN substrate is shown as a hexagon. b) SEM image of the tilted view of a device is illustrated; the yellow color indicates the deposited metal electrode on the ZnO backbone. All other device layers are labeled in the figure.

Current-voltage (I-V) characteristics of the as grown n-p heterojunctions exhibit rectifying behavior for small and large groups of nanowires or nanowalls. Multiple devices were repeatedly operated from -25 V to 27 V in order to examine the structural and functional stability of the p-n heterojunctions. Results in Figures 15a and 15b, respectively, show the I-V scans of a 5 mm x 5 mm n-p heterojunction array and a 100 μm-size array.

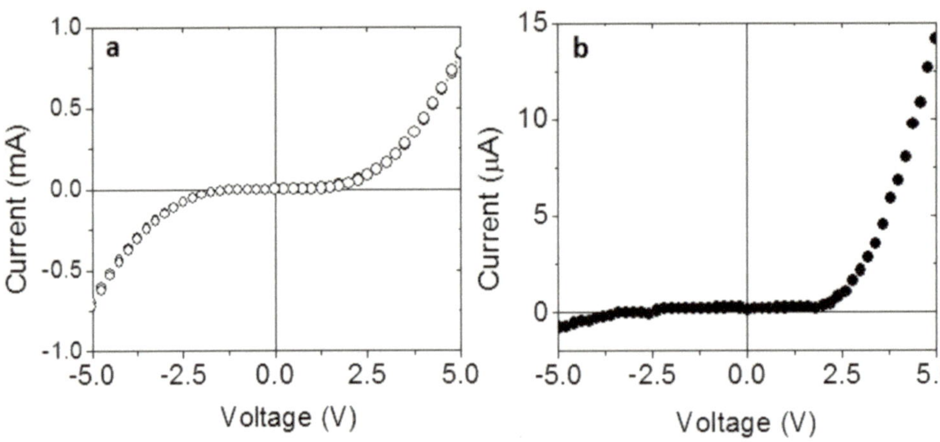

Fig. 15. I-V scans of two nanowall-LED devices with lateral dimensions of: a) 5 mm and b) 100 μm.

As seen in these figures, the injected current is substantially higher in larger sized devices, due to the larger number of involved nanowires as well as the contribution of the ZnO backbone in conducting the charges to the GaN layer. We typically observe a turn on voltages of 2.7 V to 3.8 V (± 0.2), breakdown voltages of 3V to 9V, and a diode ideality factor ranging from 3 to 3.4, which are comparable to those of free-standing heterojunction nanowires (Nikoobakht and Herzing 2010) (Tian, Zheng et al. 2007).

3.2 Electrically-driven light emission of p-n heterojunction arrays

In the fabricated devices, the dopant concentration of the GaN layer is 5 x 10^{17} cm^{-3} and the carrier concentration of the ZnO nanowires is about 1×10^{18} cm^{-3} (based on previous electrical measurements (Nikoobakht 2007)). Upon applying a DC voltage to metalized arrays of p-n heterojunctions made from ZnO nanowalls on p-GaN, individual heterojunctions in the array independently emit light, as can be seen from the electroluminescence (EL) image in Figure 16a. This image was collected from part of a device with a faceted-spiral pattern (inset) containing a millimeter-long ZnO backbone that is under an injection current of ~0.9 mA at +7V. Results show that both nanowires and nanowalls emit along their entire length in contrast to the mechanically formed p-n junctions where emission originates only from the parts of a nanowire that are in contact with the substrate (Lee, Kim et al. 2007; Zimmler, Stichtenoth et al. 2008). Another EL emission example of a nanowall-based p-n heterojunction array with lateral dimensions of 5 mm x 5 mm is illustrated in Figure 16b. The dark lines in this image are the metal contacts that block some of the emitted light. The strength of this growth technique is that it allows large scale charge injection into planar nanowire LEDs with a unique control that is difficult to achieve using other techniques.

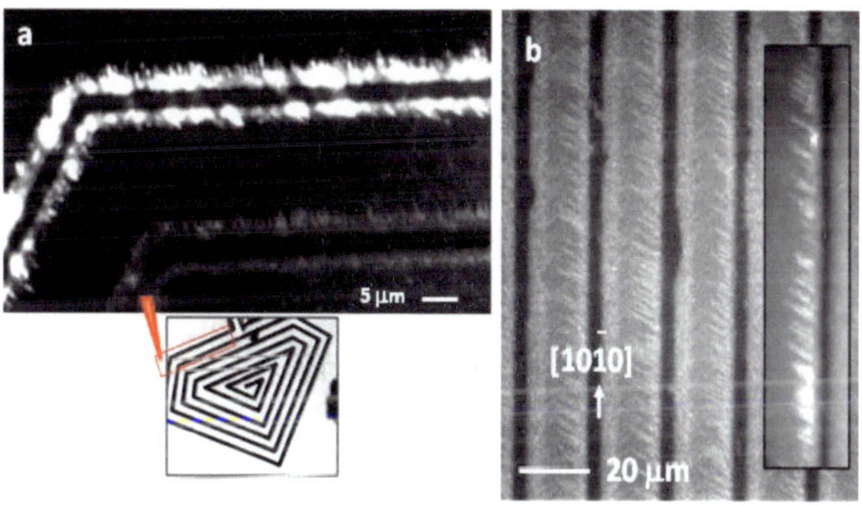

Fig. 16. a) The electroluminescence image of a group of emitting p-n heterojunctions that is part of a millimeter-long nanowall array with a ZnO backbone in the shape of a faceted spiral, which is shown in the optical image in the inset. In this example, the injected current is ~0.9 mA at +7V. b) The electroluminescence image of a 5 mm-long array of emitting nanowall p-n heterojunctions with a 20 µm-pitch array spacing. The arrays are formed along the specified direction relative to the GaN *m*-planes. The inset shows the same array, but emitting under a forward injection current of 0.02A.

In order to further probe electro-optical characteristics of these heterojunctions, the EL spectra of a number of heterojunction arrays with different shapes and sizes (from 100 μm to 5 mm) were collected at voltages from -25 V to 27 V using an optical fiber mounted above an active device. Figure 17a illustrates an example in which two spectra are collected from a large array of nanowall p-n heterojunctions (EL image shown in Figure 16b) under a forward bias from 9V to 25V. The strong excitonic emission of the device at about 390 nm agrees well with the photoluminescence (PL) of ZnO nanowalls at 380 nm (Fig. 17b), indicating that holes are injected into the n-type ZnO. This is also in agreement with the brightly emitting nanowires in Figure 16 (a-b) showing that the majority of the charge recombination occurs in the ZnO region. The strong emission and the narrow spectral bandwidth (32 nm FWHM) are indicative of a low density of interface states that depress the radiative recombination probability. The inset of Figure 16a shows the blue emission of a 5 mm^2-array of p-n heterojunctions at 20 V with an injection current of 0.017 A. The collective emission of the p-n heterojunctions is intense enough that can be observed by the unaided eye.

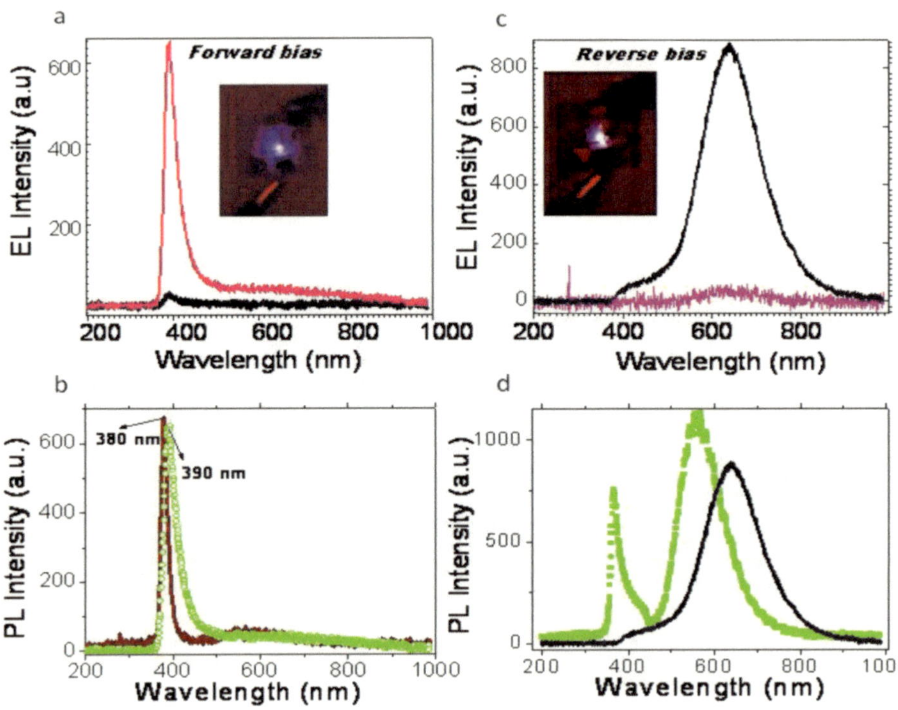

Fig. 17. a) EL spectra of a p-n heterojunction array (similar to the array in Fig.16b) with a lateral dimension of 5 mm under a forward bias of -13 V to -25 V (injection current of 0.01 A to 0.036 A). Inset shows the blue emission of the corresponding device at a forward bias of 20V collected using a digital camera after 30s of exposure time. b) EL emission of p-n heterojunctions at forward bias (circle symbol) and its good agreement with PL emission of the p-n heterojunctions (solid line). c) EL emission at 640 nm from the same array, but under a reverse bias of +7 V to 15V (0.003-0.011A injection current). Inset shows the orange emission of the same device at a reverse bias of 20 V. d) Compares EL emission at 640 nm at a reverse bias (solid line) with PL of p-GaN substrate (square symbol).

By applying reverse bias voltages above 6V, i.e., beyond the break down voltage of the examined devices, we observe a bright orange emission with a wavelength centered at 640 nm (1.95 eV) as shown in the spectra of Figure 17c and inset. This emission is different from PL emission of p-GaN (black curve-Fig. 17d), which also indicates that the charge recombination does not occur in the GaN layer. Although this is still under study, we note that the 640 nm emission intensity decreases as the surface area of the ZnO backbone-GaN decreases. This drop in intensity also coincides with a better rectifying behavior and an increase in the breakdown voltage from 4 V to more than 9 V in smaller size devices (refer to Figures 15 a-b).
Results suggest that the ZnO backbone has a contribution to this emission; nonetheless we do not observe this emission in the PL of the corresponding devices on GaN (brown curve-Fig. 17b). With regard to the electronic states involved in this emission, we tentatively attribute the EL emission under the reverse bias to a transition from an intrinsic shallow state (such as Zn_i) to an intrinsic deep state such as oxygen vacancies. The shape of the EL emission spectrum does not change by increasing the injection current in both reverse and forward biases. However, by increasing the injected current (I) the intensity of the emitted light first grows then reaches a mild saturation at higher voltages. This could be an indication of a limited charge carrier concentration at the junction, heating effects, or increase in non-radiative processes. In addition to ZnO, we observe the lateral growth for Ti_2O_3 and GaN on substrates such as Al_2O_3 (Nikoobakht 2007) and GaN, which possess hexagonal or trigonal/ rhombohedral crystal structures.

4. Conclusion

A nanodevice fabrication method is described which is based on the combination of a "bottom-up" chemical method and conventional optical lithography. Compared to other available techniques, this method controls the surface registries of nanocrystals for successive fabrication steps. The chemical method provides the horizontal growth of nanowires from individual Au nanodroplets, and their alignment is dictated by the underlying substrate. Nanowires are grown where the devices are to be fabricated, and the need for nanowire transfer or alignment is eliminated. This SVLS growth of nanowires for a wide group of materials is likely, especially III-V and II-VI semiconductors, since growth of their free-standing forms has already been reported using this process (Fortuna, Wen et al. 2008). The present technique maintains the strength of the VLS-based methods in growth of axial and core-shell heterojunctions, while allowing control of their hierarchical assembly through surface-directed processes. The flexibility in material combination and the ability to grow dense arrays of laterally grown nanowires/nanowalls could lead to the development of novel heterojunctions that are electrically addressable on relevant scales. Due to the unique control over the location and orientation of arrays of nanowires afforded by this technique, and its compatibility with conventional microfabrication methods, it is expected to impact the field of scalable nanowire-based devices by enabling the realization of numerous structural combinations and device concepts.

5. Acknowledgment

We would like to thank Dr. S. Eustis for collecting the TEM images of cross-sections of nanowires.

6. References

Bjork, M. T., et al. (2002). One-dimensional Steeplechase for Electrons Realized, *Nano Letters* Vol.2 No. (2)pp: 87-89.

Bjork, M. T., et al. (2004). Few-Electron Quantum Dots in Nanowires, *Nano Letters* Vol.4 No. (9)pp: 1621-1625.

Chen, C.-H., et al. (2009). Electroluminescence from n-ZnO nanowires/p-GaN heterostructure light-emitting diodes, *Appl. Phys. Lett.* Vol.95 No. pp: 223101.

Chen, Y., et al. (2000). Self-assembled growth of epitaxial erbium disilicide nanowires on silicon (001), *Applied Physics Letters* Vol.76 No. (26)pp: 4004-4006.

Dalui, S., et al. Electroluminescence from Solution Grown n-ZnO Nanorod/p-GaN-Heterostructured Light Emitting Diodes, *Journal of The Electrochemical Society* Vol.157 No. (5)pp: H516-H518.

Fang, F., et al. Hydrogen effects on the electroluminescence of n-ZnO nanorod/p-GaN film heterojunction light-emitting diodes, *Physical Chemistry Chemical Physics* Vol.12 No. (25)pp: 6759-6762.

Fortuna, S. A., et al. (2008). Planar GaAs Nanowires on GaAs (100) Substrates: Self-Aligned, Nearly Twin-Defect Free, and Transfer-Printable, *Nano Letters* Vol.8 No. (12)pp: 4421-4427.

Fortunato, E. M. C., et al. (2004). Wide-bandgap high-mobility ZnO thin-film transistors produced at room temperature", *App. Phys. Lett.* Vol.85 No. (54)pp: 3.

Garnett, E. C. and P. Yang (2008). Silicon Nanowire Radial p-n Junction Solar Cells, *Journal of the American Chemical Society* Vol.130 No. (29)pp: 9224-9225.

Hannon, J. B., et al. (2006). The influence of the surface migration of gold on the growth of silicon nanowires, *Nature* Vol.440 No. (7080)pp: 69-71.

Haraguchi, K., et al. (1991). GaAs p-n junction formed in quantum wire crystals, *Appl. Phys. Lett.* Vol.60 No. (6)pp: 745-747.

Huang, M. H., et al. (2001). Room-temperature ultraviolet nanowire nanolasers, *Science* Vol.292 No. (5523)pp: 1897-1899.

Huang, M. H., et al. (2001). Catalytic Growth of Zinc Oxide Nanowires by Vapor Transport, *Advanced Materials* Vol.13 No. (2)pp: 113-116.

Huang, Y., et al. (2001). Directed Assembly of One-Dimensional Nanostructures into Functional Networks, *Science* Vol.291 No. pp: 4.

ITRS (2006). International technology roadmap for semiconductors 2006 update, No.

Kelzenberg, M. D., et al. (2010). Enhanced absorption and carrier collection in Si wire arrays for photovoltaic applications, *Nat Mater* Vol.9 No. (3)pp: 239-244.

Kim, F., et al. (2001). Langmuir-Blodgett Nanorod Assembly, *J. Am. Chem. Soc.* Vol.123 No. (18)pp: 4360-4361.

Kohan, A. F., et al. (2000). First-principles study of native point defects in ZnO, *Physical Review B* Vol.61 No. (22)pp: 15019-15027.

Kozodoy, P., et al. (1998). Electrical characterization of GaN p-n junctions with and without threading dislocations, *Appl. Phys. Lett.* Vol.73 No. (7)pp: 975-977.

Law, M., et al. (2005). Nanowire dye-sensitized solar cells, *Nat Mater* Vol.4 No. (6)pp: 455-459.

Lee, S. K., et al. (2007). High-brightness gallium nitride nanowire UV-blue light emitting diodes, *Philos. Mag.* Vol.87 No. (14)pp: 2105 - 2115.

Melosh, N. A., et al. (2003). Ultrahigh-Density Nanowire Lattices and Circuits, *Science* Vol.300 No. (5616)pp: 112-115.

Morales, A. M. and C. M. Lieber (1998). A Laser Ablation Method for the Synthesis of Crystalline Semiconductor Nanowires, *Science* Vol.279 No. (5348)pp: 208-211.

Nikoobakht, B. (2007). Toward industrial-scale fabrication of nanowire-based devices, *Chem. Mater.* Vol.19 No. (22)pp: 5279-5284.

Nikoobakht, B. and A. Herzing (2010). Formation of Planar Arrays of One-dimensional p-n Heterojunctions using Surface-directed Growth of Nanowires and Nanowalls, *ACS Nano* No.

Nikoobakht, B., et al. (2004). Horizontal growth and in situ assembly of oriented zinc oxide nanowires, *Appl. Phys. Lett.* Vol.85 No. (15)pp: 3244-3246.

Nomura, K., et al. (2003). Thin-Film Transistor Fabricated in Single-Crystalline Transparent Oxide Semiconductor, *Science* Vol.300 No. (5623)pp: 1269-1272.

Ogino, T., et al. (1999). Fabrication and Integration of Nanostructures on Si Surfaces, *Acc. Chem. Res.* Vol.32 No. pp: 8.

Pan, Z. W., et al. (2002). Nanobelts of semiconducting oxides, *Science* Vol.291 No. pp: 3.

Park, W. I. and G. C. Yi (2004). Electroluminescence in n-ZnO Nanorod Arrays Vertically Grown on p-GaN, *Advanced Materials* Vol.16 No. (1)pp: 87-90.

Park, W. I. and G. C. Yi (2004). Electroluminescence in n-ZnO Nanorod Arrays Vertically Grown on p-GaN. Adv. Mater. 16: 87-90.

Pearton, S. J., et al. (2008). ZnO and Related Materials for Sensors and Light-Emitting Diodes *J. Electron. Mater.* Vol.37 No. (9)pp: 1426-1432.

Putnam, M. C., et al. Si microwire-array solar cells, *Energy & Environmental Science* Vol.3 No. (8)pp: 1037-1041.

Ragan, R., et al. (2003). Ordered arrays of rare earth silicide nanowires on Si(001), *J. Cryst. Growth* Vol.251 No. pp: 657.

Smith, P. A., et al. (2000). Electric Field Assisted Assembly and Alignment of Metallic Nanowires, *Appl. Phys. Lett.* Vol.77 No. (9)pp: 3.

Sun, X. W., et al. (2008). A ZnO Nanorod Inorganic/Organic Heterostructure Light-Emitting Diode Emitting at 342 nm, *Nano Letters* Vol.8 No. (4)pp: 1219-1223.

Sze, S. M. (1969). *Physics of Semiconductor Devices*, Wiley: New York.

Tian, B., et al. (2007). Coaxial silicon nanowires as solar cells and nanoelectronic power sources, *Nature* Vol.449 No. (7164)pp: 885-889.

Vispute, R. D., et al. (1998). Heteroepitaxy of ZnO on GaN and its implications for fabrication of hybrid optoelectronic devices. *Appl. Phys. Lett.* , AIP. 73: 348-350.

Wagner, R. S. and W. C. Ellis (1964). Vapor-Liquid-Solid Mechanism of Single Crystal Growth, *Appl. Phys. Lett.* Vol.4 No. (5)pp: 89-90.

Wang, X., et al. (2007). Direct-Current Nanogenerator Driven by Ultrasonic Waves, *Science* Vol.316 No. (5821)pp: 102-105.

Wang, Z. L. (2004). Zinc oxide nanostructures: growth, properties and applications *J. Phys.: Condens. Matter* Vol.16 No. pp: R829-R858.

Wang, Z. L., et al. (2003). Induced Growth of Asymmetric Nanocantilever Arrays on Polar Surfaces, *Phys. Rev. Lett.* Vol.91 No. (18)pp: 185502.

Wang, Z. L. and J. Song (2006). Piezoelectric Nanogenerators Based on Zinc Oxide Nanowire Arrays, *Science* Vol.312 No. (5771)pp: 242-246.

Wu, Y., et al. (2002). Block-by-Block Growth of Single-Crystalline Si/SiGe Superlattice Nanowires, *Nano Lett.* Vol.2 No. (2)pp: 83-86.

Yerushalmi, R., et al. (2007). Large scale, highly ordered assembly of nanowire parallel arrays by differential roll printing, *Applied Physics Letters* Vol.91 No. (20)pp: 203104-3.

Yu, G., et al. (2007). Large-area blown bubble films of aligned nanowires and carbon nanotubes, *Nat Nano* Vol.2 No. (6)pp: 372-377.

Zimmler, M. A., et al. (2008). Scalable Fabrication of Nanowire Photonic and Electronic Circuits Using Spin-on Glass, *Nano Lett.* Vol.8 No. (6)pp: 1695-1699.

13

Organic Surface Modification of Silicon Nanowire-Based Sensor Devices

Louis C.P.M. de Smet[1], Daniela Ullien[1], Marleen Mescher[1,2,3] and Ernst J.R. Sudhölter[1]
[1]*Delft University of Technology, Faculty of Applied Science, Department of Chemical Engineering, Nano-Organic Chemistry,*
[2]*Materials innovation institute (M2i),*
[3]*Micro Systems and Devices, Philips Research Eindhoven, Netherlands*

1. Introduction

The year 2011 marks the 10th anniversary of silicon nanowire (SiNW)-based electronic devices. Since their introduction (Cui & Lieber, 2001) SiNW-based sensor devices have gained considerable interest as a general platform for ultra-sensitive, electrical detection of biological and chemical species (Figure 1). Although SiO_x coatings can be used for the detection of protons (Cui, 2001), and gases (Wan, 2009), the specific detection of other analytes, including ions and biomolecules requires the presence of an affinity layer that interacts with the analyte of interest. Such a layer can be added on top of the nanowire by the modification of the nanowire surface. In this chapter we review the surface modification strategies that have been explored on SiNW-based devices over the past decade.

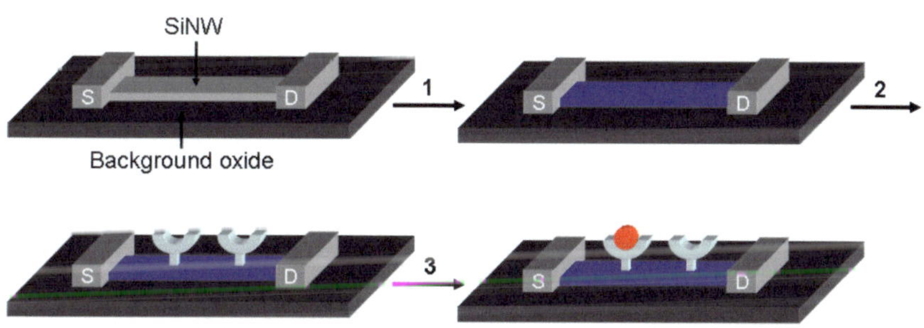

Fig. 1. Schematic representation of a SiNW-based device (top left), of which the nanowire has been chemically modified (step 1), followed by the immobilization of receptors (step 2) and the binding of an analyte (step 3). S and D stand for source and drain, respectively.

1.1 Scope and organization of this review chapter

This chapter focuses on surface modification strategies of SiNW-based devices. Our review concentrates on electrical devices that consist of in-plane orientated SiNWs, which are positioned between two ohmic contacts, often referred to as source and drain.

For an extensive overview of different materials that have been explored to prepare NW-based sensors, including silicon, we refer to the review of Ramgir et al. (Ramgir, 2010).
Several reviews on SiNW-based devices have been written with a clear focus on the chemical-vapour-deposition fabricated NWs by the Lieber and Zhou groups (Li, 2006; Patolsky, 2006b; Carlen & van den Berg, 2007) or top–down fabricated SiNW-based sensors from the Reed group (Stern, 2008). In this chapter we will discuss issues related to the modification of SiNW-based devices with organic compounds. The topics of SiO_x (Aswal, 2006) and Si-H (Aswal, 2006; Stutzmann, 2006; Hamers, 2008) surface modification with covalently linked organic monolayers have been reviewed extensively. However, the main focus of these reviews is on the modification of large planar substrates, while the focus if this chapter is on SiNW-based devices. A review by Wanekaya et al. contains a paragraph on the covalent functionalization of SiNWs with biomolecules (Wanekaya, 2006). Very recent reviews also include sections with brief schemes for SiNW functionalization (Shao, 2010; Tey, 2010).
In this chapter we aim to give an overview of all the different surface modification strategies that have been explored to modify SiNW-based devices, including non-covalent immobilization strategies. The methods can be divided into different categories (Table 1). The first one is based on silanization chemistry, i.e., the reaction between hydroxyl-terminated surfaces and organosilanes. Also other compounds like activated esters or organophosphonates have been covalently attached to SiO_x surfaces. In addition, polyelectrolytes and bilayers have been physically adsorbed onto SiO_x surfaces. The last approach follows the route of oxide removal, followed by the reaction with (ω-functionalized)-1-alkenes.

Surface	Modification Method	Type of Binding	Nature Film	# Papers (Year of First Report)	§
SiO_x	Silanization	Covalent	Monolayer	>50 (2001)	2.1.1
SiO_x	Esterification	Covalent	Monolayer	1 (2004)	2.2
SiO_x	Phosphorization	Covalent	Monolayer	1 (2008)	2.2
SiO_x	Layer-by-Layer	Electrostatic	Multilayers	1 (2010)	2.3.1
SiO_x	Lipid membrane	Electrostatic	Bilayer	2 (2009)	2.3.2
Si-H	Hydrosilylation	Covalent	Monolayer	6 (2006)	3

Table 1. Overview of different strategies used in the modification of SiNW-based devices.

2. SiO_x-covered SiNW-based sensor devices

Cleaned silicon surfaces exposed in an ambient environment such as air or solutions will oxidize spontaneously (Zhang, 2001b). The silicon oxide layer, in this chapter referred to as SiO_x –either native oxide or oxide grown under controlled conditions–, passivates the silicon surface. This allows the silicon to be used as an electronic material in processing and application environments. After fabrication, or at least after exposure to ambient environment, also SiNWs contain a SiO_x layer. Typically the thickness of this SiO_x layer on the NW is 1.5-10 nm, whereas the background oxide is usually >10 nm. It is important to note that –in terms of surface modification– one cannot discriminate between the oxide of the nanowire and background oxide; both are modified. The surface modifications described in §2 deal with SiO_x-coated SiNW devices.

2.1 Covalent functionalization

2.1.1 Silanization

The most applied method to functionalize silica surfaces is through the self-assembly of organofunctional alkoxysilane and chlorosilane molecules. While early silane-based monolayer work is based on Langmuir-Blodgett techniques, the first silane-based self-assembled monolayer was reported by Sagiv in 1980 (Sagiv, 1980). In general, a hydroxylated surface is introduced into a solution of a silane derivative in an organic solvent, but gas-phase methods have also been used (Aswal, 2006). In all cases, thorough cleaning of the substrate is a must for obtaining a clean and activated (hydroxylated) oxide layer with a high density of silanol groups on the surface (~ 10^{15} per cm^2) (Aswal, 2006). Scheme 1 gives a schematic representation of the formation of a silane monolayer onto SiO_x surfaces. It should be noted that this chemistry is not limited to SiO_x surfaces; other hydroxylated oxides surfaces can be modified as well (e.g., Al_2O_3 (Wang, 2008)).
Also in the case of the surface functionalization of the SiO_x surfaces of SiNW-based devices the most applied method is silanization. Figure 2 gives an overview of the different silane derivatives that have been used, while Table 2 gives a number of examples of surface silanization, together with the target analytes. It should be noted that on planar SiO_x substrates more compounds have been used; here we aim to give an overview for SiNW-based devices only. Table 2 summarizes the first papers on SiNW-based sensor devices and selected papers that use other silanization methods. We have categorized the references in terms of pH sensors, biosensors, cation sensors and gas sensors. This way, the cited work does not only represent the diversity of silane compounds that have been used, it also shows the variation in the further (bio)functionalization of SiNW-based devices.

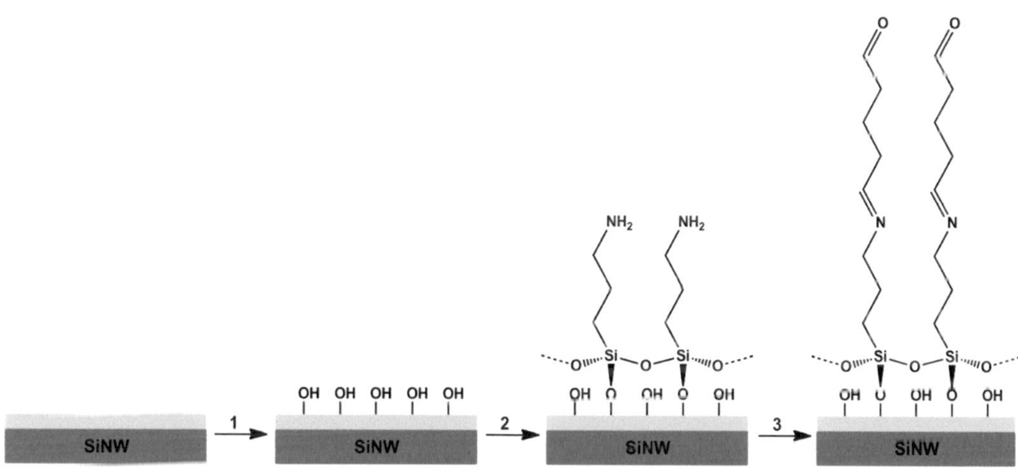

Scheme 1. Schematic representation of the 1) cleaning/hydroxylation of the SiO_x surface, 2) silanization (here: silane 1a, Figure 2), and 3) further functionalization with glutaraldehyde. It should be noted that the exact structure of the bonded silanes is arbitrary as also two or three silanol groups per attached molecule may react with the surface hydroxyl groups. Here we adapted the scheme from Aswal et al. (Aswal, 2006).

Sensor Type	Surface Modification*		Receptor	Analyte	Reference
pH	SiO_2			H^+	
	APTES (**1a**)			H^+	(Cui, 2001; Bunimovich, 2006; Chen, 2006)
	APTES (**1a**)		DNA-templated polymerization of aniline	H^+	(Ma, 2004)
Bio			Biotin	Streptavidin	(Cui, 2001)
	APTES (**1a**)		15-mer ss-DNA	Hybridization-induced charges in poly-T/poly-A 15-mer DNA	(Lin, 2007)
	APDMES (**1b**)		16-mer ss-DNA	Complementary DNA	(Bunimovich, 2006)
	MPTMS (**4a**)		12-mer ss-DNA	Single mismatch	(Li, 2004)
	APMS (**3**)		Monoclonal antibodies	PSA	(Zheng, 2005)
	APMS (**3**)		Antibody	influenza	(Patolsky, 2004)
	APMS (**3**)		Tyrosine kinase	ATP	(Wang, 2005)
	APTES (**1a**)	GA	Estrogen receptors protein	Estrogen response elements	(Zhang, 2010)
	APTES (**1a**)	GA	Anti-PSA	PSA	(Kim, 2007)
	APTES** (**1a**)	GA	Glucose oxidase	Glucose	(Wang, 2008)
	APTES (**1a**)	GA	Anti-CRP	CRP	(Lobert, 2003; Patolsky, 2006a)
	APTES (**1a**)		EDC-supported carboxy-phenylboronic acid (CPBA)	Dopamine	(Li, 2010)
	AEAPS (**2**)		Gold nanoparticles	Thiol-terminated enzyme	(Sheu, 2005)
Cations	SiO_2		Calmodulin	Ca^{2+}	(Cui, 2001)
	MPTES (**4b**)			Hg^+, Cd^+	(Luo, 2009)
	AEAPS (**2**)	GA	(phosphor)tyrosine	Ca^{2+}	(Bi, 2008b)
	APMS (**3**)		Gly-Gly-His	Cu^{2+}	(Bi, 2008a)
Gas	APTES (**1a**)			TNT	(Engel, 2010)

Table 2. Examples of surface modifications on SiO_x-covered SiNW devices and target analytes.

GA = Glutaraldehyde; EDC = 1-Ethyl-3-[3-dimethylaminopropyl]carbodiimide hydrochloride; PSA = Prostate specific antigen; ATP = Adenosine triphosphate; TNT = Trinitrotoluene; CRP = C-reactive protein; * The chemical structures of the silanes are given in Figure 2; ** The silicon nanowires are covered with a layer of 10 nm Al_2O_3, grown by atomic layer deposition.

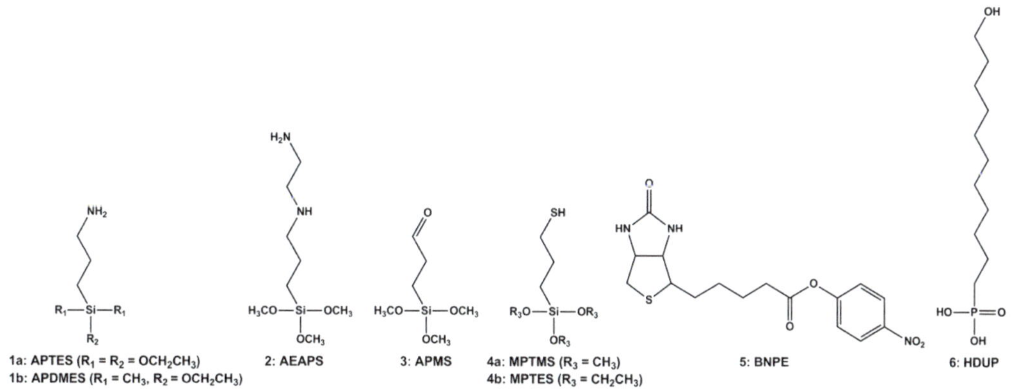

Fig. 2. Compounds that have been used in the modification on SiO_x-coated NWs in NW-based devices: 3-aminopropyltriethoxysilane (APTES, 1a), (3-aminopropyl)-dimethyl-ethoxysilane (APDMES, 1b), *N*-(2-aminoethyl)-3-aminopropyltrimethoxysilane (AEAPS, 2), 3-aldehydepropyltrimethoxysilane (APMS, 3), mercaptopropyltrimethoxysilane (MPTMS, 4a), mercaptopropyltriethoxysilane (MPTES, 4b), biotin 4-nitrophenyl ester (BNPE, 5) and 11-hydroxyundecyl-phosphonate (HUP, 6).

2.1.1.1 3-Aminopropyltriethoxysilane

The first paper of the Lieber group on SiNW-based devices shows the utilization of silane-based functionalization of SiO_x-coated nanowires (Cui, 2001). The devices were chemically modified with 3-aminopropyltriethoxysilane (APTES, Figure 2, silane **1a**). The resulting devices showed a linear source-drain conductance versus pH-response for pH values from 2 to 9. The authors explained these results by the presence of both NH_3^+ and SiOH groups, which have different acid dissociation constants. At low pH, the amino group is protonated, and acts as a positive gate, which depletes hole carriers in the p-type SiNW and decreases the conductance. At high pH, the SiOH group is deprotonated, causing an increase in conductance. Ma et al. realized pH sensitivity via the enzymatic polymerization of aniline, which was templated by DNA on APTES (Ma, 2004) (Figure 3). The direct use of the stretched, immobilized DNA molecules as templates prevents the agglomeration of the polyaniline/DNA complexes formed in solution.

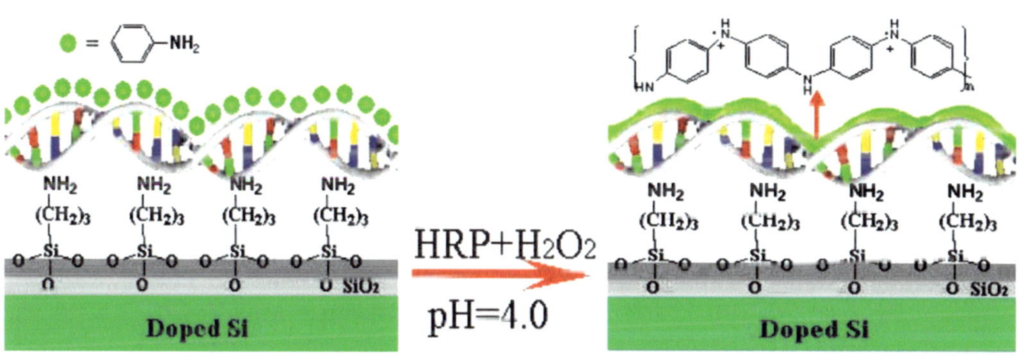

Fig. 3. Fabrication of a polyaniline nanowire immobilized on a SiO_x surface with stretched double-stranded DNA as a guiding template, resulting in pH sensitivity. HRP stands for horseradish peroxidase. Figure from (Ma, 2004) reprinted with permission from the American Chemical Society ©.

Besides its use as sensitive layer in the detection of protons, recently APTES has also been used as a sensing layer for the vapor-phase detection of 2,4,6-trinitrotoluene (TNT) (Engel, 2010). The sensing mechanism is based on the acid-base interaction between the amino-terminated nanowire surface and the polarized TNT nitro groups, which have a weak basicity. The binding of electron-deficient TNT molecules to the amino groups on the SiNW sensor surface is believed to result in the formation of charge-transfer complexes, which act as effective molecular gating elements. The conductance response of the resulting devices has a clear relation with the concentration of the explosive analytes and the device was found to be able to distinguish the target analyte from structurally related compounds. This example nicely shows that SiNW-based sensors are not only sensitive to changes in charges, but also to changes in charge density.

Given its reactivity towards aldehyde, carboxylic acid and epoxy functionalities, APTES has become the most frequently used linker compound for further (bio)functionalization of SiNWs. The activation procedure of the SiO_x nanowire surface using an oxygen plasma, followed by the immersion in a solution of APTES in ethanol and subsequent heating of the chip as reported by the Lieber group (Cui, 2001) is commonly used (Patolsky, 2006a; Wang, 2008). Alternative approaches have been reported as well, e.g., the use of UV/ozone for the surface activation, followed by the use of neat APTES (Li, 2010). Also toluene (Lobert, 2003) and acetone (Lin, 2007) have been used as a solvent in the silanization step.

Because of the possibility of hydrogen bond formation between the amine of APTES and the SiO_x surface, both head and tail groups can be oriented towards the surface, which can result in a large disorder in APTES layers (Vandenberg, 1991). Additionally, cross-linking between the alkoxy silane units may yield oligomerized silane structures on the surface, resulting in rough layers that are thicker than a monolayer. The optimal conditions for solvent-based silanization using APTES were investigated on planar surfaces (Howarter & Youngblood, 2006). The morphology and growth kinetics of APTES films deposited from solutions were found to be affected by reaction time, solution concentration, and temperature. Experiments with an APTES concentration of 1% only produced good films when the reaction was time-limited (1 h). Increasing the reaction time increased the APTES film thickness.

To overcome the problem of the disordered monolayers, Lin et al. performed a post-treatment of APTES-functionalized devices (Lin, 2007). After the APTES modification, they aligned the internal dipoles of the APTES molecules using high electric fields, thus decreasing the disorder in the monolayer. In addition, it is shown that the source-drain conductance of the APTES-functionalized devices changes upon UV illumination. The authors explained this by a strengthening effect of the internal APTES dipoles by UV illumination. Additionally, UV illumination excites extra charge carriers in the nanowire. Depending on the size of both effects, the resistance of the nanowire can be increased or decreased, which was found to be measurable with the produced devices.

A modification that often follows the attachment of APTES to the SiO_x nanowire surface is the reaction with glutaraldehyde (GA, Scheme 1, right). This linker molecule is used to form an aldehyde-terminated surface, which increases the possibilities for further reactions, including the reaction with amine groups of, e.g., anti-bodies (Kim, 2007), proteins and enzymes (Lee, 2008; Wang, 2008; Zhang, 2010). Also amino-terminated, single-stranded DNA (ss-DNA) has been attached to APTES/GA-modified surfaces to measure hybridization-induced mismatches in 15-mer ss-DNA chains (Lin, 2007). The addition of glutaraldehyde is, like that of APTES, both performed from solution (Bi, 2008b; Zhang, 2010) and from the gas phase (Wang, 2008).

Li et al. reported the functionalization of an APTES-coated thin oxide gate with a dopamine receptor (Li, 2010). Although the devices do no belong to the class of SiNW-based sensors, the CMOS open-gate Field-Effect Transistor (FET) devices have (sub)micrometer gate dimensions. They have immobilized carboxyphenylboronic acid onto the APTES layer. The boronic acid group specifically reacts with the catechol functionality of, e.g., dopamine, which was detected in the femtomolar range.

2.1.1.2 Other silane derivatives

Apart from APTES, also other amino-terminated silane derivatives have been used. For example, (3-aminopropyl)-dimethyl-ethoxysilane (APDMES, Figure 2, silane **1b**) has been used in the modification of SiNW-based devices to bind a 16-mer ss-DNA (Bunimovich, 2006). In contrast to APTES (**1a**), APDMES (**1b**) is a mono-alkoxydimethylsilane, which is a crucial difference when it comes to the monolayer quality. Although silanes with only alkoxy group can react with each other, oligomerization can not take place. Consequently, the use of mono-alkoxydimethylsilanes results in the formation of a true monolayer. Very recently this issue was addressed in a detailed study by Dorvel et al. who synthesized an amino- and an epoxy-terminated mono-alkoxydimethylsilane, which were immobilized onto planer silicon oxide surfaces (Dorvel, 2010).

In addition, *N*-(2-aminoethyl)-3-aminopropyltrimethoxysilane (AEAPS, Figure 2, silane **2**) has been applied to immobilize an amino acid (Bi, 2008b). The nanowires devices were exposed to a solution of AEAPS in water, subsequently washed and finally dried to stimulate cross-linking of the molecules. This was followed by the reaction with GA to produce an aldehyde-terminated surface. Finally, tyrosine and phosphotyrosine were cross-linked with the aldehyde group through the formation of a Schiff base, which was further reduced to a stable secondary amine using sodium cyanoborohydride ($NaBH_3CN$). The studies, however, do not contain information on potential benefits of the alternative silanization agents used.

Some other functionalities rather than amino groups have been employed in silane chemistry (Figure 2, silanes 3 and 4). Apart from APTES, the Lieber group used 3-aldehydepropyltrimethoxysilane (APMS, Figure 2, silane 3) (Zheng, 2005) to directly obtain an aldehyde-terminated surface. Subsequently, monoclonal antibodies were attached to detect prostate specific antigens (PSA). The use of APMS enabled them to make a thinner organic layer as compared to combining APTES with glutaraldehyde. As oxygen in air will slowly oxidize aldehydes to acids or peracids, it is advisable to work with freshly prepared APMS surfaces or to make use of the reaction of glutaraldehyde with an APTES surface, directly prior to further functionalization.

As a last example we mention the use of 3-mercaptopropyltriethoxysilane (MPTES, Figure 2, silane **4b**) (Luo, 2009), which was applied to detect Cd^{2+} and Hg^{2+} ions. The thiol-terminated devices showed a linear relation between the logarithm of the Cd^{2+} and Hg^{2+} concentration and the source-drain conductance. The response towards K^{+}, Na^{+}, Ca^{2+}, Ba^{2+}, and Mg^{2+} (hard Lewis acids) was considerably smaller as compared to Cd^{2+} and Hg^{2+} (soft Lewis acids). This difference can be explained by the chemical softness of thiol groups, which hardly coordinate with hard Lewis acids.

2.1.2 Alternative approaches

In this paragraph we address two different types of non silane-based compounds that have been used in the chemical modification of SiO_x surfaces of SiNW-based devices. Hahm and Lieber report on the 4-(dimethylamino)pyridine (DMAP)-catalyzed transesterification of biotin 4-nitrophenyl ester with the hydroxyl groups of SiO_x nanowire surface (Hahm & Lieber, 2004)

(Figure 4, left). Afterwards, the devices were exposed to phosphate-buffered saline (PBS) solution of avidin, which was followed by the addition of biotinylated peptide nucleic acid (PNA) capture probes. A generally benefit of the use of PNA over DNA is that it enables sensing in an ultra-low background due to the lack of charges on the peptide backbone. In the work of Hahm and Lieber the surfaces were not analyzed after each modification step. However, based on the electrical characteristics of the modified devices it can be concluded that the PNA immobilization was successful. It has to be noted though that siloxane ester bonds (Si-O-(CO)) have a limited hydrolytic stability, at least lower than in the case of siloxane bonds (Si-O-Si). As a result, this immobilization technique is not often used.
Cattani-Scholz et al. investigated hydroxyalkylphosphonate monolayers as a platform for the biofunctionalization of SiNW-based field effect sensor devices (Cattani-Scholz, 2008) (Figure 4, right). They employed a stepwise functionalization protocol using 11-hydroxyundecylphosphonate and a maleimide heterobifunctional linker system. The surface chemistry was also performed on planar surfaces, which made it possible to extensively analyze the surfaces after each step. It was found that the hydroxyalkylphosphonate monolayer effectively passivates the Si surface against electrochemical leakage current into the electrolyte solution through the interface, with maximum currents of the order of 10 pA measured at an applied voltage of 200 mV. In a comparative study it has been shown that phosphonate monolayers on the native oxide of Ti-6Al-4V are hydrolytically more stable than the corresponding siloxane monolayers (Silverman, 2005). It is also reported that the monolayer growth is not limited by the surface OH content,(Hanson, 2003) making this type of surface modification an interesting alternative for silane-based chemistry.

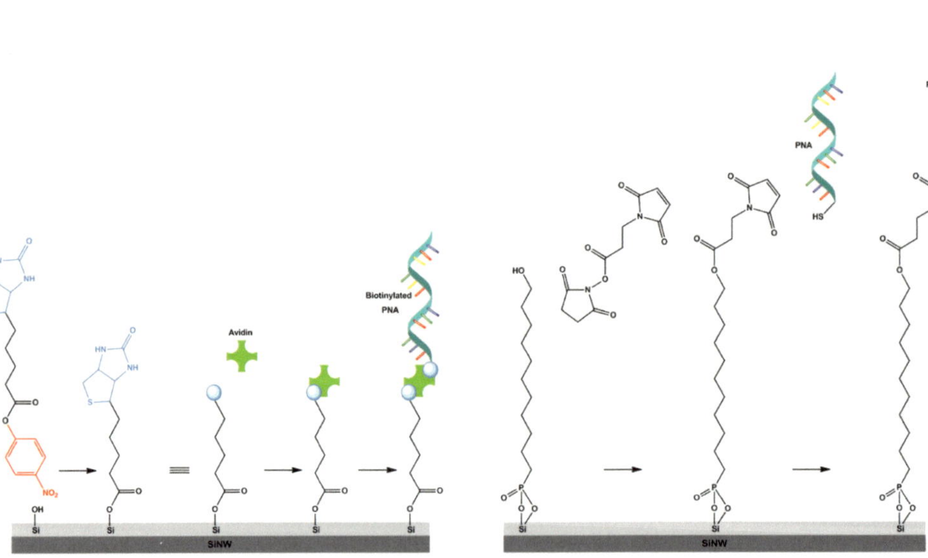

Fig. 4. A schematic representation of two different surface chemistries that have been used to attach PNA onto SiO_x: (left) the use of biotin 4-nitrophenyl ester (the blue part of the structure indicates biotin molecule, whereas the red part of the structure indicates the 4-nitrophenyl group, which is used to activate the ester), the following steps involve the use of avidin and biotinylated PNA, and (right) the formation of organophosphonate monolayer, followed by the reactions with a maleimide heterobifunctional linker and subsequently with a thiolated PNA derivative.

2.2 Non-covalent functionalization

2.2.1 Adsorption of polyelectrolytes

The negative nature of SiO_x at pH > 3.5 allows one to electrostatically immobilize positively charged polymers (polycations), also called generally polyelectrolytes (PEs). Subsequently, a negatively charged polymer (polyanion) can be adsorbed on top. Providing that each adsorption step of charged polymers leads to a charge inversion of the surface (charge overcompensation effect), the subsequent deposition finally results in a PE multilayer, stabilized by strong, multivalent electrostatic forces. This method is known as layer-by-layer (LbL) deposition (Decher, 1997). A variety of deposition methods can be used, including dip-coating, spin-coating, spray-coating and flow-based techniques.

Recently Ingebrandt and coworkers applied the LbL deposition technology onto SiNW-based devices (Vu, 2010). They immobilized polyallylamine hydrochloride (PAH: MW 15,000) onto 6 nm thick, thermally grown SiO_x via dip coating. Subsequently, polystyrenesulphonate (PSS: MW 60,000-80,000) was immobilized and this sequence was repeated several times to make a stack of 6 bilayers (Figure 5). The thickness of the resulting multilayer has not been reported. Assuming a monolayer thickness of ~ 2 nm (Decher, 1997; Poghossian, 2007) the multistack thickness would be ~ 24 nm. It has to be noted, however, that the salt concentration has an effect on the layer thickness (Decher & Schmitt, 1992), which makes it difficult to predict the thickness accurately. Although the noise level of PE-coated devices was slightly higher than that of their standard FET devices, the sensitivity for PE adsorption measurements was 3 to 4 times higher as compared to previous reports.

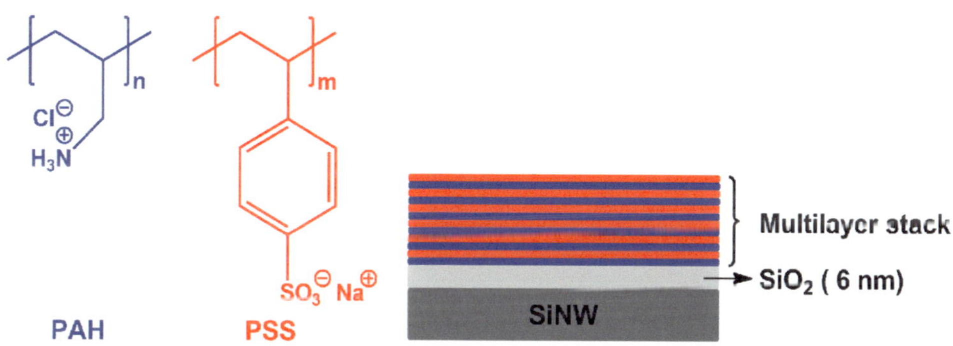

Fig. 5. Structures of the polycation polyallylamine hydrochloride (PAH) and polyanion polystyrenesulphonate (PSS) used by Ingebrandt and coworkers to prepare layer-by-layer multilayers onto SiO_x-covered SiNW-based devices.

2.2.2 Lipid membranes

Silica surfaces have also been used as a platform to immobilize model membrane systems like supported lipid bilayers, which have been extensively used in understanding the fundamental properties of biological membranes. However, there have been only a few attempts to apply lipid membranes on nanoelectronic devices. Misra et al. incorporated lipid bilayer membranes into SiNW transistors (Figure 6). In more detail, they covered the NWs and the background oxide with a continuous lipid bilayer shell, forming a barrier between the SiO_x surface and solution species (Martinez, 2009; Misra, 2009). The method is based on the fusion of spherical unilamellar vesicles onto a SiNW surface producing a planar conformal lipid bilayer coating (Huang, 2007).

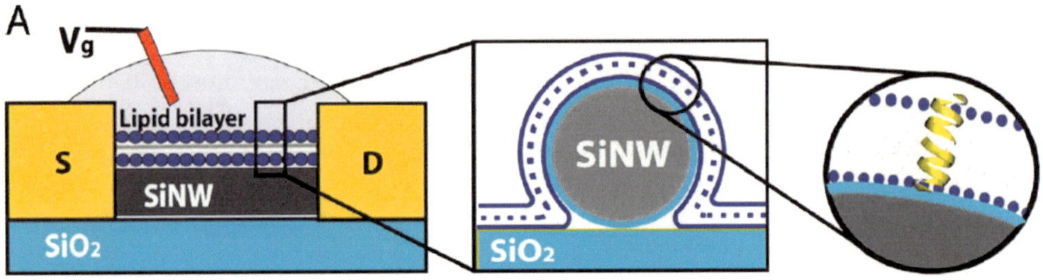

Fig. 6. Device schematics showing a SiNW connected to microfabricated source (S) and drain (D) electrodes. Although this is not clear from the figure it should be noted that the source and drain are insulated with Si_3N_4. The insets show the configuration of the lipid bilayer and a pore channel placed in the bilayer membrane. Figure from (Misra, 2009) reprinted with permission from National Academy of Sciences, USA.

Cyclic voltammetry using potassium ferrocyanide ($K_4Fe(CN)_6$) as a redox probe showed that the presence of a lipid bilayer on the NW surface reduced the limiting current by 85–95% relative to the native oxide-covered NW device (Misra, 2009). Subsequently, two different transmembrane peptides (gramicidin A and alamethicin) were incorporated in the vesicles, resulting in ion channel-containing lipid bilayers after fusion with SiNW-based devices. It was shown that the incorporation of transmembrane peptide pores enabled ionic-to electronic-signal-transduction by using voltage-gated and chemically gated ion transport through the membrane pores.

3. Oxide-free SiNW-based sensor devices

3.1 Hydrosilylation

As an alternative to silane-based chemistry, silicon substrates can also be modified using so-called hydrosilylation chemistry, e.g., the reaction between a surface Si-H and an organic compound containing a terminal unsaturated carbon-carbon bond (alkenes and alkynes) to form a Si-C linked monolayer (Scheme 2).

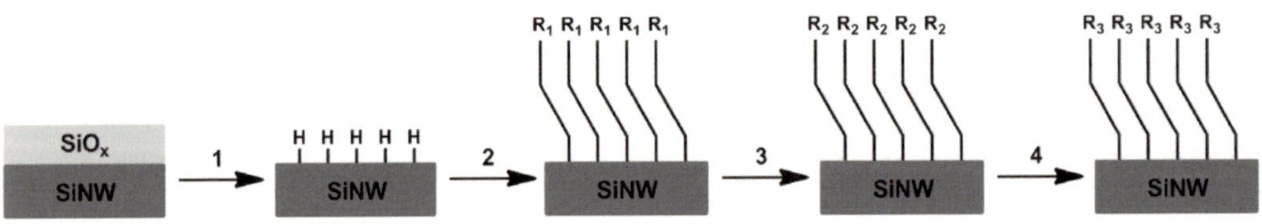

Scheme 2. Schematic representation of the 1) etching of the SiO_x surface, yielding a H-terminated surface, 2) formation of Si-C monolayers via hydrosilylation (R_1 represents a functional group that is chemically protected towards the reaction with the H-terminated Si surface, e.g., an ester or amide), 3) deprotection the functional group (represented by R_2, e.g., a carboxylic acid or amine), 4) attachment of a receptor (represented by R_3).

This work has been pioneered by Linford and Chidsey in the early 1990s (Linford & Chidsey, 1993; Linford, 1995) and this type of monolayers has continued to attract attention ever since. Hydrosilylation chemistry requires removal of SiO_x, which in most cases is achieved using wet chemistry. On planar silicon surfaces diluted, aqueous HF and NH_4F solutions are used to etch Si(100) and Si(111) substrates, respectively, although a wide variety of etching methods has become available (Zhang, 2001a). This process results in hydrogen-terminated, oxide-free silicon substrates with Si-H groups. The reaction of H-terminated silicon surfaces with alkenes and alkynes can be performed by making use of high temperatures, UV and visible light irradiation, electrochemistry, hydrosilylation catalysts (e.g., $AlCl_3$), and chemomechanical scribing, as summarized in different reviews (Buriak, 2002; Stutzmann, 2006; Hamers, 2008; Ciampi, 2010).

3.2 Organic monolayers: Si-C versus SiO_x-C

Si-C bonded monolayers have a number of advantages over silane-based monolayers on SiO_x surfaces. First, interface trap densities for Si-C bonded monolayers can be considered remarkably lower than in the case of alkyl chains organic monolayers on naturally oxidized silicon surfaces (Kar, 2001). Also, as already mentioned in §2.1.1.1, silane oligomerization can result in rough layers that are thicker than a monolayer. In contrast, the reaction between a H-terminated Si surface and alkenes or alkynes results in a true monolayer. This is related to the involved zipper mechanism, in which each binding of an alkyl chain directly generates a new radical site on the next-nearest Si surface atom (Linford, 1995). Moreover, the chemical stability of monolayers on SiO_x in aqueous media is poor since the exposure to water pilots the destruction of the monolayer through hydrolysis of Si-O bonds (Calistri-Yeh, 1996). Organic layers formed by hydrosilylation are stable in a number of environments, including air, boiling water, organic solvents, and acids (Linford, 1995).

In the field of SiNW-based devices, the application of Si-C monolayers has a number of other benefits as compared to silane-based surface modifications. First, the removal of SiO_x brings a sensing event (target/analyte interaction) closer to the conducting surface, which increases the sensitivity of the device (Bunimovich, 2006; Zhang, 2008a). Second, the selective functionalization of oxide-free SiNWs is considered to be a supplementary factor that improves the detection sensitivity by avoiding the binding competition between the SiNW and the oxide surface on the rest of the device. Third, several studies show improved electrical properties of Si-C linked monolayers over monolayers on SiO_x, in terms of the flat-band potential and source-drain conductance (Faber, 2005b; Vilan, 2010).

A crucial difference between the silane-based and alkene-based approach becomes clear from, e.g., the preparation of amine-terminated monolayers. While in the case of SiO_x unprotected amines can be used (§2.1.1), e.g., APTES, amine-containing alkenes should be chemically protected as amines can react with H-terminated Si surfaces. The *N*-tert-butoxycarbonyl (t-BOC, Figure 7, alkenes **7a** and **7b**) is a commonly used protecting group for primary amines, but other protecting groups have been used as well (Sieval, 2001a). Apart from amines, also carboxylic acid and alcohol functionalities should be protected before application in the hydrosilylation chemistry (Buriak, 2002).

Characteristic of Si-C bonded monolayers is that the molecular cross-section of the attached alkyl chain prevents reaction of all individual Si–H bonds, i.e., upon formation of a monolayer on Si(111) about 50-55% of the Si–H sites has reacted (Sieval, 2001b), while on Si(100) this is about 30-35% (Sieval, 2000a). The remaining, unreacted Si-H sites are a potential source for the formation of interface states, e.g., via oxidation, which is an unfavourable process in terms of device performance.

3.3 Hydrosilylation on SiNW-based sensor devices

The reported chemistry for functionalizing the oxide-free, H-terminated SiNW sensor devices is limited to one method, i.e., hydrosilylation catalyzed by UV irradiation. This

approach has been applied by different research groups using different etching conditions and probing different types of analytes (Table 3).
The first report on the modification of oxide-free SiNW-based devices is by Heath and coworkers (Bunimovich, 2006). The authors followed an established, multistep protocol (Bunimovich, 2004; Streifer, 2005) that starts with the hydrosilylation of H-terminated SiNWs with tert-butyl allylcarbamate (Figure 7, alkene **7a**) irradiated at 254 nm. After deprotection, the researchers obtained a positively charged amine-terminated monolayer on the SiNWs, enabling the adsorption of negatively charged, oligo ss-DNA.
Surface characterization of the nanowires embedded in a device is a challenge, since conventional analytical tools used for surface characterization require substrates on the micro-scale. For that reason a widely used strategy is to perform the applied chemistry also on planar samples.

Analyte	Receptor	Alkene*	Oxide	Etching Conditions	Reference
ss-DNA	ss-DNA	**7a**	native	2% HF_{aq} (3 s)	(Bunimovich, 2006)
K^+, Na^+	Crown Ether	**7b**	5-10 nm	1% HF_{aq} (50 s) + NH_4F_{aq} (60 s)	(Zhang, 2007)
ss-DNA	ss-PNA	**7b**	~ 5 nm	1% HF_{aq} (50 s) + NH_4F_{aq} (60 s)	(Zhang, 2008a; Zhang, 2008b)
Avidin/ Streptavidin	Biotin	**7b**	not reported	10:1 v/v 40% NH_4F_{aq}/ 49% HF_{aq} (5 s)	(Stern, 2007)
Antigen	Antibody	**7b**			
-	-	**8, 9**			

Table 3. Overview of the references that report on the chemical modification of the SiNW-based devices via hydrosilylation chemistry.

* The chemical structures of the alkenes are given in Figure 7.

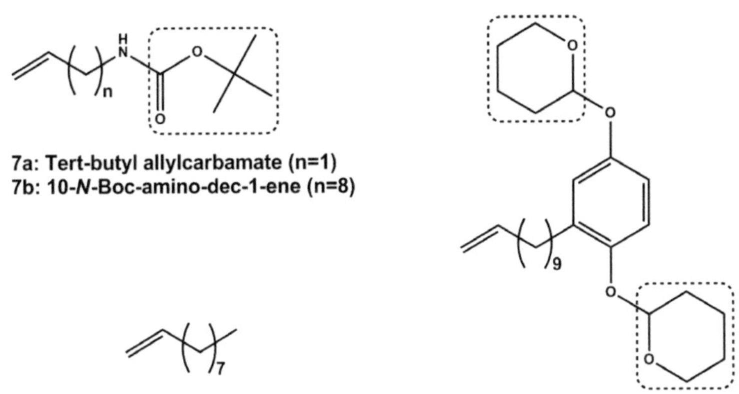

7a: Tert-butyl allylcarbamate (n=1)
7b: 10-*N*-Boc-amino-dec-1-ene (n=8)

9: 1-Decene **8: Tetrahydropyranyl-terminated alkene**

Fig. 7. Alkene derivatives that have been chemically bound to oxide-free SiNW sensor devices. The protecting groups on various alkenes are marked in dashed rectangles. For alkene 7a and 7b the protecting group is *N*-tert-butoxycarbonyl (t-BOC), for alkene 8 it is the tetrahydropyranyl (THP) group. See text for further information.

This strategy was also followed by Bunimovich et al. and planar Si(100) samples were analyzed by water contact angle measurements and X-ray Photoelectron Spectroscopy. Moreover, this interesting paper compares Si-C and SiO_x-based monolayers. It is shown that the electrical read-out of oxide-free SiNWs for sensing complementary 16-mer DNA was more profound than on modified, SiO_x-covered SiNWs.

Also Zhang et al. used hydrosilylation chemistry to coat SiNW devices with ss-DNA, although a different strategy was chosen (Zhang, 2008a; Zhang, 2008b). The modification scheme starts with the UV-initiated attachment of 10-*N*-boc-amino-dec-1-ene (Figure 7, alkene **7b**) to a device with H-terminated SiNWs. After the deprotection of the amino groups the surface was chemically modified with glutaraldehyde, allowing the covalent binding of amino-terminated compounds. Inspired by the work of others on devices with SiO_x nanowires (Strother, 2000; Hahm & Lieber, 2004; Bi, 2008a), they subsequently immobilized PNA. The benefit of this approach over the method reported by Bunimovich et al. is twofold: PNA is not only neutral (see also §2.1.2), it is also bonded covalently. Although multivalent, ionic interactions might be stronger than one single covalent bond, the latter is independent of the salt concentration, making it more robust in wide range of chemically different environments.

As for sample characterization the research group was quite successful in depicting the chemistry performed on the SiNWs. Transmission electron microscopy (TEM) was performed for a morphological study of the non-oxidized SiNW surface. From the TEM data the authors concluded that the surface was nearly uniform, oxide free and did not show visible TEM-detectable defects (Figure 8b). In addition, binding of fluorescently labeled, complementary DNA was used to demonstrate the selectivity of the non-oxidized SiNW surface chemistry and the specificity of PNA-DNA hybridization. Indeed, strong fluorescent signals on the SiNWs were obtained and the bright arrays of SiNWs were clearly visible (Figure 8c).

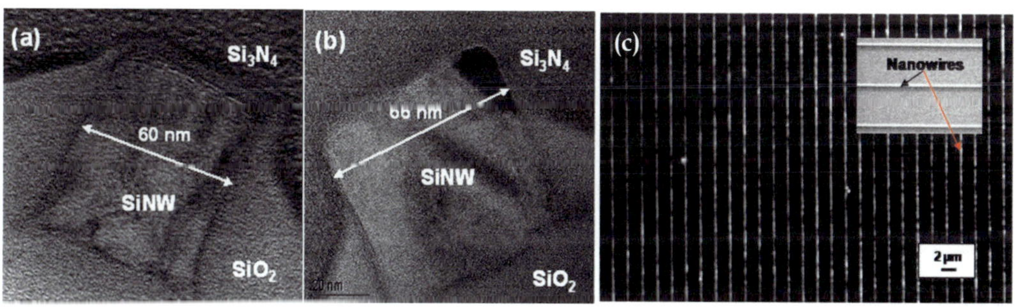

Fig. 8. Cross-sectional TEM micrographs of SiNWs before and after HF and NH_4F etching: (a) SiO_x layer coated on the SiNW surface prior to etching, (b) hydrogen-terminated SiNW after etching, (c) Fluorescent images of Cy3-labeled complementary DNA hybridized to the PNA-functionalized SiNWs. The inset shows SEM micrograph of SiNW arrays. Figure from (Zhang, 2008a) reprinted with permission from Elsevier.

Earlier work by Zhang et al. focused on the sensing of alkali metal ions by the chemical modification of SiNW-based devices with crown ethers (Zhang, 2007). Also in this case alkene **7b** was used, followed by the reaction with glutaraldehyde, resulting in an aldehyde-terminated monolayer. Subsequently, amine-terminated crown ethers were covalently attached. Based on the complexation selectivity of crown ethers to alkali metal ions, the resulting functionalized oxide-free SiNW-based devices could detect Na^+ and K^+.

The sensitivity of protein binding by oxide-free SiNW-based FETs was first reported by Stern et al. using biotin-avidin/streptavidin interactions and antibody-antigen sensing (Aswal, 2006). Three different functionalized alkene derivatives were used in the UV-initiated reaction with H-terminated SiNW devices: 10-*N*-boc-amino-dec-1-ene (**7b**), 2-[2-(undec-10-enyl)-4-(tetrahydro-2H-pyran-2-yloxy)phenoxy] tetrahydro-H-pyran containing a tetrahydropyranyl (THP) protecting group (**8**), and 1-decene (**9**). Only alkene **7b** could be realized for sensing purposes, whereas modification with alkene **8** was ineffective for sensing and alkene **9** aimed as a control. Using alkene **7b**, amino-terminated monolayers were obtained, followed by a common biotinylation protocol using N-hydroxysulfosuccinimide for capturing avidin and streptavidin. For antibody-antigen sensing the capture antibodies were bound to hydrosilylated SiNWs by N-hydroxysulfosuccinimide/ethylene dicarbodiimide coupling chemistry (Hermanson, 2008).
The strategy for protein binding using alkene **8** is different and more complicated, yet very appealing. That is because it creates electrochemically active monolayers on top of the SiNWs. Such electro-active monolayers ensure not only the monolayer formation qualitatively, but they also allow the quantification of the surface coverage. This is based on the fact that the required electrons for the reduction of the electro-active moiety can be measured. The number of electrons is directly proportional to the number of the electro-active groups present and hence to the coverage. As a first step THP-protected alkene **8** was photochemically bound to hydrogen-terminated SiNW. Removal of the protecting THP ethers under mild conditions of 1% toluene sulfonic acid (TSA) in methanol leads to the hydroquinone (Scheme 3). Cyclic voltammetry (CV) in PBS oxidized the hydroquinone to quinone. Finally, they aimed to couple thiolated biotin selectively to the thus formed quinone moieties, but this was not successful, most likely due to the degradation of the device performance that was observed upon chemical modification.

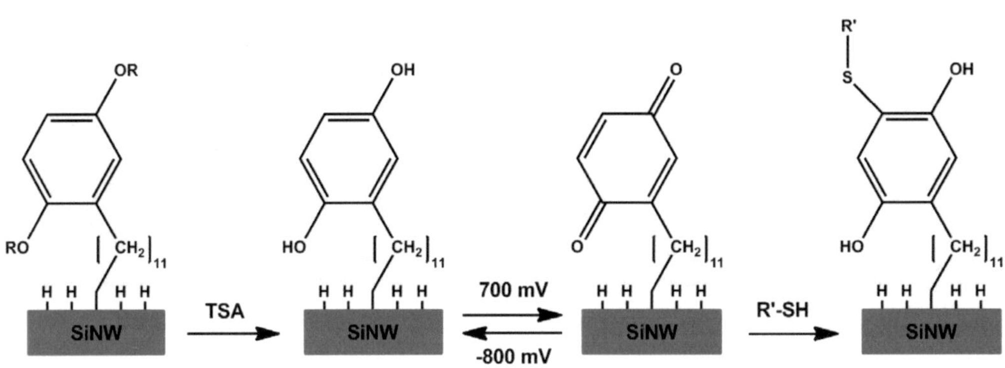

Scheme 3. The strategy reported by the Heath group (Bunimovich, 2004) and also investigated by the Reed group (Stern, 2007) for the selective immobilization of thiolated biotin (R'-SH) via a Michael addition. TSA stands for 1% toluene sulfonic acid. See text for further explanation.

While devices that were modified via UV-induced hydrosilylation using alkenes **7b** and **9** experienced increased leakage currents, the use of compound **8** led to loss of gating behavior, most likely due to the formation of redox-active surface traps. For compound **7b** the authors report a device yield as low as <2% after the deprotection. The work of Stern et al. highlights the importance of validating the device performance after chemical modification.

3.4 Organic monolayers on oxide-free SiNW devices: towards sensing

The observation of changes in device performance upon chemical modification of its sensing area has not only been reported by Stern et al., but also in a few other cases. Yet there is evidence that even non-charged molecules adsorbed on the nanowire alter the device characteristics. Haick et al. showed an improved conductance of the H-terminated SiNWs upon the chlorination/alkylation of the nanowire (Haick, 2006). They reported that methyl-terminated SiNWs exhibit higher source-drain conductance values, lower surface defect levels and higher on-off ratios than oxygenated SiNWs.

Another example of characterizing the surface passivation of a NW prior to sensing is the work of Masood et al. (Masood, 2010). An improved device behavior after hydrosilylation on SiNW with 1-octadecene has been shown by means of capacitance-voltage and current-front-gate voltage electrical measurements. Based on other studies showing that hydrogen-terminated Si(111) surfaces have the lowest reported surface recombination velocity and improved electronic performance as compared to Si(100) surfaces with SiO_2 interfaces, the authors stress the importance of using Si(111) surface to further improve the Si-NW biosensor performance.

4. Concluding remarks and outlook

In order to bind (bio)receptor molecules onto SiNW-based devices, most studies apply the silanization on the nanowire oxide layer. Clearly APTES is the most frequently used silane-based agent, enabling the subsequent immobilization of carboxylic acid- or aldehyde-terminated biomolecules. More often, further derivativization with glutaraldehyde has been applied to chemically bind amine-terminated biomolecules. A disadvantage of silane compounds is their ability to react with each other via cross-linking of the alkoxy units, resulting in rough, unordered multilayers. Based on the optimization studies on planar silica substrates that have become available (Howarter & Youngblood, 2006) and reports on the quality differences between vapour- and wet chemistry-prepared APTES layers (Fiorilli, 2008), there is room for the improvement of silane chemistry on SiNW-based devices. Phosphonate monolayers are an interesting alternative for silane-based monolayers, due to their higher hydrolytical stability and the less critical pretreatment of the oxide surface that is required.

Apart from covalent functionalization of SiNW-based sensor devices also physisorption has been studied, which enabled the detection of several compounds and ions. Covalent attachment of (bio)molecules to sensor surfaces may be appealing in terms of device stability, but some systems, e.g., membrane receptors require a more dynamic environment. It has been shown that a phospholipid monolayer can be covalently immobilized on SiO_x (Heysel, 1995) and Si-H (Charrier, 2010) surfaces, enabling the non-covalent immobilization of a second phospholipid layer. However, this strategy may hamper the fluidic properties of the bilayer, which is vital for the incorporation of biologically active structures into the membrane. The fusion of spherical unilamellar vesicles onto surfaces overcomes this problem. So far, only two different transmembrane ion channels on SiNW-based devices have been reported (Misra, 2009), but lipid bilayers provide a matrix for a virtually unlimited number of transmembrane proteins that can accommodate different functionalities.

Physisorbed PE layers are not covalently attached; nevertheless they are very stable, due to the multivalent character of the interaction. The multivalency is also beneficial regarding the detection of a PEs that adsorbs onto the SiNW-based device. LbL films of PE have been employed for various applications, including electrochemical sensing and biosensing and they have also contributed to the investigation of fundamental studies in electrochemistry (Crespilho, 2006). The work of Ingebrandt and coworkers (Vu, 2010) can now be extended to the use of biologically relevant PEs and perhaps also the incorporation of nanoparticles.

An interesting candidate to replace SiO_x layers on silicon surfaces is found in alkene-prepared organic monolayers. There are a number of benefits for monolayers on oxide-free silicon over silane-based layers: a) no risk of oligomerization, resulting in a true monolayer, b) higher chemical stability, c) selective functionalization, i.e., the background surface is not functionalized, and d) a lower density of interface traps. Although a wide variety of preparation procedures for hydrosilylation surface chemistry are available, so far only UV-initiated methods have been used in the chemical modification of oxide-free H-terminated SiNW devices. This can be explained by the fact that the research in this direction is still in its early days, and it probably will not take long before some of the other available preparation methods will be explored on SiNW-based devices. An important issue in this regard is the surface orientation of the nanowire (Bunimovich, 2004). In addition, a very good monolayer quality is needed, reducing the number of interfacial defects (Seitz, 2006). In contrast to SiO_x-covered SiNW-based devices, the insulating properties in the case of oxide-free SiNW-based devices are exclusively to be realized by the organic monolayer itself. One possibility to improve the monolayer quality is by using alkynes instead of alkenes (Sieval, 2000b; Faber, 2005a; Scheres, 2010).

To conclude, selective functionalization of the nanowires is critical to retain sensitivity as the modification of the background oxide results in a reduced sensitivity. In addition, the modification of individual nanowires that are part of a larger array is essential to perform a multi-analysis using a single device. Continued advances in the preparation of well-defined, defect-free, organic-semiconductor interfaces will further improve the performance of SiNW-based devices. The combination of an increasingly growing number of interesting (bio)receptors, sophisticated device fabrication methods, improved electrical characterization procedures and advanced organic surface modification does not only provide unlimited research opportunities, it will also contribute significantly to diverse, highly usable sensor applications in a variety of fields.

5. Acknowledgment

This research was supported by the Dutch Technology Foundation STW, the Materials innovation institute M2i and by a VENI grant from the Netherlands Organization for Scientific Research (NWO) to LCPMdS.

6. References

Aswal, D. K., S. Lenfant, D. Guerin, J. V. Yakhmi and D. Vuillaume (2006). Self assembled monolayers on silicon for molecular electronics. *Analytica Chimica Acta* 568(1-2), 84-108.

Bi, X. Y., A. Agarwal, N. Balasubramanian and K. L. Yang (2008a). Tripeptide-modified silicon nanowire based field-effect transistors as real-time copper ion sensors. *Electrochemistry Communications* 10(12), 1868-1871.

Bi, X. Y., W. L. Wong, W. J. Ji, A. Agarwal, N. Balasubramanian and K. L. Yang (2008b). Development of electrochemical calcium sensors by using silicon nanowires modified with phosphotyrosine. *Biosensors & Bioelectronics* 23(10), 1442-1448.

Bunimovich, Y. L., G. L. Ge, K. C. Beverly, R. S. Ries, L. Hood and J. R. Heath (2004). Electrochemically programmed, spatially selective biofunctionalization of silicon wires. *Langmuir* 20(24), 10630 10638.

Bunimovich, Y. L., Y. S. Shin, W. S. Yeo, M. Amori, G. Kwong and J. R. Heath (2006). Quantitative real-time measurements of DNA hybridization with alkylated nonoxidized silicon nanowires in electrolyte solution. *Journal of the American Chemical Society* 128(50), 16323-16331.

Buriak, J. M. (2002). Organometallic chemistry on silicon and germanium surfaces. *Chemical Reviews* 102(5), 1271-1308.

Calistri-Yeh, M., E. J. Kramer, R. Sharma, W. Zhao, M. H. Rafailovich, J. Sokolov and J. D. Brock (1996). Thermal stability of self-assembled monolayers from alkylchlorosilanes. *Langmuir* 12(11), 2747-2755.

Carlen, E. T. and A. van den Berg (2007). Nanowire electrochemical sensors: can we live without labels? *Lab on a Chip* 7(1), 19-23.

Cattani-Scholz, A., D. Pedone, M. Dubey, S. Neppl, B. Nickel, P. Feulner, J. Schwartz, G. Abstreiter and M. Tornow (2008). Organophosphonate-based PNA-functionalization of silicon nanowires for label-free DNA detection. *ACS Nano* 2(8), 1653-1660.

Charrier, A., T. Mischki and G. P. Lopinski (2010). Direct stabilization of a phospholipid monolayer on H-terminated silicon. *Langmuir* 26(4), 2538-2543.

Chen, Y., X. H. Wang, S. Erramilli, P. Mohanty and A. Kalinowski (2006). Silicon-based nanoelectronic field-effect pH sensor with local gate control. *Applied Physics Letters* 89(22), 223512

Ciampi, S., J. B. Harper and J. J. Gooding (2010). Wet chemical routes to the assembly of organic monolayers on silicon surfaces via the formation of Si-C bonds: surface preparation, passivation and functionalization. *Chemical Society Reviews* 39(6), 2158-2183.

Crespilho, F. N., V. Zucolotto, O. N. Oliveira Jr. and F. C. Nart (2006). Electrochemistry of Layer-by-Layer Films: a review. *Int. J. Electrochem. Sci.* 1, 194-214.

Cui, Y. and C. M. Lieber (2001). Functional nanoscale electronic devices assembled using silicon nanowire building blocks. *Science* 291(5505), 851-853.

Cui, Y., Q. Q. Wei, H. K. Park and C. M. Lieber (2001). Nanowire nanosensors for highly sensitive and selective detection of biological and chemical species. *Science* 293(5533), 1289-1292.

Decher, G. (1997). Fuzzy nanoassemblies: toward layered polymeric multicomposites. *Science* 277(5330), 1232-1237.

Decher, G. and J. Schmitt (1992). Fine tuning of the film thickness of ultrathin multilayer films composed of consecutively alternating layers of anionic and cationic polyelectrolytes. *Progr. Colloid Polym. Sci.* 89, 160-164.

Engel, Y., R. Elnathan, A. Pevzner, G. Davidi, E. Flaxer and F. Patolsky (2010). Supersensitive detection of explosives by silicon nanowire arrays. *Angewandte Chemie-International Edition* 49(38), 6830-6835.

Faber, E. J., L. C. P. M. de Smet, W. Olthuis, H. Zuilhof, E. J. R. Sudhölter, P. Bergveld and A. van den Berg (2005a). Covalently attached 1-alkynes on silicon surfaces provide superious insulators. SAFE 2005, 8th Annual Workshop on Circuits, Systems and Signal Processing, Veldhoven, the Netherlands November 17-18, 2005.

Faber, E. J., L. C. P. M. de Smet, W. Olthuis, H. Zuilhof, E. J. R. Sudhölter, P. Bergveld and A. van den Berg (2005b). Si-C linked organic monolayers on crystalline silicon surfaces as alternative gate insulators. *ChemPhysChem* 6(10), 2153-2166.

Fiorilli, S., P. Rivolo, E. Descrovi, C. Ricciardi, L. Pasquardini, L. Lunelli, L. Vanzetti, C. Pederzolli, B. Onida and E. Garrone (2008). Vapor-phase self-assembled monolayers of aminosilane on plasma-activated silicon substrates. *Journal of Colloid and Interface Science* 321(1), 235-241.

Hahm, J. and C. M. Lieber (2004). Direct ultrasensitive electrical detection of DNA and DNA sequence variations using nanowire nanosensors. *Nano Letters* 4(1), 51-54.

Haick, H., P. T. Hurley, A. I. Hochbaum, P. D. Yang and N. S. Lewis (2006). Electrical characteristics and chemical stability of non-oxidized, methyl-terminated silicon nanowires. *Journal of the American Chemical Society* 128, 8990-8991.

Hamers, R. J. (2008). Formation and characterization of organic monolayers on semiconductor surfaces. *Annual Review of Analytical Chemistry* 1, 707-736.

Hanson, E. L., J. Schwartz, B. Nickel, N. Koch and M. F. Danisman (2003). Bonding self-assembled, compact organophosphonate monolayers to the native oxide surface of silicon. *Journal of the American Chemical Society* 125(51), 16074-16080.

Hermanson, G. T. (2008). *Bioconjugate Techniques* (2nd Edition), Elsevier Science & Technology Books, ISBN 0123705010, New York, Chapter 11.

Heysel, S., H. Vogel, M. Sanger and H. Sigrist (1995). Covalent attachment of functionalized lipid bilayers to planar wave-guides for measuring protein-binding to biomimetic membranes. *Protein Science* 4(12), 2532-2544.

Howarter, J. A. and J. P. Youngblood (2006). Optimization of silica silanization by 3-aminopropyltriethoxysilane. *Langmuir* 22(26), 11142-11147.

Huang, S. C. J., A. B. Artyukhin, J. A. Martinez, D. J. Sirbuly, Y. Wang, J. W. Ju, P. Stroeve and A. Noy (2007). Formation, stability, and mobility of one-dimensional lipid Bilayers on polysilicon nanowires. *Nano Letters* 7, 3355-3359.

Kar, S., C. Miramond and D. Vuillaume (2001). Properties of electronic traps at silicon/1-octadecene interfaces. *Applied Physics Letters* 78(9), 1288-1290.

Kim, A., C. S. Ah, H. Y. Yu, J. H. Yang, I. B. Baek, C. G. Ahn, C. W. Park, M. S. Jun and S. Lee (2007). Ultrasensitive, label-free, and real-time immunodetection using silicon field-effect transistors. *Applied Physics Letters* 91(10), 103901.

Lee, M. H., K. N. Lee, S. W. Jung, W. H. Kim, K. S. Shin and W. K. Seong (2008). Quantitative measurements of C-reactive protein using silicon nanowire arrays. *International Journal of Nanomedicine* 3(1), 117-124.

Li, D. C., P. H. Yang and M. S. C. Lu (2010). CMOS open-gate ion-sensitive Field-Effect Transistors for ultrasensitive dopamine detection. *IEEE Transactions on Electron Devices* 57(10), 2761-2767.

Li, Y., F. Qian, J. Xiang and C. M. Lieber (2006). Nanowire electronic and optoelectronic devices. *Materials Today* 9(10), 18-27.

Li, Z., Y. Chen, X. Li, T. I. Kamins, K. Nauka and R. S. Williams (2004). Sequence-specific label-free DNA sensors based on silicon nanowires. *Nano Letters* 4(2), 245-247.

Lin, M. C., C. J. Chu, L. C. Tsai, H. Y. Lin, C. S. Wu, Y. P. Wu, Y. N. Wu, D. B. Shieh, Y. W. Su and C. D. Chen (2007). Control and detection of organosilane polarization on nanowire field-effect transistors. *Nano Letters* 7(12), 3656-3661.

Linford, M. R. and C. E. D. Chidsey (1993). Alkyl monolayers covalently bonded to silicon surfaces. *Journal of the American Chemical Society* 115(26), 12631-12632.

Linford, M. R., P. Fenter, P. M. Eisenberger and C. E. D. Chidsey (1995). Alkyl monolayers on silicon prepared from 1-alkenes and hydrogen-terminated silicon. *Journal of the American Chemical Society* 117(11), 3145-3155.

Lobert, P. E., D. Bourgeois, R. Pampin, A. Akheyar, L. M. Hagelsieb, D. Flandre and J. Remacle (2003). Immobilization of DNA on CMOS compatible materials. *Sensors and Actuators B-Chemical* 92(1-2), 90-97.

Luo, L. B., J. S. Jie, W. F. Zhang, Z. B. He, J. X. Wang, G. D. Yuan, W. J. Zhang, L. C. M. Wu and S. T. Lee (2009). Silicon nanowire sensors for Hg2+ and Cd2+ ions. *Applied Physics Letters* 94(19), 3.

Ma, Y., J. Zhang, G. Zhang and H. He (2004). Polyaniline nanowires on Si surfaces fabricated with DNA templates. *Journal of the American Chemical Society* 126(22), 7097-7101.

Martinez, J. A., N. Misra, Y. M. Wang, P. Stroeve, C. P. Grigoropoulos and A. Noy (2009). Highly efficient biocompatible single silicon nanowire electrodes with functional biological pore channels. *Nano Letters* 9(3), 1121-1126.

Masood, M. N., S. Chen, E. T. Carlen and A. v. d. Berg (2010). All-(111) surface silicon nanowires: selective functionalization for biosensing applications. *ACS Applied Materials & Interfaces* 2(12), 3422-3428.

Misra, N., J. A. Martinez, S. C. J. Huang, Y. M. Wang, P. Stroeve, C. P. Grigoropoulos and A. Noy (2009). Bioelectronic silicon nanowire devices using functional membrane proteins. *Proceedings of the National Academy of Sciences of the United States of America* 106(33), 13780-13784.

Patolsky, F., G. F. Zheng, O. Hayden, M. Lakadamyali, X. W. Zhuang and C. M. Lieber (2004). Electrical detection of single viruses. *Proceedings of the National Academy of Sciences of the United States of America* 101(39), 14017-14022.

Patolsky, F., G. F. Zheng and C. M. Lieber (2006a). Fabrication of silicon nanowire devices for ultrasensitive, label-free, real-time detection of biological and chemical species. *Nature Protocols* 1(4), 1711-1724.

Patolsky, F., G. F. Zheng and C. M. Lieber (2006b). Nanowire-based biosensors. *Analytical Chemistry* 78(13), 4260-4269.

Poghossian, A., M. H. Abouzar, F. Amberger, D. Mayer, Y. Han, S. Ingebrandt, A. Offenhausser and M. J. Schoning (2007). Field-effect sensors with charged macromolecules: Characterisation by capacitance-voltage, constant-capacitance, impedance spectroscopy and atomic-force microscopy methods. *Biosensors & Bioelectronics* 22(9-10), 2100-2107.

Ramgir, N. S., Y. Yang and M. Zacharias (2010). Nanowire-based sensors. *Small* 6(16), 1705-1722.

Sagiv, J. (1980). Organized monolayers by adsorption.1. Formation and structure of oleophobic mixed monolayers on solid-surfaces. *Journal of the American Chemical Society* 102(1), 92-98.

Scheres, L., M. Giesbers and H. Zuilhof (2010). Self-assembly of organic monolayers onto hydrogen-terminated silicon: 1-alkynes are better than 1-alkenes. *Langmuir* 26(13), 10924-10929.

Seitz, O., T. Bocking, A. Salomon, J. J. Gooding and D. Cahen (2006). Importance of monolayer quality for interpreting current transport through organic molecules: Alkyls on oxide-free Si. *Langmuir* 22(16), 6915-6922.

Shao, M. W., D. D. D. Ma and S. T. Lee (2010). Silicon nanowires - synthesis, properties, and applications. *European Journal of Inorganic Chemistry*(27), 4264-4278.

Sheu, J. T., C. C. Chen, P. C. Huang, Y. K. Lee and M. L. Hsu (2005). Selective deposition of gold nanoparticles on SiO2/Si nanowires for molecule detection. *Japanese Journal of Applied Physics Part 1-Regular Papers Brief Communications & Review Papers* 44(4B), 2864-2867.

Sieval, A. B., R. Linke, G. Heij, G. Meijer, H. Zuilhof and E. J. R. Sudhölter (2001a). Amino-terminated organic monolayers on hydrogen-terminated silicon surfaces. *Langmuir* 17(24), 7554-7559.

Sieval, A. B., R. Linke, H. Zuilhof and E. J. R. Sudhölter (2000a). High-quality alkyl monolayers on silicon surfaces. *Advanced Materials* 12(19), 1457-1460.

Sieval, A. B., R. Opitz, H. P. A. Maas, M. G. Schoeman, G. Meijer, F. J. Vergeldt, H. Zuilhof and E. J. R. Sudhölter (2000b). Monolayers of 1-alkynes on the H-terminated Si(100) surface. *Langmuir* 16(26), 10359-10368.

Sieval, A. B., B. van den Hout, H. Zuilhof and E. J. R. Sudhölter (2001b). Molecular modeling of covalently attached alkyl monolayers an the hydrogen-terminated Si(111) surface. *Langmuir* 17(7), 2172-2181.

Silverman, B. M., K. A. Wieghaus and J. Schwartz (2005). Comparative properties of siloxane vs phosphonate monolayers on a key titanium alloy. *Langmuir* 21(1), 225-228.

Stern, E., J. F. Klemic, D. A. Routenberg, P. N. Wyrembak, D. B. Turner-Evans, A. D. Hamilton, D. A. LaVan, T. M. Fahmy and M. A. Reed (2007). Label-free immunodetection with CMOS-compatible semiconducting nanowires. *Nature* 445(7127), 519-522.

Stern, E., A. Vacic and M. A. Reed (2008). Semiconducting nanowire field-effect transistor biomolecular sensors. *IEEE Transactions on Electron Devices* 55(11), 3119-3130.

Streifer, J. A., H. Kim, B. M. Nichols and R. J. Hamers (2005). Covalent functionalization and biomolecular recognition properties of DNA-modified silicon nanowires. *Nanotechnology* 16(9), 1868-1873.

Strother, T., R. J. Hamers and L. M. Smith (2000). Covalent attachment of oligodeoxyribonucleotides to amine-modified Si (001) surfaces. *Nucleic Acids Research* 28(18), 3535-3541.

Stutzmann, M., J. A. Garrido, M. Eickhoff and M. S. Brandt (2006). Direct biofunctionalization of semiconductors: A survey. *Physica Status Solidi a-Applications and Materials Science* 203(14), 3424-3437.

Tey, J. N., I. P. M. Wijaya, J. Wei, I. Rodriguez and S. G. Mhaisalkar (2010). Nanotubes-/nanowires-based, microfluidic-integrated transistors for detecting biomolecules. *Microfluidics and Nanofluidics* 9(6), 1185-1214.

Vandenberg, E. T., L. Bertilsson, B. Liedberg, K. Uvdal, R. Erlandsson, H. Elwing and I. Lundstrom (1991). Structure of 3-aminopropyl triethoxy silane on silicon-oxide. *Journal of Colloid and Interface Science* 147(1), 103-118.

Vilan, A., O. Yaffe, A. Biller, A. Salomon, A. Kahn and D. Cahen (2010). Molecules on Si: electronics with chemistry. *Advanced Materials* 22(2), 140-159.

Vu, X. T., R. Stockmann, B. Wolfrum, A. Offenhaesser and S. Ingebrandt (2010). Fabrication and application of a microfluidic-embedded silicon nanowire biosensor chip. *Physica Status Solidi a-Applications and Materials Science* 207(4), 850-857.

Wan, J., S. R. Deng, R. Yang, Z. Shu, B. R. Lu, S. Q. Xie, Y. F. Chen, E. Huq, R. Liu and X. P. Qu (2009). Silicon nanowire sensor for gas detection fabricated by nanoimprint on SU8/SiO2/PMMA trilayer. *Microelectronic Engineering* 86(4-6), 1238-1242.

Wanekaya, A. K., W. Chen, N. V. Myung and A. Mulchandani (2006). Nanowire-based electrochemical biosensors. *Electroanalysis* 18(6), 533-550.

Wang, W. U., C. Chen, K. H. Lin, Y. Fang and C. M. Lieber (2005). Label-free detection of small-molecule-protein interactions by using nanowire nanosensors. *Proceedings of the National Academy of Sciences of the United States of America* 102(9), 3208-3212.

Wang, X. H., Y. Chen, K. A. Gibney, S. Erramilli and P. Mohanty (2008). Silicon-based nanochannel glucose sensor. *Applied Physics Letters* 92(1), 013903.

Zhang, G. J., A. Agarwal, K. D. Buddharaju, N. Singh and Z. Q. Gao (2007). Highly sensitive sensors for alkali metal ions based on complementary-metal-oxide-semiconductor-compatible silicon nanowires. *Applied Physics Letters* 90(23), 233903.

Zhang, G. J., J. H. Chua, R. E. Chee, A. Agarwal, S. M. Wong, K. D. Buddharaju and N. Balasubramanian (2008a). Highly sensitive measurements of PNA-DNA hybridization using oxide-etched silicon nanowire biosensors. *Biosensors & Bioelectronics* 23(11), 1701-1707.

Zhang, G. J., M. J. Huang, Z. H. H. Luo, G. K. I. Tay, E. J. A. Lim, E. T. Liu and J. S. Thomsen (2010). Highly sensitive and reversible silicon nanowire biosensor to study nuclear hormone receptor protein and response element DNA interactions. *Biosensors & Bioelectronics* 26(2), 365-370.

Zhang, G. J., G. Zhang, J. H. Chua, R. E. Chee, E. H. Wong, A. Agarwal, K. D. Buddharaju, N. Singh, Z. Q. Gao and N. Balasubramanian (2008b). DNA sensing by silicon nanowire: charge layer distance dependence. *Nano Letters* 8(4), 1066-1070.

Zhang, X. G. (2001). *Electrochemistry of Silicon and its Oxide,* Kluwer Academic/ Plenum Publishers, 0-306-46541-8, New York, Chapter 2.

Zheng, G. F., F. Patolsky, Y. Cui, W. U. Wang and C. M. Lieber (2005). Multiplexed electrical detection of cancer markers with nanowire sensor arrays. *Nature Biotechnology* 23(10), 1294-1301.

14

Characterization and Application of Thermoelectric Nanowires

Huzel, D.[1], Reith, H.[1], Schmitt, M.C.[1], Picht, O.[2], Müller, S.[2], Toimil-Molares, M.E.[2] and Völklein, F.[1]
[1]*RheinMain University of Applied Sciences Wiesbaden,*
[2]*GSI Helmholtzzentrum für Schwerionenforschung GmbH, Darmstadt, Germany*

1. Introduction

In recent years, numerous studies on the electrical transport properties of semiconducting and metallic nanostructures have been performed because of their possible applications in future miniaturized devices, such as thermoelectric sensors or generators. In particular, the electrical transport properties of quasi one-dimensional structures such as nanowires were investigated intensively (Dresselhaus et al., 2003; Cornelius et al., 2006; Toimil-Molares et al., 2003). When the structure size is comparable to the electronic mean free path, finite-size effects and quantum-size effects are expected to occur (Fuchs, 1937; Sondheimer, 1952; Mayadas & Shatzkes, 1970). These effects are anticipated to affect also the thermal transport mechanisms. Besides a decrease of the thermal conductivity λ, theories predict an increase of the Seebeck coefficient S and the thermoelectric efficiency $z = S^2\sigma/\lambda$, where σ denotes the electrical conductivity, when reducing the dimensionality of the specimen (e.g. by decreasing the diameter of nanowires) (Lin et al., 2000; Zou & Balandin, 2001; Dresselhaus & Heremans, 2006). Therefore, nanowires are promising candidates for new thermoelectric devices (TED) with significantly increased thermoelectric figure of merit.
Experimental verification of these predicted effects requires the simultaneous measurement of the above mentioned transport coefficients (S, σ, λ) on single nanowires. In contrast to the extensive studies on electrical transport, only few experimental investigations of thermal conductivity and Seebeck coefficient of nanowires have been reported. For thermal conductivity experiments any thermal bypass has to be avoided, i.e., these studies must not be performed on single wires lying on a substrate but on individual suspended needles. To measure the Seebeck effect accurately, a temperature difference along the nanowire axis (between the electrical contacts) of a few Kelvin has to be established. In the case of a single nanowire contacted between two electrodes placed 10 µm apart, the temperature gradient along the wire amounts 10^5 K/m. Furthermore, the thermoelectric voltage U caused by this temperature gradient and the temperatures at the nanowire contacts (i.e. at two positions separated by only few micron) have to be measured precisely. Thus, sophisticated micro-structured devices are required for the determination of the thermoelectric transport coefficients of individual nanostructures (Shi et al., 2003; Boukai et al., 2006). Measurements of S, σ and λ have been performed on assemblies of many identical nanowires of bismuth

(Boukai et al., 2006) and silicon (Boukai et al., 2008) and on individual silicon nanowires (Li et al., 2003; Hochbaum et al., 2008).
Here, we present new methods and microchips for the measurement of electrical and thermal conductivity and Seebeck coefficient on individual nanowires prepared by two different methods: (i) wires electrochemically deposited in etched ion-track membranes (Toimil-Molares et al., 2001; Cornelius et al., 2005; Karim et al., 2006) and (ii) nanostructures fabricated by focused electron/ion beam induced deposition (FEBID/FIBID). In addition, we discuss finite element method (FEM) simulations performed to optimize the design and the fabrication process of a *z*-microchip that can be applied for the complete characterization of the thermoelectric efficiency of an individual nanowire. In addition, we have investigated the thermoelectric parameters of nanowire arrays embedded in 30 µm thick etched ion-track polymer foils. Measurements of the cross-plane thermal conductivity of template-embedded bismuth nanowires using a steady-state method are presented.

2. Fabrication of thermoelectric nanostructures by ion-track technology and FEBID/FIBID

2.1 Fabrication of nanowires by ion-track technology

The steps of nanowire fabrication by electrodeposition and ion-track technology are schematically depicted in figure 1. In a first step, a 30-60 µm thick polycarbonate foil is irradiated with a high-energetic ion beam (typical ions: Au, Pb, U; E ~ GeV) at the UNILAC accelerator of GSI. Each ion generates a damage-zone called latent track along its trajectory.

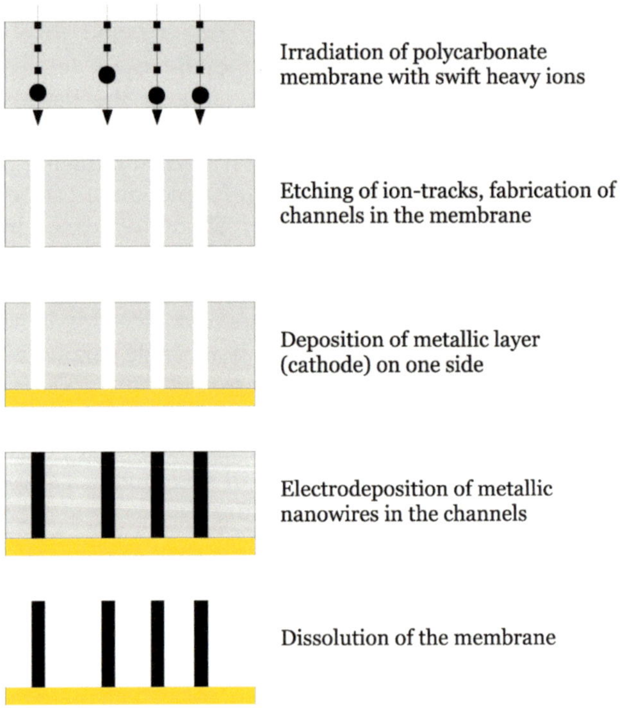

Fig. 1. Schematic of the nanowire fabrication process by electrodeposition and ion-track technology.

In a second step, the tracks are selectively etched in a 6M NaOH solution at 50 °C, leading to the formation of nearly cylindrical nanochannels, whose diameter *d* increases linearly with the etching time. After rinsing the membrane in distilled water, a Au metal layer is deposited on one membrane surface. This layer acts as cathode during the electrochemical deposition of Bi, $Bi_{1-x}Sb_x$ ($0 < x < 1$) and Bi_2Te_3 nanowires (Cornelius, 2005; Picht, 2009). The length *l* of the nanowires is limited by the thickness of the membrane (60 µm) and is controlled by the duration of the deposition process. Free-standing nanowires are created by dissolving the matrix in an organic solvent. Subsequently, the nanowires are detached from the back electrode by ultra-sonication, and nanowires with lengths ranging from a few to 60 µm are available in the suspension. The wires are then placed onto a substrate of choice (e.g. a Si wafer with a SiO_2 film) by applying few drops of the solvent containing the needles. The solvent evaporates within a few minutes leaving behind randomly distributed wires. By controlling the fabrication parameters, polycrystalline as well as single-crystalline wires can be fabricated (Toimil-Molares et al., 2001; Cornelius et al., 2005; Karim et al., 2006). Figure 2 presents a representative scanning electron microscopy (SEM) image, displaying Pt nanowires with smooth contours and narrow diameter distribution.

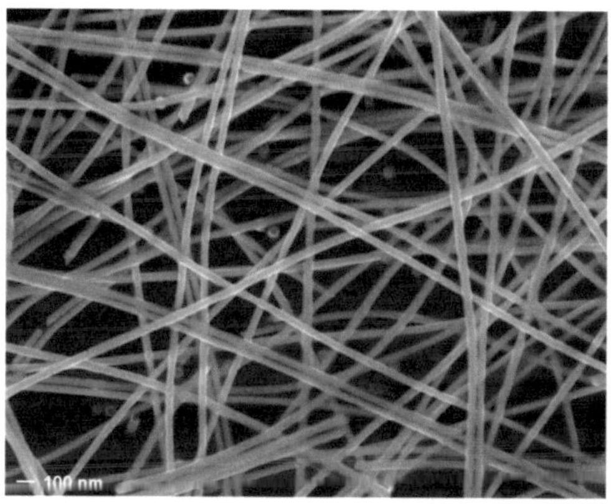

Fig. 2. SEM image of Pt nanowires after dissolving the polycarbonate foil and detachment from the back electrode.

2.2 Focused Electron/Ion Beam Induced Deposition

Focused electron/ion beam induced deposition (FEBID/FIBID) is a direct beam writing technique for the realization of nano- and microstructures. The schematic FEBID process is represented in figure 3. Precursor molecules are emitted from a gas injection system close to the focal point of an electron beam. The molecules are dissociated by primary, backscattered and secondary electrons. With a predefined pattern, the primary electron beam can be scanned over the substrate surface. Relevant process parameters for this scanning process are the distance between successive dwell points of the electron beam (pitch) and the time period during which the electron beam is held at each dwell point (dwell time). Typical pitches vary between 10 to 100 nm. The required dwell times depend on the precursor gas and the substrate material. Detailed recent reviews concerning FEBID/FIBID and related techniques are given in (Utke et al., 2008; Huth, 2011). By proper selection of the precursor gas and the process parameters, different functionalities of the resulting deposits can be obtained. FEBID-generated nano-granular films are very promising substances regarding the optimization of thermoelectric material parameters due to nano-dimensions. Quite generally, nano-granular metals can be considered as tunable model systems for studying the interplay of electronic correlation effects, quantum size effects and disorder. Particularly interesting are nano-granular FEBID/FIBID structures for which an insulator-to-metal transition is observed as a function of the inter-grain coupling strength. Valuable information about the electronic properties close to this transition can be obtained from Seebeck effect measurements.

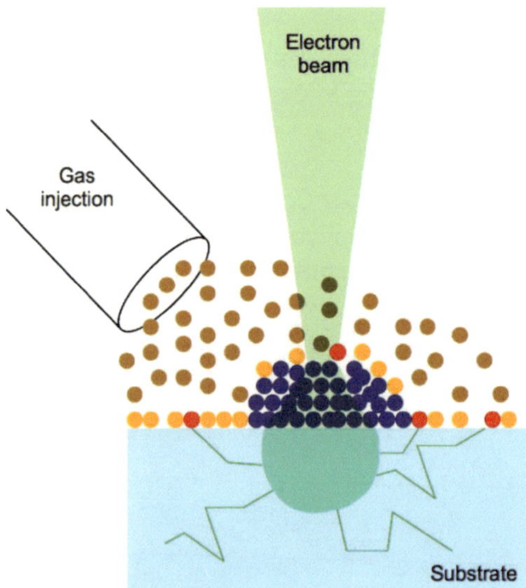

Fig. 3. Schematic representation of the FEBID process: The adsorbed precursor molecules (orange spheres) are dissociated by electron impact (red spheres) and a permanent deposit (blue spheres) is formed in the focal area of the electron beam. The green lines indicate exemplary trajectories of electrons leaving the excitation volume.

3. Electrical conductivity

3.1 Fabrication of electrical contacts

The thermoelectric characterization of nanowires and FEBID nanostructures requires their integration into specific microchips. In this nano-micro-integration process it is very important to fabricate reliable electrical contacts that do not influence the thermoelectric properties of the nanostructures (e.g. by diffusion processes). To obtain electrical contacts and measuring devices for nanostructures such as nanowires or FEBID/FIBID lines two different approaches are commonly used:

i. The nanostructure is (randomly) placed on a substrate and the measuring device is realized around the nanostructure by micropatterning using surface micromachining technologies. This approach includes the direct formation of electrical connections to the nanostructures as part of the construction of the measuring device.
ii. Specific microchips are produced (preferably) by standard semiconductor batch-process fabrication and the nanostructures are subsequently placed onto the chips by particular handling and deposition techniques. Highly resistive electrical contacts to the pre-fabricated microchips have to be improved by additional methods.

The first approach has the advantage of requiring no equipment or techniques for handling and micro-positioning of the nanostructure. On the other hand, due to the small dimension of the samples, the patterning of the measuring device has to be performed using local microscopic lithography which makes the fabrication time-consuming (Voelklein et al., 2009a, 2009b, 2010). The second approach, applying separately batch-fabricated chips and afterwards putting the nanostructures onto these devices, is associated with challenging handling and positioning procedures. One method is to place droplets of a solution containing nanowires (see section 2.1) onto the device and to repeat the procedure until one nanowire is (randomly) placed at the right position (Hochbaum et al., 2008; Shi et al., 2005; Mavrokefalos et al., 2009). Another method is the direct positioning of the sample onto the microchip using a nano-manipulator (Hochbaum et al. 2008, Shi et al., 2003). After such placing procedures usually the electrical contacts of the nanowire to the electrical connections (e.g. thin film contact pads) has to be improved by local deposition of FEBID/FIBID layers at the overlap of nanowire and contact pads.

3.1.1 Lift-off process

Electrical connections to individual nanowires fabricated according to approach i) can be realized by the lift-off technique. The preparation sequence for electrical contacts on both ends of a single nanowire is depicted in figure 4.

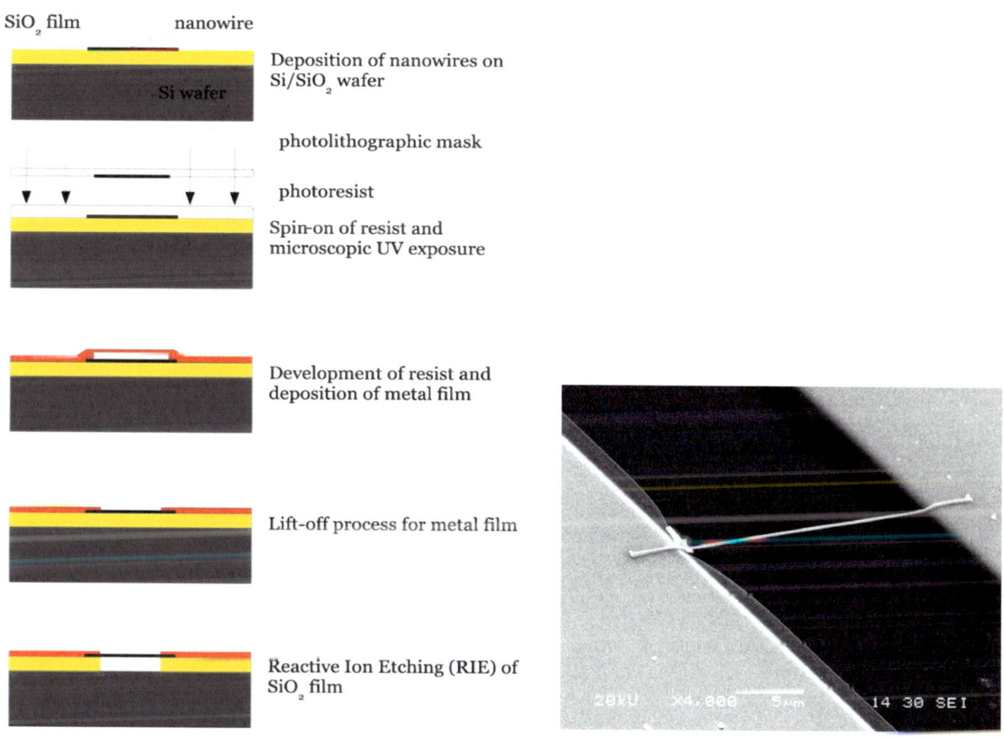

Fig. 4. left: Preparation scheme of electrical contact pads for a single nanowire performed with local microscopic photolithographic exposure of photoresist, resist development, thin-film deposition, and lift-off technique; right: SEM image of electrical contact layers above a 20 µm long suspended nanowire on a Si/SiO_2 substrate.

Nanowires are placed randomly on a (100)-oriented silicon wafer with an insulating (300 nm thick) SiO_2 layer on top. After spin-coating the photoresist on the SiO_2 surface and inspection of the nanowire positions, the resist is locally exposed to UV light at both ends of the nanowire using a photolithographic micro-mask. The exposed photoresist is then developed and removed at both ends, whereas the central area of the nanowire is still protected. Next, a thin metal contact film is deposited onto the wafer by sputtering. Finally, this film is structured by lift-off technique forming contact pads at both ends of the selected nanowire. This contacting procedure is used for all Seebeck coefficient, electrical and thermal conductivity measurements. To determine the thermal conductivity, suspended nanowires have to be prepared to avoid any thermal bypass through the substrate. Therefore, after contact preparation the SiO_2 layer below the nanowire is removed by reactive ion etching (RIE). Due to optimized process parameters (etching gas, pressure, bias voltage) the usually anisotropic RIE process is transformed in a partially isotropic one, removing the SiO_2 layer below the nanowire within a few minutes. During the etching process, the metal contact films serve as etching mask layers, and the SiO_2 film is removed exclusively in the gap area.
Long-term stability is a crucial criterion of thin film contacts on nanostructures, especially for their applications at elevated temperatures. Therefore thermal cycling tests have been performed on various film materials to study the stability of the contacts regarding their ohmic behavior. Contacts consisting of 20 nm Ti (diffusion barrier) and 150 nm Au film have shown best long-term stability without degradation due to diffusion effects.

3.1.2 Direct writing of FEBID/FIBID nanostructures on measuring devices

Figure 5 represents an example for the deposition of nanostructures on a pre-fabricated microchip according to approach ii). Here, nano-granular FEBID layers are placed by direct writing between prepared Au contact films of a sensor chip.

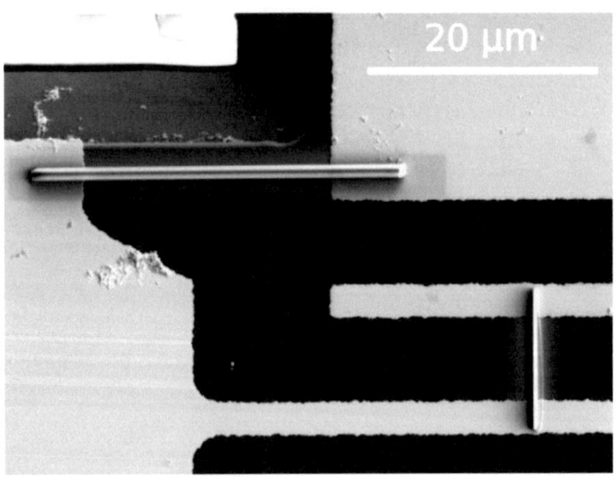

Fig. 5. Nano-granular sensor element (left-to-right structure) and reference element (top-to-bottom structure) prepared by FEBID direct writing between Au contact films using the precursor $MeCpPt(Me)_3$

A very important benefit of such FEBID techniques arises from the fact that nano-granular structures can be deposited at arbitrary positions of a planar or 3-dimensional measuring device or also of a specific application device. For example, in figure 5 the nano-granular layer is deposited on a 10 µm thin cantilever in order to investigate its resistivity as function of induced strain.

3.2 Four-point-probe measurement

Contact resistances are known to have considerable influence on the measuring results. The effect of the contact resistances on nanostructures has been investigated and eliminated by using a 4-point-probe arrangement, fabricated as previously described using the lift-off technique. The structure of the used lithographic micro-mask for 4-point measurements is presented in figure 6. The occurrence of contact resistances can be deduced by comparing the results of 2-point-probe and 4-point-probe resistivity measurements. To obtain the 2-point-probe resistivity ρ_{2p} a current flow I and the corresponding voltage drop U_{2p} between the contacts 1 and 4 are measured. In case of negligible contact resistance the resistivity ρ_{2p}, calculated according to

$$\rho_{2p} = \frac{U_{2p} A}{I(l_{12} + l_{23} + l_{34})} \tag{1}$$

represents the true resistivity of the nanowire, where A indicates its cross section area and the l_{ij} denote the lengths of the wire segments between the contact probes. For the determination of the 4-point-probe resistivity ρ_{4p} the same current flow I is supplied using contacts 1 and 4, whereas the potential difference is measured between contacts 2 and 3. Without any contact resistance, the difference between the 2-point-probe and 4-point-probe resistivity (ρ_{2p} - ρ_{4p}), related to ρ_{2p}

$$\frac{\rho_{2p} - \rho_{4p}}{\rho_{2p}} = 1 - \frac{U_{4p}(l_{12} + l_{23} + l_{34})}{U_{2p} l_{23}} \tag{2}$$

should be zero.

Fig. 6. Ti/Au film 4-point-probe contact structure (bright area). The gaps between contact probes (marked with 1 to 4) are 5 µm.

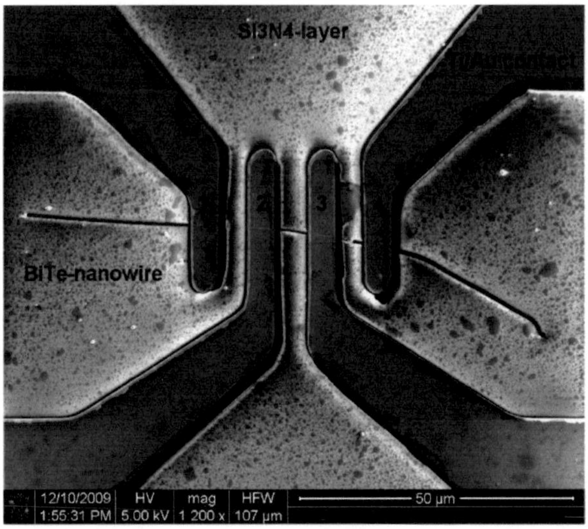

Fig. 7. SEM image of a Bi_2Te_3 nanowire with a diameter of 180 nm contacted in a 4-point-probe structure.

Figure 7 shows a SEM image of a Bi_2Te_3 nanowire contacted by a 4-point-probe arrangement. Its dimensions (length, diameter) are determined by evaluation of high-resolution SEM pictures. Table 1 lists the lengths of the nanowire segments between the contact probes and the corresponding voltage drops.

l_{12} [µm]	l_{23} [µm]	l_{34} [µm]	U_{2p} [mV]	U_{4p} [mV]
2.73	3.87	3.55	12.7	4.7

Table 1. Lengths of segments of a 180 nm diameter Bi_2Te_3 nanowire between contact probes and voltage drops U_{2p} in 2-point-probe and U_{4p} in 4-point-probe arrangement for I=0.88 µA

The relative deviation between 2-point-probe and 4-point-probe resistivity

$$\frac{\rho_{2p}-\rho_{4p}}{\rho_{2p}} = 1 - \frac{U_{4p}(l_{12}+l_{23}+l_{34})}{U_{2p}l_{23}} < 0.03 \tag{3}$$

of this nanowire is smaller than 3% and indicates that the effect of contact resistances can be neglected for our preparation technique. This may be attributed to sputter deposition of the adhesion/barrier layer (Ti) before the deposition (thermal evaporation) of the conducting Au film, since sputtered particles have a very high energy (about 100 eV) and form very intimate bindings with the nanowire surface atoms.

In addition, conventional 2-point-probe contacts can be transferred to 4-point-probe arrangements using FEBID for direct writing of potential contacts as indicated in figure 8. Furthermore, the additional application of a focused ion beam (FIB) allows controlled doping of potential barriers (Schottky barriers) which may appear between metal-semiconductor junctions. Therefore FIB can be used for reducing contact resistances, where the junction is doped with Ga until a metallic behavior is achieved (Cronin et al., 2002). In general, it is necessary to verify 2-point-probe measurement results with appropriated 4-point-probe measurements.

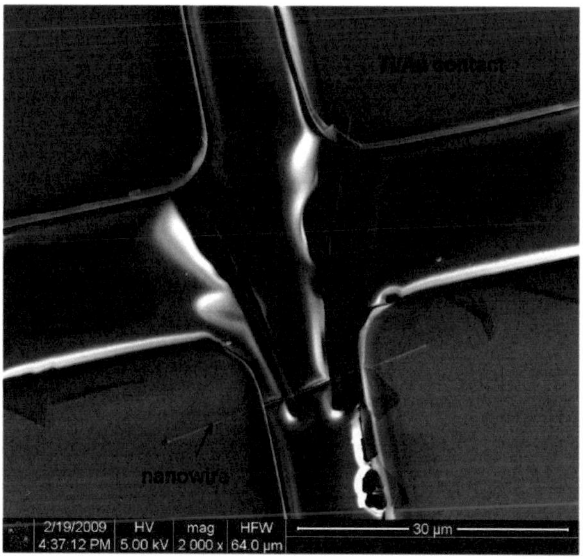

Fig. 8. SEM image of a 30 µm long Pt nanowire contacted in a 4-point-probe arrangement with the interior potential contacts written by FEBID (see labels in the picture for details).

3.3 Current-voltage characteristics

It is known that potential barriers (Schottky barriers) may occur at metal-semiconductor junctions. Current-voltage characteristics measured on single nanowires reveal the metal-like or semiconductor-like character of the contacts. Thus, a metal-like (ohmic) contact leads to a linear current-voltage curve, as shown in figure 9 for the Bi_2Te_3 nanowire with negligible contact resistance at room temperature (see equation (3)).

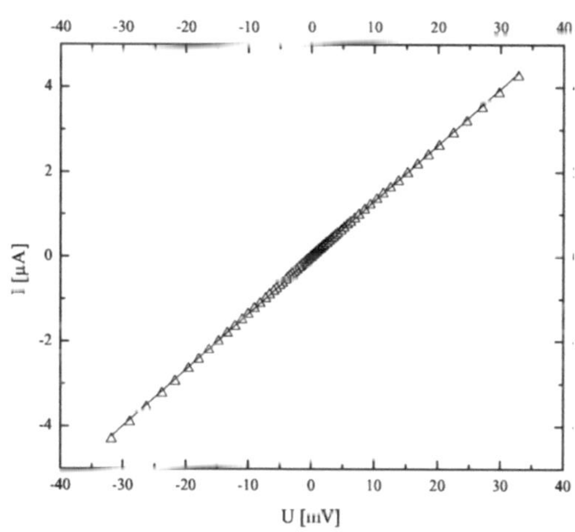

Fig. 9. Current-voltage characteristic of a Bi_2Te_3 nanowire (l = 18.4 µm , d = 180 nm) measured at room temperature

Figure 10 displays the current-voltage characteristics of an individual $Bi_{0.9}Sb_{0.1}$ nanowire contacted with Ti/Au pads measured at both room temperature and 100 K. The observed non-linear current-voltage characteristics indicate a larger barrier height at 100 K compared to room temperature. Since the Schottky barrier height is increased by decreasing carrier concentration in the semiconducting material of a metal-semiconductor contact, the current-voltage behavior can be interpreted by the semiconducting state of the $Bi_{0.9}Sb_{0.1}$ nanowire. Compared to the situation at 100 K the nanowire carrier concentration is increased at room temperature. Thus, the influence of the Schottky barrier decreases and the current-voltage characteristic becomes almost linear.

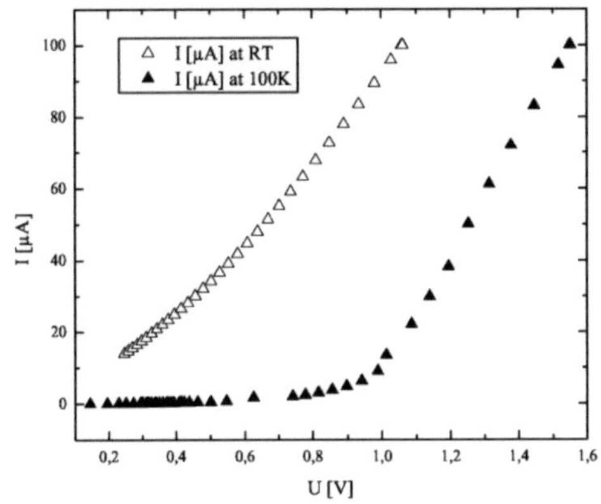

Fig. 10. Current-voltage characteristics of a $Bi_{0.9}Sb_{0.1}$ nanowire (l = 20 μm, d = 200 nm) measured at room temperature (empty triangles) and at 100 K (filled triangles)

3.4 Electrical conductivity and temperature coefficient of resistance

Since we showed that the contact resistance of the system (Au/Ti)/Bi_2Te_3 nanowire/(Au/Ti) was negligible (see section 3.2), Bi_2Te_3 nanowires displaying a linear current-voltage characteristic have been also measured in a 2-point-probe arrangement in the temperature range between 4K - 300 K. In figure 11 the resistance and electrical resistivity of a Bi_2Te_3 nanowire with diameter 180 nm and length 18.4 μm is presented. The geometrical data were obtained by SEM measurements.

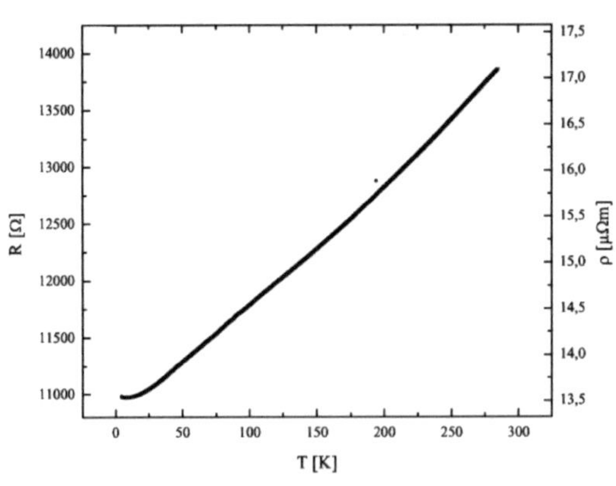

Fig. 11. Resistance/resistivity as function of temperature of a Bi_2Te_3 nanowire (d = 180 nm, l = 18.4 μm)

The measurements of the resistance as function of temperature were performed in a ^{4}He cryostat. A very small electrical power of 10^{-8} W was dissipated in the nanowire during this measurement by appropriate choice of the applied current. With respect to the high current densities generated in the nanowire, especially for low resistances (metal nanowires), preferably an AC current source and lock-in technique should be used to prevent damages of the nanowire by electromigration. The resistivity value at room temperature, $\rho \approx 17$ μΩm, is consistent with values previously reported in literature for polycrystalline Bi_2Te_3 material. The resistivity decreases with decreasing T indicating the metallic behavior of the Bi_2Te_3 nanowire.
The temperature coefficient of resistance is calculated using TCR = $[1/R_0]dR(T)/dT$ being R_0 the resistance of the wire measured at T_0 = 295 K. The slope can be obtained either by differentiation of an appropriated fit function of the resistivity data or by a partial linearization of the data points. The TCR presented in figure 12 is calculated using a polynomial fit of the resistivity data.

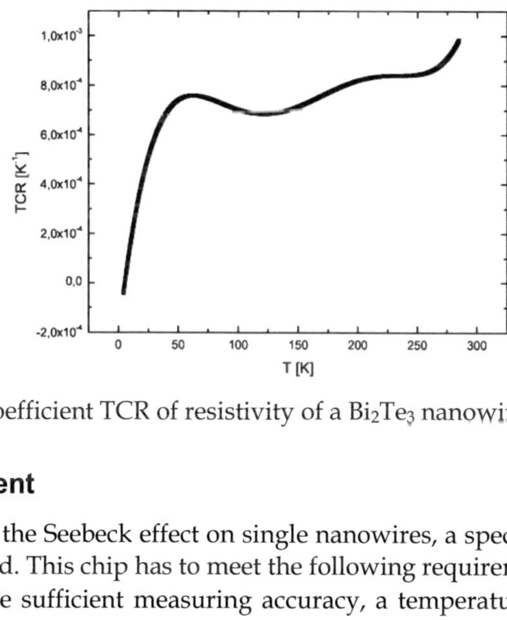

Fig. 12. Temperature coefficient TCR of resistivity of a Bi_2Te_3 nanowire.

4. Seebeck coefficient

To accurately measure the Seebeck effect on single nanowires, a specific microchip has been designed and optimized. This chip has to meet the following requirements:

- In order to achieve sufficient measuring accuracy, a temperature difference along the nanowire axis of at least 1 K should be realized. Thus, a temperature gradient of at least 10^5 K/m has to be established between the electrical contacts (assuming a contact distance of about 10 μm).
- Besides the thermoelectric voltage U caused by the temperature difference, also the temperatures at the nanowire contacts (i.e., at two positions with a distance of only a few microns) have to be measured precisely.

Micromachining and microlithographic methods were employed to realize a newly developed microchip, which meets both experimental challenges.

4.1 Experimental details

A 300 nm thin Si_3N_4 membrane is prepared by micromachining in the center of a silicon chip. After low pressure chemical vapour deposition (LPCVD) of Si_3N_4 layers on both sides of a silicon wafer, a square-shaped window is opened by photolithographic patterning and reactive ion etching (RIE) of the Si_3N_4 layer on the back side. Then, the exposed (100)-oriented silicon is removed by anisotropic etching in KOH-solution at 80 °C. The LPCVD layer is not attacked by this etchant; therefore a square-shaped Si_3N_4 membrane of 2x2 mm^2 remains at the front side of the chip. This very thin membrane with low thermal conductivity is a crucial feature of the chip with respect to realization of the above mentioned temperature gradient.

Prior to the membrane preparation, nanowires are placed on the front side Si_3N_4 layer of the (100)-oriented silicon wafer. After inspection of the wires and their positions on the Si_3N_4 surface, one pair of Ti/Au contact films is structured at the ends of a selected wire by lift-off technique (described in figure 4). An identical second pair of contact films is patterned adjacent to the first one as shown in figure 13. After preparation of these contact layers, the opening of the back side windows for anisotropic silicon etching is established by photolithography, using a double-side exposure mask aligner. Due to double-side alignment, the correct position and orientation of the back side window with respect to the nanowire and contact film positions at the wafer front side is achieved. The anisotropic etching is performed in a KOH-resistant etching box in order to protect the wafer front side against the etchant. When aligned correctly, the single nanowire is located near the centre of the membrane (or window) and the edges of the rectangular contact films are oriented parallel to the membrane edges, as shown in figure 13. Next, a Seebeck reference film (thickness 200 nm) with known Seebeck coefficient is deposited on the second pair of contact films by thermal evaporation through a metal diaphragm. Polycrystalline bismuth films have been applied as reference films, because their thermoelectric properties and Seebeck coefficients have been investigated in detail (Voelklein & Kessler, 1986). Furthermore, the Seebeck reference film can be deposited simultaneously on the bulk silicon rim of the chip, where its Seebeck coefficient can be measured with calibrated thermocouples.

Finally, a thin-film heater is deposited onto the membrane region of the chip in order to generate the required temperature difference. This heating film can be placed on the back side of the membrane without additional electrical insulation. Alternatively, it can also be deposited on the membrane front side, if the heater is separated from the contact films by a thin insulating layer. The contact films are electrically connected by wire bonding, where the bond contacts have to be realized outside the membrane region on the bulk silicon rim.

The heater generates a temperature difference ΔT at the gap between the metal contacts. The chip design, shown in figure 13, was investigated and optimized by finite element method (FEM) simulations. As demonstrated in the next section, an identical temperature difference ΔT between both pairs of contact films can be achieved. Therefore, ΔT at the nanostructure is determined by using the reference voltage U_R of the Seebeck reference film

$$U_R = (S_R - S_C)\Delta T \tag{4}$$

where S_R is the known Seebeck coefficient of the reference film ($S_R = -63$ µV/K for 200 nm thick Bi films at room temperature) and S_C is the Seebeck coefficient of the contact pads. For Ti/Au contact films, S_C amounts −0.4 µV/K at room temperature. With the measured Seebeck voltage U_N of the nanowire (and the known temperature difference) its Seebeck coefficient S_N can be calculated according to

$$S_N - S_C = \frac{U_N}{\Delta T} = \frac{U_N(S_R - S_C)}{U_R} \tag{5}$$

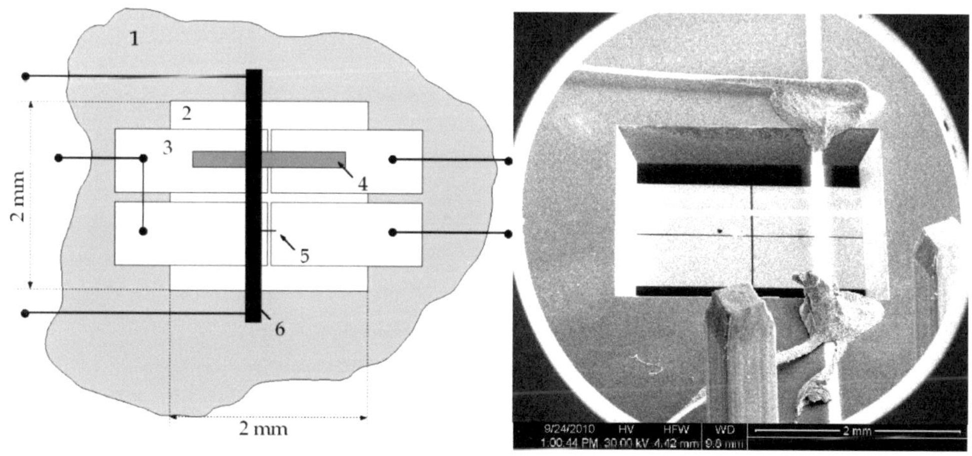

Fig. 13. left: Scheme of the microchip for Seebeck coefficient measurements with 1: surface of silicon wafer, 2: Si_3N_4 membrane, 3: Ti/Au contact pad, 4: bismuth reference layer, 5: position of the nanowire, 6: heating layer; right: SEM picture (back side view) of the microchip, showing electrical connections to the heating layer, the Ti/Au contact pads and Bi reference film.

4.2 FEM simulation

By finite element method (FEM) simulations using COMSOL Multiphysics, the temperature profile in the membrane region as a function of the heating power and the position of the thin-film heater has been investigated. The temperature difference at the gap between the contact films is the most important feature of the chip design. Since the bulk silicon rim has a very high thermal capacitance and thermal conductivity compared to the membrane, it can be considered as a heat sink with constant ambient temperature T_0. Therefore, we apply the boundary condition of constant temperature $T=T_0$ at the membrane edges. Because of the very small thickness of the membrane and of all deposited films compared with their lateral dimensions, the FEM simulation can be performed with a two-dimensional (2D) model. Figure 14 shows the various areas of the membrane region that must be distinguished because of their different thin-film stacks. By using a 2-dimensional model, all individual areas (subdomains) are modeled by a uniform model thickness d. Then the effective thermal conductivities λ_{eff} of the subdomains have to be calculated by

$$\lambda_{eff} d = \sum_i \lambda_i d_i \tag{6}$$

with λ_i denoting the thermal conductivities and d_i representing the thicknesses of the individual layers of the stacks. Radiation losses of the membrane areas are included in the calculation by Stefan-Boltzmann radiation law. With respect to thermal radiation, the subdomains are characterized by two emissivities ε_1 and ε_2 for the top and bottom surface, respectively.

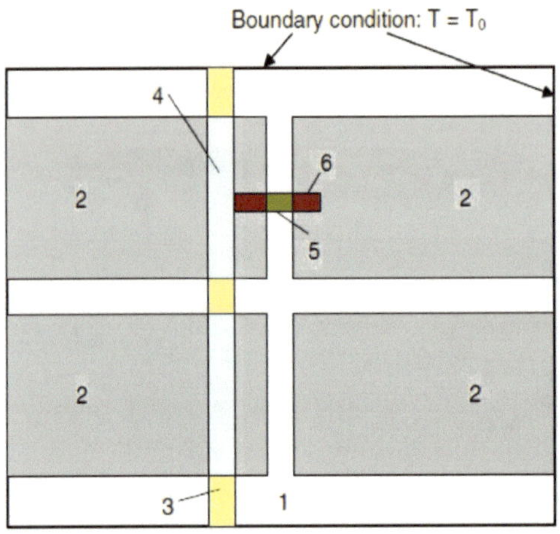

Fig. 14. FEM subdomains in the membrane region, each characterized by specific effective thermal conductivities λ_{eff} and emissivities ε_1, ε_2. 1: membrane, 2: membrane + Ti/Au film, 3: membrane + Bi heating film, 4: membrane + Ti/Au film + Bi heating film, 5: membrane + Bi reference film, 6: membrane + Ti/Au film + Bi reference film.

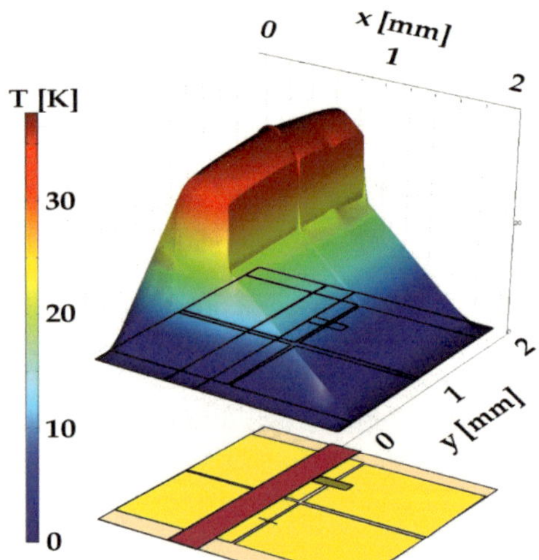

Fig. 15. Simulated three-dimensional temperature profile in the membrane region for a microchip operating in high vacuum (pressure p < 10^{-5} mbar) with a thin-film heater 0.2 mm distant from the gap and a heating power density of 28 W/mm^3, indicating temperature differences $\Delta T > 10$ K for the nanowire and reference film, respectively.

Simulations were performed for microchips applied both in high vacuum and at atmospheric pressure. Heat losses by convection can be neglected for measurements in high vacuum at pressures $p < 10^{-5}$ mbar. Convective heat losses in atmospheric air are simulated by using a heat transfer coefficient a = 6.4 W/m^{2}K. The temperature differences ΔT at the gaps between the contact pairs depend on the heating power density Q and on the distance between gap and thin-film heater. Figure 15 demonstrates the 3-dimensional temperature profile in the membrane region for a microchip with a thin-film heater placed on the back side of the membrane at a distance of 0.2 mm to the gap. A moderate heating power density Q = 28 W/mm^3 generates temperature differences of ΔT > 10 K between the nanowire contacts for operation in high vacuum as well as at atmospheric pressure. A nearly identical temperature difference can be achieved between the contact pairs for the Seebeck reference film (200 nm Bi), because of its small thermal conductance.

4.3 Seebeck coefficient results

First experimental investigations of the Seebeck coefficient were performed on nano-granular (W)-FEBID lines (Voelklein et al., 2010). Nano-granular metals may be promising thermoelectric materials because of the expected reduction of thermal conductivity. Therefore, their thermopower has been investigated with respect to thermoelectric applications. A typical FEBID layer is shown in figure 5. For Seebeck effect measurements nano-granular W-FEBID lines (length: 13 μm, thickness: 300 nm, width: 1.2 μm) are deposited between the Ti/Au contact pads of the chip. Applying a voltage of 20 mV to a thin-film heater of 50 Ω, a temperature difference of a few Kelvin is generated at the gap. By increasing the heating power, a linear increase of the thermoelectric voltage has been observed (figure 16). The Seebeck coefficient is calculated using the slope of the thermoelectric voltage as function of temperature difference. A small Seebeck coefficient $S_N - S_C = -8.1$ μV/K has been determined for nano-granular W-FEBID with a metal content of about 22 at.% at room temperature, where S_C denotes the Seebeck coefficient of the contact material (- 0.4 μV/K).

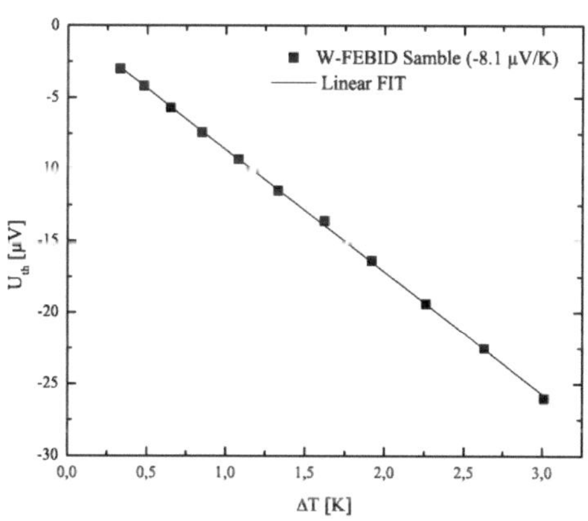

Fig. 16. Thermoelectric voltage U_{th} as function of temperature difference ΔT for a nano-granular W-FEBID layer (thickness: 300 nm, width: 1.2 μm, length: 13 μm) with a metal content of about 22 at.% at room temperature

Figure 17 shows the thermoelectric voltage U_{th} as function of the temperature difference ΔT measured on a single Bi_2Te_3 nanowire at various ambient temperatures. The Seebeck coefficient decreases with decreasing temperature in the observed temperature range. Figure 18 indicates the Seebeck coefficient of a 200 nm diameter $Bi_{0.9}Sb_{0.1}$ nanowire in a temperature range from 80 K to 300 K. The measured Seebeck coefficient of −95 ± 5 μV/K at room temperature is close to the value reported for $Bi_{1-x}Sb_x$ thin films with an antimony content of about 10 at.% (Cho et al., 2000; Liu et al., 2007), as expected for nanowires with these dimensions.

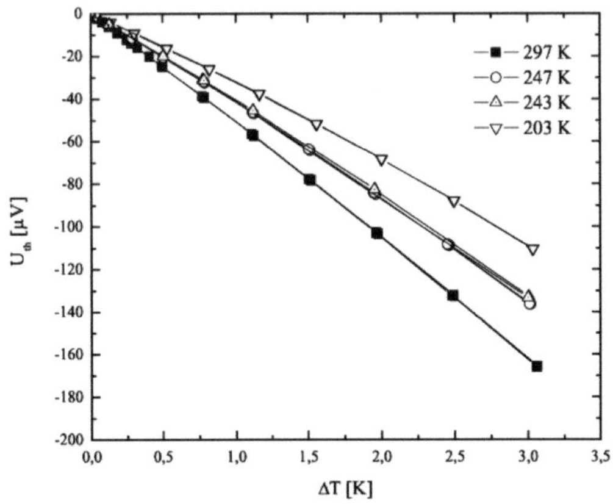

Fig. 17. Thermoelectric voltage U_{th} as function of temperature difference ΔT for a Bi_2Te_3 nanowire (*d*: 180 nm, *l*: 18 µm) at different ambient temperatures

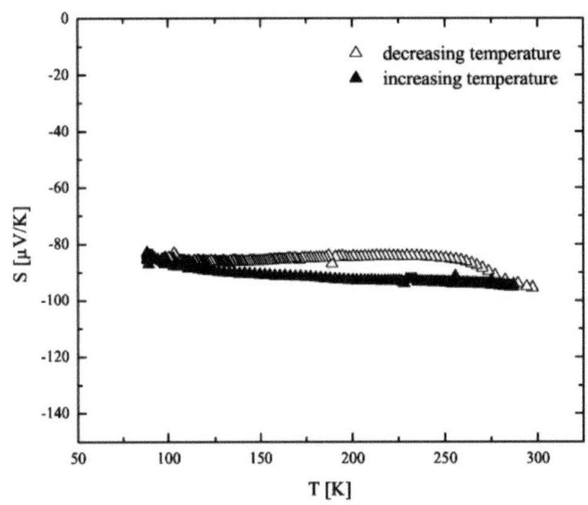

Fig. 18. Seebeck coefficient of a 200 nm diameter $Bi_{0.9}Sb_{0.1}$ nanowire in the temperature range from 80 K to 300 K. Data points indicate measured values during continuous temperature increase (filled triangles) and decrease (empty triangles).

5. Thermal conductivity

Sophisticated micro devices have been designed and applied for the measurement of thermal conductivity of nanostructures (Shi et al., 2003; Boukai & Heath, 2006; Boukai et al., 2008). Extremely fragile suspended cantilevers are prepared for measuring platforms in order to reduce the thermal conductance of the device. We present an alternative steady-state method that requires only heat sink conditions at the ohmic contacts of the nanostructure. Measurements of the thermal conductivity at the single nanowire level are important to investigate the influence of finite-size and quantum-size effects. In addition, methods to determine the thermal conductivity of nanowire arrays embedded in the polymer are relevant to explore the potential integration and application of thermoelectric nanowires in TEDs.

5.1 Steady-state method for suspended nanowires

5.1.1 Experimental details

For thermal conductivity measurements, the nanowire must be suspended to avoid thermal bypass though the substrate. The fabrication and contacting of suspended nanowires by RIE and lithography has been discussed in section 3.1.1. The nanowire endings are in intimate thermal contact with the metallic contact pads and are separated from the bulk silicon wafer only by a thin Si_3N_4 layer. During electrical heating of the nanowire by an electrical current I, most of the heating power is dissipated in the suspended part of the wire, since the resistance of this part is large compared to the resistance of the contact pads. Therefore, the ends of the nanowire can be considered as heat sinks of constant temperature T_0, as demonstrated by finite-element method (FEM) simulations (see figure 19).

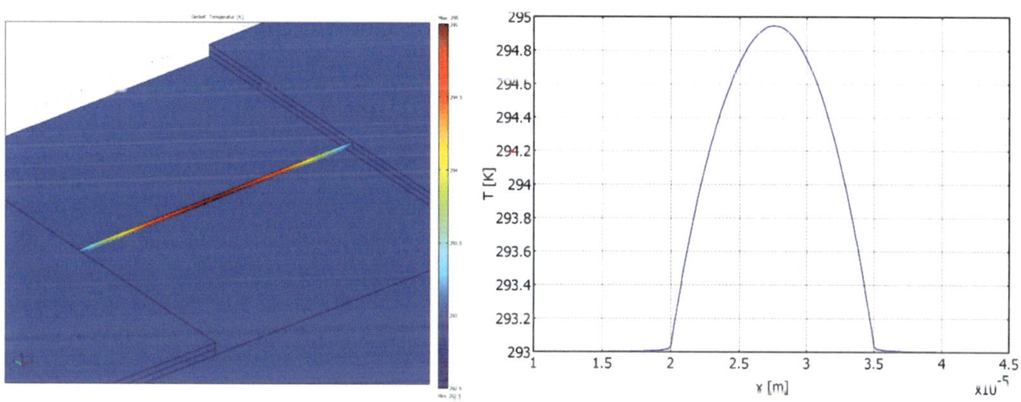

Fig. 19. Temperature profile (FEM simulation) in a suspended, electrically heated nanowire (heating power I^2R = 1 µW, wire radius = 100 nm, length = 15 µm) contacted on a Si_3N_4/Si chip (Si_3N_4-thickness amounts 0.3 µm) with Ti/Au contact layers, demonstrating heat sink conditions at the contacts.

The measuring principle and experimental setup for the thermal conductivity measurements are illustrated in figure 20. The measuring sequence starts with the determination of the nanowire resistance $R(T_0)$ at any ambient temperature T_0, using a 4-point-probe technique and a low measuring current I_0. The temperature T_0 of the setup can be varied between 80 K and 400 K, but, in general, measurements can also be performed at liquid He temperatures. Because of the low I_0, the temperature increase of the nanowire due to Joule heating can be neglected. From these measurements, the slope of the resistance-temperature curve $m = dR(T_0)/dT_0$ is determined at an arbitrary T_0. Then, a heating current I is applied leading to a temperature increase of the wire which involves a rise of the electrical resistance $\Delta R=R(T_M)-R(T_0)$, where $R(T_M)$ denotes the resistance and T_M the mean temperature of the heated nanowire. The usually applied heating currents lead to mean temperature increases $\Delta T_M = T_M-T_0$ of less than 10 K. ΔT_M can be determined experimentally by using the measured resistance increase ΔR and the slope m according to

$$\Delta T_M = \frac{\Delta R}{m} \tag{7}$$

On the other hand, ΔT_M is a function of the thermal conductivity λ of the nanowire. In order to find the dependence of ΔT_M on λ, the thermal heat equation has to be solved for the specific boundary conditions, given by the experimental setup.

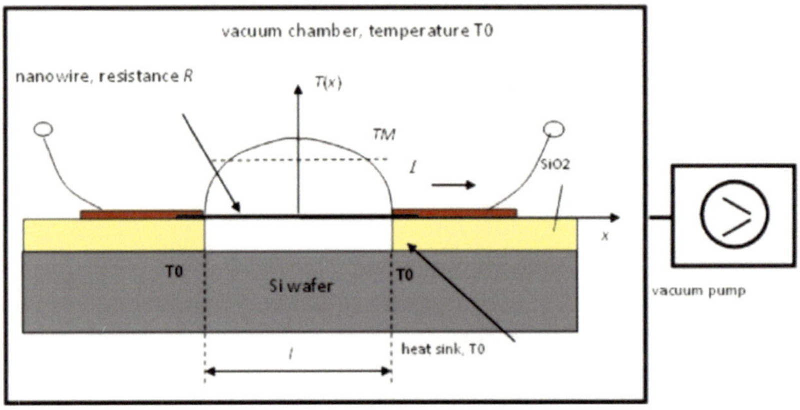

Fig. 20. Schematic cross section of the experimental setup with temperature profile T(x)

Measurements are performed in a high vacuum (pressure $p < 10^{-5}$ mbar). Therefore, convective heat losses from the heated wire to the ambient can be neglected. Furthermore, radiation losses according to the Stefan-Boltzmann law can be omitted because of the small ΔT_M and the dimensions of the suspended nanowire. With respect to the wire length of (typically) 10 µm compared to its small diameter in the order of 100 nm, the heat transport can be described using a one-dimensional model. A detailed derivation of the heat equation (Voelklein et al., 2009b) leads to

$$\Delta T_M = \frac{I^2 R}{12\lambda(A/l)} \tag{8}$$

Combining equations (7) and (8), we can calculate the thermal conductivity

$$\lambda = \frac{mI^2 Rl}{12A\Delta R} \tag{9}$$

by using the measured increase in resistance ΔR, the heating power I^2R, the slope m of the resistance-temperature curve and the geometrical dimensions (A, l) of the wire, which are determined by SEM investigations.

5.1.2 Thermal conductivity results

For a first test of measuring principle the thermal conductivity of individual polycrystalline Pt nanowires has been investigated (Voelklein et al., 2009b). The presented measurements are performed in the temperature range from 260 K - 360 K on a nanowire with 131 nm diameter and a suspended length of 14.7 µm. The values of electrical conductivity σ (figure 21) are deduced from measurements of the nanowire resistance $R(T_0)$ performed with a low

AC current of frequency 113 Hz. During these measurements a very low Joule heat of 10^{-7} W is dissipated and, therefore, the mean temperature increase ΔT_M in the nanowire remains smaller than 0.3 K. Compared to the bulk value $\sigma_{bulk} = 9.6 \times 10^6\ (\Omega m)^{-1}$ at 295 K, the electrical conductivity of the Pt nanowire is reduced by a factor of 2.5.

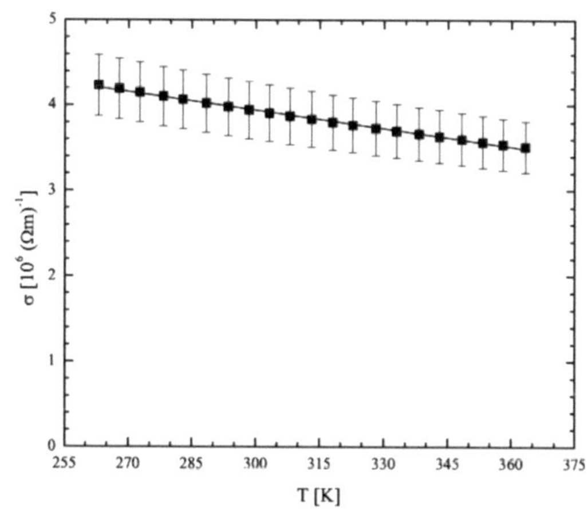

Fig. 21. Electrical conductivity σ of a Pt nanowire as function of temperature (Voelklein et al., 2009b)

This decrease is in good agreement with classical size effect theory (Sondheimer, 1952; Mayadas & Shatzkes, 1970). For thermal conductivity measurements an electrical heating power of 4×10^{-6} W was applied, causing a mean temperature increase of ΔT_M = 12.7 K. Figure 22 shows the thermal conductivity in the temperature range 260 K - 365 K. The bulk values (e.g. λ_{bulk} = 71 W/mK at room temperature) are more than three times larger than the measured thermal conductivities of the polycrystalline Pt nanowire. This is also attributed to size effects (Ouarbya et al., 1981), mainly caused by grain boundary scattering of electrons.

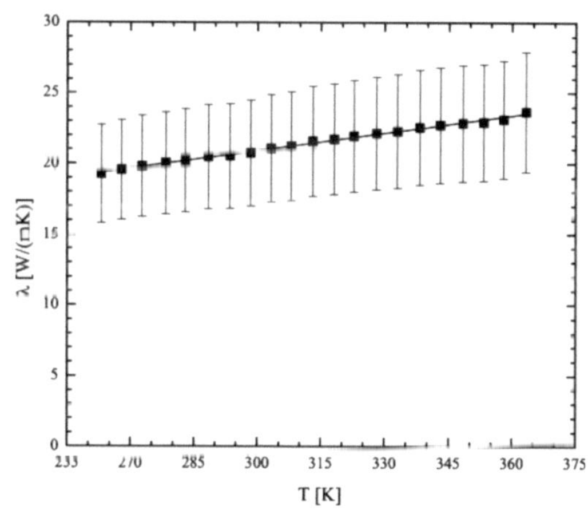

Fig. 22. Thermal conductivity λ of a Pt nanowire as function of temperature (Voelklein et al., 2009b)

5.2 Cross-plane thermal conductivity of embedded nanowires

Arrays of nanowire bundles can be applied for thermoelectric micro generators or for arrays of thermoelectric radiation sensors. Figure 23 shows the fabrication sequence of an array of thermoelectric sensors, where each sensor pixel consists of a thermocouple of p-type and n-type nanowire bundles.

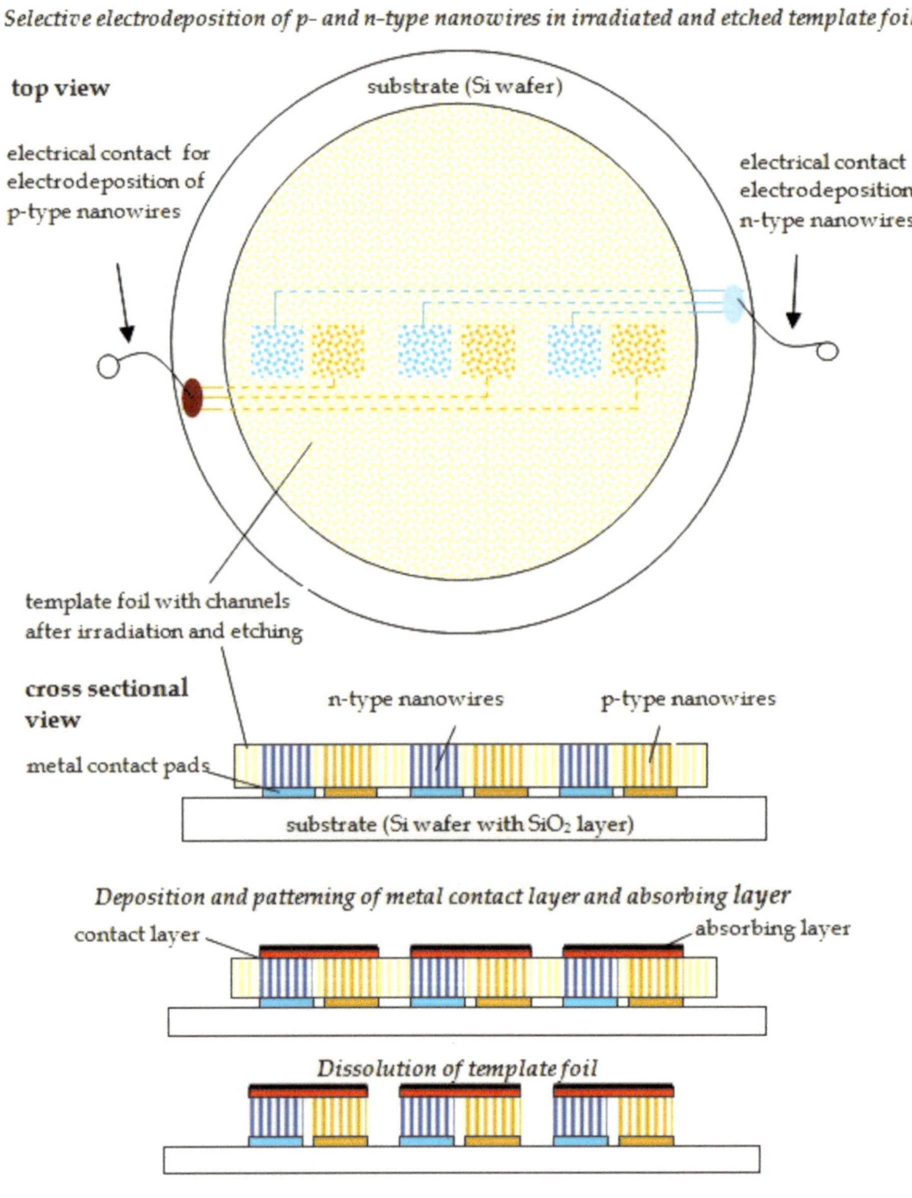

Fig. 23. Fabrication process for an array of thermoelectric sensors. top: selective electrochemical deposition of p- and n-type nanowire arrays. bottom: fabrication of the sensor pixels, each consisting of a bundle of p- and n-type nanowires.

To achieve a selective deposition of two different kinds of nanowire arrays, it is necessary to provide independent contacts to the template material. As heat sink, a substrate with high thermal conductivity is used. In particular, a silicon wafer with an electrical insulating layer (e.g. SiO_2 or Si_3N_4), on which metal contact pads were structured by microlithography (figures 23 and 24) acts as a heat sink. A polycarbonate etched ion-track membrane with pore diameters between 100 and 200 nm, is placed on such a substrate. Alternatively, thin polymer layers such as PMMA, SU8 negative resist or thick AZ photoresist can be deposited by spin coating, irradiated with heavy ions and etched in suitable solutions. Half of the metal contact pads are connected together for selected fabrication of n-type nanowire bundles in a first electrodeposition process and the second group of contact pads is used in the next electrodeposition step for the realization of p-type nanowire bundles. Then, thermocouples are formed by connecting a p- and n-type bundle by physical vapor deposition (PVD) and photolithographic patterning of a metal layer. To form a radiation sensor array these metal contact layers are coated with a broadband radiation absorbing layer with high absorption coefficient from the visible to the infrared radiation range. Finally, the template foil can be removed by dissolution (figure 24), using an organic solvent, in order to increase sensitivity and detectivity of the array pixels. By using the presented technology, a minimum size of an individual thermocouple (pixel) of about 10x20 μm^2 can be achieved. For simulation and characterization of the sensor parameters, the thermoelectric figure of merit of such template foils filled with nanowires has to be investigated.

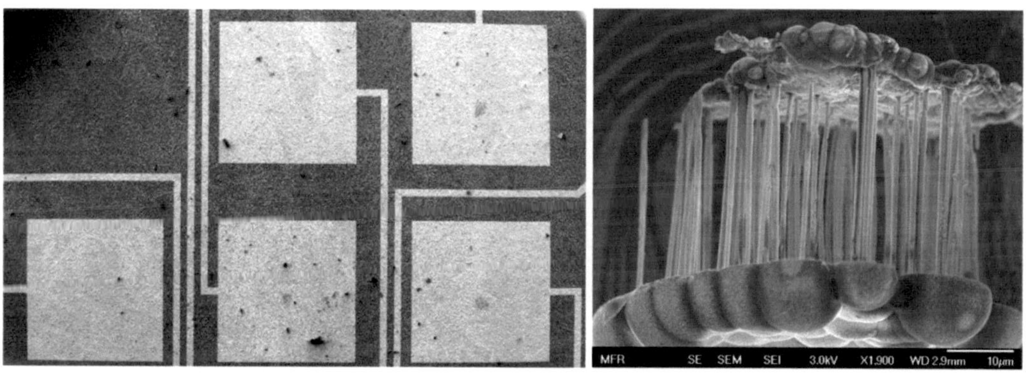

Fig. 24. left: SEM image shows the surface of an etched ion-track membrane and the Au contact pads patterned by microlithography to act as cathodes for selective electrodeposition of different materials; right: bundle of nanowires with metal contact layer after dissolution of template foil (Rauber, 2008).

The investigation of nanowire arrays requires the measurement of the cross-plane thermal conductivity of thin foils with thickness in the order of 10 μm. The measuring principle and experimental setup of the applied steady-state method is represented in figure 25. Metallic heater/thermometer films with high temperature coefficients of resistivity are deposited on the top surfaces of the samples and electrically insulated from them by thin dielectric layers (interfaces). Measurements are performed on two foils and results are compared in order to calculate the thermal conductivity of embedded nanowires: on a template foil after irradiation and etching (figure 26) and on an identical foil (same thickness, nanochannel diameters and ion-track density) filled with thermoelectric nanowires. The template foils are fixed on a substrate by an adhesive and thus are thermally separated from the substrate by an interface with unknown thermal resistance. The substrate is fixed on a heat sink with temperature T_0. On the heater/thermometer films (length l, width b) on top of the two template foils with cross-plane thermal conductivities λ_1 and λ_2, the electrical heating powers N_1 and N_2 are dissipated, respectively.

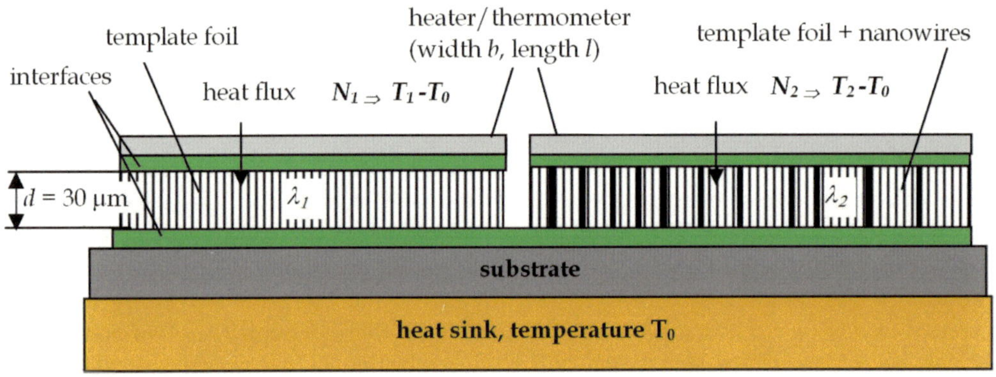

$$\frac{T_1 - T}{N_1} = R_{T1} = \left(R_{TSub} + R_{TInt} + \frac{d}{\lambda_1 bl} + \frac{d_H}{2\lambda_H bl} \right) \qquad \frac{T_2 - T_0}{N_2} = R_{T2} = \left(R_{TSub} + R_{TInt} + \frac{d}{\lambda_2 bl} + \frac{d_H}{2\lambda_H bl} \right)$$

Fig. 25. Method and experimental setup (cross-sectional view) for steady-state measurements of cross-plane thermal conductivity of thin template foils (without and with embedded nanowires).

These cause temperature increases T_1-T_0 and T_2-T_0, respectively, which are measured by the resistance increases of the heating films (thermometers). The ratios of the specific temperature increases and heating powers $(T_1-T_0)/N_1$ and $(T_2-T_0)/N_2$, respectively, represents the total thermal resistances R_{T1} and R_{T2} of the samples. Besides the thermal resistances of the templates $R_{TF1} = d/(\lambda_1 bl)$ and $R_{TF2}=d/(\lambda_2 bl)$, these total thermal resistances include the thermal resistances of the substrate R_{TSub}, of the interfaces R_{TInt} and of the heater/thermometer $R_{TH}=d_H/(2\lambda_H bl)$. Usually, the thermal resistance of the heater/thermometer is very small and can be neglected, since the thickness d_H of the used metallic heating films amounts 100 nm and its thermal conductivity λ_H is higher than 100 W/mK. For our measurements the thermal resistance of all interfaces R_{TInt} can also be neglected, since the insulation between heater and foil is realized by spin coating of a very thin resist layer and the interface between foil and substrate is formed by a thin metal-filled adhesive with high thermal conductivity. Furthermore, by proper choice of the geometrical dimensions l and b of heater and sample, the spreading thermal resistance R_{TSub} of the substrate can also be omitted. In general, the geometrical conditions $b > d$ and $b \ll l$ should be fulfilled. Then, the measured total resistance R_{T1} represents the thermal resistance R_{TF1} of the irradiated and etched polycarbonate foil and R_{T2} is equivalent to the thermal resistance R_{TF2} of the foil with embedded nanowires. Figure 26 shows a cross sectional view of a template foil after etching of nanochannels. For our measurements we used polycarbonate foils of about 30 μm thickness with an average channel diameter of 200 nm and an ion track density of $10^9/cm^2$. Consequently, the volume content of etched channels V_{Ch} in the foil amounts about 30%. Figure 26 shows the thermal conductivity λ_1 of the polycarbonate foil with nanochannels as function of temperature, which has been determined from the measured data of the thermal resistance R_{T1}, and the calculated thermal conductivity λ_{PC} of polycarbonate, assuming the nanochannels are filled with air at atmospheric pressure. Then λ_{PC} is evaluated by

$$\lambda_{PC} = \lambda_1 \frac{(V_{PC} + V_{Ch})}{V_{PC}} - \lambda_{air} \frac{V_{Ch}}{V_{PC}} \tag{10}$$

where λ_{air} indicates the thermal conductivity of air at atmospheric pressure and V_{PC} is the volume content of polycarbonate of the etched foil.

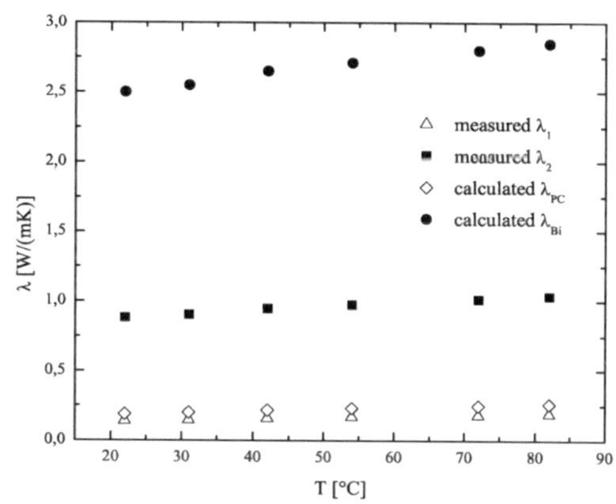

Fig. 26. left: Template foil after irradiation with heavy ions and etching of nanochannels (Spohr, 1990); right: Measured thermal conductivities λ_1 and λ_2 and calculated thermal conductivities of polycarbonate template foil λ_{PC} and embedded Bi nanowires λ_{Bi}.

Figure 26 indicates also the thermal conductivity λ_2 of such a polycarbonate foil filled with Bi nanowires. Taking into consideration the volume content of Bi nanowires of 30%, the thermal conductivity of embedded Bi nanowires λ_{Bi} (also shown in figure 26) can be calculated using

$$\lambda_{Bi} - \lambda_{air} = (\lambda_1 - \lambda_2) \frac{(V_{PC} + V_{Ch})}{V_{Ch}} \tag{11}$$

Figure 25 demonstrates the problems associated with the measurement of the cross-plane thermal conductivity. Interfaces between substrate and template foil and between template foil and heater/thermometer may have a thermal resistance that cannot be ignored in comparison to the thermal resistance of the investigated template. However, both the presented steady-state technique as well as the transient 3-Omega method have to solve this experimental challenge. One advantage of the steady-state method compared to the 3-Omega method is a much lower electrical power density required for the heater/thermometer and a less expensive/complex electrical measuring setup.

6. Microchip for the complete characterization of thermoelectric transport coefficients (σ, S, λ) on an individual nanowire (*z*-chip)

Various specific and sophisticated microchips and methods for the measurement of thermoelectric transport properties of nanostructures have been presented in the previous sections. In the ideal case, only one chip should be applied for the complete characterization of all thermoelectric transport coefficients (σ, S, λ) on an individual nanowire. This section deals with the design (simulation) and realization of a z-chip fabricated by micromachining. Figure 27 shows the scheme of the *z*-chip and the measuring principle. By microlithographic patterning and anisotropic etching in silicon a comb-like silicon cantilever array is fabricated. The cantilevers (length l = 550 µm, width b = 16 µm, thickness d = 40 µm) are separated by a distance w of about 10 µm. The top surface of the cantilevers is coated with a thin insulating Si_3N_4 film and two thin film Ti/Au contact stripes are patterned by photolithography on each cantilever. The cantilever array is fabricated by using a SOI wafer with a top silicon wafer of 40 µm thickness, an insulating SiO_2 interface and a bottom silicon wafer of 300 µm thickness. After patterning of the electrical contact stripes, the cantilever array is realized by anisotropic etching of the top wafer. Finally, the cantilevers are thermally separated from each others by anisotropic etching of a cavity in the bottom wafer, where the etching process is stopped beneath the cantilevers due to the SiO_2 interface. The thermal cross-talk between the cantilevers caused by the remaining SiO_2 interface layer (300 nm thick) is very small and can be minimized by removing the SiO_2 film using reactive ion etching (RIE).

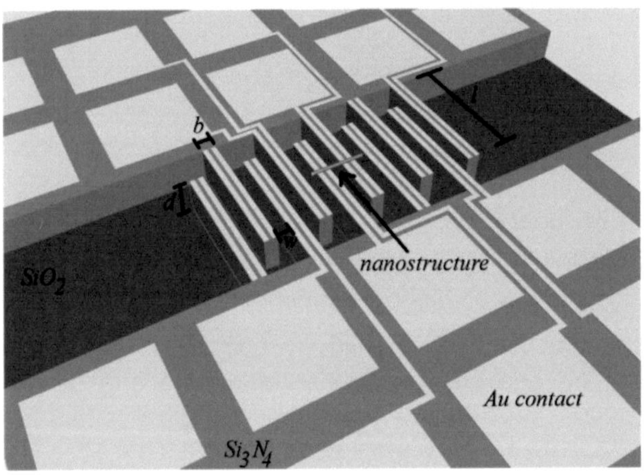

Fig. 27. Scheme and measuring principle of a cantilever array for the characterization of thermoelectric efficiency z of individual nanowires (z-chip).

One part of the bulk silicon rim of the *z*-chip is connected with a heat sink of temperature T_1, the other part with a heat sink of temperature T_0. Because of the very small thermal cross-talk between the cantilevers and the high thermal conductivity of silicon, one set of the cantilevers have a temperature close to the heat sink temperature T_1, the other half close to T_0. Thus two neighboring cantilevers possess a temperature difference which is proportional to the temperature difference of the heat sinks T_1-T_0. Figure 28 shows the temperature profiles along two neighboring cantilevers, one of them connected with a heat sink of temperature T_1=298 K and the other with a heat sink of temperature T_0=295 K. Figure 28 demonstrates that the temperature difference between the cantilevers is about 2.4 K and nearly independent of the position. Ion-track fabricated nanowires are deposited on the cantilever array (figure 29, left) using few drops of the solvent containing the wires (see section 2.1). The solvent evaporates within a few minutes leaving behind randomly distributed suspended wires (figure 29, right) fixed on neighboring cantilevers.

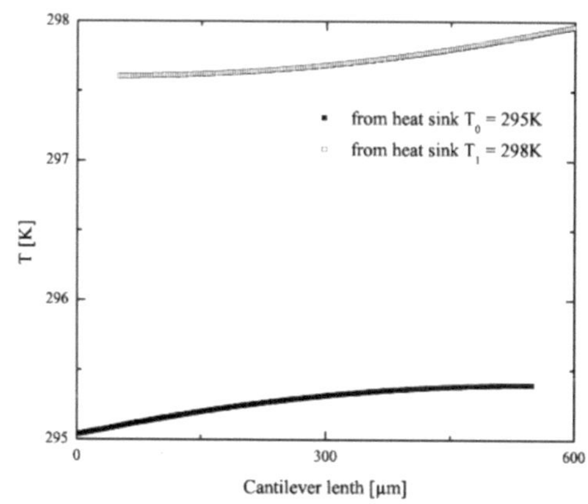

Fig. 28. Temperature profile along two neighboring cantilevers connected to heat sinks of temperatures T_1=298 K and T_0=295 K, respectively.

Measurements of electrical conductivity σ are performed using a 4-point-probe technique, where the 4 thin film Ti/Au lines of two neighboring cantilevers are applied as current and potential probes. For the measurement of the Seebeck coefficient S a defined temperature difference between the heat sinks is generated, leading to a specific lower temperature difference between neighboring cantilevers and consequently to a thermoelectric voltage, that can be measured using the Ti/Au contact lines. The temperature difference between neighboring cantilevers is a defined function of the heat sinks temperature difference (T_1-T_0) and has been determined by separate measurements, using thin film temperature sensors.

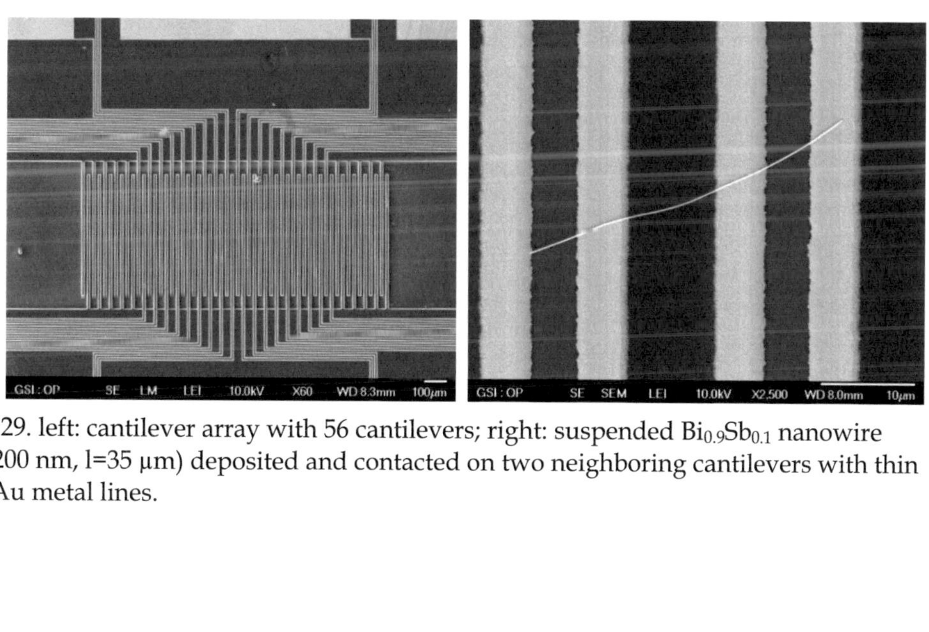

Fig. 29. left: cantilever array with 56 cantilevers; right: suspended $Bi_{0.9}Sb_{0.1}$ nanowire (d=200 nm, l=35 µm) deposited and contacted on two neighboring cantilevers with thin Ti/Au metal lines.

For measurements of thermal conductivity λ both heat sinks are kept at the same temperature. Because of their high thermal conductivity, the cantilevers act as heat sinks of identical temperature despite electrical heating of the nanowire. Therefore, the steady-state method described in section 5.1 can be applied. The 4-point-probe technique is employed for the determination of the heating power I^2R dissipated in the nanowire and the resistance change ΔR, and λ is calculated using equation (9). Thus, all transport coefficients (σ, S, λ) can be characterized on one and the same individual nanowire.

7. References

Boukai, A., Xu, K. and Heath, J.R., Size-Dependent Transport and Thermoelectric Properties of Individual Polycrystalline Bismuth Nanowires, Advanced Materials, 18, 864, 2006

Boukai, A., Bunimovich, Y., Tahir-Kheli, J., Yu, J.K., Goddard III, W.A. and Heath, J.R., Silicon Nanowires as Efficient Thermoelectric Materials, Nature, 451, 168, 2008

Cho, S.L., Kim, Y., DiVenere, A., Wong, G., Ketterson, J.B. and Meyer, J.R., Anisotropic Seebeck and Magneto-Seebeck Coefficients of Bi and Bi0.92Sb0.08 Alloy Thin Films, Journal of Applied Physics, 88, 808, 2000

Cornelius, T.W., Brotz, J., Chtanko, N., Dobrev, D., Miehe, G., Neumann, R. and Toimil-Molares, M.E., Controlled Fabrication of Poly- and Single-Crystalline Bismuth Nanowires, Nanotechnology, 16, S246, 2005

Cornelius, T.W., Toimil-Molares, M.E., Neumann, R. and Karim, S., Finite-Size Effects in the Electrical Transport Properties of Single Bismuth Nanowires, Journal of Applied Physics, 100, 114307, 2006

Cronin, S.B., Lin, Y.M., Rabin, O., Black, M.R., Ying, J.Y., Dresselhaus, M.S., Gai, P.L., Minet, J.P. and Issi, J.P., Making Electrical Contacts to Nanowires with a Thick Oxide Coating, Nanotechnology, 13, 653, 2002

Dresselhaus, M.S., Lin, Y.M., Rabin, O., Jorio, A., Souza Filho, A.G. Pimenta, M.A., Saitof, R., Samsonidze, G. and Dresselhaus, G., Nanowires and Nanotubes, Materials Science and Engineering C, 23, 129, 2003

Dresselhaus, M.S. and Heremans, J.P., Recent Developments in Low-Dimensional Thermoelectric Materials, Thermoelectric Handbook – Macro to Nano, CRC Handbook, D.M. Rowe (Ed.), Taylor & Francis, 2006

Fuchs, K., The Conductivity of Thin Metallic Films According to the Electron Theory of Metals, Mathematical Proceedings of the Cambridge Philosophical Society, 34, 100, 1937

Hochbaum, A.I., Chen, R., Delgado, R.D., Liang, W., Garnett, E.C., Najarian, M., Majumdar, A. and Yang, P., Enhanced Thermoelectric Performance of Rough Silicon Nanowires, Nature, 451, 163, 2008

Huth, M., Focused Electron Beam Induced Deposition - Principles and Applications, Proceedings of Beilstein Symposium Functional Nanosciences, Bozen, Italy (2010), to be published 2011

Karim, S., Toimil-Molares, M.E., Maurer, F., Miehe, G., Ensinger, W., Liu, J., Cornelius, T.W. and Neumann, R., Synthesis of Gold Nanowires with Controlled Crystallographic Characteristics, Applied Physics A, 84, 403, 2006

Li, D., Wu, Y., Kim, P., Shi, L., Yang, P. and Majumdar, A., Thermal conductivity of individual silicon nanowires, Appl. Phys. Letters, 83, 2934, 2003

Lin, Y.M., Sun, X. and Dresselhaus, M.S., Theoretical Investigation of Thermoelectric Transport Properties of Cylindrical Bi Nanowires, Physical Review B, 62, 4610, 2000

Liu, H.J., Song, C.M., Wu, S.T. and Li, L.F., Processing Method Dependency of Thermoelectric Properties of Bi85Sb15 Alloys in Low Temperature, Cryogenics, 47, 56, 2007

Mavrokefalos, A., Moore, A.L., Pettes, M.T. and Shi, L.,Wang, W. and Li, X., Thermoelectric and Structural Characterizations of Individual Electrodeposited Bismuth Telluride Nanowires, Journal of Applied Physics, 105, 104318, 2009

Mayadas, A.F. and Shatzkes, M., Electrical-Resistivity Model for Polycrystalline Films: the Case of Arbitrary Reflection at External Surfaces, Physical Review B, 1, 1382, 1970

Ouarbya, L., Tosser, A.J. and Tellier, C.R., Effects of Electron Scatterings on Thermal Conductivity of Thin Metal Films, Journal of Materials Science, 16, 2287, 1981

Picht O., Müller, S.; Rauber, M., Lauterbach, S., Toimil-Molares, M.E. and Neumann, R., GSI Annual Report, 446, 2009

Rauber, M., courtesy of M. Rauber, TU-Darmstadt, 2008

Shi, L., Li, D., Yu, C., Jang, W., Kim, D., Yao, Z., Kim, P. and Majumdar, A., Measuring Thermal and Thermoelectric Properties of One-Dimensional Nanostructures Using a Microfabricated Device, Journal of Heat Transfer, 125, 881, 2003

Shi, L., Yu, C. and Zhou, J., Thermal Characterization and Sensor Applications of One-Dimensional Nanostructures Employing Microelectromechanical Systems, Journal of Physical Chemistry B, 109, 22102, 2005

Sondheimer, E.H., The Mean Free Path of Electrons in Metals, Advances in Physics, 50, 499, 1952

Spohr, R., Ion Tracks and Microtechnology, Ed. by K. Bethge, Friedr. Vieweg & Sohn Verlagsgesellschaft mbH, Braunschweig, 1990

Toimil-Molares, M.E., Buschmann, V., Dobrev, D., Neumann, R., Scholz, R., Schuchert, I.U. and Vetter, J., Single-Crystalline Copper Nanowires Produced by Electrochemical Deposition in Polymeric Ion Track Membranes, Advanced Materials, 13, 62, 2001

Toimil-Molares, M.E., Hohberger, E.M., Schaeflein, C., Blick, R.H., Neumann, R. and Trautmann, C., Electrical Characterization of Electrochemically Grown Single Copper Nanowires, Applied Physics Letters, 82, 2139, 2003

Utke, I., Hoffmann, P. and Melngailis, J., Gas-Assisted Focused Electron Beam and Ion Beam Processing and Fabrication, Journal of Vacuum Science & Technology B, 26, 1197, 2008

Voelklein, F. and Kessler, E., Analysis of the Lattice Thermal Conductivity of Thin Films by Means of a Modified Mayadas-Shatzkes Model: The Case of Bismuth films, Thin Solid Films, 142, 169, 1986

Voelklein, F., Schmitt, M.C., Cornelius, T.W., Picht, O., Müller, S. and Neumann, R., Microchip for the Measurement of Seebeck Coefficients of Single Nanowires, Journal of Electronic Materials, 38, 1109, 2009

Voelklein, F., Reith, H., Cornelius, T.W., Rauber, M. and Neumann, R., The Experimental Investigation of Thermal Conductivity and the Wiedemann-Franz Law for Single Metallic Nanowires, Nanotechnology, 20, 325706, 2009

Voelklein, F., Reith, H., Schmitt, M.C., Huth, M., Rauber, M. and Neumann, R., Microchips for the Investigation of Thermal and Electrical Properties of Individual Nanowires, Journal of Electronic Materials, 39, 1950, 2010

Zou, J. and Balandin, A., Phonon Heat Conduction in a Semiconductor Nanowire, Journal of Applied Physics, 89, 2932, 2001

15

Silicon-Based Nanowire MOSFETs: From Process and Device Physics to Simulation and Modeling

Jin He, Haijun Lou, Lining Zhang and Mansun Chan
Peking University Shenzhen SOC Key Laboratory,
Institute of PKU-HKUST of IER Department of ECE,
Hong Kong University of Science and Technology, Hong Kong,
P.R.China

1. Introduction

Over the past few years tremendous progress has been made on the process, application, device physics and compact modeling of nanowire MOSFETs. We would like to review the above aspects of silicon-based nanowire, focusing on its crucial compact models and the circuit performance demonstration based on our group research work and understanding on nanowire MOSFET progress.

Nanowire MOSFETs are recognized as one of the most promising candidates to extend Moore's law into nanoelectronics era. Both the top-down (Singh et al., 2008) and bottom-up (Lu & Lieber, 2006) approaches are widely studied to prepare ultra small nanowire. With bottom-up method, nanowires are generally synthesized by using metal nanoclusters as catalysts via a vapor-liquid-solid process (Lu & Lieber, 2006). After growth nanowires are transferred to silicon substrate to form FET structure. With top-down technique there are various fabrication approaches, such as hard mask trimming, etching in H_2 ambient and stress limited oxidation (Singh et al., 2008). 5-nm gate length device has been demonstrated (Liow et al., 2008). Nanowire MOSFETs prepared with both methods find application in logic circuits (Singh et al., 2008), memory (Singh et al., 2008) and sensors (Stern et al., 2008).

Due to the quantum confinement in cross section of nanowire MOSFETs, especially of nanowires with diameter smaller than 15nm, electron mobility behaves differently from its bulk counterpart. Phonon-limited electron mobility decreases with reducing the wire size (Kotlyar et al., 2004) while total electron mobility is enhanced due to volume inversion at high transverse field (Jin et al., 2007). Although whether or not ballistic transport can occur in the silicon nanowire MOSFETs with ultrasmall channel length is disputable (Ferry et al., 2008), it deserves our attention. In the ballistic transport regime, carrier scattering in the device channel is totally suppressed. The study of ballistic transport in nanowire MOSFETs provides the upper limit to their performances. Under extreme scaling of nanowire MOSFETs, the atoms in nanowire cross section are countable. It is believed that the change in bandstructure of one dimensional nanowire influences the device performances (Neophytou et al., 2008). The above mentioned phenomenon are studied and simulated with various numerical approaches and also need to be accounted for in the advanced compact models.

Compact models for silicon nanowire MOSFETs have been developed for design and simulation of nanowire-based circuits. They also provide guides for optimal device design. Natural length theory of nanowire MOSFETs sets the criterion of optimizing device parameters, e.g. the radius, oxide thickness to maintain electrical performance. Superior to Plummer's work, Taur's scaling theory (Yu et al., 2008) takes into account the lateral distribution of gate oxide field and allows consideration of relatively thicker high-k dielectric. Core models of undoped nanowire MOSFETs where quantum effect is not significant are well established (Iñíguez et al., 2005; Bian et al., 2007; He J. et al., 2007). By rigorously solving Poisson's equation in polar coordinate, the electrostatic potential distribution in cross section of nanowire is obtained. Combined with Pao-Sah's drift-diffusion formula, current characteristics of long channel ideal nanowire MOSFETs are derived. Following Ward's channel charge partition scheme, terminal capacitance models are obtained. Various forms of this core model exist, such as charge-based (Iñíguez et al., 2005), potential-based (Bian et al., 2007) and carrier-based (He J. et al., 2007) model. Each has its own advantages. Potential-based model is further used to explain the dynamic depletion effect in nanowire (Zhang L. et al., 2009a), while charge-based model has been extended to cover the doped nanowire MOSFETs (Yang et al., 2008). In order to accurately reproduce the electric performances of nanowire MOSFETs, advanced effect models are integrated into the core model frame, e.g. short channel effects , quantum confinement effects , velocity saturation effect (Zhang L. et al., 2009b), etc. A preliminary compact model for silicon nanowire MOSFETs is presented in (Yang et al., 2008) where several advanced effect models are included as optional modules. Another capacitance based analytic model of ballistic silicon nanowire is also given in (Wang, 2005).
Working out an analytic model covering ballistic and diffusive transport and also realizing transition between both (Michetti et al.,2009) is another challenging task. With the above mentioned preliminary compact model for silicon nanowire MOSFET implemented into circuit simulations by Verilog-A, several representative logic circuits are simulated (Yang et al., 2008). Current status process, device physics, simulation and modeling of silicon-based namowire MOSFETs is reviewed, and the circuit performance is also analyzed. Moreover, the future possible trend of nanowire MOSET is finally outlined in this chapter.

2. Nanowire fabrication process

Silicon nanowire (SiNW) transistors have shown promising potential to revolutionize the applications of electronic, optical, chemical and biological devices (Black, 2005 ; Barrelet et al.,2004; Ramanathan et al., 2005; Hood et al., 2004). The conventional approach for the fabrication of silicon nanowires is a bottom-up approach from one of many pathways ranging from chemistry, laser-assisted or e-beam directed patterning (Cui et al., 2001a) in a controllable fashion down to sub 10nm diameter in width. One of the many established methods is by nanocluster assisted vapor-liquid-solid (VLS) growth mechanism in which metal nanoclusters mediate the nanowire growth. A more subtle approach of the bottom-up method for the patterning of silicon nanowires is by laser-assisted catalytic growth (Zhang, Y. F. et al., 1998). A bottom-up formation approach is advantageous for creating small silicon nanowires. However, the orientation growth of the wires is a major issue and the repeatability for device use is highly challenging. The formation of silicon nanowires from the top-patterning takes a completely different philosophy from the chemical or laser-assisted method. It relies on the grid formation on standard mask and transformation onto the silicon wafer surface. Due to the limitations of photo-lithography, mostly top-down controlled patterning methods have been developed by successive stress-limited oxidation. One possible method is by thermal oxidation of silicon pre-cursor under refinement (Liu et al., 1993). The starting Si columns are first patterned by Reactive-Ion-Based techniques to refine very small dimensions of silicon pattern, allowing for a series of thermal oxidation with controlled temperature and time. Due to the accumulated stress, the oxidation rate becomes extremely slow and results in the formation of silicon nanowires. Silicon pillars down to 2nm in diameter have been successfully fabricated by stress-limited oxidation (Liu et al., 1994; Kedzierski et al., 1997).

2.1 Single nanowire fabrication

2.1.1 Top-down method

The formation of Silicon Nanowire from the top-patterning takes a completely different philosophy than the chemical or laser assisted method (Kedzierski et al., 1997). It relies on the grid formation on standard mask and transformation onto the silicon wafer surface. Unfortunately, the common challenge of photo-lithography is the definition limit due to the physical obstruction of visible light. Numerous methods are proposed to stretch the top-down patterning limit since silicon nanowires are widely studied. One possible method is by thermal oxidation of silicon pre-cursor under refinement as shown in Figure 1. A small silicon is firstly patterned by Reactive-Ion-Based techniques to refine very small dimensions of silicon pattern allowing for a series of thermal oxidation with controlled temperature and exposure time. Due to the surrounding stress developed at the silicon outer wall, the inside of silicon gradually decreases and results in trends of silicon nanowires.

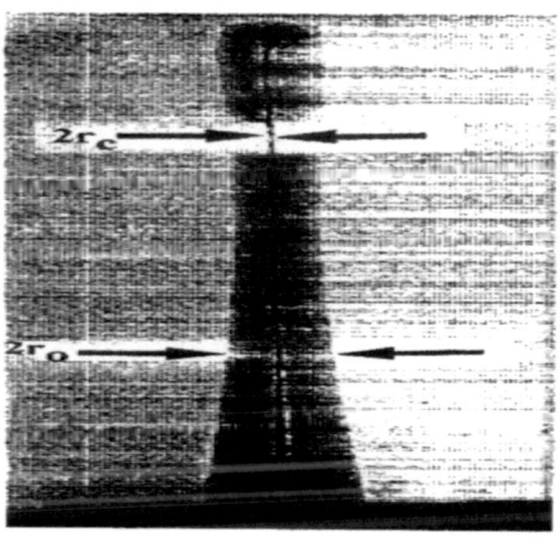

Fig. 1. Stress limited oxidation Silicon Nanowire (From Liu H. et al. Appl. Phys. Lett., Vol. 64, No.11, (Mar 1994))

This method is also applied onto Silicon-On-Insulator based materials (Black, 2005) to promote better stress limiting effect such that more rounded silicon cores are formed as shown in Figure 2. The extra buried oxide layer under the device silicon acts as a relieve site for oxidation flow promotion. In other words, the oxidation rate around the silicon is isotropic and hence the stress is evenly distributed along the surface. As a result, better control rates on the silicon cross-section can be achieved.

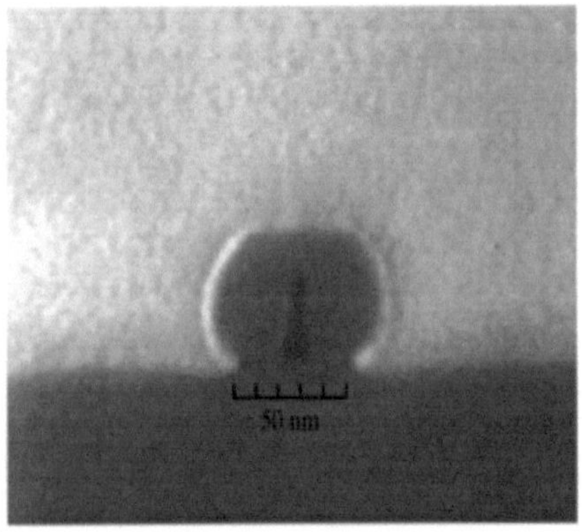

Fig. 2. Silicon-On-Insulator based Silicon Nanowire (From Kedzierski J., J. Vac. Sci. Technol. B, Vol. 15, No. 6, (July 1997))

2.1.2 Bottom-up method

The fabrication of silcon nanowire comes from many pathways ranging from chemistry, laser assisted or e-beam directed patterning (Cui et al., 2001a) in a controllable fashion down to sub 10nm diameter in width. One out of the many established method is by nanocluster assisted vapor-liquid-solid (VLS) growth mechanism in which metal nanoclusters mediate the Nanowire growth. The size of the metal catalysts determines the diameter of the Nos implying that with a narrow size distribution could be obtained by exploiting well-defined catalysts. High single-crystallinity configuration is reached by the chemical controlled method; however, the drawback is an uncontrolled growth orientation which deters its motivation for transistor application use.

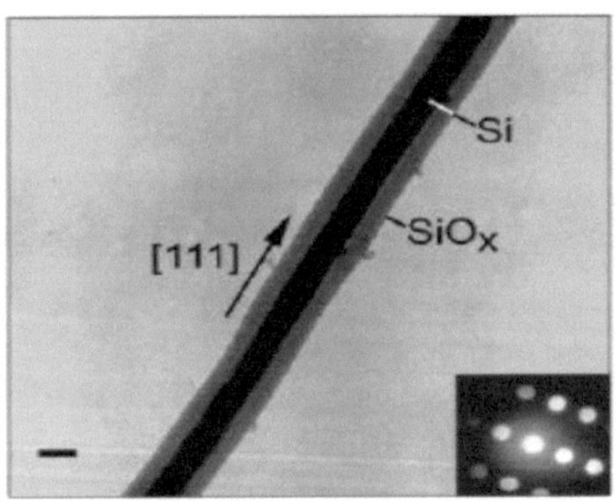

Fig. 3. Silicon Nanowire prepared by laser ablation method (From Alfredo M. Morales, *Science* 279, 208 (1998))

A more subtle approach of the bottom-up method as shown in Figure 3 for Silicon Nanowire patterning is by laser assisted catalytic growth (Zhang, Y. F. et al., 1998). The energy and pulsed controlled laser beam with sub-UV wavelength range is capable of cutting tiny target Silicon element site followed by a high temperature vapor condensation with the addition of metallic nano chemical cluster. The ablated silicon-metallic served as a growing site for silicon reaction growth in the liquid state under extremely high temperature. After cooling, the grown silicon solidities and results in strands of nanowires with better crystallinity and straightness. The above mentioned bottom-up formation approach is advantageous in creating small dimensions silicon nanowires (Figure 4). However, the orientation growth of the wires is a major issue and the repeatability for device use is highly challenging.

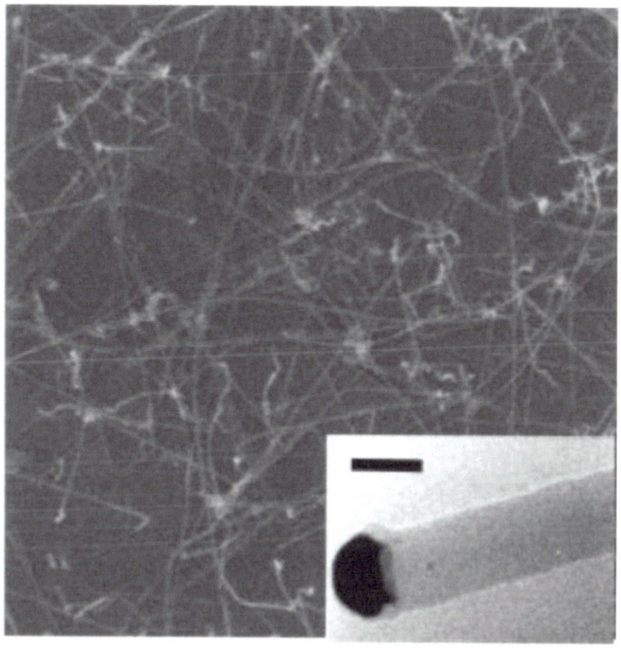

Fig. 4. Diameter controlled synthesis by single metallic crystal. The scale bar is 20 nm. (From Cui Y., Applied Physics Letters Volume 78, Number 15)

2.2 Multiple nanowire

The nanowires discussed above are single wires with localized shrinkage at a direct point of location (Black, 2005). As mentioned above, the excitement of multiple wires is the capability to increase the current throughput due to the carrier transport limitation for each single wire. The fabrication process for multiple wires is much complicated due to the replicating challenge for multiple patterns, or the ability to define grids in the multiple directions. A horizontal pitched silicon nanowire is shown by applying self-alignment of diblock copolymer films on an already lithographically defined mask (Figure 5). The interaction between the photo resist and co-reacting polymer aids to sub-divide the nano patterns into more and refined line structures (Bera et al., 2008). The promotion of multiple wires in the vertically direction is also proposed but with much greater challenge. One difficulty of vertical wires is the ability to control the vertical pitches. This problem can be resolved by utilizing the crystallinity orientation of silicon lattice structure along different direction. The existence of KOH etch is known for highly crystallinity oriented disruption. By splitting the crystal orientation, different etch planes are exposed and therefore oxidation of the structure will separate the silicon into different regions as shown in Figure 6. It is seen the above method has limitation to double silicon nanowires. To extend the number of small wires, another approach is proposed by utilizing the Si/SiGe technology (Namatsu et al., 1997). A stacked heterogeneous layer of mega structure is used to generate different silicon sites. The pattern precursor is then subjected to conventional oxidation process which than leaving multiple wires behind. This is currently the most reliable repeating process with virtually no limit on the number of stacked cores to be formed (Figure 7).

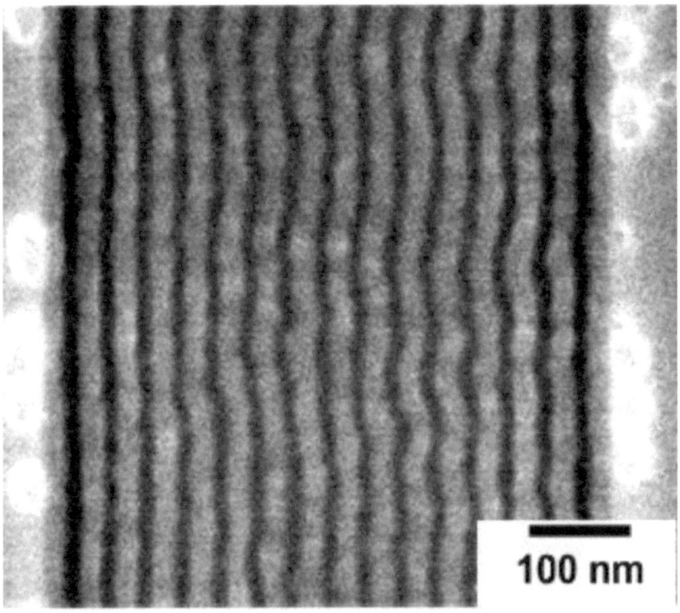

Fig. 5. Horizontal Multiple Silicon Nanowires. (From, C. T. Black, Applied Physics Letters 87, 163116 2005)

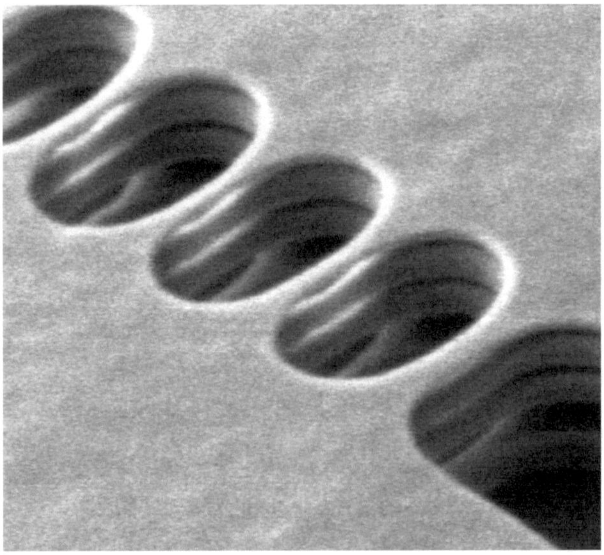

Fig. 6. SiGe based vertical multiple Silicon Nanowire (From L. K. Bera, IEEE IEDM '06, pp. 551-554)

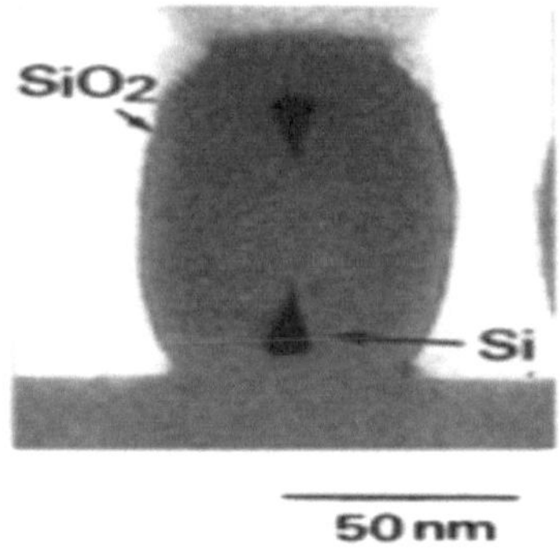

Fig. 7. Twin KOH based Multiple Silicon Nanowire (From Hideo Namatsu, J. Vac. Sci. Technol. B, Vol.15, No.5)

2.3 Vertically stacked nanowire process

The advantage of multiple wires is the capability to increase the current throughput with an increasing number of conducting channels. However, horizontal stacking increases the integration density and vertical pillars are hard to integrate into a traditional planar technology. A vertically stacked silicon nanowire concept has been introduced resulting in an increase in the drive current without impacting the density of integration (Fang et al., 2007).
Fabricating horizontal SiNWs using self-limiting oxidation has also been proposed (Liu et al., 2007). Silicon lines are defined on silicon-on-insulator with an electron beam lithography followed by metal liftoff process and silicon plasma etching. Low temperature oxidation is then used to shrink these lines to a sub-10 nm diameter. But the resulting geometry and shape of the nanowire are difficult to control and need sophisticated equipments and process control. A more controllable method has been demonstrated to fabricate horizontal nanowires using a multi-material system such as Si/SiGe/Ge/SiGe stacks (Bera et al., 2006; Dupre et al., 2008) but the process requires careful handling of the hetero-material interface. Among the methods to form vertical nanowires, the Bosch process plus stress-limited oxidation is the simplest in terms of the equipment needed.

2.3.1 Bosch process plus stress-limited oxidation approach and advantage

Figure 8 shows the key steps of the process to form vertically stacked nanowires. First, the Bosch Process is repeated by ICP to form a scallop pattern along the sidewall of a tall silicon ridge. Then stress-limited oxidation trims down the narrow region at the silicon ridge to form stacked nanowires (Ng.et al.,2007, 2009).
The active area is defined by a high resolution photo lithography on an oxide hardmask. The Bosch cycle technique (Chang et al., 2005) was employed to create a periodic sidewall profile by balancing the ICP etch and passivation step in each cycle. After the first etching phase, the patterned bulk-Si is first masked by a C_4F_8-based resistant polymer before the subsequent SF_6 plasma etches in the unprotected region. ICP etch is proceeded for 7 seconds under C_4F_8 (15sccm), SF_6 (45sccm), O_2 (5sccm) at 600W. The passivation cycle continues for 5.2 seconds under C_4F_8 (75sccm) at 200W. The scallop pattern that results from this process is shown in Figure 9(a). The widths of the ridges and troughs are controlled by the initial lithography and the etch time versus the passivation time in each cycle. After forming the

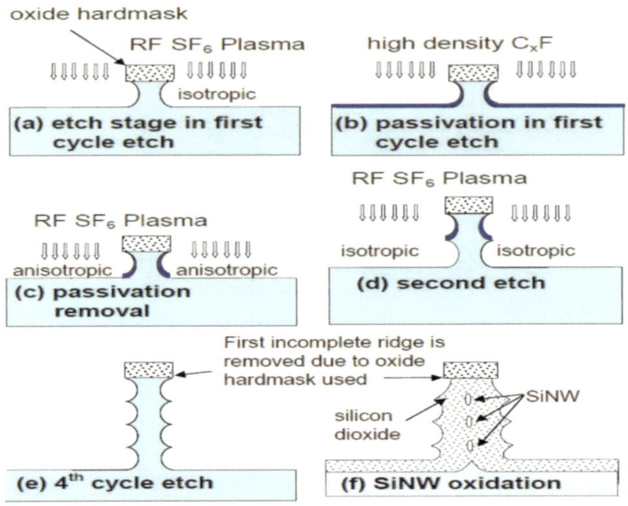

Fig. 8. Nanowire formation by Bosch etching and stress-limited oxidation (From R. M. Y. Ng, IEEE Elect. Dev. Lett , Vol. 30, No. 5, 2009)

scallop pattern, dry oxidation at 1000ºC is performed to consume the narrow region of the silicon ridge and isolate the nanowires. The evolution of the process with time is shown in the SEM images in Figure 9(b)-(d). The oxidation is self-limiting due to the stress resulting from the volume expansion during the formation of the silicon dioxide as shown in Figure 9. It is noted that for an initial mask width of 100nm, the entire silicon ridge is consumed. For an initial mask width of 200nm and 300nm, the oxidation is self-limiting as shown in Figure 10. A reduction of oxide thickness (negative oxidation rate) is observed after prolonged oxidation due to the oxide densification effect. It should be noted that the oxide hardmask facilitates the removal of the incomplete ridge at the top of the silicon nanowire stack as illustrated in Figure 8(e) and (f) so that the dimensions of the stacked nanowires become more uniform.

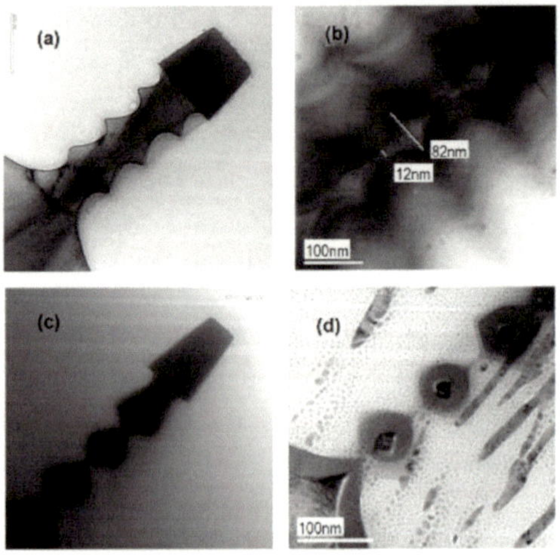

Fig. 9. TEM images showing the time evolution of the nanowire geometry (a) right after the Bosch etching and after (b) 30mins (c) 60mins and (d) 120mins of oxidation (From R. M. Y. Ng, IEEE Elect. Dev. Lett , Vol. 30, No. 5)

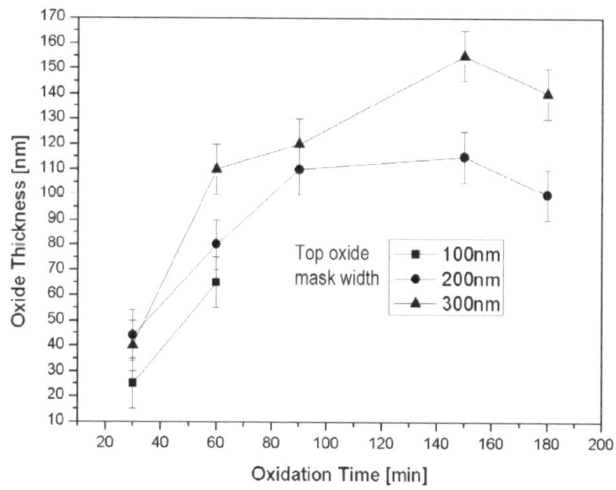

Fig. 10. Sidewall oxidation rate at 1000oC in oxygen ambient with different top oxide hardmask widths (From R. M. Y. Ng, IEEE Elect. Dev. Lett , Vol. 30, No. 5)

By controlling the initial width of the wide and narrow regions of the silicon ridges the vertical separation between the wider regions, elliptical and even triangular nanowires can be formed. To ensure the nanowires can be separated, the width of the narrow regions of the ridge should be less than 200nm. Due to the curvature effect, oxidation is in general slower at the narrow regions. As a result, wide regions should be at least 100nm wider than the narrow regions to form the residual nanowires after prolonged oxidation. Despite the different crystal orientation, the oxidation rate after the nanowire separation is quite uniform in all directions regardless of initial shape and dimension. This observation is due to the nature of the stress-limited process rather than the reaction rate limited process.

Therefore, vertically stacked SiNWs have been successfully fabricated using Inductive Coupled Plasma etching followed by stress-limited oxidation through a careful design process and optimization. Since the SiNW is trimmed from the Si wafers, the single crystal property of the SiNW is maintained when observed from the atomic plane, showing the existence of the crystallographic alignment. The shape and size of the nanowires can be controlled by varying the process conditions.

3. Silicon-based nanowire MOSFETs modeling and simulation

3.1 Core model for undoped nanowire

As a result of the interest in implementing SRG MOSFET to extend the scaling of CMOS technology, physics-based models are important in electrical circuit simulators to predict the performance of circuits when these devices are in use. It is well known that a complete surrounding-gate MOSFET model should not only be used to predict the SRG MOSFET current-voltage characteristics, but also be used in the calculation of terminal charges and various capacitances in the large signal and small signal simulations.

Here, an analytic carrier-based terminal charge and capacitance-voltage compact model for the long channel undoped surrounding-gate MOSFETs is also developed directly from both the current continuity principle and channel charge partition scheme based on our previous theoretical results (He, J. et al.,2004, 2006a). The analytic model is based on the exact device

physics and covers all regions of SRG MOSFET operation, from the sub-threshold to the strong inversion and from the linear to the saturation. Terminal charges and trans-capacitances of SRG MOSFETs are calculated with the newly developed model and further compared with the three-dimensional (3-D) numerical simulation.
The coordinate system and energy band used in this work is shown in Figure 11 with r representing the radial distance from the centre of the silicon film and $r = R$ giving the silicon film radius. It is also assumed that the quasi-Fermi level is constant in the radial direction, so that the current flows only along the channel (y direction). The energy levels are referenced to the electron quasi-Fermi level of the source end since there is no body contact in the undoped SRG MOSFETs.

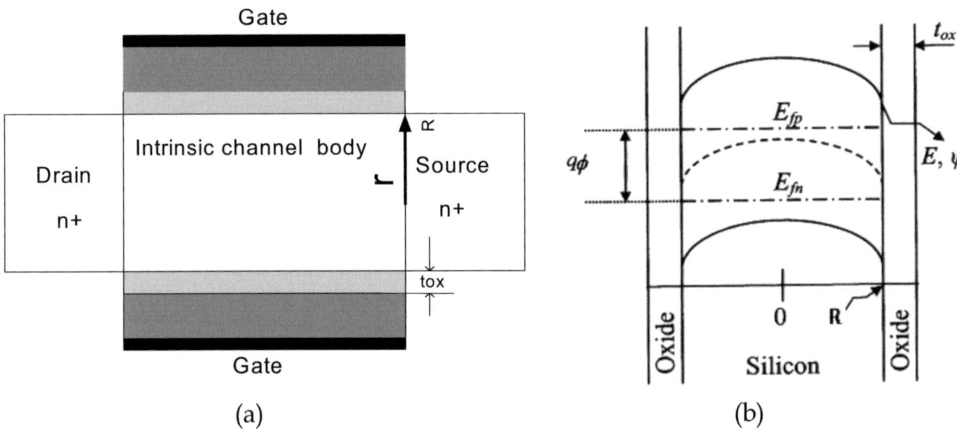

Fig. 11. The coordinate system(a) and energy band(b) used in this work. (From He J. IEEE TED, Vol. 54, No. 6, June 2007)

Following the basic device physics of SRG MOSFET (Jiménez et al., 2004; Iñíguez et al., 2005; Moldovan et al., 2007; He, J. et al.,2004, 2006a) under the Gradual-Channel–Approximation (GCA), the solution to the1-D Poisson–Boltzmann equation is valid in terms of the carrier concentration:

$$V_{gs} - \Delta\phi_i - V_{ch} = \frac{1}{\beta}\ln\left(\frac{n_0}{n_i}\right) - \frac{2}{\beta}\ln\left[1 - \frac{R^2}{8L_i^2}\frac{n_0}{n_i}\right] + \frac{R^2\varepsilon_{si}\ln\left[1+\frac{t_{ox}}{R}\right]}{2\beta L_i^2\varepsilon_{ox}}\frac{\frac{n_0}{n_i}}{\left[1-\frac{R^2}{8L_i^2}\frac{n_0}{n_i}\right]} \tag{1}$$

where all symbols have their common physics meanings:
$1/\beta = kT/q$ is the thermal voltage.
n_i is the silicon intrinsic carrier concentration ($1.14\times10^{10}cm^{-3}$ at room temperature).
n_0 is the induced carrier concentration at the reference coordinate point (at the centre of the silicon film in this study).
$\Delta\phi_i$ is the work function difference between the gate and the channel silicon body.
V_{ch} is the quasi-Fermi-potential with $V_{ch} = 0$ at the source end and $V_{ch} = V_{ds}$ at the drain end.
L_i^{-2} $=q^2ni/kT\varepsilon_{si}$ is the reciprocal of the square of the intrinsic silicon Debye length.

Based on Poisson's equation solution, the total inversion charge density is expressed as:

$$Q_i = \frac{R\varepsilon_{si}}{2\beta L_i^2} \frac{\frac{n_0}{n_i}}{\left[1 - \frac{R^2}{8L_i^2}\frac{n_0}{n_i}\right]} \quad (2)$$

Following Pao-Sah current formulation (Pao & Sah, 1966), the drain current can be written as

$$I_{DS} = \mu_0 \frac{W}{L} \int_0^{V_{DS}} Q_i(V)dV = \mu_0 \frac{W}{L} \int_{n_{0S}}^{n_{0D}} Q_i(n_0)\frac{dV}{dn_0}dn_0 \quad (3)$$

where n_{0s} and n_{0d} are the solutions of Eq.(1) corresponding to $V_{ch} = V_s = 0$ and $V_{ch} = V_{ds}$, respectively. μ_0 is the effective mobility, assumed constant here. W and L are the SRG MOSFET channel width and length, respectively.
Substituting n_0 with dV_{ch} / dy and the inversion charge expression in Eq. (2) into (3) and performing analytical integration, the carrier-based drain current is obtained:

$$I_{ds} = \mu_0 \frac{2\pi\varepsilon_{si}}{L}\left(\frac{2}{\beta}\right)^2 F[n_0]\Big|_{n_{0d}}^{n_{0s}} \quad (4)$$

where

$$F[n_0] = \left[\ln\left[1 - \frac{R^2}{8L_i^2}\frac{n_0}{n_i}\right] + 2\left(1 - \frac{R^2}{8L_i^2}\frac{n_0}{n_i}\right)^{-1} + \theta\left(\left(1 - \frac{R^2}{8L_i^2}\frac{n_0}{n_i}\right)^{-2} - 2\left(1 - \frac{R^2}{8L_i^2}\frac{n_0}{n_i}\right)^{-1}\right)\right] \quad (5)$$

with $\theta = \frac{2\varepsilon_{si}}{\varepsilon_{ox}} \ln\left[1 + \frac{t_{ox}}{R}\right]$

In order to simplify the derivation, we define a normalized carrier concentration $\rho = 1 - \frac{R^2}{8L_i^2}\frac{n_0}{n_i}$. As a result, (5) is simplified to $F(\rho) = \ln\rho + 2\rho^{-1} + \theta\left(\rho^{-2} - 2\rho^{-1}\right)$.

Based on a similar procedure, the analytic source and drain terminal charges are obtained and the final results are given as follows

$$Q_s = \frac{8\varepsilon_{si}\pi L}{\beta} \frac{F(\rho_d)G(\rho)\Big|_{\rho_d}^{\rho_s} + M(\rho)\Big|_{\rho_d}^{\rho_s}}{\left[\Gamma(\rho)\Big|_{\rho_d}^{\rho_s}\right]^2} \quad (6a)$$

$$Q_d = \frac{8\varepsilon_{si}\pi L}{\beta} \frac{F(\rho_s)G(\rho)\Big|_{\rho_s}^{\rho_d} + M(\rho)\Big|_{\rho_s}^{\rho_d}}{\left[F(\rho)\Big|_{\rho_s}^{\rho_d}\right]^2} \quad (6b)$$

with

$$M(\rho)=\left(\frac{4}{3}\rho^{-3}-\frac{5}{2}\rho^{-2}-\rho^{-1}+\frac{1-3\rho}{\rho^{2}}\ln\rho-\frac{1}{2}(\ln\rho)^{2}\right)$$
$$+\theta\left(\frac{3}{2}\rho^{-4}-\frac{43}{9}\rho^{-3}+\frac{9}{2}\rho^{-2}+\frac{2\text{-}6\rho+6\rho^{2}}{3\rho^{3}}\ln\rho\right) \tag{6c}$$
$$+2\theta^{2}\left(\frac{1}{5}\rho^{-5}-\rho^{-4}+\frac{5}{3}\rho^{-3}-\rho^{-2}\right)$$

and

$$G(\rho)=\rho^{-2}-3\rho^{-1}-\ln\rho+2\theta\left(\frac{1}{3}\rho^{-3}-\rho^{-2}+\rho^{-1}\right) \tag{6d}$$

From Eqs.6 (a-d), all three terminal charges of SRG MOSFETs can be analytically calculated from the normalized electron concentration ρ_s and ρ_d from the solutions of (1) corresponding to $V_{ch}=0$ and $V_{ch}=V_{ds}$, respectively.
The SRG MOSFET is essentially a three terminal device, a nine-capacitance matrix is written as

$$C_{ij}=\begin{cases}\dfrac{\partial Q_i}{\partial V_j} & \text{i}=\text{j}\\ -\dfrac{\partial Q_i}{\partial V_j} & \text{i}\neq\text{j}\end{cases}=\begin{bmatrix}C_{ss} & C_{sd} & C_{sg}\\ C_{ds} & C_{dd} & C_{dg}\\ C_{gs} & C_{gd} & C_{gg}\end{bmatrix} \tag{7}$$

With the analytical expressions of the three terminal charges formulated, all capacitances are derived as a function of the carrier concentration in the silicon centre n_{0s} and n_{0d} by means of the series law of calculus. The analytical expressions of all nine trans-capacitances can be simplified by using the dependency of the trans-capacitance matrix:

$$C_{gg}=C_{sg}+C_{dg}=C_{gs}+C_{gd} \tag{8a}$$

$$C_{ss}=C_{sd}+C_{sg}=C_{gs}+C_{ds} \tag{8b}$$

$$C_{dd}=C_{sd}+C_{gd}=C_{ds}+C_{dg} \tag{8c}$$

The details of these independent trans-capacitances are displayed in (He J. et al., 2007). Using the derived analytic expressions of the carrier concentrations, terminal charges, and the trans-capacitances, all current-voltage and capacitance-voltage characteristics of a long channel surrounding-gate MOSFET device can be predicted and analyzed with respect to different geometrical parameters and bias operation conditions in Figure 12-18. The detailed discussion is included in (He J. et al., 2007). The validity of the analytical solutions is confirmed by comparing model predictions with simulation data obtained using the 3-D numerical solvers. Through the results and discussion, we get that the explicit expressions to the terminal charges and trans-capacitance not only lead to a more clear understanding of surrounding-gate MOSFET device physics, but also provide a better infrastructure to develop a complete carrier-based model for the surrounding-gate MOSFET based circuit simulation.

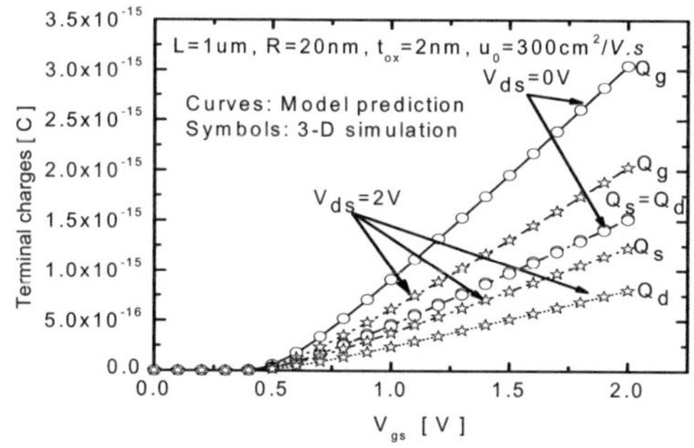

Fig. 12. Terminal charge versus gate voltage for different V_{ds}, compared with the numerical 3-D simulation (symbols) for a long channel undoped surrounding-gate MOSFET with L=1um, tox=2nm, R=20nm, $u_0 = 300cm^2 / Vs$, and W= $2\pi R$. (From He J. IEEE TED, Vol. 54, No. 6, June 2007)

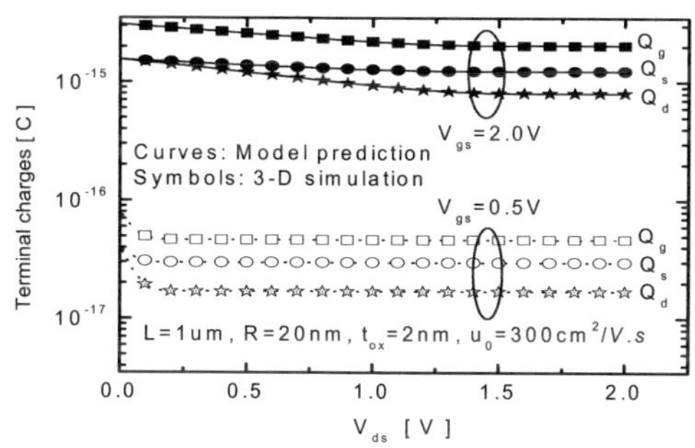

Fig. 13. Terminal charge versus V_{ds} for different V_{gs} (From He J. IEEE TED, Vol. 54, No. 6, June 2007)

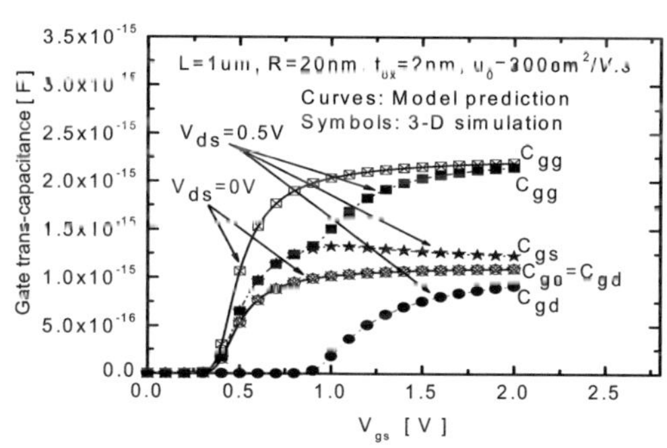

Fig. 14. Gate trans-capacitance versus gate voltage for different V_{ds} (He J. From He J., IEEE TED, Vol. 54, No. 6, June 2007)

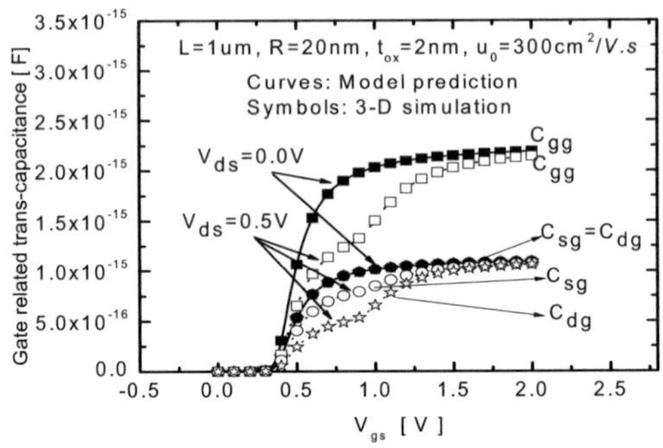

Fig. 15. Gate voltage related trans-capacitances versus V_{ds} (From He J. IEEE TED, Vol. 54, No. 6, June 2007)

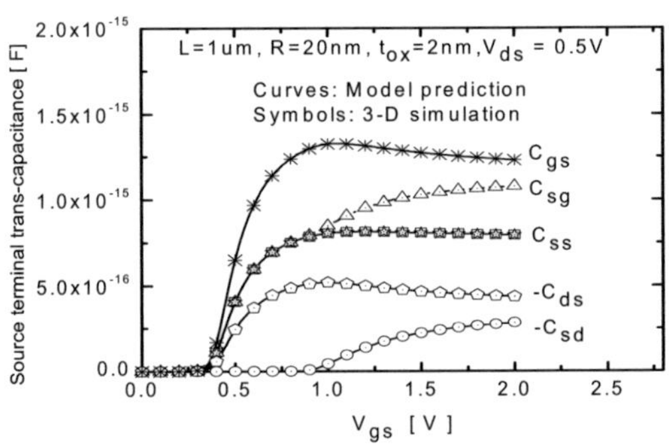

Fig. 16. Source terminal trans-capacitances versus V_{gs} (From He J. IEEE TED, Vol. 54, No. 6, June 2007)

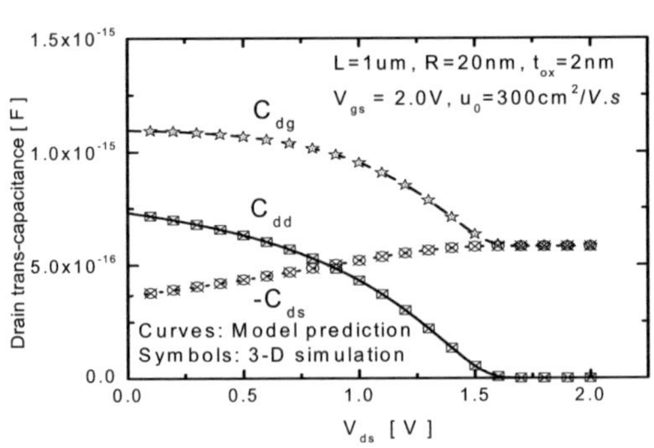

Fig. 17. Drain terminal trans-capacitances versus V_{ds} (From He J. IEEE TED, Vol. 54, No. 6, June 2007)

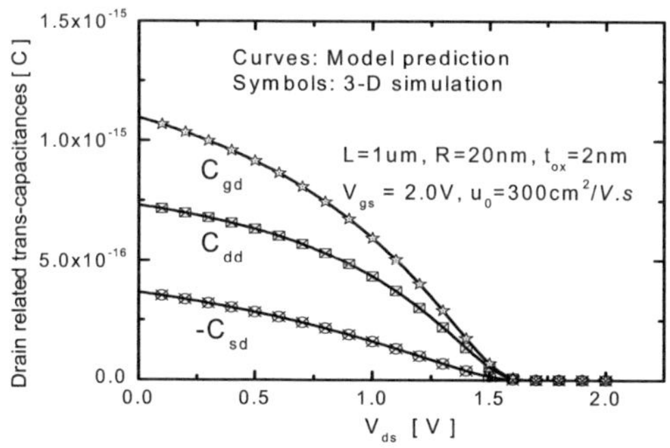

Fig. 18. Drain related trans-capacitances versus V_{ds} (From He J. IEEE TED, Vol. 54, No. 6, June 2007)

A basic feature of the above modeling works is to use an undoped (or lightly doped) body assumption to sustain the theory results. The idea of an undoped body, sometimes referred to as "intrinsic channel", is expected to have special advantages such as the low leakage current, free-statistic dopant fluctuation, and improved short-channel effects. However, the practical SRG MOSFET is always a doped body structure due to a small order unintentional doping ($10^{12}cm^{-3} \sim 10^{15}cm^{-3}$) during the real fabrication process. Thus, the undoped body is only an ideal approximation for light and low-doped case in all non-classical CMOS device. The practical SRG device may be designed as a fully depleted MOSFET in order to take the advantages of the undoped body via the low body concentration process and materials. Since the dopant concentration does not only change the surface potential magnitude, but also strongly changes the device sub-threshold slope, an analytic doped SRG MOSFET model is highly desirable for the circuit design and performance test. So a compact model for doping nanowire is showed next.

3.2 Unified core model for dopings in nanowire

Following the carrier-based approach (He, J. et al., 2006a, 2006b) and using a superposition principle, an approximate carrier-based compact model for the fully depleted SRG MOSFETs with a finite doping body is approximately developed directly from both the Poisson equation solution and the Pao-Sah current formulation. The standard cylindrical coordinate formulation of the Poisson Boltzmann equation in a doped SRG MOSFET is written as:

$$\frac{d^2\phi}{dr^2}+\frac{1}{r}\frac{d\phi}{dr}=\frac{qN_a}{\varepsilon_{si}}\left[1+\left(\frac{n_i}{N_a}\right)^2\exp(\beta(\phi-V))\right] \quad (9)$$

Where n_i and N_a are the intrinsic silicon concentration and the body doping concentration in the silicon film with the unit of cm^{-3}, respectively ε_{si} and ϕ are the silicon dielectric constant and the electrostatic potential in *Volt* respectively. r is the cylindrical coordinate in cm along the radius direction of the silicon film. $1/\beta$ And V are the thermal voltage and the quasi-Fermi-potential in Volt.

The Boltzmann statistics can be expressed as:

$$n = \frac{n_i^2}{N_a}\exp(\beta(\phi - V)) \tag{10a}$$

and

$$n_0 = \frac{n_i^2}{N_a}\exp(\beta(\phi_0 - V)) \tag{10b}$$

Where n_0 and ϕ_0 are the induced electron concentration in cm^{-3} and the electrostatic potential in V at $r = 0$. In a SRG MOSFET, the silicon radius center where the electric field is always zero is chosen as coordinate reference point.
The Poisson equation solution in terms of the carrier concentration can be written as (He, J. et al., 2006a, 2006b)

$$\phi_{sI} - \phi_{0I} = \frac{2}{\beta}\ln\left[1 - \frac{R^2}{8L_D^2}\frac{n_0}{N_a}\right] \tag{11}$$

Where ϕ_{SI} and ϕ_{0I} are the silicon surface and centric potentials in *Volt* contributed by the induced electron charge. $L_D = \sqrt{\varepsilon_{si}kT/q^2N_a}$ is the Debye length in cm of silicon film with doping concentration of N_a.
Similarly, we obtain the inversion density from the carrier-based Poisson equation solution in (He, J. et al., 2006a, 2006b)

$$Q_I = \frac{\varepsilon_{si}R}{2\beta L_D^2}\frac{n_0 / N_a}{1 - \frac{R^2}{8L_D^2}\frac{n_0}{N_a}} \tag{12}$$

If only the dopant is considered in (9) and the fully depletion approximation is used, the Poisson equation solution in the SRG can be written as

$$Q_b = \sqrt{q\varepsilon_{si}N_a(\phi_s - \phi_0)} \tag{13}$$

$$\phi_{sB} - \phi_{0B} = \frac{q_b R}{4\varepsilon_{si}} \tag{14}$$

where Q_b is the depleted charge density in $C \cdot cm^{-2}$ in the silicon film contributed by the doping atom. ϕ_{sB} and ϕ_{0B} are the silicon surface and centric potentials in V contributed by the depletion charge density, respectively.
thus $\phi_s = \phi_{sI} + \phi_{sB}$

$$\phi_s = \phi_0 + \sqrt{\frac{qN_aR^2}{16\varepsilon_{si}}(\phi_s - \phi_0)} - \frac{2}{\beta}\ln\left[1 - \frac{R^2}{8L_D^2}\frac{n_0}{N_a}\right] \tag{15}$$

and (4) plus (5):

$$Q_{tot} = Q_b + Q_I = \sqrt{q\varepsilon_{si}N_a(\phi_s - \phi_0)} + \frac{\varepsilon_{si}R}{2\beta L_D^2}\frac{n_0 / N_a}{1 - \frac{R^2}{8L_D^2}\frac{n_0}{N_a}} \tag{16}$$

where ϕ_s and Q_{tot} are the total silicon surface potential in V and the total charge in $C \cdot cm^{-2}$ contributed by the induced mobile electron charges and the depleted charges, respectively. Through (15), we can get the surface potential in terms of the carrier concentration.

$$\phi_s = \phi_0 + \left[\sqrt{\alpha^2 / 4 - \frac{2}{\beta}\ln\left[1 - \frac{R^2}{8L_D^2}\frac{n_0}{N_a}\right]} + \alpha / 2\right]^2 \tag{17}$$

where $\alpha^2 = \frac{qN_aR^2}{16\varepsilon_{si}}$, and the complete analytic solution of the Poisson equation is obtained in terms of the carrier concentration:

$$\begin{aligned} V_{gs} - \Delta\varphi - V - 2\phi_F &= \frac{1}{\beta}\ln\left(\frac{n_0}{N_a}\right) + \left[\sqrt{\alpha^2 / 4 - \frac{2}{\beta}\ln\left[1 - \frac{R^2}{8L_D^2}\frac{n_0}{N_a}\right]} + \frac{\alpha}{2}\right]^2 \\ &+ \gamma\left[\sqrt{\alpha^2 / 4 - \frac{2}{\beta}\ln\left[1 - \frac{R^2}{8L_D^2}\frac{n_0}{N_a}\right]} + \frac{\alpha}{2}\right] + \frac{\varepsilon_{si}R}{2\beta C_{OX}L_D^2}\frac{n_0 / N_a}{1 - \frac{R^2}{8L_D^2}\frac{n_0}{N_a}} \end{aligned} \tag{18}$$

where $\gamma = \frac{\sqrt{2q\varepsilon_{si}N_a}}{C_{ox}}$ is the bulk bias factor.

(18) gives the closed form expression of electron concentration at the silicon film center as a function of bias conditions, dopant concentration, and geometry sizes.
The drain current is written as:

$$I_{DS} = \mu\frac{W}{L}\int_0^{V_{DS}} q_{I}dV = \mu\frac{W}{L}\int_{n_{0S}}^{n_{0D}} q_I(n_0)\frac{dV}{dn_0}dn_0 \tag{19}$$

Where n_{0S} and n_{0D} are solutions of (18) corresponding to $V = 0$ being the source end voltage and $V = V_{ds}$ being the drain end voltage, respectively. μ is the effective channel mobility of the SRG MOSFET in $cm^2 / V.s$. L and $W = 2\pi R$ are the effective channel length and width of the SRG MOSFET in cm, respectively. Note that the dV / dy can also be expressed as a function of n_0 by differentiating (18). Substituting these factors into (19), integrating can be performed approximately yet analytically to yield:

$$I_{ds} = \mu\frac{W}{L}\int_0^{V_{DS}} q_I dV = \frac{4\pi\mu\varepsilon_{si}}{L}\left(\frac{2}{\beta}\right)^2 f(n_0)\Big|_{n_{0d}}^{n_{0s}} \tag{20}$$

where

$$f(n_0)=\frac{1}{2}\ln\left[1-\frac{R^2}{8L_D^2}\frac{n_0}{N_a}\right]+\frac{\left[\varepsilon_{ox}-2\varepsilon_{si}\ln\left(1+\frac{t_{ox}}{R}\right)\right]}{\varepsilon_{ox}\left[1-\frac{R^2}{8L_D^2}\frac{n_0}{N_a}\right]}+\frac{\varepsilon_{si}\ln\left(1+\frac{t_{ox}}{R}\right)}{\varepsilon_{ox}\left[1-\frac{R^2}{8L_D^2}\frac{n_0}{N_a}\right]^2}$$
$$+\frac{\beta}{2}(\alpha+\gamma)\sqrt{\alpha^2/4-\frac{2}{\beta}\ln\left[1-\frac{R^2}{8L_D^2}\frac{n_0}{N_a}\right]} \tag{21}$$

The fully depleted SRG MOSFET characteristics for all operation regions can be predicted from this compact yet continuous, analytic model. The characteristics of surface potential and centric potential for different doped concentrations from the intrinsic undoped assumption are showed in Figure 19 and 20, and their relative error is also predicted by the analytic model compared with the 3-D numerical simulation. It is found that the relative error increases with the increase of the body concentration for the given geometry parameters, when the body doping concentration is $1e16cm^{-3}$ the relative error is within the order of 1e-3, but the relative error increases to the order of 1e-2 for the doping concentration up to $1e17cm^{-3}$. Figure 21 and 22 show the comparison of I_{ds} versus V_{GS} and V_{DS} between the analytic solution (curves) and the 3-D simulation (symbols) for the different doping concentration. The SRG MOSFET current predicted by the analytic model shows a good match with the 3-D numerical simulation from the sub-threshold region to the strong inversion region for most doping concentrations.

We should point out that some second-order physics effects of the nanoscale SRG MOSFETs such as short-channel effects, the drain induced barrier lowering effect, and the more important inversion layer quantum effect (QME) are important for the analysis of nanoscale SRG MOSFET although they are ignored in above study to give a clear presentation of the SRG MOSFET charge and capacitance model. For example, the QME drives the peaks of electron concentration away from the interface between the gate oxide and the silicon film,

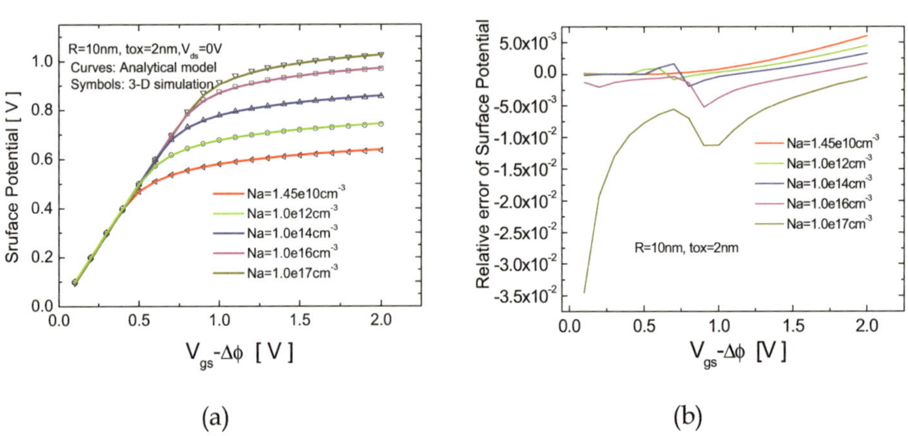

Fig. 19. (a) the comparison of the surface potential versus gate voltage curves for five different doped concentrations and (b) the relative error of the surface potential. (From He J. Semicond. Sci. Technol. 22 (2007) 671–677)

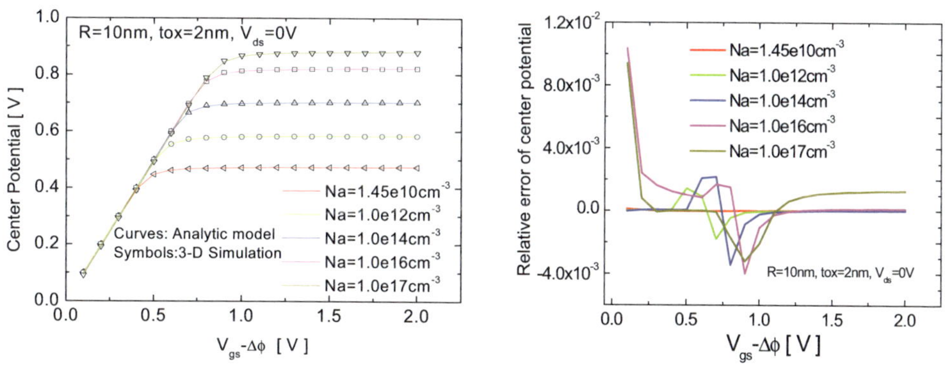

(a) (b)

Fig. 20. (a) the comparison of the centric potential versus gate voltage curves and (b) the resultant relative error(from He J. Semicond. Sci. Technol. 22 (2007) 671–677)

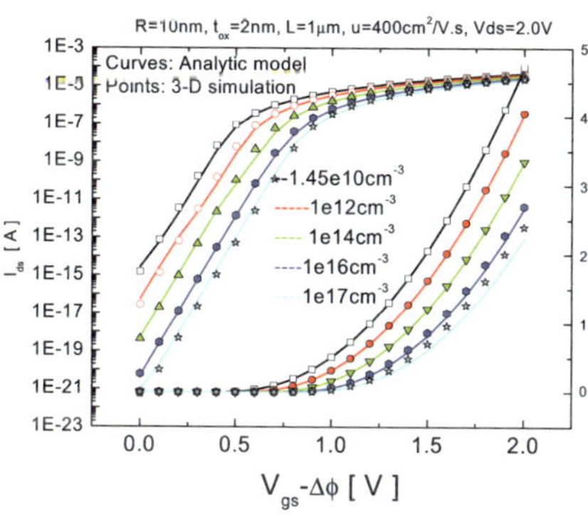

Fig. 21. I_{ds} versus V_{GS} and compared with the 3-D numerical simulation (from He J. Semicond. Sci. Technol. 22 (2007) 671–677)

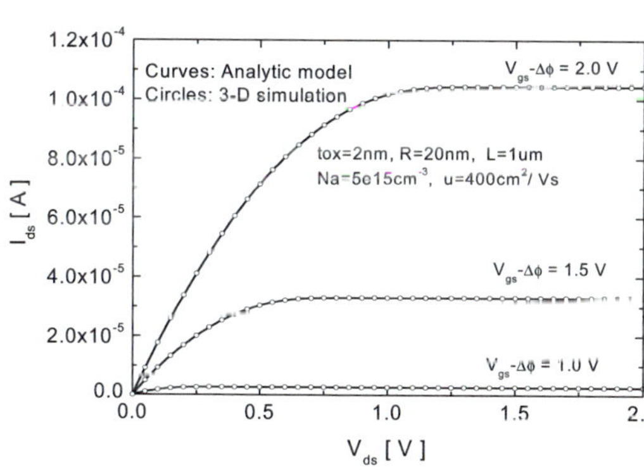

Fig. 22. I_{ds} versus V_{DS} and compared with the 3-D numerical simulation (from He J. Semicond. Sci. Technol. 22 (2007) 671–677)

and pushes the peaks toward the centre of the silicon channel. As a result, the QME causes an increase in the threshold voltage and degradation in the sub-threshold slope, especially with the reduction of the silicon film radius. In addition, the short-channel effects also lead to the threshold voltage reduction and sub-threshold slope degradation with the reduction of the SRG MOSFET channel length. Next, we will introduce second-order physics effects in the compact model of the nanoscale SRG MOSFETs.

3.3 Unified model for nanowire with advanced effects

Starting from Poisson's equation solution, an accurate inversion charge (Qin) equation is obtained for the long channel SNWTs with arbitrary doped bodies. Moreover, some advanced physical effects such as short channel effects (SCEs), quantum mechanical effects (QMEs), high field mobility degradation and velocity saturation have all been incorporated into the model.

3.3.1 Modeling short channel effects

For the transistor with short channel length, the voltage at the drain terminal has significant impact on the channel potential, known as SCEs. In generally, SCEs induce threshold voltage roll-off, subthreshold slope degradation, drain induced barrier lowering. In addition, channel length modulation and carrier velocity saturation and overshoot become important at short channel length as well. Here, SCEs are modeled following the BSIM5 approach. Assume the surface potential in a short channel device can be expressed as:

$$\phi_s(y) = \phi_{s0} + \phi_{s,3D}(y) \tag{22}$$

where ϕ_{s0} is the solution of the 2D Poisson's equation. After substituting (22) into (9) and considering boundary condition, we got

$$\phi_S(y) = \phi_{S0} + \frac{(v_{ds} + v_{bi} - \phi_{s0})\sinh[y/\lambda] + (v_{bi} - \phi_{s0})\sinh\left[(L-y)/\lambda\right]}{\sin[L/\lambda]} \tag{23}$$

Its minimum value, which determines the threshold voltage at a low v_{ds}, is given by:

$$\phi_{s,\min} = \phi_{s0} + \frac{(v_{ds} + 2v_{bi} - 2\phi_{s0})\sinh[L/2\lambda]}{\sin[L/\lambda]} \tag{24}$$

where $\lambda = \sqrt{t_{oxeff} \cdot \frac{R}{2} \cdot \frac{\varepsilon_{si}}{\varepsilon_{ox}} + \frac{1}{4}R^2}$ is the natural length of SNWT by assuming the highest leakage path lies at the center of the channel. Combining ϕ_{s0} with equation of boundary condition, we get a new surface potential solution in terms of inversion charge:

$$q_{dep} + q_{in} = v_{gs} - \Delta\phi - \alpha \cdot \phi_s + f_{SCE} \cdot (v_{ds} + 2v_{bi}) \tag{25a}$$

and a inversion charge solution can be obtained:

$$\frac{v_{gs} - V_{th}}{\alpha} - v_{ch} = \frac{q_{in}}{\alpha} + \ln q_{in} + \ln(1 + H \cdot q_{in}) \tag{25b}$$

where $\alpha = 1 + 2 \cdot f_{SCE}$ is the subthreshold slope (SS);
$f_{SCE} = 1/\left[2\cosh(L/2/\lambda) - 2\right]$ is the SCEs factor;
$v_{th} = v_{th0} + \Delta v_{th,VOL} + \Delta v_{th,SCE}$ is the new threshold voltage;
$\Delta v_{th,SCE} = f_{SCE} \cdot \left[2 \cdot \left(v_{th,Long} - q_{dep} - v_{bi}\right) - v_{ds}\right]$ is the threshold voltage roll-off induced by the SCEs. A drain current expression can further be derived using (25b):

$$I_{ds} = \frac{2\pi R \mu_{eff} C_{oxeff}}{\beta^2 L_{eff}} \left[f(q_d) - f(q_s)\right] \tag{26a}$$

with

$$f(q_{in}) = -\frac{q_{in}^{\ 2}}{2\alpha} - 2q_{in} + \frac{1}{H}\ln(1 + H \cdot q_{in}) \tag{26b}$$

Comparison with numerical simulation results shown in Figure 23(a) shows the correctness of the threshold voltage roll-off and DIBL described by the proposed SCEs model. Figure 23(b) shows that the transfer characteristics with subthreshold slope degradation as predicted in the proposed the short channel model. The strong inversion drain current is not affected by the SCEs, which is expected.

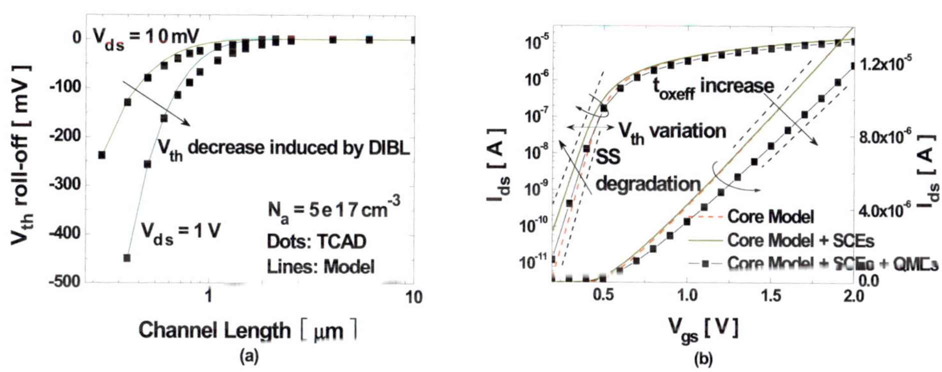

Fig. 23. (a) the threshold voltage roll-off versus channel length and (b) transfer characteristics in the short channel model (From Yang J. IEEE TED, Vol. 55, No. 11, November 2008)

3.3.2 Modeling quantum-mechanical (QM) effects and other advanced physical effects

In highly scaled device, the QM confinement of the carrier in the thin silicon channel is significant. As in bulk MOSFETs, the large vertical electric field leads to a strong band bending and carrier confinement at the surface, known as EC as shown in Figure 24. The deviation of the location of peak carrier concentration from the surface decreases the channel carrier concentration and leads a decrease in the gate capacitance (Francis et al., 1992):

$$\Delta t_{ox} = \Delta = a \cdot \left(\frac{\hbar^2}{2qm_e E_{avg}}\right)^{1/3} \tag{27a}$$

$$C_{oxeff} = \varepsilon_{ox}\left[(R - \Delta/3) \cdot \ln\left((R + t_{ox})/(R - \Delta/3)\right)\right]^{-1} \tag{27b}$$

where m_e is the effective mass of the electron corresponding to the lowest electric sub-band, and $E_{avg} = C_{ox}\left(q_{dep} + \overline{q}_{in}/3\right)/\beta\varepsilon_{si}$ is the average surface field. Besides the EC, there is a strong carrier confinement in nanoscale SNWTs even at low electric fields in the channel. It is because the carriers are confined in a rectangular well formed by the gate insulator around, known as SC. In SC and EC, the conduction band split into several subbands as shown in Figure 24. Since the carriers stays at the subband with the lower energy first, the reduction of the amount of carriers can be modeled by the widening the effective band-gap (Arora, 1993) and replace v_{ch} in (25b) by

$$v_{ch} \rightarrow v_{ch} + \frac{E_0}{q} = v_{ch} + \lambda_q \frac{1}{q}\left(\frac{\hbar^2\pi^2}{4m_e R^2}\right) \tag{28}$$

The simulation results shown in Figure 23(b) illustrates the model can predicted the threshold voltage roll off and gate capacitance degradation induced by QME correctly.

The modeling of other advanced physical effects such as poly-depletion effects and mobility degradation of this work are imported from the BSIM5 approach. Velocity saturation, velocity overshoot and ballistic transport (source-end velocity limit) is handled in a unified way using the saturation charge concept:

$$q_{insat} = \frac{q_{in}}{1 + 2\beta n v_{sat} L_{eff} / \mu_{eff}\left(2n + k_{sativ} q_{in}\right)} \tag{29a}$$

$$q_{ineff} = \sqrt[m]{q_{insat}{}^m + q_{in}{}^m} \tag{29b}$$

where n and m are the only parameters. The effective inversion charge q_{ineff} at the source and drain ends are substituted into the final current expression (26) to calculate the device output current. A complete ballistic transport model is not considered in this work, as ballistic transport will not be significant until the channel length is scaled to less than 10nm (Duan et al., 2001). The dc characteristics for a small size SNWT predicted by the complete compact model is demonstrated in Figure 25.

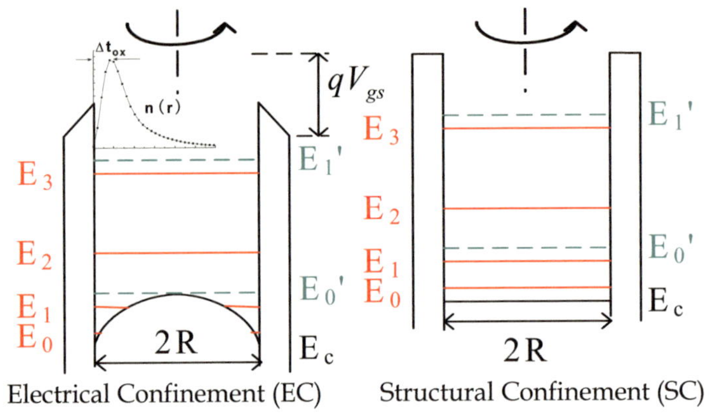

Fig. 24. Energy-band diagrams showing the carrier confinement and the quantization of electronic energy levels in the small size channel due to electrical confinement and structural confinement (From IEEE TED, Vol. 55, No. 11, November 2008)

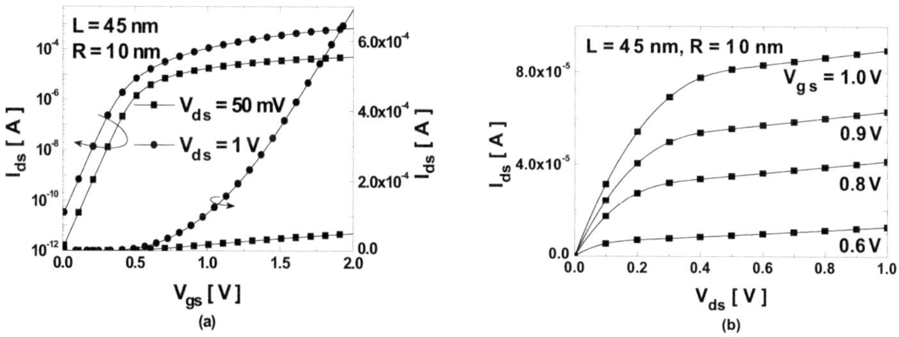

Fig. 25. (a) Ids-Vgs and (b) Ids-Vds characteristics predicted by the complete compact model for a small size SNWT (From IEEE TED, Vol. 55, No. 11, November 2008)

3.4 Ge/Si core/shell nanowire MOSFETs modeling

As one promising molecular device, Ge/Si core/shell nanowire MOSFET has been intensively studied experimentally and theoretically (Xiang et al., 2006; Lu et al., 2005; Fan et al., 2008; Musin & Wang, 2005; Liang et al., 2007; He, Y. et al., 2008) in recent years. Such a device utilizes the bandstructure engineering (Xiang et al., 2006), strain effect (He, Y. et al., 2008) and ballistic transport of 1-dimensional (1D) hole gas (Liang et al., 2007), resulting in very high carrier mobility and excellent performance. At the same time, an analytic compact model is also highly desirable for both physical insight into and performance evaluation of the core/shell based NWFET devices. However, it has not been available so far due to the complexity in analytically modeling the corresponding heterostructure, strain effect and ballistic transport. Nevertheless, some theoretical progresses on non-classical MOSFETs have provided the base for us to develop a primary analytical electrostatic potential model.

Here, starting from Poisson-Boltzmann equation in the Ge/Si core/shell NWFETs, the classical analytic expressions of electrostatic potential and charges in the semiconductor layers are derived out under the gradual channel approximation.

The schematic diagram and corresponding coordinate of a long channel Ge/Si core/shell NWFET are shown in Figure 26, where z is the channel direction, r is the radial direction, t_{ox} is the thickness of gate dielectric, r_0 and R are the radii of germanium core and silicon shell, respectively.

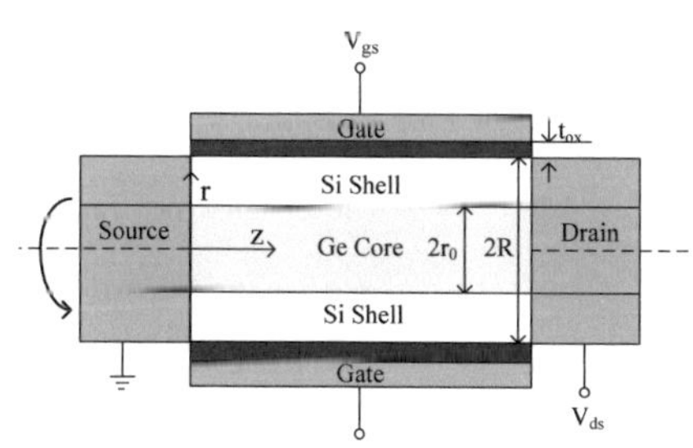

Fig. 26. Shematic diagrame of a Ge/Si core/shell NWFET

The assumed condition in the model derivation include: an abrupt heterojunction between germanium and silicon, an intrinsic or lightly doped body and a long device channel. Quantum confinement and strain effect are neglected for simplicity. Under gradual channel approximation (GCA), the Poisson-Boltzmann equations in Ge/Si core/shell NWFETs are written as the following forms with only hole term considered:

$$\frac{1}{r}\frac{d}{dr}\left(r\frac{d\phi}{dr}\right)=-\frac{qn_{iGe}}{\varepsilon_{Ge}}\exp\left(\frac{V_{bi}-\phi+V_{ch}}{V_t}\right),0\le r\le r_0 \tag{30}$$

$$\frac{1}{r}\frac{d}{dr}\left(r\frac{d\phi}{dr}\right)=-\frac{qn_{isi}}{\varepsilon_{si}}\exp\left(\frac{V_{ch}-\phi}{V_t}\right),r_0<r\le R \tag{31}$$

where ϕ is the electrostatic potential, V_{bi} is the self-built voltage of intrinsic Ge/Si heterojunction, V_{ch} is hole quasi-Fermi potential, q is the electronic charge, V_t is the thermal potential, $n_{iGe}(n_{isi})$ and $\varepsilon_{Ge}(\varepsilon_{si})$ are the intrinsic carrier concentration and permittivity of germanium (silicon), respectively. The Ge/Si core/shell NWFET behaves like a depletion p-channel device due to the band offset between intrinsic germanium and silicon. Hole accumulation in both germanium core and silicon shell are handled while electron is neglected. Also notice that V_{ch} is assumed to be constant along the radius according to quasi-equilibrium approximation, being a negative value for p type device.
For simplicity, a normalized formulation of (31) is used to find its solution in the shell region, the detailed process is showed in (Zhang L. et al., 2008). (31) in the silicon shell layer is solved concisely as:

$$\phi_{si}(r)=V_{ch}-V_t\ln(2A^2L_{si}^2/r^2)-V_t\ln\left[\beta_{si}(r/r_0)^A\right]+2V_t\ln\left[1-\beta_{si}(r/r_0)^A\right] \tag{32}$$

where β_{si} and A are two intermediary parameters to be decided by the boundary condition, L_{si} is the Debye length of intrinsic silicon $L_{si}=\sqrt{\varepsilon_{si}V_t/(qn_{isi})}$.
And then the radial electric field in the silicon shell layer is obtained:

$$E_{si}(r)=-\frac{d\phi_{si}}{dr}=\frac{V_t}{r}\left[A-2+2A\frac{\beta_{si}(r/r_0)^A}{1-\beta_{si}(r/r_0)^A}\right] \tag{33}$$

Similarly, electrostatic potential and electric field in the germanium core layer are given at A=2:

$$\phi_{Ge}(r)=V_{bi}+V_{ch}-V_t\ln(8L_{Ge}{}^2/r_0^2)-V_t\ln(\beta_{Ge})+2V_t\ln[1-\beta_{Ge}(r/r_0)^2] \tag{34}$$

and

$$E_{Ge}(r)=-\frac{d\phi_{Ge}}{dr}=\frac{4V_t}{r_0}\frac{\beta_{Ge}(r/r_0)}{1-\beta_{Ge}(r/r_0)^2} \tag{35}$$

where L_{Ge} is the Debye length of intrinsic germanium $L_{Ge}=\sqrt{\varepsilon_{Ge}V_t/(qn_{iGe})}$. Also the intermediary parameter β_{Ge} is evaluated from the boundary condition.

At the interface between silicon and gate oxide layer, surface potential and electric filed are obtained according to the above equation:

$$\phi_s = V_{ch} + V_t \ln(2A^2 \frac{L_{si}^2}{R^2}) + V_t \ln\left[\beta_{si}(\frac{R}{r_0})^A\right] - 2V_t \ln\left[1-\beta_{si}(\frac{R}{r_0})^A\right] \tag{36}$$

$$E_s = \frac{V_t}{R}\left[(A-2) + 2A\frac{\beta_{si}(R/r_0)^A}{1-\beta_{si}(R/r_0)^A}\right] \tag{37}$$

Following Gauss's Law and electric flux continuity condition, we get the final input voltage equation:

$$\begin{aligned}\frac{V_{ch}-(V_{gs}-\Delta\phi)}{V_t} &= \frac{\varepsilon_{si}}{\varepsilon_{ox}}\ln(1+\frac{t_{ox}}{R})\left[(A-2)+2A\frac{\beta_{si}(R/r_0)^A}{1-\beta_{si}(R/r_0)^A}\right]+ \\ &+\ln(\frac{2A^2L_{si}^2}{R^2}) + \ln\left[\beta_{si}(\frac{R}{r_0})^A\right] - 2\ln\left[1-\beta_{si}(\frac{R}{r_0})^A\right]\end{aligned} \tag{38}$$

Similar to the existing compact models for bulk and other non-planar MOSFET, the input voltage equation and its solution is the base for further developing drain current models and advanced effect models. The detailed solving procedure for (38) is described as follows. At the interface between germanium core layer and silicon shell layer both the potential and electric flux need to be continuous. So the parameters A , β_{si} and β_{Ge} are arrived.

For the studied Ge/Si core/shell structure, it is verified that A is always larger than 2 with gate voltage increasing for the Ge/Si core/shell structure. On the other hand, A has an upper limit set which is independent of the bias condition (Figure 27). And the three intermediary parameters are obtained simultaneously by the shooting secant method. After that, interface potential ϕ_{r0} and centric potential ϕ_0 are obtained directly by substituting $r = r_0$ and $r = 0$ into (34), respectively:

$$\phi_{r_0} = V_{bi} + V_{ch} - V_t \ln(8L_{Ge}{}^2 / r_0^2) - V_t \ln(\beta_{Ge}) + 2V_t \ln[1-\beta_{Ge}] \tag{39}$$

$$\phi_0 = V_{bi} + V_{ch} - V_t \ln(8L_{Ge}{}^2 / r_0^2) - V_t \ln(\beta_{Ge}) \tag{40}$$

Hole charge density in germanium core and silicon shell layers are directly obtained through Gauss's law:

$$Q_{mGe} = 4\varepsilon_{Ge}\frac{V_t}{r_0}\frac{\beta_{Ge}}{1-\beta_{Ge}} \tag{41a}$$

$$Q_{moi} = 2A\varepsilon_{si}\frac{V_t}{R}\left[\frac{1}{1-\beta_{si}(R/r_0)^A} - \frac{1}{1-\beta_{si}}\right] \tag{41b}$$

respectively.

The total hole charge density in the channel is also obtained as:

$$-Q_g = \frac{\varepsilon_{si} V_t}{R}\left[(A-2) + 2A\frac{\beta_{si}(R/r_0)^A}{1-\beta_{si}(R/r_0)^A}\right] \quad (41c)$$

Note that Q_{msi} represents the charge density per unit area at the interface between the silicon shell and oxide layer while Q_{mGe} denotes that at the interface between the germanium core layer and silicon shell layer. As a result, gate capacitance of the core/shell structure is obtained by differentiating gate charge Q_g with respect to the gate voltage as follows:

$$C_g = \frac{dQ_g}{dV_{gs}} = C_{ox}\left[1 - \frac{d\phi_s}{dV_{gs}}\right] = C_{ox}\left[1 - \frac{d\phi_s}{dA} / \frac{dV_{gs}}{dA}\right] \quad (42)$$

where $d\phi_s / dA$ and dV_{gs} / dA are obtained from (36) and (38), respectively.

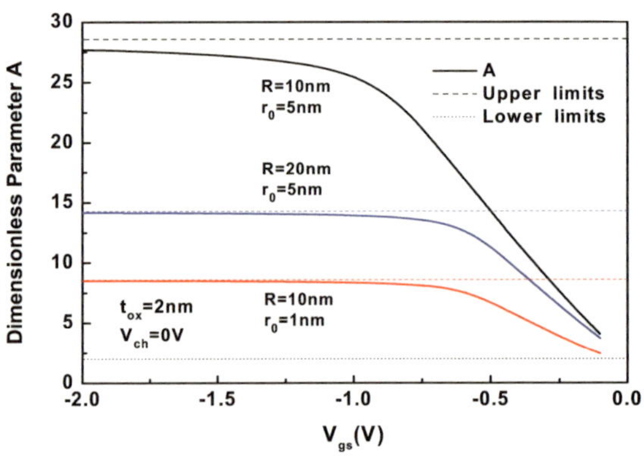

Fig. 27. The parameter A versus V_{gs} for different Ge/Si core/shell structures (Zhang L. IEEE TED, Vol. 55, No. 11, Nov. 2008)

In summary, the analytical electrostatic potential equation and its solution are presented above. In this section the analytic potential model is verified through numerical simulation. The physical parameters of silicon and germanium in both the analytical model and numerical simulator are taken from (Sze, 1981). The spatial potential, electric field and hole charge density are obtained under classical device physics through intermediary parameter A, β_{si} and β_{Ge} for the given gate voltage, quasi-Fermi potential and structure parameters. So once the three intermediary parameters are solved for the given bias and device geometry, the characteristics of the Ge/Si core/shell structure are directly obtained form the analytic model prediction. The potential, electric field and charge density solution are illustrated in Figure 28-32 for one geometry configuration: $R = 10nm$, $r_0 = 5nm$, and $t_{ox} = 2nm$ in comparison with numerical simulation results. The results show that excellent agreements are observed between the analytic model and the numerical simulation in all these figures and the Ge/Si core/shell heterostucture NWFET exhibits its unique characteristics, much different from a common SRG MOSFET. A more detailed discussion is displayed in (Zhang L. et al., 2008).

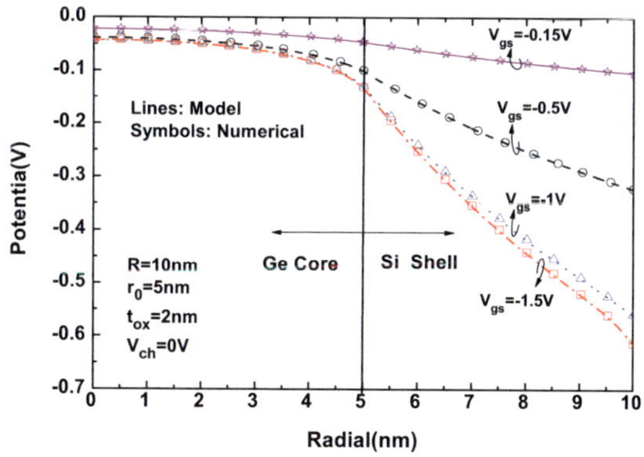

Fig. 28. Potential distribution along the radial of a Ge/Si core/shell structure. The lines denote results obtained from the analytic model and the symbols are results from numerical simulation (Zhang L. IEEE TED, Vol. 55, No. 11, Nov. 2008)

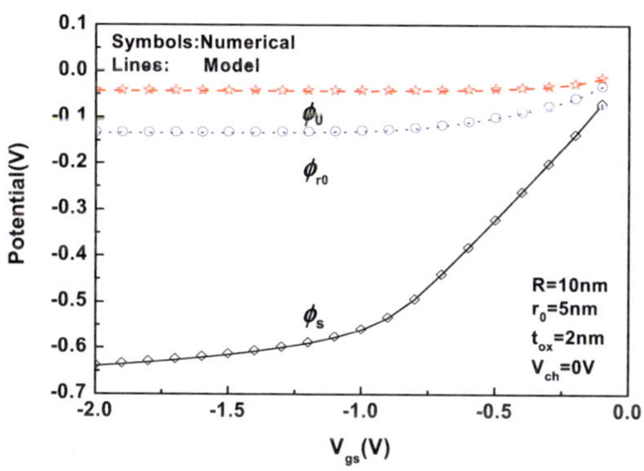

Fig. 29. ϕ_s, ϕ_{r_0} and ϕ_0 of a Ge/Si core/shell structure as functions of V_{gs} obtained form the analytic model (curves) in comparison with numerical simulation (symbols) (Zhang L. IEEE TED, Vol. 55, No. 11, Nov. 2008)

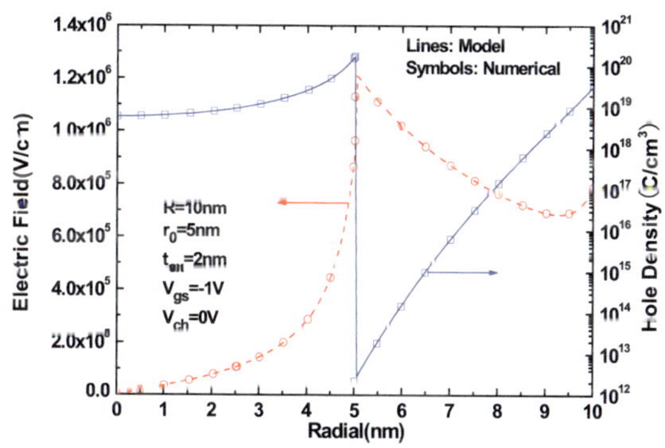

Fig. 30. Electric field and hole density distributions along the radial of a Ge/Si core/shell structure at $V_{gs} = -1V$, obtained from the analytic model (lines) in comparison with numerical simulation (symbols) (Zhang L. IEEE TED, Vol. 55, No. 11, Nov. 2008)

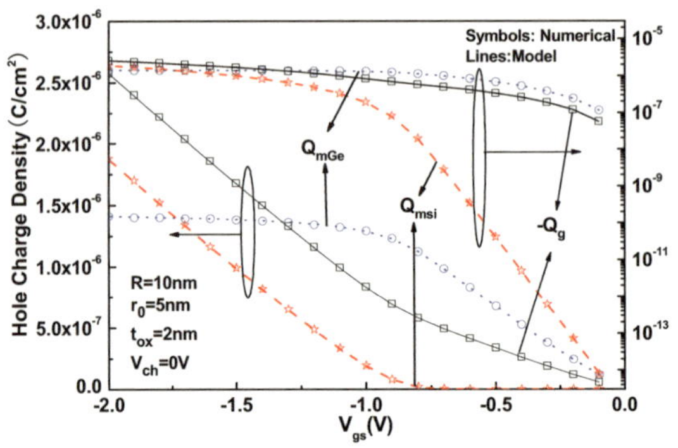

Fig. 31. Charge density Q_{mGe}, Q_{msi} and $-Q_g$ of a Ge/Si core/shell structure as functions of V_{gs} both in linear and logarithm coordinates, obtained from the analytic model (lines) in comparison with numerical simulation (symbols). (Zhang L. IEEE TED, Vol. 55, No. 11, Nov. 2008)

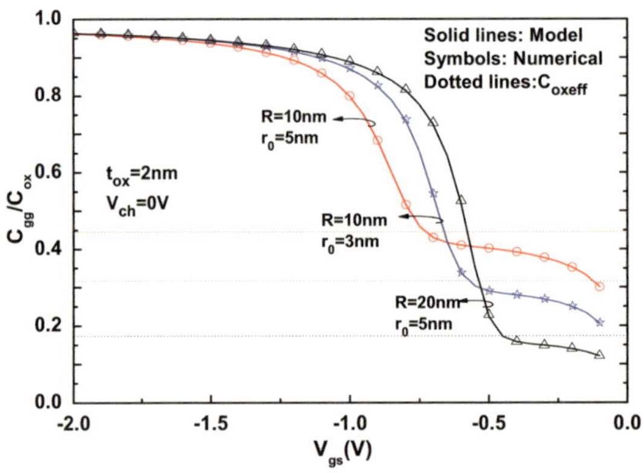

Fig. 32. Normalized gate capacitance C_{gg} / C_{ox} curves as a function of V_{gs} from the analytic model (solid lines) in comparison with numerical simulation (symbols). Dotted lines denote three corresponding C_{oxeff} (Zhang L. IEEE TED, Vol. 55, No. 11, Nov. 2008)

4. Nanowire–based circuit simulation

Paralleling the advance of process technology to fabricate SNWT (Cui et al., 2001c; Duan et al., 2001), compact models for the SNWTs have also been developed for circuit simulations in recent years. A general model for doped SNWTs may be very useful for device scientists to optimize the device structure and for circuit designers to evaluate the performance of the SNWT circuits. Here, a design oriented compact model for SNWTs, applicable for a wide range of doping concentrations (i.e. from 10^{10} to 10^{19}cm-3) and geometrical dimensions is presented. Starting from Poisson's equation solution, an accurate inversion charge (Qin) equation is obtained for the long channel SNWTs with arbitrary doped bodies. Then a charge based drain current (Ids) expressions is derived. Transconductance (gm), output conductance (gds), terminal charges (Q) and capacitance (Cij) are all derived analytically and verified by TCAD simulation (Synopsys, 2005). And then this model is also implemented in the circuit simulator by the Verilog-A language and its application in circuit simulations is also demonstrated.

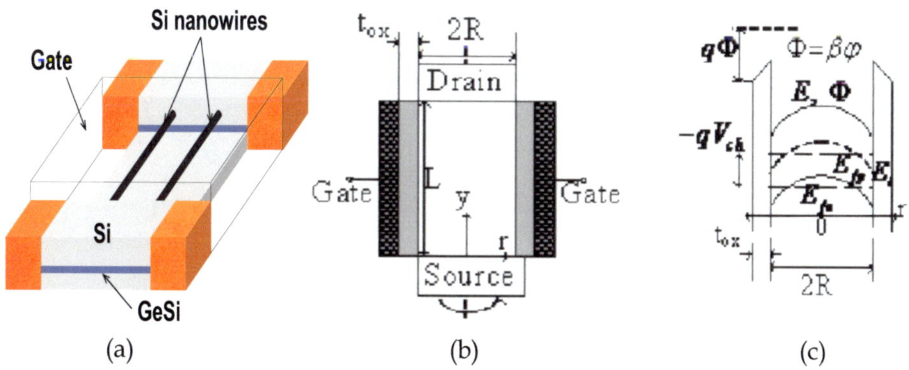

Fig. 33. (a) Stereoscopic schematic, (b) cross-section schematic, and (c) energy band diagram of an N-type SNWT (From Yang J., IEEE TED, Vol. 55, No. 11, Nov. 2008)

The device structure, coordinate system, and corresponding energy band diagram of a doped SNWT studied in this work are shown in Figure 33 (a), (b) and (c), respectively. For an n-type device, the hole Fermi level Efp remains constant along the channel if the current of majority carriers is neglected. Consequently, Efp can be defined as the energy reference level in this work. The 3D Poisson-Boltzmann equation in the SNWT structure is written as

$$\frac{1}{r}\frac{d}{dr}\left(r\frac{d\phi}{dr}\right)+\frac{d^2\phi}{dy^2}=\frac{1}{L_D{}^2}\left[1+e^{\phi-v_{ch}-2\phi_f}\right] \tag{43}$$

where $\phi_f=\ln(N_a/n_i)$ is the Fermi-potential, v_{ch} is the quasi-Fermi potential, ϕ is the channel potential, and $L_D{}^2=kT\varepsilon_{si}/q^2N_a$ is the Debye's length of the silicon body with doping N_u. All other variables have there usual meanings. All the potentials in this work are normalized by thermal voltage β, and the charges by β/C_{ox}, unless otherwise specified. $C_{ox}=\dfrac{\varepsilon_{ox}}{R\ln(1+t_{ox}/R)}$ is the effective oxide capacitance.

This 3D problem is simplified into two separate parts along the vertical and current flow directions. And we get the relationship:

$$v_{gs}-v_{th0}\quad\Delta v_{th,VOL}-v_{ch}=q_{in}+\ln q_{in}+\ln(1+H\cdot q_{in}) \tag{44}$$

where $v_{th0}=\Delta\phi+2\phi_f+q_{dep}-\ln\dfrac{4\varepsilon_{si}}{RC_{ox}}q_{dep}$ is the transistor threshold voltage similarly to bulk MOSFET; $\Delta v_{th,VOL}=-\ln\dfrac{1}{2q_{dep}}\left[1-\exp\left(-\dfrac{C_{ox}R}{2\varepsilon_{si}}q_{dep}\right)\right]$ is the extra part of threshold voltage induced by the special geometric structure of SNWT and $q_{in}=q^2n_0R/2kTC_{ox}$ is given as the total mobile charge sheet density in the channel.

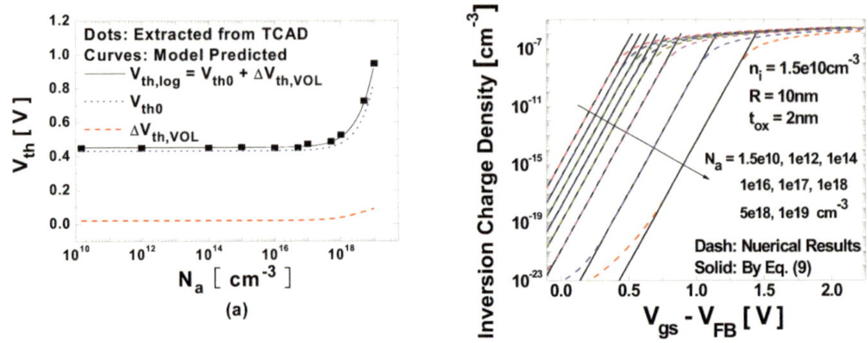

(a) (b)

Fig. 34. (a) Vth model is compared with TCAD extracted by second derivative method in a long channel SNWT (From Yang J., IEEE TED, Vol. 55, No. 11, Nov. 2008)

Through the above equation, the validity of the threshold voltage modeling is verified, see Figure 34， and the characteristics of inversion charge with different condition are got and shown in Figure 35.

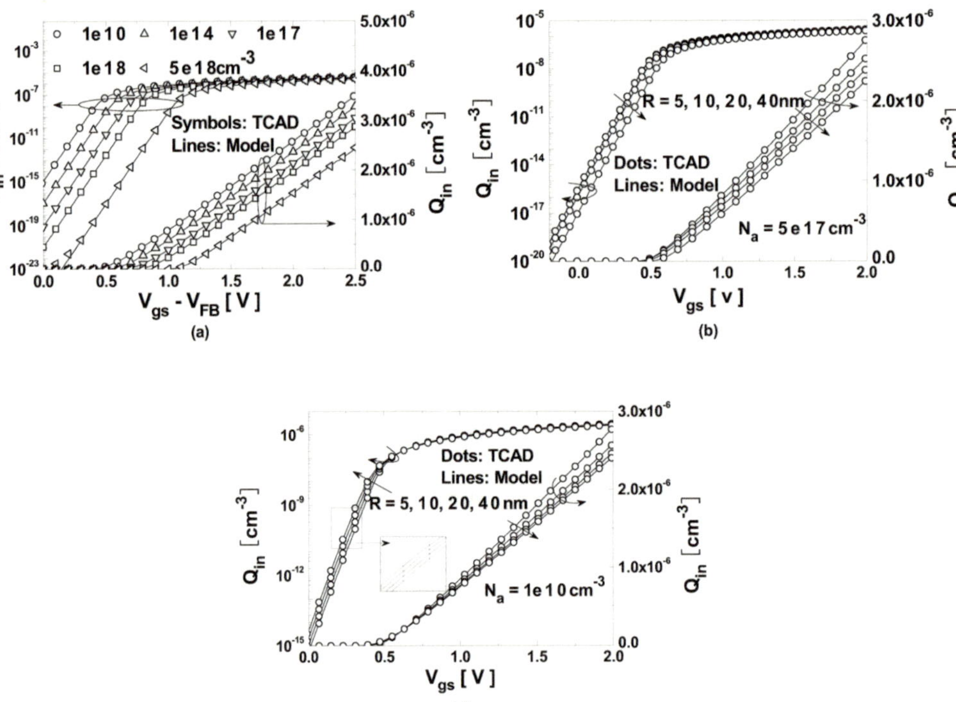

Fig. 35. Inversion charge equation verification for long channel SNWTs (a) with doping variation, (b) with geometric size variation in doped case and (c) with geometric size variation in undoped case (From Yang J., IEEE TED, Vol. 55, No. 11, Nov. 2008)

The analytical drain current expression is obtained as:

$$I_{ds} = \frac{2\pi R \mu_{eff} C_{ox}}{\beta^2 L_{eff}} \left[f(q_d) - f(q_s) \right] \tag{45}$$

where μ_{eff} is the effective mobility in the channel; V_{ds} is the quasi-Fermi potential in the drain terminals; q_d and q_s are normalized inversion charge per unit gate area at the source and drain terminals with $f(q_{in}) = -\frac{1}{2} q_{in}^2 - 2q_{in} + \frac{1}{H} \ln(1 + H \cdot q_{in})$

Transconductance and output conductance can be derived analytically from the expressions of drain current:

$$g_m = \frac{\partial I_{ds}}{\partial V_{gs}} \Big|_{V_{ds}} = \frac{2\pi R \mu_{eff} C_{ox}}{\beta L_{eff}} (q_s - q_d) \tag{46a}$$

$$g_{ds} = \frac{\partial I_{ds}}{\partial V_{ds}} \Big|_{V_{gs}} = \frac{2\pi R \mu_{eff} C_{ox}}{\beta L_{eff}} q_d \tag{46b}$$

Analytical expressions for inversion charges at each terminal are desired for efficient transient circuit simulation. For a SNWT, there are three terminal charges, associating with gate, source, and drain, respectively: the gate charge Qg can be computed by integrating the channel charge density along the channel; The drain and source charges, denoted as Qd and Qs, can be derived analytically by using the Ward-Dutton linear-charge-partition method:

$$Q_g = \frac{2\pi R C_{ox}}{\beta^2} \int_0^{V_{ds}} q_{in} dy \tag{47a}$$

$$Q_d = -\frac{2\pi R C_{ox}}{\beta^2} \int_0^{V_{do}} \frac{y}{L} q_{in} dy \tag{47b}$$

$$Q_s = -Q_g - Q_d \tag{47c}$$

where y/L are obtained from the current continuity condition

$$\frac{y}{L} = \frac{f(q_{in}) - f(q_s)}{f(q_d) - f(q_s)} \tag{48}$$

The model predicted I-V characteristics are verified by comparing with TCAD simulation under various biasing voltages, a wide range of doping concentrations and geometric dimensions as shown in Figure 36. The error between the proposed model and the 3-D numerical simulation is less than 5% with devices with intrinsic to heavily doped body with a doping of 10^{19} cm-3. The numerical simulation results demonstrate the accuracy of the proposed model from the linear to saturation and from the sub-threshold to strong inversion regions.

In contrast to digital circuits, analog design focus on the first derivatives such as gm/Ids and Cij, which are shown in Figure 37. At Vds=0, Cgs=Cgd and Csg=Cdg in the figures indicate that the developed model has inherent the source/drain symmetry characteristic, which is important for analog and RF applications. Note that the verifications of the core I-V and C-V models are done without any fitting parameters.

The compact SNWT model has been implemented in a commercial circuit simulator ADS2006A by the Verilog-A language to demonstrate practicality of the model. This implementation includes a procedure to produce an initial guess and convergent correction methods to help the nonlinear equation (25b) to calculate the inversion charge. DC, ac, and transient simulations of many sample circuits are performed. Figure 38 shows the transient analysis of a 21-stage SNWT ring oscillator by this model. The waveforms of three successive stages are plotted in the figure. Compared with a ring oscillator constructed by DG MOSFETs (Taur et al., 2004), the oscillator frequency in this work is much higher, illustrating a smaller transistor input capacitances relative to the higher current drive in SNWT device. In the case of high parasitic capacitance, parallel connection of several SNWTs can be used.

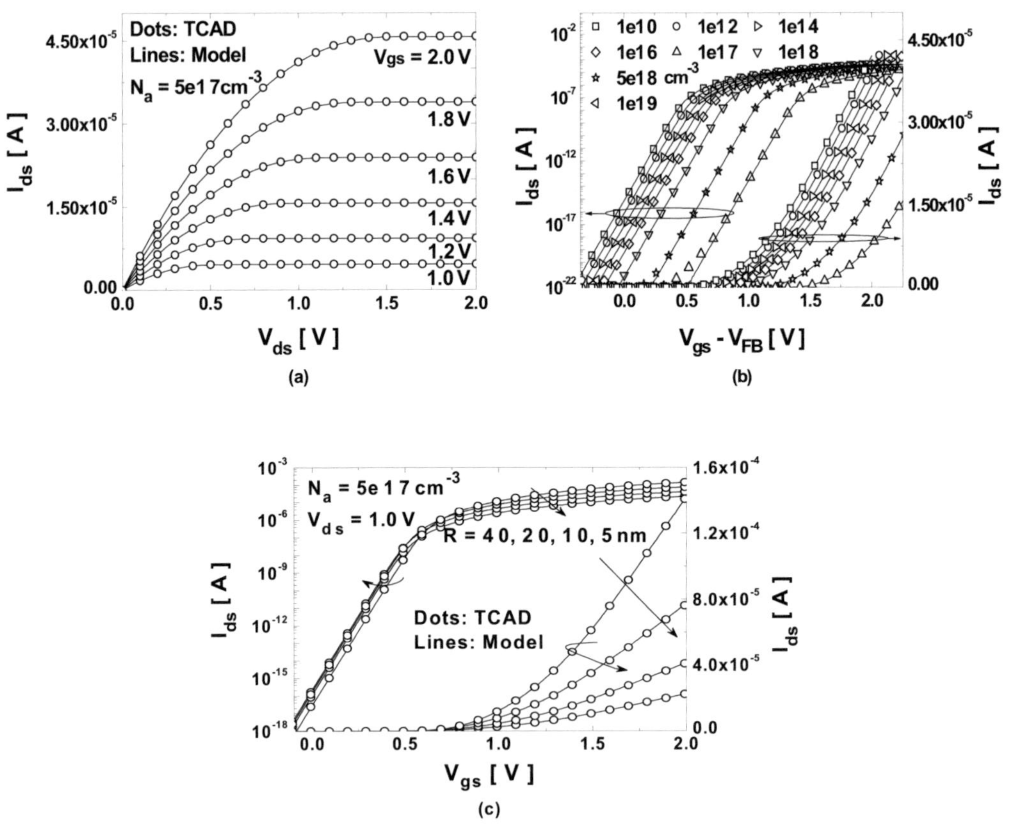

Fig. 36. dc I-V characteristics verification for long channel SNWTs (a) Ids-Vds (b) Ids-Vgs with doping variation (Symbols: TCAD, Lines: Model) and (c) Ids-Vgs with geometric size variation. (From Yang J., IEEE TED, Vol. 55, No. 11, Nov. 2008)

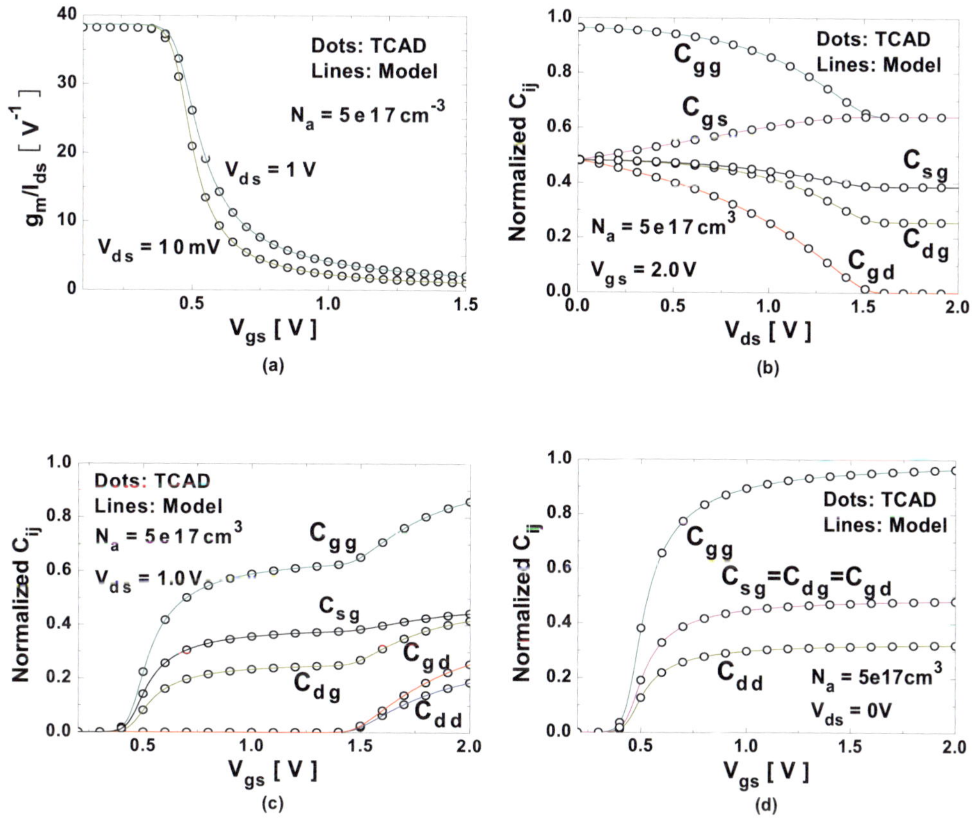

Fig. 37. ac characteristics verification of the model for a long channel SNWT (a) gm efficiency, (b) C-Vds, (c) C-Vgs (at Vds≠0) and (d) C-Vgs (at Vds=0) (From Yang J., IEEE TED, Vol. 55, No. 11, Nov. 2008)

5. Conclusion

In this chapter, after a brief outline of the silicon-based nanowire fabrication technology, we discussed simulation and compact model of nanowire MOSFET following the device physics understanding and exploring. In the developed compact model, the short channel effect, quantum mechanic effect and other are analyzed and the predicted results are also compared with the 3-D numerical simulation. Based on such a verified physics based SPICE model, the nanowire based circuit performance is simulated and demonstrated

6. Acknowledgment

This work is supported by the Fundamental Research Project of Shenzhen Science & Technology Foundation (JC200903160353A), the Shenzhen Science & Technology Foundation (CXB201005250031A), and the Industry, Education and Academy Cooperation Program of Guangdong (2009B090300318).

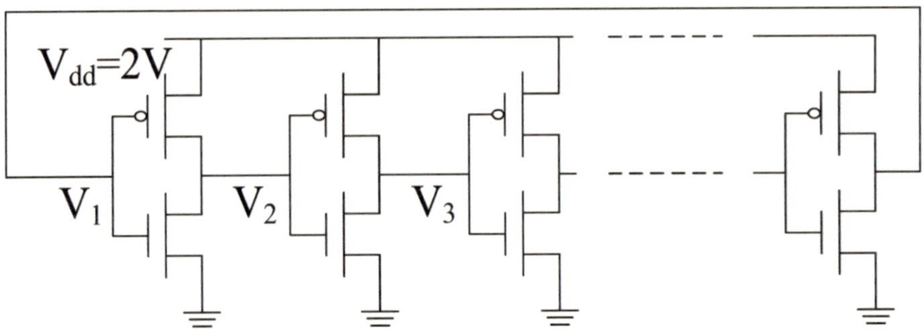

(a)

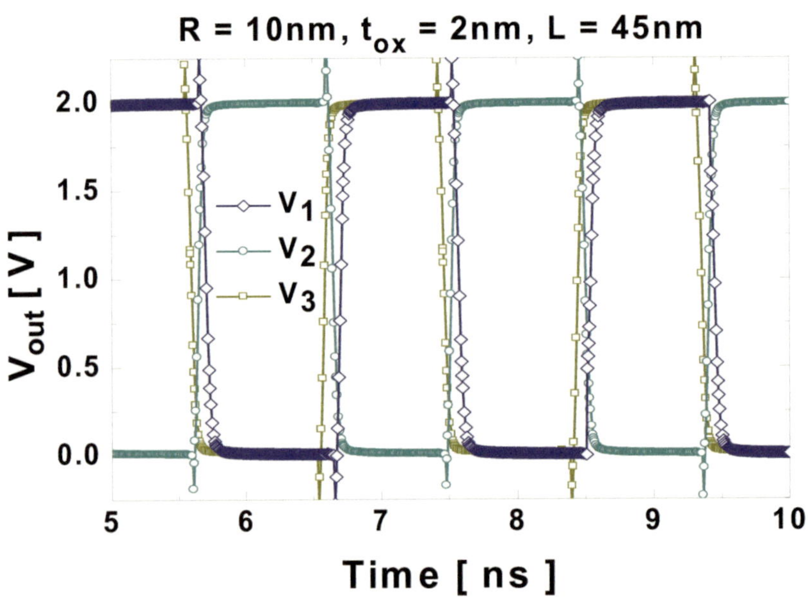

(b)

Fig. 38. (a) Circuit schematic of a 21-stage SNWT ring oscillator, and (b) the waveforms of three successive outputs in a transient analysis by ADS2006A (From Yang J., IEEE TED, Vol. 55, No. 11, Nov. 2008)

7. References

Arora, N., MOSFET models for VLSI circuit simulation: theory and practice, *Springer-Verlag New York Inc.*, ISBN 0387823956, Secaucus, NJ, USA

Barrelet, C. J., Greytak, A. N., & Lieber, C. M., (2004). Nanowire Photonic Circuits Elements, *Nano Letters*, vol. 4, no. 10, (September, 2004), pp.1981-1985, ISSN 10.1021

Bera, L. K., Nguyen, H. S., Singh, N., Liow, T. Y., Huang, D. X., Hoe, K. M., Tung, C. H., Fang, W. W., Rustagi, S. C., Jiang, Y., Lo, G. Q., Balasubramanian, N., & Kwong, D. L., Three Dimensionally Stacked SiGe Nanowire Array and Gate-All-Around p-

MOSFETs, *IEEE International Electron Devices Meeting, 2006*, pp. 551-554, ISBN 1-4244-0438-X, San Francisco, CA, USA, Dec 11-13, 2006

Black, C. T., (2005). Self-aligned self assembly of multi-nanowire silicon field effect transistors, *Appl. Phys. Lett.* Vol. 87, No.16, (October 2005), pp.163116-163118, ISSN 0003-6951

Bian, W., He, J., Tao, Y., Fang, M., & Feng, J., (2007). An analytic potential-based model for undoped nanoscale surrounding-gate MOSFETs, *IEEE Transactions on Electron Devices*, Vol. 54, No.9, (Sept. 2007), pp. 2293-2303, ISSN 0018-9383

Chang, C., Wang, Y. F., Kanamori, Y., Shih, J. J., Kawai, Y., Lee, C. K., Wu, K. C., & Esashi, M., (2005). Etching submicrometer trenches by using the Bosch process and its application to the fabrication of antireflection structures, *J. Micromech. Microeng.*, Vol. 15, No. 3, (March 2005), pp.580-585, ISSN 0960-1317

Cui, Y. & Lieber, C. M., (2001). Functional Nanoscale Electronic Devices Assembled Using Silicon Nanowire Building Blocks, *Science*, Vol.291, No.5505, (February 2001), pp.851-853, ISSN 1095-9203

Cui, Y., Lauhon, L., Gudiksen, M., Wang, J., & Lieber, C., (2001). Diameter-controlled synthesis of single-crystal silicon nanowires, *Applied Physics Letters*, Vol. 78, No.15, (APRIL 2001), pp. 2214-2216, ISSN 1077-3118

Cui, Y., Wei, Q., Park, H., & Lieber, C. M., (2001). Nanowire nanosensors for highly sensitive and selective detection of biological and chemical species, *Science*, Vol. 293, No. 5533, (August 2001), pp. 1289–1292, ISSN 1095-9203

Duan, X., Huang, Y., Cui, Y., Wang, J., & Lieber, C. M., (2001). Indium phosphide nanowires as building blocks for nanoscale electronic and optoelectronic devices, *Nature*, Vol. 409, (January 2001), pp. 66–69, ISSN 0028-0836

Dupre, C., et. al. 15nm-diameter 3D Stacked Nanowires with Independent Gates Operation: ΦFET, *IEEE International Electron Devices Meeting, 2008*, pp. 749-752, ISBN 978-1-4244-2377-4, San Francisco, CA, USA, Dec15-17, 2008

Fang, W. W., Singh, N., Bera, L. K., Nguyen, H. S., Rustagi, S. C., Lo, G. Q., Balasubramanian, N., & Kwong, D. L., (2007). "Vertically stacked SiGe nanowire array channel CMOS transistors", *IEEE Electron Device Letters*, Vol. 28, No. 3, (March 2007), pp. 211-213, ISSN 0741-3106

Fan, Z., Ho, J. C.. Jacobson, Z. A., Razavi, H.. & Javey, A., (2008). Large-scale, heterogeneous integration of nanowire arrays for image sensor circuitry, *Proc. Nat. Acad. Sci.*, Vol.105, No.32, (Aug., 2008), pp.11066 -11070, ISSN 11066-11070

Ferry, D., Gilbert, M., & Akis, R., (2008). Some considerations on nanowires in nanoelectronics, *IEEE Transactions on Electron Devices*, Vol. 55, No.11, (Nov. 2008), pp. 2820-2826, ISSN 0018-9383

Francis, P., Terao, A., Flandre, D., & Van de Wiele, F., (1992). Characteristics of nMOS/GAA (Gate-All-Around) transistors near threshold, *Microelectronic Engineering*, Vol. 19, No. 1-4, (August 2002), pp. 815-818, ISSN 0167 9317(92)90551-2

He, J., & Chan, M., physics based analytical solution to undoped cylindrical surrounding-gate (SRG) MOSFETs. *15th IEEE Int. Conf. on Devices, Circuits and Systems*, pp. 26-28, ISBN 0-7803-8777-5, Dominican Republic, Nov.3-5, 2004

He, J, Zhang, X., Zhang, G., Chan, M., & Wang, Y., (2006). A carrier-based DCIV model for long channel undoped cylindrical surrounding-gate MOSFETs, *Solid-State Electronics*, Vol. 50, No. 3, (March, 2006), pp. 416-421, ISSN 0038-1101

He, J., Zhang, X., Zhang, G., Chan, M., & Wang, Y., A complete carrier-based non-charge-sheet analytic theory for nano-scale undoped surrounding-gate MOSFETs. *Proceedings of ISQED'2006*, pp.115-120, ISBN 0-7695-2523-7, San Jose, Costa Rica, Mar. 28, 2006

He, J., Bian, W., Tao, Y., Yang, S., & Tang, X., (2007). Analytic carrier-based charge and capacitance model for long-channel undoped surrounding-gate MOSFETs, *IEEE Transactions on Electron Devices*, Vol. 54, No.6, (June 2007), pp. 1478-1485, ISSN 0018-9383

He, Y., Fan, C., Zhao, Y., Du, G., Liu, X., & Han, R., Impact of inhomogeneous strain on the valence band structure of Ge-Si core-shell nanowires, *Proc. Int. Conf. on Simulation of Semiconductor Process and Devices*, pp. 121-124, ISBN 978-1-4244-1753-7 Hakone, Sept. 9-11, 2008

Hood, L., Heath, J. R., Phelps, M. E., & Lin, B., (2004). Systems Biology and New Technologies Enable Predictive and Preventative Medicine, *Science*, Vol. 306, No.5696, (October 2004), pp.640-643, ISSN 1095-9203

Iñíguez, B., Jiménez, D., Riog, J., Hamid, H.A., Marsal, L. F., & Pallarès, J., (2005). Explicit continuous model for long channel undoped surrounding-gate MOSFETs, *IEEE Trans on Electron Devices*, Vol.52, No.8, (Aug. 2005), pp. 1868-1873, ISSN 0018-9383

Jiménez, D., Sáenz, J. J., Iñíguez, B., Suñé, J., Marsal, L. F., & Pallarès, J., (2004). Continuous analytical current-voltage model for surrounding-gate MOSFETs, *IEEE Electron Device Letters*, Vol. 25, No.8, (Aug. 2004), pp. 571-573, ISSN 0741-3106

Jin, S., Fischetti, M., & Tang, T., (2007). Modeling of electron mobility in gated silicon nanowires at room temperature: Surface roughness scattering, dielectric screening, and band nonparabolicity, *Journal of Applied Physics*, Vol. 102, No.8, (October 2007), pp. 083715-083728, ISSN 0021-8979

Kedzierski, J., Bokor, J., & Kisielowski, C., (1997). Fabrication of planar silicon nanowires on silicon-on-insulator using stress limited oxidation, *J. Vac. Sci. Technol. B*, Vol. 15, No. 6, (July 1997), pp.2825-2828, ISSN 0734-211X

Kotlyar, R., Obradovic, B., Matagne, P., Stettler, M., & Giles, M., (2004), Assessment of room-temperature phonon-limited mobility in gated silicon nanowires, *Applied Physics Letters*, Vol. 84, No.25, (June 2004), pp. 5270-5272, ISSN 0003-6951

Liang, G., Xiang, J., Kharche, N., Klimeck, G., Lieber, C. M., & Lundstrom, M., (2007). Performance Analysis of a Ge/Si Core/Shell Nanowire Field-Effect Transistor, *Nano Letter*, Vol. 7, No.3, (January, 2007), pp. 642-646

Liow, T., Tan, K., Lee, R., Zhu, M., Tan, B., Samudra, G., Balasubramanian, N., & Yeo, Y., (2008), "5 nm gate length Nanowire-FETs and planar UTB-FETs with pure germanium source/drain stressors and laser-free Melt-Enhanced Dopant (MeltED) diffusion and activation technique," *Symposium on VLSI Technology*, pp. 36-37, ISBN 978-1-4244-1802-2, Honolulu, HI, June 17-19, 2008

Liu, H., Biegelsen, D. K., Johnson, N. M., Ponce, F. A., & Peace, R. F. W., (1993). Self-limiting oxidation of Si nanowires, *J. Vac. Sci. Technol. B*, Vol. 11, No. 6, (Nov 1993), pp.2532-2537, ISSN 1071-1023

Liu, H., Biegelsen, D. K., Ponce, F. A., Johnson, N. M., & Peace, R. F. W., (1994). Self-limiting oxidation for fabricating sub-5nm silicon nanowires, *Applied Physics Letters*, Vol. 64, No.11, (Mar 1994), pp.1383-1385, ISSN 0003-6951

Lu, W., Xiang, J., Timko, B. P., Wu, Y., & Lieber, C. M., (2005), One-dimensional hole gas in germanium/silicon nanowire heterostructures, *Proc. Nat. Acad. Sci.*, Vol.102, No.29, (July, 2005), pp.10046 -10051, ISSN 10046-10051

Lu, W. & Lieber, C., (2006), Semiconductor nanowires, *Journal of Physics D: Applied Physics*, vol. 39, NO. 21, (June, 2006), pp. R387, ISSN 0022-3727

Michetti, P., Mugnaini, G., & Iannaccone, G., (2009). Analytical model of nanowire FETs in a partially ballistic or dissipative transport regime, *IEEE Transactions on Electron Devices*, Vol. 56, No.7, (July 2009), pp. 1402-1410, ISSN 0018-9383

Moldovan, O., Iñíguez, B., Jiménez, D., & Riog, J., (2007). Analytical Charge and Capacitance Models of Undoped Cylindrical Surrounding-Gate MOSFETs, *IEEE Trans on Electron Devices*, Vol.54, No.1, (January, 2007), pp.162-165, ISSN 0018-9383

Musin, R. N. & Wang X. Q., (2005). Structural and electronic properties of epitaxial core-shell nanowire heterostructures, *Physics Review B*, Vol.71, No.15, (April 2005), pp. 15538-15541, ISSN 1098-0121

Morales, A. & Lieber, C., A laser ablation method for the synthesis of crystalline semiconductor nanowires, (1998). *Science*, Vol. 279, No.208, (JANUARY 1998), pp. 208-211, ISSN 1095-9203

Namatsu, H., Horiguchi, S., Nagase, M., & Kurihara, K., (1997). "Fabrication of one-dimensional nanowires structures utilizing crystallographic orientation in silicon and their conductance characteristics", *J. Vac. Sci. Technol. B*, Vol.15, No.5, (July 1997), pp.1688-1696, ISSN 0734-211X

Neophytou, N., Paul, A., Lundstrom, M., & Klimeck G., (2008), Bandstructure effects in silicon nanowire electron transport, *IEEE Transactions on Electron Devices*, Vol. 55, No.6, (June 2008), pp. 1286-1297, ISSN 0018-9383

Ng, R. M. Y., Wang, T., Liu, F., Zuo, X., He, J. & Chan, M., (2009). Vertically Stacked Silicon Nanowire Transistors Fabricated by Inductive Plasma Etching and Stress Limited Oxidation, *IEEE Electron Device Letters*, Vol. 30, No. 5, (May 2009), pp. 520-522, ISSN 0741-3106

Ng, R. M. Y., Wang, T., & Chan, M., A New Approach to Fabricate Vertically Stacked Single-Crystalline Silicon Nanowires, *Proceedings of 2007 IEEE EDSSC*, pp. 133-136, ISBN 978-1-4244-0637-1, Tainan, Taiwan, China, December 20-22, 2007

Pao, H. C. & Sah, C. T., (1966). Effects of diffusion current on characteristics of metal-oxide (insulator)-semiconductor transistors, *Solid-State Electronics*, Vol. 9, No.10, (September 2002), pp. 927-937, ISSN 0038-1101(66)90068-2

Ramanathan, K., Bangar, M. A., Yun, M., Chen, W., Myung, N. V., & Mulchandani, A., (2005). Bioaffinity Sensing Using Biologically Functionalized Conducting Polymer Nanowire, *J. AM. CHEM. SOC.*, Vol. 127, No.2, (December, 2004), pp.496-497, ISSN 10.1021

Singh, N., Buddharaju, K. D., Manhas, S. K., Agarwal, A., Rustagi, S. C., Lo, G. Q., Balasubramanian, N., & Kwong, D. L., (2008), Si, SiGe Nanowire Devices by Top-Down Technology and Their Applications, *IEEE TRANSACTIONS ON ELECTRON DEVICES*, VOL. 55, NO. 11, (July, 2008), pp. 3107- 3118, ISSN 0018-9383

Stern, E., Vacic, A., & Reed, M., (2008), Semiconducting nanowire field-effect transistor biomolecular sensors, *IEEE Transactions on Electron Devices*, Vol. 55, No.11, (Nov. 2008), pp. 3119-3130, ISSN 0018-9383

Synopsys, TCAD Sentaurus Device User's Manual , Mountain View, CA, 2005

Sze, S.M., Physics of Semiconductor Devices, 2nd ed. *Wiley*, ISBN 978-0-471-14323-9, New York, USA

Taur, Y., Liang, X., Wang, W., & Lu, H., (2004). A continuous, analytic drain-current model for DG MOSFETs, *IEEE Electron Device Letter*, Vol. 25, No. 2, (Feb. 2004), pp. 107–109, ISSN 0741-3106

Xiang, J., Lu, W., Hu, Y., Yan, H., & Lieber, C. M., (2006). Ge/Si nanowire heterostructures as high performance field-effect transistors, *Nature*, Vol.441, No.25, (May 2006), pp.489-493, ISSN 0028-0836

Yang, J., He, J., Liu, F., Zhang, L., Zhang, X., & Chan M., (2008). A compact model of silicon-based nanowire MOSFETs for circuit simulation and design, *IEEE Transactions on Electron Devices*, Vol. 55, No.11, (Nov. 2008), pp. 2898-2906, ISSN 0018-9383

Yu, B., Wang, L., Yuan, Y., Asbeck, P., & Taur, Y., (2008). Scaling of nanowire transistors, *IEEE Transactions on Electron Devices*, Vol. 55, No.11, (Nov. 2008), pp. 2846-2858, ISSN 0018-9383

Zhang L., He J., Zhang J., Liu F., Fu Y., Song Y., & Zhang X., (2008). An Analytic Model for Nanowire MOSFETs With Ge/Si Core/Shell Structure," IEEE Transactions on Electron Devices , Vol. 55, No.11, (Nov. 2008), pp. 2907-2917, ISSN 0018-9383

Zhang. L., Zhang. J., Liu. F., Chen. L., Xu. Y., Zhou. W., He, F., An analytic channel potential based model for dynamic depletion surrounding-gate mosfets with arbitrary doping level, 1st Asia Symposium on Quality Electronic Design, ASQED 2009, pp. 131-135, ISBN 978-1-4244-4952-1, Kuala Lumpur, Malaysia, 15-16 July 2009

Zhang, L., Guan, Y., Zhou, W., Chen, L., Xu, Y., & He, J., (2009). A carrier-based analytic drain current model incorporating velocity saturation for undoped surrounding-gate MOSFETs, *Semiconductor Science and Technology*, Vol. 24, No. 11, (Oct. 2009), pp. 115003-115010, ISSN 0268-1242

Zhang, Y. F., Tang, Y. H., Wang, N., Yu, D. P., Lee, C. S., Bello, I., & Lee, S. T., (1998). Silicon nanowires prepared by laser ablation at high temperature, *Applied Physics Letters*, Vol.72, No.15, (Apr 1998), pp.1835-1837, ISSN 0003-6951

Wang, J., "Device Physics and Simulation of Silicon Nanowire Transistors", PhD thesis, Purdue University, 2005

Part 4

Nanowire Fabrication

16

Obtaining Nanowires under Conditions of Electrodischarge Treatment

Dikusar Alexandr
Shevchenko State University, Tiraspol,
Pridnestrovie Institute of Applied Physics,
Academy of Science of Moldova, Chishinau,
Republic of Moldova

1. Introduction

At present various methods for obtaining nanowires and nanotubes are known using different materials. Nevertheless, the list of these methods grows constantly. This may be accounted for by the fact that, on the one hand, new methods for developing nanomaterials appear using both the technology of bottom-up and top-down. On the other hand, it becomes clear that nanowires and nanotubes can be manufactured using the methods and technologies that are known for a long time under certain conditions.

One such method is the electrodischarge treatment that is the basis for the electrodischarge machining, the method that was proposed more than 50 years ago by the spouses B.R. Lazarenko and N.I. Lazarenko .

This work describes the peculiarities of application of the electrodischarge machining-electrodischarge doping. The conditions for manufacturing nanowires are described along with certain mechanical properties of the surfaces that are developed by introducing the nanowires into the surface layer composition.

2. Electrodischarge machining (EDM) and its technological use

When a certain value of a critical voltage U_{cr} is applied across the interelectrode gap (IEG) that consists of two electrodes and is filled with a dielectric liquid (kerosene or deionized water) the electrical breakdown of the gap (i.e., the formation of the electroconducting region in this medium) is registered. The order of lifetime in this region is ~ 10^{-7} s (Fig. 1)

$$U_{cr} = l\, E_{cr}\,,$$

where E_{cr} is the critical value of the field intensity that induces the gap breakdown (a discharge); l is the distance between the electrodes.

Since both electrodes in the considered situation have a natural roughness, E_{cr} will be reached firstly at the points with a minimum interelectrode distance l_{min}.

The electron flow that forms on the cathode, evaporates and ionizes the liquid due to its motion to the counter electrode. By the moment the electron avalanche reaches the anode, this flow turns out to be separated from the environment (the liquid) by the vapor-gas-plasma cover. After the IEG breakdown the discharge channel tends to be wider and a shock wave is followed by forcing out the liquid in the radial direction with respect to the discharge channel axis. High pressure forms at a front of the shock wave. A certain part of the electric energy introduced into the IEG is transformed by the shock wave into the mechanical work of compression in the working medium. The channel radius is generally less than 10^{-1} mm, the duration of this part of the discharge is short, i.e., within a few microseconds the front moves away for such distances that the energy gain becomes insufficient to ionize the substance.

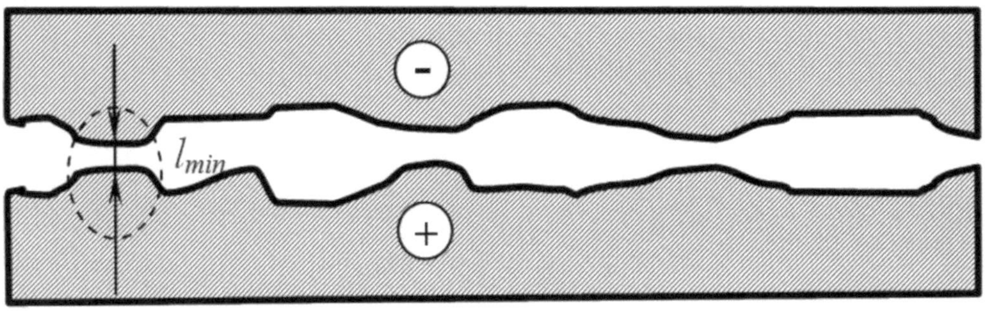

Fig. 1. Scheme of the electric discharge formation.

The EDM is usually characterized by the pulsed supply of the voltage, and during a single pulse the applied voltage changes from ~200 to 23 – 25 V, while the lifetime of the plasma channel that arises is up to 200 μs. Moreover, during the time of 10^{-6} - 10^{-7} s an abrupt increase in the electric current occurs, and the expanding front of the discharge wave increases the radius of the discharge channel. The energy densities within a single pulse reach 3 J/mm^2. The situation after the breakdown is referred to as *a spark form of a discharge*. It is characterized by the times of ~10^{-8} – 10^{-7} s, the current densities of 10^6 – 10^7 A/cm^2 and temperatures of 10^4 – 10^5 K. The high local temperature in the discharge channel ensures a possibility of phase transitions across both electrodes, since the obtained temperatures may exceed not only melting but also boiling temperatures.

The removal of the material from the surfaces of both electrodes results from the spark discharge. It concerns the anode in a greater degree, since the cathode melting, as a rule, takes more time versus that of the anode melting. The reason for this is that the electrons have a higher mobility and, hence, reaching a high temperature followed by melting and evaporation of part of the surface starts from the initial period of a pulse (during a few microseconds). The less mobile ions, unlike electrons, ensure the phase transitions across the cathode with a time delay.

After-the-breakdown stage is characterized by a sharp collapse of the discharge plasma channel and the formation of a gaseous bubble. The melted parts of the surface are removed from the surface of the electrodes and are transferred (in a solid state form) into the liquid. The radius of the formed cavities depends on the energy of a single pulse and ranges from 1μm to ~100 μm.

The rate of erosion is determined by the volume of a sum of cavities that are removed from the surface per the unit of time. The volume of a single dimple determines also the roughness of the surface after the treatment, which is formed by the overlap of single dimples. The erosion

causes an increase in the value of the local IEG (Fig. 1) and a transition of the discharge to another IEG point. In other words, the considered form of a discharge is a certain form of a non-stationary discharge and a local melting (and evaporation) of the electrode material is the basis of the electroerosion method of treatment that is most popular today.
The electroerosion treatment is performed under the pulse conditions. A pulse generator supplies the currents with several tens of amperes at a regular frequency in the range from the units to hundreds of kHz. An ejection of the melt from the zone of a spark discharge can occur both at the moment of the pulse supply and after its termination. Various hypotheses exist to account for the mechanism of the material removal from the zone of treatment, namely:

- A single ejection of the melt from the erosion dimple at a minimum pressure in the vapor-gas bubble that resulted from a single discharge;
- An ejection of the melt affected by the ponderomotive forces (a current pulse generates a strong magnetic field); the interaction between the vortex current and the magnetic field (that induced the latter) leads to arising the electrodynamic forces;
- Due to the presence of the pressure of the vapors of the materials evaporated from the surface;
- The emission of the products of destruction during the electroerosion treatment of brittle materials that results from the nonuniform thermal expansion of the material and arising thermal strains in the latter.

It is obvious that the EDM real process occurs under the conditions of a simultaneous effect of several factors that determine both the destruction and the emission of the destruction products from the discharge zone.
At present, the EDM serves the following purposes: a 3D copying, producing holes (including those of irregular shapes), treatment and a complicated-profile cutting using an electrode-wire, and the combined treatment (electroerosion polishing), etc. One form of the EDM is an electrospark doping (ESD) which is a process based on a polar transfer of the anode material onto the cathode under the conditions of a spark discharge in a gaseous phase

3. ESD – pulsed air arc deposition

Under the ESD conditions, both electrodes are eroded during the discharge pulse. For the case of the ESD, the anode is less than a cathode, and the cathode surface is treated by the anode (i.e., the anode material is transferred onto the cathode surface).
The basis of this process, just as that of the EDM, is the local melting (evaporation) of the anode material. However, since the transfer occurs in air medium, the surface coating always contains oxides, nitrides, carbides, etc.
The advantages of the EDS are the following:

- The possibility of using different materials in order to change the properties of a surface layer and participation of the interelectrode medium allow one to extensively modify the surface properties and to obtain hard, wear resistant, temperature-resistant, corrosion-resistant, antifriction, and decorative coatings, along with the repair and reconditioning the auto-workpieces;
- The method is simple for implementation and is comparatively cheap;
- The deposited layer has a strong cohesion with the substrate;
- The preliminary surface preparation is unnecessary.

At the same time, a relatively high roughness of the manufactured surface and the restrictions that result from the impossibility to produce fairly thick layers, restrict its more extended application.
In order to carry out a discharge in gaseous media, the RC-generators of pulses are commonly used (Fig. 2). A capacitor is charged using the current source through the ballast resistance R. As the electrodes TE (a vibrating TE is used in this method) and P (TE is the electrode-tool and P is the workpiece) move to contact, a breakdown of a gaseous gap occurs at a certain l_{min} length. Because of the polar effect, the transfer of the eroded material mostly from the anode onto cathode involves the formation of a site with certain physicochemical properties across the latter.

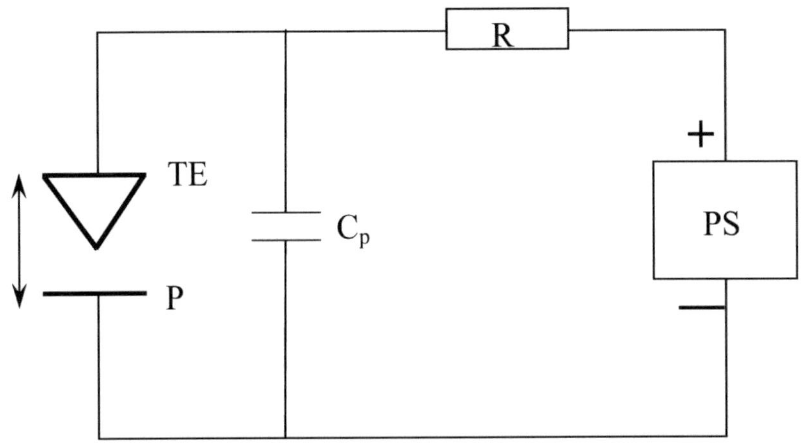

Fig. 2. Scheme of the simplest RC-generator.

As a rule, the surface layer of the cathode changes its composition and structure due to the ESD. The characteristics of this layer can be varied within the wide range due to a selection of the electrode material, composition of the interelectrode medium, parameters of the pulse discharges and other conditions when forming a layer on the cathode. It is obvious that the ESD ensures wide possibilities for creation of the working areas with specified operational characteristics.
The amount of the anode material that is transferred during a single act of the discharge is small. Thus, for a hard alloy (titanium-tungsten) at the discharge energy of 1 J it is 2 – 3 10^{-6} g. Therefore, in order to form a layer of a required thickness across the cathode, both a periodic commutation between the anode and cathode and a displacement (scanning) of the anode along the treated surface or a displacement of the cathode with respect to the fixed anode are necessary.
The periodic commutation of the cathode with anode is performed using various facilities, e.g., special vibrators, rotating disks or discs with the electrodes in the form of plates or small wires located along its perimeter which contact the cathode due to vibration, and the vertical feed of the automatic controller.
The ESD versions were developed to form the layer and perform the polar transfer using a powder material that was introduced into IEG . Here, the ESD advantages of using the compact electrodes are combined with the wide possibilities of the powder methods for the coating deposition.

3.1 Dynamics of transfer of the electrode materials at EDS

In the case of the compact materials used as the anode, the most popular variant of treatment is one at which the commutation between the anode and cathode is possible due to vibration. The processes are performed at U ~15--200 V, the pulse duration is in the range from the tens of microseconds to milliseconds, the frequency of vibration is of 50--300 Hz, and the amplitude is up to 0.2—0,5 mm. The breakdown of the interelectrode gap at the indicated voltages can occur at the distances that equal ~ 0.01 - 10 μm. Taking the frequency and amplitude of vibration into account, the time of passing the indicated distances by the anode is from several to the tens of microseconds. Hence, the discharge can occur completely in the gaseous phase and it can stop upon the contact of the electrodes. At U < 100 V the discharge develops and terminates actually upon the contact of the electrodes.

In 10^{-7} - 10^{-8} s after the breakdown and the beginning of formation of a discharge channel (a plasma jet of the discharge), the evaporation from the surface of the electrodes in the form of jets and vapors and the ejection of the liquid phase by means of dispersion starts. Since these phenomena take place in a fairly small interelectrode gap that in addition decreases continuously, favorable conditions are created for the transfer of the flow of energy to the counter electrodes.

Upon the current pulse of certain duration, the electrodes manage to approach each other almost to a full contact before the discharge termination. But the full contact apparently fails to occur between the anode and cathode, since the pressure of the vapors of the metals in the evaporation zone can reach 10^8 Pa which exceeds considerably the pressure that is developed by the electromagnetic system of the vibrator in the contact zone.

The liquid volumes of the approached anode and cathode are exposed to the effect of several forces: a hydrodynamic pressure of the flames, a gas-kinetic pressure of the discharge channel, and the electrodynamic force, etc. The volumes of the liquid metal are distorted under the total effect of those forces and eject from the dimple. Since this takes place during the contact, the integration of the liquid phases of the electrode materials occurs along with their convective mixing.

Due to the polar effect and the aforementioned factors, the quantity of the liquid phase across the anode must be substantially higher compared to that of the cathode and, hence, the surface layer that was formed on the cathode must consist mainly of the anode material. But the convective mixing is responsible for the fact that a fairly great amount of the material of the cathode is also distributed in this layer. In addition, it is noteworthy that the treatment takes place in a gaseous medium that comprises the elements which can form the chemical compounds (oxides and nitrides) that determine the surface layer.

3.2 Effect of various conditions on the formation of a surface layer on the cathode

The formation of the ESD surface layer is performed by a successive local exposure to the pulse discharge of all sites of the treated cathode. As a rule, the required characteristics of the layer can be achieved by a repeated travel of the anode over one site of the cathode. In most case, in order to obtain a uniform layer over the entire treated surface, a constant shift of the anode with respect to the zone of interaction of the discharge with the cathode is necessary. This shift is usually selected experimentally.

The quantity of the transferred material onto the cathode is generally fixed in the form of a change in the cathode weight. The weight change in the cathode during 1 min upon the treatment of 1 cm^2 of the surface is generally referred to as a specific gain. It is actually a characteristic of the intensity of the ESD process. A detailed study of the formation of the layer across the cathode and the anode erosion under different conditions of treatment showed that the effect of the following factors is most significant, namely, that of the parameters of the pulse discharge, the duration of the treatment, the nature of the electrode materials, the interelectrode medium and a form of the anode motion with respect to the cathode.

The dynamics of the formation of the surface layers is characterized by the fact that the intensity of the anode material transferred to the cathode is found to be the highest at the initial moments of the process, but then it decreases. Eventually, the weight gain of the cathode may change for the inverse process, i.e., its erosion ("a negative gain"). The combination of these two processes, at a fairly high share of the latter, may lead to a restriction in the thickness of the coatings that is really observed in many cases. Usually, in the range of the discharge energies of 0.1 - 3 J the treatment of a 1 cm^2 surface during 0.5 - 2 min yields a maximum (or close to it) value of the cathode gain.

There are various ways to increase the rates of deposition and thicknesses of the deposited layers. One such is the use of the rotating electrode instead of the vibratiory one. In the latter case, a position of the discharge channel and a zone of interaction of the electrodes during the discharge, shift in the direction of motion of the anode and the erosion trace widens along the cathode surface. This leads to the change of the thermal mode of the treatment which, as a result, affects the obtained thickness of the coating. In this case, the thicknesses of the coatings may reach 1 - 2 mm, which significantly exceeds the thicknesses developed under conditions of the vibratory electrode. However, the surface roughness may also increase.

The restriction of the thicknesses of the coatings results from the dynamics of changes in the values of the remaining strains in the manufactured coatings. The obtained results show that with an increase in the specific duration of the alloying, the level of the stretching remaining strains in the developed layers increases. However, definite values of a specific duration of the alloying exist for each kind of material at which the maximum level of the stretching remaining strains is observed.

Study of the effect of a composition of the gaseous environment showed that its change not only allows one to control the deposition rate, but also the composition and structure of the developed layers. This was manifested most vividly during the treatment in the reducing media (hydrogen and argon).

In the process of developing the surface layer on the cathode, between the liquid phases of the electrode material there occurs interaction which leads to establishing chemical bonds between the components and to the formation of intermetallic compounds and alloys, as well as the development of the process of self- and heterodiffusion.

Under these situations of the materials interaction the processes of crystallization, mass transfer and other phenomena occur under extremely nonequilibrium conditions that result in formation of nano- and fine-crystalline structures up to the formation of the amorphous structures.

4. The ESD by the Al-Sn alloy electrode-tool

4.1 The ESD installation and the TE material

The power for the ESD was supplied using an ALIER-31 installation (SCINTI, Moldova). The ALIER type electrospark power supplies are successfully used for various types of the electrospark deposition of coatings. A specific feature of this installation is that the frequency of the generated pulses is not directly related to the TE vibration frequency; it is set independently. Thus, it depends on the pulse energy. Table 1 lists the parameters of the technological pulses of the ALIER-31 generator.

In this study, we used all seven modes of the ALIER-31 installation (Table 1) at a constant treatment time of 1 min; the frequency of the set pulses corresponded to that shown in Fig. 3. This was achieved by means of a special frequency regulator (energy coefficient). Here, it should be taken into account that, since we used an installation with a manual TE, the real number of pulses in a gap depended on the operator's hand and on the conditions in it. The number of pulses in a gap can be assumed to be 0.6 - 0.9 from the values corresponding to those presented in Fig. 3.

As a TE, we used the rods from a specially prepared Al–Sn alloy doped with copper and titanium (~1 wt % Cu and 0.1 wt % Ti) with a diameter of ~8 mm.

No.	Mode	Pulse duration, μs (±10%)	Pulse current amplitude, A (± 20%)	Pulse energy, J
1	1	16	125	0.036
2	2	31	125	0.07
3	3	62	175	0.2
4	4	125	175	0.39
5	5	250	175	0.79
6	6	500	175	1.58
7	7	1000	175	3.15

Table 1. Parameters of the technological pulses of the generator of the ALIER-31 installation

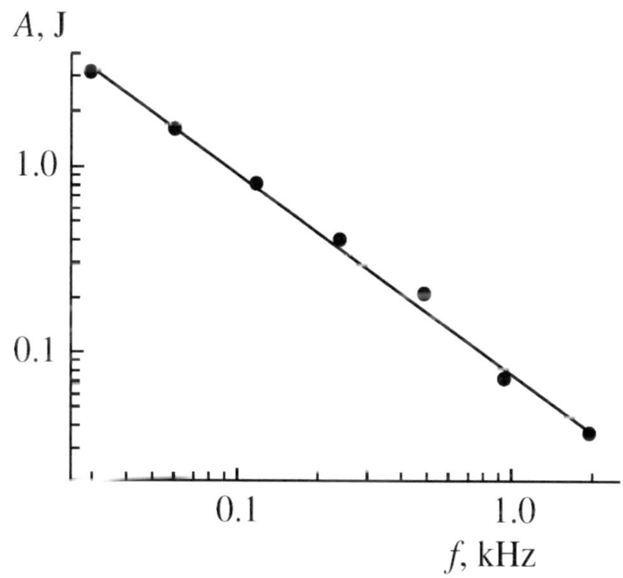

Fig. 3. Frequencies and values of the pulse energies in various modes of the electrodischarge treatment.

The alloy of the required chemical composition was melted in a graphite melting pot in the inductor of a V4I10U high-frequency installation; then, it was poured into a specially prepared chill mould with a size of 8.50 mm. The procedure of the manufacturing consisted of the following operations: (a) preparation of the working mixture; (b) melting in an induction furnace; (c) pouring the melt into a chill mold; and (d) topping, clearing, and turning.
In order to obtain an alloy with a preset composition, we used pure aluminum and tin. The doping components were introduced in the form of foundry alloys (50% Al + 50% Cu and 90% Al + 10% Ti). The working mixture was calculated with respect to the mean content of the elements: 20 wt % Sn, 1 wt % Cu, 0.1 wt % Ti, and the rest was Al. As a sample, we used a D1 aluminum alloy (State Standard GOST 4784). Manual treatment was carried out. The TE and the sample were weighed before and after the treatment in each experiment. Their surface (before and after the treatment) was studied by means of a scanning electron microscopy (a TESCAN scanning electron microscope with an INCA Energy EDX detachable device for the element analysis of the surface (Oxford, Great Britain)).

4.2 Composition and structure of the TE applied

Figure 4 shows a diagram of the state for the Al–Sn binary system. One can see that at room temperature (up to the tin melting point of 228° C), the material used as a TE (AlSn20) must be an aluminum matrix with tin metal dispersed in it. This is confirmed by the results of the scanning electron microscopy and the EDX element analysis (Fig. 5), as well as by the sample surface scanning with the simultaneous determination of the aluminum and tin (Fig. 6). One can see that the TE is really an aluminum matrix with tin particles with a size of 3 – 5 µm dispersed in it (Fig. 6).

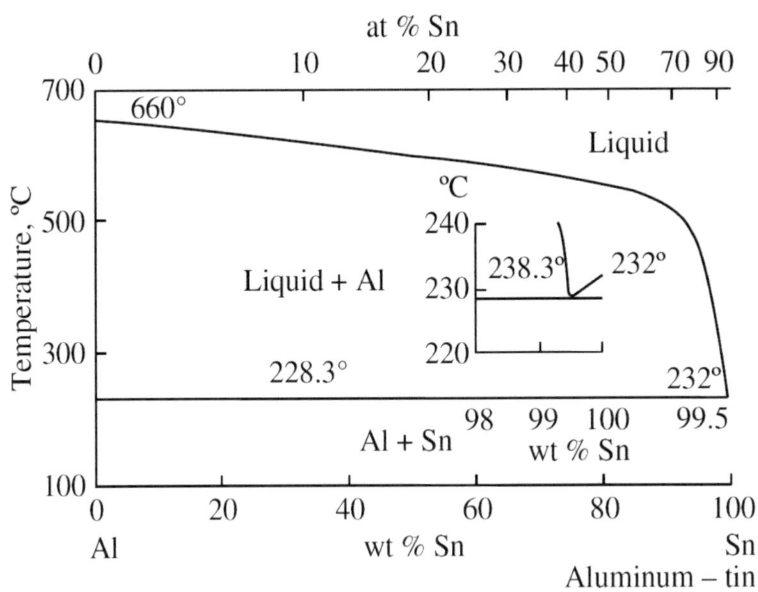

Fig. 4. State diagram for the aluminum–tin system.

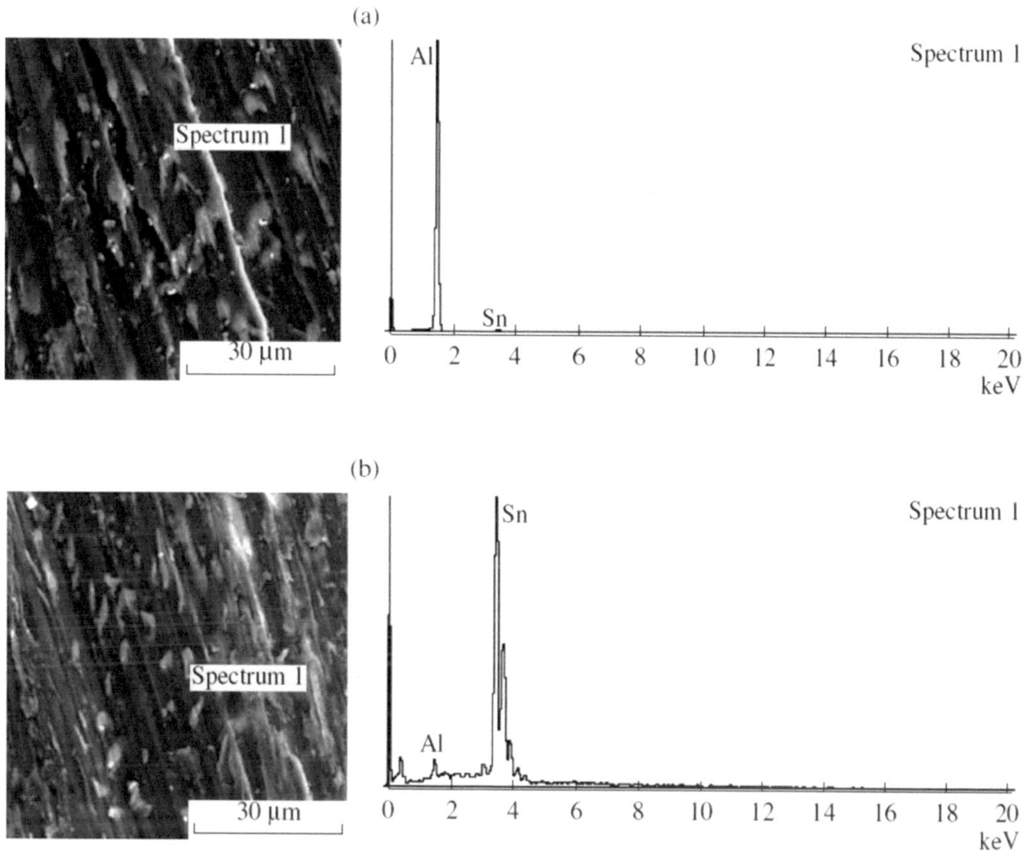

Fig. 5. SEM images of the TE surface and the EDX spectra of the matrix (a) and the dispersed tin (b). The crosses in the microphotographs show the places of the local EDX analysis.

4.3 Mass transfer effects

Figure 7 shows changes in the weight in the process of the experiments both for the sample and for the TE. It is seen that, at relatively low pulse energies (and, respectively, high frequencies, see Fig. 3), a loss in the sample mass occurs, which increases as the pulse energy grows. This is region I in Fig.7. In this region, a loss in the TE mass is also observed; however, it is relatively small. It is obvious that region I is the region characteristic of electroerosion treatment in an air medium, which is characterized by the removal of substance from the work piece surface (that increases with the growing pulse energy) and by the low wear of the TE. As the pulse energy increased, the situation varied significantly. At relatively high energies, a polar transfer on the sample surface was observed; it was accompanied by weight gain. The transfer coefficient (the ratio of the gained weight to the weight of the substance removed from the TE) was ~0.3. This region (II in Fig. 7) is the ESD classical variant with respect to the TE material used. In the intermediate region, considerable wear of the TE was observed at relatively low material transfer on the sample surface (Fig. 7).

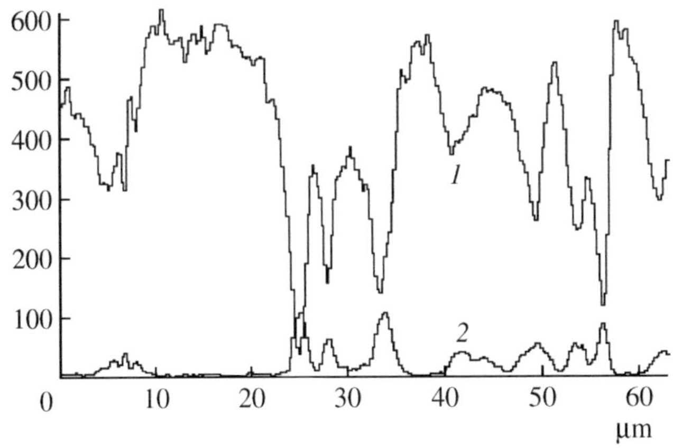

Fig. 6. Distribution of the aluminum (1) and tin (2) over the TE surface. The levels of the EDX spectrum are given in relative units.

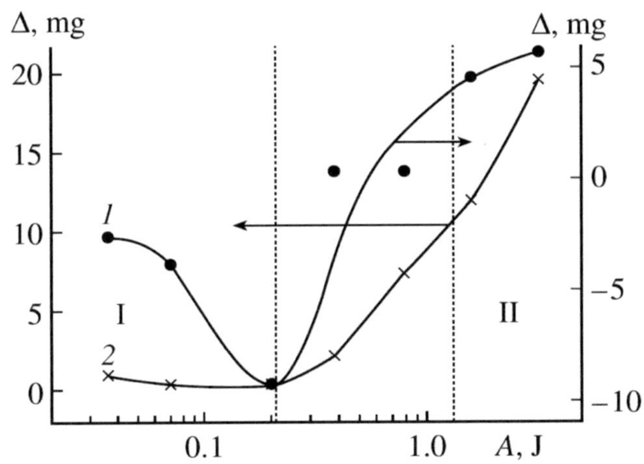

Fig. 7. Influence of the pulse energy on the variation of the weight of the sample (1) and the TE (2).

4.4 Morphology and composition of the surface

Figure 8 shows the surfaces of the sample and the TE after treatment in the modes corresponding to region I in Fig. 7 and to the intermediate region. It is clear from the results of element analysis (Table 2) that they are a mixture of oxides and nitrides of aluminum (the sample) as well as of oxides of aluminum, tin, and copper (the TE). The presented photographs of the surface and the analysis results are typical for treatment in modes 1–5. Here, the composition of the surfaces could change insignificantly. In particular, on the sample surfaces in modes 2–5, there was tin (a trace amount); sometimes, copper. The composition of the TE and the sample comprised carbon. This can be caused by the fact that the TE melting was performed in a graphite melting pot. The morphology of the surfaces obtained in these conditions is characteristic of the fritted surfaces.

Element	Al	Sn	O	N	C
wt %	50.0 ± 4.8	34.8 ± 2.6	7.7 ± 1.5	2.2 ± 0.6	5.3 ± 2.6

Table 2. Composition of the sample's surfaces after treatment in modes 6 and 7 (mean values)

Element	Al	Sn	O	N	C
wt %	19.1	61.5	11.5	4.9	3.0

Table 3. Element composition of the wire on the TE after treatment in mode 6 (see Fig. 9a)

The given results confirm the results of the measurements of the loss in mass; it follows from the latter that region I is the region of the electroerosion treatment in an air medium. A similar situation is observed for the samples treated in modes 4 and 5 (see Table 1, the region of intermediate modes shown in Fig. 7).

As to region II in Fig.7, alongside with the composition and morphology of the surface characteristic of region I and the intermediate region, on the surfaces of the samples and the TE treated in these modes, specific structures in the form of wires with a diameter less than 1 μm are registered (Fig. 9). Figure 10 shows nanowires at higher magnification. It is seen that their diameter ranges within ~200–600 nm. Table 3 presents the composition of the samples surface after treatment in these conditions. One can see that, in comparison with the other modes, the tin concentration on the surfaces of the samples sharply increases (Fig. 11). The analysis of the wire's composition shows that this is tin for the most part (Table 3, Fig. 9a). Table 3 gives the results of the element analysis of a large fragment of the wire on the TE; it follows from the data that the tin concentration in the wires is more than 60 wt % (also see the EDX spectrum in Fig.9a).

However, it is impossible to answer the question whether they consist exclusively of tin (its oxides) or comprise aluminum and its oxides as well (yet in a less amount than tin), because the analysis technique applied records the ratio of the components in volumes that exceed the volume of only the wires, and it partially includes the volumes of the surface layers on which these wires locate.

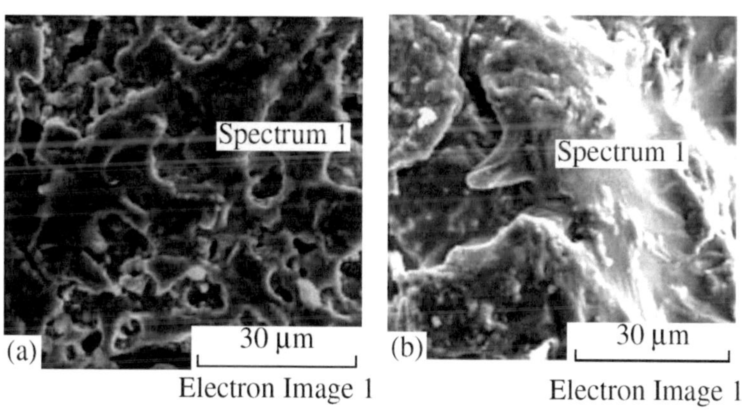

Fig. 8. Morphology of the surface of the sample (a) and the TE (b) after treatment in modes 1 (a) and 4 (b).

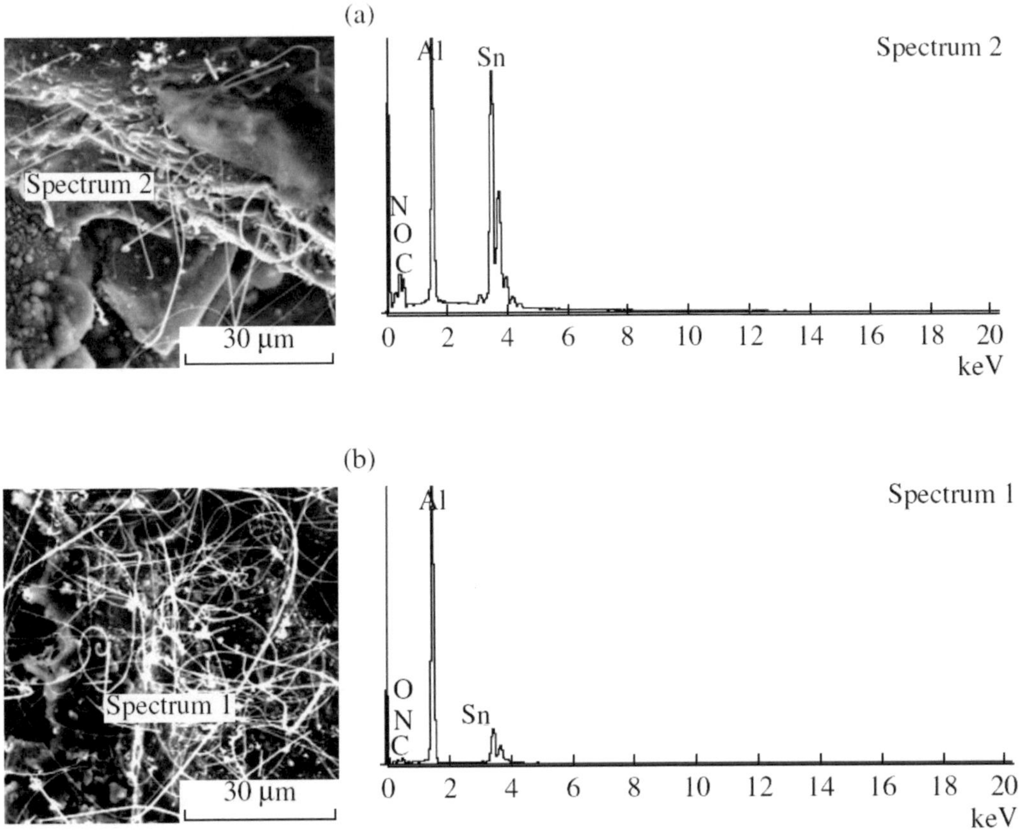

Fig. 9. Morphology of the surface and the EDX spectra of the TE (a) and the sample (b) after treatment in modes 6 (a) and 7 (b).

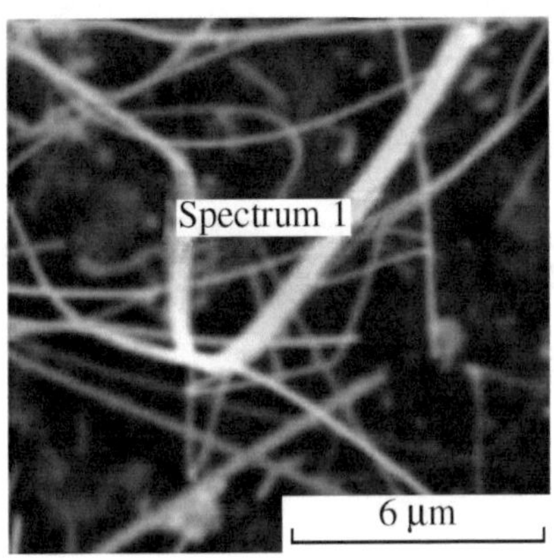

Fig. 10. Surface of the sample after treatment in mode 7 (fragment of Fig. 9b).

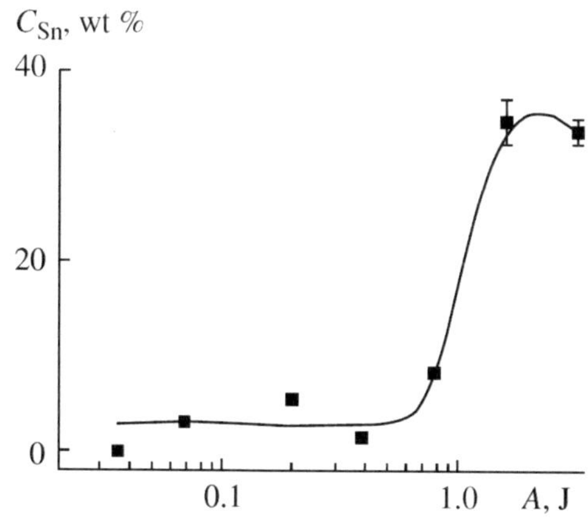

Fig. 11. Influence of the pulse energy on the tin concentration on the sample surfaces.

4.5 Physical bases of obtaining the nanowires.

The cause for the formation of the nanowires is apparently the specific character of the Al–Sn state diagram at temperatures that are higher than the tin melting point (228 °C) but lower than the melting point of the Al–Sn alloy (~655° C for the AlSn20 alloy) (Fig. 4). In this case, the system represents melted particles of dispersed tin being in a solid matrix of aluminum (Fig. 4 - 6). The transfer of these particles to the interelectrode gap occurs due to the ponderomotive forces that deform the surface of a melted drop if the surface tension force of the melt–air system is sufficiently low for the melted particles. It is known, that minimal values of the surface tension forces are observed precisely for the tin melt–air system. It is obvious that, since the surface tension forces of the melted particles of tin (or tin partially enriched in aluminum and their oxides are low, the ponderomotive forces, which appear due to the presence of a field in the gap, exceed them so that, as a result, wires with a diameter of ~1 μm and less are formed (Fig 9, 10).

The effects of this kind must be observed not only for the Al–Sn systems but also for any other systems that, at certain temperatures, are a system of melted particles in a solid matrix, for example, the Al–Pb system.

As to the possible practical applications that follow from the results of the experiments that have been described, we should mention the possibility of increasing the concentration of a low-melting component in the near-surface layer under the ESD conditions with the use of a TE of this kind (Fig. 11).

5. Mechanical properties of the surfaces obtained after the ESD by the Al-Sn alloy

5.1 Wear resistance of the coatings after the ESD by the Al-Sn alloy

The study of the mechanical properties of the surfaces developed with the formation of nanowires from the easily melted component was carried out using a see-saw friction machine (Institute of the Applied Physics, Moldova Academy of Sciences). The rectangular 3×5×55-mm plates from aluminum alloy D1 (State Standard GOST 4784) were used as the specimens to be studied (the surface with a smaller size contacted a counterbody). The counterbody was a rectangular 3×25×30-mm plate from the hardened steel St 45 with

microhardness of 650 ± 50 kgf/mm^2. The surface of a smaller size contacted with the surface under study, thus, the area of a contact was 9 mm^2 and the counterbody was fixed at an angle of 90° with respect to the exposed specimens.
The counterbody performed a reciprocable movement relatively the specimen under study at a rate of 45 double movements a minute. The length of the working surface that contacted the conterbody was 48 mm. The initial microhardness of the unfinished surface of the alloy D1 was about 100 kgf/mm^2. A vaseline oil was used during the testing friction process.
The testing comprised two stages, since, as a rule, the surfaces with high roughness result from the ESD. At the first stage during 10 h the surface under study and the counterbody were exposed to a preliminary wear-in. It consisted in a successive increase in the load from 20 to 100 kgf/cm^2; at the beginning of the work it was every 1 h of work and then every 2 h of work. At the second stage, the major testing process took place; it was performed for 20 h at a load of 100 kgf/cm^2.
Prior to the testing, the roughness of the specimen under study was measured after the ESD (R_a^0). Similar measurements were taken after the stage of the preliminary wear-in (R_a^I), as well as after the termination of the testing (R_a^{II}). The weight loss measurements were performed both for the counterbody (ΔU^{cb}) and for the tested specimens (ΔU). The latter was a sum of losses after the preliminary wear-in and after the main tests at a fixed load. A degree of the wear was estimated both in the absolute (ΔU, ΔU^{cb}) and in the relative ($\Delta U^{cb}/\Delta U$) values.
The ESD of the tested specimens was carried out using an ALIER-31 installation (SCINTI, Moldova) in the modes of 4 and 6. A wear degree was estimated compared both to the unfinished and the treated surfaces, but with the (aluminum) electrode which lacked the easily melted component (Sn). The use of the aluminum electrode as the TE, appeared to be possible (provided the treated specimens gained in weight) only in mode 4 (in more intense modes, the welding of the TE to the specimen was registered and the EDS was impossible). Therefore, the results of comparison of AlSn20 used as the TE both in mode 4 and mode 6 are listed in Table 4).

№	Material TE/P	Mode	R_a^0, µm	R_a^I, µm	R_a^{II}, µm	ΔU, mg	ΔU^{cb}, mg	$\Delta U^{cb}/\Delta U$
1	-/Al	-	0.72±0.08	0.32±0.21	0.25±0.15	137.3	0.4	$2\cdot10^{-2}$
2	-/Al	-	0.52±0.25	0.17±0.08	0.20±0.10	114.3	4.3	
3	Al/Al	4	26.0±4.6	11.0±1.7	7.6±1.7	22.4	1.5	$7\cdot10^{-2}$
4	Al-Sn/Al	4	13.1±2.4	6.5±0.9	5.9±1.0	12.1	9.35	0.8
5	Al-Sn/Al	4	14.2±2.6	7.5±2.0	7.8±1.3	2.6	5.95	2.3
6	Al-Sn/Al	4	13.1±1.2	7.6±1.6	7.8±1.0	5.1	5.95	1.2
7	Al-Sn/Al	4	10.7±1.6	8.5±1.3	8.8±1.0	0.6	6.7	11.2
8	Al-Sn/Al	4	15.1±3.4	8.6±1.0	5.8±1.1	7.6	32.0	4.2
9	Al-Sn/Al	4	18.9±1.5	10.9±1.5	7.5±1.5	9.1	37.0	4.1
10	Al-Sn/Al	6	17.9±1.5	12.9±2.2	12.1±2.0	4.1	50.8	12.4
11	Al-Sn/Steel	4	5.90±0.85	2.92±0.79	2.62±0.61	3.3	2.85	0.9
12	Al-Sn/Al	6	10.7±2.4	9.4±3.2	10.5±2.6	0.3	1.03	3.4
13	Al-Sn/Al	6	7.3±1.2	7.3±1.5	7.0±1.0	0.25	0.9	3.6

Table 4. Results of the wear tests of the surfaces

The ESD process was performed manually. As was aforementioned. the ESD indices depend strongly on the method for the coating deposition. Two methods were used in the considered experiments. The first experiment consisted in the successive coating of the entire treated surface during 4 min.. i.e.. first. one part. then another. etc. In the second case. the whole surface was coated during 1 min and then the following treatment was performed. The overall time of treatment was 4 min.

The surface roughness after the ESD in the first variant of treatment was significantly lower (R_a = 12.8 ± 1.3 μm). In the second case it was 17.3 ± 2.0 μm. and during the treatment using the Al TE it was 26.0 ± 4.5 μm (test 3. Table 4).

The EDS was performed not only on the specimens from the aluminum alloys D1. but also on steel St 45. In this case the roughness was considerably lower (R_a = 5. 90 ± 0. 85 μm (test 11. Table 4).

In the process of the preliminary wear-in. R_a decreased and. as a rule. the changes in the roughness in the process of the testing were not registered (Table 4). In certain experiments. during the treatment in mode 6. i.e.. under the conditions when the maximum roughness of the treated surface was observed. the latter was exposed to a mechanical treatment (polishing) before the testing. About a 0.2 mm-alloyed layer was removed (tests 12. 13. Table 4). In this case the roughness of the surface remained unchanged both in the preliminary wear-in and the main processes.

The results of Table 4 show that unlike the untinished surface and the surface treated only by Al for which $\Delta U^{cb}/\Delta U << 1$. $\Delta U^{cb}/\Delta U > 1$ resulted from the treatment by the AlSn20 alloy and only for the surfaces developed in mode 6 (i.e.. according to the results with an increased Sn concentration in a coating presented in part 4). the wear of the counterbody is higher by more than an order of magnitude than that of the tested surface (Table 5. Figs. 12. 13).

Even after the mechanical removal of a part of the treated surface (tests 12. 13. Table 4) this value is 3.5 ± 0.1. i.e.. the counterbody wear exceeds the wear of the tested surface (0.25--0.3 mg during the entire testing time) by a few times.

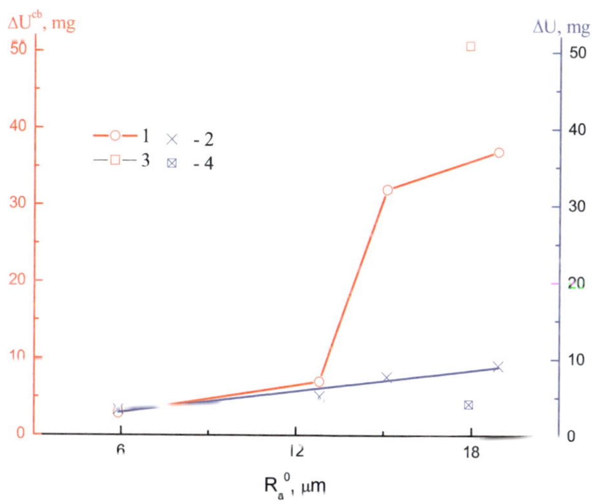

Fig. 12. Dependence of the absolute wear of the counterbody (1) and the treated surface (2) on the surface roughness. developed after the ESD in modes 4 and 6 (3 is the counerbody wear. 4 is the wear of the treated surface).

It is evident that the wear of both the counterbody and that of the treated surface depend on its roughness (Fig. 12). However. upon the treatment by the Al electrode (without Sn) even in the presence of a considerably greater initial roughness of the specimen surface (prior to the tests) (test 3. Table 4) the opposite picture is observed. according to which the wear of the tested surface exceeds that of the counterbody by more than an order of magnitude (Fig. 12). It is clear that the only reason for the increased wear resistance of the ESD treated surfaces is the Sn-dopant introduced into their composition and which under the indicated conditions passes onto the treated surface also in the form of nanowires.

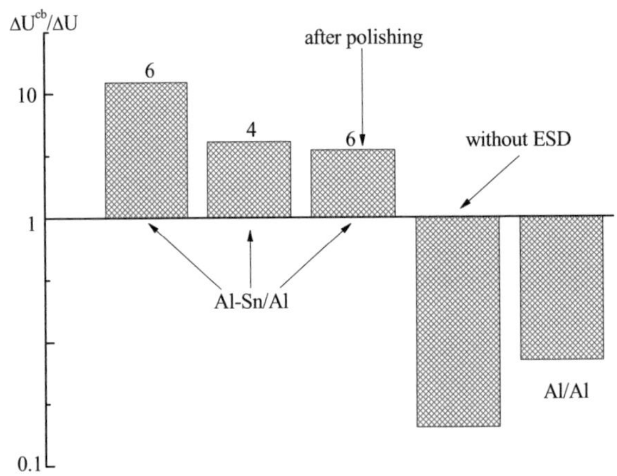

Fig. 13. Relative wear of the surfaces.

5.2 On the causes of the increase in wear resistance of the surfaces treated by the Al-Sn alloy

It seems evident that the increase in the wear resistance of the surfaces that have been treated by the Al-Sn alloy results precisely from tin presence in the TE. since the ESD using pure Al fails to lead to such effects. However. the Sn hardness is very low and the enhanced wear resistance can occur only due to the formation of certain compounds of Sn (or Al) in the surface layer. The appropriate oxides (Al_2O_3 and SnO_2) are characterized by a considerably higher hardness. However. the element analysis of the coatings prior to and after the wear tests showed that the oxygen content in the surface layer (in at. %) is considerably lower than a total amount of Al and Sn (on the average it was ~12--20 at. % of oxygen. while the amount of Al and Sn is always above 50 at. %). This is indicative of the fact that. if the above oxides are present. their concentration is relatively low.

Special XRD measurements. prior to and after the tests for wear showed that the coatings are amorphous (or the concentration of the crystalline formations in the coatings was out of the bounds of the XRD measurements). In other words. no relevant oxides. carbides. or nitrides. etc.. were found in the coatings.

Figure 14 shows the surface morphology and the EDX-spectrum of the surface after the wear tests (test 12. Table 4). A peculiarity of this test was that after the ESD the surface (prior to the wear tests) was exposed to a mechanic polishing with an ~200 µm-thick layer being removed. This has led to an abrupt decrease in an absolute wear. while a relative wear of the counterbody also increased by several times compared to the surface wear that was subjected to the ESD (Table 4).

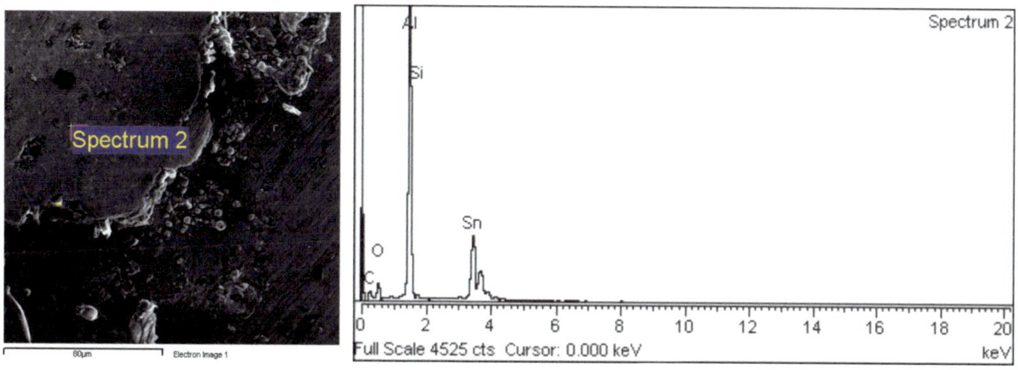

Fig. 14. SEM images of the specimen surface after the treatment and testing (test 12). On the right-hand side the spectrum of the EDX analysis is shown at a point that corresponds to the upper left-hand corner of Spectrum 2.

As is seen from the results in Figs. 14 and 15. in the area of the contact with the counterbody (where the element composition was measured) (Fig. 14). the Sn content is fairly high (wt %): Al is 39.2. Sn is 36.9. and O is 12.8. The test results showed that the counterbody wear (with no changes in the roughness both after the preliminary wear and fundamental tests) exceeded by 3.5 times the wear of the coating (Table 4).

It is of special interest that a classical method could be used to measure microhardness in the contact areas (both after the polishing and testing). It was similar to the microhardness of substrate (~100 kgf/mm^2). In other words. it seems evident that the only cause for the wear resistance increase in the coating compared to the counterbody is the presence of the Sn (or SnO_2) particles in it with such a degree of dispersion which fails to be registered by the classical XRD method. The most probable cause is the formation of the SnO_2 (or some other tin compounds) nanowires in the coating. Since they cannot form a uniform coating. their local effect facilitates the counterbody wear. but they cannot be detected simultaneously using the classical identification methods.

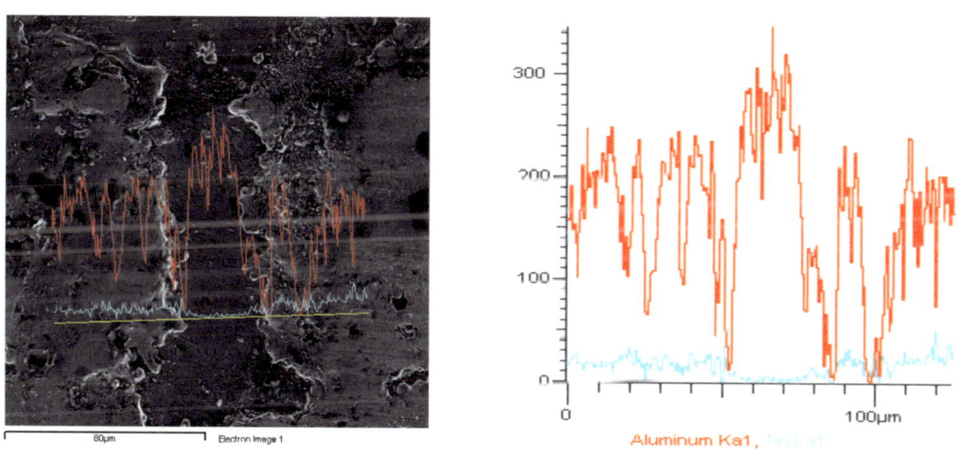

Fig. 15. Distribution of aluminum and tin on the treated and tested (test 12) surface. The EDX results are presented in the arbitrary units. On the left-hand side are the SEM images of the analyzed surface.

6. Conclusions

The presented results have shown that the conditions of the electrodischarge treatment using the tool-electrodes that represent a mechanical mixture of an easily melting component in the matrix from a component with higher temperature of melting. facilitate the formation of nanowires from the compounds of an easily melting component (including the element of the environment). It is obvious. that these possibilities do not confine only to describing the ESD using the Al-Sn electrode. The method under consideration for obtaining the nanowires may be further developed both in the direction of a research of new double systems for the manufacturing of electrodes and in the direction of the work under conditions of the controlled medium where the electrical discharge occurs.

This work shows that the functional properties of the coatings obtained using the above method can have an extensive practical use. It is clear that not only the mechanical properties of these coatings may be of interest. The research in this direction is at the initial stage and the spheres of efficient application for such coatings have yet to be studied.

7. Acknowledgements

This study was performed under the financial support of the Moldova Academy of Sciences (the project: Electrophysicochemical Methods of New Materials and Coatings with Improved Functional Properties Manufacturing and Treatment) and Shevchenko Pridnestrov'e State University.

The author thanks V.I. Yurchenko and V.M. Fomichev for the assistance in performing the EDS process. and also V.I. Agafii and V.I. Petrenko for carrying out the mechanical tests.

8. References

Lazarenko. B.R.. Lazarenko N.I. (1964) *Electrospark Machining of Metals*. B.R. Lazarenko (Ed.). Consulting Bureau. New York. USA.

DiBitonto. D.D.. Eubank. P.T.. Patel. M.R.. Barruffet. M.A. (1989) Theoretical Models of Electrical Discharge Machining Process. I. A Simple Cathode Erosion Model. *J. Appl. Phys*. Vol. 66. No 9 pp. 4095-4103; Patel. M..R.. Barruffet. M.A.. Eubank. P.T.. II. The Anode Erosion Model. pp. 4104-4111.

Samsonov. G.V.. Verchoturov. A.D.. Bovkun. G.A.. Sitchev. V.S. (1976) *Spark Discharge Doping*. Nauka. Kiev. USSR.

Gitlevich. A.E.. Mihailov. V.V.. Parkansky. N.Y.. Revuzky. V.N. (1985) *Electric Spark Alloying of Metal Surfaces*. Stiintsa. Kishinev. USSR

Parkansky. N.Y.. Boxman. R.L.. Goldsmith. S. (1993) Development and Application of Pulsed-Air-Arc Deposition. *Surf. Coat. Techn*. Vol. 6. pp. 268-273.

Yurchenko. V.I.. Yurchenko. E.V.. Fomichev. V.M.. Baranov. S.A.. Dikusar. A.I. (2009) Obtaining of Nanowires in Conditions of Electrodischarge Treatment with Al-Sn Alloy. *Surf. Engineering Appl. Electrochem*. Vol. 45. No 4. pp.259-264.

Vol. A.E. (1959) *Structure and Properties of Binary Metal Systems*. N.V.Ageev (Ed.) Vol. 1. Fizmatgiz. Moscow. USSR.

The Selective Growth of Silicon Nanowires and Their Optical Activation

Lingling Ren, Hongmei Li and Liandi Ma
National Institute of Metrology, Beijing, China

1. Introduction

1.1 The selective growth of silicon nanowires via vapor-liquid-solid mechanism

Compared with bulk semiconductors, 1D semiconducting nanowires possess some very unique properties such as quantum confinement effects, surface sensitivity, intrinsically miniaturized dimensions, and low leakage currents which make them attractive as building blocks for functional nanosystems and next generation electronics (Khanal et al., 2007; Andersen et al., 2007; Tilke et al., 2003, as cited in Wu, 2008; Huang et al., 2010). This can be inferred from the sharply increasing number of publications in this field. Figure 1.1 shows the number of publications on nanowires or nanowhiskers by year, determined from a CAS SciFinder search (Wang et al., 2006). Especially for Silicon Nanowires, there are more than 700 articles published in 2008, which is twice the number published in 2005 (Schmidt et al., 2010).

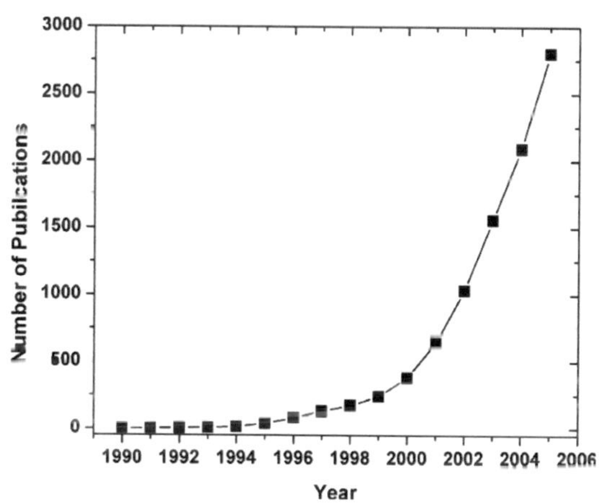

Fig. 1.1 Number of publications on Nanowires or Nanowhiskers by year, determined from CAS SciFinder search.

The earliest silicon wires were produced in the late 1950s as silicon whiskers (Treuting & Arnold, 1957, as cited in Schmidt et al., 2010). Nowadays, the term whisker has been almost displaced by the term wire and nanowire. Rodlike crystals with a diameter of less than 100nm will be referred to as nanowires while the term wire is used to the rodlike crystals of larger diameters (more than 100nm).

1.2 Silicon nanowire synthesis based on Vapor-Liquid-Solid (VLS) mechanism

1.2.1 VLS mechanism

Over the past several years, great efforts have been placed on the bulk synthesis of one-dimensional nanoscale materials, and various synthesis methods, such as chemical vapor deposition (CVD) (Amelinckx et al., 1994, as cited in Huang et al., 2010), arc discharge, laser ablation (Thess et al., 1994; Morales & Lieber, 1998; Yu et al., 1998a; Yudasaka et al., 1997, as cited in Huang et al., 2010), template-assisted growth (Dai et al., 1995; Han et al. 1997a, 1997b, 1998, 1999; Zhang et al., 2000, 2001, as cited in Huang et al., 2010), physical evaporation (PE) (Yu et al., 1997b; Zhang et al., 1999; Zhang et al., 2000, as cited in Huang et al., 2010) and lithography (Giovine et al., 2001, as cited in Huang et al., 2010) have been exploited. The most prominent method for silicon nanowire synthesis is the VLS growth mechanism, which is firstly proposed in March 1964 by Wagner and Ellis.

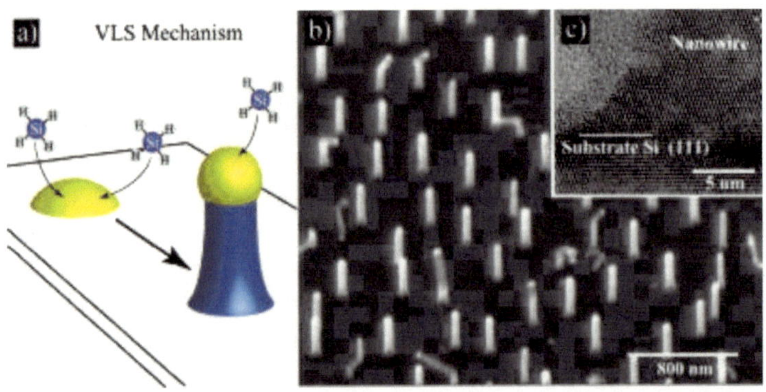

Fig. 1.2 (a) Schematics of the vapor-liquid-solid growth mechanism. (b) Scanning electron micrograph of epitaxially grown Si nanowires on Si <111>. (c) Transmission electron micrograph of the interface region between Si nanowire and substrate. Note the epitaxy and the curved shape of the nanowire flank.

The name VLS mechanism reflects the pathway of Si, which coming from the vapor phase diffuses through the liquid droplet and ends up as a solid Si wire. The VLS mechanism represents the core of silicon wire research, though it does not only work for silicon wire but also for a much broader range of wire materials, such as Ge (Wang et al., 2006) and other Ⅲ-Ⅴ nanowires (Mandl et al., 2011). The VLS growth process can be summarized in the following four steps (Givargizov, 1975, as cited in Schmidt et al., 2010): (1) mass transport of SiH_4 from the gas phase to the Au surface; (2) reaction of SiH_4 on the Au surface; (3) diffusion of Si through the Au–Si eutectic liquid phase; (4) crystallization of Si from the supersaturated Au–Si eutectic liquid. The VLS mechanism can best be explained on the basis of Au catalyzed Si wire growth on silicon substrates by means of chemical vapor deposition (CVD) using a gaseous silicon precursor such as silane. Different metal-Si alloy system posses characteristic phase diagrams, which will be elaborated in section 1.2.1.1. The Au-Si binary phase diagram possesses a melting point of the Au-Si alloy strongly depends on composition. If heating Au in the presence of a sufficient amount of Si, considering e.g. a Au film on a Si substrate, to temperatures above 363 °C, the melting point of Au-Si alloy of 19 atom % Si and 81 atom % Au, will result in the formation of liquid Au-Si alloy droplets as schematically depicted in Figure 1.2a. When a gaseous silicon precursor such as silane, SiH_4, covered these Au-Si alloy droplets, the SiH_4 will be catalyzed to solid Si in the droplet

interface. A continuous supply silicon precursor leads Si consequently to the growth of wires with a Au-Si droplet at their tip. Figure 1.2b is an example of Au-catalyzed Si nanowires grown homoepitaxially on a <111> substrate via the VLS-mechanism and Figure 1.2c is the transmission electron micrograph proves the epitaxial relation between nanowire and substrate (Schmidt, 2005, as cited in Schmidt et al., 2010).
As discussed above, the VLS nanowire growth mechanism is merely deduced from the fact that these nanowires generally have alloy droplets on their tips, while the direct evidence, however, is still lacking. A better understanding of the nanowire growth process in the vapor phase is necessary to pin down the growth mechanism and to be able to rationally control their compositions, sizes, crystal structures, and growth directions. P. Yang group (Wu et al., 2001) reported the real-time observation of semiconductor nanowire growth in an in-situ high temperature transmission electron microscope (TEM).

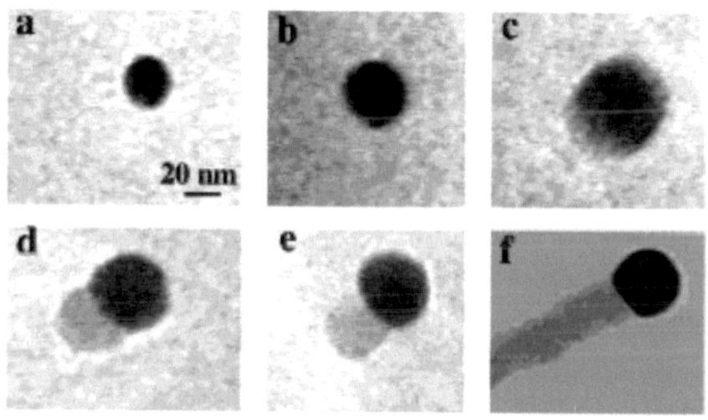

Fig. 1.3 In situ TEM images recorded during the process of nanowire growth. (a) Au nanoclusters in solid state at 500 ° C; (b) alloying initiates at 800 ° C, at this stage Au exists in mostly solid state; (c) liquid Au/Ge alloy; (d) the nucleation of Ge nanocrystal on the alloy surface; (e)-(f) Ge nanocrystal elongates with further Ge condensation and eventually a wire forms. (Wu et al., 2001)

Figure 1.3a-f shows a sequence of TEM images during the growth of a Ge nanowire in situ. This real-time observation of the nanowire growth directly mirrors the proposed VLS mechanism in Figure 1.4a. We have examined over 50 individual Au clusters during the in situ catalytic nanowire growth. In general, three stages (I-III) could be clearly identified.

i. *Alloying process (Figure 1.3a-c).* Au clusters remain in the solid state up to our maximum experimental temperature 900 °C if there is no Ge vapor condensation. This is confirmed by selected area electron diffraction on the pure Au clusters. With increasing amount of Ge vapor condensation and dissolution, Ge and Au form an alloy and liquefy. The volume of the alloy droplets increases, and the elemental contrast decreases (due to dilution of the heavy metal Au with the lighter element Ge) while the alloy composition crosses sequentially, from left to right, a biphasic region (solid Au and Au/Ge liquid alloy) and a singlephase region (liquid). This alloying process can be depicted as an isothermal line in the Au-Ge phase diagram (Figure 1.4b).
ii. *Nucleation (Figure 1.3d,e).* Once the composition of the alloy crosses the second liquidus line, it enters another biphasic region (Au/Ge alloy and Ge crystal). This is where nanowire nucleation starts. Knowing the alloy volume change, we estimate that the nucleation generally occurs at Ge weight percentage of 50-60%. This value differs from the composition calculated from the equilibrium phase diagram which indicates the

first precipitation of Ge crystal should occur at 40% Ge (weight) and 800°C. This difference indicates that the nucleation indeed occurs in a supersaturated alloy liquid.

iii. *Axial growth (Figure 1.3d-f).* Once the Ge nanocrystal nucleates at the liquid/solid interface, further condensation/ dissolution of Ge vapor into the system will increase the amount of Ge crystal precipitation from the alloy. This can be readily accounted for, using the famous lever rule of phase diagram. The incoming Ge species prefer to diffuse to and condense at the existing solid/liquid interface, primarily due to the fact that less energy will be involved with the crystal step growth as compared with secondary nucleation events in a finite volume. Consequently, secondary nucleation events are efficiently suppressed, and no new solid/liquid interface will be created. The existing interface will then be pushed forward (or backward) to form a nanowire (Figures 1.3f, 1.4b). After the system cools, the alloy droplets solidify on the nanowire tips. Their compositions were analyzed with energy-dispersive X-ray spectroscopy (EDAX), and it was found that the weight percentage of Ge matches qualitatively well with the estimated alloy composition at which first Ge nanocrystal nucleates.

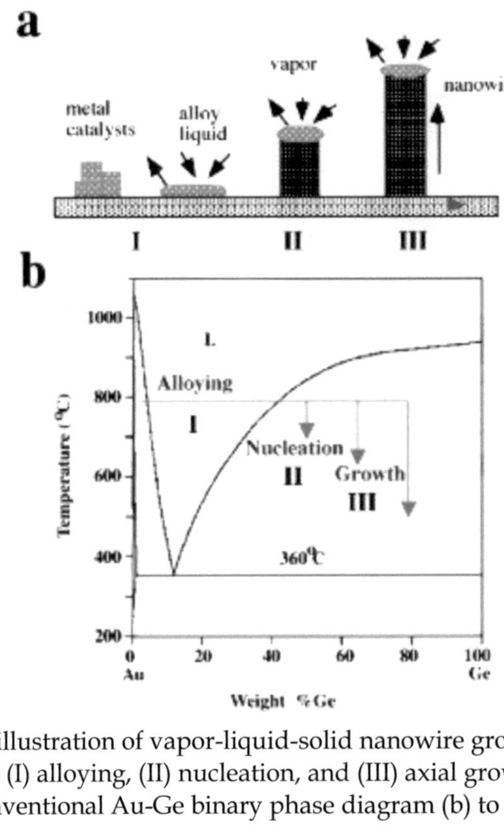

Fig. 1.4 (a) Schematic illustration of vapor-liquid-solid nanowire growth mechanism including three stages (I) alloying, (II) nucleation, and (III) axial growth. The three stages are projected onto the conventional Au-Ge binary phase diagram (b) to show the compositional and phase evolution during the nanowire growth process. (Wu et al., 2001)

The direct observation of nanowire growth unambiguously confirms the validity of vapor-liquid-solid crystal growth mechanism at the nanometer scale and should allow us to rationally control the nanowire growth which is critical for their potential implementation into the nanoscale electronic and optoelectronic devices.

1.2.1.1 Phase diagrams of different catalyst-Si alloy system

The most remarkable feature of the VLS growth mechanism, however, is its universality. VLS growth works well for a multitude of catalyst and wire materials and, regarding Si wire growth, over a size range of at least 5 orders of magnitude; from wire diameters of just a few nanometers up to several hundred micrometers.

The characteristic of VLS mechanism is the application of metal catalyst, which is generally selected by phase diagrams. Figure 1.5 is showing phase diagrams of different metal-Si alloy system. To formulate the requirement on the catalyst-Si binary phase diagram in a more abstract way, Si wire growth requires a nonhorizontal phase boundary over which one can push the catalyst-Si system to enforce the precipitation of a Si rich solid. The catalyst materials are classified into three different categories by the phase diagrams of metal-Si system: Type-A, Type-B, Type-C (Bootsma & Gassen, 1971, as cited in Schmidt et al., 2010), as shown in figure 1.6.

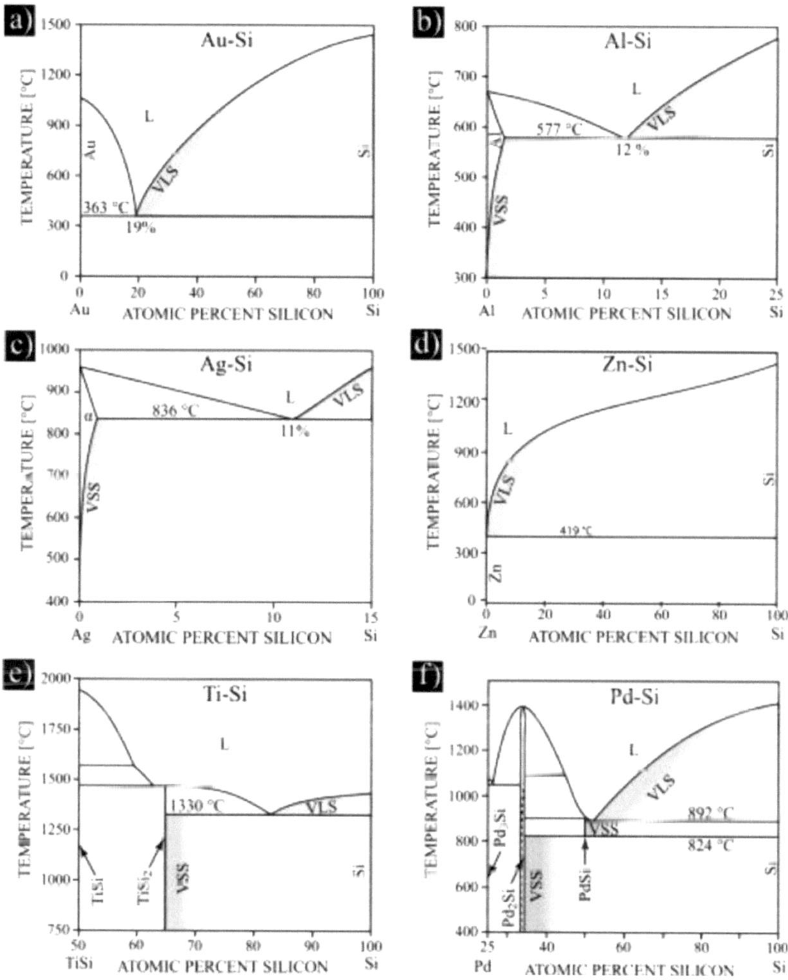

Fig. 1.5 Schematic phase diagrams of different metal-Si systems. (a) Au-Si, (b) Al-Si, (c) Ag-Si, (d) Zn-Si, (e) Ti-Si, (f) Pd-Si (Weber, 2002, as cited in Schmidt et al., 2010).

Type-A catalysts are the Au-like metals. Their phase diagram is of the simple eutectic type; that is, it is dominated by a single eutectic point. This eutectic point is located at a Si composition of more than 10 atom % Si. Furthermore, type-A catalysts do not possess any metal-silicide phases. There are only three type-A metals: Al, Ag, and Au.

Au is the most convenient and effective catalyst due to its nontoxicity, chemical stability and availability. Au is available not only by evaporation, but by Au colloid nanoparticles. And Au doesn't oxidize in air to make an in situ deposition unnecessary, which is a decisive advantage for the pregrowth sample preparation. The Au-Si phase diagram is of the simple eutectic type, with its dominant feature being a eutectic point at a composition of about 19 atom % Si, indicating that Si likes to mix with Au. The eutectic temperature is 363 °C, a quite remarkable reduction of melting temperature, which is 700 K lower than the melting point of pure Au and about 1050 K below the melting point of pure Si. The phase within the V-shaped region, visible in Figure 1.5a, is the liquid phase, the actual composition of which depends on the amount of Si supplied. For Au-Si alloy droplets on a Si substrate, Si is abundant, and the composition of such Au-Si droplets is therefore given by the position of the liquidus line on the Si side, i.e. the phase boundary on the right-hand side (rhs) of the liquid phase. If such Au-Si droplets on a Si substrate, held at temperatures above the eutectic temperature, are exposed to a Si precursor such as silane, SiH_4, silane molecules will crack at the surface of the droplets, thereby supplying additional Si to the droplet. This additional Si supply causes an increase of the Si concentration in the droplet to a value greater than the equilibrium concentration. Considering the Au-Si phase diagram shown in Figure 1.5a, this means that, by switching on the silane, the Au-Si droplet system is pushed over the liquidus line; and the only way for the droplet to reduce the Si concentration is to precipitate a Si-rich solid. In general, the composition of such a Si-rich solid would be given by the nearest phase boundary on the Si side of the liquidus. Consequently, the droplet precipitates Si, which with time results in the growth of a wire (Schmidt et al., 2010).

The drawback for Au is the contamination of the nanowires (Allen et al., 2008; Perea et al., 2006; Shchetinin et al., 1991) in the semiconductor industry. Because it is associated with deep-level defects in Si, leading to strongly enhanced carrier recombination. Metal impurities in semiconductors are generally known to affect the charge carrier lifetimes by facilitating charge carrier recombination. The recombination rate critically depends on the energy difference between the impurity level or levels and the band gap middle; the closer the impurity level is to the band gap middle, the more efficient it is as a recombination center. The use of metals with impurity levels close to the band gap middle, so-called deep levels, is therefore to be avoided.

Al, another Type-A catalyst, shows the closest similarity of Al-Si binary phase diagram (see Figure 1.5b) with Au-Si, excluding that the eutectic point of the Al-Si system is located at a higher temperature (577°C) and at a slightly lower Si concentration (12 atom%). So Al could catalyze the VLS growth of Si nanowires undoubtly. Osada et al. (Osada et al., 1979, as cited in Schmidt et al., 2010) demonstrated Al catalyzed VLS growth of crystalline Si wires in a CVD process using silane and applying temperatures of 580-700°C. Compared with Au, Al catalyst shows the great advantage that Al does not create deep level defects. The major drawback of the use of Al, however, is its oxygen sensitivity. Oxidation of the Al catalyst particle has to be prevented during the whole processing sequence, which clearly limits the usability of Al.

Silver is the second nongold, type-A catalyst. The Ag-Si system (see Figure 1.5c) possesses a single eutectic point (at 11 atom % Si and 836 °C) (Weber, 2002, as cited in Schmidt et al., 2010). Due to the high eutectic temperature, high process temperatures at around 1000 °C are required for Ag-catalyzed VLS growth of Si wires. So the high process temperature becomes the disadvantage of Ag catalyst, because the Ag catalyst did not evaporate completely under these conditions, as the vapor pressure of Ag reaches a value close to 10^{-2} mbar at 1000 °C, which is about 3 orders of magnitude larger than that of Au (Geiger et al., 1987, as cited in Schmidt et al., 2010).

Type-B catalysts are the low Si solubility metals. Their phase diagrams also show a single dominant eutectic point but no silicide phases. In contrast to the type-A catalysts, the eutectic point is located at much lower Si concentrations, less than 1 atom % Si. Typical type-B catalysts are In, Ga, or Zn. The Zn-Si binary phase diagram is dominated by a single eutectic point at 420 °C and 0.02 atom % Si, shown in Figure 1.5d, and despite its high vapor pressure of 0.2 mbar at 420 °C, Zn has proven to be an effective catalyst material for VLS growth (Chung et al., 2000). However, the impurity levels of Zn in Si are basically as detrimental as those of Au in the view of the electronic properties. The only advantage of using Zn is that a potential Zn contamination of wafers or equipment can be removed more easily than a potential Au contamination.

Ga or In as catalyst of VLS growth appears to be much more attractive than that of Zn from a vapor pressure point of view. At 500 °C, the vapor pressure of In is below 10^{-7} mbar, and the vapor pressure of Ga is even lower: 10^{-10} mbar (Schmidt et al., 2010). Moreover, In and Ga would also be attractive from an electronics point, as both would induce a p-type doping of the wires. In terms of phase diagrams, Ga and In show great similarities. The Si concentrations at the eutectic point (smaller 0.01 atom %) as well as the eutectic temperatures (Ga, 30°C; In, 156°C) are very low in both cases, and any reasonable CVD growth temperatures will be way above the respective eutectic temperature. Therefore In or Ga can be expected to produce similar Si nanowire results (Givargizov & Sheftal, 1971, as cited in Schmidt et al., 2010).

Type-C catalysts are the silicide forming metals. Their phase diagram indicates the presence of one or more silicide phases. In addition, the lowest eutectic temperature is higher than 800°C. Typical type-C catalysts are Cu, Pt, or Ti. Here is showing based on Ti catalyst. Figure 1.5e schematically depicts the Si-rich half of the Ti-Si phase diagram. As indicated therein, Ti-Si possesses a eutectic point at 1330°C adjoining the pure Si side of the phase diagram, whose liquidus can be used for Si wire growth via the VLS mechanism. At growth temperatures below 1330°C, growth should theoretically proceed via the phase that at this temperature is neighboring the pure Si side. As one can see in Figure 1.5e, this would be $TiSi_2$. Considering growth at 1000°C and starting from a Ti particle, this Ti particle will first transform into Ti_5Si_3 and then into Ti_5Si_4, which becomes TiSi, which will finally transform into $TiSi_2$. Only once this transformation process is completed can Si wire growth start.

To explain the one-dimensional growth of Si nanowires prepared via VLS, the generation of invisible charged nanoparticles during VLS is experimentally confirmed. In an effort to confirm whether charged silicon nanoparticles were also generated during the synthesis of Si nanowires by VLS, a differential mobility analyzer (DMA) combined with a Faraday cup electrometer (FCE) was connected to an atmospheric-pressure CVD reactor under typical conditions for Si nanowire growth, as shown in Figure 1.7 (Kim et al., 2010).

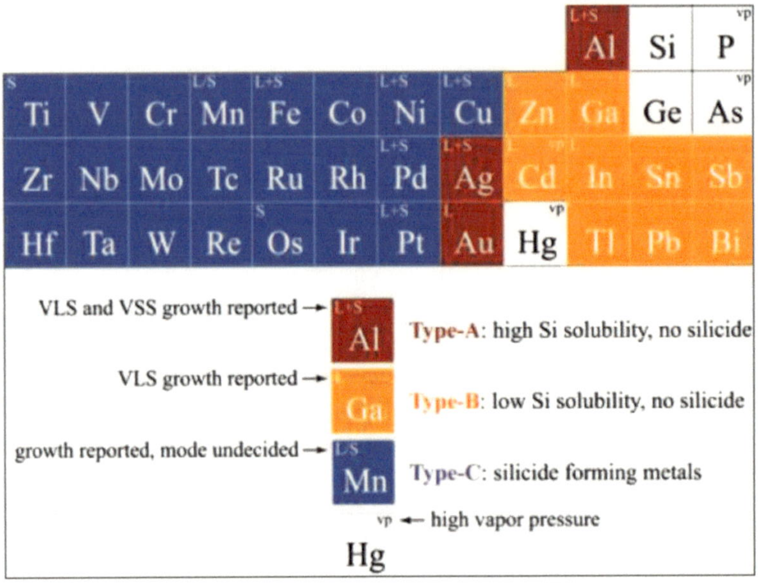

Fig. 1.6 Periodic table with potential catalyst metals classified according to their phase diagram. Type-A: phase diagram dominated by a eutectic point at a Si concentration >10%; no metal-silicide phase present. Type-B: phase diagram dominated by a eutectic point at a Si concentration <1%; no metal-silicide phases present. Type-C: phase diagram with one or more metal-silicide phases; eutectic points located at temperatures above 800 ° C. Elements marked with superscript ion have a vapor pressure of more than 0.01 mbar at 300 ° C (Bootsma & Gassen, 1971, as cited in Schmidt et al., 2010).

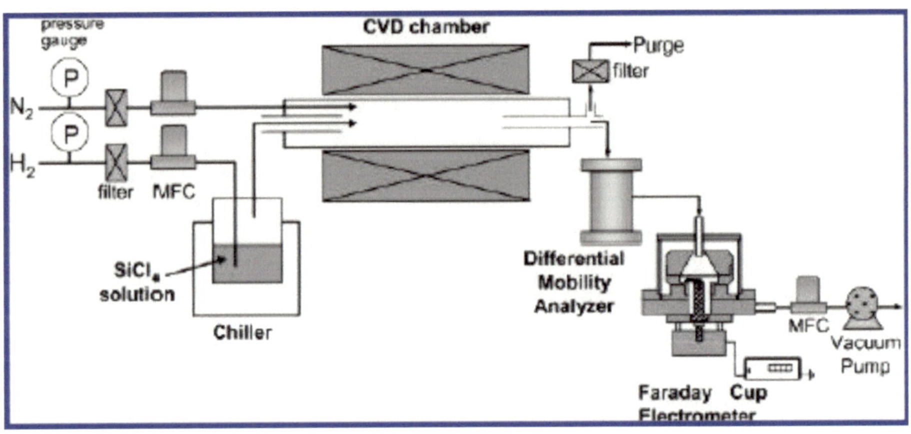

Fig. 1.7 Schematic of experimental setup for the measurement of charged nanoparticles generated during theVLS process (Kim et al., 2010).

To investigate the effect of the reactor temperature on the deposition of Si nanowires and on the formation of nanoparticles in the gas phase, the reactor temperature was varied from 900 to 1000 °C at a $SiCl_4/H_2$ molar ratio of 0.1 and a hydrogen flow rate of 5 sccm, with the total flow rate of nitrogen and hydrogen fixed at 1000 sccm, and in situ measurements of charged nanoparticles were carried out using a DMA-FCE system. The results show that as the reactor temperature was increased, Si nanowire growth is enhanced with increasing reactor temperatures within the examined temperature range, and the size distribution of both positively and negatively charged nanoparticles shifted to smaller particle sizes and the number concentration of charged nanoparticles increased. At reactor temperatures above 900°C, the number concentration of charged nanoparticles smaller than ~30 nm increased drastically. These results indicate that the decomposition of $SiCl_4$ increased with increasing reactor temperature. Thus, nucleation and charging of nanoparticles in the gas phase are enhanced with increasing reactor temperature due to enhanced thermal decomposition of $SiCl_4$.

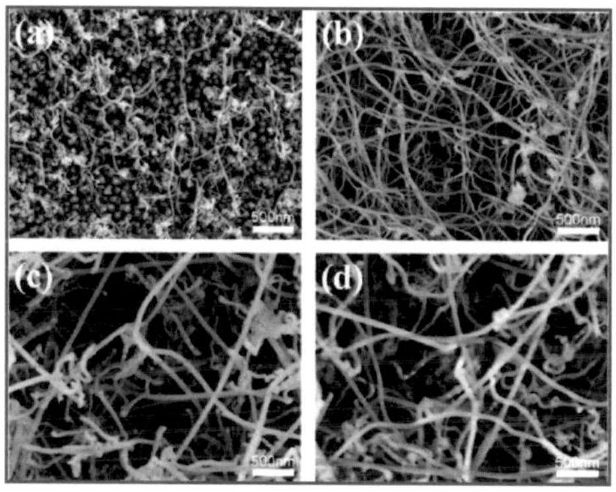

Fig. 1.8 FESEM images of Si nanowires at $SiCl_4/H_2$ molar ratios of (a) 0.05, (b) 0.1, (C) 0.15, and (d) 0.2 at a hydrogen flow rate of 5 sccm and a reactor temperature of 975° C (the scale bar is 500 nm) (Kim et al., 2010).

To examine the effect of the molar ratio of $SiCl_4/H_2$ on both Si nanowire growth and the size distribution of charged nanoparticles, the molar ratio of $SiCl_4/H_2$ was varied at a fixed hydrogen flow rate of 5 sccm and a reactor temperature of 975°C. Figure 1.8 shows the FESEM images of nanowires formed at different molar ratios of $SiCl_4/H_2$. As the ratio of $SiCl_4/H_2$ was increased, both the diameter and the length of the nanowires increased. As the ratio of $SiCl_4/H_2$ was increased from 0.05 to 0.1, the diameter increased from 24 to 35 nm, and the length and the density of the nanowires were also markedly increased. As the ratio $SiCl_4/H_2$ was further increased to 0.15 and 0.20, the diameter drastically increased to 61 and 65 nm, respectively. This result indicates that the ratio of $SiCl_4/H_2$ is an important parameter controlling the diameter and length of Si nanowires. Therefore, the size of the gold nanoparticles is not the only parameter that determines the diameter of the Si nanowires that grow on them.
Figure 1.9 shows the particle size distribution of positively (Figure 9a) and negatively (Figure 9b) charged nanoparticles generated at various $SiCl_4/H_2$ ratios. The size distribution of positively and negatively charged nanoparticles generated at a $SiCl_4/H_2$ ratio of 0.05 has a peak at 14.6 nm. As the ratio of $SiCl_4/H_2$ increased, the size distribution of both positively and negatively charged nanoparticles shifted to larger particle sizes and the number concentration of charged nanoparticles increased drastically. However, the peak diameter did not change considerably at a $SiCl_4/H_2$ ratio higher than 0.1. These results show that the ratio of $SiCl_4/H_2$ not only affects the growth behavior of Si nanowires but also affects the size distribution of charged nanoparticles.

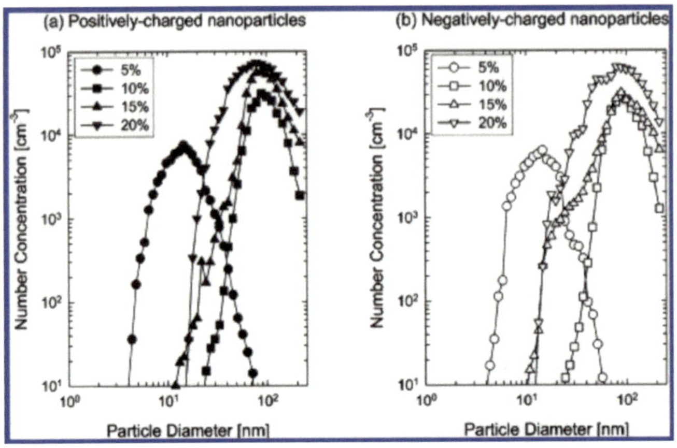

Fig. 1.9 Size distribution of (a) positively and (b) negatively charged nanoparticles at various molar ratios of $SiCl_4/H_2$ at a hydrogen flow rate of 5 sccm and a reactor temperature of 975 °C (Kim et al., 2010).

Similiarly, the hydrogen flow rate has a sensitive effect on the generation of charged nanoparticles as well as the growth behavior of Si nanowires. As the hydrogen flow rate was increased from 5 to 10 sccm, the number concentration of charged nanoparticles increased, but it decreased at the hydrogen flow rate of 15 sccm. Especially at a hydrogen flow rate of 15 sccm, where Si nanowires did not grow, charged nanoparticles below 40 nm were not detected.

The formation mechanism of charged silicon nanoparticles generated during Si nanowire synthesis by VLS could be deduced that gas phase nuclei of silicon would be formed first and then these nuclei undergo surface ionization on any surface, such as thevquartz tube of the reactor (Kim et al., 2009b). The surface ionization of gas phase nuclei is very similar to the electrostatic charging or triboelectric charging that is experienced commonly in everyday life. The electrostatic charging shows that charging can occur even at room temperature. it is more probable that the charged nanoparticles, instead of individual atoms or molecules, produced in the VLS process, should interact with the catalytic gold particles to produce the Si nanowires. The electrostatic attraction between the charged nanoparticles and the conducting gold particles would be much stronger than that between the charged nanoparticles and the insulating native oxide of the silicon substrate. The nanoparticles carried by the gas flow have difficulty in landing on a surface because of the levitation force: the gas flow velocity is zero on the surface and increases away from the surface. For charged nanoparticles to land on any surface, the electrostatic attraction between charged nanoparticles and the surface should overcome this levitation force. It appears that the electrostatic attraction between charged nanoparticles and the conducting gold particles is larger than the levitation force, whereas the electrostatic attraction between charged nanoparticles and the native silicon oxide of the substrate or the silicon surface of nanowires is less than the levitation force. This might be why charged nanoparticles land preferentially on the conducting gold particles, leading to nanowire growth. When the rod-shaped silicon is formed with a gold nanoparticle at its tip, the charged silicon nanoparticles would be attracted to the gold tip more preferentially than to the side of the silicon rod, considering that the charged nanoparticles should have higher electrostatic attraction energy with the conducting gold nanoparticles than with the side of the semiconducting silicon rod.

1.2.2 The selective growth of SiNWs based on VLS mechanism

1.2.2.1 SiNW diameter controlling via VLS mechanism

Precise control of the SiNW diameter strongly affects the electrical and optical properties of the nanowires (Brus, 1994). The diameter of each Si nanowire is largely determinded by the size of the catalyst particle and growth conditions. In general, Au colloids are used to define the diameter and position of the SiNWs. Au colloids are ideal seeds for controlling the SiNW diameter: They act as the seeding metal for nanowire growth by the VLS process, and Au colloids may be synthesized or obtained commercially with relatively narrow size distributions. Since each colloid seeds the growth for one nanowire, aligned nanowires can be grown with narrow size distributions approaching those of the seed particles. Hence, by seeding wire growth with colloids of different average size, the average diameter of the SiNW arrays could be precisely controlled (Cui et al., 2001; Wu et al., 2004).

Using a thin polyelectrolyte layer, gold colloids are electrostatically attracted to and immobilized on the substrate to act as seeds for Si nanowires grown using the VLS-CVD method. The diameter of the colloids precisely controls the nanowire diameter. The colloid solution concentration controls the density of growth. Microcontact printing of the polyelectrolyte layer confines wire growth to patterned regions (Hochbaum et al., 2005). As seen in Figure 1.10b, d, and f. Size distributions of both colloids and nanowires were determined from TEM micrographs. SiNWs grown from Au colloids of 50 (56 ± 5.0), 30 (30 ±3.3), and 20 (20 ± 2.1) nm diameters were 93±7.4, 43± 4.4, and 39 ± 3.7 nm in diameter, respectively (Hochbaum et al., 2005).

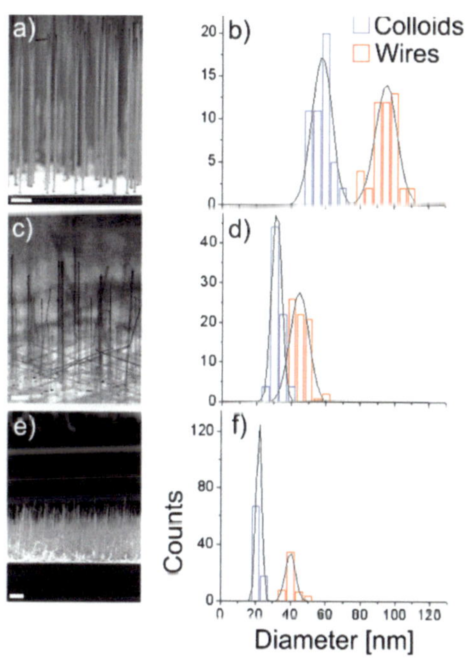

Fig. 1.10 (a, c, e) Cross-sectional images of nanowire samples grown from 50, 30, and 20 nm (nominally) Au colloids, respectively. Scale bars are 1 μm. (b, d, f) Size distributions of Au colloids and resulting SiNW diameters. (g) High-resolution transmission electron microscopy image of a single crystalline Si nanowire. Scale bar is 3 nm (Hochbaum et al., 2005).

The density of nanowire growth is also critical to device function. By varying the concentration of the seeding solution (using 50 nm Au colloids as example here), we were able to control the seeding density on the substrate surface. The graph in Figure 1.11 shows the relationship between nanowire growth density, as determined from SEM images, and dilution of the gold colloid stock solution. Wires were seeded with densities ranging over an order of magnitude, from ~0.1-1.8 wires/μm^2. In general, a good 1-to-1 nanoparticle/ nanowire ratio can be achieved, although it was observed that optimal growth conditions varied slightly with nanowire seeding density.

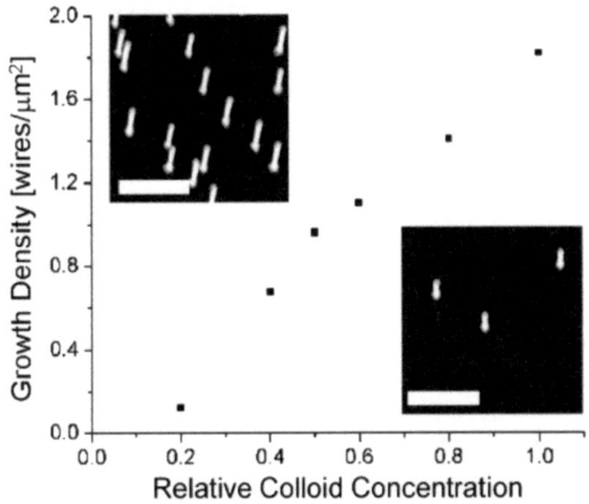

Fig. 1.11 SiNW growth density dependence on the relative concentration of the seeding solution. All colloid solutions were diluted from the same stock solution. Insets show typical nanowire growth at 4/5 (top left) and 2/5 dilution (bottom right). Scale bars are 1μm (Hochbaum et al., 2005).

1.2.2.2 Spatial controlling via VLS mechanism

For both phenomenological studies and the implementation of practical applications of SiNWs, fabrication of regular arrays of Si wires, with precise control of the crystallographic orientation, dimension, and density will be of great value. For laser interference lithography, the ultimate achievable resolution equals to one-fourth of the laser wavelength (Solak, 2006).To overcome the limitation in achieving reliable diameter reduction by reactive ion etching (RIE) for the smaller dimensions, template-based methods offer access to a wide variety of nanowires with a broad diversity in composition and shape. As template material, most commonly substrates such as track-etched polycarbonate membranes (Azarian et al., 2009), mesoporous silica (Petkov et al., 2007) and porous anodic alumina membranes (Xiang et al., 2008; Gao et al., 2002) are used.

A porous anodic alumina (PAA) is a material characterized by a honeycomb pattern of nanometer-sized pores with uniform diameter and spacin (Masuda & Fukuda, 1995). The pores are formed during the electrochemical anodization of aluminum under controlled conditions; different self-ordering conditions yield pores of different diameters and spacing (Nielsch et al., 2002). They give easy access to self-organized pores with high aspect ratios and tunable pore sizes (Gao et al. 2002; Masuda & Fukuda et al., 1995; Krishnan & Thompson, 2007; Nielsch et al., 2000). These pores are then used as a molding system for the synthesis of the desired nanostructures (Xiang et al., 2010).

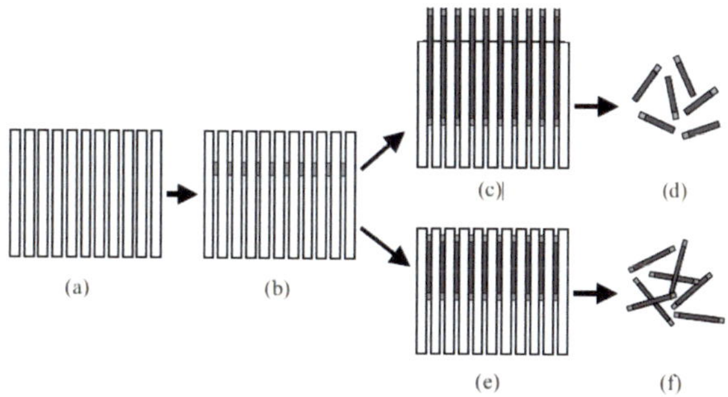

Fig. 1.12 Schematic diagram of the fabrication method used for the PAA template-based growth of SiNWs. (a) Cross-section of PAA, (b) electro-deposition of Au within pores, (c) VLS growth of SiNWs out of membrane surface, (d) SiNW removal by mechanical agitation, (e) VLS growth of SiNWs within membrane, and (f) SiNW removal by wet etching of membrane (Bogart et al., 2005).

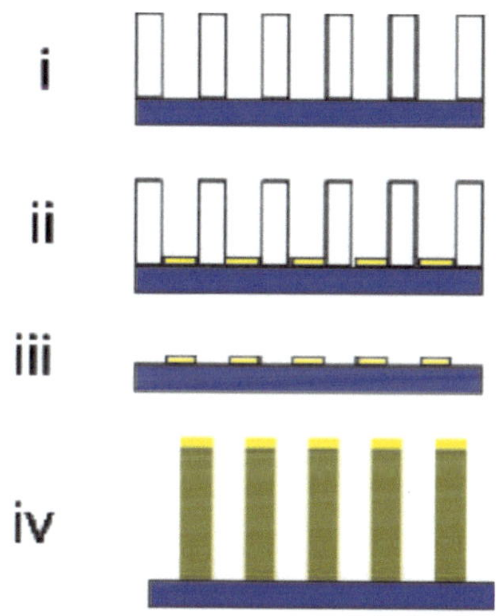

Fig. 1.13 Schematic diagram of the fabrication method used for the PAA template-based growth of SiNWs. (i) Transfer of the PAA mask on Si (111), (ii) evaporation of Au through the PAA pores, (iii) PAA removal, and (iv) growth of SiNWs from the patterned substrate.

The VLS growth of SiNWs inside the pores of an unsupported PAA template using SiH_4 as the precursor gas has been completed (Bogart et al., 2005; Lew & Redwing, 2003; Zhang et al., 2001), as shown in Figure 1.12. Briefly, gold is electro-deposited inside a free-standing PAA membrane, and the Au loaded template is placed inside the VLS growth furnace where SiNWs with uniform diameter are grown inside the pores of the alumina film. Subsequently, the template is dissolved (Bogart et al., 2005). The disadvantage of this process is that produced SiNWs are unsupported. In order to overcome the production of unsupported nanowires, due to the preventing their integration into functional devices (Lombardi et al., 2006), evaporation of Au through the PAA pores have been applied to overcome the dispesrion of SiNWs after removal of PPA membrane, as shown in Figure 1.13 (Lombardi et al., 2006).

Figure 1.14 shows the ordered pattern of gold nanodots after the thermal evaporation step and PAA mask removal. The size and arrangement of the gold clusters closely match those of the PAA pores. The average diameter of the dots is 53 nm with a standard deviation of 7.2%. The SiNW density could be controlled by simply changing the acid etch of the PAA film yielding pores up to required diameter while keeping the distance between pores unchanged. As a result, the gold nanoparticle arrays are arranged in a highly ordered configuration reflecting the hexagonal pattern of the nanopores of the PAA mask. SiNWs with controlled dimensions and spacings are grown in a rather inexpensive manner with this approach over large areas (~7 mm × ~7 mm). Additionally, this fabrication method is capable of producing SiNWs with a packing density as high as $6 \times 10^9/cm^2$, which could be achieved otherwise only with lithographic techniques, entailing serious limits for mass scale production of these nanomaterials.

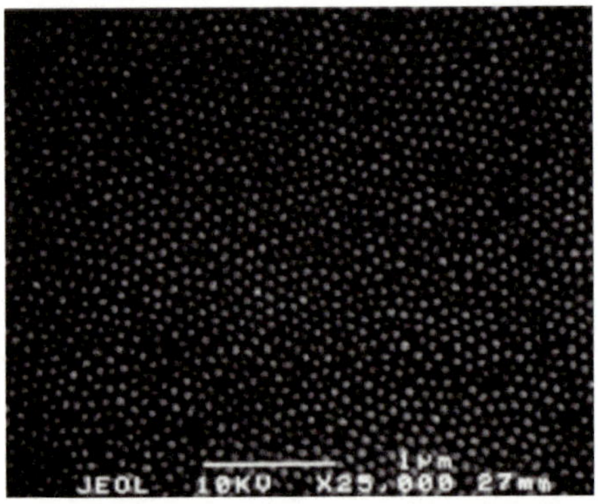

Fig. 1.14 Ordered arrays of Au nanodots evaporated through the nanochannels of the PAA mask. The clusters have an average diameter of 53 nm and a dot-to-dot distance of 100 nm. The height of the dots is 5 nm.

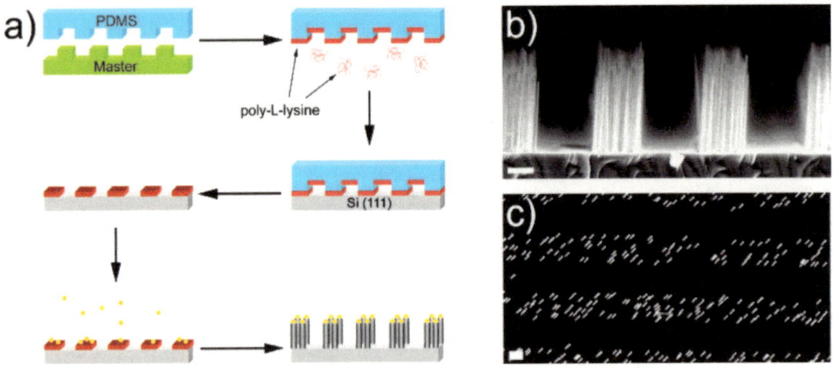

Fig. 1.15 (a) Schematic of PDMS patterning of Au colloids. Briefly, a PDMS stamp is molded to the relief pattern of a photoresist master. After curing the polymer, the stamp is removed from the master and "inked" with a solution of poly-L-lysine. The stamp pattern is transferred to the Si (111) substrate, which is then immersed in the Au colloid solution. The colloid-patterned substrate is grown using the conventional VLS-CVD synthesis, resulting in a corresponding pattern of SiNW arrays. (b) Cross-sectional SEM image of PDMS patterned SiNW growth, and (c) plane-view SEM image of the same. Scale bars are 1 μm (Hochbaum et al., 2005).

A poly(dimethylsiloxane) (PMDS) stamp was made using a photoresist master of 2 μm lines with 2 μm separation, and the pattern was transferred to the substrate by previously established techniques (Hochbaum et al., 2005) (Figure 1.15a). The stamp was "inked" with the poly-L-lysine solution by the same method described above for deposition on the Si substrates. The polymer pattern was transferred to a Si wafer by placing the stamp on the substrate and heating at 70 °C for 5 min. The patterned substrate was immersed in the 50 nm Au colloid solution for a short time, such that colloids only adhered to the polyelectrolyte and not the bare Si. Nanowires were synthesized on these substrates by the same VLS method. The resulting growth, seen in Figure 1.15b and c, is strictly confined to the regions of poly-L-lysine deposition. The plane-view SEM image (Figure 15c) shows a small part of the pattern, which is consistent over several square millimeters - the extent of the stamped area that was immersed in the colloid solution. Thus, spatial control over SiNW growth was achieved by patterning regions of seed particles using microcontact printing.

1.3 Conclusion

The VLS mechanism is a well established growth technology for SiNWs. Based on the VLS mechanism, the diameter, the density and the special arrays of SiNWs can be well controlled by careful selections of catalysts, the templates and the growth conditions. The future challenge is the controllable growth of SiNWs matching with the devices requirements, such as single SiNW growth in an ordered place, the uncontaminated SiNWs.

2. Optical activation of Si and Ge nanowires codoping with Er rare earth by sol-gel methods

2.1 Introduction

The discovery of the strong Stokes shift between absorption and emission in Si nanostructures suggested that emission could be due to an interfacial radiative emission center (Song & Bao, 1997; Klimov et al., 1998; Iacona et al., 2000, as cited in Priolo et al., 2001). Moreover, the quantum confinement picture and the interfacial state model have been reconciled by demonstrating that a Si=O double bond introduces size dependent levels (both for electrons and for holes) within the gap (Wolkin et al., 1999). According to this picture it is the radiative recombination of this electron-hole pair trapped at the Si=O double bond, the process responsible for the observed emission. This emission, however, is still nc size dependent, due to quantum confinement effects, thus explaining the observed blueshift. migration and trapping of excitons within nanostructures have been studied in both porous Si and Si nc formed by ion implantation (Pavesi, 1996; Linnros et al., 1999, as cited in Priolo et al., 2001). But Si nanocrystals emit light at room temperature in the range 700–1100 nm (Priolo et al., 2001, as cited in Priolo et al., 2001).

In order to realize the Si based light emitter in the technologically important 1.5 μm range (Ennen et al., 1983, as cited in Suh et al., 2005), the Er-doping of Si has attracted great attention because of its promising future in the development of light-emitting diodes and lasers operating at a wavelength of 1.54 μm, which coincides with the absorption minimum of optical fibers (Park et al., 2005). The Er doping of Si nanocrystals (ncs) holds some promise for efficiently generating light emission, since Si nanocrystals in the presence of Er act as efficient sensitizers for Er ions (Polman, 1997; Seo & Shin, 2001; Schmidt et al., 2002; Iacona et al., 2002, as cited in Park et al., 2005).

2.2 Optical activation of Er-doped silicon nanocrystals

Er is an important rare earth material for optoelectronic devices due to its luminescence wavelength at about 1540 nm, which coincides with a minimum loss in optical fibers. Er-doped nanocrystals of indirect-gap semiconductors like Si are being widely studied as they would open new possibilities for applications in optoelectronics and microelectronics, with the main advantage of being compatible with actual device technology. As shown (Coffa et al., 1994, as cited in Park et al., 2005) in the Auger excitation model, the electron–hole pairs are bound to Er-related states below the conduction band in Si ncs. The exactions can then recombine and thereby excite the Er ions with an excess energy to the difference between the bound state and the conduction band in Si ncs. When Er is introduced in the sample the Er–nc interaction is particularly strong. In this case the excited nc preferentially transfers its energy to the Er ion. The Er ion is excited in a high energy state and decays very rapidly in the first excited state ($^4I_{13/2}$). At this stage its energy is too small to be transferred back to the nc and the Er remains excited until the radiative emission occurs at 1.54 μm while the nc luminescence at around 800 nm is totally quenched (Kenyon et al., 1994; Fujii et al., 1997, 1998; Franzo\` et al., 1999; Chryssou et al., 1999; Kenyon et al., 2000; Shin et al., 2000a; Franzo\` et al., 2000; Kik et al., 2000, as cited in Priolo et al., 2001). Therefore in the presence of Er ions a transfer of the energy among Si nc is thought to be less probable (Priolo et al., 2001).
Excellent optical properties were obtained by using silicon-rich silicon oxide (SRSO), which consists of Si ncs embedded inside a SiO_2 matrix (Kenyon et al., 1994; Fujii et al., 1997; Shin et al., 1998, as cited in Suh et al., 2005). By now, optical gain (Han et al., 2002, as cited in Suh et al., 2005) as well as efficient light-emitting diodes (Franzó et al., 2002, as cited in Suh et al., 2005) have been demonstrated using Er-doped SRSO. Theoretical calculation (Nishio et al., 2003) also showed that the radiative recombination rate for Amorphous Si quantum dots (a-Si QDs) is higher by two or three orders of magnitude than that for Si ncs indicating that better performance can be obtained when a 1.54-μm light source fabricated using an Er-doped a-Si QD structure is employed. However, isolation of Si ncs inside the SiO_2 matrix makes current injection into SRSO difficult. Thus, SRSO-based LEDs generally require either very high voltages (Franzó et al., 2002, as cited in Suh et al., 2005) or very thin SRSO layers that can limit the light output (Irrera et al., 2002, as cited in Suh et al., 2005). Furthermore, excitation occurs via impact excitation by energetic carriers, which raises questions about the long-term reliability of such devices.

2.3 Optical activation of Er-doped silicon nanowires

Compared with the isolation of Si nc, SiNWs have diameters in the range of 10-100 nm and with lengths exceeding 1 μm. Thus, Er-doped SiNWs may provide a very high areal density of Er^{3+} ions in the case of Er-doping. The ease of charge injection into, and transport along the SiNWs are expected to improve the photoluminescence (PL) intensity. Si/Er core-shell nanowires with erbium enriched at the surface (Wang & Coffer, 2002) were fabricated successfully via a VLS process, which is modified by passing the He through a bubbler (heated to ~144 °C) containing $Er(tmhd)_3$ after SiH_4 flow. This process results in introducing Er to the surface of SiNWs, as shown in Figure 2.1.
Detectable erbium emission at room temperature in a crystalline semiconductor is often difficult to achieve. But the anticipated Er^{3+} with Er^{3+} luminescence near 1.54 μm, associated with the ($^4I_{13/2}$) → ($^4I_{15/2}$) transition, is observed upon excitation at 488 nm, as shown in Figure 2.2. However, it is regretted that the Photoluminescence (PL) intensity is very weak.

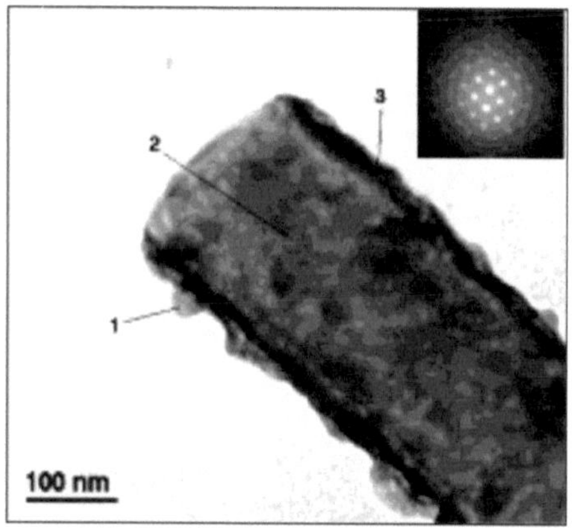

Fig. 2.1 A typical TEM image of a surface Er enriched Si wire (JEOL JEM-3010). Inset: SAED pattern from the center of the wire. EDX analysis for the marked three areas are presented: Si 96% (area 2), erbium concentration ~12% (area 3), erbium 53% and silicon 47% (area 1).

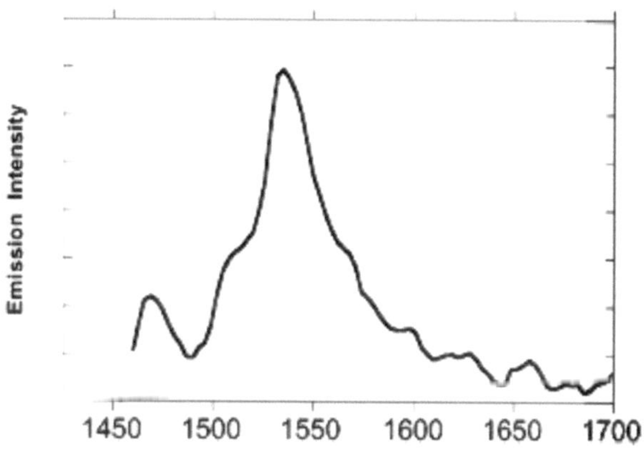

Fig. 2.2 Room-temperature photoluminescence (PL) spectrum of Er surface-enriched Si nanowires after a vacuum anneal, demonstrating the near-IR emission near 1540 nm (λ_{ex}) 488 nm).

2.3.1 Improvement of PL intensity of Er-doped silicon nanowires

2.3.1.1 Silicon nanowires coating with Er-doped SiO_2 derived from sol-gel solutions

As above mentioned, the optimum location for Er^{3+} is not inside Si, but in the nanometer-thin oxide shell right next to Si (Shin et al., 2000b; Kimura et al., 2003; Stepikhova et al., 1997, as cited in Suh et al., 2005). This suggests that rather than trying to dope Si-NWs directly, it would be preferable to coat the Si-NWs with high-quality Er-doped silica in order to optically activate the Si-NWs. Er coated SiNWs are achieved by using sol-gel derived Er-doped silica. The sol-gel technique is a low-temperature route widely employed to prepare thin film for integrated optic devices, because it can offer the homogeneous thin films at molecular scale and control of chemical purity. In this system, the pre-produced SiNWs are embedded inside an Er-doped silicon oxide film.

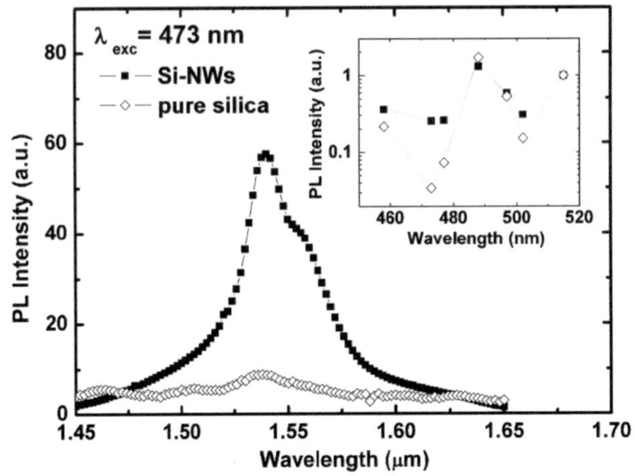

Fig. 2.3 Room temperature PL spectra of the Si-NWs and pure Er-doped silica film, pumped with the wavelength of 473 nm. The inset shows the wavelength dependence of the PL peak intensities (Suh et al., 2005).

Figure 2.3 shows the room temperature PL spectra of the Si-NWs and the pure Er-doped silica thin film, pumped with the 473 nm line of an Ar laser. The 473 nm line was chosen because it is absorbed only by Si-NWs and not directly by Er^{3+} ions. Hardly any luminescence from the pure silica film is not observed. However, the pure silica film does show Er^{3+} luminescence when Er^{3+} ions can directly absorb the pump photons, as shown in the inset. The strong Er^{3+} luminescence at 1.54 μm from the Si-NWs indicates energy transfer from carriers in Si-NWs to Er^{3+} ions.

Figure 2.4 shows the effect of temperature on the integrated Er^{3+} PL intensities. The pure silica film was pumped using the 488 nm line of an Ar laser in order to directly excite the Er^{3+} ions. On the other hand, the Si-NWs were pumped using the 477 nm line of an Ar laser, which is not absorbed optically by Er^{3+} ions, in order to probe only those Er^{3+} ions that can be excited via carriers. The temperature dependence of Er^{3+} luminescence lifetimes are nearly identical, as is shown in the inset. The Er^{3+} luminescence lifetime from the pure silica film decreases from 13.6 to 11 ms as the temperature is raised from 25 to 300 K, while that from the Si-NWs decreases from 8.3 to 6.9 ms. Such complete suppression of thermal quenching of Er^{3+} luminescence cannot be due to any quantum effects of the large diameter of SiNWs (Suh et al., 2005). Given the identical temperature dependence of the Er^{3+} luminescence lifetime, the reduction of the Er^{3+} luminescence lifetime by Si-NWs is attributed to the effect of increased effective refractive index (Snoeks et al., 1995) rather than any degradation of the quality of sol-gel derived silica film. Thus, it could be ensured that the Er^{3+} ions excited via carriers generated in Si-NWs are actually in the sol-gel derived Er-doped silica film that is coating the Si-NWs (Shin et al., 2000b; Kimura et al., 2003; Stepikhova et al., 1997), indicating that Si-NWs, even without quantum confinement effects, are much more effective for exciting Er^{3+} ions in an oxide layer on their surface. This conclusion is supported by the low-temperature PL spectra, as shown in Figure 2.5. The $^4I_{13/2}\rightarrow{}^4I_{15/2}$ transition responsible for the 1.54 μm Er^{3+} luminescence is a parity-forbidden transition that occurs in part due to the effects of the crystal field, and the exact shape and position of the luminescence spectra depend on the chemical and structural environment of Er (Stepikhova et al., 1997).

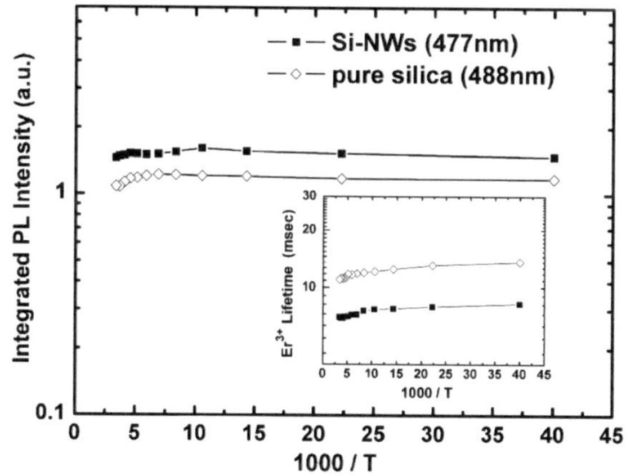

Fig. 2.4 The temperature dependence of Er^{3+} integrated PL intensities, showing the complete suppression of thermal quenching. The inset shows the temperature dependence of the luminescence lifetimes.

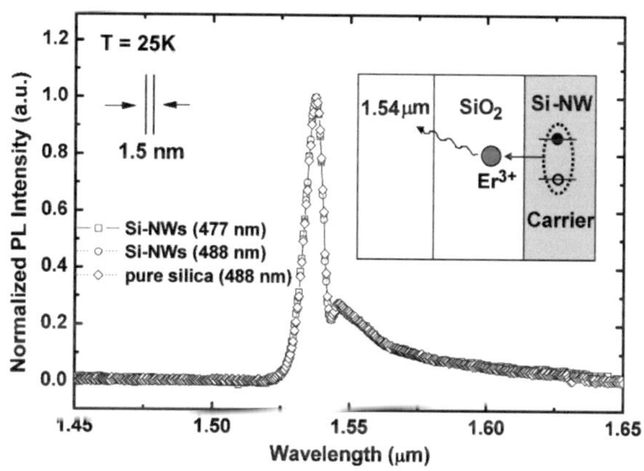

Fig. 2.5 Normalized PL spectra of pure Er-doped silica (488 nm), with Si-NWs (488 nm) and with Si-NWs (477 nm) at 25 K. The spectra are completely idenctical within the spectral resolution, indicating that the luminescent Er^{3+} ions are all in silica. The inset shows the schematic description of the proposed Er^{3+} luminescence.

Therefore, the results indicate that by using sol-gel derived Er-doped silica to optically activate Si-NWs, we can simultaneously achieve the ease of carrier injection, high carriermediated Er^{3+} excitation efficiency, and high Er^{3+} luminescence efficiency in a thick, easily produced film with a very high areal density of Er^{3+} ions, thus providing a promising material platform for Si photonics. The effectiveness of Si NWs in enhancing the Er^{3+} luminescence can be increased by simply increasing the density of Si-NWs. Furthermore, because the growth of Si-NWs and its coating by sol-gel derived Er-doped silica are performed separately, the two processes can be optimized separately.

2.3.1.2 Silicon nanowires coating with Er-doped SiO_2 and Al_2O_3 derived from sol-gel solutions

SiO_2 derived from sol-gel was used as a precursor to disperse Er^{3+} ions. But the film is uncontinuous and the erbium ions gather to clusters, which may induce the concentration quenching of the photoluminescence (PL) intensity. In fact, the concentration quenching at high concentration of Er should be avoided to realize the strong luminescence of the devices. Al_2O_3 is another promising material for high quality, low loss waveguide fabrication. Optical doping of thin Al_2O_3 films by Er ion implantation (van den Hoven et al., 1993, 1996) and by sol-gel technology (Feofilov, 1998; Patra, 2004) have been reported. Not only because the relatively high refraction index of Al_2O_3 (Al_2O_3, n=1.64 is higher than SiO_2 (n=1.45)) (Wang & Lei, 2005) is expected to improve the optical transfer efficiency, but the similarity in valency and lattice constants between Al_2O_3 and Er_2O_3 may enable disperse high concentrations optical doping with Er (van den Hoven et al., 1993, 1996; Feofilov, 1998). Furthermore, Aluminium oxide possesses the low thermal expansion coefficient (Jaymes et al., 1996), high chemical durability, and good mechanical property, which are benefit to Al_2O_3 film derived from sol-gel solutions on Si wafer substrate. It is very important for the application of the optical devices (Jimenez de Castro et al., 2000; Wang et al., 2004a, 2004b; Armelao et al., 2005). In order to prevent the clustering of erbium ions in the silica network, Al ions could be added into the silicon oxide structure or completely aluminum oxide structure. Therefore, the optically harmful Er clustering could be prevented by the selective coordination of Al^{3+} around the Er^{3+} ions (Patra, 2004) allowing for a homogeneous dispersion of Er ions in the silicon oxide structure.

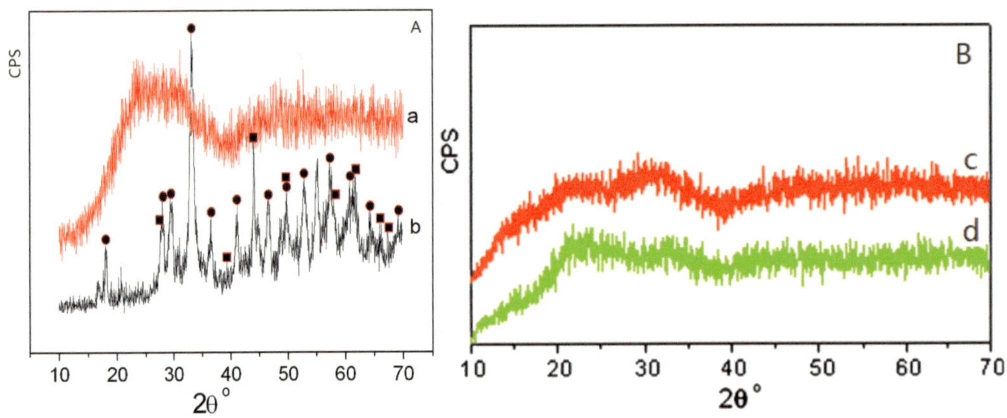

Fig. 2.6 X-ray diffraction patterns of (A-a (Ren, 2008) and B-c (Ren et al., 2007a)) Er-doped SiO_2 derived from sol-gel heat-treated at 750°C(Er/Si = 0.05), (A-b (Ren, 2008)) Er-doped Al_2O_3 derived from sol-gel heat-treated at 750°C (Er/Al = 0.10), and (B-d) (Ren et al., 2007a) Er-doped Si and Al oxides derived from sol-gel heat-treated at 750°C. The hollow circle (●) represents $Er_3Al_5O_{12}$ peaks; the solid circle (■) represents $ErAlO_3$ peaks.

In order to understand the relationship between the PL properties and the structures of Er-doped Al_2O_3, Si and Al oxides complex (SiAlO) and SiO_2 sol-gel films, the powder XRD patterns of heat-treated sol-gels at 750 °C with compositions of Er-doped SiO_2 (A-a and B-c), Er-doped SiAlO (B-d) and Er-doped Al_2O_3 (A-b) are tested, as shown in Figure 2.6 (Ren et al., 2007a, 2008). Figure 2.6 (A-a and B-c) shows the XRD pattern of Er-doped Si oxides. It shows the amorphous structure due to the short range ordering of Si network (Stepikhova et al., 2004). As shown in Figure 2.6 (B-d), the Er-doped Si–Al oxides also show an amorphous structure, indicating that the incorporation of Al into the Si network does not change the structures. However, it is expected that a more homogeneous incorporation of Er ions into

Al-Si network compared to the Si network is likely by the selective coordination of Al^{3+} around the Er^{3+} ions. However, the $Er_3Al_5O_{12}$ (PDFN 78-1451) phase and $ErAlO_3$ (PDFN 24-0396) phase are observed as shown in Fig. 2.6(A-b), which is consistent with Tanner's report (Kenyon, 2002) where the mixture of $Er_3Al_5O_{12}$ and $ErAlO_3$ are achieved for concentrations at or below 10% Er doped into Al_2O_3 due to the valence match between the rare-earth ions (Er^{3+}) and the substituted cation (Al^{3+}). As a result, it allows the incorporation of Er^{3+} into the Al_2O_3 lattice (Hochbaum et al., 2005) despite of the large size difference between Er^{3+} (0.89 nm) and the Al^{3+} (0.53 nm) (Tanner et al., 2004). It indicates that Er ions are homogeneous in Al_2O_3 even at high concentrations.

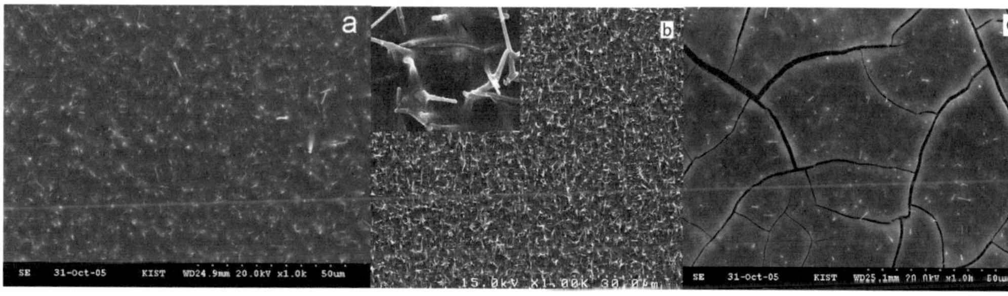

Fig. 2.7 SEM images of (a) SiNW-ErAlO (Ren, 2008), (b) SiNW-ErSiAlO (Si/Al=3/1) (Ren et al., 2007a) and (b) SiNW-ErSiO (Ren, 2008).

Figure 2.7 shows the SEM images of SiNW-ErAlO (a), SiNW-ErSiAlO (b) and SiNW-ErSiO (c) films. As shown in Figure 2.7, SiNW-ErAlO (a) and SiNW-ErSiAlO (b) film are continuous without cracks but numerous splits emerge in SiNW-ErSiO (c) film. A zoomed image further shows that the sol-gel solution has completely penetrated the SiNWs arrays, and had formed a crack-free, integrative film with tight SiNWs. The higher quality of film derived from Al-Si solution is due to the comparable thermal expansion coefficient of the film over that of the Si substrate. The thermal expansion coefficient of SiO_2, Al_2O_3 and Si is 66 x 10^{-6}/K, 23 x 10^{-6}/K and 23 x 10^{-6}/K, respectively. Thus, the cracks in the Er-Si oxide film are due to large differences in the thermal expansion coefficient between the SiO_2 and the Si binary system. In addition, an addition of Al to the Si-oxide while maintaining its amorphous structures matches the thermal expansion coefficient of the film (55 x 10^{-6}/K, according to the rule of mixture) to the Si substrate and suppressed the formation of cracks during the heat treatment. Thus, the reason of the high quality of SiNW-ErAlO and SiNW-ErSiAlO films are those the thermal expansion coefficient of the film over that of Si substrate matched well whereas the cracks in SiNW-ErSiO film are due to large differences in thermal expansion coefficient between the SiO_2 and the Si binary system.

In order to further understand the relationship between PL properties and the structures of Er-doped Al_2O_3 and SiO_2 sol-gel films, the TEM of heat treated sol-gels at 750°C are tested. Figure 2.8 was TEM images of Er-doped SiO_2 and Er-doped Al_2O_3. As shown in the inset of Figure 2.8a and Figure 2.8b, the Er-doped SiO_2 is amorphous due to its out-of-order structure in long range but the Er-doped Al_2O_3 is crystal structure because the optically active Er^{3+} ion readily substitutes for aluminium ions occupying octahedral sites in alumina (Kenyon, 2002), which is consistent with the XRD results. As shown in Figure 2.8a and 2.8b, the cluster sizes of Er-doped SiO_2 (about 300 nm) are much larger than that of Er-doped Al_2O_3 (≤100 nm), indicating that the dispersion of Al_2O_3 is better than that of SiO_2.

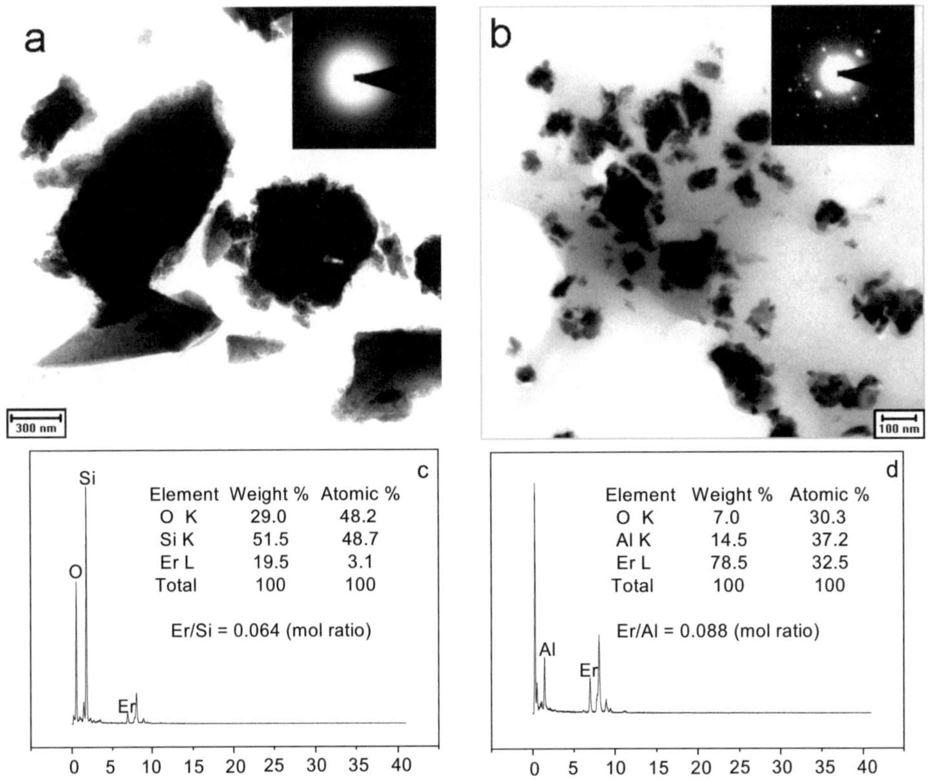

Fig. 2.8 TEM images and EDX of (a, c) Er-doped SiO_2 derived from sol-gel heat-treated at 750°C, inset is the selected area electron diffraction (SAED) pattern of the particles; (b, d) Er-doped Al_2O_3 derived from sol-gel heat-treated at 750°C, inset is the SAED pattern of the particles (Ren, 2008).

Furthermore, the EDX results also show that Er ions are dispersed in the SiO_2 and Al_2O_3, as shown in Figure 8c and 8d, the mol ratios of Er/Si and Er/Al are 0.064 and 0.088 respectively, which are close to the original mol ratios. As we have discussed that the crystal mixture of $Er_3Al_5O_{12}$ and $ErAlO_3$ created when Er ions are doped into Al_2O_3, but the Er-doped SiO_2 is amorphous, and TEM images indicate that Er-doped Al_2O_3 clusters are much smaller than that of Er-doped SiO_2 clusters. So it could be deduced that the dispersion of Er ions at high concentration in Al_2O_3 is better than that in SiO_2.

Figure 2.9 shows the PL spectra of SiNW-ErSiO, SiNW-ErSiAlO and SiNW-ErAlO at the room temperature, pumped with the 477 nm line of an Ar laser. The 477 nm line was chosen because it is absorbed only by SiNWs and not directly by Er^{3+} ions (Suh et al., 2005), which is also confirmed by our result, as shown in Figure 2.9A-a. No peak is observed in Figure 2.9A-a when only Er-doped SiO_2, Er-doped SiAlO or Er-doped Al_2O_3 without SiNWs is pumped with the 477 nm line. This ensures that Er^{3+} excitation occurs via carriers only, and represents an accurate simulation of the situation under current injection. However we observed strong Er^{3+} luminescence at 1534nm (the PL intensity is 66.3 a.u.) from the SiNW-ErSiO even with the high Er concentration (Er/Si = 0.05, concentration 9 at.%), indicating that SiNWs provide a very high areal density of Er and energy transfer from carriers in

SiNWs to Er^{3+} ions dispersed in SiO_2, as shown in Figure 2.9A-b. It can be seen that the Full Width at Half Maximum (FWHM) of the luminescence is about 40.5 nm. This emission band is very typical and is attributed to the Stark splitting of Er^{3+} embedded in an amorphous structure plus small additional inhomogeneous and homogeneous broadening (Polman et al., 1991). Similarly, a strong Er luminescence was observed 1.534 μm from the SiNWs coated with Er-doped Si Al oxide, as shown in Figure 2.9B-d, indicating an energy transfer from carriers in the SiNWs to the Er^{3+} ions. In addition, it can be seen that the FWHM (full width at half maximum) of the luminescence is approximately 56 nm. This emission band structure is especially typical, and may be attributed to the Stark splitting of Er^{3+} embedded in the amorphous structure. It could also be attributed to the presence of many different environments for Er^{3+} ions in the binary Si–Al oxides. The broadening of the spectra suggests a wider and homogeneous distribution of Er^{3+} sites in the matrix. As a result, the PL intensities of SiNW-ErSiO and SiNW-ErSiAlO are similar except that the film quality of SiNW-ErSiAlO id better than that of SiNW-ErSiO after addition of aluminum sol-gel into that of Silicon sol-gel.

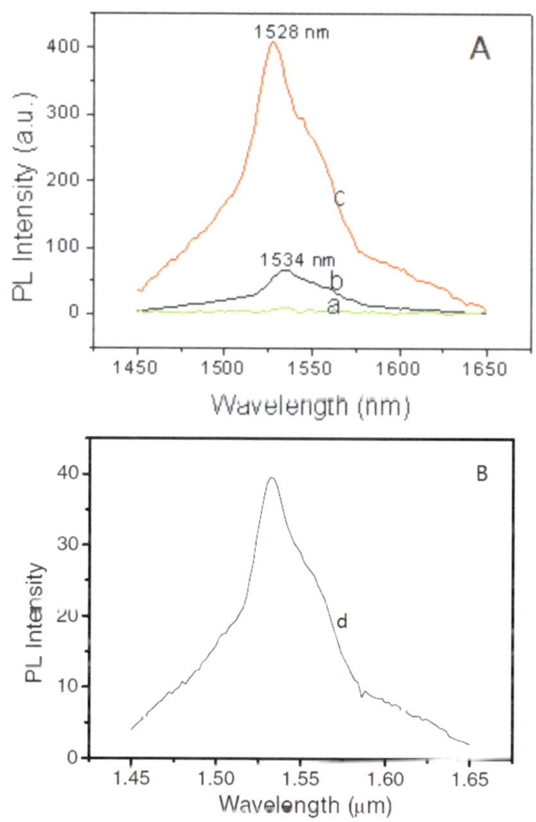

Fig. 2.9 Room temperature PL spectra of (A-a) (Ren, 2008) Er-doped SiO_2 (Er/Si = 0.05), Er-doped Si–Al oxides (Er/(Si+Al) = 0.05, Si/Al = 3:1) and Er-doped Al_2O_3 (Er/Al = 0.10) without SiNWs, (A-b) (Ren, 2008) SiNW-ErSiO (Er/Si = 0.05), (A-c) (Ren, 2008) SiNW-ErAlO (Er/Al = 0.10) and (B-d) (Ren et al., 2007a) SiNW-ErSiAlO (Er/(Si+Al)=0.05, Si/Al = 3:1) films heat-treated at 750°C, wave guide with 477 nm excitation from Ar+ laser.

However, the PL spectrum of SiNW-ErAlO is shown in Figure 2.9A-c. Stronger Er^{3+} luminescence (the PL intensity is about 407.6 a.u.) at 1528 nm from the SiNW-ErAlO is observed, which is approximate six times higher than that of SiNW-ErSiO, indicating the concentration quenching is not obvious even at so high Er concentration (Er/Al = 0.1), and there is more energy transfer from carriers in SiNWs to Er^{3+} ions in Al_2O_3 medium. Perhaps the reasons include two aspects. One is that Er ions with high concentration in Al_2O_3 disperse better than Er ions in SiO_2, which excludes the concentration quenching. The other is that Al_2O_3 has an even higher refraction index, so it couples more efficiently to semiconductor laser materials. And waveguides with smaller dimensions and tighter optical modes can be produced. Moreover, an obvious blue shift of luminescence peak occurs in Figure 2.9A-c and the FWHM is about 46.9 nm, which is wider than that of SiNWs coating with Er-doped SiO_2. It is known that the main reason for the broadening is the local crystal field symmetry at the rare earth ion site and the luminescence spectra of Er-doped nanoparticles are depended on the host structure (Patra, 2004). The XRD results and SAED patterns have shown that the crystal mixture of $Er_3Al_5O_{12}$ and $ErAlO_3$ created when Er ions are doped into Al_2O_3, but the Er-doped SiO_2 is amorphous, which change the local structure around Er^{3+} ions (d'Acapito et al., 2001). Perhaps these are the reasons of the blue shift of luminescence peak and broadening of FWHM.

2.3.2 Controlled PL of Er-doped silicon nanowires

As mentioned, there is a fixed method for preparing Er-doped SiNWs as following processes. Firstly, sputtered Au nanoparticles were employed as the catalyst and the NWs' growth proceeded through a vapor-liquid-solid (VLS) crystal growth mechanism. Then, the long SiNWs were dipped into Sol-gel solutions by surface coating with Er-doped silica or other materials. That is because the optimum locations for Er ions are not inside Si, but in the nanometer thin oxide shell right next to Si (Kimura et al., 2003; Shin et al., 2000b; Stepikhova et al., 1997). As a result, rather than trying to dope SiNWs directly, it would be preferable to coat the SiNWs with high quality, Er-doped oxides by sol-gel in order to optically activate the SiNWs. Lastly, the Er-doped SiNWs derived from sol-gel solution are sintered.

Controlled growth of well-ordered Er-doped SiNWs is crucial, and would eliminate much of the processing associated with device fabrication. Selective growth of SiNWs via a VLS mechanism were achieved using an ion implantation mask, (Sood et al., 2006) lithographically defined regions of SiNW growth by thin film evaporation (Gangloff et al., 2004; Islam et al., 2004) and seeding colloids (Hochbaum et al., 2005). These methods employ expensive or complex processing techniques, but are unfit for the controlled growth of SiNWs and selective coating of Er-doped solutions at the same time.

However, combined sol-gel techniques and gold colloids, a simple technique to achieve patterned growth of SiNWs coating with Er-doped aluminous film is presented. The advantage of this method is that the growth of patterned SiNWs, the doping of Er ions and the sintered process are completed by one step. The Er-doped aluminium gels are calcined to be powders when SiNWs are grown by a VLS mechanism, removing sintered process, so the energy consume is reduced considerable (Ren et al., 2007). SiNWs grew from a solution-based precursor containing ASB ($Al(O\text{-}sec\text{-}C_4H_9)_3$), hydrogen tetrachloroaurate ($HAuCl_4$) and Erbium by VLS mechanism for optical activation. In this process, the functional materials, Au nanoparticles and Er ions, are all evenly dispersed in the sol-gel solution at the same time. Furthermore, ASB was selected as the precursor to disperse Er ions because of its relatively high refraction index (n =1.64) waveguides compared with SiO_2 (n = 1.45), leading to efficient pumping and amplification (Wang & Lei, 2005), lower thermal expansion coefficient (Jaymes et al., 1996). And aluminum could also be either a network former, supplying non-bridging oxygen bonds, or an octahedral-coordinated network modifier (Ryu & Kim, 1995). Moreover, the solution-based precursor allows locating gold/Aluminium gel on any well-defined position on substrate by shadow mask, printing, or soft lithography methods, which could facilitate the integration of SiNWs for characterization and devices. Micropatterns of GeNWs have been grown in a high yield on sol-gel prepared gold/silica substrate by using shadow mask with the help of an additive agent to keep a continuous film (Pan et al., 2005).

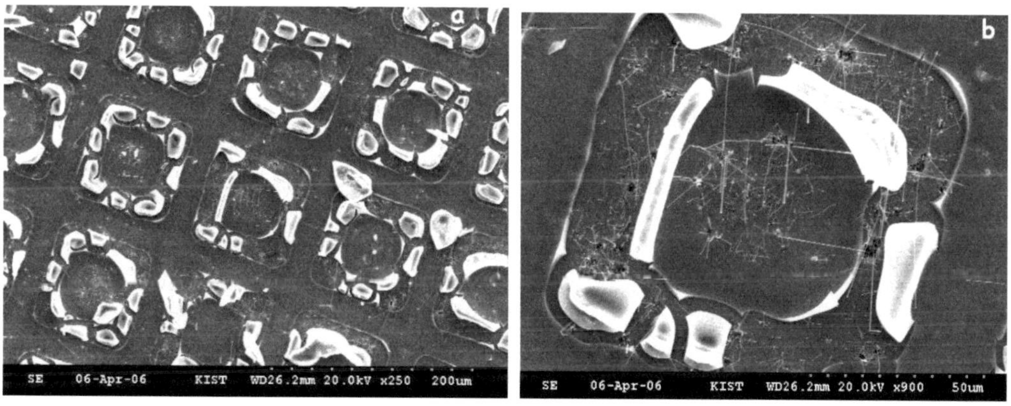

Fig. 2.10 SEM images of Si nanowires patterns grown from sol-gel prepared gold/Er-doped aluminous film by using TEM grids as the masks (a) the ordered arrays of Si nanowires (b) High-magnification image showing no unwanted nanowires grown in the interspaces (Ren et al., 2007b).

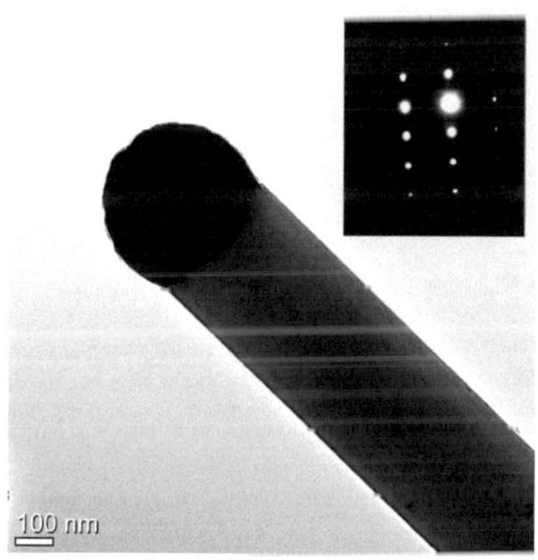

Fig. 2.11 High-magnification TEM image of an individual Si nanowire grown from sol-gel prepared gold/Er-doped aluminous film with a hemispherical Au nanoparticle located at the tip (Ren et al., 2007b).

Figure 2.10 is a SEM image of SiNW pattern produced by using a G200 copper TEM grid as the mask. As shown in Figure 2.10(a), the ordered arrays are obtained. And there is no unwanted nanowires grown in the interspaces except the obtained nanowires in the square area as shown in Figure 2.10(b), indicating that the spatial resolution of the patterns can be well controlled by the resolution of the mask used. Moreover, the SiNWs are tipped with an Au nanoparticle (determined by energy-dispersive X-ray (EDX) spectroscopy) with diameter comparable to the connected nanowire (Figure 2.11), thus showing a typical VLS growth feature. This result suggests that the growth of SiNWs occurs through an Au particle-catalyzed VLS growth mechanism even the gold nanoparticles dispersed in sol-gel solution. In addition, the dense of nanowires could be controlled by the amount of gold nanoparticles, and the thickness of the gel film could also be adjusted through changing the ASB concentration. Moreover, other techniques such as soft lithography (Kind et al., 2000; Huang et al., 2000) and direct printing (Cassell et al., 2001) could also be used to print the gold-containing solution to a desired position on the substrate surface, which might facilitate the integration of Si nanowires into devices.

Figure 2.12 is the XRD pattern of the Er -doped aluminous sol-gel materials sintered at 900°C. The characteristic XRD pattern of $Er_3Al_5O_{12}$ phase (PDF 78 1451) is observed, indicating that the Al^{3+} cations are substituted by the Er^{3+} rare-earth ions sintered at 900°C due to the valence match between the Er^{3+} ions and the substituted Al^{3+} ions, which is consistent with Tanner's report (Tanner et al., 2004) where the mixture of $Er_3Al_5O_{12}$ and $ErAlO_3$ are achieved for concentrations at or below 10% Er doped into Al_2O_3 due to the valence match between the rare-earth ions (Er^{3+}) and the substituted cation (Al^{3+}). As a result, it allows the incorporation of Er^{3+} into the Al_2O_3 lattice (Hochbaum et al., 2005) despite of the large size difference between Er^{3+} (0.89 nm) and the Al^{3+} (0.53 nm) (Tanner et al., 2004)]. Moreover, the θ-Al_2O_3 phase (PDF 35 0121) were observed, which is consistent with former reports (Wang et al., 2004a, 2004b) that when the sintered temperature is at 900°C, the γ-Al_2O_3 transfer to θ-Al_2O_3 phase. In addition, the small peaks of the Er_2O_3 (PDF 77 0464) phase are also observed.

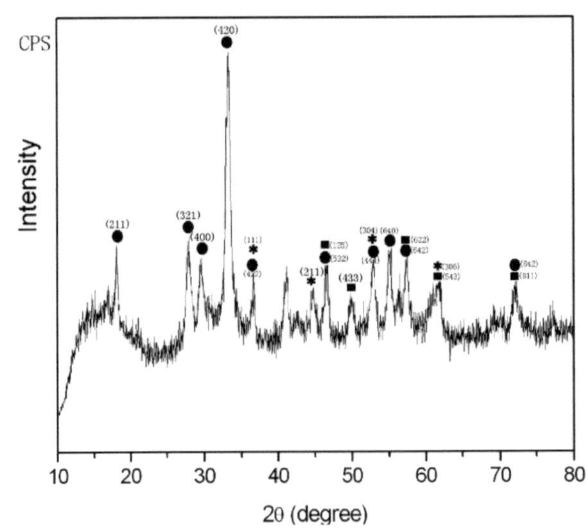

Fig. 2.12 X-ray diffraction patterns of Er-doped aluminous film sintered at 900 °C. The solid circle (•) represents $Er_3Al_5O_{12}$ peaks; the solid square (■) represents Er_2O_3 peaks; the star (*) represents θ-Al_2O_3 peaks (Ren et al., 2007b).

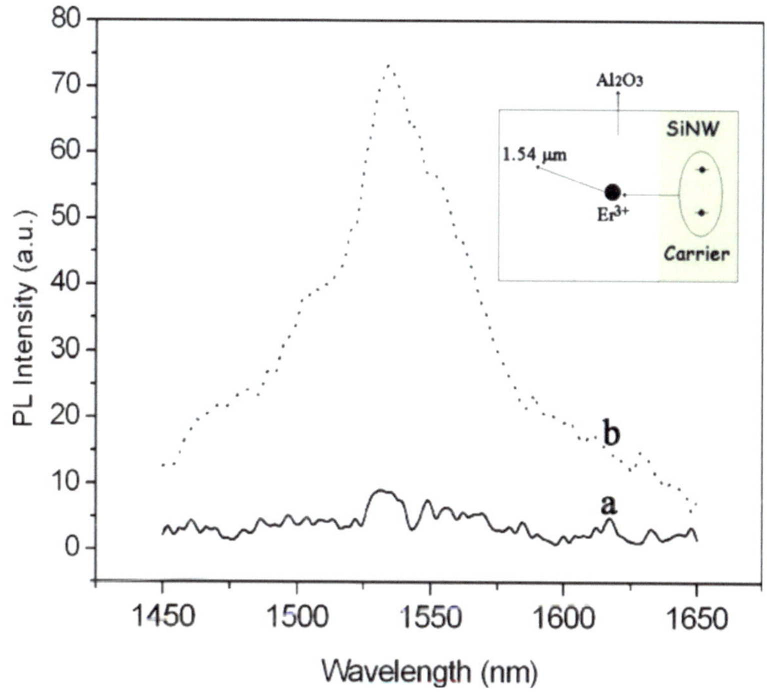

Fig. 2.13 Room temperature PL spectra of SiNWs grown from sol-gel prepared gold/Er-doped aluminous film at 900 °C, wave guide with 477 nm excitation from Ar^+ laser. (a) no SiNWs (b) with SiNWs. The inset shows the schematic process of carrier transfer from SiNW to Er ion (Ren et al., 2007b).

Figure 2.13 shows the room temperature PL spectra of Si nanowires patterned grown on sol-gel prepared gold/Er-doped aluminous film, pumped with the 477 nm line of an Ar laser at room temperature. The 477 nm line was chosen because it is absorbed only by NWs and not directly by Er ions (Suh et al., 2005). This ensures that Er^{3+} excitation occurs via carriers only, and represents an accurate simulation of the situation under current injection, which has been confirmed by our result, as shown in Figure 2.13a. No peak is observed in Figure 13a when only Er-doped Al_2O_3 without SiNWs is pumped with the 477 nm line. This ensures that Er^{3+} excitation occurs via carriers only, and represents an accurate simulation of the situation under current injection. Nevertheless, we observed strong Er^{3+} luminescence at 1534 nm (the PL intensity is 75 a.u.) from the SiNWs coating with Er-doped Al_2O_3 as shown in Figure 2.13b, indicating that energy transfer from carriers in SiNWs to Er^{3+} ions dispersed in Al_2O_3. The schematic process is shown in the inset, which suggests that SiNWs are more effective for exciting Er ions in an oxide layer on their surface.

However, when the Si nanowires patterned grown on sol-gel prepared gold/Er-doped aluminous film are excited by the 488 nm line of an Ar laser at room temperature, reflecting the ${}^4I_{15/2} \rightarrow {}^4F_{7/2}$ and ${}^4I_{15/2} \rightarrow {}^4H_{11/2}$, the PL intensity (about 550 a.u.) is improved largely as shown in Figure 2.14. It could be explained that the 488 nm line of an Ar laser could excite not only Si nanowires, but also the Er^{3+} ions directly. The schematic process is shown in the inset. As a result, the excited Er concentrations are increased, indicating that the optically excited and carrier-excited Er ions are all possible.

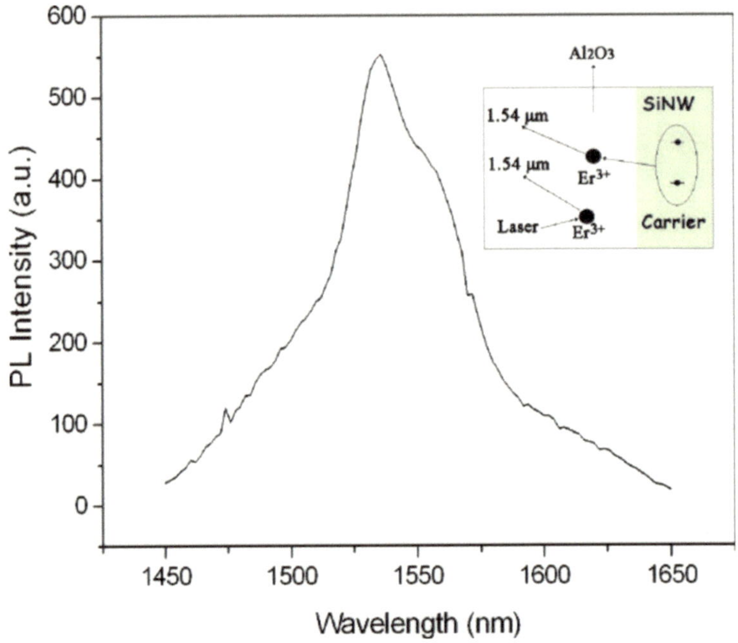

Fig. 2.14 Room temperature PL spectra of SiNWs grown from sol-gel prepared gold/Er-doped aluminous film at 900 °C, wave guide with 488 nm excitation from Ar^+ laser. The inset shows the schematic process of carrier transfer from SiNW to Er ion and excited Er ion directly (Ren et al., 2007b).

3. Other nanowires for PL, such as Ge nanowires

Ge nanowires (GeNWs) is another promising nanowires due to its high carrier mobility and a small band gap (Gu et al., 2001; Wang & Dai, 2002). Germanium is also an important semiconducting electronic material with indirect band gap. Moreover, the direct-gap (0.88 eV) is close to its indirect gap (0.75 eV) in Ge, predicting that quantum confinement effects would appear more pronounced in Ge than Si. The optical activation of GeNWs doping with Er ions (Er-GeNWs) instead of SiNWs will also be achieved and the PL intensity at 1540 nm should be improved largely. But a major limitation is cooperative interactions, such as up conversion and fast energy migration between Er^{3+} ions, limiting emission efficiency at high Er concentration (Snoeks et.al.,1997). Ytterbium is widely used to improve the PL intensity of Er ions because Yb is a well-known sensitizer for the Er emission at 1540 nm due to fast energy transfer from the Yb ions to Er centers and no up conversion in Yb-Yb, and absorption band of Yb ions cover a broad extent from 850 nm to 1000 nm (Kozanecki et al., 1999, 2001).

Thus coating the SiNWs/GeNWs with high quality, Er/Er:Yb doped oxides by sol-gel are produced in order to optically activate the SiNWs/GeNWs (Ren et al., 2007c). Figure 3.1 shows the scanning electron microscope (SEM) image of as-grown SiNWs (a), GeNWs (b) for further coating sol-gel solutions as substrate and Er:Yb-SiNWs (c) and Er:Yb-GeNWs (d). As shown in Figure 3.1(a), the dense arrays of straight SiNWs by VLS are well aligned. The SiNWs lie at an angle of about 60° from the surface normal, which are about 100 nm in diameter and 10μm in length. Similarly, we also obtained GeNWs with smaller than 100 nm in diameter and 10 μm in length in Figure 3.1(b). As shown in Figure 3.1(c) and 3.1(d), both films are continuous without any cracks. The sol-gel solution has completely penetrated into the SiNWs and GeNWs arrays, forming a crack-free, integrative film tight of NWs, even after sintered at 750°C, the film still keeps continuous and the NWs can also be seen well, indicating that Er and Yb ions dispersed in the oxide film derived from sol-gel solutions are continuous to coat the SiNWs/GeNWs with high quality. These are important for the applications of planar devices.

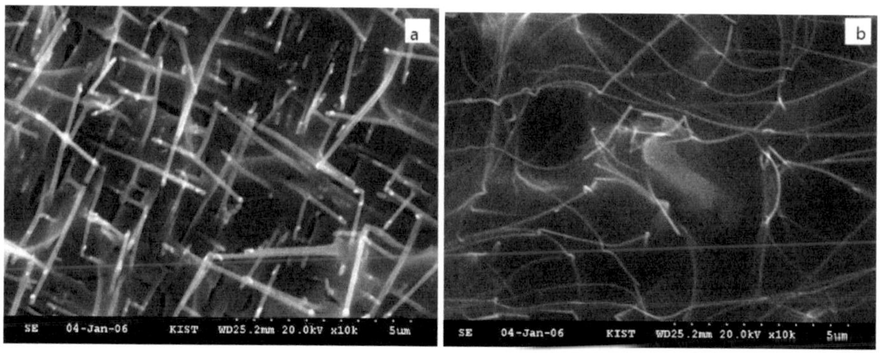

Fig. 3.1 SEM image of (a) Er:Yb-SiNWs, (b) Er:Yb-GeNWs. Er:Yb codoped Si-Al oxides coating on NWs films heated at 750°Cderived from the sol-gel solution (Si/Al = 3:1) (Ren et al., 2007c).

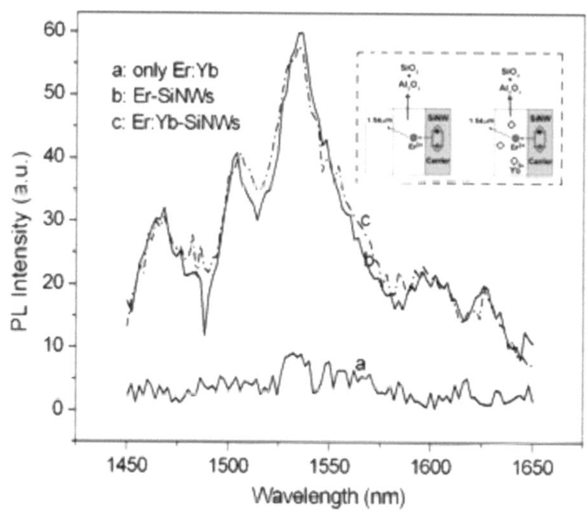

Fig. 3.2 Room temperature PL spectra of Er:Yb codoped Si-Al oxides coating SiNWs film heat-treated at 750°C, wave guide with 477 nm excitation from an Ar$^+$ laser. (Si/Al = 3:1) (a) Er:Yb codoped Si-Al oxides without SiNWs, (b) Er-SiNWs, (c) Er:Yb-SiNWs, the inset is the corresponding schematic route of optical activation of SiNWs (Ren et al., 2007c).

Figure 3.2 shows the room temperature PL spectra of Er-SiNWs and Er:Yb-SiNWs sintered at 750°C, pumped with the 477 nm line of an Ar laser. The 477 nm line was chosen because it is absorbed only by NWs and not directly by Er and Yb ions (Suh et al., 2005). This ensures that Er^{3+} excitation occurs via carriers only, and represents an accurate simulation of the situation under current injection. As shown in Figure 3.2(a), hardly any luminescence from the Er or Er:Yb codoped SiO_2-Al_2O_3 without SiNWs is observed, which is consistent with the previous report (Suh et al., 2005) that either Yb or Er ions are not excited by this 477 nm wavelength laser, or there is no energy transfer to Er ions. However, a strong Er^{3+} luminescence (PL intensity is about 60 a.u.) at 1534 nm from Er-SiNWs was observed as shown in Figure 3.2(b), indicating SiNWs are excited by 477 nm wavelength and energy transfer from carriers in SiNWs to Er^{3+} ions. Unimaginably, the PL intensity (57.8 a.u. at 1534 nm) of Er:Yb-SiNWs was affected weakly after codoping Yb with Er,[15] as shown in Figure 2.16(c), indicating that energy transfer from carriers in SiNWs not to Yb ions but to Er ions only. This result further confirms that the pumped energy of 477 nm wavelengths is directly absorbed by SiNWs, not by Er or Yb ions, suggesting that the Yb ions are not sensitizer for the Er emission at 1534 nm after SiNWs take part in the optical excitation and energy transfer.

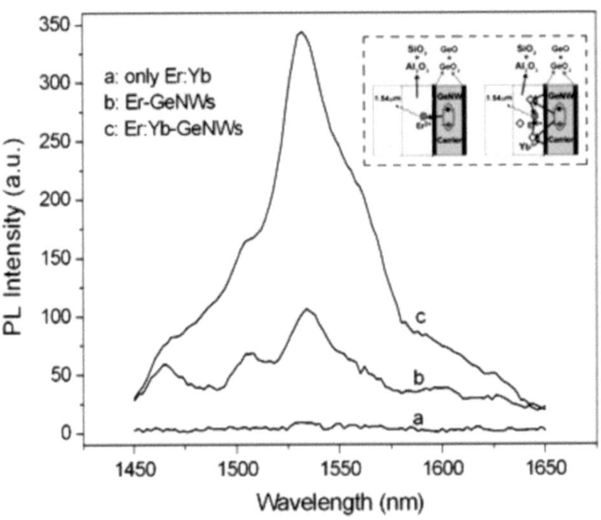

Fig. 3.3 Room temperature PL spectra of Er:Yb codoped Si-Al oxides coating GeNWs film heat-treated at 750°C, wave guide with 477 nm excitation from an Ar^+ laser. (Si/Al = 3:1) (a) Er:Yb codoped Si-Al oxides without GeNWs, (b) Er-GeNWs, (c) Er:Yb-GeNWs, the inset is the corresponding schematic route of optical activation of GeNWs (Ren et al., 2007c).

Figure 3.3 shows the room temperature PL spectra of Er-GeNWs and Er:Yb-GeNWs. The same result with Figure 3.2(a) is shown in Figure 3.3(a) that few luminescence from the Er or Er:Yb codoped SiO_2-Al_2O_3 without GeNWs was observed. But strong Er^{3+} luminescence (PL intensity is about 106.5 a.u.) at 1534 nm from Er-GeNWs is observed as shown in Figure 3.3(b), which is approximate two times higher than that of Er-SiNWs in Figure 3.2(b), indicating more energy transfer from carriers in GeNWs to Er^{3+} ions than that of in SiNWs. Furthermore, the PL intensity of Er^{3+} (about 343.6 a.u.) at 1532 nm from Er:Yb-GeNWs as shown in Figure 3.3(c) increased three times higher than that of only Er-GeNWs in Figure 3.3(b) after codoping Yb ions with Er ions, and a blue shift of the luminescence peak occurs as the Yb ions were added, suggesting that energy transfer from carriers in GeNWs to Yb ions and Yb ions are the sensitizer for the Er luminescence at 1534 nm after GeNWs participate in the luminescence excitation.

The reasons of the PL distinction between Figure 3.2 and Figure 3.3 are deduced. One is perhaps due to the quantum effect of SiNWs and GeNWs. As shown in Figure 3.1, the diameter of the GeNWs is smaller than SiNWs. As we known, the optical properties of a quantum-confined system strongly depend on its size. Another reason is perhaps due to the indirect band gap nature of Si and Ge semiconductors. In Figure 3.2, photons are absorbed by the SiNWs and promote an electron from conduction band (CB) to the valence band (VB). Then the recombination of the electron with a hole in the valence band gives the typical Si nc light emission at ~0.8 μm (corresponding to 1.5 eV) (Khriachtchev & Räsänen, 2005). Since the 1.5 eV couples well with the $^4I_{9/2}$ level of the Er manifold, in presence of Er, the energy can be transferred to the Er ions to excite it (Franzo` et al., 2000), as shown in Figure 3.4A ((a)-(e)). From this level, a rapid relaxation occurs to the $^4I_{11/2}$ level with the subsequent emission of 0.98 μm photons or with a relaxation to the metastable level $^4I_{13/2}$ and emission of photons at 1534 nm. For GeNWs, the recombination of the electrons with holes in the valence band gives the Ge nc light emission at the range of 0.9-2.3 eV (Kanemitsu et al., 1992; de Azevedo et al., 2005). Other wavelength besides 1.5 eV, such as 0.9 eV, could also couple well with the metastable $^4I_{13/2}$ level of the Er manifold and emission of photons at 1534 nm (as shown in Figure 3.4B). These different wavelengths from GeNWs could excite more Er ions than single wavelength (1.5 eV) from SiNWs. As a result, the PL intensity of Er-GeNWs is much higher than that of Er-SiNWs (as shown in Figure 3.2(b) and Figure 3.3(b)).

The optical properties of Er codoped with Yb further proved that the indirect gap nature of Si and Ge semiconductors decides the energy transfer between semiconductor and rare earth ions. If the nanowire diameter decides the optical property, the PL intensity of Er:Yb-SiNWs should improved than that of only Er-SiNWs due to the sensitizer of Yb to Er. Actually, the Yb ions have no any effects on the PL intensity of Er:Yb-SiNWs, because the 1.5 eV wavelengths from excited SiNWs cannot be absorbed by Yb ions, but only by Er ions. As a result, the Yb ions cannot be excited and the PL intensity of Er:Yb-SiNWs didn't change compared with that of Er-SiNWs after Yb ions doped with Er ions coating on the SiNWs surface.

However, the Yb ions have great effects on that of Er:Yb-GeNWs. GeNWs emit not only 0.9 eV and 1.5 eV wavelengths to couple well with the metastable $^4I_{13/2}$ level of the Er ions directly, but also 1.3 eV wavelengths, coupling well with the excited state of Yb^{3+}-$^2F_{5/2}$ states. So the excited energy from Yb^{3+} can be transferred resonantly to the $^4I_{11/2}$ level of Er^{3+}. This excitation process is shown in Figure 3.4C. Moreover, Ge is well known to oxidize to form GeO and GeO_2 in air on the GeNW surfaces (Tabet et al., 1999; Wang et al., 2004). These oxidations can be excited to emit a luminescence around 3.1 - 4.1 eV (Oku et al., 2000). Theses lights cannot be absorbed by Er ions but can be absorbed by Yb ions (Kozanecki et al., 1999). As a result of photon absorption, the Yb ion becomes excited its 2+ state, leaving a hole, h_{VB}, in the valence band, which most probably is localized near Yb^{2+} ion. In the subsequent, Yb^{2+} + h_{VB} recombination process Yb^{3+} is left in its excited $^2F_{5/2}$ state, from which energy then can be transferred resonantly to the $^2I_{11/2}$ level of Er^{3+} (Song et al., 2006). Hence, the PL intensity of Er:Yb-GeNWs is the total intensities of the three processes and it is higher than that of without Yb codoping. The blue shift of the luminescence peak further indicates that Yb ions take part in the energy transfer during the optical activation of Er:Yb-GeNWs. In summary, the optical activations of SiNWs and GeNWs are mainly dependent on the indirect gap nature of Si and Ge semiconductors. Stronger optical activation of Er:Yb-GeNWs is achieved.

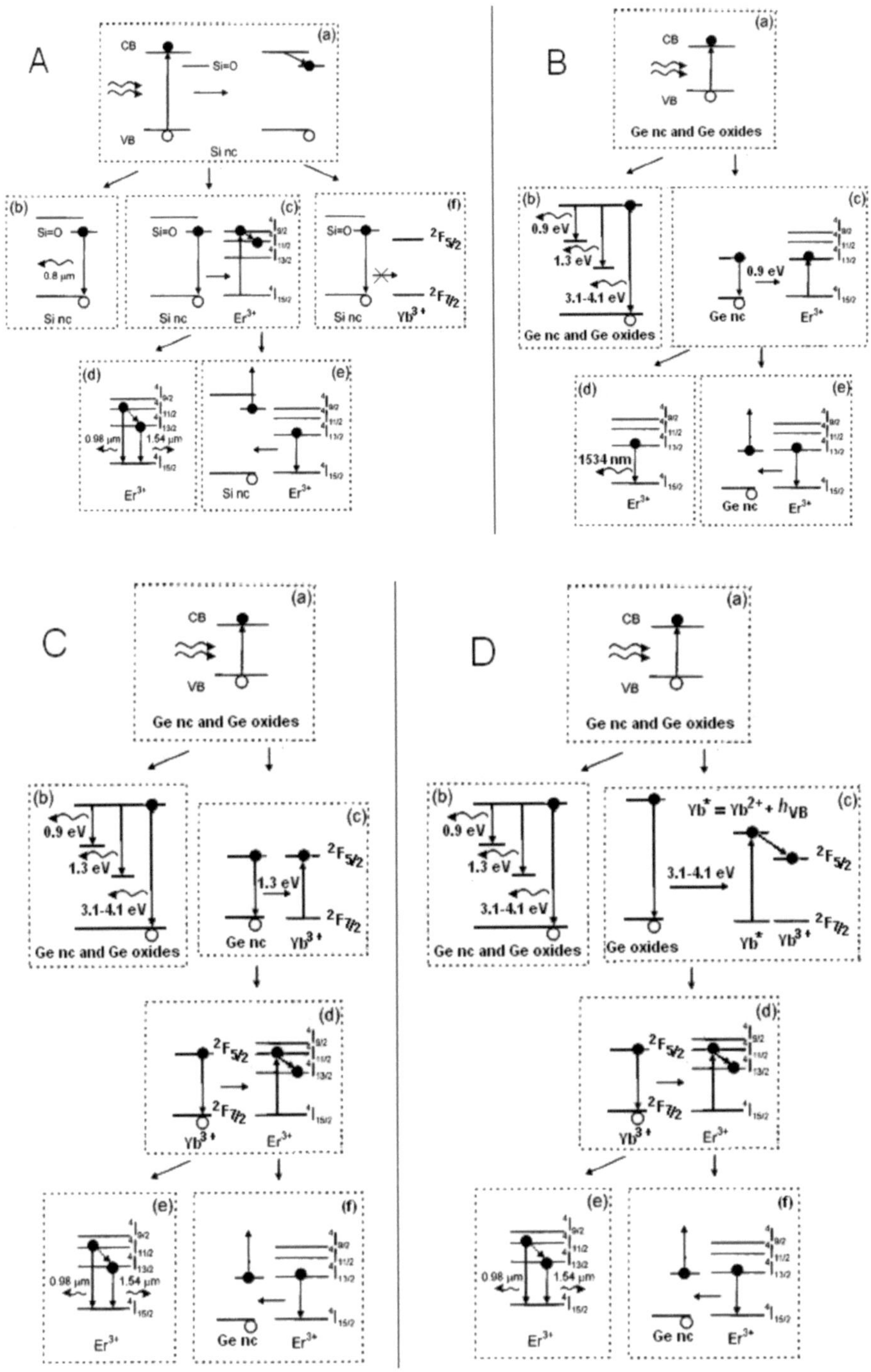

Fig. 3.4 Schematic of the mechanisms of optical activation of Er-doped and Er:Yb-codoped SiNWs and GeNWs. (A) Er-doped and Er:Yb-codoped SiNWs; (B) Er-doped GeNWs; (C) and (D) Er:Yb-codoped GeNWs (Ren et al., 2007c).

4. Conclusion

Optical activations of Er doped Si ncs, SiNWs and GeNWs have been observed, but the PL intensity of Si ncs is weaker than those of SiNWs and GeNWs. So much progress has been made to improve the photoluminescence (PL) intensity; however, a challenging problem in Er-doped semiconductors is that only a small fraction of the excitation energy is transferred from ncs to Er ions. One of the major limitations is that isolated ncs cannot be efficiently addressed electrically. On this point, the successful optical activation of Er-doped SiNWs (Er-SiNWs) has been achieved. More important that the optimum location for Er^{3+} is not inside Si, but in the nanometer-thin oxide shell right next to Si. This gives us simple ways to dope the NWs with additional technologies, such as sol-gel method. Thus, this process avoids the fabrication of SiNWs with Er ions directly. The long length of SiNWs exceeding 1 μm overcomes the isolation of Si nc dispersed in the SiO_2 and Al_2O_3 matrix to provide a very high areal density of Er ions. Moreover, the controlled growth of well-ordered Er-doped SiNWs has been achieved. The growth of patterned SiNWs, the doping of Er ions and the sintered process are completed in one step. The Er-doped aluminium gels are calcined to be powders when SiNWs are grown by a VLS mechanism, removing sintered process, so the energy consume is reduced. Similarly, the GeNWs posses the same result. Moreover, the rare earth Yb may improve the PL intensity of Er-doped GeNWs additionally.

In the future, the additional technologies for Er doping are required more attentions. For sol-gel method, the sol-gel film quality is important, such as no cracks, homogeneous, no concentration quench, and so on.The challenging researches focus on how to further improve the film quality and fine control of the film thickness to match well with the actual requirements. In addition, the SiNW diameter also has great affection on the PL intensity. The controlled growth of Er-doped SiNWs with stable diameter and arrays are also good challenge.

5. Reference

d'Acapito, F., Mobilio, S., Gastaldo, P., Barbier, D., Santos,Luis F., Martins, O. & Almeida, R. M. *J. Non-Cryst. Solids*, Vol. 293-295, (2001). pp. (118-120).

Allen, J. E., Hemesath, E. R., Perea, D. E., Lensch-Falk, J. L., Li, Z. Y., Yin, F., Gass, M. H., Wang, P., Bleloch, A. L., Palmer, R. E. & Lauhon, L. J. (2008). High-resolution detection of Au catalyst atoms in Si nanowires, *Nat. Nanotechnol.*, Vol. 3, (Feb. 2008). pp. (168-173), ISSN: 1748-3387.

Amelinckx, S. Zhang, X. B. Bernaerts, D. Bernaerts, D. Zhang, X. F.;Ivanov, V. Nagy, J. B. (1994). A Formation Mechanism for Catalytically Grown Helix-Shaped Graphite Nanotubes, Science, Vol. 265, Issue 5172, (July 1994). pp. (635-639), ISSN: 0036-8075.

Andersen, S. E. S., Brenner, R., Wellard, C. J., Yang, C., Hopf, T., Escott, C. C., Clark, R. G., Dzurak, A. S., Jamieson, D. N. & Hollenberg, C. L. (2007). Charge State Control and Relaxation in an Atomically Doped Silicon Device, *Nano Lett.*, Vol. 7, Issue 7, (July 2007). pp. (2000-2003), ISSN: 1530-6984.

Armelao, L., Gross, S., Obetti, G., Tondello, E. (2005). Er^{3+}-doped SiO_2–Al_2O_3 thin films prepared by the sol-gel route, *Surf. Coating Tech*, Vol. 190, Issue 2-3, (2005). pp. (218-222), ISSN: 0257-8972.

Azarian, A., Iraji Zad, A., Dolati, A. & Mahdavi, S. M. (2009). Field emission of Co nanowires in polycarbonate template, *Thin Solid Films,* Vol. 517, Issue 5, (Jan. 2009). pp. (1736–1739), ISSN: 0040-6090.

de Azevedo, W.M., da Silva Jr, E.F., de Vasconcelos, E.A., Boudinov, H. (2005). Visible photoluminescence from Ge nanoclusters implanted in nanoporous aluminum oxide films, Microelectronics Journal, Vol. 36, Issue 11, (Nov. 2005). pp. (992-994), ISSN: 0026-2692.

Bogart, T. E., Dey, S., Lew, K.-K., Mohney, S. E. & Redwing, J. M. (2005). Diameter-Controlled Synthesis of Silicon Nanowires Using Nanoporous Alumina Membranes, *AdV. Mater.,* Vol. 17, Issue 1, (Jan. 2005). pp. (114-117), ISSN: 1521-4095.

Bootsma, G. A. & Gassen, H. J. (1971). A quantitative study on the growth of silicon whiskers from silane and germanium whiskers from germane, *J. Cryst. Growth,* Vol. 10, Issue 3, (Aug. 1971). pp. (223-234), ISSN: 0022-0248.

Brus, L. (1994). Luminescence of Silicon Materials: Chains, Sheets, Nanocrystals, Nanowires, Microcrystals, and Porous Silicon, *J. Phys. Chem.,* Vol. 98, Issue 14, (Apr. 1994). pp. (3575-3581).

Cassell, A. M., Verma, S., Delzeit, L., Meyyappan, M. & Han, J. (2001). Combinatorial Optimization of Heterogeneous Catalysts Used in the Growth of Carbon Nanotubes, *Langmuir,* Vol. 17, Issue 2, (Jan. 2001). pp. (260-264), ISSN: 0743-7463.

Chryssou, C. E., Kenyon, A. J., Iwayama, T. S., Pitt, C. W. & Hole, D. E. (1999). Evidence of energy coupling between Si nanocrystals and Er^{3+} in ion-implanted silica thin films, *Appl. Phys. Lett.,* Vol. 75, Issue 14, (1999). pp. (2011-2014), ISSN: 0003-6951.

Chung, S. W., Yu, J. Y. & Heath, J. R. (2000). Silicon nanowire devices, *Appl. Phys. Lett.,* Vol. 76, Issue 15, (Apr. 2000). pp. (2068-2070), ISSN: 0003-6951.

Coffa, S., Franzo` ,G., Priolo, F., Polman, A. & Serna, R. Temperature dependence and quenching processes of the intra-4f luminescence of Er in crystalline Si. *Phys. Rev., B,* Vol. 49, Issue 23 (1994). pp. (16313-16320), ISSN: 1098-0121.

Cui, Y., Lauhon, L. J., Gudiksen, M. S., Wang, J. & Lieber, C. M. (2001). Diameter-controlled synthesis of single-crystal silicon nanowires, *Appl. Phys. Lett.,* Vol. 78, Issue 15, (Apr. 2001). pp. (2214-2216), ISSN: 0003-6951..

Dai, H. J., Wong, E. W., Lu, Y. Z., Fan, S. S. & Lieber, C. M. (1995). Synthesis and characterization of carbide nanorods, *Nature,* Vol. 375, (June 1995). pp. (769-772), ISSN: 0028-0836.

Ennen, H., Schneider, J., Pomrenke, G. & Axmann, A. (1983). 1.54-μm luminescence of erbium-implanted III–V semiconductors and silicon, *Appl. Phys. Lett.,* Vol. 43, Issue10, (1983). pp. (943-946), ISSN: 0003-6951.

Feofilov, S.P., Kulinkin, A.B., Kutsenko, A.B. & Zakharchenya, R.I. (1998). Selective laser spectroscopy of RE^{3+} and Mn^{4+} in sol-gel technique produced Al_2O_3, *Journal of Luminescence,* Vol. 76&77, (1998). pp. (217-220), ISSN: 0022-2313.

Franzo`, G., Vinciguerra, V. & Priolo, F. (1999). The excitation mechanism of rare-earth ions in silicon nanocrystals, *Appl. Phys. A: Mater. Sci. Process.,* Vol. 69, Issue 1, (Jul. 1999). pp. (3-12), ISSN (printed): 0947-8396.

Franzo`, G., Pacifici, D., Vinciguerra, V., Priolo, F. & Iacona, F. (2000). Er^{3+} ions–Si nanocrystals interactions and their effects on the luminescence properties. *Appl. Phys. Lett.,* Vol. 76, Issue 16, (2000). pp. (2167-2179), ISSN: 0003-6951.

Franzó, G., Irrera, A., Moreira, E. C., Miritello, M., Iacona, F., Sanfilippo, D., Di Stefano, G., Fallica, P. G. & Priolo, F. (2002). Electroluminescence of silicon nanocrystals in MOS structures, *Appl. Phys. A: Mater. Sci. Process.*, Vol.74, Issue 1, (Jan. 2002). pp. (1-5), ISSN (printed): 0947-8396.

Franzo` , G., Boninelli, S., Pacifici, D., Priolo, F., Iacona, F. & Bongiorno, C. (2003). Sensitizing properties of amorphous Si clusters on the 1.54-μm luminescence of Er in Si-rich SiO_2, Appl. Phys. Lett., Vol. 82, Issue 22 , (2003). pp. (3871-3875), ISSN: 0003-6951.

Fujii, M., Yoshida, M., Kanzawa, Y., Hayashi, S. & Yamamoto, K. (1997). 1.54 μm photoluminescence of Er^{3+} doped into SiO_2 films containing Si nanocrystals: Evidence for energy transfer from Si nanocrystals to Er^{3+}, *Appl. Phys. Lett.*, Vol. 71, Issue 9, (1997). pp. (1198-1200) , ISSN: 0003-6951.

Fujii, M., Yoshida, M., Hayashi, S. & Yamamoto, K. (1998). Photoluminescence from SiO_2 films containing Si nanocrystals and Er: Effects of nanocrystalline size on the photoluminescence efficiency of Er^{3+}, *J. Appl. Phys.*, Vol. 84, Issue 8, (1998). pp. (4525-4531).

Gangloff, L., Minoux, E., Teo, K. B. K., Vincent, P., Semet, V., Binh, V. T., Yang, M. H., Lacerda, R. G., Bu, I. Y. Y., Pirio, G., Schnell, J. P., Pribat, D., Hasko, D. G., Amaratunga, G. A. J., Milne, W. I. & Legagneux, P. (2004). Self-Aligned, Gated Arrays of Individual Nanotube and Nanowire Emitters, *Nano Lett.*, Vol.4, Issue 9, (Sep. 2004). pp. (1575-1579), ISSN: 1530-6984.

Gao, T., Meng, G., Wang, Y., Sun, S. & Zhang, L. (2002). Electrochemical synthesis of copper nanowires, *J. Phys.: Condens. Matter*, Vol. 14, Issue 3, (Jan. 2002). pp. (355–363), ISSN: 0953-8984.

Geiger, F., Busse, C. A. & Loehrke, R. I. (1987). The vapor pressure of indium, silver, gallium, copper, tin, and gold between 0.1 and 3.0 bar, *Int. J. Thermophys.*, Vol. 8, Issue 4, (1987). pp. (425-436), ISSN: 0195-928X.

Giovine, E., Notargiacomo, A., Di Gaspare, L., Palange, E., Evangelisti, F., Leoni, R., Castellano, G., Torrioli, G. & Foglietti, V. (2001). Investigation of SiGe-heterostructure nanowires, *Nanotechnology*, Vol. 12, Issue 2, (June 2001). pp. (132-135), ISSN: 0957-4484.

Givargizov, E.I. (1975). Fundamental aspects of VLS growth, *J. Crystal Growth*, Vol. 31, (Dec. 1975). pp. (20-30), ISSN: 0022-0248.

Givargizov, E. I.; Sheftal, N. N. Morphology of silicon whiskers grown by the VLS-technique, *J. Cryst. Growth*, Vol. 9, (May 1971). pp. (326-329), ISSN: 0022-0248.

Gu, G., Burghard, M., Kim, G. T., Du¨sberg, G. S., Chiu, P. W., Krstic, V., Roth, S. & Han, W. Q. (2001). Growth and electrical transport of germanium nanowires, *J. Appl. Phys.*, Vol. 90, Issue 11, (Dec. 2001). pp. (5747-5752), ISSN:0021-8979.

Han, H-S., Seo, S-Y., Shin, J. H. & Park, N. (2002). Coefficient determination related to optical gain in erbium-doped silicon-rich silicon oxide waveguide amplifier. *Appl. Phys. Lett.*, Vol. 81, Issue 20, (2002). pp. (3720-3722), ISSN: 0003-6951.

Han, W. Q., Fan, S. S., Li, Q. Q. & Gu, B. L. (1997). Synthesis of silicon nitride nanorods using carbon nanotube as a template, *Appl. Phys. Lett.*, Vol. 71, Issue 16, (Oct. 1997a). pp. (2271-2273), ISSN: 0003-6951.

Han, W. Q., Fan, S. S., Li, Q. Q., Liang, W.J., Gu, B. L. & Yu, D. P. (1997). Continuous synthesis and characterization of silicon carbide nanorods, *Chem. Phys. Lett.*, Vol. 265, Issue 3-5, (Feb.1997b). pp. (374-378), ISSN: 0009-2614.

Han, W.Q., Bando, Y. Kurashima, K. & Sato, T. Synthesis of boron nitride nanotubes from carbon nanotubes by a substitution reaction, *Appl. Phys. Lett.*, Vol. 73, Issue 21, (Nov. 1998). pp. (3085-3087), ISSN: 0003-6951.

Han, W. Q., Philipp, K. R., Ernst, F. & Ru¨ hle, M. (1999). Formation of (BN)xCy and BN Nanotubes Filled with Boron Carbide Nanowires, *Chem. Mater.*, Vol. 11, Issue 12, (Dec. 1999). pp. (3620-3623), ISSN: 0897-4756.

Hochbaum, A. I., Fan, R., He, R. & Yang, P. (2005). Controlled Growth of Si Nanowire Arrays for Device Integration, *Nano Lett.*, Vol. 5, Issue 3, (Mar. 2005). pp. (457-460), ISSN (printed): 1530-6984.

van den Hoven, G. N., Snoeks, E., Polman, A., van Uffelen, J. W. M., Oei,Y. S. & Smit, M. K. (1993). Photoluminescence characterization of Er–implanted Al_2O_3 films, *Appl. Phys. Lett.*, Vol. 62, Issue 24, (Jun 1993). pp. (3065-3067), ISSN: 0003-6951.

van den Hoven, G. N., Snoeks, E., Polman, A., van Dam, C., van Uffelen, J. W. M. & Smit, M. K. (1996). Upconversion in Er–implanted Al_2O_3 waveguides, *J. Appl. Phys.*, Vol. 79, Issue 3, (1996). pp. (1258-1266), ISSN: 0021-8979.

Huang, J., Chiam, S. Y., Tan, H. H., Wang, S. & Chim W. K. Fabrication of Silicon Nanowires with Precise Diameter Control Using Metal Nanodot Arrays as a Hard Mask Blocking Material in Chemical Etching, *Chem. Mater.*, Vol. 22, (2010). pp. (4111–4116), ISSN 0897-4756.

Huang, S. M., Mau, A. M. H., Turney, T. W., White, P. A. & Dai, L. M. (2000). Patterned Growth of Well-Aligned Carbon Nanotubes: A Soft-Lithographic Approach, *J. Phys. Chem. B*, Vol. 104, Issue 10, (March 2000). pp. (2193-2196), ISSN: 1520-6106.

Iacona, F., Franzo`, G. & Spinella, C. (2000). Correlation between luminescence and structural properties of Si nanocrystals, *J. Appl. Phys.*, Vol. 87, Issue 3, (Feb. 2000). pp. (1295-2303), ISSN: 0021-8979.

Iacona, F., Pacifici, D., Irrera, A., Miritello, M., Franzo` , G., Priolo, F., Sanfilippo, D., DiStefano, G. & Fallica, P.G. (2002). Electroluminescence at 1.54 μm in Er-doped Si nanocluster-based devices, *Appl. Phys. Lett.*, Vol. 81, Issue 17 , (2002). pp. (3242-3244), ISSN: 0003-6951.

Irrera, A., Pacifici, D., Miritello, M., Franzó, G., Priolo, F., Iacona, F., Di Stefano, G. & Fallica, P. G. (2002). Excitation and de-excitation properties of silicon quantum dots under electrical pumping, *Appl. Phys. Lett.*, Vol. 81, Issue 10, (2002). pp. (1866-1868), ISSN: 0003-6951.

Islam, M., Sharma, S., Kamins, T. I. & Williams, R. S. (2004). Ultrahigh-Density Semiconductor Nano-bridges Formed Between two Semiconductor Surfaces, *Nanotechnology*, Vol. 15, Issue 5 (May 2004). pp. (L5–L8), ISSN 0957-4484.

Jaymes, I., Douy, A., Massiot, D. & Coutres, J. P. (1996). Evolution of the Si environment in mullite solid solution by 29Si MAS-NMR spectroscopy, *J. Non-Cryst. Solids*, Vol. 204, Issue 2, (1996). pp. (125-134), ISSN: 0022-3093.

Jimenez de Castro, M., Serna, R., Chaos, J. A., Afonso, C. N. & Hodgson, E. R. (2000). Influence of defects on the photoluminescence of pulsed-laser deposited Er-doped amorphous Al_2O3 films, *Nucl. Instrum. Methods B*, Vol. 166-167, (2000). pp. (793-797), ISSN: 0168-583X.

Kanemitsu, Y., Uto, H., Masumoto, Y. & Maeda, Y. (1992). On the origin of visible photoluminescence in nanometer-size Ge crystallites, *Appl. Phys. Lett.*, Vol. 61, Issue 18, (Nov. 1992). pp. (2187-2189), ISSN:0003-6951.

Kenyon, A. J., Trwoga, P. F., Federighi, M. & Pitt, C. W. (1994). Optical properties of PECVD erbium-doped silicon-rich silica: evidence for energy transfer between silicon microclusters and erbium ions, *J. Phys.: Condens. Matter*, Vol. 6, Issue 21, (1994). pp. (L319), ISSN 0953-8984.

Kenyon, A. J., Chryssou, C. E. & Pitt, C. W. (2000). Indirect excitation of 1.5 μm emission from Er3+ in silicon-rich silica, *Appl. Phys. Lett.*, Vol. 76, Issue 6, (2000). pp. (688-690), ISSN: 0003-6951.

Kenyon, A. J. (2002). Recent developments in rare-earth doped materials for optoelectronics, *Progress in Quantum Electronics*, Vol. 26, Issue 4-5, (2002). pp. (225-284), ISSN: 0079-672).

Khanal, D. R., Yim, J. W. L., Walukiewicz, W. & Wu, J. (2007). Effects of Quantum Confinement on the Doping Limit of Semiconductor Nanowires, *Nano Lett.*, Vol. 7, Issue 5, (May 2007). pp. (1186 -1190), ISSN: 1530-6984.

Khriachtchev, L. & Räsänen, M. (2005). Free-standing silica film containing Si nanocrystals: Photoluminescence, Raman scattering, optical waveguiding, and laser-induced thermal effects, *Appl. Phys. Lett.*, Vol. 86, Issue 14, (Apr. 2005). pp. (141911-141913), ISSN:0003-6951.

Kik, P. G., Brongersma, M. L. & Polman, A. (2000). Strong exciton-erbium coupling in Si nanocrystal-doped SiO_2, *Appl. Phys. Lett.*, Vol. 76, Issue 17, (2000). pp. (2325-2327) , ISSN: 0003-6951.

Kim, C. S., Chung, Y. B., Youn, W. K. & Hwang, N. M. (2009). Generation of charged nanoparticles during the synthesis of carbon nanotubes by chemical vapor deposition, *Carbon*, Vol. 47, Issue 10, (Aug. 2009b). pp. (2511-2518), ISSN: 0008-6223.

Kim, C. S., Kwak, I. J., Choi, K. J., Park, J. G. & Hwang, N. M. (2010). Generation of Charged Nanoparticles During the Synthesis of Silicon Nanowires by Chemical Vapor Deposition, *J. Phys. Chem. C*, Vol. 114, Issue 8, (Mar. 2010). pp. (3390– 3395), ISSN: 1932-7447.

Kimura, T., Isshiki, H., Ide, S., Shimizu, T. & Ishida, T. (2003). Suppression of Auger deexcitation and temperature quenching of the Er-related 1.54 μm emission with an ultrathin oxide interlayer in an Er/ SiO_2/Si structure, *J. Appl. Phys.*, Vol. 93, Issue 5, (2003). pp. (2595-2601), ISSN:0021-8979.

Kind, H., Bonard, J., Forro, L., Nilsson, L. & Schlapbach, L. (2000). Printing Gel-like Catalysts for the Directed Growth of Multiwall Carbon Nanotubes, *Langmuir*, Vol. 16, Issue 17, (August 2000). pp. (6877-6883), ISSN: 0743-7463.

Klimov, V. I., Schwarz, C., McBranch, D. & White, C. W. (1998). Initial carrier relaxation dynamics in ion-implanted Si nanocrystals: Femtosecond transient absorption study, *Appl. Phys. Lett.*, Vol. 73, Issue 18, (Nov. 1998). pp (2603-2605), ISSN. 0003-6951.

Kozanecki, A., Homewood, K. & Sealy, B. J. *J. Appl. Phys.*, Vol. 75, (1999). pp. (793-795).

Kozanecki, A., Sealy, B.J., Homewood, K., Ledain, S., Jantsch, W. & Kuritsyn, D. (2001). Sensitization of the 1.54 μm luminescence of Er3+ in SiO_2 films by Yb and Si-nanocrystals. Materials Science and Engineering B, Vol. 81, Issue 1-3, (Apr. 2001). pp. (23-28). ISSN: 0921-5107.

Krishnan, R. & Thompson, C. V. (2007). Monodomain High-Aspect-Ratio 2D and 3D Ordered Porous Alumina Structures with Independently Controlled Pore Spacing and Diameter, *Adv. Mater.,* Vol. 19, Issue 7, (Apr. 2007). pp. (988–992), ISSN: 1521-4095.

Lew, K. K. & Redwing, J. M. (2003). Growth characteristics of silicon nanowires synthesized by vapor–liquid–solid growth in nanoporous alumina templates, *J. Cryst. Growth,* Vol. 254, Issue 1-2, (June 2003). pp. (14-22), ISSN: 0022-0248.

Linnros, J., Lalic, N., Galeckas, A. & Grivickas, V. (1999). Analysis of the stretched exponential photoluminescence decay from nanometer-sized silicon crystals in SiO_2, *J. Appl. Phys.*, Vol. 86, Issue 11, (1999). pp. (6128-6134), ISSN: 0021-8979.

Lombardi, I., Hochbaum, A. I., Yang, P., Carraro, C. & Maboudian, R. (2006). Synthesis of High Density, Size-Controlled Si Nanowire Arrays via Porous Anodic Alumina Mask, *Chem. Mater.,* Vol. 18, Issue 4, (Feb. 2006). pp. (988-991), ISSN: 0897-4756.

Moore, G. E. *Electronics,* Vol. 38, (1965). pp. (84-90).

Mandl, B., Stangl, J., Hilner, E., Zakharov, A. A., Hillerich, K., Dey, A. W., Samuelson, L., Bauer, G., Deppert, K. & Mikkelsen, A. (2010). Growth Mechanism of Self-Catalyzed Group III−V Nanowires, *Nano Lett.,* Vol. 10 Issue 11, (Dec. 2010). pp. (4443–4449), ISSN (printed): 1530-6984.

Masuda, H. & Fukuda, K. (1995). Ordered Metal Nanohole Arrays Made by a Two-Step Replication of Honeycomb Structures of Anodic Alumina, *Science,* Vol. 268, Issue 5216, (Jun. 1995). pp. (1466–1468), ISSN 0036-8075 (print).

Morales, A. M. & Lieber, C. M. (1998). A Laser Ablation Method for the Synthesis of Crystalline Semiconductor Nanowires, *Science,* Vol. 279, Issue 5348, (Jan. 1998). pp. (208-211), ISSN 0036-8075 (print).

Nielsch, K., Mu¨ ller, F., Li, A.-P. & Go¨ sele, U. (2000). Uniform Nickel Deposition into Ordered Alumina Pores by Pulsed Electrodeposition, *Adv. Mater.,* Vol. 12, Issue 8, (Apr. 2000). pp. (582–586), ISSN: 1616-301X (Print).

Nielsch, K., Choi, J., Schwirn, K., Wehrspohn, R. B. & Gosele, U. (2002). Self-ordering Regimes of Porous Alumina: The 10 Porosity Rule, *Nano Lett.,* Vol. 2, Issue 7, (Jul. 2002). pp. (677-680), ISSN (printed): 1530-6984.

Nishio, K., Koga, J., Yamaguchi, T. & Yonezawa, F. (2003). Theoretical study of light-emission properties of amorphous silicon quantum dots, *Phys. Rev. B,* Vol. 67, Issue 19, (2003). pp. (195304-195308), ISSN: 1098-0121,).

Oku, T., Nakayama, T., Kuno, M., Nozue, Y., Wallenberg, L. R., Niihara, K. & Suganuma, K. (2000). Formation and photoluminescence of Ge and Si nanoparticles encapsulated in oxide layers, Materials Science and Engineering B, Vol. 74, Issue 1-3, (May 2000). pp. (242-247), ISSN: 0921-5107.

Osada, Y., Nakayama, H., Shindo, M., Odaka, T. & Ogata, Y. (1979). Growth and Structure of Silicon Fibers, *J, Electrochem. Soc.,* Vol. 126, Issue 1, (Jan. 1979). pp. (31-36), ISSN: 0013-4651 (print).

Pan, A. W., Dai, S. & Lowndes, D. H. (2005). Straight single-crystalline germanium nanowires and their patterns grown on sol–gel prepared gold/silica substrates, *Solid State Communications,* Vol. 134, Issue 4, (April 2005). pp. (251-255), ISSN: 0038-1098.

Park, N.-M., Kim, T.-Y., Kim, S. H., Sung, G. Y., Cho, K. S., Shin, J. H., Kim, B.-H., Park, S.-J., Lee, J.-K. & Nastasi, M. (2005). Luminescence of Er-doped amorphous silicon quantum dots, *Thin Solid Films* Vol. 475, (2005). pp. (231– 234), ISSN: 0040-6090.

Patra, A. (2004). Study of photoluminescence properties of Er3+ ions in SiO_2–GeO_2 and Al_2O_3 nanoparticles, *Solid State Communications,* Vol. 132, Issue 5, (2004). pp. (299-303), ISSN: 0038-1098.

Pavesi, L. (1996). Influence of dispersive exciton motion on the recombination dynamics in porous silicon, *J. Appl. Phys.,* Vol. 80, Issue 1, (1996). pp. (216-225), ISSN: 0021-8979.

Perea, D. E., Allen, J. E., May, S. J., Wessels, B. W., Seidman, D. N. & Lauhon, L. J. (2006). Three-Dimensional Nanoscale Composition Mapping of Semiconductor Nanowires, *Nano Lett.,* Vol. 6, Issue 2, (Feb. 2006). pp. (181-185), ISSN (printed): 1530-6984.

Petkov, N., Platschek, B., Morris, M. A., Holmes, J. D. & Bein, T. (2007). Oriented Growth of Metal and Semiconductor Nanostructures within Aligned Mesoporous Channels, *Chem. Mater.,* Vol. 19, Issue 6, (Mar. 2007). pp. (1376–1381), ISSN 0897-4756 (print).

Polman, A., Jacobson, D. C., Eaglesham, D. J., Kiatler, R. C. & Poate, J. M. *J. Appl. Phys.* Vol. 70, (1991). pp. (3778-3781).

Polman, A. Erbium implanted thin film photonic materials. (1997). *J. Appl. Phys.,* Vol. 82, Issue 1, (1997). pp. (1-39), ISSN: 0021-8979.

Priolo, F., Franzo, G., Pacifici, D. & Vinciguerra, V. (2001). Role of the energy transfer in the optical properties of undoped and Er-doped interacting Si nanocrystals, *J. Appl. Phys.,* Vol. 89, Issue 1, (Jan. 2001). pp. (264 272), ISSN: 0021-8979.

Ren, L. (2008). Optical Activation of Si Nanowires Coating with Er-Doped SiO2 and Al2O3 Derived from Sol-Gel Solutions, *Journal of Nanoelectronics and Optoelectronics,* Vol. 3, Issue 2, (Jul. 2008). pp. (190-194), ISSN: 1555-130X (Print).

Ren, L., Jeung, W. Y., Han, H. C., Suh, K. Shin, J. H. & Choi, H. J. (2007a). Optical activation of Si nanowires by Er3+ doped binary Si–Al oxides films derived from sol–gel solutions, *Optical Materials,* Vol. 30, Issue 3, (Nov. 2007a). pp. (497-501), ISSN: 0925-3467.

Ren, L., Jeung, W. Y. & Choi, H. J. (2007b). Optical activation of patterned Si nanowires grown from sol–gel prepared gold/Er-doped aluminous film, *Applied Surface Science,* Vol. 253, Issue 22, (Sep. 2007b). pp. (8885-8888), ISSN: 0169-4332.

Ren, L. Jeung, W. Y. Han, H. C. & Choi, H. J. (2007c). Optical Activation of Si and Ge Nanowires Codoping with Er:Yb Rare Earth by Sol–Gel Methods, *Journal of Nanoelectronics and Optoelectronics,* Vol. 2, Issue 2, (Agu. 2007c). pp. (191-196) , ISSN: 1555-130X (Print).

Ryu, C. K. & Kim, K. (1995). Fabrication of highly concentrated Er3+ doped aluminosilicate films via sol-gel processing. *Appl. Phys. Lett.,* Vol. 66, Issue 19, (May 1995). pp. (2496-2498), ISSN:0003-6951.

Schmidt, M., Heitmann, J., Scholz, R. & M. Zacharias, (2002). Bright luminescence from erbium doped nc-Si/SiO_2 superlattices. *J. Non-Cryst. Solids,* Vol. 299-302, Part 1, (Jan. 2002). pp. (678-682), (ISSN: 0022-3093).

Schmidt, V.; Senz, S.; Go¨sele, U., (2005). UHV chemical vapour deposition of silicon nanowires, *Zeitschrift für Metallkunde,* Vol. 96, No. 5, (2005). pp. (427-428), ISSN: 0044-3093.

Schmidt, V, Wittemann, J. V. & Go¨sele, U. (2010). Growth, Thermodynamics, and Electrical Properties of Silicon Nanowires, *Chem. Rev.,* Vol. 110, Issue 1, (Jan. 2010). pp. (361-388), ISSN: 0009-2665 (print).

Shchetinin, A. A., Nebol'sin, V. A., Kozenkov, O. D., Tatarenkov, A. F. & Dunaev, A. I. N. E. P. (1991). Further studies on the role of neighboring group participation in carbonyl substitution reactions of group 6 metal carboxylates, *Inorg. Mater.*, Vol. 30, ISSUE 5, (Mar.1991). pp. (1137-1142), ISSN: 0020-1669 (print).

Seo, S.-Y. & Shin, J. H. Exciton–erbium coupling and the excitation dynamics of Er3+ in erbium-doped silicon-rich silicon oxide, *Appl. Phys. Lett.*, Vol. 78, Issue18, (2001). pp. (2709-2712). (ISSN:0003-6951).

Solak, H. H. (2006). Nanolithography with coherent extreme ultraviolet light, *J. Phys. D: Appl. Phys.*, Vol. 39, Issue 9, (May 2006). pp. (R171-R171), ISSN 0022-3727 (Print).

Shin, J. H., Kim, M-J., Seo, S-Y. & Lee, C. (1998). Composition dependence of room temperature 1.54 μm Er^{3+} luminescence from erbium-doped silicon:oxygen thin films deposited by electron cyclotron resonance plasma enhanced chemical vapor deposition. *Appl. Phys. Lett.*, Vol. 72, Issue 9, (1998). pp. (1092-1094), ISSN: 0003-695).

Shin, J. H., Seo, S.-Y., Kim, S. & Bishop, S. G. (2000). Photoluminescence excitation spectroscopy of erbium-doped silicon-rich silicon oxide. *Appl. Phys. Lett.*, Vol. 76, Issue 15, (2000). pp. (1999-2001), ISSN: 0003-6951.

Shin, J. H., Jhe, J-H., Seo, S-Y., Ha, Y. H. & Moon, D. W. (2000). Er–carrier interaction and its effects on the Er3+ luminescence of erbium-doped Si/ SiO_2 superlattices, *Appl. Phys.Lett.*, Vol. 76, Issue 24, (Jun 2000). pp. (3567-3569), ISSN: 0003-6951.

Snoeks, E., Lagendijk, A. & Polman, A. (1995). Measuring and Modifying the Spontaneous Emission Rate of Erbium near an Interface, *Phys. Rev. Lett.*, Vol. 74, Issue 13, (1995). pp. (2459-2462), ISSN:0031-9007.

Snoeks, E., van den Hoven, G. N., Polman, A., Hendriksen, B., Diemeer, M. B. J. & Priolo, F. (1997). Cooperative upconversion in erbium-implanted soda-lime silicate glass optical waveguides, *J. Opt. Soc. Am. B*, Vol. 12, Issue 8, (1997). pp. (1468-1474), ISSN: 0740-3224.

Song, H. & Bao, X. (1997). Visible photoluminescence from silicon-ion-implanted SiO_2sfilm and its multiple mechanisms, *Phys. Rev. B*, Vol. 55, Issue 11, (Mar. 1997). pp. (6988-6993), ISSN: 1098-0121.

Song, Q., Li, C.-R., Li, J.-Y., Ding, W.-Y., Li, S.-F., Xu, J., Deng, X.-L. & Song, C.-L. (2006). Photoluminescence properties of the Yb:Er co-doped Al_2O_3 thin film fabricated by microwave ECR plasma source enhanced RF magnetron sputtering, *Optical Materials*, Vol. 28, Issue 12, (Sep. 2006). pp. (1344-1349), ISSN: 0925-3467.

Sood, D. K., Sekhar, P. K. & Bhansali, S. (2006). Ion implantation based selective synthesis of silica nanowires on silicon wafers, *Appl. Phys. Lett.*, Vol. 88, Issue 14, (Apr. 2006). pp. (143110-143112), ISSN: 0003-6951.

Stepikhova, M., Jantsch, W., Kocher, G., Palmetshofer, L., Shoisswohl, M. & Von Bardeleben, H. J. (1997). Direct excitation spectroscopy of Er centers in porous silicon, *Appl. Phys.Lett.*, Vol. 71, Issue 20, (Nov 1997). pp. (2975-2977), ISSN:0003-6951.

Stepikhova, M. V., Cerqueira, M. F., Losurdo, M., Giangregorio, M. M., Alves, E., Monteiro, T. & Soares, M. J. *Phys. Solid State,* Vol. 46, (2004). pp. (113-116), ISSN: 1063-7834.

Suh, K., Shin, J. H., Park, O.-H., Bae, B.-S., Lee, J.-C. & Choi, H.-J. (2005). Optical activation of Si nanowires using Er-doped, sol-gel derived silica, *Appl. Phys. Lett.*, Vol. 86, (2005). pp. (053101-1 - 053101-3), ISSN 0003-6951.

Tabet, N., Al-Sadah, J. & Salim, M. (1999). Growth of oxide layer on germanium (011) substrate under dry and wet atmospheres, Surf. ReV. Lett., Vol. 6, Issue 6, (Dec. 1999). pp. (1053-1060), ISSN: 0218-625X.

Tanner, P.A., Wong, K.L. & Liang, Y. (2004). Multiple phase production on doping Er^{3+} into α-Al_2O_3, *Chem. Phys. Lett.*, Vol. 399, Issue 1-3, (Nov. 2004). pp. (15-18), ISSN: 0009-2614.

Thess, A., Lee, R., Nikolaev, P., Dai, H., Petit, P., Robert, J., Xu, C., Lee, Y. H., Kim, S. G., Rinzier, A. G., Colbert, D. T., Scuseria, G. E., Tomanek, D., Fischer, J. E. & Smalley, R. E. (1996). Crystalline Ropes of Metallic Carbon Nanotubes, *Science,* Vol. 273, issue 5274, (Jul. 1996). pp. (483-487), ISSN 0036-8075 (print).

Tilke, A. T., Simmel, F. C., Lorenz, H., Blick, R. H. & Kotthaus, J. P. (2003). Quantum interference in a one-dimensional silicon nanowire, *Phys. Rev., B,* Vol. 68, Issue 7, (Agu. 2003). pp. (075311-1 - 075311-6), ISSN 1050-2947 (Print).

Treuting, R. G. & Arnold, S. M. (1957). Orientation habits of metal whiskers, *Acta Met.,* Vol. 5, Issue 10, (Oct. 1957). pp. (598-598), ISSN: 0412-1961.

Wang, D. & Dai, H. (2002). Low-Temperature Synthesis of Single-Crystal Germanium Nanowires by Chemical Vapor Deposition, *Angew. Chem. Int. Ed.,* Vol. 41, Issue 24, (Dec. 2002). pp. (4783-4786), ISSN:1521-3773.

Wang, D., Chang, Y.-L., Wang, Q., Cao, J., Farmer, D. B., Gordon, R. G. & H. Dai, (2004). Surface Chemistry and Electrical Properties of Germanium Nanowires, J. Am. Chem. Soc., Vol. 126, Issue 37, (Sep. 2004). pp. (11602-11611), ISSN: 0002-7863.

Wang, F., Dong, A., Sun, J., Tang, R., Yu, H. & Buhro, W. E. (2006). Solution–Liquid–Solid Growth of Semiconductor Nanowires, *Inorg. Chem.,* Vol. 45, Issue 19, (Sep. 2006). pp. (7511-7521), ISSN 0020-1669 (printable version).

Wang, X. J., Lei, M. K., Yang, T. & Wang, H. (2004). Phase structure and photoluminescence properties of Er^{3+}-doped Al_2O_3 powders prepared by the sol–gel method, *Opt. Mater.,* Vol. 26, Issue 3, (2004). pp. (247-252), ISSN: 0925-3467.

Wang, X. J., Lei, M. K., Yang, T. & Cao, B. S. (2004). Coherent effect of Er^{3+}–Yb^{3+} co-doping on enhanced photoluminescence properties of Al_2O_3 powders by the sol–gel method, *Opt. Mater.,* Vol. 26, Issue 3, (2004). pp. (253-259), ISSN: 0925-3467.

Wang, X. J. & Lei, M. K. (2005). Preparation and photoluminescence of Er^{3+}-doped Al_2O_3 films by sol–gel method, *Thin Solid Films,* Vol. 476, Issue 1, (2005). pp. (41-45), ISSN: 0040-6090.

Wang, Z. & Coffer, J. L. (2002). Erbium Surface-Enriched Silicon Nanowires, *Nano Lett.,* Vol. 2, No. 11, (2002). pp. (1303-1305), ISSN (printed): 1530-6984.

Weber, L. (2002). Equilibrium solid solubility of silicon in silver, *Metall. Mater. Trans. A,* Vol. 33, Issue 4, (Apr. 2002). pp. (1145-1150), ISSN: 1073-5623 (print version).

Wolkin, M. V., Jorne, J., Fauchet, P. M., Allan, G. & Delerue, C. (1999). Electronic States and Luminescence in Porous Silicon Quantum Dots: The Role of Oxygen, *Phys. Rev. Lett.,* Vol. 82, Issue 1, (1999). pp. (197-200), ISSN: 0031-9007.

Wu, X., Kulkarni, J. S., Collins, G., Petkov, N., Alme´cija, D. Boland, J. J., Erts, D. & Holmes J. D. (2008). Synthesis and Electrical and Mechanical Properties of Silicon and Germanium Nanowires, *Chem. Mater.,* Vol. 20,Issue 19, (Oct. 2008). pp. (5954 - 5967), ISSN 0897-4756.

Wu, Y. & Yang, P. (2001). Direct Observation of Vapor-Liquid-Solid Nanowire Growth, *J. Am. Chem. Soc.,* Vol. 123, Issue 13, (Apr. 2001). pp. (3165-3166), ISSN 0002-7863.

Wu, Y., Cui, Y., Huynh, L., Barrelet, C. J., Bell, D. C. & Lieber, C. M. (2004). Controlled Growth and Structures of Molecular-Scale Silicon Nanowires, *Nano Lett.,* Vol. 4, Issue 3, (Feb. 2004). pp. (433-436) , ISSN (printed): 1530-6984.

Xiang, Y., Lee, W., Nielsch, K., Abstreiter, G. & Fontcuberta i Morral, A. (2008). Microstructured horizontal alumina pore arrays as growth templates for large area few and single nanowire devices, *Phys. Status Solidi RRL*, Vol. 2, Issue 2, (Jan. 2008). pp. (59–61), ISSN: 1862-6270.

Xiang, Y., Keilbach, A., Codinachs, Lia M., Nielsch, K., Abstreiter, G., Fontcuberta i Morral, A. & Bein, T. (2010). Multiple Nanowire Species Synthesized on a Single Chip by Selectively Addressable Horizontal Nanochannels, *Nano Lett.*, Vol. 10, (Mar. 2010). pp. (1341–1346), ISSN (printed): 1530-6984.

Yu, D. P., Lee, C. S., Bello, I., Sun, X. S., Tang, Y. H., Zhou, G. W., Bai, Z. G., Zhang, Z. & Feng, S. Q. (1998). Synthesis of nano-scale silicon wires by excimer laser ablation at high temperature, *Solid State Commun.*, Vol. 105, Issue 6, (Feb. 1998a). pp. (403-407), ISSN: 0038-1098.

Yu, D. P., Bai, Z. G., Ding, Y., Hang, Q. L., Zhang, H. Z., Wang, J. J., Zou, Y. H., Qian, W., Xiong, G. C., Zhou, H. T. & Feng, S.Q. Nanoscale silicon wires synthesized using simple physical evaporation, *Appl. Phys. Lett.*, Vol. 72, Issue 26, (Apr. 1998b). pp. (3458-3460), ISSN 0003-6951 (print).

Yu, J. Y., Chung, S. W. & Heath, J. R. (2000). Silicon Nanowires: Preparation, Device Fabrication, and Transport Properties, *J. Phys. Chem. B*, Vol. 104, (Nov. 2000). pp. (11864-11870), ISSN 1089-5647.

Yudasaka, M., Komatsu, T., Ichihashi, T. & Iijima, S. (1997). Single-wall carbon nanotube formation by laser ablation using double-targets of carbon and metal, *Chem. Phys. Lett.*, Vol. 278, No. 1-3, (Oct. 1997). pp. (102-106), ISSN: 0009-2614.

Zhang, H. Z., Kong, Y. C., Wang,Y. Z., Du, X., Bai, Z. G., Wang, J. J., Yu, D. P., Ding, Y., Hang, Q. L. & Feng, S. Q. (1999). Ga_2O_3 nanowires prepared by physical evaporation, *Solid State Commun.*, Vol. 109, No. 11, (Mar. 1999). pp. (677-680), ISSN: 0038-1098.

Zhang, Y. F., Tang,Y. H., Lam C., Wang, N., Lee, C. S., Bello, I. & Lee, S. T. (2000). Bulk-quantity Si nanowires synthesized by SiO sublimation, *J. Cryst. Growth*, Vol. 212, No. 1-2, (Apr. 2000). pp. (115-118), ISSN: 0022-0248.

Zhang, Y. J., Zhu, J., Zhang, Q., Yan, Y. J., Wang, N. L. & Zhang, X. (2000). Synthesis of GeO_2 nanorods by carbon nanotubes template, *Chem. Phys. Lett.*, Vol. 317, No. 3-5, (Feb. 2000). pp. (504-509), ISSN: 0009-2614.

Zhang, Y. J., Liu, J., He, R. R., Zhang, Q., Zhang, X. & Zhu, J. (2001). Synthesis of Aluminum Nitride Nanowires from Carbon Nanotubes, *Chem. Mater.*, Vol. 13, No. 11, (Sep. 2001). pp. (3899-3905), ISSN 0897-4756.

Zhang, X.-Y., Zhang, L.-D., Meng, G.-W., Li, G.-H., Phillip, N.-Y. J. & Phillip, F. (2001). Synthesis of Ordered Single Crystal Silicon Nanowire Arrays， AdV. Mater., Vol. 13, No. 16, (Aug. 2001). pp. (1238-1241), ISSN: 1521-4095.

18

Nano-Scale Measurements of Dopants in Individual Silicon Nanowires Using Kelvin Probe Force Microscopy

Elad Koren[1], Jonathan E. Allen[2], Uri Givan[2], Noel Berkovitch[1], Eric R. Hemesath[2], Lincoln J. Lauhon[2] and Yossi Rosenwaks[1]
[1]*School of Electrical Engineering, Tel-Aviv University,*
[2]*Dept. of Materials Science and Engineering, Northwestern university,*
[1]*Israel*
[2]*USA*

1. Introduction

Semiconductor nanowires are one of the most promising building blocks for near future nano-electronics because they provide a new route to continuing miniaturization as well as a wealth of opportunities in nanoscale science and technology. As the synthesis of new compositionally and morphologically diverse nanowires expands, and the expectations for reliable nanowire devices increase, challenges in nanoscale metrology assume increasing importance. The chemical composition of an individual semiconductor nanowire is of great importance to nanowire electronic properties and device performance, but it is not easy to determine. At least three major challenges in nanowire synthesis and device fabrication are intimately related to the challenge of analyzing nanowire composition: (1) controlling the concentration of intentional dopants; (2) identifying unintentional doping, and (3) electrical junction formation. There are many nanowire systems in which the majority carrier type is controlled by intentional doping (Haraguchi, Katsuyama et al. 1992), but in all cases the actual dopant concentration, its spatial distribution, and the fraction of active dopants are unclear. The origins of dopant compensation (Wang, Lew et al. 2005), including the presence of residual catalyst atoms in nanowires grown by the vapor-liquid-solid (VLS) method (Wagner and Ellis 1964), are therefore speculative. In some material systems, and particularly in quantum dots (Erwin, Zu et al. 2005), impurity doping is known to be difficult. This challenge is compounded by a lack of appropriate characterization tools, and issues such as dopant segregation are generally addressed by indirect means
Individual nanowires have already been configured as field-effect transistors (Cui and Lieber 2001), photodetectors (Wang, Gudiksen et al. 2001), and bio/chemical sensors (Patolsky, Zheng et al. 2006). The continued advances of these nanometer-scale devices depend critically on knowledge of their atomic-scale structures (Castell, Muller et al. 2003), as compositional fluctuations as small as a single dopant atom can affect a device's performance. Additionally, the important challenge of doping atoms into the 'bulk' of nanowires and nanocrystals, while avoiding surface segregation, further emphasizes the need for 3D composition characterization

of these nanostructures. It should be also emphasized that the performance of almost all current transistor technology depends critically on controlling the concentration and location of dopants to create abrupt homojunctions in Si. Such junctions have been realized in Si nanowires synthesized using the VLS growth mechanism by *in situ* doping, (Gudiksen, Lauhon et al. 2002; Yang, Barrelet et al. 2006) but unintentional surface doping caused by vapor-solid (VS) deposition on the sides of the nanowires (Tutuc, Chu et al. 2006) during growth can complicate the formation of strictly axial homojunctions. Surface doping has been seen in phosphorous-doped Ge nanowires by electrical characterization (Tutuc, Chu et al. 2006) and atom probe tomography, (Perea, Wijaya et al. 2008; Perea, Hernesath et al. 2009) and in doped Si nanowires by Raman spectroscopy (Imamura, Kawashima et al. 2008) and scanning photocurrent microscopy (Allen, Perea et al. 2009). Surface doping and other inhomogeneities are of utmost importance because the success or failure of an integrated circuit is rapidly approaching the point at which a few misplaced atoms can ruin a device. It is highly desirable to determine the atomic scale composition and dopant distribution of individual nanowires with precision, and this is the main objective of this research. As will be show in this review, of our recent work (Koren et al. 2009, 2010, 2011), this can be accomplished by employing combined current-voltage and Kelvin probe force microscopy (KPFM) measurements.

In particular, it shows that KPFM can be use for both longitudinal and radial dopant distribution profiling for *n*-type Si nanowires, grown by the VLS method. The presented results show, that along with a longitudinal non-uniformity, there is also an exponential decrease in dopant concentration across the nanowire radius with a difference of ~ 2 orders of magnitude between the nanowire surface and its core. In addition, the radial dopant profile is consistent with a diffusion coefficient of $D\sim1\text{x}10^{-19}$ m^2s^{-1}. This relatively unexpected high diffusivity will be discussed further in this chapter. In addition, low temperature annealing treatment (460 ^{o}C) was used to achieve a uniform dopant distribution. The low temperature treatment was essential to prevent the diffusion of the Au catalyst into the nanowire, which might affect the opto-electronic properties of the Si (Grimmeiss 1977; Queisser 1978).

2. Experimental

2.1 Device fabrication

For all electrical measurements conducted in this chapter, electrical contacts were fabricated through a series of lithography steps. In chronological order, the process for nanowire device fabrication was as follows. First, heavily doped Si (100) substrates with a thin dielectric layer of 200 nm Si_3N_4 were patterned with both large electrodes and a grid used for nanowire location via photolithography and subsequent metal evaporation. Figure 1 presents a typical photolithography pattern for substrate preparation. The metal deposition for the patterned grid and large electrodes consisted of 5 nm of Cr followed by 80 nm of Au. Nanowires were suspended in isopropyl-alcohol solution by low-power sonication of the growth substrate. The nanowires in solution were then drop-cast on the prepared substrates and imaged using an optical microscope. The images were used to establish the position of the nanowires with respect to the patterned grid in the middle of the large contacts so that a high accuracy mask could be designed for electrodes. Then, small electrodes were design by CAD software and written by electron beam lithography (EBL). The electron beam resist was poly-methyl-methacrylate (PMMA), which was deposited by standard spin coating technique. After developing the EBL pattern, a wet chemical etch with a buffer HF solution was used to remove the native oxide from the nanowires surface, while the residual polymer resist was removed using a standard plasma clean step. Ohmic contacts were fabricated by evaporating Ni immediately after etching.

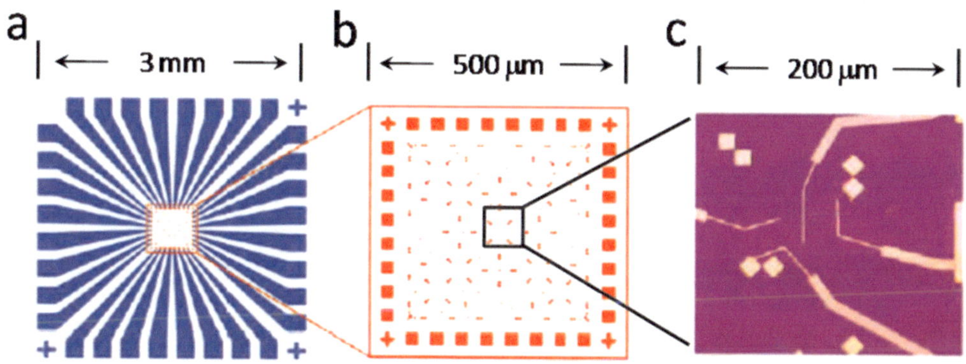

Fig. 1. **(a)** Photolithography pattern for typical device substrate. **(b)** Patterned grid for nanowire location and contacts writing. **(c)** Complete devices after EBL. Taken from (Allen 2008).

2.1.1 Nanowire growth

The Si nanowires used in this study (data summarized in table 1) were grown in a custom-built chemical vapor-deposition reactor designed to provide a large dynamic range in the process variables that most influence nanowire morphology and composition. Silane (SiH_4) has been used as reactant and dilute phosphine (PH_3) has been used as dopant. The background gas composition can be tuned continuously from pure Ar, N_2 and H_2. The ratio of Si to P (in the gas phase) can be varied from 10^6:1 to 10^2:1. Using commercially available colloidal Au nanoparticles, nanowire diameters from 5-100 nm can be grown, and the length is controlled by the growth time. The exact nanowire growth conditions in each part are summarized in the following table:

Part	Au (nm)	Gas phase Dopant:Si	T (°C)	P (Torr)	Carrier gas	Flow rates (SCCM) Carrier/dopant/SiH_4	Time (min)
3	50	1:1500	460	40	N_2	43.3/6.67/2	15
3, 4	50	1:500	460	40	N_2	30/20/2	15
5	80	1:500	460	40	N_2	30/20/2	15

Table 1. Growth conditions for the nanowires used in each part.

2.2 Kelvin probe force microscopy

KPFM measures contact potential difference (defined below) at the nanoscale. It is based on the conventional atomic force microscope (AFM) invented in 1985 (Binnig, Quate et al. 1986). KPFM was used for the first time (Martin, Abraham et al. 1988);(Weaver and

Abraham 1991) to measure charges and potential variation on metal surfaces. In this section the basic setup of the KPFM used for our measurements is described.

KPFM measures the contact potential difference (CPD) between a probing tip and a sample. The CPD represents the difference between the tip work-function (ϕ_t) and sample work-function (ϕ_s) and is defined as:

$$CPD = -\frac{1}{q}(\phi_t - \phi_s) \tag{1}$$

The sample work function is a sum of the intrinsic work function and any additional surface potential, thus measuring CPD across a sample surface gives local changes in the surface potential. The basic principle of the measurement is schematically described for a metallic tip and sample in Fig. 2. When the tip and the sample are not connected their Local-vacuum-levels (LVLs) are aligned, but there is a difference in their Fermi levels. Upon electrical connection, electrons flow from the material with the lower work function to the one with a higher work function (in this case from the sample to the tip) as shown in Fig. 2 (b). This process continues until the Fermi levels are aligned; the two materials are now charged and there is a difference in their LVLs. Due to the charging of the tip and the sample, an electrostatic force develops. This force can be nullified by applying an additional bias between the tip and the sample.

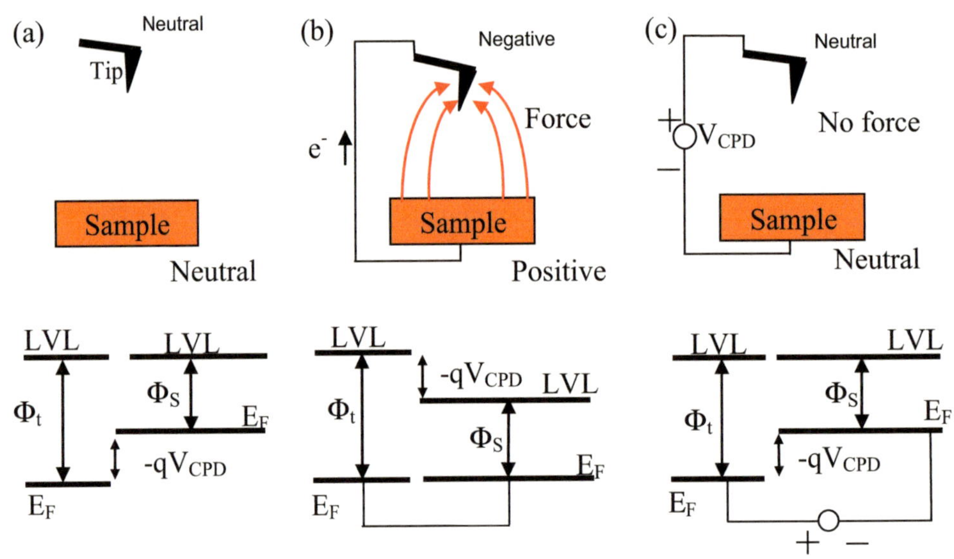

Fig. 2. Basic measurement setup of contact potential difference (CPD).

The magnitude of this bias is the CPD as seen in Fig. 2 (c). The nullifying procedure is based on applying a bias either to the tip or to the sample. The CPD measurement is conducted either in non-contact, or lift modes of the AFM. It is well known that in these modes the forces between the tip and the sample are relatively small, thus modulation (AC) techniques are used in order to increase the force sensitivity. A typical AC measurement is conducted in the following way. An AC bias at a frequency ω is applied between the tip and the sample. It can be shown (Martin, Abraham et al. 1988) that the force component at this frequency is proportional to the CPD and therefore, can be nullified using a feedback loop whose input is the ω component of the force. The most naive way to derive this force is to treat the tip-sample system as a parallel plate capacitor with one plate as the tip apex, and the other as the sample underneath it (Weaver and Abraham 1991). Under this assumption the force which is just the derivative of the electrostatic energy with respect to the tip-sample separation (z), is given by:

$$F = -\left.\frac{\partial U}{\partial z}\right|_Q = -\frac{1}{2}V^2\frac{\partial C}{\partial z} \propto -\frac{1}{2}\frac{V^2}{z^2} \tag{2}$$

where the electrostatic energy (U) is given for a parallel plate capacitor configuration by:

$$U = \tfrac{1}{2}CV^2 \tag{3}$$

with C the tip sample capacitance, and V the potential difference between the AFM tip and the sample. Using the following expression for the potential difference:

$$V = CPD - V_{DC} + V_{AC}\sin(\omega \cdot t) \tag{4}$$

and inserting it in Eq. 2 gives:

$$F \propto \frac{1}{2z^2}\left[(CPD - V_{DC})^2 + (CPD - V_{DC})V_{AC}\sin(\omega \cdot t) + V_{AC}^2\sin(\omega \cdot t)^2\right] \tag{5}$$

Thus the force at frequency ω is indeed proportional to the CPD.

In this work, KPFM measurements were conducted in Lift mode by a commercial AFM system (Dimension 3100 *Veeco, Inc.*). On the first pass, topography is measured in standard Tapping mode. For imaging, the probing tip is driven by a piezoelectric element at a frequency slightly below its first resonance. On the second pass, the mechanical drive to the piezo is turned off, and an alternating voltage, $V_{AC}\,sin(\omega t)$ at the resonance frequency is applied to the cantilever in order to induce an electrostatic force between the tip and the sample. The force at the frequency ω is given by:

$$F_\omega \propto \frac{\partial C}{\partial Z}\left[(V_{CPD} - V_{DC})V_{AC}\sin(\omega \cdot t)\right] \tag{6}$$

Hence the CPD between the tip and the sample surface is measured by nullifying the output signal of a lock-in amplifier, which measures the electrostatic force at the frequency ω.

2.3 Measurements setup

The measurements are conducted in a controlled nitrogen environment glove box (less than ~2 ppm H_2O). Voltages and currents are supplied and measured using Agilent 4155c SPA and a lock in amplifier which are connected to the sample through BNC cables. The samples were placed on a ceramic chip carrier and a wire bonder was used to connect the chip's Gate/Source/Drain contacts to the chip carrier's contacts. The ceramic chip carrier was placed in a chip socket which in turn is connected to the outer cables and the SPA (Fig. 3). This setup enables simultaneous KPFM and current-voltage measurements.

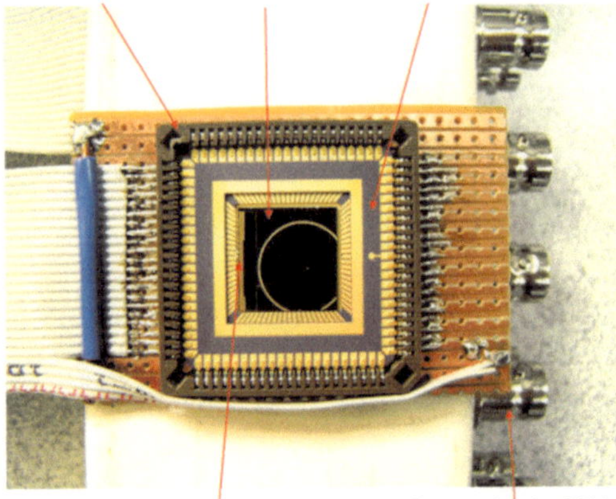

Fig. 3. The custom made electronics connections to the nanowire chip

3. Non-uniform dopant distribution along Si nanowires revealed by Kelvin probe force microscopy and scanning photocurrent microscopy

The future of semiconductor nanowires as building blocks for high-performance nanoscale devices is promising. However, it remains an important challenge to synthesize materials with sufficient control over composition and doping to compete with existing commercial devices. For example, the performance of existing CMOS technology depends critically on the ability to control the location of dopants to create abrupt homojunctions in Si. Such junctions have been realized in Si nanowires synthesized using the VLS growth mechanism by in-situ doping (Gudiksen, Lauhon et al. 2002; Yang, Barrelet et al. 2006), but unintentional surface doping caused by vapor-solid deposition on the sides of the nanowires during growth can complicate the formation of strictly axial homojunctions. Evidence of surface doping has been seen in phosphorous-doped Ge nanowires by electrical characterization (Tutuc, Chu et al. 2006) and atom probe tomography (Perea, Hernesath et al. 2009), and in boron doped Si nanowires by Raman spectroscopy (Imamura, Kawashima et al. 2008). In each of these cases, the nanowires were highly tapered, which is indicative of significant Vapor-solid deposition during synthesis. Even in the absence of tapering, it was shown (Allen, Perea et al. 2009) through scanning photocurrent microscopy that surface doping can occur for phosphorous-doped Si nanowires, resulting in a nonuniform doping profile.

KPFM and scanning photocurrent microscopy (SPCM) were used to measure the doping distribution along single phosphorous doped silicon nanowire grown by the VLS method. A non-linear potential drop along biased silicon nanowires is detected both by measuring the surface potential directly via KPFM and by integrating the photocurrent measured by SPCM (Allen, Perea et al. 2009). These variations in the potential and field are further analyzed to extract the longitudinal dopant distribution along an individual silicon nanowire. The results show a very good agreement between the two methods to quantitatively detect potential, field and doping variations along doped silicon nanowires. Considering that KPFM measures the surface potential whereas SPCM samples the nanowire volume, the correlation between these two methods can improve our understanding regarding the role of surfaces vs. bulk in nanowires electrical properties.

Since the KPFM measures the surface potential profile, it is used as a "four-point-probe" where the scanning tip replaces the two inner contacts of the conventional setup. Although KPFM measures the potential at the surface, in the case of heavily doped wires used here, this potential can be considered, to a very good approximation, as the bulk nanowire potential. The current through the nanowire is measured and can be used to calculate the local effective doping, $N_D(x)$ along the Si nanowire:

$$j / E(x) = \sigma = q\mu N_D(x) \tag{7}$$

where j is the current density, $E(x)$ is the local electric field, σ is the electrical conductivity, q is the elementary charge and m is the mobility for electrons. $E(x)$ is simply obtained by differentiating the measured surface potential along the nanowires. Figure 4(b) shows the expected potential profile along a biased Si nanowire for the case of uniform doping, in contrast to the case of gradually decreasing doping (Fig. 4(c)). Figure 4(d,e) shows the KPFM-measured surface potential profiles along an n-Si nanowire under several applied potentials (-3V to 3V with 1V steps). The voltage was applied by grounding one contact and biasing the other, while the back gate electrode was grounded throughout the whole measurement. The potential peaks are due to the applied bias at the left and the right metal contacts. The black dashed lines represent a linear potential drop and therefore help to visualize the deviation of the measured potential profiles from the case of a uniform doping concentration. The perpendicular doted lines (blue) represent the edges of the metal contacts. When applying the bias to the contact closer to the gold catalyst end, the surface potential profile is concave (Fig. 4(d)), while biasing the other contact the profile is convex (Fig. 4(e)). Nevertheless, the absolute averaged electric field (the average was done for four measurements where the applied bias was 1V and -1V, each time on a different contact; the error represent the standard deviation) is always larger near the gold catalyst (Fig. 5(a)), and therefore such a potential curvature is expected. For this reason, the observed field gradient is due to intrinsic variations within the nanowire itself rather than effects related to the metal electrodes. Furthermore, the field gradient is too small to cause significant variations in the local carrier mobility, so we assume a constant mobility of ~150 $cm^2V^{-1}s^{-1}$ (Ben G.Streetman 1999) consistent with electrons in bulk silicon doped to similar levels. This assumption, together with the measured potential, allows one to extract the longitudinal dopant distribution along the Si nanowire as explained above.

Figure 5(b) shows the averaged doping profile extracted by applying 1V and -1V, each time for a different contact. The doping profile increases monotonically along the 8μm long nanowire. The rate of increase is greater when the PH_3 partial pressure is increased (Allen, Perea et al. 2009). In all cases, the doping is higher further from the catalyst side. This is consistent with the fact that this end of the wire was exposed a longer time to the PH_3 gas.

Next, the potential profiles obtained by the SPCM were used for direct comparison with the KPFM data. Since the SPCM measurements were conducted on two-terminal devices, extraction of the absolute potential profiles from the SPCM-measured photocurrent required knowledge of a proportionality factor, representing the potential drop over the contacts, which can be determined by four-probe measurements (Allen, Perea et al. 2009). Therefore, we have scaled and offset the integrated photocurrent profiles to fit the measured KPFM potential profiles, as the latter represent the actual nanowire potential. The scaling factor was 1.5±0.1 for all curves. Figure 6 shows the normalized integrated-photocurrent profiles measured by the SPCM together with the KPFM measured potential profiles for the same wire under the same applied potentials (1V, -1V for each one of the contacts). The perpendicular doted lines (blue) represent the edges of the metal contacts. Our results show an excellent agreement between these two methods, and confirm the validity of the potential measurements by the SPCM method.

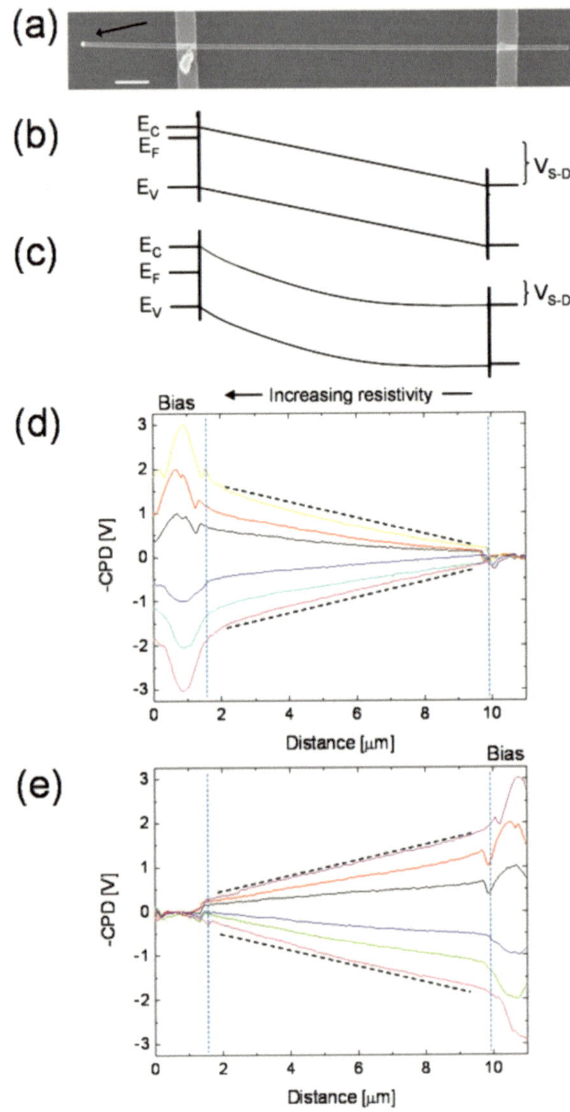

Fig. 4. **(a)** HRSEM image of a typical nanowire device used in our measurements. The Au catalyst is marked by an arrow, contacts are made out of Ni, scale bar is 1μm; schematic band-diagram for the case of a uniformly **(b)** and non-uniformly **(c)** doped Si nanowires under an applied bias. -CPD measured profiles along *n*-Si nanowire biased at different voltages (-3V to 3V with 1V steps), applied to the gold catalyst end **(d),** and to the other side **(e)**. The perpendicular doted lines (blue) represent the edges of the metal contacts. The dashed lines (black) are a guide to the eye to show the deviation from a linear potential profile in the case of a uniform doping distribution.

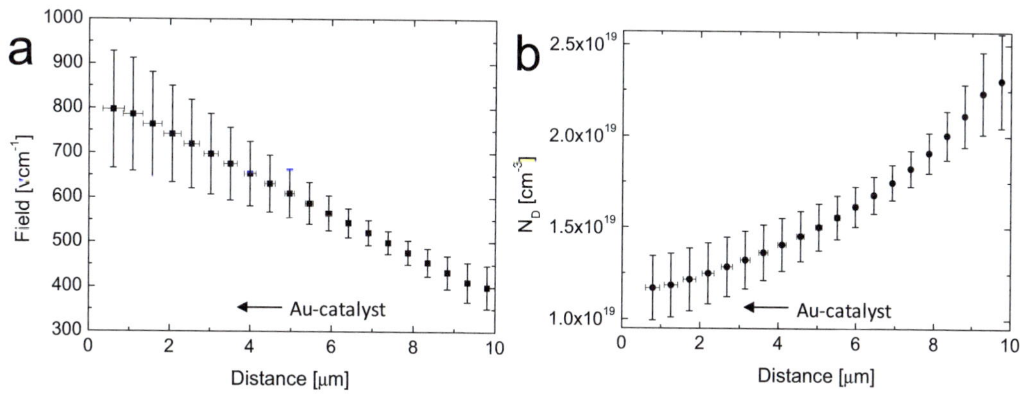

Fig. 5. Averaged electric field **(a)** and doping **(b)** profiles extracted from the measured KPFM profiles along biased *n*-Si nanowire (the average is between four cases where the applied bias was 1V and -1V, each time for a different contact. The error bars represent the standard deviation of the four measurements).

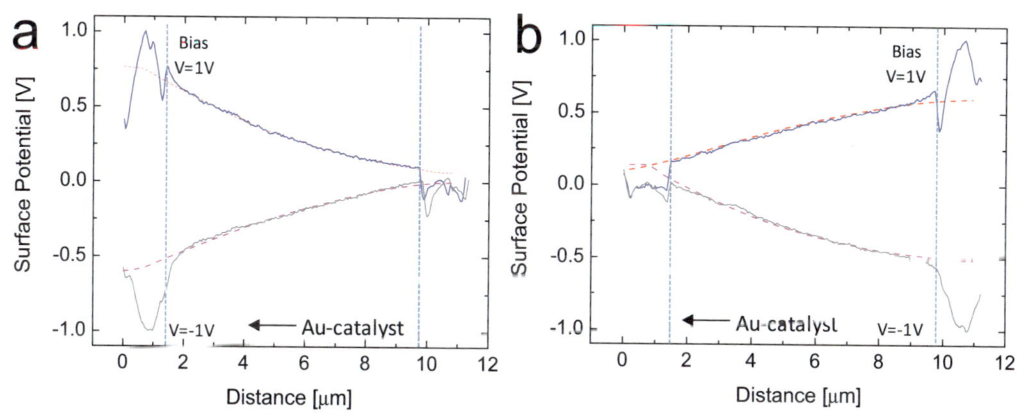

Fig. 6. Surface potential (solid lines) and normalized integrated-photocurrent (dashed lines) profiles along biased *n*-Si nanowire when the bias is applied to the gold catalyst end **(a)** and to the other contact **(b)**. The perpendicular doted lines (blue) represent the edges of the metal contacts.

There are several consequences of the observed non-uniform axial doping. As pointed out by others (Tutuc, Chu et al. 2006; Kempa, Tian et al. 2008), the synthesis of complex axial doping profiles is complicated by unintentional radial doping profiles. Surface doping may also diminish the influence of surface states on conductivity (Kimukin, Islam et al. 2006, Seo, Sharma et al. 2006) by narrowing depletion layers. Even in the absence of surface states, a radially non-uniform doping profile will create a radially non-uniform potential and could produce carrier confinement at or away from the surface. Moreover, considering that KPFM measures the surface potential whereas SPCM the nanowire volume, the excellent agreement between the two methods and the fact that the surface is more heavily doped leads to the conclusion that most of the current-flow takes place between the outer shell of the nanowire and its core. Measurements of the radial doping profile (presented in the next part) are needed to understand how great is the influence of this surface doping on operating devices.

In summary, we have measured the variations of potential and electric field along biased *n*-Si nanowires by KPFM and SPCM, respectively. The potential profiles indicate a non-uniform doping along VLS based *n*-Si nanowires as recently reported (Allen, Perea et al. 2009). This non-uniformity occurs as a result of the exposure of the nanowire surface to PH_3 gas throughout the growth. In addition we have shown good correlation between KPFM and SPCM to quantitatively detect potential, field and doping variations within doped Si nanowires.

4. Measurement of radial dopant distribution and diffusion in individual Si nanowire

There are very few methods that can be used to measure the dopant concentration and distribution in a single nanowire. Radial dopant profiling in silicon nanowires using capacitance-voltage measurements was recently demonstrated (Garnett, Tseng et al. 2009) where the nanowire doping was controlled by post-growth BCl_3 gas diffusion. They have found that the dopant concentration decreased by almost two orders of magnitude over a radial distance of 25 nanometers from the wire circumference to its center. 3-D atom probe tomography was used (Perea, Hernesath et al. 2009) to directly measure the dopant concentration in individual tapered Ge nanowires. They have found that differences in precursor decomposition rates between the liquid catalyst and the solid nanowire surface result in a radial dopant decrease of one and a half orders of magnitude for a radial distance of 10 nanometers.

Here, we quantitatively demonstrate the important role of dopant diffusion taking place during VLS growth of individual silicon nanowires. We have used KPFM to measure the active radial dopant distribution within a single untapered *n*-type silicon nanowire. It was found that the dopant concentration decreases by almost two orders of magnitude from the wire surface to its core. Moreover the profile is consistent with dopant diffusion during the growth with a diffusion coefficient of $D=1x10^{-19}$ m^2s^{-1} which is much larger than the expected value for this temperature in bulk silicon. This implies that P diffusion during the VLS growth of silicon nanowires is very significant and subsequent thermal annealing must be used when a homogenous dopant distribution is required.

Before etching the wires, the active dopant concentration at the nanowire surface was determined to be $\sim 1x10^{20}$ cm^{-3} by measuring the potential drop along a biased nanowire using KPFM and assuming electron mobility of $\sim$ 150 $cm^2V^{-1}s^{-1}$ as described in the previous section.

The method for measuring the radial dopant distribution by KPFM is described schematically in Figure 7(a) in both cross sections and top views. It is based on surface etching of a portion of a contacted nanowire (Allen, Hemesath et al. 2009), followed by measurement of the potential difference between the etched and the unetched areas using KPFM. This process is repeated several times to gradually remove material, and the surface potential difference (between the etched and unetched parts) is measured for a number of nanowire radii. The radial dopant distribution is then obtained by fitting the measured potentials with a 3D solution of Poisson equation as explained in more details below.

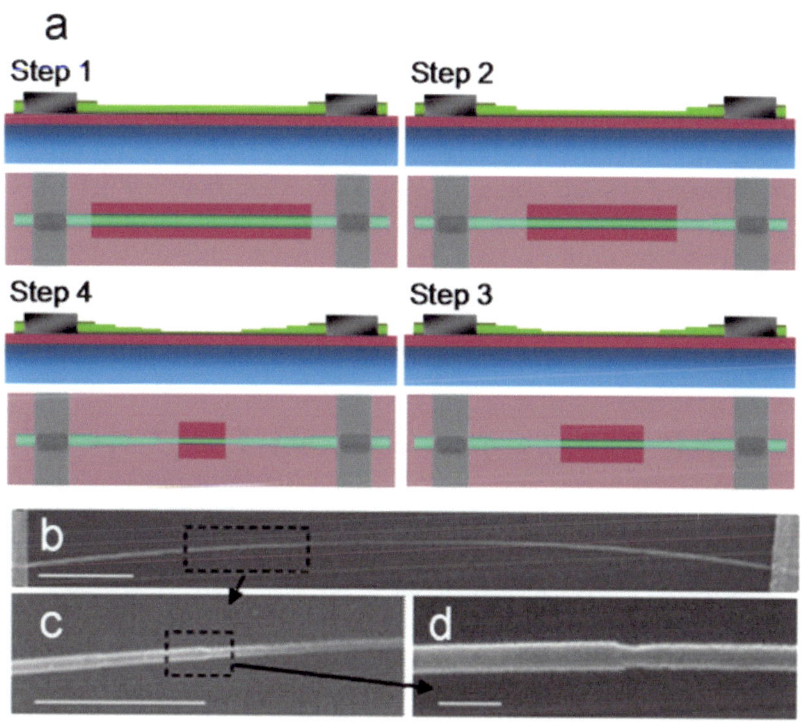

Fig. 7. (a), schematic illustration of four etching steps and their resulting nanowire structures; in each step a different size of window is opened over the nanowire channel by electron-beam lithography and few nm of the nanowire surface are etched. (b), HRSEM image of the nanowire after five etching steps, scale bar is 1μm. (c), magnified HRSEM image showing three steps over the nanowire, scale bar is 1μm. (d), magnified HRSEM image on a single step, scale bar is 100nm.

The nanowire selective etching was carried out by coating the sample with a thick layer of PMMA and defining windows by EBL in the center of the contacted wire device. The PMMA is protecting the metal contacts and ensuring that the nanowire is etched only in the exposed region defined by the window as demonstrated in Figure 7. Wet chemical etching was carried out by the following procedure: After a short oxygen plasma clean, the native oxide was removed using a 2-second buffered HF etch, and a few nanometers (5-7nm) of the silicon surface were subsequently removed by a 20-second NH_4F etch. Four sequential wet chemical etching steps were applied, where in each step a smaller window was defined and opened; the result was a stairs case-like structure of the nanowire, as schematically depicted in Figure 7. An HRSEM image of three etching steps (Fig. 7 (c)) and a magnified image of a single step (Fig. 7(d)) show parts of the etched nanowire. We note that, following each etching step, the nanowire surface becomes rougher resulting in a larger distribution of nanowire diameters; having a larger number of nanowire radii enables the determination of the radial dopant distribution with a better depth resolution, as explained in the next sections.

Figure 8 shows the measured topography (a), and -CPD (b) of a single silicon nanowire following four etching steps. For convenience we present -CPD, because it expresses the surface potential of the nanowire. The observed change in -CPD as a function of the etching-depth indicates a change in the Fermi level (E_F) relative to the sample local vacuum level (E_{LVL}) as shown in Figure 8 (c). We attribute this change to a decreasing doping level from the nanowire shell to its core. The observed correlation between the topography and -CPD cannot be due to the changes in the scanning tip height as the nanowire is etched. Such tip convolution effects were analyzed by us in the past (Strassburg, Boag et al. 2005) and have shown that such small (<60mV) CPD differences between neighboring areas (in this case the nanowire and the substrate) cannot account for the CPD changes measured here.

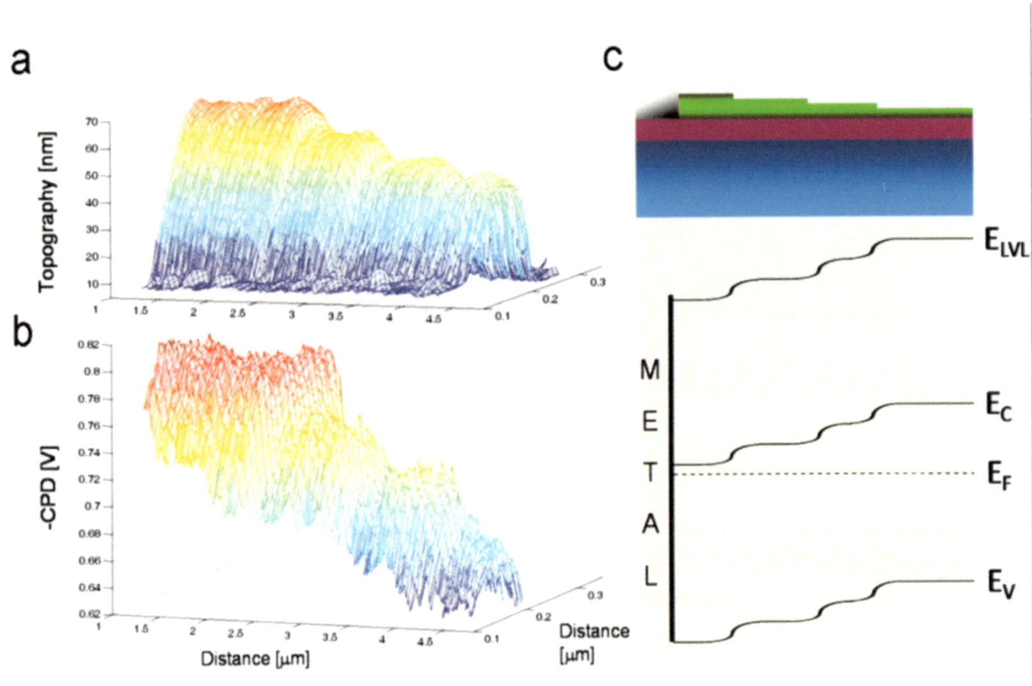

Fig. 8. **(a)**, 3D AFM image showing four steps in topography, each step is ~ 7 nm. **(b)**, 3D - CPD image showing four steps in -CPD, each step is ~ 15 mV. **(c)**, schematic band diagram of the etched silicon nanowire.

To obtain the radial dopant distribution, both the topography and the -CPD values from all the measured points of the top 10 nm etched nanowire area were averaged in order to minimize the effect of the nanowire curvature. Subsequently, a plot of the averaged surface potential as a function of the radial etching distance was obtained (symbols Fig. 9); the measured profile was averaged over three different wires measured at the higher doped side. The solid lines in Figure 9 are surface potential profiles calculated as explained in the following paragraph.

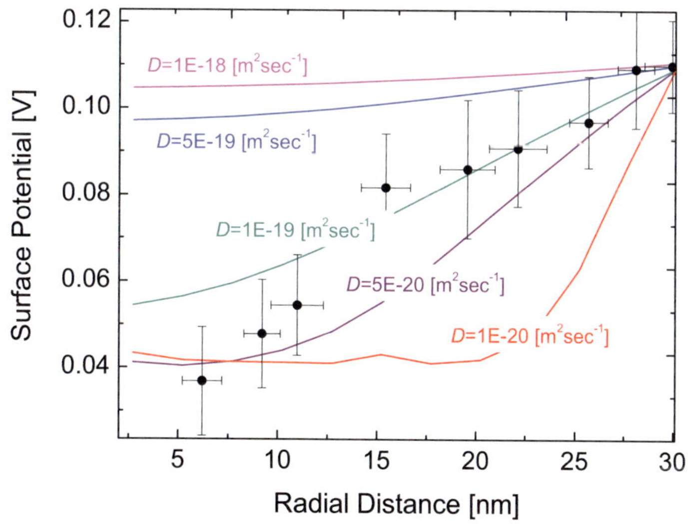

Fig. 9. Simulated (solid lines) and measured (symbols) surface potential profiles. Simulation was carried out for different doping profiles; each profile was calculated for a different diffusion coefficient. The measured surface potential profile was averaged over three different samples; shows a relatively good agreement with the calculated doping profile for $D \sim 1\text{x}10^{-19}$ m^2s^{-1} (until a radial distance of ~12nm). The error bars in the measured profile represent the standard deviation in the KPFM measurement for all the three nanowires.

As mentioned above, during the VLS growth dopants are incorporated into the wire both through the gold catalyst (causing a uniform dopant concentration, *C (t=0)*) and through the surface by uncatalyzed decomposition. Given the higher surface doping and the growth temperature of 460°C it is reasonable to assume that the dopant profile is governed by P diffusion into the nanowire. Thus, the time-dependent diffusion equation with cylindrical symmetry was solved to obtain the radial dopant profile of the nanowire (Crank 1975):

$$C(r,t)=k\left(t-\frac{a^2-r^2}{4D_{Growth}}\right)+\frac{2k}{aD_{Growth}}\sum_{n=1}^{\infty}\exp\left(-D_{Growth}\alpha_n^2 t\right)\frac{J_0(r\alpha_n)}{\alpha_n^3 J_1(a\alpha_n)}+C\Big|_{t=0} \quad (8)$$

The diffusion equation was solved for different diffusion coefficients with a linear increasing concentration of P (with the growth time) on the surface of the nanowire, accounting for the evolving surface doping during the VLS growth. The linear increasing surface doping was determined from the longitudinal dopant distribution along the nanowire and the growth rate of the nanowire; both were assumed to be approximately linear. Thus, a surface doping rate (*dC/dt*) of ~ $1\text{x}10^{17}$ $cm^{-3}s^{-1}$, was estimated by the product between the growth rate of the nanowires (*dZ/dt* ~ $1\text{x}10^{-6}$ $cm\,s^{-1}$) and the surface doping distribution along the nanowires (*dC/dZ* ~ $1\text{x}10^{23}$ cm^{-4}), where C is the P concentration on the nanowire surface; *t* is time and Z is the longitudinal distance along the nanowire. The background concentration of P inside the nanowire which was assumed to be homogeneous and caused as a result of intentional doping through the Au catalyst was determined to be $1\text{x}10^{18}cm^{-3}$ in order to be self consistence both with the P surface concentration ($1\text{x}10^{20}$ cm^{-3}) and with the extracted doping profile.

These profiles are then used as an input to the 3D Poisson equation solved using the *Sentaurus TCAD* device simulator *(Synopsys inc.)* to obtain the wire surface potential as a function of the etching height. This solver is a 3D finite element simulator which solves the Poisson equation coupled with the continuity equation for holes and electrons for the desired doping profile. The model follows the real device geometry including the topography steps resulted from the applied etching. Correlation between the radial distance and the surface potential was carried out by taking the potential and height values from the surface of each stair. Figure 9 shows both the simulated and the averaged-measured surface potential profiles. The calculated doping profile for D = 1x10^{-19} m^{2}s^{-1} (Fig. 10(a)) leads to a surface potential profile which is in agreement with the measured profile up to a radial distance of ~12nm.

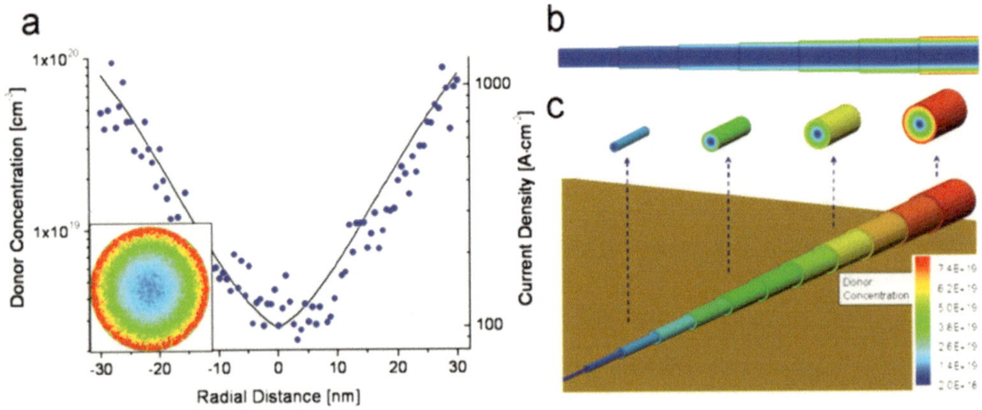

Fig. 10. **(a)**, calculated radial doping profile for D = 1x10^{-19} m^{2}s^{-1} (solid line) and simulated current density (symbols in the graph and inset image) through the nanowire under an applied bias of 0.1V. P concentration decreases exponentially from ~1x10^{20} cm^{-3} at the nanowire surface to ~2x10^{18} cm^{-3} at the nanowire core. The current density is higher (> order of magnitude) close the nanowire surface in agreement with the doping profile. Calculated 2D **(b)** and 3D **(c)** of the dopants distribution in the nanowire for D = 1x10^{-19} m^{2}s^{-1} following several etching steps (a matched 3D distribution of the potential was used to extract the radial surface potential profile).

Calculated 2D and 3D of the dopants distribution in the nanowire following several etching steps for D = 1x10^{-19} m^{2}s^{-1} are presented in Figure 10(b) and 10(c), respectively. These dopant profiles are used for calculating the current density through the nanowire under an applied bias voltage of 0.1V between the two contacts. The resulting non-uniform current flow (symbols and inset image) is shown in Figure 10(a) together with the dopant distribution profile (solid line). The current density is larger by more than an order of magnitude at the nanowire circumference compared to its interior part.
Following etching of more than 15nm, the measured surface potential profile deviates (downward bending) from the calculated dopant diffusion profile indicating a lower Fermi energy with respect to the conduction band minimum, E_c. This deviation can be due to negatively charged surface states which induce upward band bending at the wire circumference. This was simulated by including a negative surface charge

(N_{SC}) of = 5x10^{11} cm^{-2}, which has been estimated for Si nanowires by several groups (Bjork, Schmid et al. 2009; Garnett, Tseng et al. 2009). Figure 11(a) shows the simulated (solid lines) surface potentials for a diffusion constant of D =1x10^{-19} m^{2}s^{-1} including negative surface charge (N_{SC}) of 0, 5x10^{11} and 1x10^{12} cm^{-2} together with the average-measured potential profile (symbols). As the nanowire is etched, with a concurrent decrease in its surface doping concentration, the existence of a negative surface charge results in larger surface band-bending (Fig. 11(b)).

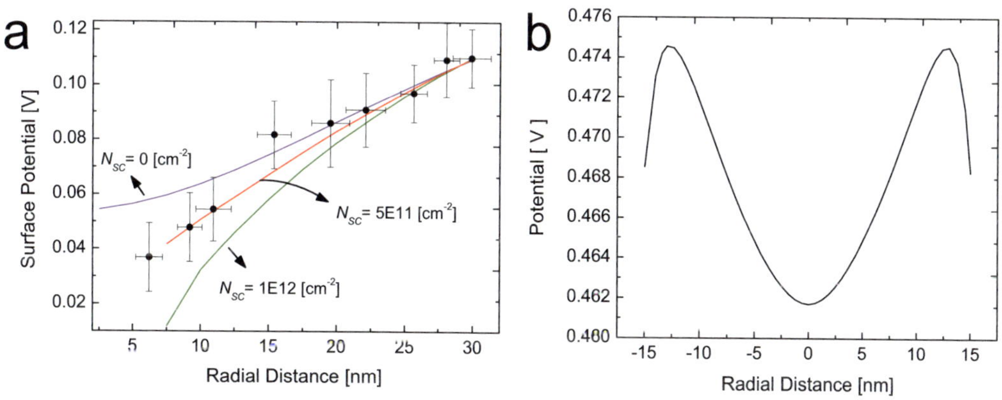

Fig. 11. **(a)** Simulated (solid lines) surface potential profiles with negative surface charge of 0, 5x10^{11} cm^{-2} and 1x10^{12} cm^{-2} and the averaged measured profile (symbols). The error bars in the measured profile represent the standard deviation in the KPFM measurement for all the three nanowires. The averaged measured profile shows a relatively good agreement with the simulated profile for N_{SC} = 5x10^{11} cm^{-2}. **(b)** Simulated surface potential band-bending of the etched nanowire with the presents of negative surface charge of N_{SC} = 5x10^{11} cm^{-2}.

The extracted value for the diffusion coefficient, D~1x10^{-19} m^{2}s^{-1}, is considerably larger from any value extrapolated from high temperature data (F.F.Y.Wang 1981), and to the best of our knowledge there are no reports of P diffusion in bulk silicon with conditions resembling the VLS growth. However, many works have shown that P diffusion is remarkably enhanced under several conditions. For example, the diffusion is enhanced when the surface concentration of P exceeds a value of n_s ~1x10^{20} cm^{-3} (Matsumot.S, Yoshida et al. 1974; F.F.Y.Wang 1981; Gorban and Gorodokin 1988) (as in the present study). This high surface concentration leads to an apparent "kink" in the dopant profile which is a result of dissociation of P$^+$V$^=$ ion pairs, causing an enhanced diffusion in the form of a "tail" (Schwettmann and Kendall 1972). It was shown (Fair and Tsai 1977) that this effect increases the diffusivity at 600°C by four orders of magnitude relative to the value extrapolated from high temperatures; they have also shown that this enhancement will be even larger at lower temperatures. It was reported (Shibayama, Masaki et al. 1976) that Boron and Arsenic diffusivities can reach values of ~ 1x10^{-20} m^{2}s^{-1} in the temperature range of 500-800°C and with no sign for temperature dependence. They associated this enhancement to a generation of excess vacancies at low temperatures. Figure 12 summarizes most of the literature reported data of P low temperature diffusion for the case of high surface concentration in comparison with the our result. The figure shows that the maximum predicted diffusivity, based on the above, for our growth temperature of

460ºC is ~ 1x10^{-21} m^{2}s^{-1} (see our linear fit in Fig. 12) which is still smaller than our extracted value. This may be attributed to: a) generation of excess vacancies at the nanowires surface, b) relatively short time and range of the dopants diffusion during growth (10-15 minutes and 30nm, respectively) in nanowires compared to bulk; and c) to the very large surface to volume ratio present in nanowires (Fig. 13d).

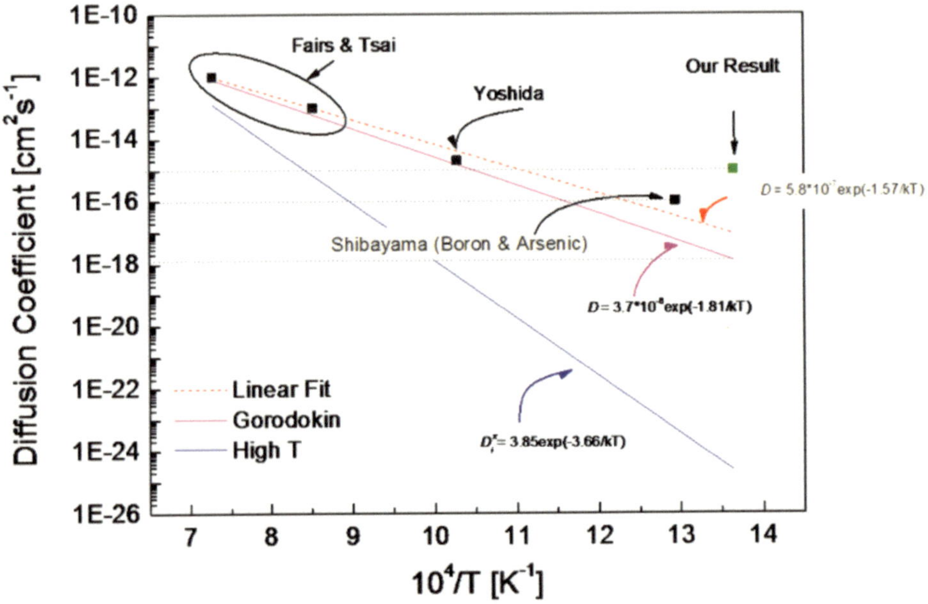

Fig. 12. Summary of the diffusion coefficients from several literature sources including, high surface concentration and low temperature reports together with high temperature data and our result. The red dashed line is a linear fit which made by us in order to predict what should be the diffusivity of P at 460C. It is related to the data from the works of Fair & Tsai, Yoshida, and Shibayama. This linear fit agrees with the following formula: 5.8x10^{-7}exp (-1.57/kT).

Furthermore, the kink observed in our KPFM measured surface potential profiles (see individual wire profiles, Fig. 13) is in agreement with P diffusion in bulk silicon and may imply that similar diffusion mechanisms exist in nanowires (Tsai 1969). However more dopant distribution measurements are required in order to obtain the P diffusion mechanism in silicon nanowires.

In summary, KPFM was used to map the radial active dopant distribution within a single *n*-type silicon nanowire grown by the VLS method. Results show a radial decrease in the dopant concentration from the surface toward the wire core, with a change of almost two orders of magnitude even when there is no indication of taper. Furthermore, the diffusion coefficient of the dopant P atoms in the silicon nanowires was estimated to be D ~1x10^{-19} m^{2}s^{-1}. The latter does confirm the diffusion of P from the vapor phase into the silicon nanowire during its growth, in the VLS method. Finally, the simulated non-uniform current density through the nanowire demonstrates the implications of such non homogenous dopant distribution on the performance of nanowire based electrical devices.

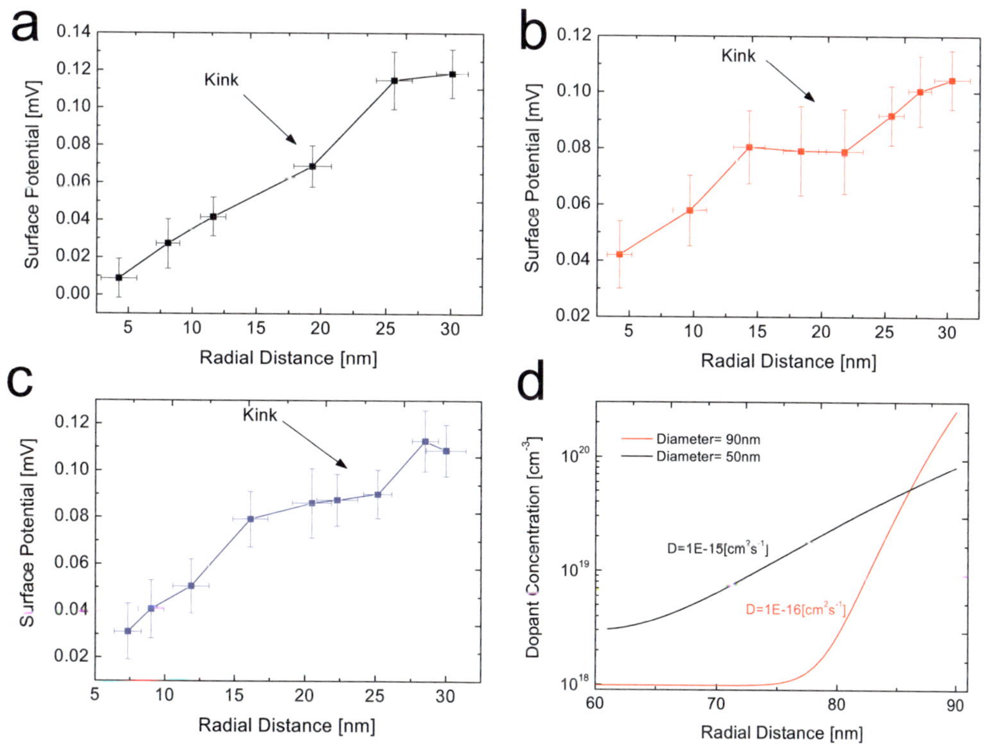

Fig. 13. **(a, b, c)** Individual measured surface potential profiles for three nanowires from the same growth show a kink in the potential profile in a radial distance of ~ 23nm. The same kink has been reported for the case of P diffusion in bulk silicon. The error bars represent the standard deviation in the KPFM measurement. **(d)** Measured radial dopant concentration profiles and diffusivities for nanowires with different diameters, showing higher diffusivity for lower diameter.

5. Obtaining uniform dopant distribution in Si nanowires

As presented in parts 3-4, vapour-solid deposition during nanowire growth can lead to a highly non-uniform distribution of dopants both along the nanowire and across its radii, even in the absence of tapering. In this section, the high diffusivity of P dopants that occurs in nanowires was exploited to reduce the radial non-uniformity by employing thermal annealing following nanowire growth. In addition, no measurable diffusion of the Au catalyst was observed, as probably would occur at higher temperature treatment (Hannon, Kodambaka et al. 2006).

Thermal annealing of the nanowires (prior to their separation from the substrate by sonication i.e. before device fabrication) was carried out at 460 °C in forming gas (90% N_2 and 10% H_2) for one hour using a commercial *AS-MICRO* system of *ANNEALSYS* Company. The radial dopant distribution for both as-grown and annealed nanowires was measured by etching a section of the nanowire surface, followed by the measurement of the potential difference between the etched and the unetched areas using KPFM (as described in the previous part).

Figure 14a presents the simulated (solid lines) and measured (symbols) surface potential profiles for both as-grown and annealed nanowires. The plotted profiles of measurements are averages from 5 different wires, measured at the end that is more highly doped; the error bars represent the standard deviation in the measurement for all the averaged profiles. The simulated potential profiles in Figure 14a are produced after first calculating the dopant profile. As mentioned earlier, during the VLS growth dopants are incorporated into the wire both through the gold catalyst (causing a uniform dopant concentration, $C\ (t=0)$) and through the surface by uncatalyzed decomposition. The time-dependent diffusion equation (eq. 8) was solved to obtain the radial dopant profile inside the as-grown nanowires. The calculated profile assuming D_{Growth}= $1\text{x}10^{-16}$ cm^2s^{-1} (green solid line in Figure 14b) is in good correspondence with the measured surface potential profile (empty symbols in Figure 14a).

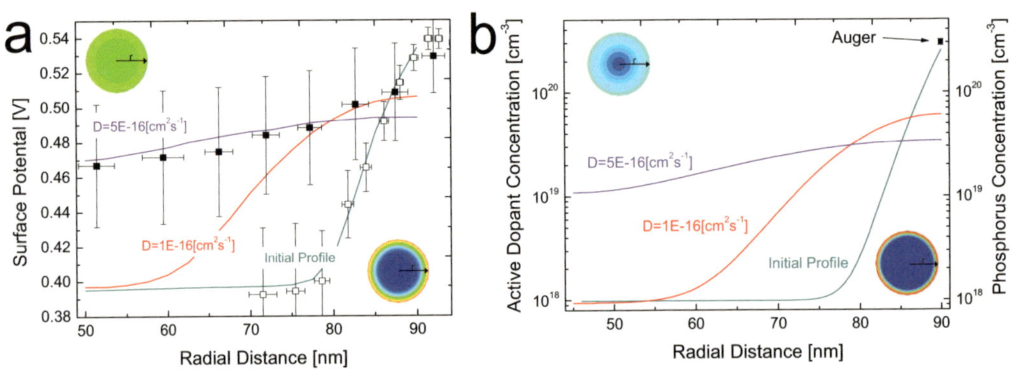

Fig. 14. **(a)** simulated (solid lines) and measured (symbols) radial potential profiles for the as-grown (initial profile) and annealed nanowires. **(b)** calculated radial active dopants concentration profiles (left scale) of the as-grown (initial profile) and annealed nanowires and Auger spectroscopy measurements (right scale) of the nanowire P surface concentration (symbol) before and after the anneal. The insets are the 2D potential **(a)** and concentration **(b)** profiles for the as-grown (right sides) and for the annealed (left sides) nanowires.

The same procedure was used to simulate the radial surface potential profile for the annealed nanowires following a calculation of the dopant diffusion. Thus, we have used the initial dopant distribution, $f(r)$, (established by equation 8 for the as-grown nanowires) to solve the diffusion equation for impermeable cylinder (Crank 1975) (to simulate the annealing treatment):

$$C(r,t)=\frac{2}{a^2}\left\{\int_0^a rf(r)dr+\sum_{n=1}^{\infty}\exp\left(-D_{anneal}\alpha_n^2 t\right)\times\frac{J_0(r\alpha_n)}{J_0^2(a\alpha_n)}\int_0^a rf(r)J_0(\alpha_n r)dr\right\} \tag{9}$$

Here, C is the dopant concentration, r is the radial distance from the middle of the nanowire, $f(r)$ is the initial dopant distribution, α_n's are the roots of: $J_1(a\alpha_n)=0$, a is the nanowire radius and D_{anneal} is the diffusion coefficient during the annealing. Figure 14a, b present surface potential and dopant concentration profiles, respectively, for two different diffusion coefficients after 1 hour of annealing. Using the calculated profile assuming D_{anneal}= $5\text{x}10^{-16}$ cm^2s^{-1} (purple solid line in Figure 14b), we get good agreement with the measured surface potential profile (filled symbols in Figure 14a). The insets in figure 14a, b are the 2D potential and concentration profiles, respectively, for the as-grown (right side) and for the annealed (left side) nanowires.

We note the relatively constant, high measured surface potential of the nanowires implies that a high surface dopant concentration remained after annealing. In order to explore how the annealing process affected the surface P concentration and the fraction of active dopants, we conducted nano-scale Auger measurements to evaluate the P concentration (for Auger spectroscopy: Si nanowires were deposited onto a GaAs wafer, measured by a PHI 700Xi Scanning Auger nanoprobe). Comparing the Auger transitions of Si and P obtained in locations comparable to where the KPFM measurements were done revealed a similar surface dopant concentration (symbol in Fig. 14b) of 0.51% ±0.03% (or $2.55x10^{20}$±$0.15x10^{20}$ cm^{-3}, taking the Si solid concentration to be $5x10^{22}$ cm^{-3}) before and after annealing. The high surface concentration may result not only from direct surface doping, but also from surface segregation of P atoms, which has been predicted to occur in Si nanowires (Fernandez-Serra, Adessi et al. 2006; Peelaers, Partoens et al. 2006) and has been extensively studied in thin films (Nutzel and Abstreiter 1996; Thompson and Jernigan 2007). Specifically, the segregation of P to the Si-SiO interface (Lau, Mader et al. 1989). The stability of this interfacial excess upon annealing, has been observed before in Si thin films (Schwarz, Barton et al. 1981). The high dopant concentration measured in the annealed nanowires is consistent with the expectation of surface segregation.

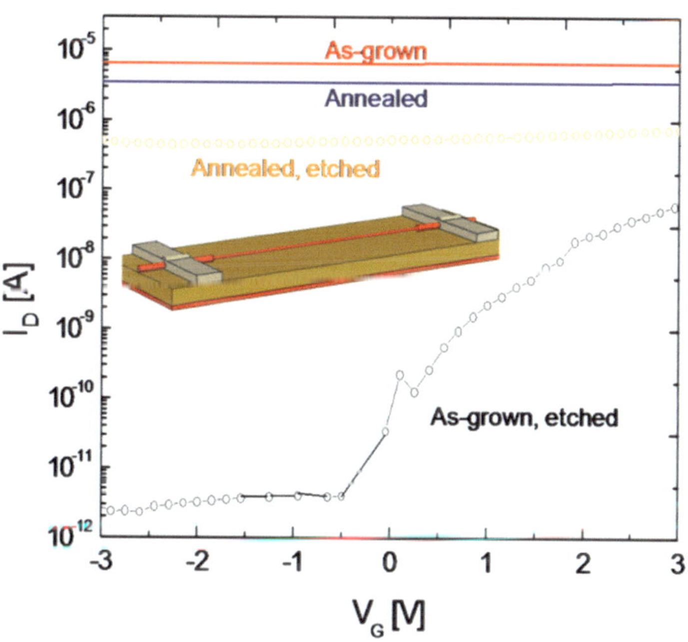

Fig. 15. I_D-V_G current-voltage measurements for as-grown and annealed nanowires before (solid lines) and after (symbols) 20nm etch of the Si surface. The inset is a schematic illustration of the measured etched device.

The redistribution of dopants during annealing was confirmed by current-voltage measurements of two-terminal devices. To highlight the influence of the surface dopants, 20 nm was removed from the surface of both as-grown and annealed nanowires in a single step, and the characteristics were compared with unetched nanowires (Fig. 15). Etched, as-grown nanowire devices showed rectifying behavior in agreement with previous work (Allen, Perea et al. 2009), and their conductivity was strongly influenced by a back-gate voltage. However, the conductivity of the annealed nanowires was only weakly affected by the back gate voltage both before and after the etching, as expected for higher doped nanowires. This behavior further confirms that the annealing treatment encouraged the diffusion of P atoms from the nanowires surface to core, resulting in a uniform dopant distribution with $N_D \sim 1\text{x}10^{19}$ cm^{-3}. Importantly, TEM images show no evidence of Au diffusion after the annealing (Fig. 16c, d), indicating the appropriateness of the low temperature annealing.

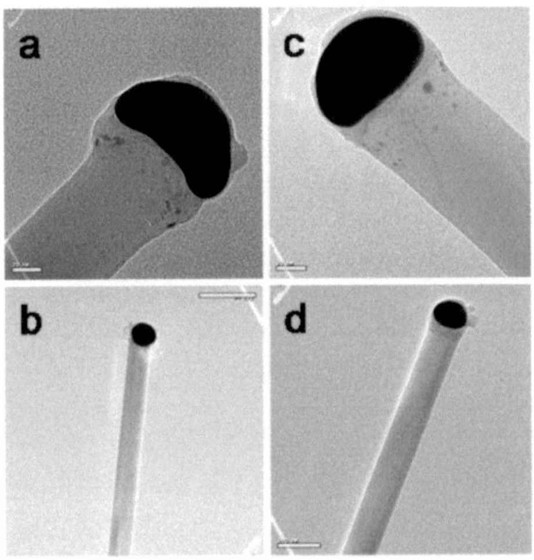

Fig. 16. TEM images of the as-grown **(a, b)** and annealed **(c, d)** nanowires. Scale bars are: 20nm, 200nm, 20nm and 100nm for **(a)**, **(b)**, **(c)**, and **(d)**, respectively.

In summary, a promising approach to obtain uniform radial dopant concentration profiles within P doped silicon nanowires, grown by the VLS process, was demonstrated. In particular, it was shown that low temperature annealing induced the diffusion of P atoms from the enriched nanowire surface to core, with a diffusion coefficient of $\sim 5\text{x}10^{-16}$ cm^{2}s^{-1}, resulting in a more uniform dopant distribution with $N_D \sim 1\text{x}10^{19}$ cm^{-3}, and with no sign of Au diffusion into the nanowires or on the surface.

6. Acknowledgment

The authors would like to thank Dr. John Hammond and Dennis Paul from PHI electronics for their help with scanning Auger measurements. This research was generously supported by Grant No. 2008140 from the United States - Israel Binational Science Foundation [BSF].

7. References

Allen, J. E. (2008). Thesis (Ph.D.) "Investigation of carrier transport in semiconductor nanowires by scanning probe techniques." Materials science and engineering department, *Northwestern University*, USA.

Allen, J. E., E. R. Hemesath, et al. (2009). "Scanning Photocurrent Microscopy Analysis of Si Nanowire Field-Effect Transistors Fabricated by Surface Etching of the Channel." *Nano Letters* 9(5): 1903-1908.

Allen, J. E., D. E. Perea, et al. (2009). "Nonuniform Nanowire Doping Profiles Revealed by Quantitative Scanning Photocurrent Microscopy." *Advanced Materials* 21(30): 3067.

Ben G.Streetman, S. Banerjee. (Prentice-Hall, Englewood Cliffs, 1999). "*Solid State Electronic Devices*" 5th ed: 99.

Binnig, G., C. F. Quate, et al. (1986). "Atomic Force Microscope." *Physical Review Letters* 56(9): 930-933.

Bjork, M. T., H. Schmid, et al. (2009). "Donor deactivation in silicon nanostructures." *Nature Nanotechnology* 4(2): 103-107.

Castell, M. R., D. A. Muller, et al. (2003). "Dopant mapping for the nanotechnology age." *Nature Materials* 2(3): 129-131.

Crank, J. (1975). "*The Mathematics of Diffusion*", 2nd ed.; Oxford Univ. Press: Oxford, 1975; p. 75.

Cui, Y. and C. M. Lieber (2001). "Functional nanoscale electronic devices assembled using silicon nanowire building blocks." *Science* 291(5505): 851-853.

Erwin, S. C., L. J. Zu, et al. (2005). "Doping semiconductor nanocrystals." *Nature* 436(7047): 91-94.

F.F.Y.Wang (1981). "*Materials Processing Theory and Practices*"; North Holland: Amsterdam, New York, Oxford, 1981; Vol. 2, p 342.

Fair, R. B. and J. C. C. Tsai (1977). "A Quantitative Model for the Diffusion of Phosphorus in Silicon and the Emitter Dip Effect." *Journal of the Electrochemical Society* 124(7): 1107-1118.

Fernandez-Serra, M. V., C. Adessi, et al. (2006). "Surface segregation and backscattering in doped silicon nanowires." *Physical Review Letters* 96(16).

Garnett, E. C., Y. C. Tseng, et al. (2009). "Dopant profiling and surface analysis of silicon nanowires using capacitance-voltage measurements." *Nature Nanotechnology* 4(5): 311-314.

Gorban, A. N. and V. A. Gorodokin (1988). "Low-Temperature Diffusion of Phosphorus in Silicon." *Russian Physics Journal* 31(9): 56-60.

Grimmeiss, H. G. (1977). "Deep Level Impurities in Semiconductors." *Annual Review of Materials Science* 7: 341-376.

Gudiksen, M. S., L. J. Lauhon, et al. (2002). "Growth of nanowire superlattice structures for nanoscale photonics and electronics." *Nature* 415(6872): 617-620.

Hannon, J. B., S. Kodambaka, et al. (2006). "The influence of the surface migration of gold on the growth of silicon nanowires." *Nature* 440(7080): 69-71.

Haraguchi, K., T. Katsuyama, et al. (1992). "Gaas P-N-Junction Formed in Quantum Wire Crystals." *Applied Physics Letters* 60(6): 745-747.

Imamura, G., T. Kawashima, et al. (2008). "Distribution of active impurities in single silicon nanowires." *Nano Letters* 8(9): 2620-2624.

Kempa, T. J., B. Tian, et al. (2008). "Single and Tandem Axial p-i-n Nanowire Photovoltaic Devices." *Nano Letters* 8(10): 3456-3460.

Kimukin, I., M. S. Islam, et al. (2006). "Surface depletion thickness of p-doped silicon nanowires grown using metal-catalysed chemical vapour deposition." *Nanotechnology* 17(11): S240-S245.

Koren, E., Rosenwaks, Y., et al. (2009). "Nonuniform doping distribution along silicon nanowires measured by Kelvin probe force microscopy and scanning photocurrent microscopy." *Applied Physics Letters* 95(9): 092105.

Koren, E., Berkovitch, N., et al. (2010). "Measurement of Active Dopant Distribution and Diffusion in Individual Silicon Nanowires." *Nano Letters* 10(4): 1163–1167.

Koren, E., Hyun, J. K., et al. (2011). " Obtaining Uniform Dopant Distributions in VLS-Grown Si Nanowires." *Nano Letters* 11(1): 183–187.

Lau, F., L. Mader, et al. (1989). "A Model for Phosphorus Segregation at the Silicon Silicon Dioxide Interface." *Applied Physics a-Materials Science & Processing* 49(6): 671-675.

Martin, Y., D. W. Abraham, et al. (1988). "High-resolution capacitance measurement and potentiometry by force microscopy." *Applied Physics Letters* 52(13): 1103-1105.

Matsumot.S, M. Yoshida, et al. (1974). "Confirmation of Surface Effect Upon Phosphorus Diffusion into Silicon." *Japanese Journal of Applied Physics* 13(11): 1899-1900.

Nutzel, J. F. and G. Abstreiter (1996). "Segregation and diffusion on semiconductor surfaces." *Physical Review B* 53(20): 13551-13558.

Patolsky, F., G. F. Zheng, et al. (2006). "Nanowire-based biosensors." *Analytical Chemistry* 78(13): 4260-4269.

Peelaers, H., B. Partoens, et al. (2006). "Formation and segregation energies of B and P doped and BP codoped silicon nanowires." *Nano Letters* 6(12): 2781-2784.

Perea, D. E., E. R. Hernesath, et al. (2009). "Direct measurement of dopant distribution in an individual vapour-liquid-solid nanowire." *Nature Nanotechnology* 4(5): 315-319.

Perea, D. E., E. Wijaya, et al. (2008). "Tomographic analysis of dilute impurities in semiconductor nanostructures." *Journal of Solid State Chemistry* 181(7): 1642-1649.

Queisser, H. J. (1978). "Recombination at Deep Traps." *Solid-State Electronics* 21(11-1): 1495-1503.

Schwarz, S. A., R. W. Barton, et al. (1981). "Studies of Phosphorus Pile-up at the Si-Sio2 Interface Using Auger Sputter Profiling." *Journal of the Electrochemical Society* 128(5): 1101-1106.

Schwettmann, F. N. and D. L. Kendall (1972). "On the nature of the kink in the carrier profile for phosphorus-diffused layers in silicon." *Applied Physics Letters* 21(1): 2-4.

Seo, K. I., S. Sharma, et al. (2006). "Surface charge density of unpassivated and passivated metal-catalyzed silicon nanowires." *Electrochemical and Solid State Letters* 9(3): G69-G72.

Shibayama, H., H. Masaki, et al. (1976). "Emitter Dip Effect by Low-Temperature Heat-Treatment of Arsenic-Diffused Layer." *Journal of the Electrochemical Society* 123(5): 742-747.

Strassburg, E., A. Boag, et al. (2005). "Reconstruction of electrostatic force microscopy images." *Review of Scientific Instruments* 76(8).

Thompson, P. E. and G. G. Jernigan (2007). "Determination of the surface segregation ratio of P in Si(100) during solid-source molecular beam epitaxial growth." *Semiconductor Science and Technology* 22(1): S80-S83.

Tsai, J. C. C. (1969). "Shallow phosphorus diffusion profiles in silicon." *Proceedings of the IEEE* 57(9): 1499-1506.

Tutuc, E., J. O. Chu, et al. (2006). "Doping of germanium nanowires grown in presence of PH[sub 3]." *Applied Physics Letters* 89(26): 263101.

Wagner, R. S. and W. C. Ellis (1964). "Vapor-liquid-solid mechanism of single crystal growth." *Applied Physics Letters* 4(5): 89-90.

Wang, J. F., M. S. Gudiksen, et al. (2001). "Highly polarized photoluminescence and photodetection from single indium phosphide nanowires." *Science* 293(5534): 1455.

Wang, Y. F., K. K. Lew, et al. (2005). "Use of phosphine as an n-type dopant source for vapor-liquid-solid growth of silicon nanowires." *Nano Letters* 5(11): 2139-2143.

Weaver, J. M. R. and D. W. Abraham (1991). "High-Resolution Atomic Force Microscopy Potentiometry." *Journal of Vacuum Science & Technology B* 9(3): 1559-1561.

Yang, C., C. J. Barrelet, et al. (2006). "Single p-type/intrinsic/n-type silicon nanowires as nanoscale avalanche photodetectors." *Nano Letters* 6(12): 2929-2934.

19

Fabrication of Conducting Polymer Nanowires

WooSeok Choi, Taechang An and Geunbae Lim
Pohang University of Science and Technology
Korea

1. Introduction

The advancement of nanotechnology provides opportunities for fabrication of nanoscale materials and higher performance devices using nanomaterials with high precision. Currently, various nanomaterials and nanostructures in the range of 1 to 100 nm have been produced by chemical and physical methods. Among various nanostructured materials, one-dimensional (1D) materials, such as nanowires, nanotubes, nanorods, and nanobelts, have potential applications in nanoscale electronics (Cui & Lieber, 2001), optoelectronics (Duan et al., 2001), photonics (Gudiksen et al., 2002; Huang et al., 2001), sensors (Cui et al., 2001), and solar cells (Law et al., 2005) due to their unique electrical, chemical, and optical properties (Li et al., 2006; Thelander et al., 2006; Wang et al., 2008). Nanowires are useful in chemical or biological sensors for detecting single molecules because they have a high surface-to-volume ratio and a highly sensitive 1D nanostructure that gives rise to large conductivity change associated with binding molecules (Cui et al., 2001; Ramanathan et al., 2005).

Conducting polymers, such as polypyrrole, polyaniline, polythiophene, and their derivatives, are promising materials for synthesis of nanostructured materials and devices (Langea et al., 2008; Malinauskas et al., 2005). Compared with other materials, conducting polymers have some unique electrical, chemical, and optical properties because of their conjugated structures, and they are easily synthesized using chemical or electrochemical synthetic methods at room temperatures with low cost (Aleshin, 2006; Briseno et al., 2008; Guimard et al., 2007; Xia et al., 2010). Conducting polymers have electrical and optical properties similar to those of metals and semiconductors, while maintaining the flexibility and properties commonly associated with conventional polymer substances (Dai et al., 2002; Heeger, 2002; Shirakawa, 2002). For example, the electrical conductivity of these polymers can be adjusted from an insulator to traditional metals by varying the species and concentrations of doping ions. Undoped conducting polymers with conductivities of 10^{-10} to 10^{-5} S cm^{-1} can be changed into semiconducting or metallic materials with conductivities of 1 to 10^{4} S cm^{-1} through a chemical or electrochemical doping process (MacDiarmid, 2002). Also, optical absorption bands and mechanical volume of conducting polymers can be changed by entrapped doping ions. Therefore, they have been used for various applications such as electronic devices (Hashizume, 2006), optoelectronic devices (Noy et al., 2002), actuators (Berdichevsky & Lo, 2006), transistors (Alam et al., 2005), and chemical sensors (Bangar et al., 2009; García-Aljaro et al., 2010).

In recent years, 1D conducting polymer nanostructures have been demonstrated to have improved performance with low dimensionality. Many different fabrication methods have been applied to making conducting polymer nanowires. These methods include e-beam lithography, focused ion-beam etching, dip-pen lithography, electro-spinning, DNA scaffolding, mechanical break junction, hydrodynamically focused stream, and nanopore template. In this chapter, we will introduce various fabrication methods for conducting polymer nanowires. This chapter is describes four methods: the template method, electro-spinning method, nanolithography method, and microfluidic method.

2. Template method

The template method has been widely used to fabricate 1D conducting polymer nanostructures such as nanowires and nanotubes because of its simplicity, versatility, and controllability (Cho & Lee, 2007; Tran et al., 2009). Generally, the template method has been classified into a hard-template method, which uses a nanopore film such as anodic aluminium oxide (AAO), and a soft-template method, which uses the self-assembly ability of molecules such as surfactants and DNA.

2.1 Hard template

In the early 1990s, Martin and co-workers introduced the fabrication of various conducting polymer nanowires such as polypyrrole, polyaniline, poly(3-methylthiophene), and polyacetylene using a nanoporous polycarbonate and aluminium oxide template. Figure 1 shows SEM images of these nanowires and their templates (Cai et al., 1991; Cai & Martin, 1989; Liang & Martin, 1990; Martin et al., 1993; Parthasarathy & Martin,1994). These polymers are electrochemically or chemically polymerized from corresponding monomers using an oxidizing agent (Martin, 1994). This process is composed of simple steps: 1) fill nanopores of the membrane with monomers, 2) polymerize the monomers inside the nanopores, and 3) remove the nanopore template to obtain pure polymers (Tran et al., 2009).

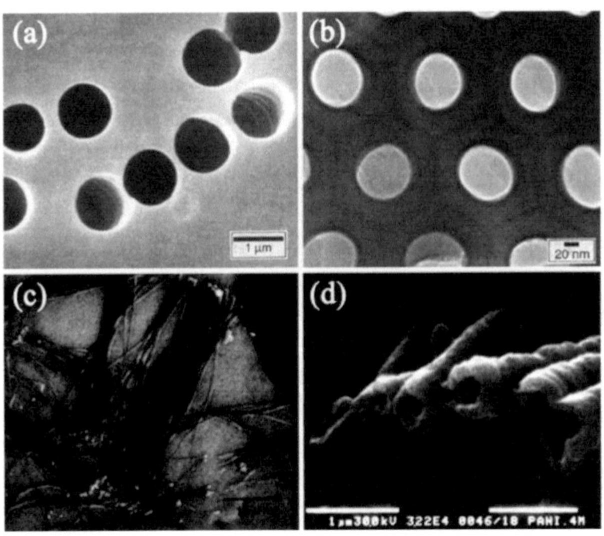

Fig. 1. Conducting polymer nanowires and template from early research by Martin and co-workers. SEM images of (a) polycarbonate and (b) the AAO template (Martin, 1994) and of (c) polyaniline (Partharathy & Martin, 1994) and (d) polyacetylene nanowires (Liang & Martin, 1990).

Later, a method was sought to control the composition or properties of conducting polymer nanowires. Jèrôm et al. (1999) fabricated high-aspect-ratio polypyrrole nanowires 600 nm in diameter and 300 μm in length using a two-step electrochemical method. After electro-polymerization of polyacrylate films on the cathode, polypyrrole was synthesized by chronopotentiometry under a constant current of 0.5 mA/cm^2 in a dimethylformamide solution. Fu et al. (2001) synthesized polythiophene nanowires on flexible gold film by electro-polymerization of thiophene in boron trifluoride diethyl etherate solution using an alumina membrane. The length and diameter of various conducting polymer nanowires synthesized using AAO templates can be controlled by polymerization time and current (Joo et al., 2005; 2003; Park et al., 2005; Xiao et al., 2007). The electrical properties of conducting polymers were controlled through various synthetic conditions, such as doping level, dopant, and template-dissolving solvents (Kim et al., 2005). Li et al. (2001; 2002) copolymerized pyrrole/aniline and pyrrole/thiophene composite nanowires using AAO. The diameter and length could be controlled by the shape of the nanopores of the AAO membrane. The composition of nanowires can be controlled by electrochemical polymerization potentials, and it can be estimated indirectly using cyclic voltammetry and IR spectroscopy. Wang et al. (2005) fabricated polypyrrole/carbon nanotube composite nanowires to improve the electrical and mechanical properties of polymers. Electro-polymerization of polypyrrole was achieved using carboxylated CNT (carbon-nanotube) dopants in nanoporous alumina membranes. Figure 2 shows a SEM image of various conducting polymer nanowires using a hard template.

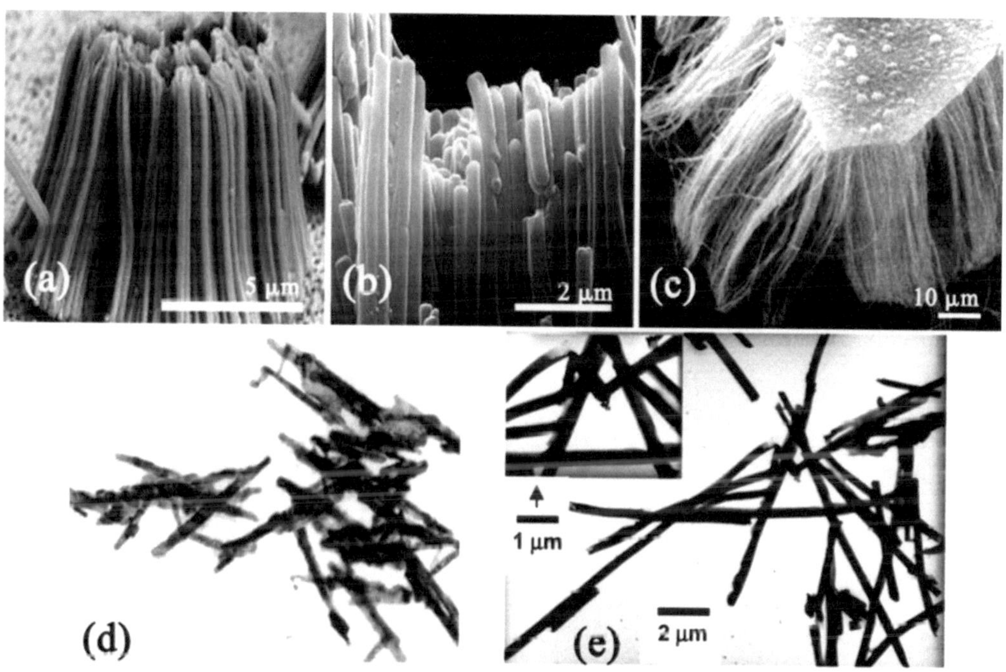

Fig. 2. SEM Images of (a) poly(3-methylthiophene) nanowires, (b) polypyrrole nanowires (Joo et al., 2003) and (c) polythiophene nanowires on flexible substrate (Fu et al., 2001). TEM images of (d) polypyrrole/polyaniline copolymer nanowires (Li et al., 2002) and (e) polypyrrole-carbon nanotube composite nanowires (Wang et al., 2005).

The fabrication method using a hard template provides a straightforward system to synthesize conducting polymer nanowires. The diameter and length of nanowires are controllable by adjusting the pore size, thickness of the membrane, and polymerization conditions. On the other hand, this method requires a post-synthesis task to obtain pure polymer nanowires. In some cases, polymer nanowires can be destroyed or damaged because the template removal process entails harsh conditions. AAO, one of the most widely used nano templates, is a representative example.

2.2 Soft template

The soft-template method which is also called the self-assembly method typically uses surfactants or DNA as templates. This method has the advantage of a simple fabrication process, and the template removal process is achieved under mild conditions or is not required. Thus, it is possible to avoid the damage that occurs during the removal of a hard template.

Surfactants offer a representative soft template because of their self-assembly ability. Wan et al. (1998) accidentally discovered that polyaniline nanowires could be synthesized by *in-situ* doping polymerization in the presence of β-naphthalene sulfonic acid (β-NSA) as the dopant without the need to use any membrane as a hard template. This method does not require the removal of a β-NSA template after polymerization because the membrane is a dopant of polyaniline. Wei et al. (2002) reported that the diameter of polyaniline nanowires can be controlled by adjusting the ratio of β-NSA to aniline monomer concentration. Zhang et al. (2004; 2006) reported various polypyrrole nanostructures in the presence of various surfactants (anionic, cationic, or non-ionic surfactants) with various oxidizing agents. Various polypyrrole nanostructures can be synthesized according to the selection of surfactants and oxidizing agents. Especially, it is possible to obtain nanowire structures in the presence of long-chain cationic surfactants, such as cetyltrimethylammonium bromide, dodeyltrimethylammonium bromide, and anions of the oxidizing agent of ammonium persulfate. Li & Zhang et al. (2004) reported dendritic polyaniline nanowires with diameters between 60 and 90 nm by chemical oxidative polymerization in the presence of a special surfactant gel, which consisted of hexadecyltrimethylammonium chloride ($C_{16}TMA$), acetic acid, aniline, and water at -7°C. Figure 3 shows examples of conducting polymer nanowires fabricated using surfactant as a template.

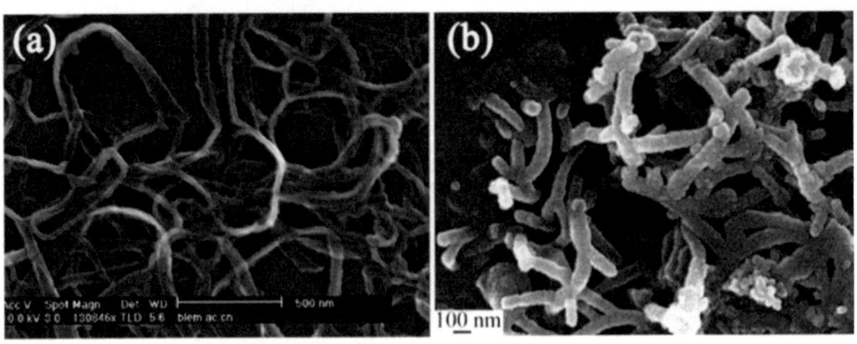

Fig. 3. SEM Images of conducting polymer nanowires using surfactant as a template; (a) polypyrrole nanowires (Zhang et al., 2006); (b) dendric polyaniline nanowires (Li & Zhang, 2004).

DNA molecules also provide attractive soft templates for nanowire fabrication because they are chemically robust and are able to react with monomers without obvious structural requirements or functional group adjacency (Moon et al., 2010; Xia et al., 2010). Ma et al. (2004) immobilized stretched double stranded λ-DNA on a thermally oxidized silicon chip using the molecular combining method. The DNA templates were incubated in aniline monomer solutions (19 mM, pH = 4.0), and the phosphate groups in the DNA templates and protonated aniline monomers were organized by electrostatic interactions. The aligned aniline monomers were enzymatically polymerized by adding horseradish peroxidase and H_2O_2 successively. Figure 4 shows the fabrication process of polyaniline/DNA nanowires introduced by Ma et al. (2004). Polypyrrole nanowires were chemically polymerized on mica with $FeCl_3$ oxidant using DNA as a template by Dong et al. (2007). Moon et al. (2010) also chemically polymerized polypyrrole nanowires using DNA as a template on a (3-aminopropyl)triethoxysilane modified silicon wafer with ammonium persulfate as an oxidant. Hassanien et al. (2010) reported that polyindole nanowires were synthesized on a λ-DNA template by chemical oxidation of indole using $FeCl_3$ as an oxidant. Figure 3 shows AFM images of conducting polymer nanowires using DNA as a template.

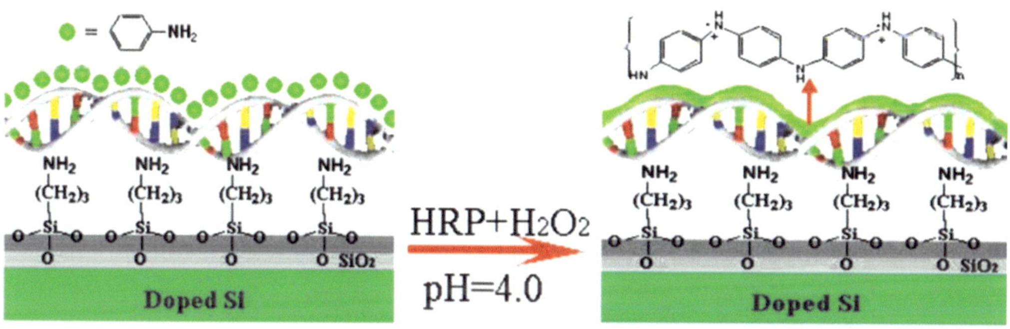

Fig. 4. Fabrication of a polyaniline nanowire immobilized on a Si surface with stretched double stranded DNA as a guiding template (Ma et al., 2004).

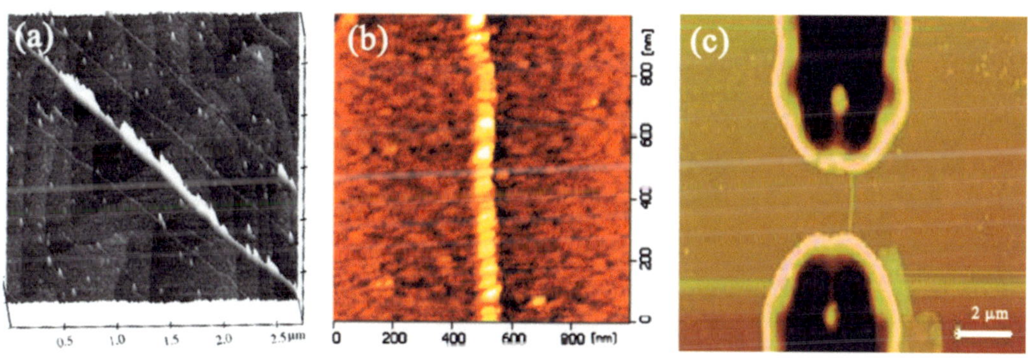

Fig. 5. AFM images of DNA/conducting polymer nanowires; (a) DNA/polypyrrole nanowire. The highlighted part is a DNA/polypyrrole nanowire and the others are bare DNA-scaffold molecules (Dong et al., 2007); (b) DNA/polypyrrole nanowire (Moon et al., 2010); (c) DNA/polyindole nanowires aligned between Au electrodes (Hassanien et al., 2010).

3. Electro-spinning

Electro-spinning has been recognized as one of the most efficient techniques for the fabrication of micro- or nanoscale polymer fibers. Electro-spinning occurs with the development of a jet when the repulsion forces of a charged solution overcome the surface tension of the solution under a high electrostatic field. When the jet solidifies, polymer nanofibers remain (Huang et al., 2003; Reneker & Chun, 1996).

In recent years, it has been reported that various polymers including conducting polymers have been successfully electrospun into polymer nanofibers. MacDiarmid et al. (2001) reported fabrication of polyanlin nanowires with sub-100 nm diameters doped with *d,l* camphorsulfonic acid (PAn.HCSA) as a blend in polyethylene oxide (PEO) based on their previous research (Norris et al., 2000). Zhou et al. (2003) reported an electrospun polyaniline/polyethylene oxide blend nanowire with diameters below 30 nm with optimized process parameters. Chronakis et al. (2006) reported electrospun polypyrrole/PEO nanofibers with diameters in the rage of about 70–300 nm with improved electrical conductivity using the functional doping agent di(2-ethylhexyl) sulfosuccinate sodium salt (NaDEHS). Ju et al. (2007) reported an electro-spinning method for polypyrrole/sulfonated-poly(styrene-ethylene-butylenes-styrene) composite nanofibers with an average diameter of about 300 nm and a uniform and smooth appearance. Choi et al. (2010) reported a method of fabricating electrospun poly(3,4-ethylenedioxythiophene) (PEDOT): poly(styrenesulfonate) (PSS)/PVP nanofibers for a chemical vapor sensor, and Laforgue & Robitaille (2010) introduced a method for fabrication of PEDOT nanofibers by a combination of electro-spinning and vapor-phase polymerization. Figure 6 shows SEM images of conducting polymer nanowires fabricated using electro-spinning.

Fig. 6. SEM Image of (a) polyaniline fibers with an average diameter of 139 nm, (MacDiarmid et al., 2001), (b) polypyrrole/PEO composite nanofibers (the scale bar is 1 μm), (Chronakis et al. 2006), (c) PEDOT:PSS/PVP nanofibers (Choi et al., 2010), and (d) PEDOT nanofibers fabricated using a combination of electro-spinning and vapor-phase polymerization (the scale bar is 5 μm) (Laforgue & Robitaille, 2010).

Electro-spinning is a very effective method for fabrication of conducting polymer nanowires; however, it is limited in its application for various electronic devices owing to the nonwoven fiber shape and the difficulty of controlling positioning. Kameoka et al. (2003) presented a method for the controlled deposition of oriented nanowires with the rotation of collectors using a chopper motor. Although this method is not limited by nonwoven shape, it still has limits compared with other methods.

4. Nanolithography

Nanowire sensors are among the most promising applications because of their impressive sensitivities to detect nanomolar concentrations of DNA (Kemp et al., 2007) and ~10 fM concentrations of micro-RNAs (Fan et al., 2007). To detect such small concentrations of analytes, electrode–wire–electrode assemblies are required (Thapa et al., 2009). The aforementioned fabrication methods can yield uniform nanowires with a high throughput, but they require the elaborate post-synthesis task of positioning the nanowires with the desired precision and electrical contact properties to create one-dimensional nanostructures. The simplest nanowire-positioning technique uses dielectrophoresis (Dan et al., 2007) and self-assembly at the air/water interface (Xu et al., 2009). On the other hand, nanolithography makes it possible to synthesize the nanowires in the desired position. This method allows simultaneous synthesis and positioning of the conducting polymer nanowires.

4.1 Dip-pen nanolithography

Dip-pen nanolithography (DPN), in which an atomic force microscope (AFM) tip transfers alkane thiol to a gold surface, is one of the best-known nanolithography techniques. It is possible to directly write on the desired position without a photomask using an AFM tip as a "pen" (Piner et al., 1999). Lim & Mirkin (2002) synthesized conducting polymer nanowires using DPN using self-doped sulfonated polyaniline (SPAN) and doped polypyrrole as "ink" on a positively charged surface using 2% (trimethoxysilyl)propyldiethylenetriamine (DETA) (Fig. 7). Because SPAN and doped polypyrrole are negatively charged, an electrostatic

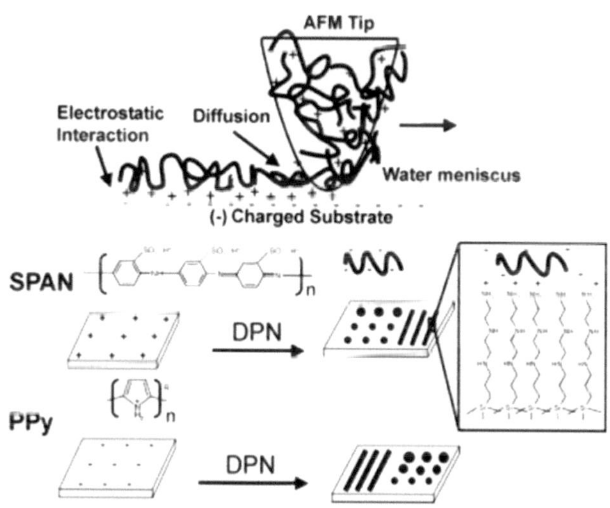

Fig. 7. Schematic representation of dip-pen nanolithography for charged conducting polymers (Lim & Mirkin, 2002).

interaction makes it possible to draw conducting polymer nanowires using DPN. It is possible to control the width of the polymer pattern owing to the linear dependence on the root of contact time (Fig. 8). Maynor et al. (2002) reported the PEDOT line deposition on an anodic silicon surface simultaneous with oxidation. Fig. 9 shows conducting polymer lines drawn using DPN.

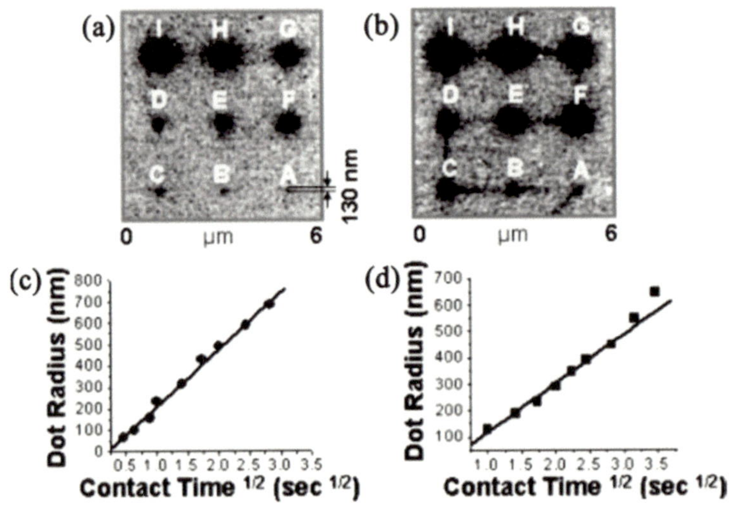

Fig. 8. Diffusion properties of conducting polymers on modified silicon substrates. Lateral force microscopy (LFM) images of (a) SPAN dots and (b) polypyrrole dots as a function of tip-substrate contact time. (c) Plot of the feature radius as a function of $t^{1/2}$ for SPAN and (d) polypyrrole (Lim & Mirkin, 2002).

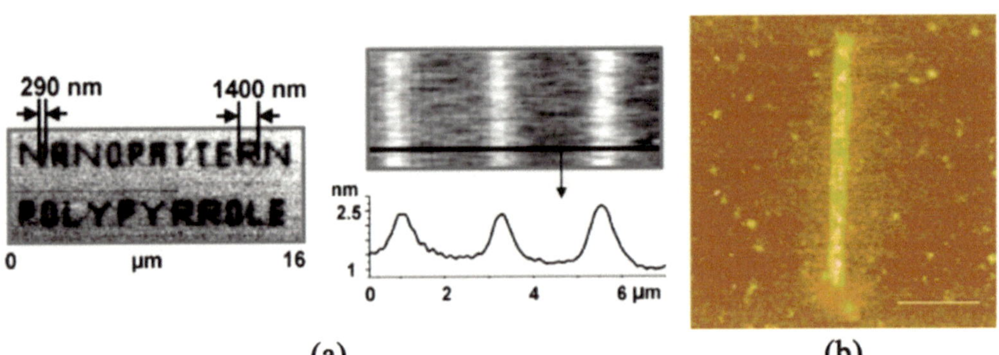

Fig. 9. (a) LFM (left) and topography (right-top) images of polypyrrole lines and the cross-sectional profile (right-bottom) (Lim & Mirkin, 2002). (b) PEDOT line patterned on SiO_2. Polymer line width: 200 nm; scale bar: 1 µm (Maynor et al., 2002).

4.2 e-beam lithography

Yun et al. (2004) and Ramanathan et al. (2004) reported a method of fabricating an array of individually addressable conducting polymer nanowires using e-beam lithography. Figure 10 shows schematic diagram and SEM image of a conducting polymer nanowire using e-beam lithography. They patterned nano channel arrays for the electro-deposition of wires between the electrodes using e-beam lithography. Monomer solutions, such as pyrrole (0.06 M in 0.01 M KCl) or aniline (0.1 M in 0.01 M KCl), were placed on the nanochannel between the electrodes, and the electro-polymerization was generated by applying an electric current. Because the electro-polymerization occurs only in nanochannels, it is possible to control various properties of the nanowires such as dimensions, position, alignment, and chemical composition. This technique is similar to the hard-template method in that nanowire synthesis occurs in nanochannels. The hard-template method is suitable for fabricating nanowires with high density within definite area, whereas e-beam lithography is suitable for fabricating a single nanowire at a desired position, such as in an electrode–wire–electrode structure.

Conducting-polymer fabrication methods based on nanolithography such as Dip-pen and e-beam are attractive because it is possible to synthesize conducting polymers in a desired position to avoid the post-synthesis positioning task. However, they have obvious limitations, such as high cost and low yield, which are characteristic of nanolithgraphy.

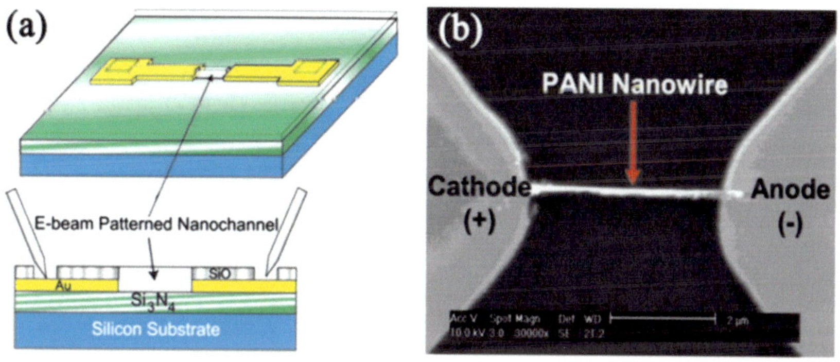

Fig. 10. (a) Schematics diagram of a structure used for the electrochemical wire growth (Yun et al., 2004) and (b) SEM image of a 100 nm width and 4 µm length polyaniline nanowire (Ramanathan et al., 2004).

5. Fabrication using microfluidics

Nanolithography is an attractive technique because nanowires with a diameter of several tens of nanometers can be reproducibly obtained at a desired position. However, nanolithography is not suitable for commercialization because of a low yield and high cost. Fabrication methods using microfluidics make it possible not only to synthesize nanowires in the desired position but also to produce nanowires at a low cost. Limitations such as reproducibility and a slightly larger diameter of nanowires still exist.

5.1 Hydrodynamic focusing

Hou et al. (2008) introduced a fabrication method that uses the characteristics of laminar flow in micro-channels. In the microfluidic system, a low Reynolds number causes a small diffusion layer between miscible liquids. The microfluidic device reported by Hou et al. was made up of polydimethylsiloxane micro-channels and a platinum electrode array on a glass substrate (Fig. 11a). The pyrrole solution (0.1 M pyrrole, 0.1 M LiClO, and 1.0 mM HCl) was inserted into the centre of the microchannel at a constant flow rate, while the distilled water comprised a sheath flow at a variable flow rate. Figure 11b shows the fluorescence images of hydrodynamically focused pyrrole solutions containing a fluorescent component (10 nM 5-carboxyfluorescein). The two-step sheath flow created a compressed monomer layer in the centre of the microchannel, and this layer flowed across the platinum electrode arrays. The thickness of the compressed layer was controlled by changing the flow rate of the sheath flow. Applying constant current between the electrodes, a conducting polymer micropattern was electropolymerized in a focused monomer stream as a dynamic template (Fig. 11c). The width of the polypyrrole wire could be controlled by the width of the focused stream, the gap between electrodes, and the electro-polymerization time.

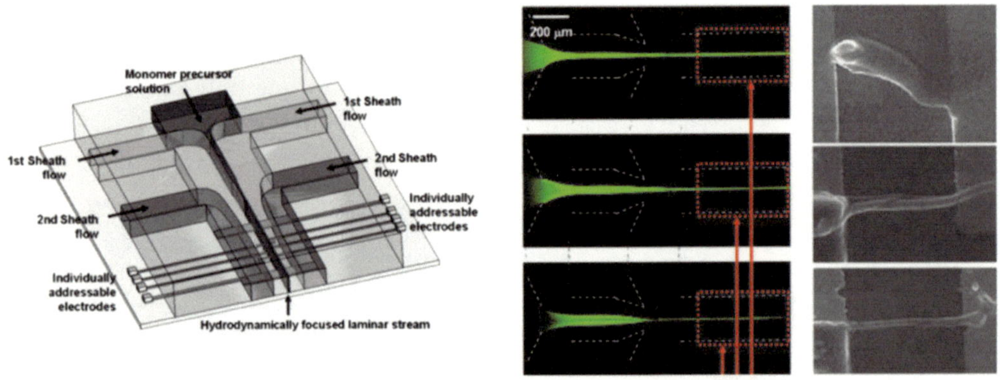

(a) (b) (c)

Fig. 11. (a) Microfluidic device for a hydrodynamically focused laminar stream. (b) Fluorescence images of hydrodynamically focused streams with various widths in a microfluidic device. (c) SEM images of various widths of polypyrrole; 5-, 2-, and 1-µm width (Hou et al., 2008).

Strictly speaking, it is hard produce a nanowire because the smallest published width of a conducting polymer wire is just 1 µm. However, this technique shows the potential to fabricate conducting polymer nanowires by controlling the polymerization conditions. Moreover, the conducting polymer can be synthesized with low cost and high throughput by applying an electric field to the desired electrodes. For this reason, we introduce a hydrodynamic focused fabrication method in this chapter.

5.2 Dielectrophoresis

Choi et al. (2009) electrochemically synthesized CNT–conducting polymer nanowires using dielectrophoresis. CNTs and monomer precursors were gathered between the electrodes where the electric-field gradient is greater due to their higher conductivity compared with the surrounding medium (Fig. 12a). After the suspension was partially removed, the remaining suspension was compressed, and it formed a concave meniscus with evaporation due to surface tension (Fig. 12b). As a result, the electric current was concentrated through the compressed CNTs and monomers. Gathered CNTs generated dynamic templates for electro-polymerization of conducting polymer nanowires. Because nanowires are synthesized between the electrodes to which an electric field is applied, it is possible to individually address a conducting polymer nanowire array to a single chip. Figures 13 shows SEM images of various conducting polymer nanowires on a single chip. These CNT–

conducting polymer nanowires are a few hundred nanometers in diameter and a few micrometers in length.

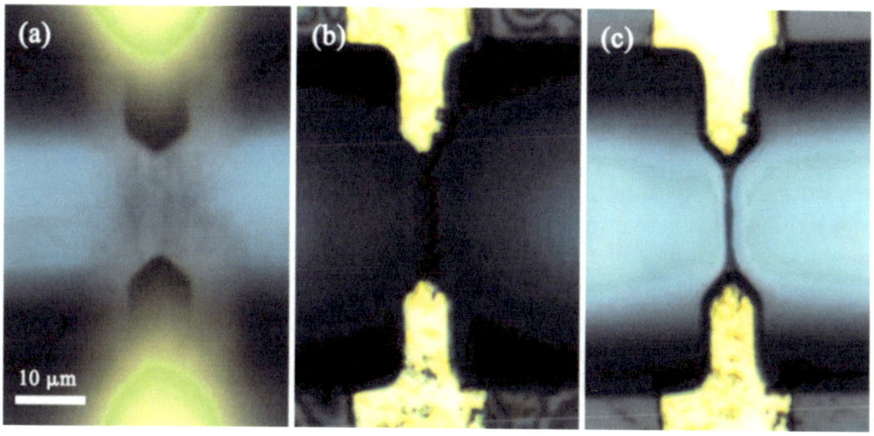

Fig. 12. Microscope images of the CNT-conducting polymer nanowire fabrication process using dielectrophoresis. (a) Attraction of the CNT and monomer molecules between electrodes with an AC electric field; (b) Compression of the CNT and monomer by suspension evaporation; (c) A CNT-conducting polymer nanowire synthesized between electrodes.

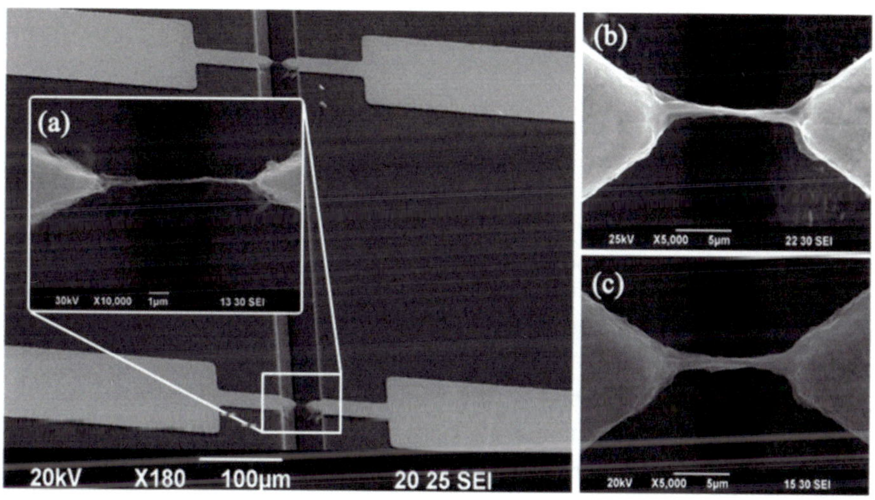

Fig. 13. SEM images of a CNT-conducting polymer composite nanowire array on a single chip. (a) polypyrrole nanowire; (b) polyaniline nanowire; (c) polythiophene nanowire (Choi et al., 2009).

6. Conclusion

This chapter has provided a brief overview of the methods of fabricating conducting polymer nanowires. Since Martin and co-workers first reported conducting polymer nanowires, which were synthesized using nanopore templates, various fabrication methods have been developed. Template methods consist of a simple synthesis process to produce nanowires and can control the size of nanowires by adjusting the nanopore of templates. Electro-spinning is a method to obtain a mass quantity of nanowires with a nonwoven fiber structure using a high electric field. Template and electro-spinning methods are possible options to fabricate nanowires with high yield and low cost; however, they require a post-synthesis task to address the desired nanowire position in some applications that need a single nanowire. In the case of nanolithography, wire synthesis and positioning occur simultaneously because polymerization occurs according to the nano pattern drawn by e-beam or dip-pen nanolithography. This method can control the properties of each individual nanowire; however, it is expensive, and a long production time is required to synthesize a large quantity of nanowires. In the case of the method using microfluidics, polymerization occurs in compressed monomer precursors through hydrodynamic focusing or dielectrophoresis. This method allows for the control of the properties of each individual nanowire in a similar manner to nanolithography. Compared with nanolithography, microfluidics has advantages such as low cost and high yield along with limitations such as size control and reproducibility.

Although many techniques have been developed to fabricate conducting polymer nanowires, still some limitations remain such as size control, low yield, high cost, or long production time due to post-synthesis tasks. Further research is needed to overcome these limitations and develop applications for commercialization.

7. Acknowledgment

This work was supported by Mid-career Research program through NRF grant funded by the MEST (No. 2009-0085377) and the World Class University program through the National Research Foundation of Korea funded by the Ministry of Education, Science and Technology (R31-2008-000-10105-0).

8. References

Alam, M. M., Wang, J., Guo, Y., Lee, S. P. & Tseng, H. R. (2005). Electrolyte gated transistors based on conducting polymer nanowire junction arrays, *J. Phys. Chem. B* 109: 12777.

Aleshin, A. N. (2006). Polymer nanofibers and nanotubes: Charge transport and device applications, *Adv. Mater.* 18: 17–27.

Bangar, M. A., Shirale, D. J., Chen, W., Myung, N. V. & Mulchandani, A. (2009). Single conducting polymer nanowire chemiresistive label-free immunosensor for cancer biomarker, *Anal. Chem.* 81(6): 2168–2175.

Berdichevsky, Y. & Lo, Y.-H. (2006). Polypyrrole nanowire actuators, *Adv. Mater.* 18: 122–125.

Briseno, A. L., Mannsfeld, S. C. B., Jenekhe, S. A., Bao, Z. & Xia, Y. (2008). Introducing organic nanowire transistors, *Mater. Today* 11(4): 38–47.

Cai, Z., Lei, J., Liang, W., Menon, V. & Martin, C. R. (1991). Molecular and supermolecular origins of enhanced electric conductivity in template-synthesized polyheterocyclic fibrils. 1. Supermolecular effects, *Chem. Mater.* 3(5): 960–967.

Cai, Z. & Martin, C. R. (1989). Electronically conductive polymer fibers with mesoscopic diameters show enhanced electronic conductivities, *J. Am. Chem. Soc.* 111(11): 4138–4139.

Cho, S. I. & Lee, S. B. (2007). Fast electrochemistry of conductive polymer nanotubes: Synthesis, mechanism, and application, *Accounts Chem. Res.* 41(6): 699–707.

Choi, J., Lee, J., Choi, J., Jung, D. & Shim, S. E. (2010). Electrospun PEDOT:PSS/PVP nanofibers as the chemiresistor in chemical vapour sensing, *Synth. Met.* 160: 1415–1421.

Choi, W., An, T. & Lim, G. (2009). Fabrication of conducting polymer nanowire sensor array, *Sensors*, IEEE, New Zealand, pp. 1151–1153.

Chronakis, I. S., Grapenson, S. & Jakob, A. (2006). Conductive polypyrrole nanofibers vis electrospinning: Electrical and morphological properties, *Polymer* 47(5): 1597–1603.

Cui, Y. & Lieber, C. M. (2001). Functional nanoscale electronic devices assembled using silicon nanowire building blocks, *Science* 291(5505): 851–853.

Cui, Y., Wei, Q., Park, H. & Lieber, C. M. (2001). Nanowire nanosensors for highly sensitive and selective detection of biological and chemical species, *Science* 293(5533): 1289–1292.

Dai, L., Soundarrajan, P. & Kim, T. (2002). Sensors and sensor arrays based on conjugated polymers and carbon nanotubes, *Pure Appl. chem.* 74(9): 1753–1772.

Dan, Y., Cao, Y., Mallouk, T. E., Johnson, A. T. & Evoy, S. (2007). Dielectrophoretically assembled polymer nanowires for gas sensinig, *Sens. Actuator A* 125: 55–59.

Dong, L., Hollis, T., Fishwick, S., Connolly, B. A., Wright, N. G., Horrocks, B. R. & Houlton, A. (2007). Synthesis, manipulation and conductivity of supramolecular polymer nanowires, *Chem. Eur. J.* 13(3): 822–828.

Duan, X., Huang, Y., Cui, Y., Wang, J. & Lieber, C. M. (2001). Indium phosphide nanowires as building blocks for nanoscale electronic and optoelectronic devices, *Nature* 409(6816): 66–69.

Fan, Y., Chen, X., Trigg, A. D., hang Tung, C., Kong, J. & Gao, Z. (2007). Detection of microRNAs using target-guided formation of conducting polymer nanowires in nanogaps, *J. Am. Chem. Soc.* 129(17): 5437–5443.

Fu, M., Zhu, Y., Tan, R. & Shi, G. (2001). Aligned polythiophene micro- and nanotubules, *Adv. Mater.* 13: 1874–1877.

García-Aljaro, C., Bangar, M. A., Baldrich, E., noz, F. J. M. & Mulchandani, A. (2010). Conducting polymer nanowire-based chemiresistive biosensor for the detection of bacterial spores, *Biosens. Bioelectron.* 25: 2309–2312.

Gudiksen, M. S., Lauhon, L. J., Wang, J., Smith, D. C. & Lieber, C. M. (2002). Growth of nanowire superlattice structures for nanoscale photonics and electronics, *Nature* 415(6872): 617–620.

Guimard, N. K., Gomez, N. & Schmidt, C. E. (2007). Conducting polymers in biomedical engineering, *Prog. Polym. Sci.* 32: 876–921.

Hashizume, T. (2006). Propertiy of individual conducting polymer nanowires: conductance and fet devices, *APS Meeting Abstracts* p. 7002.

Hassanien, R., Al-Hinai, M., Al-Said, S. A. F., Little, R., Ŝiller, L., Wright, N. G., Houlton, A. & Horrocks, B. R. (2010). Preparation and characterization of conductive and phhotoluminescent dna-templated polyindole nanowires, *ACS Nano* 4(4): 2149–2159.

Heeger, A. J. (2002). Semiconducting and metallic polymers: the forth generation of polymeric metals, *Synth. Met.* 125: 23–42.

Hou, S., Wang, S., Yu, Z., Zhu, N., Liu, K., Sun, J., Lin, W.-Y., Shen, C.-F., Fang, X. & Tseng, H.-R. (2008). A hydrodynamically focused stream as a dynamic template for site-specific electrochemical micropatterning of conducting polymers, *Angew. Chem.* 120(6): 1088–1091.

Huang, M. H., Mao, S., Feick, H., Yan, H., Wu, Y., Kind, H., Weber, E., Russo, R. & Yang, P. (2001). Room-temperature ultraviolet nanowire nanolasers, *Science* 292(5523): 1897–1899.

Huang, Z.-M., Zhang, Y.-Z., Kotaki, M. & Ramakrishna, S. (2003). A review on polymer nanofibers by electrospinning and their applications in nanocomposites, *Compos. Sci. Technol.* 63(15): 2223–2253.

Jèrôme, C., Labaye, D., I.Bodart & Jèrôme, R. (1999). Electrosynthesis of polyacrylic/polypyrrole compostes: formation of polypyttole wires, *Synth. Met.* 101: 3–4.

Joo, J., Kim, B. H., Parka, D. H., Kim, H. S., Seo, D. S., Shim, J. H., Lee, S. J., Ryu, K. S., Kim, K., Jin, J. I., Lee, T. J. & Lee, C. (2005). Fabrication and applications of conducting polymer nanotube, nanowire, nanohole, and double wall nanotube, *Synth. Met.* 153: 313–316.

Joo, J., Park, K. T., Kim, B. H., Kim, M. S., Lee, S. Y., Jeong, C. K., Lee, J. K., Park, D. H., Yi, W. K., Lee, S. H. & Ryu, K. S. (2003). Conducting polymer nanotube and nanowire synthesized by using nanoporous template: Synthesis, characteristics, and application, *Synth. Met.* 135-136: 7–9.

Ju, Y.-W., Park, J.-H., Jung, H.-R. & Lee, W.-J. (2007). Electrochemical properties of polypyrrole/sulfonated SEBS composite nanofibers prepared by electrospinning, *Electrochim. Acta* 52(14): 4841–4847.

Kameoka, J., Orth, R., Yang, Y., Czaplewski, D., Mathers, R., Coates, G. W. & Craighead, H. G. (2003). A scanning tip electrospinning source for deposition of oriented nanofibers, *Nanotechnology* 14(10): 1124.

Kemp, N. T., McGrouther, D., Cochrane, J. W. & Newbury, R. (2007). Bridging the gap: Polymer nanowire devices, *Adv. Mater.* 19(18): 2634–2638.

Kim, B. H., Park, D. H., Joo, J., Yu, S. & Lee, S. H. (2005). Synthesis, characteristics, and field emission of doped and de-doped polypyrrole, polyaniline, poly(3,4-ethylenedioxythiophene) nanotubes and nanowires, *Synth. Met.* 150: 279–284.

Laforgue, A. & Robitaille, L. (2010). Production of conductive PEDOT nanofibers by the combination of electrospinning and vapor-phase polymerization, *Macromolecules* 43(9): 4194–4200.

Langea, U., Roznyatovskayal, N. V. & Mirsky, V. M. (2008). Conducting polymers in chemical sensors and arrays, *Anal. Chim. Acta* 614(1): 1–26.

Law, M., Greene, L. E., Johnson, J. C., Saykally, R. & Yang, P. (2005). Nanowire dye-sensitized solar cells, *Nat. Mater.* 4(6): 455–459.

Li, G. & Zhang, Z. (2004). Synthesis of dendritic polyaniline nanofibers in a surfactant gel, *Macromolecules* 37: 2683–2685.

Li, X., Lu, M. & Li, H. (2001). Preparation and characterization of pyrrole/aniline copolymer nanofibrils using the template-synthesis method, *J. Appl. Polym. Sci.* 81: 3002–3007.

Li, X., Lu, M. & Li, H. (2002). Electrochemical copolymerization of pyrrole and thiophene nanofibrils using template-synthesis method, *J. Appl. Polym. Sci.* 86: 2403–2407.

Li, Y., Qian, F., Xiang, J. & Lieber, C. M. (2006). Nanowire electronic and optoelectronic devices, *Mater. Today* 9(10): 18–27.

Liang, W. & Martin, C. R. 1990. Template-synthesized polyacetylene fibrils show enhanced supermolecular order, *J. Am. Chem. Soc.* 112(26): 9666–9668.

Lim, J.-H. & Mirkin, C. A. (2002). Electrostatically driven dip-pen nanolithography of conducting polymers, *Adv. Mater.* 14(20): 1474–1477.

Ma, Y., Zhang, J., Zhang, G. & He, H. (2004). Polyaniline nanowires on Si surfaces fabricated with dna templates, *J. Am. Chem. Soc.* 126(22): 7097–7101.

MacDiarmid, A. G. (2002). Synthetic metals: a novel role for organic polymers, *Synth. Met.* 125: 11–22.

MacDiarmid, A. G., Jones, W. E., Norris, I. D., Gao, J., Johnson, A. T., Pinto, N. J., Hone, J., Han, B., Ko, F. K., Okuzaki, H. & Llaguno, M. (2001). Electrostatically-generated nanofibers of electronic polymers, *Synth. Met.* 119: 27–30.

Malinauskas, A., Malinauskiene, J. & Ramanavičius, A. (2005). Conducting polymer-based nanostructurized materials: Electrochemical aspects, *Nanotechnology* 16(10): R51–R62.

Martin, C. R. (1994). Nanomaterials: A membraine-based synthetic approach, *Science* 266: 1961–1966.

Martin, C. R., Parthasarathy, R. & Menon, V. (1993). Template synthesis of electronically conductive polymers - A new route for achieving higher electronic conductivities, *Synth. Met.* 55: 1165–1170.

Maynor, B. W., Filocamo, S. F., Grinstaff, M. W. & Liu, J. (2002). Direct-writing of polymer nanostructures: Poly(thiophene) nanowires on semiconducting and insulating surfaces, *J. Am. Chem. Soc.* 124(4): 522–523.

Moon, H. K., Kim, H. J., Kim, N.-H. & Roh, Y. (2010). Fabrication of highly uniform conductive polypyrrole nanowires with dna template, *J. Nanosci. Nanotechnol.* 10: 3180–3184.

Norris, I. D., Shaker, M. M., Ko, F. K. & MacDiarmid, A. G. (2000). Electrostatic fabrication of ultrafine conducting fibers: polyaniline/polyethylene oxide blends, *Synth. Met.* 114: 109–114.

Noy, A., Miller, A. E., Klare, J. E., Weeks, B. L., Woods, B. W. & DeYoreo, J. J. (2002). Fabrication of luminescent nanostructures and polymer nanowires using dip-pen nanolithography, *Nano Lett.* 2(2): 109–112.

Park, D. H., Kim, B. H., Jang, M. K., ans S. J. Lee, K. Y. B. & Joo, J. (2005). Synthesis and characterization of polythiophene and poly (3-methylthiophene) nanotubes and nanowires, *Synth. Met.* 153: 341–344.

Parthasarathy, R. V. & Martin, C. R. (1994). Template-synthesized polyaniline microtubules, *Chem. Mater.* 6: 1627–1632.

Piner, R. D., Zhu, J., Xu, F., Hong, S. & Markin, C. A. (1999). "Dip-pen" nanolithography, *Science* 283(29): 661–662.

Ramanathan, K., Bangar, M. A., Yun, M., Chen, W., Mulchandani, A. & Myung, N. V. (2004). Individually addressable conducting polymer nanowires array, *Nano Lett.* 4(7): 1237–1239.

Ramanathan, K., Bangar, M. A., Yun, M., Chen, W., Myung, N. V. & Mulchandani, A. (2005). Bioaffinity sensing using biologically functionalized conducting-polymer nanowire, *J. Am. Chem. Soc.* 127(2): 496–497.

Reneker, D. H. & Chun, I. (1996). Nanometre diameter fibres of polymer, produced by electrospinning, *Nanotechnology* 7: 216–223.

Shirakawa, H. (2002). The discovery of polyacetylene film the dawining of an era of conducting polymers, *Synth. Met.* 125: 3–10.

Thapa, P. S., Yu, D. J., Wicksted, J. P., Hadwiger, J. A., Barisci, J. N., Baughman, R. H. & Flanders, B. N. (2009). Directional growth of polypyrrole and polythiophene wires, *Appl. Phys. Lett* 94: 033104.

Thelander, C., Agarwal, P., Brongersman, S., Eymery, J., Feiner, L. F., Forchel, A., Scheffler, M., Riess, W., Ohlsson, B. J., Gösele, U. & Samuelson, L. (2006). Nanowire-based one-dimensional electronics, *Mater. Today* 9(10): 28–35.

Tran, H. D., Li, D. & Kaner, R. B. (2009). One-dimensional conducting polymer nanostructures: Bulk synthesis and applications, *Adv. Mater.* 21: 1487–1499.

Wan, M. X., Shen, Y. Q. & Huang, J. (1998). Chinese patent no.98109916.5.

Wang, J., Dai, J. & Yarlagadda, T. (2005). Carbon nanotube-conducting-polymer composite nanowires, *Langmuir* 21(1): 9–12.

Wang, N., Cai, Y. & Zhang, R. Q. (2008). Growth of nanowires, *Mater. Sci. Eng. R-Rep.* 60: 1–51.

Wei, Z., Zhang, Z. & Wan, M. (2002). Formation mechanism of self-assembled polyaniline micro/nanotubes, *Langmuir* 18: 917–921.

Xia, L., Wei, Z. & Wan, M. (2010). Conducting polymer nanostructures and their application in biosensors, *J. Colloid Interface Sci.* 341: 1–11.

Xiao, R., Cho, S. I., Liu, R. & Lee, S. B. (2007). Controlled electrochemical synthesis of conductive polymer nanotube structures, *J. Am. Chem. Soc.* 129: 4483–4489.

Xu, J., Jiang, Y., Yang, Y. & Yu, J. (2009). Self-assembly of conducting polymer nanowires at air-water interface and its application for gas sensors, *Mater. Sci. Eng. B* 157: 87–92.

Yun, M., Myung, N. V., Vasquez, R. P., Lee, C., Menke, E. & Penner, R. M. (2004). Electrochemically grown wires for individually addressable sensor arrays, *Nano Lett.* 4(3): 419–422.

Zhang, X., Zhang, J., Liu, Z. & Robinson, C. (2004). Inorganic/organic mesostructure directed synthesis of wire/ribbon-like polypyrrole nanostructures, *Chem. Commun.* (16): 1852–1853.

Zhang, X., Zhang, J., Song, W. & Kiu, Z. (2006). Controllable synthesis of conducting polypyrrole nanostructures, *J. Phys. Chem. B.* 110(3): 1158–1165.

Zhou, Y., Freitag, M., Hone, J., Staii, C., Johnson, A. T., Pinto, N. J. & MacDiarmid, A. G. (2003). Fabrication and electrical characterization of polyaniline-based nanofibers with diameter below 30 , *Appl. Phys. Lett.* 83(18): 3800–3802.

Fabrication and Characterization of Copper Nanowires

Hardev Singh Virk
Director Research, Nanotechnology Laboratory,
DAV Institute of Engineering and Technology,
India

1. Introduction

Nanowires are especially attractive for nanoscience studies as well as for nanotechnology applications. Nanowires, compared to other low dimensional systems, have two quantum confined directions, while still leaving one unconfined direction for electrical conduction. This allows nanowires to be used in applications where electrical conduction, rather than tunneling transport, is required. Because of their unique density of electronic states, nanowires in the limit of small diameters are expected to exhibit significantly different optical, electrical and magnetic properties from their bulk 3D crystalline counterparts. The increased surface area, very high density of electronic states, enhanced exciton binding energy, diameter-dependent band gap, and increased surface scattering for electrons and phonons are just some of the ways in which nanowires differ from their corresponding bulk materials. Synthesis, characterization and application of nanowires and nanotubes comprise a significant aspect of today's endeavor in nanotechnology. During recent years, nanowires and nanorods of metallic and semi-conducting materials have drawn a lot of research interest because of their potential applications in diverse fields, for example, nanoelectronics, opto-electronics and sensors (Sarkar et al. 2007; Ratner & Ratner, 2003; Nalwa & Bandhopadhaya, 2003; Dresselhaus et al. 2003).

Many studies have focused on the fabrication of copper nanowires (Cao & Liu, 2008; Sun Shin et al. 2009; Ingunta et al. 2008; Fang et al. 2007; Motoyama et al. 2005), because of their potential applications in the micro/nanoelectronics industry and, in particular, for interconnection in electronic circuits. Copper is one of the most important metals in modern electronic technology. Many methods have been developed for the fabrication of copper nanowires but template synthesis is considered to be the most suitable and useful for growth of nanowires.

Template synthesis by electrochemical deposition route is easy, low-cost as well as less cumbersome compared to other fabrication techniques (Sarkar et al. 2007), namely, pulsed laser deposition (PLD), vapour-liquid-solid (VLS) method and chemical vapour deposition (CVD). Another advantage of the electrochemical deposition technique is the possibility of fabricating multi-layered structures within nanowires. By varying the cathodic potentials in the electrolyte which contains two different kinds of ions, different metal layers can be controllably deposited. Electrochemical cell used in electrodeposition of copper into pores of anodic alumina template was fabricated in our laboratory. Morphology of electrodeposited copper nanowires has been studied using Field Emission Scanning Electron Microscopy (FESEM) and crystal structure by XRD analysis. The diameter of nanowires generally depends upon the pore size of template. Anodic alumina discs of 200 nm and polymer membranes of 100 nm pore diameter were selected for this purpose.

2. Template synthesis of nanowires

Template-based growth is a versatile method of synthesis of metallic and semiconductor nanowires. In template-assisted synthesis of nanostructures, the chemical stability and mechanical properties of the template, as well as the diameter, uniformity and density of the pores are important characteristics to consider. Templates frequently used for nanowire synthesis include anodic alumina (Al_2O_3), nano-channel glass, ion track-etched polymers and mica films. Porous anodic alumina templates are produced by anodizing pure Al films in various acids, for example, oxalic acid is most commonly used. Under carefully chosen anodization conditions, the resulting oxide film possesses a regular hexagonal array of parallel and nearly cylindrical channels, as shown in Fig. 2(a). The self-organization of the pore structure in an anodic alumina template involves two coupled processes: pore formation with uniform diameters and pore ordering. Depending on the anodization conditions, the pore diameter can be systematically varied from < 10nm up to 200nm with a pore density in the range of 10^9 to 10^{11} pores/cm^2 (Dingle et al., 1969).

Template materials must meet certain requirements (Cao & Liu, 2008). First, the template materials must be compatible with the processing conditions. For example, an electrical insulator is required for a template to be used in electrochemical deposition. Template materials should be chemically and thermally inert during the synthesis. Secondly, depositing materials or solution must wet the internal pore walls. Thirdly, for synthesis of nanowires, the deposition should start from the bottom of the template and proceed upwards to the other side. This is known as bottom up technique in nanotechnology.

Template-based synthesis offers many advantages over other methods of synthesis (Lai & Riley, 2008): (1) It is performed under mild conditions rather than requiring high temperatures, high vacuum or expensive instrumentation; (2) templated electrodeposition has a relatively high growth rate; (3) the morphology of deposited materials depends on the shape of template pores; (4) the dimensions of the materials obtained can be tuned by tuning of the template pore size; (5) two or more components can be easily deposited into the membrane sequentially to form multi-segmented materials or hetero-junctions.

3. Materials and methods

The electrodeposition technique used in our experiment (Virk et al., 2010) is similar in principle to that used for the electroplating process. Commercial anodic alumina membranes (AAM) (anodisc 25 made by Whatman) having an average pore diameter of 200 nm, a nominal thickness of 60 μm and a pore density of 10^9 pores/ cm^2, were used as templates. A second set of polymer membrane (Sterlitech USA) of 100 nm pore diameter was selected for the sake of comparison. To achieve uniform deposition of nanowires, templates were cleaned in the ultrasonic bath for 10 minutes. The electrochemical cell, fabricated in our laboratory using Perspex sheets, was washed in double distilled water. A copper rod of 0.8 cm diameter was used as a sacrificial electrode (anode). The cathode consists of copper foil attached to alumina disc by an adhesive tape of good conductivity. Prior to the electro-deposition process, a thin film of copper (0.5 μm) was sputtered onto one side of alumina disc. This metal layer along with adhesive copper tape provides a stable substrate (cathode) for the growth of nanowires. Figure 1(a) illustrates the scheme of this process.

Polymer membranes can be prepared by irradiation of polycarbonate foils using heavy ion beams (Toimil Molares, 2001). Author has prepared polymer templates, called Ion Track Filters (Virk & Kaur, 1998), using Makrofol N and Kapton after irradiation at the UNILAC (Universal Linear Accelerator) of GSI, Darmstadt, with highly charged heavy ions having kinetic energies in the GeV range and fluences between 10^6 and 10^{10} ions/cm^2. Due to energy loss through interaction with the target electrons, each ion creates along its trajectory a cylindrical damage zone, a few nanometers in diameter. The damaged material can selectively be removed by chemical etching, resulting in pores of cylindrical geometry. Composition, concentration, and temperature of the etching solution determine the size and geometry of the resulting pores, the pore diameter increasing linearly with the etching time. A 6 N NaOH solution containing 10% methanol at T = 50 ^{0}C was used for etching to produce pore diameters between 50 and 200 nm by varying time of etching. A thin gold film was sputtered onto one side of the membrane using Jeol sputter and reinforced by copper foil attached by an adhesive tape of good conductivity to obtain a stable substrate. This serves as a cathode suitable for the growth of the nanowires in polymer template in our two-electrode electrochemical cell. A schematic diagram of the polymer template synthesis process is illustrated in Figure 1(b).

The electrolyte used had a composition of 20 gm/100ml $CuSO_4.5H_2O$ + 25% of dilute H_2SO_4 at room temperature. A high concentration of CuSO4 was used to supply a sufficiently large number of ions inside the pores during the deposition. Sulfuric acid was added to increase the conductivity of the solution and to lower the cathode over-voltage. The electrodeposition was performed at room temperature of 30 ^{0}C. The low overvoltages avoided side reactions such as hydrogen evolution. The inter-electode distance was kept 0.5 cm and a current of 2mA was applied for 10 minutes using a regulated power supply. Electrodeposition of copper nanowires depends on many factors, namely, inter-electrode spacing, electrolyte composition, temperature and pH value, current density and time of deposition. The influence of current density, temperature and type of electrolyte on the crystallinity of copper nanowires has been reported elsewhere (Toimil Molares et al., 2001). We studied the effect of current density on electrodeposition of copper nanowires in our experiment.

After the electrodeposition was over, copper foil with template-grown nanowires was divided into two parts. One part was kept for study of I-V characteristics in-situ using Dual Source Meter (Keithley Model 4200 SCS) with platinum probes for contacts. The other part was kept immersed in 1 M NaOH for 1 hour in a beaker to dissolve alumina template. The copper nanowires were liberated from the host matrix, washed in distilled water and dried in an oven at 50^{0}C for 30 minutes. The cleaned and dried nanowires were mounted on aluminium stubs with the help of double adhesive tape. Field Emission Scanning Electron Microscope (FESEM, Hitachi S-4300) was used to record cross-sectional and lateral views of grown nanowires at an accelerating voltage of 15kV using different magnifications. X-ray Diffraction studies were carried out at Sophisticated Analytical Instruments Facility (SAIF) set up by Punjab University, Chandigarh using X' Pert PRO (PANanalytical, Netherlands) using Cu Kα radiation.

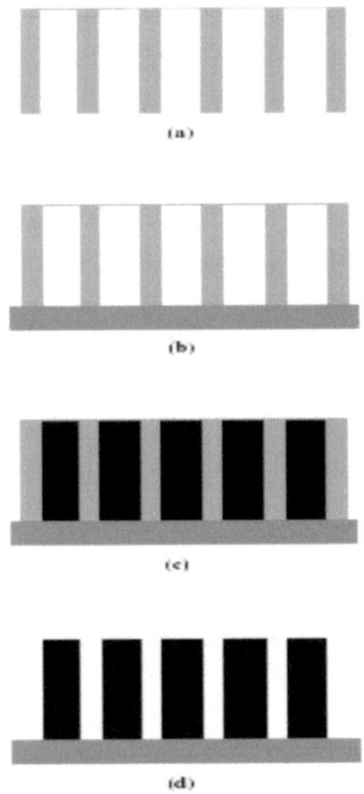

Fig. 1. (a). A schematic diagram of the template synthesis process (Gao et al., 2002):
(a) Anodic alumina template, (b) copper sputtered alumina template, (c) electrodeposited copper nanowires, and (d) copper nanowires after removal of anodic alumina template.

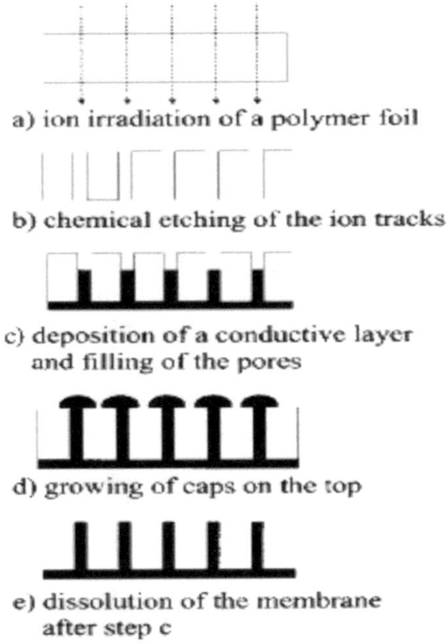

Fig. 1. (b). Scheme of the polymer template synthesis (Toimil Molares et al., 2001).

4. Characterization of copper nanowires

4.1 AFM, SEM and FESEM analysis

Commercial available templates were examined before their use using Atomic Force Microscope (NT-MDT PR 400 Model) installed in our laboratory and Scanning Electron Microscope (Jeol, JSM 6100) facility of Punjab University, Chandigarh. Atomic force microscopic technique (Menon, 2003) shows the two dimensional surface topology of the anodic alumina template with pores regularly arranged on the surface (Fig. 2a). The pores appear nearly at the centre of each hexagonal cell. After gold sputtering, using Jeol sputter JFC 1100, SEM micrograph (Fig. 2b) shows the geometrical pattern of pores on the alumina surface of anodisc.

Copper nanowires liberated from AAM were examined under SEM and FESEM under different magnifications. Two sets of templates were used for growth of copper nanowires. In one set, current density was changed intermittently which resulted in non-uniform growth of nanowires. Figure 3 represents the cross-sectional view of copper nanowires of 200nm diameter grown in alumina template. Figure 4(a) shows the SEM image of copper nanowires array in lateral view, grown under constant current conditions. Figure 4(b) represents the FESEM image of copper nanowires fabricated under transient current conditions. Overdeposition of copper is clearly visible towards the tip of nanowires resulting in capping effect. Nanawires are quite uniform with diameter in the range of 200 nm but they are not perfect cylinders. It has been reported (Schonenberger et al., 1997) that pore diameters of commercially available templates vary over a large range. The aspect ratio, that is, the ratio of length to diameter, is on the order of 300.

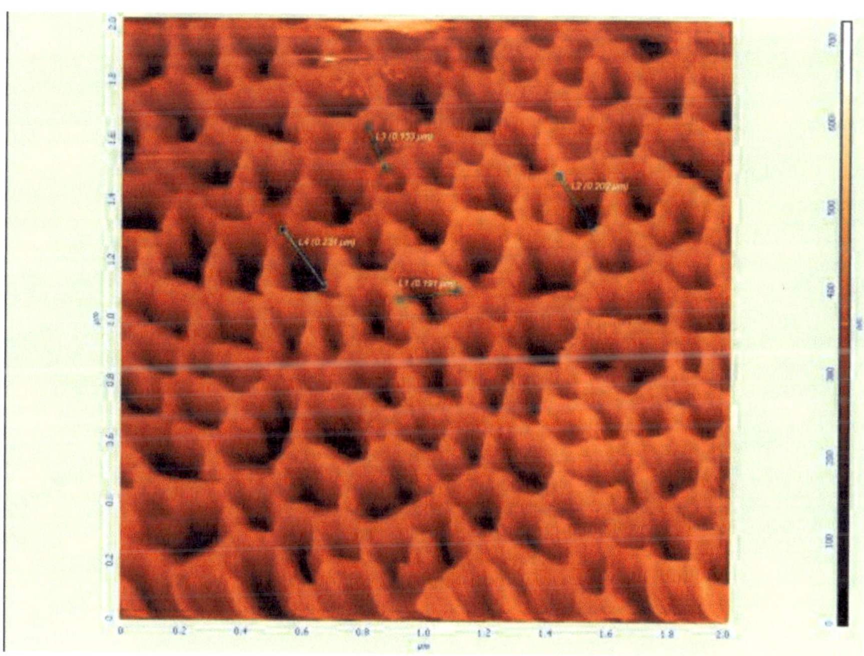

Fig. 2. (a) AFM image of hexagonal pores of anodic alumina template

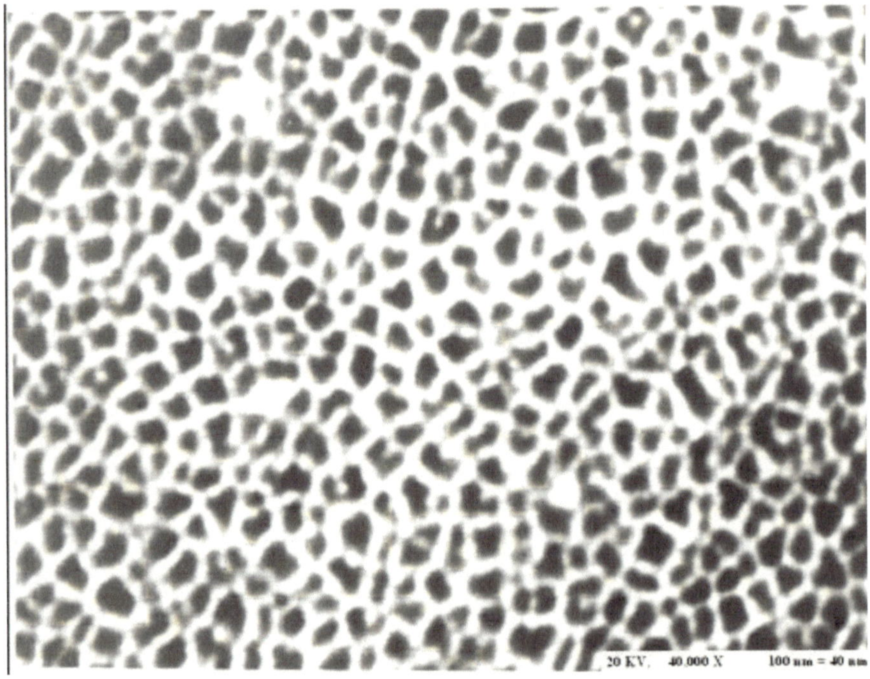

Fig. 2. (b) SEM image of anodic alumina template pores

Fig. 3. SEM image of copper nanowires (cross-sectional view, 200 nm dia.)

Fig. 4. (a) SEM image of copper nanowires fabricated under constant current

Fig. 4. (b) SEM image of copper nanowires showing capping effect

Experiment was repeated using polycarbonate membrane with pore diameter of 100 nm as a template and keeping the other conditions identical. The polymer template was dissolved in dichloromethane to liberate copper nanowires from the host matrix. The rest of the procedure is same. Instead of nanowires, we observed under FESEM the exotic patterns in the form of microflowers (Fig. 5) having their petals in nanometer dimension and copper buds (Fig. 6) leading to mushroom effect. Similar results with exotic patterns were reported in our earlier experiment (Virk et al. 2010).

There is as yet no specific theory to explain exotic patterns developed during electrodeposition of copper in anodic alumina or polymer templates. A speculative explanation (Gao et al., 2002) is provided on the basis of overdeposition. During the growth of copper nanowires in the template pores, the current remains nearly stable until the wires arrive at the template surface. If the electro-deposition process is not stopped at this stage, the current keeps on rising very gradually leading to overdeposition of copper. Flower like morphologies of metal overdeposits have been attributed to the changes in hydrodynamic conditions due to excessive hydrogen evolution during electrodeposition process (Kumar et al., 2008).

Fig. 5. FESEM micrographs showing flower patterns grown in polymer templates

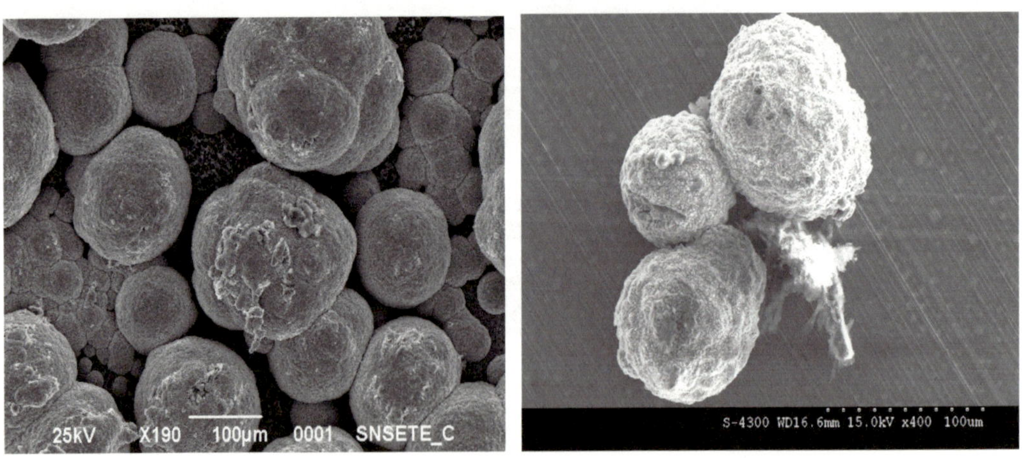

Fig. 6. FESEM micrographs showing copper buds grown in polymer templates

Fig. 7. SEM micrograph of pyramid shaped polycrystalline copper crystals

We repeated the experiment for 20 nm pore diameter polycarbonate template. The template was not coated with a conducting layer during electrodeposition. It resulted in failure to grow nanowires but the failure of experiment proved to be a blessing in disguise. Instead of copper nanowires, we observed growth of double pyramid shaped copper crystals (Fig. 7). We could not find evidence for this phenomenon in literature. It is anticipated that copper ions from the electrolyte do not enter template pores due to poor conductivity but get deposited on the cathode surface in the form of polycrystalline crystals.

4.2 X-ray and EDAX analysis

The characterization techniques that are commonly used to study the crystal structure and chemical composition of nanowires include X-ray diffraction and X-ray energy dispersion analysis (EDAX). Both these techniques have been employed in our analysis. The crystal structure of the double pyramid shaped copper crystals has been determined using X-ray diffraction analysis. XRD spectrum (Fig. 8) shows two prominent peaks corresponding to $2\theta = 43.4610$ and 50.5803, with d spacing = 2.082 and 1.804, respectively. These peaks reveal the polycrystalline nature of copper crystals, indicating that preferred growth direction of crystals is the (200) plane. Template based synthesis of single crystal copper nanowires has been reported in literature (Toimil Molares, 2001; Gao et al., 2002; Mingliang et al., 2003) with preferred growth direction along (111) plane, but to the best of our knowledge, there is hardly any report for copper nanowire arrays or copper crystals with a (200) preferred orientation.

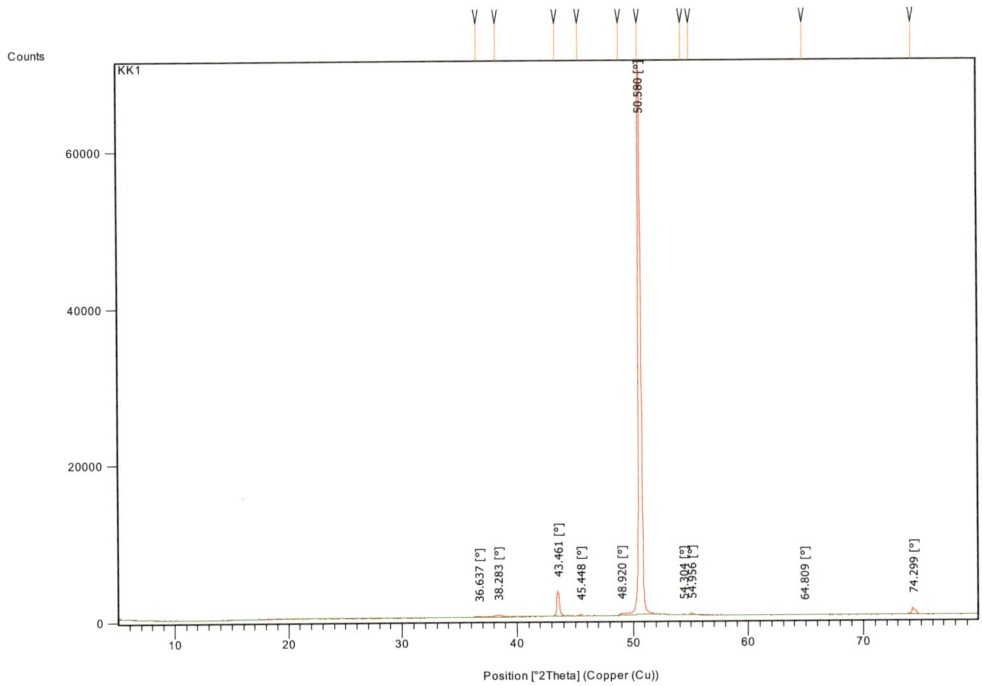

Fig. 8. XRD spectrum of pyramid shaped polycrystalline copper crystals

The crystallographic structure of copper nanowire arrays was investigated by X-ray diffraction analysis (XRD). For sake of comparison, XRD spectrum of Cu foil used as a substrate was also recorded (Fig. 9). XRD diffractograms were obtained in the 2θ range from 10^0 to 80^0 with a step of 0.02^0, using the Cu Kα radiation source of λ = 1.5406 Å. XRD spectrum (Table 1) shows three prominent peaks corresponding to 2θ = 43.5966, 50.8127 and 74.4331, with d spacing = 2.074, 1.80 and 1.27, and corresponding Miller indices, (111), (200) and (220), respectively. All peaks can be attributed to the cubic form of metallic copper (Ingunta et al., 2008). XRD spectrum of copper nanowires (Fig. 10) shows some interesting results. There are in all 8 peaks in the spectrum; with 2 additional peaks at 2θ = 37.0062 and 54.9761, which are of negligible intensity and may be ignored. Three main peaks are also there as in Fig. 9 but two of them split into double and triple peaks (Table 2), which may be attributed to X-ray scattering at the substrate. These peaks reveal the polycrystalline nature of copper nanowires, the most prominent peak at 2θ = 50.9870, indicating that the preferred growth direction of nanowires is the (200) plane. Due to polycrystalline nature of copper nanowires, the most prominent peak at 2θ = 43.5966 (Fig. 9) shifts to 2θ = 50.9870 (Fig. 10). Template based synthesis of single crystal copper nanowires have also been reported in literature (Gao et al., 2002; Mingliang et al., 2003) with preferred growth direction along (111) plane.

The average size *D* of the crystalline grains in the Cu nanowires is calculated using the Debye Scherrer's formula (Cullity, 1956): $D = 0.9\ \lambda\ /\ \beta \cos\theta$, where λ=1.5406 Å is the wavelength of the X-ray radiation used, β is the full width at half maximum (FWHM) of the diffraction peak (0.1224), K, shape factor is assumed to be 0.9 and θ is the Bragg diffraction angle of the most prominent XRD peak. Substituting appropriate values in the formula, the crystallite size value of Cu nanowires comes out to be 1.22 nm. However, the value of crystallite size calculated for Cu foil is exact double, of the order of 2.44 nm.

Pos. [°2Th.]	WHM [°2Th.]	d-spacing [Å]	Rel. Int. [%]	Area [cts*°2Th.]
43.5966	0.0612	2.07438	100.00	847.65
50.8127	0.0816	1.79542	48.53	548.43
74.4331	0.1428	1.27358	11.94	236.07

Table 1. XRD spectrum peaks data of copper film

Pos. [°2Th.]	FWHM [°2Th.]	d-spacing [Å]	Rel. Int. [%]	Area [cts*°2Th.]
37.0062	0.3346	2.42724	2.29	76.37
43.4706	0.0669	2.08010	33.59	223.69
43.8561	0.2509	2.06270	93.99	2347.28
50.5881	0.0816	1.80286	74.62	819.15
50.9870	0.1224	1.78969	100.00	1646.70
51.1172	0.1224	1.78544	87.56	1441.89
54.9761	0.4080	1.66889	0.75	41.24
74.4238	0.4080	1.27372	5.90	324.08

Table 2. XRD spectrum peaks data of copper nanowires

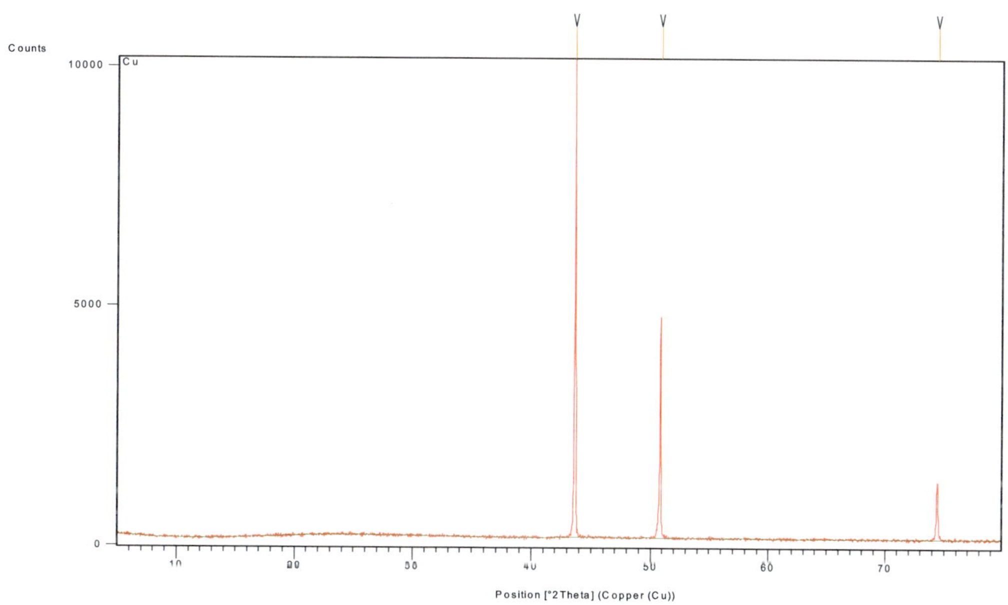

Fig. 9. XRD spectrum of Copper film serving as a substrate

Energy dispersive X-ray analysis (EDAX) of Cu nanowires was carried out at FESEM facility of CSIO, Chandigarh to determine chemical composition of nanowires. The spectrum (Fig. 11) reveals 3 peaks of copper with 100% pure copper content and no traces of any impurity in Cu nanowires. It also establishes that multiple XRD peaks are not due to any impurity but due to polycrystalline nature of Cu nanowires.

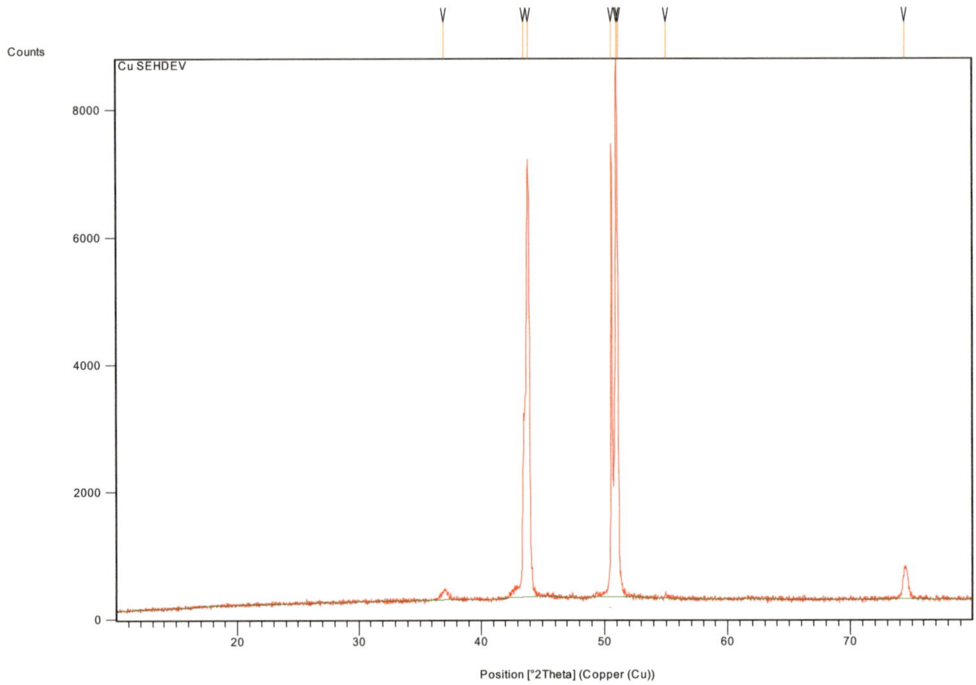

Fig. 10. XRD spectrum of Copper nanowires of 200 nm diameter

4.3 I-V Characteristics of copper nanowires

I-V properties of aligned copper nanowires have been studied using a current- sensing AFM (Cao et al., 2006). Electronic transport through nanocontacts has been an active research area. The ultimate aim for nanowires is to find applications in the nanoelectronic devices. How can a copper nanowire produce a nonlinear I–V curve? The simplest possibility for observing such a phenomenon is generation of a tunnelling barrier at the wire–lead junction whose effect gradually collapses as a function of increasing bias voltage (Mehrez & Guo, 2004). The nonlinear curves of Cu nanowire arrays may be caused by the existence of impurities (such as oxide) near the wire–lead contact region. Nonlinear phenomena of silver wire and gold wire have also been observed in air (Mehrez & Guo, 2004; Wildoer et al., 1998). It has been demonstrated that the nonlinear I–V characteristic is the basis of functional electronic devices (Itakura et al., 1999).

I-V characteristics of copper nanowires were recorded in-situ, as grown in pores of anodic alumina template, using Dual Source Meter (Keithley Model 4200 SCS) with platinum probes for contacts. The combination of copper nanowires on alumina, an insulator, results in the formation of a strange device. I-V plot (Fig. 12) shows some interesting features of a resonating tunneling diode in the forward bias mode but nothing special in the reverse bias mode. The offset in I-V plot around zero voltage may be due to slight non-ohmic characteristic of the contact, or due to quantum confinement behaviour of electrons traversing through copper nanowires.

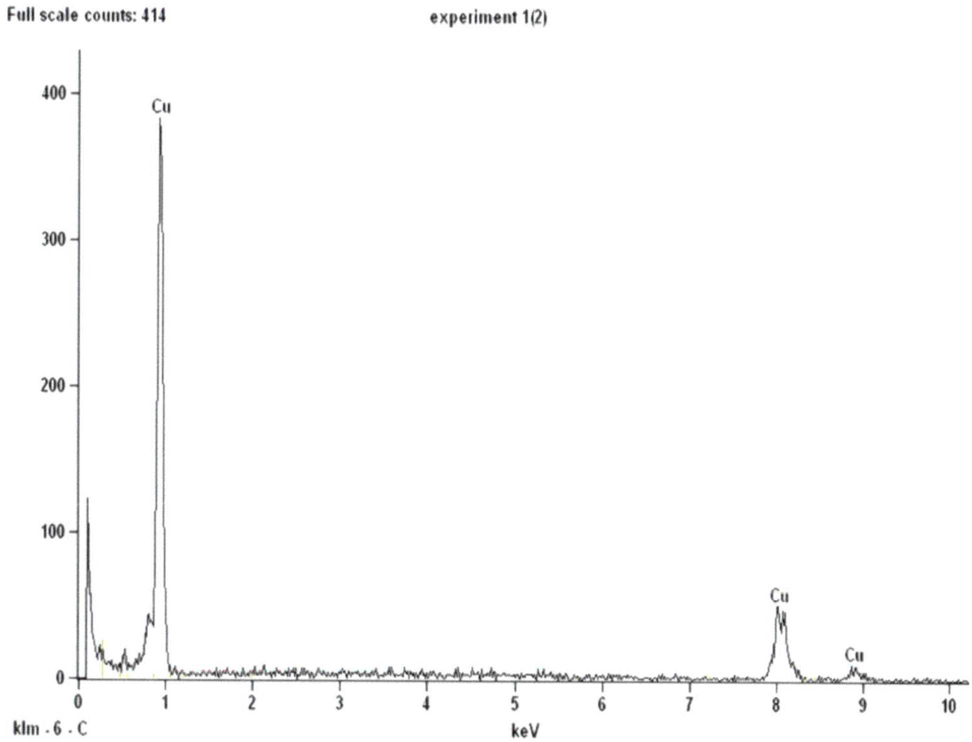

Live Time: 100.0 sec.

Quantitative Results for: experiment 1(2)

Element Line	Weight %	Weight % Error	Atom %	Atom % Error
Cu K	100.00	+/- 7.49	100.00	+/- 7.49
Total	100.00		100.00	

Fig. 11. EDAX spectrum and elemental composition of Copper nanowires

5. Conclusions

Our investigations confirm that electrodeposition of copper nanowires in anodic alumina is the simplest route to nanotechnology. The copper nanowires reveal effect of high current density resulting in overdeposition in the form of capped growth, and not as perfect cylinders. The aspect ratio is very high, of the order of 300. XRD analysis shows polycrystalline nature of nanowires and pyramid shaped copper crystals with preferred growth direction in the (200) plane. The crystallite size of nanocrystals in copper nanowires is determined to be 1.22 nm. Overdeposition results in growth of copper buds and beautiful flower patterns. I-V characteristics do not conform to normal p-n junction behaviour and need further investigation. The nonlinear I–V characteristic of the as-synthesized copper nanowire arrays suggests the presence of a potential barrier. Due to high aspect ratio, copper nanowires may be used as field emitters.

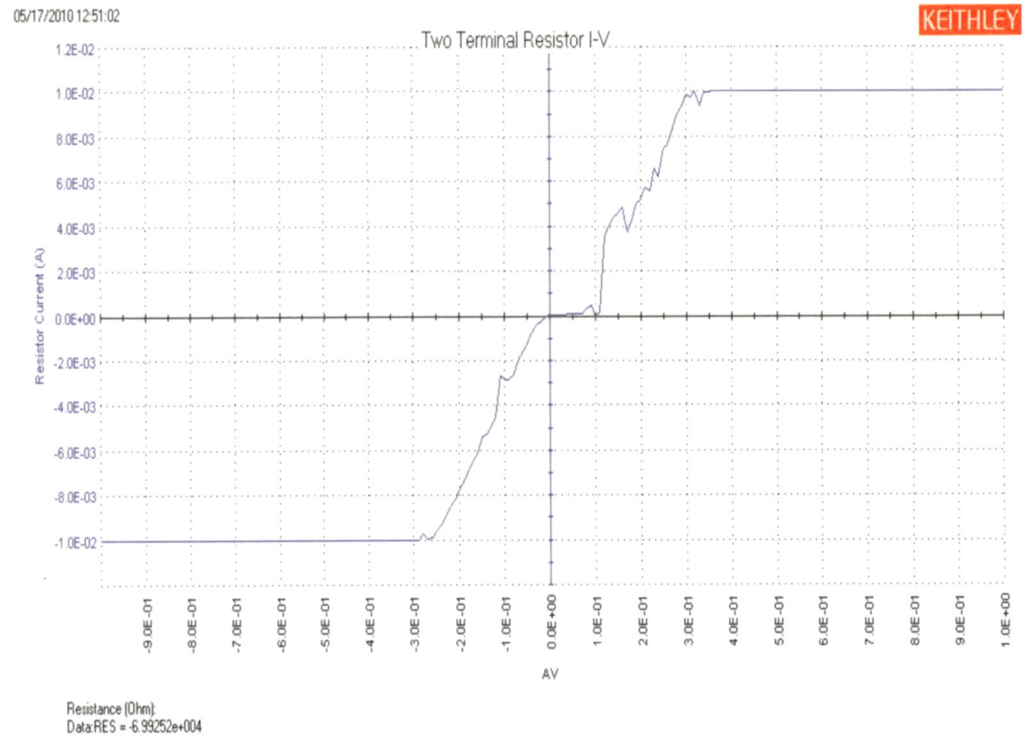

Fig.12. I-V characteristics of copper nanowires grown in-situ in anodic alumina template

Copper oxide nanowire arrays have already found applications in gas sensing, field emission and photovoltaic devices. A recent study (Rathmell et al., 2010) has established that copper nanowires could revolutionize the development and production of low-cost flexible displays, light emitting diodes and thin film solar cells. Copper is 1000 times more abundant than indium or silver, and is 100 times less expensive. As a consequence, films of copper nanowires represent a low-cost alternative to silver nanowires or ITO for use as a transparent electrode.

6. Acknowledgements

The authors are thankful to the Principal, DAV Institute of Engineering & Technology, Jalandhar and DAV College Managing Committee, New Delhi for providing research grant to set up Research Centre and Nanotechnology Laboratory. Author wishes to record his appreciation for research assistants and technical staff of Nanotechnology Laboratory in collection of data. He is thankful to Dr Lalit M. Bharadwaj and his team at CSIO, Chandigarh for providing FESEM research facility.

7. References

Cao, D. & Liu, D. (2008). Template-based synthesis of nanorod, nanowire, and nanotube arrays. *Adv. in Colloid and Interface Sci.*, Vol. 136, pp. 45- 64.

Cao, H., Wang, L., Qiu, Y. & Zhang, L. (2006). Synthesis and I-V properties of aligned copper nanowires. *Nanotechnology*, Vol.17, pp. 1736-1739.

Cullity, D.B. (1956). *Elements of X-ray Diffraction,* Addison-Wesley Inc., Massachusetts, USA.

Diggle, J.W., Downie, T.C. & Goulding, C.W. (1969). Anodic oxide films on aluminium. *Chem. Rev.,* Vol. 69, pp. 365-405.

Dresselhaus, M.S. et al. (2002). Nanowires and nanotubes. *Mater. Sci. & Engg. C,* Vol.1008, pp. 1- 12.

Fang, C., Foca, E., Xu, S., Carstensen, S.J. & Foll, H. (2007). Deep silicon macropores filled with copper by electrodeposition. *J. Electrochem. Soc.,* Vol.154, pp. D45-D49.

Gao, T., Meng, G., Wang, Y., Sun, S. & Zhang, L. (2002). Electrochemical synthesis of copper nanowires. *J. Phys.: Condens. Matter.,* Vol.14, pp. 355-363.

Ingunta, R., Piazza, S. & Sunseri, C. (2008). Novel procedure for the template synthesis of metal nanostructures. *Electrochem. Commun.,* Vol.10, pp. 506-509.

Itakura, K., Yuki, Y., Kurokawa, S., Yasuda, H. & Sakai, A. (1999). Bias dependence of the conductance of Au nanocontacts. *Phys. Rev. B,* Vol.60, pp. 11163- 11170.

Kumar, S., Kumar, V., Sharma, M.L. & Chakarvarti, S.K. (2008). Electrochemical synthesis of metallic micro-rose having petals in nanometer dimensions. *Superlattices & Microstructures,* Vol. 43, pp. 324-329.

Lai, M. & Riley, D.J. (2008). Templated electrosynthesis of nanomaterials and porous structures. *J. Colloid and Interface Sci.,* Vol.323, pp.203-212.

Mehrez, H. & Guo, H. (2004). Nanowires and Nanotubes--Materials, Properties and Devices, In: *Metal and Semiconductor Nanowires* (Vol. I). Z. L. Wang (Ed.), pp. 95–124, Tsinghua University Press, Beijing, China.

Menon, L. (2003). Synthesis of Nanowires using Porous Alumina, In: *Quantum Dots and Nanowires.* Hari Singh Nalwa and S. Bandhopadhaya (Eds.), pp. 142-187, American Scientific Publishers, Cal. USA.

Mingliang, T., Wang, J., Kurtz, J., Mallouk, T.E. & Chan, M.H.W. (2003). Electrochemical Growth of Single-Crystal Metal Nanowires via a Two-Dimensional Nucleation and Growth Mechanism. *Nano Letters,* Vol.3, pp. 919-923.

Motoyama, M., Fakunaka, Y., Sakka, T., Ogata, Y. and Kikuchi, S. (2005). Electrochemical processing of Cu and Ni nanowire arrays. *J. Electroanal. Chem.,* Vol.584, p. 84.

Nalwa, H.S. & Bandhopadhaya, S. (Eds). (2003). *Quantum Dots and Nanowires,* American Scientific Publishers, Cal., USA. pp. 142-187.

Rathmell, A.R., Bergin, S.M., Hua, Y.L., Li, Z-Y & Wiley, B.J. (2010). The growth mechanism of copper nanowires and their properties in flexible, transparent conducting films. *Adv. Materials,* Vol. 22, pp. 3558-3563.

Ratner, M., & Ratner, D. (2003). *Nanotechnology: A gentle introduction to the next big idea.* Pearson Education Publication, London.

Sarkar, J., Khan G.G. & Basumallick, A. (2007). Nanowires: properties, applications and synthesis via porous anodic aluminium oxide template. *Bull. Mater. Sci.* Vol.30, No.4, pp. 271-290.

Schonenberger, C., Vander Zande, B.M.I., Fokkink, L.G.J., Henry, M., Schmid, C., Kniger, M., Bachtold, A., Huher, R., Birk, H. & Stairfer, V. (1997). Template synthesis of nanowires in porous polycarbonate membranes, Electrochemistry and morphology. *J. Phys. Chem. B,* Vol. 101, pp. 5497-5505.

Sun Shin, H., Song, J.Y. & Yu, J. (2009). Template-assisted electrochemical synthesis of cuprous oxide nanowires. *Materials Letters,* Vol.63, pp. 397-399.

Toimil Molares, M.E., Buschmann, V., Dobrev, D., Neumann, R., Scholz, R., Schuchert, I.U. & Vetter, J. (2001). Single-crystalline copper nanowires produced by electrochemical deposition in polymeric ion track membranes. *Adv. Mater.*, Vol.13, pp. 62-65.

Virk, H.S., Baloria, V. & Kishore,K. (2010). Fabrication of Copper nanowires by electro-deposition using anodic alumina and polymer templates. *J. Nano Res.*, Vol.10, pp. 63-67.

Virk, H.S. & Amrita Kaur, S. (1998). Ion Track Filters: Properties, Development and Applications. *Curr. Sci.*, Vol.75, No.8, pp. 765-770.

Wildoer, J.W.G., Venema, L.C., Rinzler, A.G., Smalley, R.E. & Dekker, C. (1998). Electronic structure of atomically resolved Carbon nanotubes. *Nature*, 391, pp. 59-62.

21

Laser Interference Lithography for Fabricating Nanowires and Nanoribbons

Joong-Mok Park[1], Wai Leung[1], Kristen Constant[1], Sumit Chaudhary[1], Tae-Geun Kim[2] and Kai-Ming Ho[1]
[1]*Ames laboratory and Iowa State University, Ames, Iowa,*
[2]*Korea Univerisy, Seoul,*
[1]*USA*
[2]*Korea*

1. Introduction

Nanowires have been extensively studied for the last decade due to their superior electrical, optical, and mechanical properties when compared they are bulk. Their interesting properties are due to their simple nature (Wu B. et al., 2005). Various fabrication methods of nanowires have been in development including nanoimprint (Guo L., 2007; Chen L. et al., 2007), side electroplating (Shankar & Raychaudhuri, 2005; Xiang et al., 2008), self assembly (Pauzauskie & Yang, 2006), and stencil lithography (Vazquez-Mena et al., 2008) etc.

We have developed a simple, cost efficient and mass producible fabrication method for metal nanowires. First, a polymer mask is fabricated with a photosensitive material, a photoresist, using laser interference holography. Then metal is coated by physical vapor deposition (Xue et al., 2008; Kang and Guo, 2007; Kang et al., 2008), either in single or multi-stage depositions. After chemically removing the photoresist, parts of the metal nanowires remained on the substrate as ordered structures and parts are detached from the substrate which can be recovered for further analysis. This technique yields high quality nanowires (which are few cm long) either in ordered or free standing form. In this chapter, nanowires are defined as being straight as-deposited whereas nanoribbons are freestanding and are curved compared to their original configuration. The dimensions of the nanowires can be controlled by tailoring dimensions of the polymer patterns and deposition conditions such as angle and thickness.

Their structural, electrical, and optical properties are characterized by scanning electron microscopy (SEM), transmission electron microscopy (TEM), atomic force microscopy (AFM), and four point probe resistivity measurements. This fabrication method can be used for most metals (even semiconductor or insulators) to form nanowires and nanoribbons.

2. Experiment details of fabricating nanowires

The fabrication of nanowires is a two-step process. The first step is making polymer templates using two- or multiple-beam interference. Polymer templates can have a 1-dimensional grating or 2-dimensional square structures. The templates are made on a transparent substrate (glass, sapphire, ITO-coated glass) or silicon wafer depending on the application. ITO-coated glass is used for electroplating metal nanowires. Glass is used when high optical transmission is required. The second step is depositing metallic thin films on top of the polymer patterns and on the substrate between the polymer channels. Because the polymer structures are well defined, large areas with rectangular cross section, long metallic nanowires are made easily. The resulting wires are cm long, few hundred nanometer wide and less than 100 nanometer thick.

2.1 Interference holography

Interference between two coherent beams is well known in classical optics (Hecht, 1987). The typical configuration of interference holography is splitting a coherent beam into two then overlapping the beams to make interference pattern as Fig. (1)a (Guo H. et al., 2007). The drawback of this setup is that the two beam intensities are close to each other and the incidence angles of the two beams are adjusted separately. When incidence angle needs to be changed, both mirrors have to be adjusted.

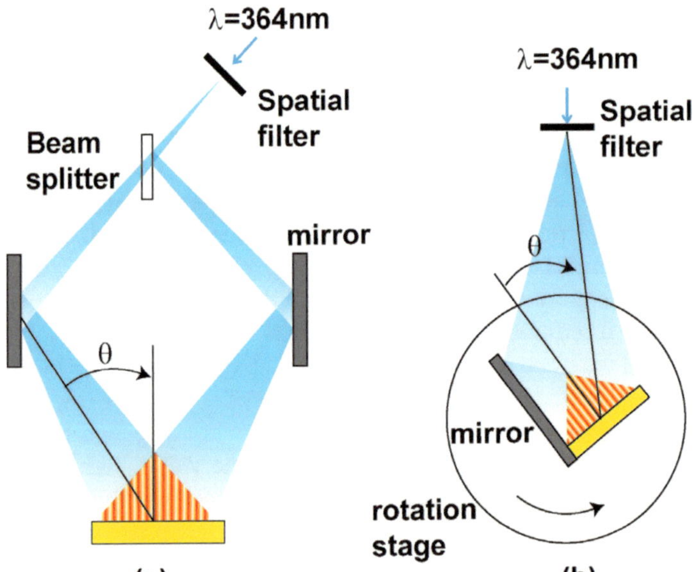

Fig. 1. Schematic diagrams of (a) two-beam interference holography with a beam splitter and (b) with Lloyd's mirror configuration.

A Lloyd's mirror setup can avoid this by using single mirror mounted perpendicular to the sample as Fig. (1)b (Onoa et al., 2005). Half of the beam is reflected from the mirror and the other half is the original beam. Because the reflecting mirror is attached to the sample, the incidence angles can be change together by rotating the sample-mirror stage. Also, mechanical rigidity reduces vibrations between interfering beams, resulting well-defined patterns even for long exposure times.

The pitch (or period, center to center distance), Λ, of the interference intensity is a function of wavelength (λ) and incidence angle (θ) as Eq. (1).

$$\Lambda = \frac{\lambda}{2\sin\theta} \tag{1}$$

Ideally, the pitch can be the half of the wavelength of the laser when the incidence angle is 90 degrees. The smallest pitch is typically larger than that because the intensity decreases at high incidence angles. Also mechanical vibration between two beams, air turbulence, laser stability, dust particles and imperfect optical components limits the pitch. There is no upper limit in pitch when incidence angle approaches to zero. The sample size is determined by the interfering area. In case of a small incidence angle, the overall interference area is reduced in horizontal direction (perpendicular to the grating direction). This is because the beam size coming from the mirror is proportional to sin(θ). Another factor determining sample size is the coherence length of the laser. A typical gas laser (Ar-ion or He-Cd) has a fairly large coherence length (> 30 cm). So the typical exposed size (less than 10 cm) is not limited by that.

A Lloyd's mirror configuration is as followings. An Ar-ion laser with UV prism (Coherent Inc.) is used to generate coherent light of 364 nm wavelength. A spatial filter, 10x ultraviolet objective lens and 10 μm diameter pinhole, expands the beam and makes it homogeneous. The distance between the spatial filter and sample is 2 m, which is far enough to ignore beam divergence and results in a large exposed area. The pitch can be adjusted by the incidence angle by rotating the stage.

The polymer templates with photo sensitive material (photoresist) are made by following process. Substrates (typically glass, Si wafer, or ITO coated glass) are prebaked 150 °C for 10 min to remove any moisture remaining after standard cleaning process. Then an adhesion promoter MCC primer (Microchem Inc.) is spin coating and baked at 120 °C for 2 minutes on hot plate. An adhesion promoter is used to enhance a bonding between photoresist and substrate. AZ HiR 1075 photoresist (AZ-EM electronics) is spin coated on top of prebaked promoter at 4000 rpm for 60 sec. The thickness of photoresist is controlled by adding thinner to the solution or by increasing the spin speed. The spin-coated photoresist is prebaked at 60 °C for 30 min in oven to remove solvent in the photoresist. Then the sample is mounted on a rotating sample stage and exposed. A typical dose is about 200 mJ (total exposure time is 30 sec). The exposed sample is baked 110 °C for 60 sec on hot plate and then developed in MIF 300 developer (AZ-EM electronics) for 60 sec. The developed sample is rinsed with de-ionized water and dried with blowing nitrogen. The thickness of photoresist including the adhesion promoter is 750 nm measured from a cross sectional image taken with JEOL 840 scanning electron microscope.

The intensity of the interference of two coherent beams at the sample surface is the time average of the magnitude of two interfering electric fields intensity squared as Eq. (2).

$$I = \left\langle (E_1 + E_2)^2 \right\rangle = I_0\left[1 + \cos(2kx\sin\theta)\right] = I_0\left[1 + \cos\left(\frac{2\pi x}{\Lambda}\right)\right] \tag{2}$$

The normalized intensity is illustrated in Fig. (2)a with the response line of the photoresist. The area exposed at intensity above the response line is completely removed and the area below this line remains when using a positive type photoresist. The patterns after developing are long straight lines with rectangular cross section as in Fig. (2)b. The width of individual photoresist is about half of the pitch for optical dose. It can be controlled by the exposure time and laser intensity (i.e. dose). The dose can increase or decrease the amplitude of the intensity curve in Fig (2)a whereas the pitch can be controlled by incidence angle and wavelength of the laser. Controlling width of the photoresist by dose is not practical because the nonlinear response of the photoresist at high or low dose.

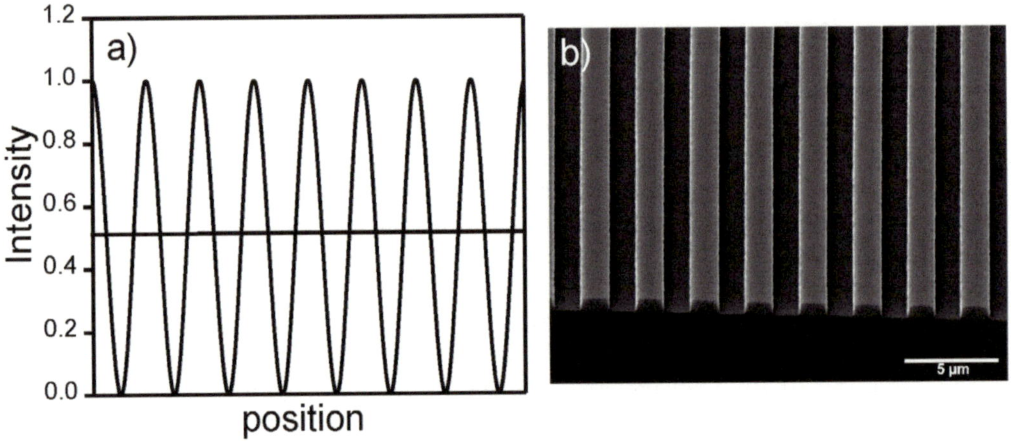

Fig. 2. (a) Normalized intensity of two beam interference and (b) SEM image of the photoresist grating made with two beam interference holography.

2.2 Shadow deposition of metals

After the polymer template is made with a photoresist, metal is deposited on the template with e-beam evaporation to form long nanowires. The distance between the sample and evaporation source is about 1m, so the deposition is approximately collimated. A quartz crystal monitor is used to monitor the deposition rate and thickness during the deposition. The deposition rate is maintained at about 1 Å/sec. The pressure is below 10^{-6} Torr during the deposition. When deposited at normal angle, a thin metal layer is coated on top of the photoresist and also on the substrate in the channels between the photoresist bars as shown in Fig. (3)a and Fig. (4)a. If desired, deposition on the substrate can be avoided by oblique angle deposition as Fig. (3)b (Bai et al., 2007; Chen L., 2007; Chen Y. and Glidman, 2008).

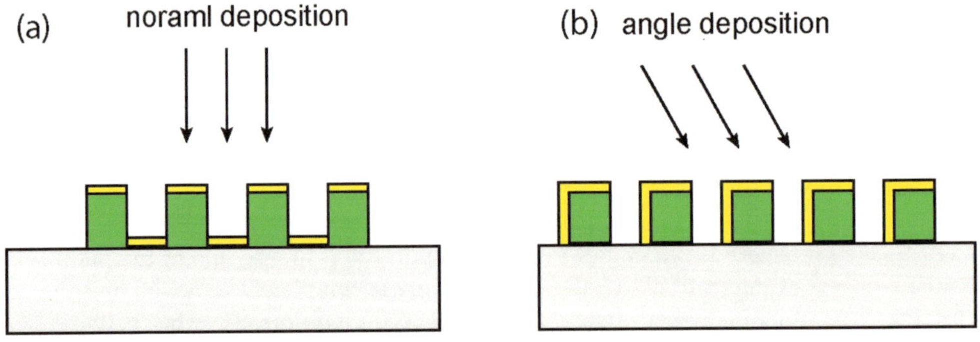

Fig. 3. Schematic diagram of two e-beam deposition methods (a) normal angle and (b) oblique angle.

In this case, only the sidewalls and top of the photoresist are deposited by an oblique deposition angle. Partial deposition on the substrate is also possible. The deposition width on the substrate depends on the thickness and width of the photoresist and angle of deposition. The pitch, width, and thickness of the deposited materials can be controlled by the dimensions of the polymer template and metal deposition rate and time.

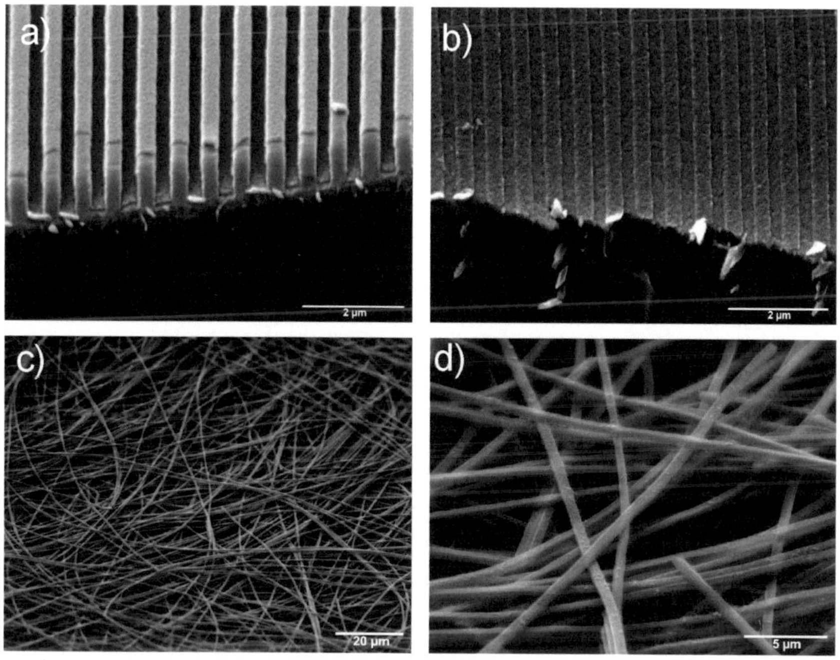

Fig. 4. SEM images of nanowires: (a) Al nanowires on the ITO coated glass substrate at normal angle deposition, (b) Al nanowires after photoresist removal. (c, d) Collected Ti nanoribbons from the IPA solution.

After deposition, the photoresist is removed by immersion in a photoresist stripper, Remover PG (Microchem Inc.), at room temperature for 30 min and the samples are dried with N_2 gas after rinsing with IPA (isopropyl alcohol). The metal nanowires deposited on the substrate between the channels of the photoresist remain even after photoresist removal as in Fig. (4)b where as thin metallic films deposited on top of the pattern are detached during rinsing process. The detached materials can be recovered as nanoribbons which have the same thickness and roughly same width of patterns on the substrate.
Various metals Ti, Au, Ag, Cu, Al, Pb are used as deposited materials. Pb has granular structures rather than uniform film. Other metals have uniform films on the photoresist or substrate.

2.2 Phase mask and electroplating

To fabricate large open area metal mesh structures with a single electro-deposition, a pattern is made from the interference of three diffracted beams after passing through the phase mask as in Fig. (5). The interference of three coherent beams is the superposition of three plane waves as described in Eq. (3) (De and Sevigny, 1967; Farhoud et al., 1999):

$$I = I_0\left[3 + 4\cos(kx\sin\theta + \Delta\phi)\cos\left(kz(1-\cos\theta)\right) + 2\cos(2kx\sin\theta)\right] \quad (3)$$

where I_0 is the incident laser beam intensity, $\Delta\phi=\phi_1-\phi_0$ is the phase difference between the 1st and the 0th order beam, θ is the first-order diffraction angle from the normal.

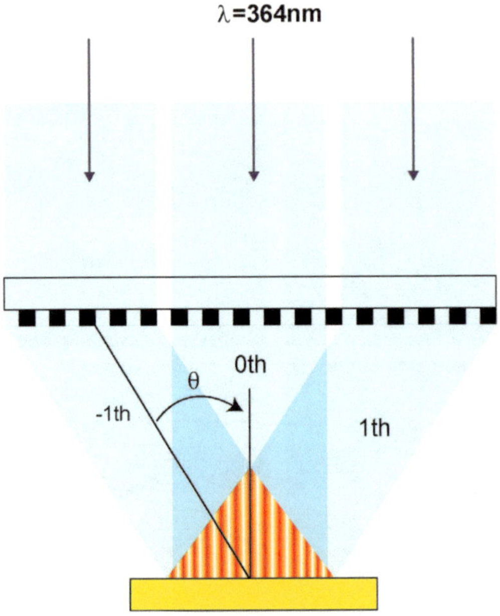

Fig. 5. Three beam interference with a diffractive phase mask.

To make metallic structures having circular or square holes, the photoresist pattern was made on indium tin oxide (ITO) coated glass to facilitate the electro-deposition of metals. Cu is electroplated by applying 50 Hz AC of 10 mA current. Forward biases of +5 V for 10 msec and reverse -3 V for 5 msec are applied in one period. AC pulse reverse plating helps initial nucleation of Cu and enhances uniform film growth. The electro-deposition is stopped before overfilling the channels. Then the photoresist is removed by immersion in stripper, Remover PG, for 30 minutes.

For 2 dimensional structures, the photoresist is exposed twice with a 90 degree rotation between exposures. The double exposed photoresist of two-beam interference is a 2 dimensional array of cylindrical pillars (Fernandez et al., 1997). Figure (6)a shows the 2D intensity map for double exposure with 90 degree rotations with phase mask. The intensity across line B and C are shown in Fig (6)b. Electroplated Cu is the inverse structure of the photoresist when electroplating is stopped before overfilling the photoresist. The final Cu has circular holes after polymer removal as shown in Fig. (6)c without a phase mask, whereas a metallic mesh structure having large open square holes is made with a diffractive phase mask as in Fig. (6)d. The phase mask is made with photoresist on glass using 2-beam interference holography. The pitch of the phase mask grating was chosen as 700 nm to have only zero- and first-order diffractions. The Cu structures made with the phase mask are narrower than those made with conventional two beam holography and are close to square in cross section as shown in Fig. (6)d.

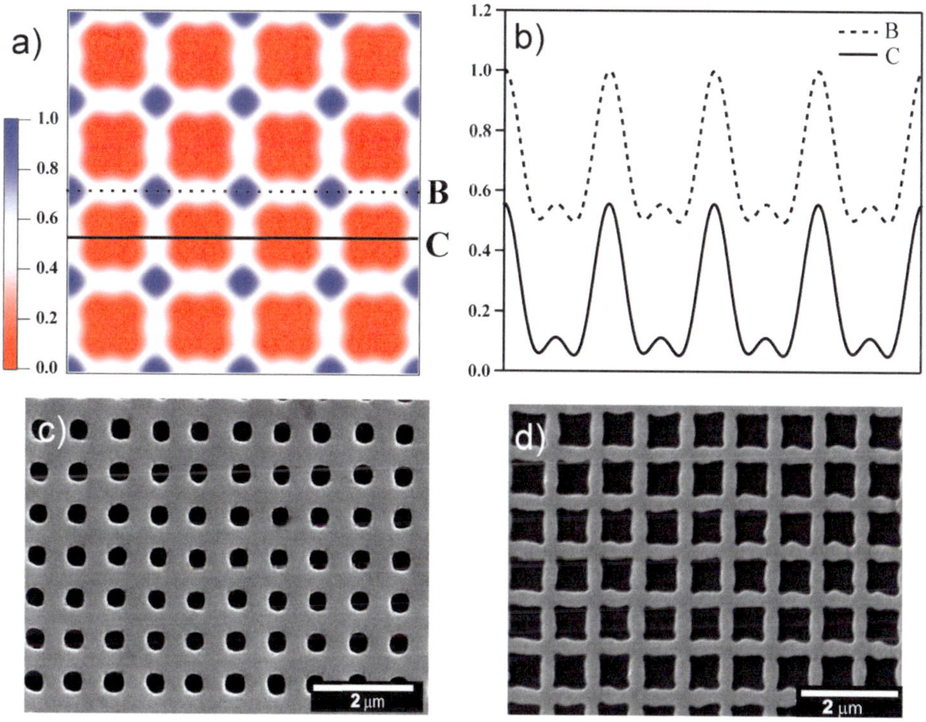

Fig. 6. (a) Intensity map of three beam interference on the substrate with double exposed with 90 degree rotation. (b) Intensity along the line B, C. (c) SEM image of Cu with two beam interference (d) with three beam interference after photoresist removal.

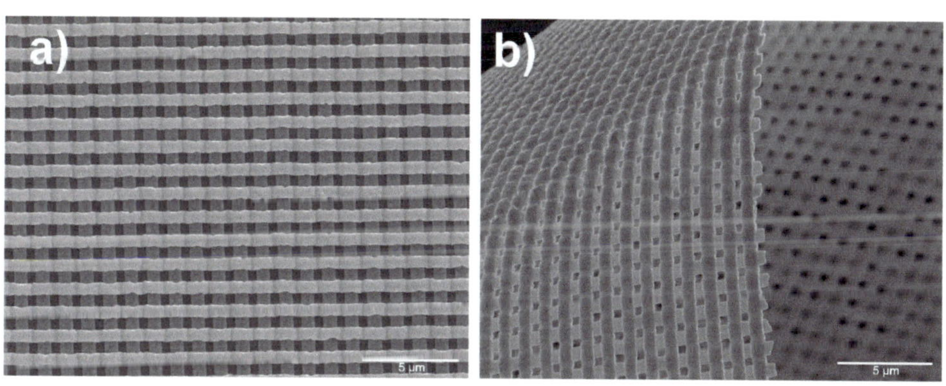

Fig. 7. (a) Cu structure electroplated between photoresist patterning. (b) Free standing Cu mesh peeled from the substrate.

An alternative method of making square metal structures is by two separate electro-depositions between exposures. The first Cu grating fills the channels of a photoresist grating made by two-beam interference. Then the second layer of photoresist is patterned perpendicular to the first Cu grating and the second Cu grating fills the channels of the photoresist perpendicular to the first layer. After removing photoresist, the remaining Cu mesh has square holes as in Fig. (7)a and can be easily detached from ITO glass making free-standing metallic mesh structures as in Fig. (7)b. Because each layer has 50% open area, meshes made by two electro-depositions have about 25% open area.

3. Results

The micro structural properties, especially grain size and surface morphology, of nanowires were characterized with TEM, AFM, and SEM. Electrical conductivity was measured with a 4 point probe. Optical transmission of nanowire grating was also measured.

3.1 Physical properties of nanowires

To examine the grain structure of nanoribbons, TEM images were taken. Nanoribbons suspended in isopropyl alcohol were ultrasonicated for 1 min to break them into small pieces. The alcohol and broken nanoribbons were dropped onto carbon grids and vacuum assist dried. The average grain sizes of Ti and Al nanoribbons were about 10-20 nm by TEM as in Fig. (8)a,b. Au appears to have a larger grain size, about 50nm as Fig. (8)c. The grain

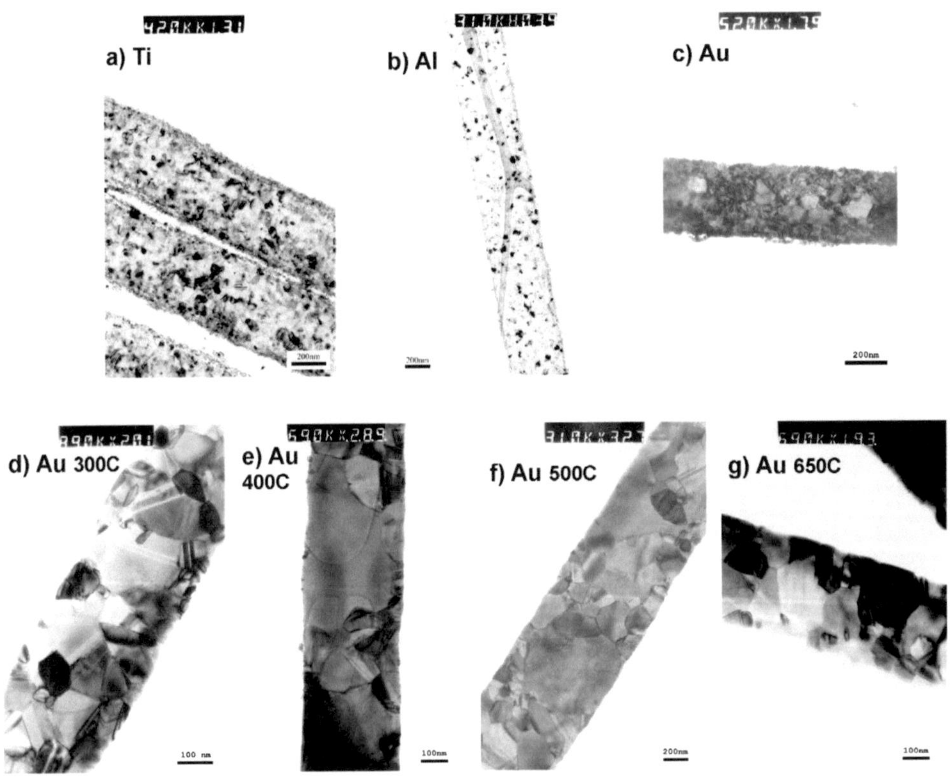

Fig. 8. Bright field TEM image of (a) Ti ribbon of 500 nm x 100nm (b) Al ribbon of 500 nm x 100 nm (c) Au 400nm x 80 nm as grown. Au ribbons annealed at different temperatures (d) 300 °C for 100 hours (e) 400 °C for 48 hours (f) 500 °C for 48 hours (d) and in situ annealed at 650 °C for 2 hours.

size depends on various factors in the deposition process (materials, impurities, temperature, substrate etc). Thermal annealing can increase the grain size and will increase the electrical conductivity of the nanoribbons but there are some limitations to grain growth related to the small dimensions of the nanoribbons. In case of Ti and Al, there was not substantial grain growth even after annealing at 600 °C for 2 hours (the melting temperatures is 660 °C for Al and 1668 °C for Ti). Grain growth in Au nanoribbons was more substantial when annealed at temperatures well below the melting temperature (1063 °C) as Fig. (8)d-g. Minimal grain growth of Al and Ti are due to the native oxides on the surface.

Average grain size of thin film after thermal annealing is about the same as thickness of nanowires (Thompson, 2000). Once large grains formed and further grain growing did not appear even after 48 hours annealing. The average grain size of annealed Au nanoribbons is approximately same as the thickness of nanoribbons.

AFM was used to study the surface topology of nanowires. To determine the dimensions and surface roughness, AFM images were taken of Au and Al nanowires made on Si substrate. AFM images show the rectangular shapes of 500 nm wide and 100 nm thick Al wires as in Fig. (9)a and 500 nm wide and 80 nm thick Au wires as in Fig. (9)c. The average surface roughness of the metal wires are about 2.5 nm for Al wires as in Fig. (9)b and 1.0 nm for Au wires as in Fig. (9)d.

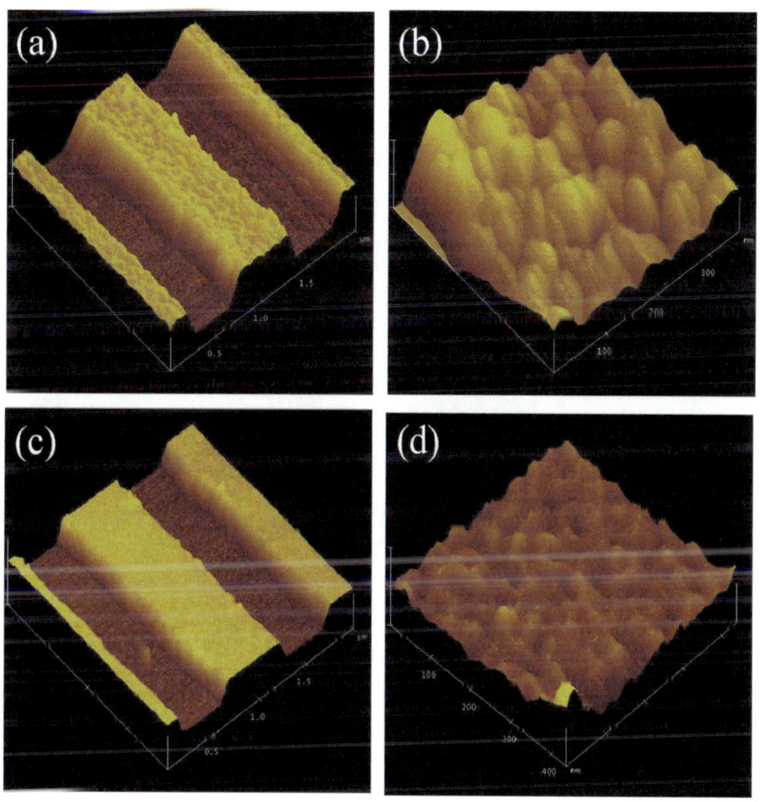

Fig. 9. (a) AFM image of Al wires of 500 nm wide and 100 nm thick on the Si substrate after removal of photoresist and (b) surface image. (c) Au wires of 500 nm width and 80 nm thick on the Si substrate (d) and surface image.

Nanowires on a Si substrate have a rectangular cross section. But after annealing their upper corners were rounded. Unlike suspended by carbon grid, metal atoms migrate into Si wafer during annealing causing the reflow and turn rectangular shape into dull round corners at the edges of nanowires.

3.1 Electric conductivity of nanowires

Electric resistivity of metal nanowires has been measured by various authors (Wu Y. et al., 2004; Walton et al, 2007; Peng et al., 2008; Sun et al., 2009). In this study, electric resistivity of Au and Al nanoribbons were measured with 4 point probe methods (4PP). The probes made of 4 Al fingers on oxidized silicon wafers using a photolithography with Cr mask. The distance between the two inner probes was 50 μm. The nanoribbons in isopropyl alcohol were ultrasonicated for 1 min to get a homogenous dispersion. Transferring the nanoribbons on these patterns could not be accomplished by drop casting from a dispersion in isopropyl alcohol because the nanoribbons have tendency to agglomerate and do not establish an electrical path between probes. A nanoribbon network between probes is created by filtering the nanoribbons dispersed in isopropyl alcohol through a cellulose filter of 0.22 μm pore size (Millipore). A continuous network of nanoribbons was obtained by vacuum filtration through the filter. The nanoribbons on the filter are stamped to the 4PP pattern under a pressure of 6.9 kPa over 12 hours then the filter is dissolved with acetone. Au ribbons has a tendency for folding more easily than Al ribbons. It is not successful to get fairly large straight Au ribbons enough to measure resistivity. So Al ribbons are used for resistivity measurement.

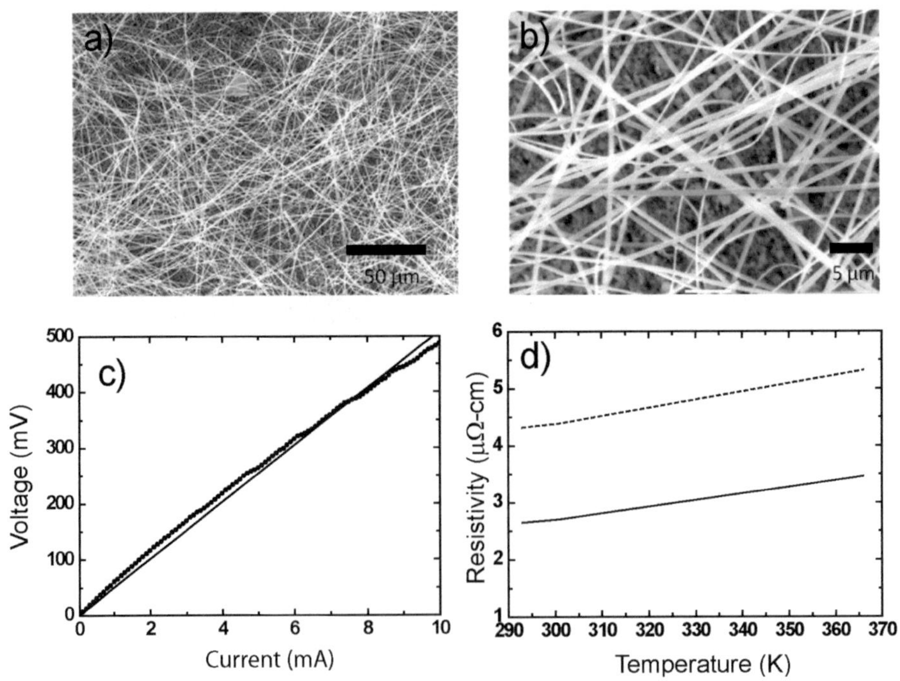

Fig. 10. (a,b) SEM image of Al ribbons transferred on cellulous filter. (c) I-V curve of a single Al ribbon (dot) with fitted line (d) temperature dependency of resistivity of a single Al nanoribbon (dashed) and Al bulk (line) from 293 K to 366 K.

Figure (10)a,b shows the network of Al nanoribbon (300 nm width and 150 nm thickness) on a cellulose filter obtained after vacuum filtration. The figure illustrates that the network is dense enough to be suitable for making transparent electrodes from nanoribbons for organic light emitting devices (LEDs) and photovoltaic applications.
The I-V curve as shown in Fig. (10)c exhibits ohmic electrical behavior of the Al nanoribbon, with resistance ~ 50 Ω. The resistivity of an individual nanoribbon, ρ was calculated from: ρ= AR/L where A is the cross sectional area of the nanoribbon and L is the electrically isolated length. Taking the width ~300 nm, thickness ~ 150 nm and length ~50 μm, the resistivity was calculated to be ~ 4.32 ± 0.12 μΩ-cm, which is higher than the bulk resistivity value ~ 2.65 μΩ-cm (Lide, 1997). Figure (10)d shows the temperature dependence of resistance of a single nanoribbon. The resistance is found to increase with temperature for both Al bulk and nanoribbons, however slope (dR/dT) is slightly different for the two cases. The resistivity of nanowires increases when the electron mean free path is larger than dimension of nanowires due to surface scattering and grain boundary scattering of electrons (Mani et al., 2006). Additionally, Al has a thin (~10 nm) native oxide layer on the surface. Thin insulating aluminum oxide layer can not be ignored in thin films and nanowires. The temperature dependence of resistivity can be explained by electron-phonon scattering (Bid et al., 2006). The dR/dT of nanowires is smaller than that of bulk because of the Debye temperature change.

3.2 Optical transmission of nanowires grating

Specular transmission data was collected for Al nanowire grating made on a glass substrate with fiber coupled spectrometer (PCI S2000, Ocean optics) for transmission measurement in the visible to near IR spectrum range (400 nm to 1000 nm) as Fig. (11). Al nanowires have period of 700 nm and thickness of 150 nm made on 200 μm thick glass substrate for transmission measurements. Narrow 100 nm width of Al nanowires are made with three angle deposition with shadow mask lithography (Park et al., 2010). Wavelengths greater than ~700 nm, which is the period of the grating, have higher transmission also there is a minima around 530nm. This is the second order diffraction of the glass-nanowire interface. Multiplying the pitch by the refractive index of glass, 1.52, and dividing by 2 yields 532 nm.

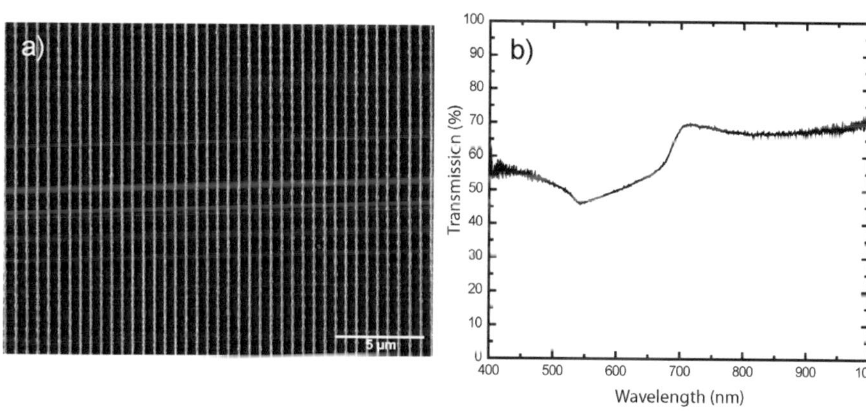

Fig. 11. (a) SEM image of Al grating of 700 nm pitch. Individual Al nanowires have 150 nm height and 100 nm width. (b) Transmission spectra of Al nanowire grating made on glass substrate.

Metallic nanowires having high transmission in visible spectrum range with high electric conductivity can replace indium tin oxide (ITO) as transparent conducting electrodes.

4. Conclusion

In summary, we fabricated metallic nanowires and nanoribbons by e-beam assisted metal depositions and electroplating with photoresist patterns made by interference holography of multiple beams. This fabrication method can produce well-defined, rectangular cross-sectioned nanowires with dimensions of a few cm long and submicron width on substrates as well as free-standing forms. Nanowires made on substrates can be further processed to make desired cross-mesh structures. Suspended nanoribbons in the IPA can be further processed or stamped as random meshes. Structural properties of metallic nanowires and nanoribbons are characterized using SEM, AFM, and TEM. Electrical resistivity of Al nanoribbons is also measured with 4PP. Potential applications of one-dimensional nanostructures are electrodes for cross-bar electronic devices like liquid crystals, and transparent electrodes for light-emitting-diodes and photovoltaic cells.
Cu metallic mesh structures are fabricated by electrochemical deposition on photoresist templates. The metallic mesh made with diffractive phase mask has large open area than that made with two beam interference. Also, the metal mesh is flexible and peeled off from the substrate to make free standing. This method is applicable to the mass production of high quality metallic nanostructures and applications requiring large areas without the need for clean room or e-beam systems.

5. Acknowledgment

This work is supported by the Division of Materials Sciences and Engineering, Basic Energy Sciences, US Department of Energy. The Ames Laboratory is operated by Iowa State University for the Office of Science, U.S. Department of Energy under Contract DE-AC02-07CH11358. This work also was supported by a Korea Research Foundation grant funded by the Korean Government (MOEHRD) (KRF-2008-D00074).

6. References

Bai, John G; Chang, Cheng-Ling; Chung, Je-Hyung and Lee, Kyong-Hoon (2007). Shadow edge lithography for nanoscale patterning and manufacturing, *Nanotechnology*, Vol. 18, (2007), pp. 405307 (8pp).

Bid, Aveek; Bora, Achyut and Raychaudhuri, A. K. (2006). Temperature dependence of resistivity of metallic nanowires of diameter >15nm: Applicability of Bloch-Gruneisen theorem, *Phys. Rev. B*, Vol. 74, (2006), pp. 035426 (8pp).

Chen, Lei; Wang, Jian; Frank, Jim; Deng, Walters, Xuegong; Buonanno, Mike; Tai, Stephen and Liu, Xiaoming (2007). Large flexible nanowire grid visible polarizer made by nanoimprint lithography, *Appl. Phys. Lett.*, Vol. 90, (2007), pp. 063111 (3pp).

Chen, Yu and Goldman, A. M. (2008). A simple approach to the formation of ultranarrow metal wires, *J. Appl. Phys.*, Vol. 103, (2008), pp. 054312 (4pp).

De, Manoranjan and Sevigny, Landre (1967). Three-Beam Holographic Interferometry, Appl. Opt., Vol. 6, (1967), pp. 1665-1671.

Farhoud, Maya; Ferrera, Juan; Lochtefeld, Anthony J.; Murphy, T. E.; Schattenburg, Mark L.; Carter, J.; Ross, C. A. and Smith, Henry I. (1999). Fabrication of 200 nm period nanomagnet arrays using interference lithography and a negative resist, *J. Vac. Sci. Technol. B* Vol. 17, (1999), 3182-3185.

Fernandez, A.; Decker, J. Y.; Herman, S. M.; Phillion, D. W.; Sweeney, D. W. and Perry M. D. (1997). Methods for fabricating arrays of holes using interference lithography, *J. Vac. Sci. Technol. B* Vol. 15.6., (1997), pp. 2439-2443.

Guo, H. C.; Nau, D.; Radke, A.; Zhang, X. P.; Stodolka, J.; Yang, X. L.; Tijhodeev, S. G.; Gippius, N. A.; Giessen H. (2005). Large-area metallic photonic crystal fabrication with interference lithography and dry etching, *App. Phys. B*, Vol. 81, (2005), pp. 271 -275.

Guo, L. Jay (2007). Nanoimprint Lithography: Methods and Material Requirements, *Adv. Mater.*, Vol. 19, (2007), pp. 495–513.

Hecht, Eugene (1987). *Optics* (2nd Edition), Addison-Wesley, ISBN 0-201-11609-X, Reading, Massachusetts.

Kang, Myung-Gyu and Guo, L. Jay (2007). Nanoimprinted Semitransparent Metal Electrodes and Their Application in Organic Light-Emitting Diodes, *Adv. Mater.*, Vol. 19, (2007), pp. 1391-1396.

Kang, Myung-Gyu; Kim, Myung-Su, Kim, Jinsang and Guo, L. Jay (2008). Organic Solar Cells Using Nanoimprint Transparent Metal Electrodes, *Adv. Mater.*, Vol. 20, (2008), pp. 1-6.

Lide, R. David (1997). *CRC Handbook of Chemistry and Physics* (87th Edition), Tayler and Francis, ISBN 978-0849304873, Boca Raton, FL.

Mani, Sathya; Saif, Taher andHan, Jong H. (2006). Effect of Annealing on the Conductivity of Electroless Deposited Ni Nanowires and Films, *IEEE Trans. on Nanotechnology*, Vol. 5, No. 2, (2006), pp. 138-141.

Onoa, G Bibiana; O'Reilly, Thomas B; Walsh, Michael E and Smith, Henry I (2005). Bulk production of singly dispersed carbon nanotubes with prescribed lengths, *Nanotechnology*, Vol. 16, No.12, (2005), pp. 2799–2803

Park, Joong-Mok; Nalwa, Kanwar Singh; Leung, Wai; Constant, Kristen; Chaudhary, Sumit and Ho, Kai-Ming (2010). Fabrication of metallic nanowires and nanoribbons using laser interference lithography and shadow lithography, *Nanotechnology*, Vol. 21, (2010), pp. 215301 (6pp).

Pauzauskie, Peter J. & Yang, Peidong (2006). Nanowire Photonics, *Materials today*, Vol.9, No. 10, (2006), pp. 36-45.

Peng, Yong; Cullis, Tony and Inkson, Beverley (2008). Accurate electrical testing of individual gold nanowires by *in situ* scanning electron microscope nanomanipulators, *Appl. Phys. Lett.*, Vol. 93, (2008), pp. 183112 (3pp).

Shankar, K. Shantha and Raychaudhuri, A.K. (2005). Fabrication of nanowires of multicomponent oxides: Review of recent advances, *Materials Science and Engineering C*, Vol. 25, (2005), pp. 738-751.

Sun, Tik; Tao, Ba; Warren, Andrew P.; Barmak, Katayun; Toney, Michael F.; Peale, Robert E. and Coffey, Kevin R. (2009). Dominant role of grain boundary scattering in the resistivity of nanometric Cu film, *Phys. Rev. B*, Vol. 79, (2009), pp. 041402 (4pp).

Thompson, C. V. (2000). Structure Evolution during Processing Polycrystalline Films, *Annu. Rev. Mater. Sci.*, Vol. 30, (2000), pp. 159–190.

Vazquez-Mena, O.; Savu, V.; Sidler, K.; Villanueva, G.; Boogaart, M. A. F. van den ; Brugger J. (2008). Sub-100 nm-scale Aluminum Nanowires by Stencil Lithography:

Fabrication and Characterization, *Proceedings of the 3rd IEEE Int. Conf. on Nano/Micro Engineered and Molecular Systems,* January 6-9, 2008, Sanya, China, pp. 807-811.

Walton, A S; Allen, C S; Critchley, K; Gorzny, M L; McKendry, J E; Brydson, R M D; Hickey, B J and Evans, S D (2007). Four-probe electrical transport measurements on individual metallic nanowires, *Nanotechnology*, Vol. 18, No. 40, (2007), pp. 065204 (6pp).

Wu, Bin; Heidelberg, Andreas and Boland, John J. (2005). Mechanical properties of ultra-high strength gold nanowires, *Nature Materials,* Vol. 4, (2005), pp. 525-529

Wu, Yue; Xiang, Jie; Yang, Chen; Lu, Wei & Lieber, Charles M. (2004). Single-crystal metallic nanowires and metal/semiconductor nanowire heterostructures, *Nature,* Vol. 430, No.1, (2004), pp. 61- 65.

Xiang, Chenxiang; Kung, Sheng-Chin; Taggart, David K.; Yang, Fan; Thompson, Michael A.; Guell, Aleix G.; Yang, Yongan and Penner, Reginald M. (2008). Lithographically Patterned Nanowire Electrodeposition: A Method for Patterning Electrically Continuous Metal Nanowires on Dielectrics, *ACS Nano,* Vol. 2, No. 9, (August 2008), pp. 1939-1949.

Xue, Mianqi; Yang, Yanlian and Cao, Tingbing (2008). Well-Positioned Metallic Nanostructures Fabricated by Nanotransfer Edge Printing, *Adv. Mater.*, Vol. 20, (2008), pp. 596-600.

Part 5

Nanowire Characterization

22

Growth and Characterisation of Ge Nanowires by Chemical Vapour Deposition

Chuanbo Li[1], Hiroshi Mizuta[2] and Shunri Oda[1]
[1]*Quantum Nanoelectronics Research Center, Tokyo Institute of Technology, Tokyo,*
[2]*School of Electronics and Computer Science,*
The University of Southampton, Highfield, Southampton,
[1]*Japan*
[2]*UK*

1. Introduction

One-dimensional Ge/Si nanostructures have attracted much attention because of their potential applications in the design of novel electronic (Fan et al, 2006; Xiang et al, 2006; Liang et al, 2007; Hu et al, 2007), photonic (Holmes et al, 2000), battery(Chan et al 2008a; Chan et al 2008b) and sensing devices (Cui et al, 2001). For examples, due to their high mobility of electrons and holes, Ge nanowires show their promising application in high-speed field-effect transistor (Xiang et al, 2006; Liang et al, 2007). And, Ge nanowires are potentially useful for high-speed quantum computing due to the long decoherence time because of their predominance of spin-zero (Hu et al, 2007; Tyryshkin et al, 2006) nuclei and the advantage of a large excitonic Bohr radius in Ge (24.3 nm) allowing for quantum confinement to be observed in relatively large structures (Sun et al, 2007; Maeda et al, 1991) and at high temperatures. In order to realize these applications, controllable and high quality nanowires growths are important. Much attention has been paid to the growth of Ge nanowires(Lauhon et al, 2002; Greytak et al 2004; Kodambaka et al 2007).In most cases, the nanowires are grown in chemical vapour deposition (CVD) method visa vapor liquid solid (VLS) mechanism (He et al, 2006; Kamins et al, 2004; Fuhrmann et al, 2005; Allen et al,2008; Hannon et al, 2006), in which the formation of Ge-Au eutectic plays the most important role in the synthesis of nanowires. It has been clarified that the VLS mechanism dominates the axial growth of Ge nanowires. However, the radial growth of nanowires is easily ignored and there are only few papers to talk about it. And, the impacts of the surface conditions on the VLS (vapor-liquid-solid) synthesis of Ge nanowires are rarely discussed. Properly understanding it could help us engineer the growth of nanowires.

In this chapter, the growth of Ge nanowires via CVD method will be discussed in detail. Vapor-solid-solid (VSS) growth mechanism is supposed to dominate the nanowire growth in a radial direction, which contributes to an increase in the diameter of nanowire. After the Au catalysts on the tip are consumed over a long growth time, the nanowire with a rough surface will be grown due to the selective VSS radial growth. The impacts of surface condition on the growth of Ge nanowires on Si (100) substrate will also be observed. On the SiO_2-terminated Si substrate, high-density Ge nanowires can easily be grown. However, it is more complex for the growth of Ge nanowires on H-terminated Si substrate. Due to Si migration and formation of a native SiO_2 overlayer on the catalyst surface, it retards the growth of Ge nanowires. After removing this overlayer in the HF solution, high-density and well-ordered Ge nanowires are grown. Ge nanowires cross vertically and form two sets parallel nanowires array. It is found that nanowires grew along <110> directions.

Nanowires incorporating heterostructures offer numerous potential applications in nanoelectronics and photonics devices because of their narrow diameter allowing for efficient strain relaxation and the formation of epitaxial structures from lattice-mismatched materials. In this chapter, the growth of Si-Ge and Ge-Si nanowire heterostructures will also be discussed via chemical vapour deposition. Due to the influence of interface energy, differing topographies of the heterostructures were observed. On initially grown Ge nanowires, numerous Si nanowire branches were grown near the tip due to Au migration. However, on initially grown Si nanowires, high-density Ge nanodots were observed.

Here, we will also demonstrate the well position-controllable Ge nanowires grown on the SiO_2 substrate by combining top-down and bottom-up methods. To realize the practical applications of Ge nanowire on a large scale, one of the key challenges is to develop a convenient and parallel method to align bottom-up nanowires into complex patterns or structures. A "pick and place" method is most widely used for integrating nanowires. However, it lacks control of precision and repeatability, and easily induce contamination in the wires. It is expected to selectively grow nanowires directly onto desired areas of the substrate and *in situ* fabricate the nanowire devices. By optimizing the EBL process, Au patterns with a diameter of 10 nm were prepared by the lift-off method. The growth of Ge nanowires can be precisely controlled by adjusting the location of catalysts, which offers the possibility of in situ fabrication of nanowire devices.

2. VSS radial growth of Ge nanowires

All Ge nanowires were grown by using a low pressure CVD method at 300 - 350 °C with 10% GeH_4 precursors (in an atmosphere of hydrogen) under the total pressure of 5 torr. Au catalysts layers were evaporated by electron beam evaporation at room temperature on SiO_2 and H-terminated substrates.

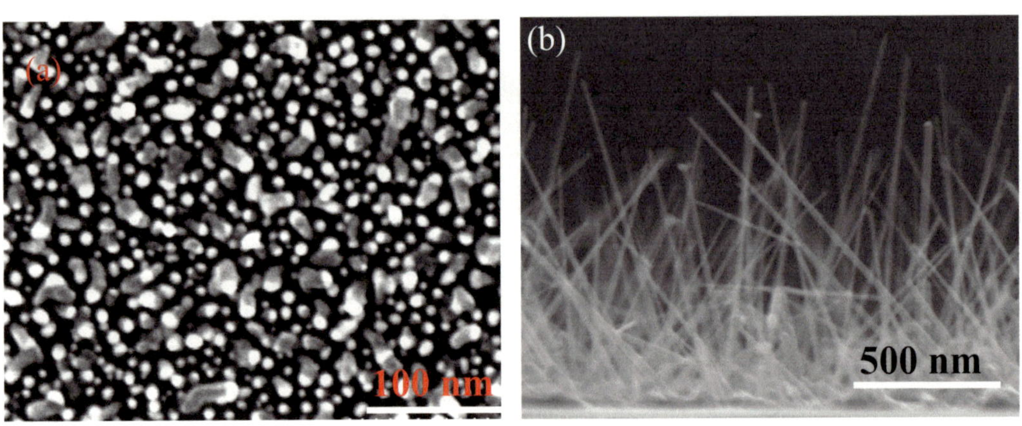

Fig. 1. (a), A SEM image of Ge nanowire at the initial growth stage grown at 300 °C ; (b), A SEM image of Ge nanowires grown on the SiO_2 substrate at 300 °C for 20 min. The nanowires have a length of 1 μm and a diameter of 15~25 nm. (Li et al, 2009a). ©American Institute of Physics.

At the initial growth stage, Ge nanowires only have a diameter of 5~10 nm, which are consistent with that of Au catalysts as deposited (Fig. 1(a)) (Li et al, 2008). However, with increasing the growth time, both the diameter and length of nanowires increase. As the scanning electron microscope (SEM) results shown in Fig. 1(b), the nanowires of 15~25 nm in diameter and 1 μm in length were grown at 300 °C for 20 min. Compared with the fast VLS growth rate in the axial direction, the growth rate in the radial direction, which contributes to an increase in the diameter of the nanowires, is much slower. There must be some other growth mechanisms instead of VLS to contribute to the nanowire growth in the radial direction. According to our control experiment results on the SiO_2 substrate, the chemical vapor deposition without the assistance of Au catalysts can be ignored. So we believe that the vapor-solid-solid (VSS) (Kodambaka et al 2007) growth mechanism should dominate the growth of nanowires in the radial direction.

As J. B. Hannon and his coworkers (Hannon et al, 2006) investigated before, there is gold migration on the surface of nanowires during the growth of nanowires. Some Au atoms will cover, wet and smooth the nanowire sidewalls. This gold migration on the nanowire surface will consume the Au droplets on the tip of nanowires, make the droplet size smaller and induce non-uniform diameter with the thicker bottom and thinner top as shown in Fig. 1(b). Clearly, some Au catalysts exist on the Ge nanowire surface. And the Au-Ge alloy will be formed on the Ge nanowire surface at growth temperature (300 °C).

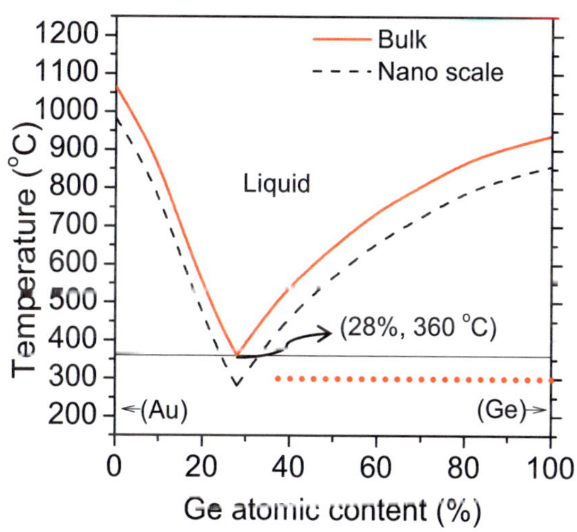

Fig. 2. A Au-Ge binary alloy diagram. Solid and dashed lines are the liquid lines for bulk material and nanoscale material respectively. Due to the size effects of nanoscale Au particles and nanowires, the liquid line in the binary alloy diagram has a downshift and Au-Ge alloy will have a lower eutectic temperature (Li et al, 2009a). ©American Institute of Physics.

However, compared with a large number of Ge atoms, the Au atoms on the surface of nanowires are much less. And the Au content in the Au-Ge binary alloy phase diagram (Kodambaka et al 2007) should be lower than their eutectic point (28%). Therefore the melting point of Au-Ge alloys on the nanowire surface is higher than both their eutectic temperature and growth temperature (300 °C) even though the melting point of the Au-Ge alloy has a little downshift due to the size effect of the nanoscale Au catalyst (Adhikari et al,2006; Buffat et al,1976) as shown in Fig.2. And the Au-Ge alloy at the nanowire surface is in solid state at 300 °C located somewhere on the dotted line in the phase diagram as shown in Fig.2. So Ge atoms can only deposit on the Ge nanowire surface in the VSS mode with the assistance of Au catalyst to contribute to the increase of the nanowires diameter.

Fig. 3. SEM images of Ge nanowires grown at 300 °C for 60 min on SiO_2 substrate. (a) the top part and (b) the middle part of Ge nanowires; the nanowires have the length of over 3 μm and the diameters of over 30 nm. The nanowires with the Au catalyst on the tip as indicated by solid arrows exhibit a smooth surface, and the VLS axial growth is not hindered; For the nanowires without the Au catalyst on the tip as indicated by the dotted arrows, a rough surfaces are shown due to the selective VSS radial growth. (Li et al, 2009a). © American Institute of Physics.

This hypothesis can be proven by conducting the nanowire growth for a longer period. With further increasing the growth time of Ge nanowires, the Au droplets will be consumed eventually to cover the surface of Ge nanowires, and the VLS growth in the axial direction will stop at the same time. However, the VSS growth in the radial direction does not stop yet since some Au catalysts still covers the nanowire surface. So the diameter of the nanowires still increases due to the VSS growth. At the same time, the surface area of nanowires will increase too, and the Au catalysts covering the nanowire surface become thinner and thinner since there is no Au atom supplied from the tip of nanowire, which migrates and covers the nanowire surface (Hannon et al, 2006). Finally, with an increase in diameter, Au catalysts cannot cover the whole nanowire surface even in an atomic layer thickness and cover only in patches. Ge atoms could be adsorbed only on the area covered with Au catalysts in the VSS mode, which contributes to the radial growth. However the radial growth does not progress in the area without the Au catalysts. So this selective VSS radial growth induces the rough nanowire surface over a longer growth time as indicated by dotted arrows shown in Fig. 3(a) and 3(b). Clearly, for the nanowire with the Au catalysts on the tip as the solid arrow indicated in Fig. 3(a), it exhibits a smooth surface because of the Au migration and wetting. And the axial growth continues. But, once the Au catalyst on the tip is consumed, due to the effect of selective VSS radial growth as mentioned above, the nanowires begin to show a rough surface and a taper shape as indicated by the dotted arrows in Fig. 3.

3. The influences of surface conditions on the growth of germanium nanowires

From our experimental results, we found (Li et al, 2008) that the growth of Ge nanowires was quite sensitive to the surface condition of the substrate. A clear understanding of this phenomenon could help us engineer the growth of nanowires.

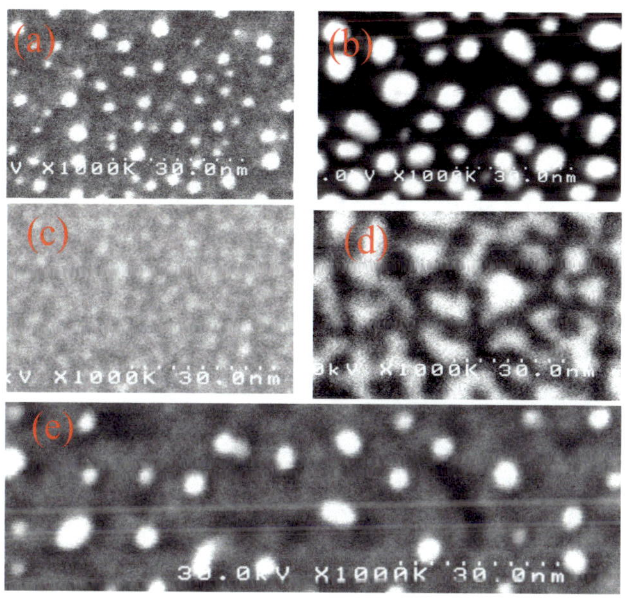

Fig. 4. SEM images of Au catalysts evaporated on SiO_2-terminated Si substrate with a thickness of 0.1 nm (a) and 1 nm (b); Au catalysts evaporated on H-terminated Si substrate with a thickness of 0.1 nm (c) and 1 nm (d); SEM image of 1 nm-thick Au evaporated on H-terminated Si after HF treatment for 2 minutes (e). (Li et al, 2008). © American Institute of Physics.

Au catalyst layers with a thickness of 0.1 nm and 1nm were evaporated by using electron beam evaporation at room temperature on two different substrates, SiO_2-terminated (a few nanometers native SiO_2 layer or 170 nm thermal oxidized SiO_2) and H-terminated (the wafers were dipped in 1.5% HF solution for 1 minute and immediately loaded in the chamber for EB evaporation) Si (100) substrates. In order to satisfy the rules of minimum surface energy, the evaporated Au layer shows different topography at different substrate surface conditions as shown in Fig. 4(a)-4(d). On the SiO_2-terminated substrate, because condensed Au adatoms are bound more tightly to each other than to the substrate[13], these atoms encounter other atoms, nucleate and agglomerate to form stable islands of 2~10 nm in diameter (Fig. 4(a), 4(b)). A thicker Au layer will result in the formation of bigger Au islands. On the other hand, on the H-terminated Si substrate, the Au catalyst prefers to deposit in the Stranski–Krastanov (SK) mode to form small Au islands with a wetting layer as shown in Fig.4(c) and 4(d).

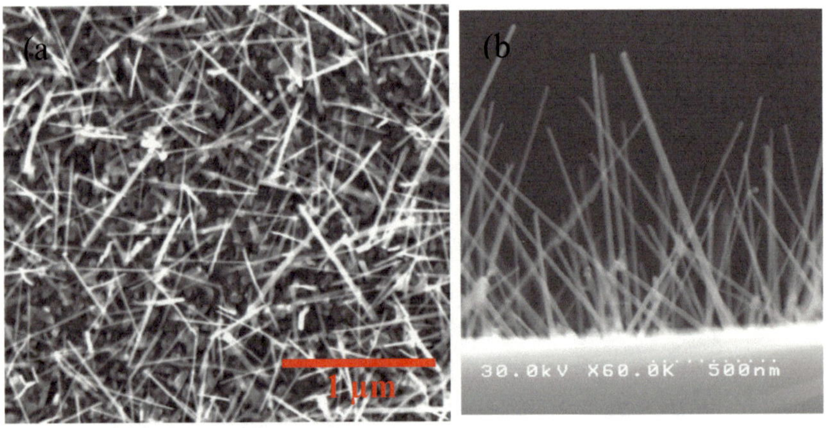

Fig. 5. (a)A top-view SEM image of high-density Ge nanowires grown on 1 nm-thick Au catalyst evaporated on the SiO_2-terminated Si substrate. (b), a cross-sectional SEM image. (Li et al, 2008). © American Institute of Physics

On the SiO_2-terminated Si substrate, without any pre-treatment to the catalysts, high-density Ge nanowires of 5~20 nm in diameter were grown on the Au catalysts with a thickness of 1 nm (Fig.5) and 0.1 nm. Both high-resolution transmission electron microscopy (TEM) and X-ray diffraction (XRD) results indicate high-quality and single-crystalline Ge nanowires with the cubic diamond structure. It was found that the thickness of the SiO_2 layer, either a few-nanometer-thick native oxide layer or a thicker thermally-oxidized SiO_2, has no influence on the growth of Ge nanowires.

However, very few Ge nanowires were grown on the H-terminated Si substrate under the same growth conditions (Fig.6). Even after high temperature pre-annealing (650 ^oC in vacuum) to dewet the Au wetting layer, it did not have any significant influence on the growth of Ge nanowires.

While observing topography of the Au catalysts evaporated on the H-terminated substrate by SEM, we found that the contrast between Au metal and Si substrate at edge area was smaller, and the brightness of Au metal in the SEM image got weaker compared with that evaporated on the SiO_2 substrate. It is likely that Au catalysts are covered by some insulating layers. So maybe the formation of a very thin Si oxide overlayer due to the gold catalyzed migration of Si through the gold film (Hiraki et al,1971, Lay, 1983; Hiraki 1984; Jagannathana et al 2006) retards the growth of Ge nanowires.

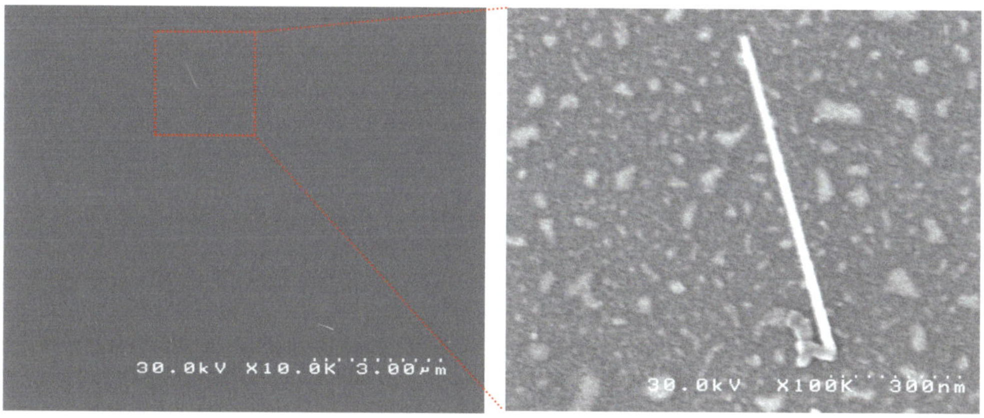

Fig. 6. A top-view SEM image of Ge nanowires grown on the H-terminated Si substrate. The inset shows a blow-up of one wire and it exhibits the same diameter as that grown on the SiO_2-terminated substrate (Li et al, 2008). © American Institute of Physics

In order to verify this hypothesis, Au catalysts evaporated on the H-terminated substrates were dipped into 1.5% HF solution for 2 minutes, and then immediately loaded into the growth chamber to grow Ge nanowires. On 1-nm-thick Au catalysts substrate, high-density and well-ordered Ge nanowires were grown as shown in Fig. 7(a). Obviously, it has a great improvement for Ge nanowires growth by removing this overlayer and it confirms that the SiO_2 overlayer formed on the catalyst surface prevents the growth of Ge nanowires. Fig. 4(e) shows the SEM image of 1-nm-thick Au catalysts after HF treatment. Clear Au dots can be found after removing the SiO_2 overlayer. In the case of 0.1-nm-thick Au catalysts, even after HF treatment, no nanowire was grown. We did not find any Au dots after HF treatment in the SEM image since it still remains at the wetting layer stage. So there are no nucleation centres for Ge atoms to grow in the axial direction. And the GeH_4 precursors will be further decomposed and deposited to form a film since the AuSi eutectic alloy exits on the whole surface. Compared with the random-directed Ge nanowires grown on the SiO_2-terminated substrate as shown in Fig.5, Ge nanowires grown on the H-terminated Si (100) substrate show a more ordered structure. Almost all Ge nanowires cross vertically and form two sets parallel nanowires array. It should be point out that each set of Ge nanowires has around 45° angle with the [110] cleavage direction as labelled in Fig. 7(a). Till now, only two growth directions, [110] and [111] were found for Ge nanowires grown on the Si substrate. Based on this finding and our experiment results, we believe that Ge nanowires are grown preferably in the [101], [011], [-101] and [0-11] directions as the four dashed arrows shown in Fig. 7(b), in which these four growth directions are crossed vertically and have a 45° degree with [110] and [-110] cleavage directions. It agrees well with the experiment results as shown in Fig. 7(a). The two dashed-dotted arrows in Fig. 7(b) are [110] and [-110] cleavage directions for (100) wafer as labelled in Fig. 7(a).

During the growth of Ge nanowires on Au catalysts evaporated on the H-terminated Si substrate after HF treatment, GeH_4 will decompose in the AuSi eutectic alloy located at the interface between the single-crystalline Si surface and Au layer and then supersaturated Ge atoms will separate out to grow in the axial direction. This ordered structure at the interface can induce the Ge nanowires grow in only the <110> directions. This capability to control the nanowire growth direction would be very important to engineer the transport characteristics for electrons and holes individually to obtain the best switching performance of a complementary logic circuit configuration[19]. And also it is quite attractive for future large-scale nanowire integration.

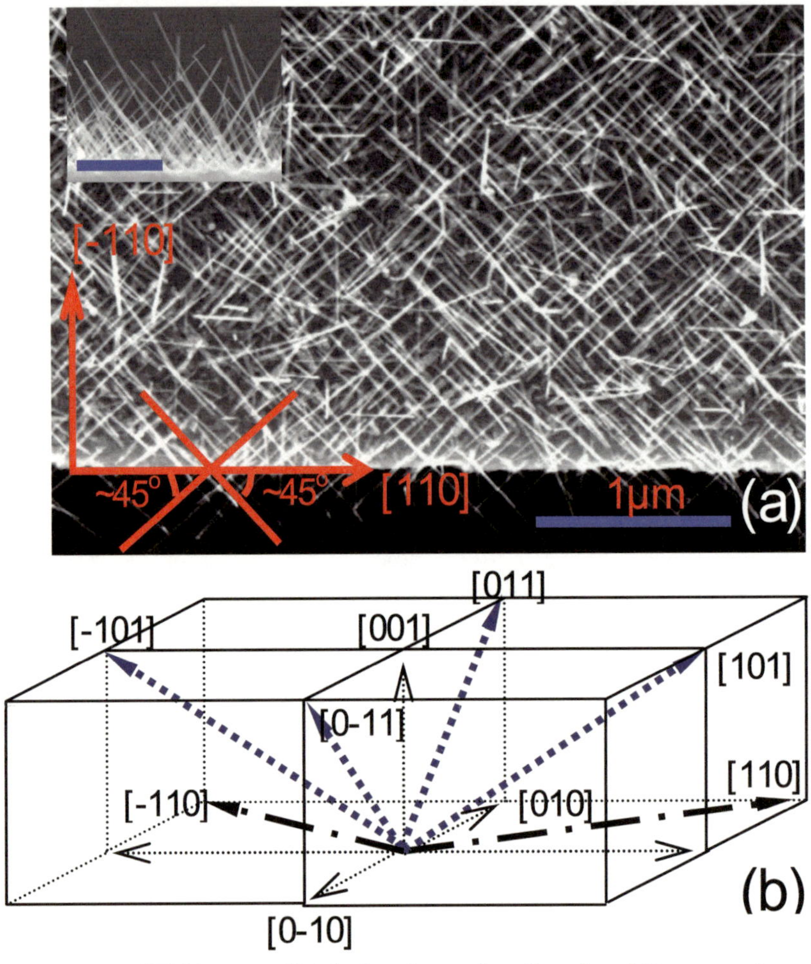

Fig. 7. (a), A top-view SEM image of high density and well-ordered Ge nanowires grown on 1-nm-thick Au catalysts evaporated on the H-terminated Si (100) after dipped into the HF solution to remove the SiO_2 overlayer. The bottom edge is the [110] cleavage direction. The inset shows its side-view SEM image and the scale bar is 500 nm; (b), A 3D schematic of the growth directions of Ge nanowires on Si (100) wafer. Four dashed arrows directed to [101], [011], [-101] and [0-11] are the Ge nanowires growth directions. (Li et al, 2008). © American Institute of Physics

In the case of Au catalysts evaporated on the SiO_2-terminated substrate, strong bonding between Si and O atoms may prevent the migration of Si atoms to the surface and the formation the SiO_2 overlayer. Therefore Ge nanowires can be grown without HF treatment.

Figures 8(a) and 8(b) show the SEM images of Ge nanowires grown at 350 **°C** in the same conditions on the H-terminated Si and the SiO_2-terminated Si substrates, respectively. The nanowires showed different density, shape and size on these two substrates. On the H-terminated substrate, similarly with that grown at 300 °C as discussed above, only low-density Ge nanowires can be grown because the SiO_2 overlayer covering the Au catalysts surface hinders the growth of Ge nanowire,. At high temperature, the growth speed increases, and thicker and longer nanowires can be grown. However, on the SiO_2-terminated substrate, thicker and shorter nanowires were grown. Compared with longer nanowires with a smooth surface as shown in Fig. 8(a), the difference between these nanowires should result from the different Au atoms supply. In the case of low-density Ge nanowires growth, there is enough Au atoms supply, which migrate from the other areas and contribute to the VLS growth over a longer period. However, on the SiO_2-terminated substrate, the Au catalyst droplets will supply for more nanowires growth and are easily consumed. After there remains no Au supply for the axial direction growth, thicker nanowires with a rough surface are grown. These facts evidence the above-mentioned explanation.

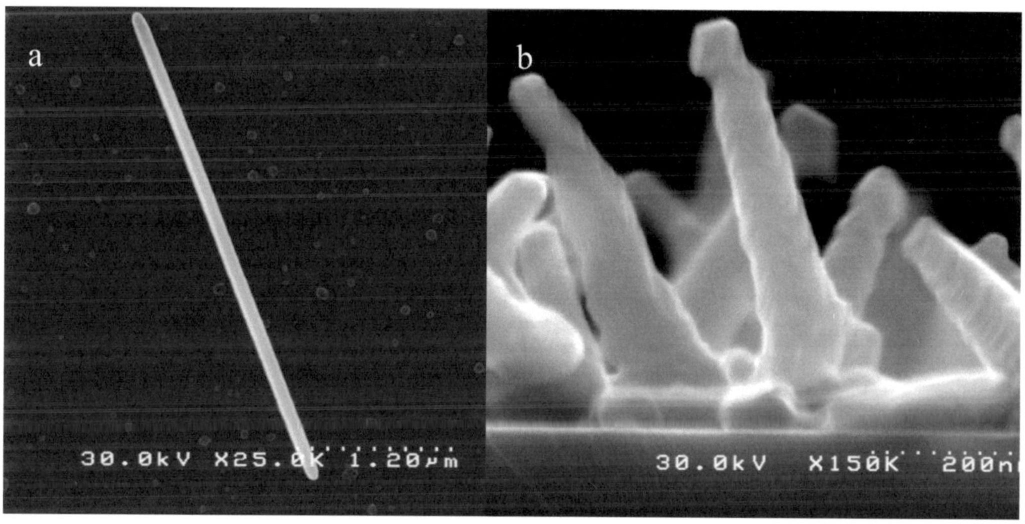

Fig. 8. SEM images of Ge nanowires grown at 350 °C for 20 minutes on H-terminated (a) and SiO_2-terminated (b) Si substrate, respectively.

4. Growth of Ge-Si nanowire heterostructures

Nanowire heterostructures offer numerous potential applications in nanoelectronics and photonics devices because their small diameter enables efficient strain relaxation and the formation of epitaxial structures from lattice-mismatched materials (Dick et al, 2007a; Ertekin et al, 2005; Kastnery & Gosele 2004; Gudiksen et al, 2002). Enormous effort has been made to the growth of Ge-Si nanowires due to their low cost and unrivaled compatibilities with the main-stream ultra-large-scale Si integration technology (Lu et al 2005; Liang et al 2007; Hu et al, 2007; Hu et al 2008). Some heterostuctures, such as Ge/Si core/shell nanowires (Xiang et al, 2006;Lauhon et al, 2002; Ben & Patolsky, 2010)and Si/SiGe nanowires superlattice (Wu et al, 2002; Clark et al, 2008) have been grown via the VLS mechanism.

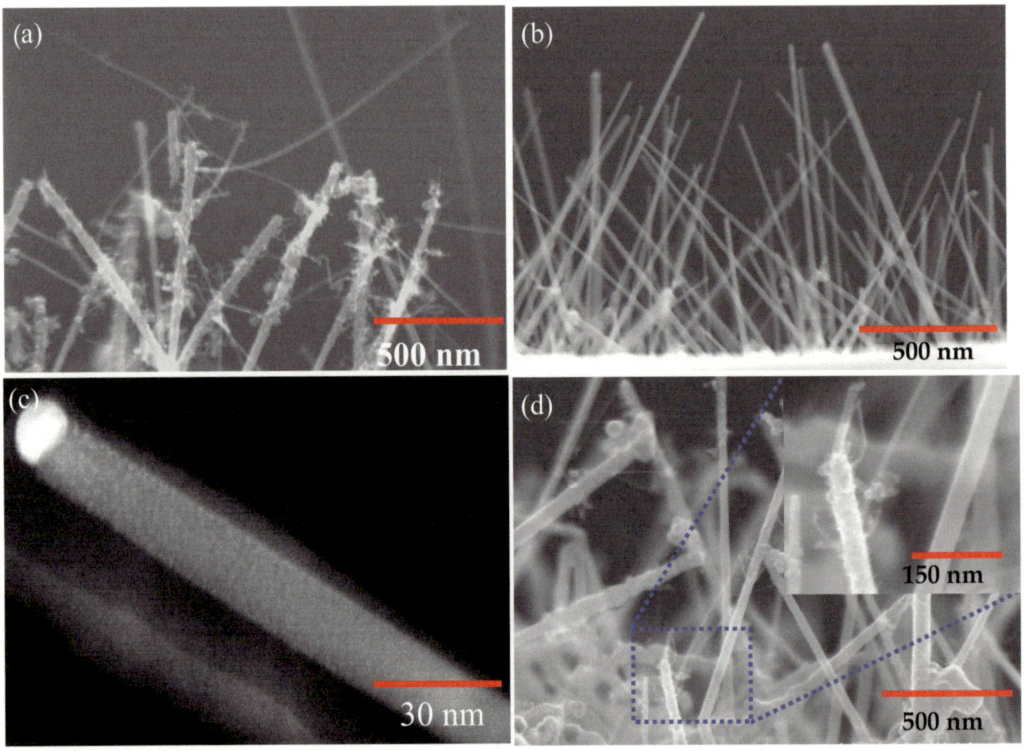

Fig. 9. (a), A SEM image of branched Si/Ge nanowire heterosturctures grown at 350 °C on Ge nanowire substrate in the continuous mode (top part of the nanowires); (b), A Cross-sectional SEM image of Ge nanowires grown on SiO_2 substrate at 300 °C for 20 min; (c), A dark field TEM image of Ge nanowires before the growth of Si nanowire branches. The Au catalysts locate at the tips of nanowires and many small Au particles on the top parts of nanowires can be observed; (d), A SEM image of bottom part of Ge nanowires and inset shows the enlarged SEM image in the dotted square. (Li et al, 2011). © Elsevier.

Branched nanowires offer another approach to increase structural complexity thus enabling greater functionalities (Wang et al, 2004). They lead to more potential applications in nanoelectronic devices by increasing the number of connection points and providing a means for parallel connectivity and interconnection of functional elements (Dick et al, 2007b). Moreover, hierarchically heterobranching and hyperbranching nanowire structures will potentially promote applications in solar energy harvesting, sensor and so forth (Bierman & Jin, 2009). A clear understanding of the growth mechanism of branched nanowires would enable to control and design the complex heterostructures which meet specific requirements of challenging applications. The growth of homogenous Si-Si nanowire branches has been demonstrated to date (Wang et al, 2004; Doerk et al, 2008). Here the growth of Si-Ge and Ge-Si nanowire heterostructures is also presented.

Si nanowires were grown at 350 °C for 20 min by using the Si_2H_6 (3.3 sccm) and H_2 (50 sccm) at 1 Torr. There is a 40 min waiting time to stabilize the growth temperature in a vacuum condition by heating (from 300 °C to 350 °C) or cooling (350 °C to 300 °C) the growth chamber between the growth intervals. Fig. 9(a) shows the SEM image of Ge-Si nanowire heterojunctions grown in a continuous mode. Compared with the smooth surface of Ge nanowire as the SEM image shown in Fig. 9(b), many nanowire branches were grown on the surfaces of Ge nanowire stems. It was confirmed that the branches on Ge nanowire stems were Si nanowires by the small angle XRD spectrum as shown in Fig. 10. Compared with the XRD spectrum of Ge nanowires without any branches (dotted line), an extra set of peaks were observed in the XRD spectrum of branched Ge nanowires (solid line). The simulation results by Philips X'pert high-score software are shown in table 1. Clearly, the peaks at 27.4112°, 45.3028 °, 53.7354 °, 65.9223 °, 72.8923 ° and 83.8253 ° match well for both samples and are related to Ge (111), (220), (311), (400), (331) and (422) plane respectively of Ge nanowires. The other set of peaks at 28.4247°, 47.2729° and 56.1197° originate from the Si (111), (220) and (311) planes respectively.

Position (° 2Θ)	FWHM (° 2 Θ)	d-spacing (Å)	Relativeintensity (%)	Chemical Formula
27.5006	0.1574	3.24344	100.00	Ge
28.4247	0.3936	3.14007	22.97	Si
45.5304	0.1574	1.99231	51.46	Ge
47.2729	0.5510	1.92286	8.86	Si
53.8240	0.2362	1.70327	27.82	Ge
56.1197	0.4723	1.63892	3.25	Si
66.1912	0.6298	1.41187	4.03	Ge
72.9023	0.5510	1.29757	7.45	Ge
83.9378	0.5760	1.15189	7.19	Ge

Table 1. Peak lists of solid line in Fig. 10 and its analysis results (Li et al, 2011). © Elsevier.

It is interesting that the Si nanowire branches were grown on the Ge nanowire surface rather than on their tips. Since the VLS nanowire growth is sensitive to Au catalysts, there must be some Au catalysts on the surface of Ge nanowires to contribute to the growth of Si nanowire branches. It has been reported that there is an Au migration on the nanowire surface during their growth (Kodambaka et al, 2007; Hannon et al, 2006; Gosele, 2006) and during the high-temperature growth interval (Doerk et al, 2008). From the dark field transmission electron microscope (TEM) image of Ge nanowires as shown in Fig. 9(c), a rough surface with numerous white spots was observed on the top parts of Ge nanowires. It is clear that the big white spot located at the tip of nanowire is the Au catalyst. The energy-dispersive-spectrometry analysis indicates that the small white dots on the top parts of nanowires are the surface-migrated Au particles. The density and the size of Au particles reduce with increase of distance from the tips of Ge nanowires and no Au particle is found on the bottom of the nanowires. This distribution of Au particles agrees well with that of Si nanowire branches grown on the Ge nanowire surface as SEM image shown in Fig. 9(d).

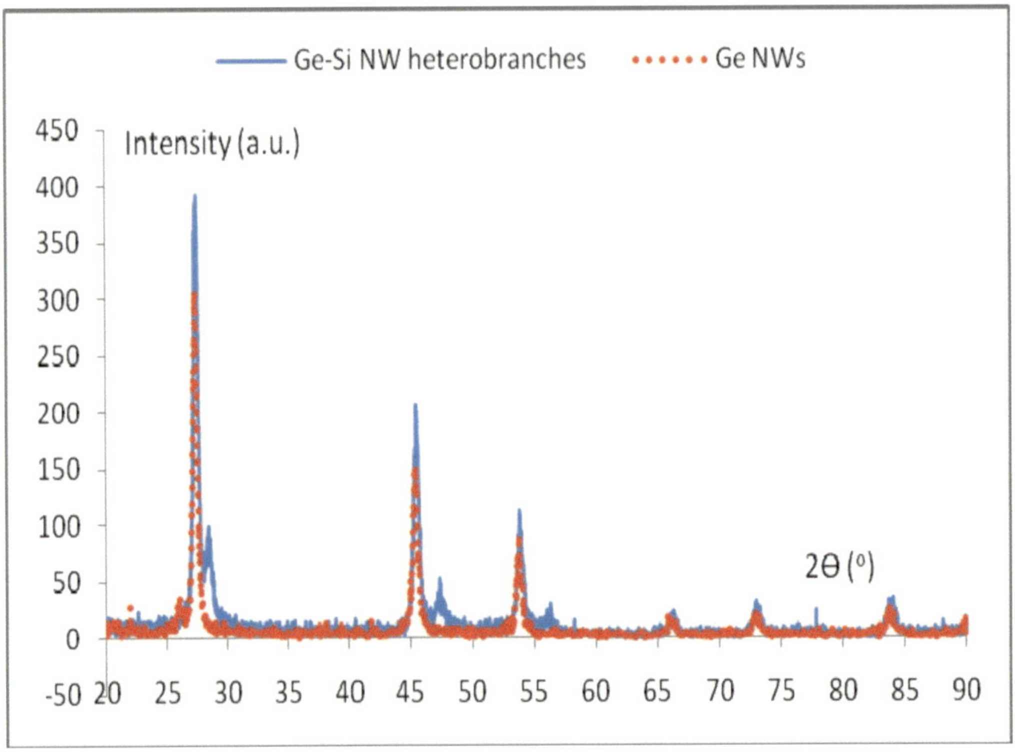

Fig. 10. Small angle XRD spectrum of Ge nanowires with and without branches. (Li et al, 2011). © Elsevier.

Clearly, these self-seeded Au particles formed by the surface migration, play an important role in controlling the growth of Si nanowire branches. The branches were only grown on the top parts of nanowire stems. With increasing distance from the tips of nanowires, the less Au catalyst exists on the surface, resulting in fewer Si nanowire branches due to the less nuclear centers for their growth. The enlarged SEM image in the inset of Fig. 9(d) shows a typical example where the branches only appear on the top parts of stems. On the short Ge nanowire stems, many Si branches were grown on its top. However, on the surface of its neighbors, as the bottom parts of long nanowires, no Si branch was found. The growth rate of nanowire branches is lower than that grown on planar substrate. Shorter and thinner nanowire branches were grown at the same condition due to the shortage of Au catalysts.

The experiments indicate that the growth mode has an important influence on the growth of nanowire branches. If Ge nanowires were kept in the air for a few days and then were reloaded into the chamber for Si nanowire branch growing in a separate mode, the Si heterostructures tend to only grow on the tips of Ge nanowires as shown in Fig. 11. The growth rate of Si nanowire branches greatly reduced compared with that grown in the continuous mode as shown in Fig. 9(a). The longer the Ge nanowires were put in air, the fewer Si nanowire branches were grown (Fig. 11(a) and 11(b)). We believe that GeO_2 formed on the nanowire surface due to the oxidization of Ge atom migrated to the surface covers the small Au catalysts dots on the top parts of Ge nanowire and retards the growth of Si nanowire branches.

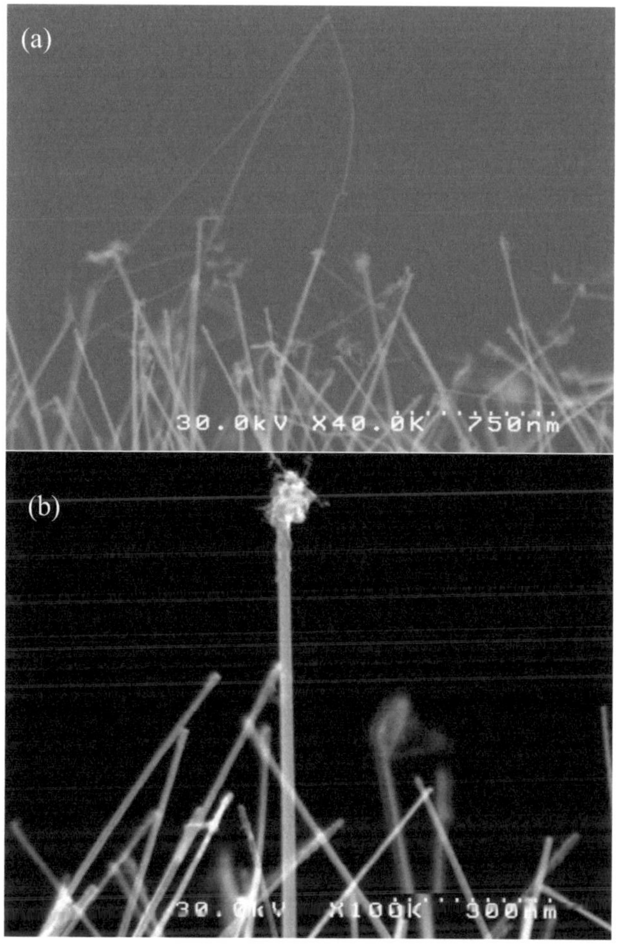

Fig. 11. SEM images of Si nanowire branches grown at 350 °C on Ge nanowire substrates in the separate mode. Ge nanowires were firstly grown at 300 °C and were put in air for one week (a) and three weeks (b) and then were reloaded in the chamber for growth of Si nanowire branches. (Li et al, 2011). © Elsevier.

As a comparison, Ge heterostructures on Si nanowire stems were also observed. After the synthesis of Si nanowires at 350 °C, the chamber was cooled down to 300 °C in 40 min, and then GeH_4 was introduced into the growth chamber to grow Ge heterostructures. Interestingly, a different topography of heterostructures was exhibited. As shown in Fig. 12(a), Si nanowire stems have a clean surface before the growth of Ge heterostructure. However, many Ge nanodots instead of branches were observed on the surface of Si nanowire stems as shown in Fig. 12(b) and 12(c). K. A. Dick and his coworkers (Dick et al, 2007a) have discussed that different morphologies of crawl or nanowire heterosturctures were grown in compound nanowire systems depending on the growth condition and interface energy between these materials. Similar effects were observed in our experiment. In order to satisfy the minimum system energy, high density Ge quantum dots other than branches were grown on the surface of Si nanowire stems.

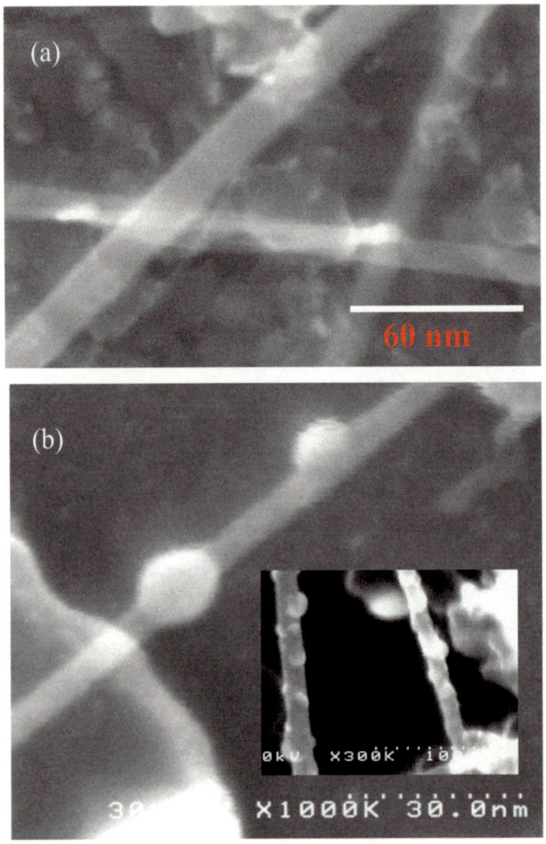

Fig. 12. (a), SEM image of Si nanowires grown on SiO_2 substrate at 350 °C for 20 min; (b) and (c), SEM images of Ge nanodots heterostructure grown at 300 °C on Si nanowire substrate in the continuous mode. (Li et al, 2011). © Elsevier.

At the same condition, the chemical vapor deposition of Ge atom without the assistance of Au catalysts can be ignored at 300 °C according to our control experiment results on SiO_2 substrate. The nanodot heterostructures are also catalyzed by the Au catalyst on the Si nanowire surface. The Ge nanodots have a diameter of 2-10 nm, which is much smaller than those grown on the Si substrate in the Stranski-Krastanov mode (Li et al, 2004a; Li et al, 2004b). These small size nanodot herterostructures will have much potential application when applied in quantum computation and optical communication.

5. Position-controllable single Ge nanowire array

To realize the applications of Ge nanowires on a large scale, one of the key challenges is to develop a convenient and parallel method to align the bottom-up nanowires into the complex patterns or structures (Li et al, 2007). Recently, a "pick and place" method is most widely used for the integrations of nanowires. However, these processes lack control of in precision and repeatability, and easily induce contamination in the wires (Conley et al, 2005). It is expected to selectively grow nanowires directly onto the desired areas of the substrate and *in situ* fabricate the nanowire devices (Li et al, 2004c). In the vapor-liquid-solid chemical vapor deposition (CVD) process, gold catalysts initiate and guide the growth of nanowires (He et al, 2006; Kamins et al, 2004; Wu et al, 2002; Kodambaka et al, 2007). Hence, precise control the location of nanowires relies on the capability to control the location of Au clusters (Lombardi et al, 2006; Fuhrmann et al, 2005). In this book, we also demonstrate the well position-controllable Ge nanowires growth on SiO_2 substrate by combining top-down and bottom-up methods.

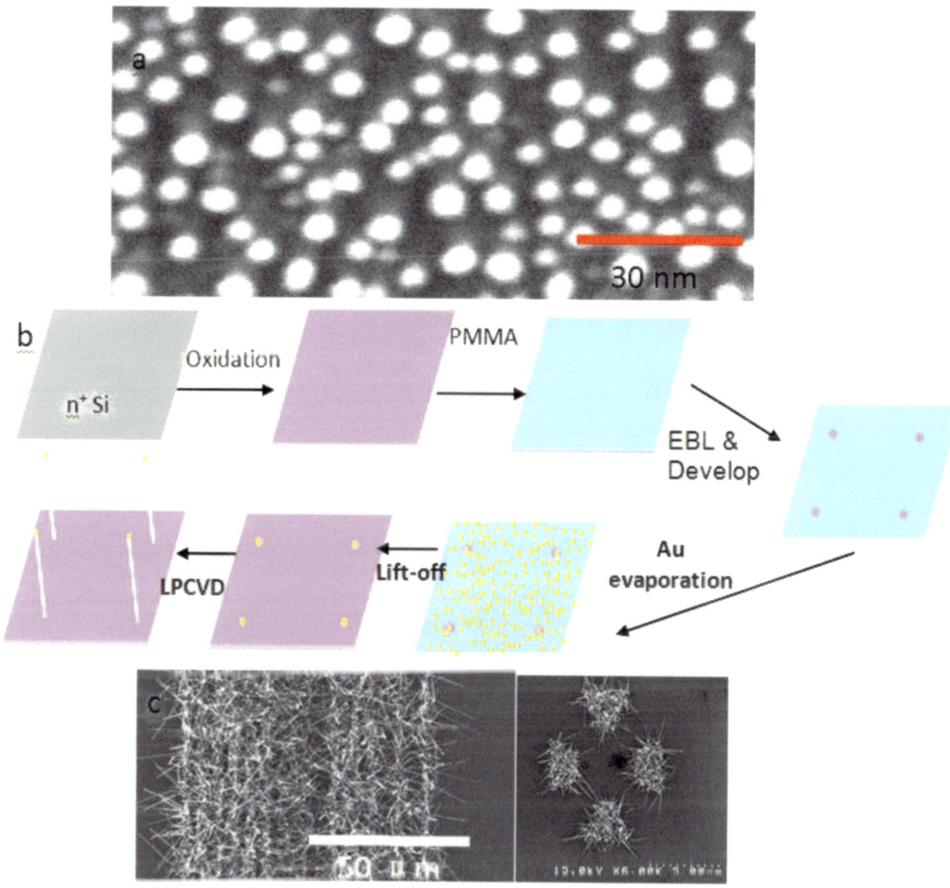

Fig. 13. (a), A SEM image of Au catalysts evaporated on SiO_2 substrate. The Au dots have a diameter of around 3~8 nm; (b), The schematic of process for the patterning Au catalysts and the growth of Ge nanowires; (c), SEM images of Ge nanowires patterns grown on the Au catalysts patterns. The wires are selectively grown on the areas with Au catalysts. (Li et al, 2009b). © Japan Society of Applied Physics.

Due to the stronger bond between the condensing Au adatoms than to the substrate (Venables, 1986), the Au atoms agglomerate to form the high-density islands with the diameter of 3~8 nm as shown in Fig. 13(a). The well-ordered Au catalysts were prepared by using a JEOL JBX-5FE electron-beam lithography (EBL) system with a beam voltage of 50 kV and a beam diameter of 8 nm and a lift-off method.

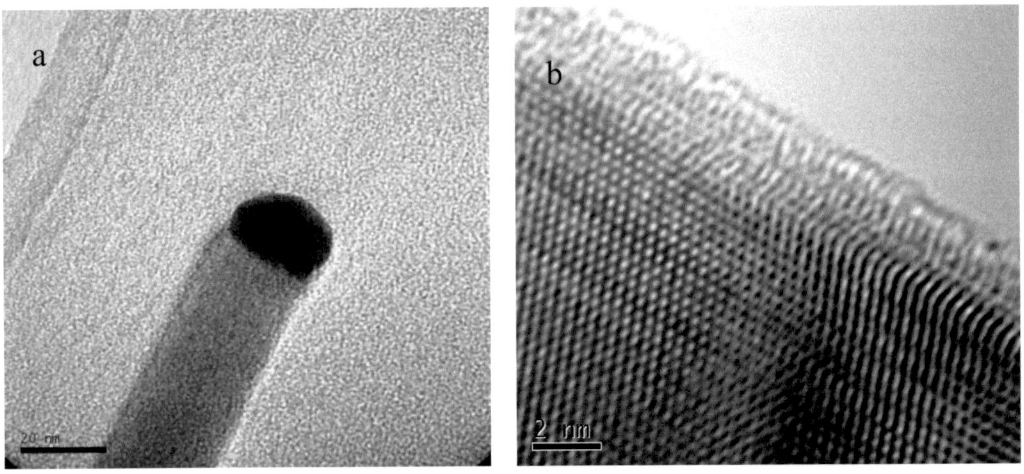

Fig. 14. (a), The TEM of Ge nanowires grown at 300 °C. The dark part on the tip of the nanowire is the Au catalyst; (b), the High-resolution TEM image of Ge nanowires showing the single-crystal structure; (Li et al, 2009b). © Japan Society of Applied Physics.

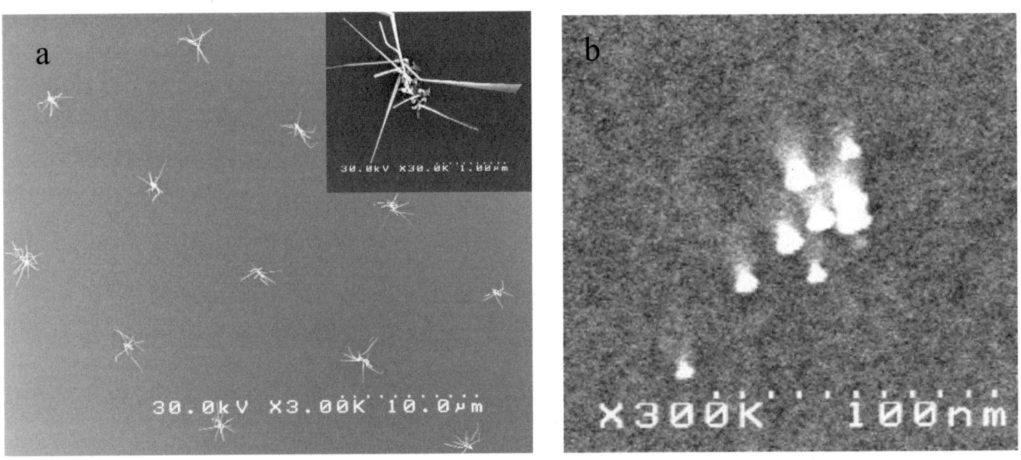

Fig. 15. (a), the SEM images of Ge nanowires array with a space distance of 10 µm; The inset is the enlarged image of one pattern of Ge nanowires. The scale bar in the inset is 1 µm. And (b) shows one of the Au patterns containing several Au dots. The scale bar in the inset is 100 nm. (Li et al, 2009b). © Japan Society of Applied Physics.

The schematic of the process for patterning the Au catalysts and the growth of Ge nanowires is shown in Fig. 13(b). After the n^+ Si(111) wafer was oxidized at 1100 ^oC for 1 hour, the positive-type diluted (50%) polymethyl methacrylate (PMMA) EB resist with a thickness of 40 nm was coated at 8000 rpm and was prebaked at 170 ^oC for 30 minutes. Then, the sample was exposed using the EBL under a dose condition of 400 $\mu C/cm^2$ and were developed by xylene for 60 seconds followed by a rinse with iso-propanol for 30 seconds. After the evaporation of Au particles and lift-off, Au pattern was formed. Finally, Ge nanowires were grown on the Au pattern by LPCVD.

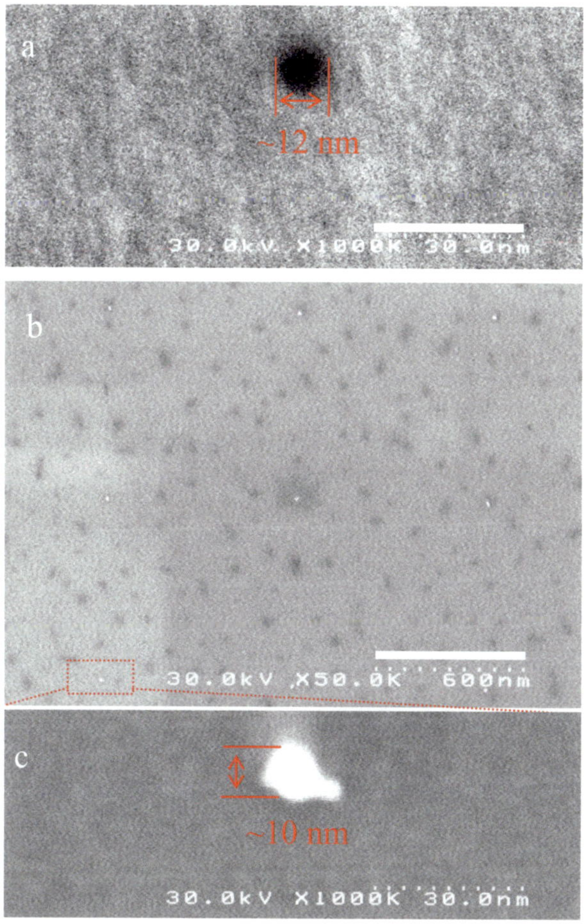

Fig. 16. (a), SEM image of one hole formed by EBL. The black hole area with a diameter of 12 nm is the SiO_2 substrate and the gray area is PMMA layer; (b), SEM image of the well-ordered Au dots array with a space of 700 nm. The white dots in the image are Au catalysts; and (c), the enlarged image of one of the Au dots in fig. 16(b). (Li et al, 2009b). © Japan Society of Applied Physics.

Fig. 13(c) shows a SEM image of high density Ge nanowire strips grown on patterned Au catalysts. Ge nanowires with a diameter of 5~20 nm can be grown selectively on the area with Au catalysts. A TEM of Ge nanowires is shown in Fig. 14(a). Almost all of the nanowires are defect free over the whole length that could be observed. The dark parts on the tips of nanowires in Fig. 14(a) are Au catalysts. The high-resolution TEM image in Fig. 14(b) reveal the high-quality single-crystalline Ge nanowires with a lattice constant of 0.565 nm, which is in excellent agreement with the diamond crystal structure known for Ge. The X-ray diffraction result (dotted line in Fig. 10) also indicates that the crystal structure of grown Ge nanowires is cubic diamond structure according to the typical six peaks. And two peaks related to the Au FCC structure were also found.

By reducing the size of each Au pattern to contain only a few Au particles (Fig. 15(b)), the Ge nanowire array with several nanowires in each area was obtained (Fig. 15). The inset of

Fig. 15 shows the enlarged image of one of the patterns. Around 10 nanowires were grown in each area.
In order to realize the alignment of single Ge nanowire, the Au catalyst array with only one Au dot in each area is needed. One way to settle this problem is to reduce the size of pattern. In the lift-off process, the thickness of the resist plays an important role in controlling the size and shape of the design. The thinner the resist is, the easier it is to control the size of pattern. In the case of Au film with only a thickness of about 0.5 nm, very thin resist can be applied. By reducing the PMMA resist down to 40 nm and optimizing the relationship between the electron dose and develop time in EBL process, a very small hole with the diameter of around 12 nm can be obtained as shown in Fig. 16(a), which is almost the same size as the beam spot of EBL. In Fig. 16(a), the black hole is the SiO_2 substrate and the gray area is PMMA. By choosing a suitable density of the Au particles to make sure that one Au dot could be deposited in this kind of hole, the well-ordered single Au dot array with a space of 700 nm was prepared as shown in Fig. 16(b). According to the enlarged image of one pattern as shown in Fig. 16(c) we can find each Au dot has a diameter of around 10 nm.

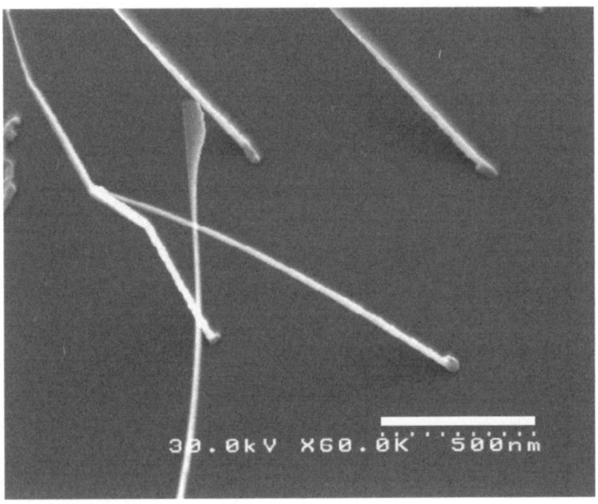

Fig. 17. A SEM image of single Ge nanowire array grown on the well-ordered Au pattern. (Li et al, 2009b). © Japan Society of Applied Physics.

On such well-ordered Au catalyst array, the single Ge nanowire arrays with a space of 700 nm were grown. The SEM image in Fig.17 shows that only one Ge nanowire is grown on each Au dot. The Au catalysts initiate and guide the growth of Ge nanowires. And the wires are located precisely where the Au catalysts sit, which may offer the possibility of *in situ* fabrication of large-scale nanowire devices.

6. Conclusion

In conclusion, the radial growth of Ge nanowire, the influence of surface condition on the growth of Ge nanowire, the growth of Ge-Si nanowire heterostructures and the patterned Ge nanowire were discussed in this chapter. The VSS growth mechanism dominates the nanowire growth in the radial direction and contributes to the increase of the diameter of nanowires. Due to the Au migration on the surface, the Au-Ge alloys on the nanowire surface are in solid state at the growth temperature. And this VSS mechanism contributes to the radial growth. Once the Au catalysts on the tip are consumed, the nanowires with the rough surface will be grown because of the selective-area VSS radial growth.

High-density Ge nanowires can be easily grown on SiO_2-terminated Si (100) substrate. However, the SiO_2 overlayer formed on the surface of Au catalysts evaporated on H-terminated Si substrate prevents the growth of Ge nanowires. After removed this SiO_2 overlayer by HF solution, high-density and well-ordered Ge nanowire can be obtained. Nanowires can be grown orderly along the [101], [011], [-101] and [0-11] directions.

The Au particles on the nanowire surfaces due to the surface migration play an important role in the controlling the growth of nanowire heterosturcture. Different topographies of heterostructure were observed. On the Ge nanowire stems, many Si nanowire branches were observed. Their distribution agrees well with that of Au particles on the Ge nanowire surface. However, the high-density Ge nanodots instead of branches prefer to grow on the Si nanowire stems.

Ge nanowires can be grown selectively on the area with Au catalysts. Single Ge nanowire array with a space distance of 700 nm was grown by LPCVD method on the patterned Au catalyst substrate.

7. Acknowledgement

This work is partially supported by SORST-JST (Japan Science and Technology) and Japan society for the promotion of science (JSPS).

8. References

Adhikari, H.; Marshall, A.F.; Chidsey, C. E. D.; McIntyre,P. C. (2006). Germanium nanowire epitaxy: Shape and orientation control. *Nano lett.* Vol. 6, pp.318-323.

Allen, J. E.; Hemesath, E. R.; Perea, D. E.; Lensch-falk, J. L.; Li, Z.Y.; Lin, F.; Gass, M. H.; Wang, P.; Bleloch, A. L.; Palmer, R. E.; Lauhon, L. J.(2008). High-resolution detection of Au catalyst atoms in Si nanowires. *Nature Nanotechnol.*,Vol.3, pp.168-173.

Ben-Ishai, M.; Patolsky, F. (2010). A Route to High-Quality Crystalline Coaxial Core/ Multishell Ge@Si(GeSi)n and Si@(GeSi)n Nanowire Heterostructures, Adv. Mater. Vol.22,902-906.

Bierman, M. J.; Jin, S.; (2009). Potential Applications of Hierarchical Branching Nanowires in Solar Energy Conversion, *Energy Environ Sci.*, Vol. 2, pp.1050 1059.

Buffat, P.; Borel, J-P. (1976). Size effect on the melting temperature of gold particles. *Phys. Rev. A* Vol.13, pp. 2287–2298

Chan, C. K.; Peng, H.; Liu, G.; McIlwrath, K.; Zhang, X. F.; Huggins, R. A.; Cui, Y. (2008a). High-performance lithium battery anodes using Si nanowires. *Nature Nanotech.*, Vol. 3, pp. 31-35.

Chan, C. K.; Zhang, X. F.; Cui, Y. (2008b) High Capacity Li Ion Battery Anodes Using Ge Nanowires. *Nano Lett.*, Vol. 8, pp.307-309.

Clark, T. E.; Nimmatoori, P.; Lew, K.; Pan, L.; Redwing, J. M., Dickey, E. C. (2008). Diameter Dependent Growth Rate and Interfacial Abruptness in Vapor-Liquid-Solid Si/Si1-xGex Heterostructure Nanowires. *Nano Lett.* Vol.8, pp.1246-1252.

Conley, J. F.; Stecker, J. L.; Ono, Y. (2005). Directed integration of ZnO nanobridge devices on a Si substrate *Appl. Phys. Lett.*, Vol.87, p.223114.

Cui, Y.; Wei, Q.; Park, H.; Lieber C. M., (2001). Nanowire Nanosensors for Highly Sensitive and Selective Detection of Biological and Chemical Species.*Science*, Vol.293, pp.1289-1292.

Dick, K. A.; Deppert, K.; Karlsson, L. S.; Larsson, M. W.; Seifert, W.; Wallenberg, L. R.; Samuelson, L. (2007b).Directed growth of branched nanowire structures.*MRS Bulletin*, Vol.32, pp.127-133

Dick, K. A.; Kodambaka, S.; Reuter, M. C. ; Deppert, K.; Samuelson, L.; Seifert, W.; Wallenberg, L. R.; Ross, F. M. (2007a). The Morphology of Axial and Branched Nanowire Heterostructures. *Nano Lett.* Vol.7, pp. 1817-1822.

Doerk, G. S.; Ferralis, N.; Carraro, C.; Maboudian, R. (2008). Growth of Branching Si Nanowires Seeded by Au-Si Surface Migration. *J. Mater. Chem.* Vol.18, pp.5376-5381.

Ertekin, E.; Greaney, P. A.; Chrzan, D. C.; Sands, T. D. (2005). Equilibrium limits of coherency in strained nanowire heterostructures. J. *Appl. Phys.* Vol. 97, p.114325.

Fan, H. J.; Werner, P.; Zacharias, M. (2006).Semiconductor Nanowires: From Self-Organization to Patterned Growth. *Small*, Vol. 2, pp.700-717.

Fuhrmann, B.; Leipner, H. S.; Hoche, H. R.; Schubert, L.; Werner, P.; Gosele, U.(2005). Ordered Arrays of Si Nanowires Produced by Nanosphere Lithography and Molecular Beam Epitaxy. *Nano Lett.*, Vol.5, pp.2524-27.

Gosele U. (2006). How clean is too clean? *Nature*, Vol. 440, pp.34-35.

Greytak,A. B.; Lauhon,L. J.; Gudiksen M. S.; and Lieber,C. M. (2004) Growth and transport properties of complementary germanium nanowire field-effect transistors. *Appl. Phys. Lett.*,Vol. 84, pp.4176-78.

Gudiksen, M. S.; Lauhon, L. J.; Wang, J.; Smith, D. C.; Lieber, C. M. (2002). Growth of nanowire superlattice structures for nanoscale photonics and electronics. *Nature* Vol. 415, pp. 617-620.

Hannon, J. B.; Kodambaka1, S.; Ross, F. M.; Tromp, R. M.(2006). The influence of the surface migration of gold on the growth of Si nanowires. *Nature*, Vol. 440, pp. 69-71.

He, J. H.; Hsu, J. H.; Wang, C. W.; Lin, H. N.; Chen, L. J.; Zhong, L.; Wang, Z. L.2006. Pattern and Feature Designed Growth of ZnO Nanowire Arrays for Vertical Devices. *J. Phys. Chem. B*, Vol.110, pp.50-53.

Hiraki, A.; E. Lugujjo, Nicolet M.-A.; Mayer, J. W. (1971). Low-temperature migration of Si through metal films importance of Si-metal interface.*Physica Status Solidi a*, Vol.7, pp.401-406.

Hiraki, A. (1984).Low temperature reactions at Si/metal interfaces; What is going on at the interfaces? *Surf. Sci. Rep.*, Vol.3, pp. 357-412.

Holmes, J. D.; Johnston, K. P.; Doty, R. C.; Korgel, B. A. (2000). Control of Thickness and Orientation of Solution-Grown Si Nanowires. *Science*, Vol. 287, pp.1471-1473.

Hu, Y. J.; Churchill, H. O. H.; Reilly, D. J.; Xiang, J.; Lieber, C. M.; Marcus, C. M. (2007). A Ge/Si heterostructure nanowire-based double quantum dot with integrated charge sensor*Nature Nanotechnol.*, Vol.2, pp. 622-625 .

Hu, Y.; Xiang, J. ; Liang, G.; Yan, H.; Lieber, C. M. (2008). Sub-100 Nanometer Channel Length Ge/Si Nanowire Transistors with Potential for 2 THz Switching Speed. *Nano Lett.* Vol.8, pp. 925-930.

Jagannathana, H.; Nishi, Y.; Reuter,M.; Copel, M.; Tutuc, E.; Guhab, S.; Pezzia, R. P. (2006). Effect of oxide overlayer formation on the growth of gold catalyzed epitaxial Si nanowire. *Appl. Phys. Lett.*, Vol. 88, p.03113.

Kastnery, G.; Gosele, U. (2004). Stress and dislocations at cross-sectional heterojunctions in a cylindrical nanowire. *[Quick Edit]Philos. Mag.* Vol.84, pp.3803-3824.

Kamins, T. I.; Li, X.; Williams, R. S.; Liu, X. (2004). Growth and Structure of Chemically Vapor Deposited Ge Nanowires on Si Substrates. *Nano Lett.*, Vol. 4, 503-506.

Kodambaka, S.; Tersoff, J.; Reuter, M. C.; Ross, F. M.(2007). Germanium Nanowire Growth Below the Eutectic Temperature.*Science*, Vol. 316, pp.729-732.

Lauhon, L. J.; Gudiksen, M. S.; Wang D.; and Lieber, C. M. (2002). Epitaxial core–shell and core– multishell nanowire heterostructures. *Nature*, Vol. 420, pp. 57-61.

Lay G. Le (1983).Physics and electronics of the noble-metal/elemental-semiconductor interface formation: A status report. *Surf. Sci.* , Vol.132, pp.169-204.

Li, C. B.; Cheng, B.W.; Mao, R.W.; Zuo, Y.H.; Shi , W.H.; Huang, C. J.; Luo, L.P.; Yu, J.Z.; Wang, Q.M. (2004b). Novel room temperature photoluminescence of Ge/Si islands in multilayer structure grown on Si-on-insulator substrate, *Thin Solid Films*, Vol.467, pp.197-200.

Li, C. B.; Mao, R. W.; Zuo, Y. H.; Zhao, L.; Shi, W. H.; Luo, L. P.; Cheng, B. W.; Yu, J. Z.; Wang, Q. M. (2004a). 1.55 mm Ge islands resonant-cavity-enhanced detector with high-reflectivity bottom mirror, *Appl. Phys. Lett.* Vol.85, pp. 2697-2699.

Li, C. B. ; Usami, K.; Mizuta, H.; Oda. S. (2009a). Vapor-solid-solid radial growth of Ge nanowires. *J. Appl. Phys.* Vol.106, P.046102

Li,C. B.; Usami, K.; Muraki, T.; Mizuta, H.; and Oda S., (2008). The impacts of surface conditions on the vapor-liquid-solid growth of germanium nanowires on Si (100) substrate. *Appl. Phys. Lett.* Vol. 93, p. 041917.

Li, C. B.; Usami, K.; Mizuta, H.; Oda, S. (2011).Growth of Ge-Si nanowires heterostructure via chemical vapour deposition, *Thin solid films*, doi:10.1016/j.tsf.2011.02.005.

Li, C. B.; Usami, K.; Yamahata, G.; Tsuchiya, Y.; Mizuta, H.; Oda, S. (2009b).Position-controllable Ge Nanowires growth on Patterned Au Catalyst Substrate. *Applied Physics Express*, Vol.2, p.015004

Li, J.Y.; Lu, C. G.; Maynor, B.; Huang, S. M.; Liu, J. (2004c). Controlled growth of long GaN nanowires from catalyst patterns fabricated by "dip-pen" nanolithographic techniques, *Chem. Mater.*, Vol. 16, pp.1633-1636.

Li, S.; Liu, N. Y.; Chan-Park, M. B.; Yan, Y. H. and Zhang Q. (2007). Aligned single-walled carbon nanotube patterns with nanoscale width, micron-scale length and controllable pitch. *Nanotechnology*, Vol.18, p.455302.

Liang, G. C.; Xiang, J.; Kharche, N.; Klimeck, G.; Charles, M.; Lieber, C. M.; Lundstrom, M.(2007). Performance Analysis of a Ge/Si Core/Shell Nanowire Field-Effect Transistor.*Nano Lett.*, Vol. 7, pp. 642-646.

Lombardi, I.; Hochbaum, A. I.; Yang, P. D.; Carraro, C.; Maboudian, R. (2006), Synthesis of high density, size-controlled Si nanowire arrays via porous anodic alumina mask, *Chem. Mater.*, Vol.18, 988-991.

Lu, W.; Xiang, J.; Timko, B. P.; Wu, Y.; Lieber, C. M. (2005). One-dimensional hole gas in germanium/Si nanowire heterostructures, Proc. Natl. Acad. Sci. Vol.102, pp.10046.

Sun, X. H.; Calebotta, G.; Yu, B.; Selvaduray G.; Meyyappan, M.;(2007). Synthesis of germanium nanowires-on-insulator (GeNOI) catalyzed by indium or antimony. *J. Vac. Sci. Technol. B,* Vol. 25, 415-420.

Tyryshkin, A. M.; Lyon, S. A.; Schenkel, T.; Bokor, J. ; Chu, J.; Jantsch, W.; Schaffler, F. ; Truitt, J. L.; Coppersmith S. N.; Eriksson, M. A.. (2006). Electron spin coherence in Si,*Physica* E, Vol. 35, 257-263

Venables, J. A., (1986). Nucleation and growth process in thin film formation, *J. Vac. Sci. Technol. B,* Vol.4, pp. 870-873.

Wang, D.; Qian, F.; Yang, C.; Zhong, Z.; Lieber, C. M. (2004). Rational Growth of Branched and Hyperbranched Nanowire Structures. *Nano Lett.* Vol.4, pp. 871-874.

Wu, Y. Y.; Fan, R.; Yang, P. D. (2002). Block-by-Block Growth of Single-Crystalline Si/SiGe Superlattice Nanowires.*Nano Lett.* Vol.2, pp.83-86.

Xiang, J.; Lu, W.; Hu, Y. J.; Wu, Y.; Yan, H.; Lieber, C. M. (2006). Ge/Si nanowire heterostructures as high performance field-effect transistors. *Nature,* Vol.441, pp. 489-493.

Y. Maeda, N. Tsukamoto, Y. Yazawa, Y. Kanemitsu and Y. Masumoto, (1991). Visible photoluminescence of Ge microcrystals embedded in SiO_2 glassy matrices. *Appl. Phys. Lett.*, Vol. 59, pp. 3168-3170

Niobates Nanowires: Synthesis, Characterization and Applications

Rachel Grange[1], Fabrizia Dutto[2] and Aleksandra Radenovic[2]
[1]Friedrich Schiller University Jena
[2]Ecole Polytechnique Fédérale de Lausanne
[1]Germany
[2]Switzerland

1. Introduction

Perovskite oxides such as alkaline niobates crystal possess many interesting properties including piezoelectricity, pyroelectricity, electro-optic and nonlinear optical response (Bhalla et al., 2000). The most common alkaline niobates material is lithium niobate ($LiNbO_3$). Since the discovery of $LiNbO_3$ ferroelectricity (Matthias & Remeika, 1949), its properties are widely exploited by electronic devices particularly in telecom applications (Wooten et al., 2000). Those devices are made from bulk or thin films material and serves as sensors, actuators, detectors or filters. Potassium niobate ($KNbO_3$) is most known for its large nonlinear coefficients ideal for wavelength conversion like second-harmonic generation (SHG), sum frequency mixing, as material in optical parametric oscillator, or lead-free piezoceramics (Saito et al., 2004). Sodium niobate ($NaNbO_3$), less studied than $LiNbO_3$ and $KNbO_3$, also belongs to the alkaline niobates. Generally associated with potassium, $NaNbO_3$ is a very promising lead-free piezoelectric ceramics (Guo et al., 2004).

Besides bulk and thin films structures, zero- (0D) and one-dimensional (1D) alkaline niobates nanostructures were synthesized recently to combine the dimensional confinement with the other known properties of perovskite materials. Different synthesis routes have been explored to obtain 0D nanoparticles or nanoflakes from alkaline niobates such as mechano-chemical milling (Kong et al., 2008; Schwesyg et al., 2007), nonaqueous route (Niederberger et al., 2004), sol-gel method (L. H. Wang et al., 2007) or hydrothermal route (An et al., 2002). Almost simultaneously, anisotropic alkaline niobates 1D structure were synthesized with various methods such as template assisted pyrolysis resulting in regular arrays of tubes (Zhao et al., 2005), solution-phase synthesis resulting in rod-like structures (Wood et al., 2008), or hydrothermal route giving free-standing nanowires with high aspect ratio (Magrez et al., 2006).

Up to now, the nanomaterials properties have been well characterized using standard materials sciences methods like X-ray diffraction (XRD), scanning electron (SEM) or transmission electron (TEM) microscopy. However, nonlinear optical or electro-optic properties have been rarely studied. Moreover, few applications have used these types of nanowires while combining the various physical properties of perovskite alkaline materials and the anisotropic shape at the nanoscale level. Nanometric SHG light probe manipulated by optical tweezers and capable of guiding light has been already demonstrated (Nakayama

et al., 2007), as well as localized SHG light source in optofluidics environment (Grange et al., 2009).
In this chapter, we focus on the synthesis, optical characterization and applications of Li-, Na-, $KNbO_3$ nanowires. Some crystal and second-order optical properties of the investigated alkaline niobates are summarized in Table 1. We will describe hydrothermal and molten salt synthesis. We will measure some optical and physical properties of these nanowires. Then we propose the use of plasmonics gold nanoshells to enhance the SHG signal and for best biocompatibility with biological applications. Finally we show interesting applications using optical tweezers and microfluidics chips by combining dielectrophoresis and SHG.

Material	Crystal system	d_{eff} (pm/V)	References
$LiNbO_3$	Trigonal	[2-34.4]	(Boyd, 2008; Fluck & Gunter, 2000; Shoji et al., 2002; Sutherland, 2003; Weber, 2003)
$KNbO_3$	Orthorhombic	[10.8-19.6]	(Fluck & Gunter, 2000; Shoji et al., 2002; Weber, 2003)
$NaNbO_3$	Orthorhombic	[0.8-4.5]	(Johnston et al., 2010; Ke et al., 2008)

Table 1. Typical crystal system at room temperature and its corresponding second-order susceptibility tensor elements (d_{eff}) range.

2. Synthesis methods

Several research groups developed different chemical synthetic approaches to fabricate crystalline alkaline niobates nanowires such as exploiting sol-gel route (Pribosic et al., 2005), hydrothermal route (An et al., 2002; Magrez et al., 2006; H. F. Shi et al., 2009; G. Z. Wang, Selbach et al., 2009; G. Z. Wang, Yu et al., 2009; Wu et al., 2010) and molten salt synthesis (MSS) (Li et al., 2009; Santulli et al., 2010).
Among the three studied alkaline niobate materials, $LiNbO_3$ nanowires synthesis is the most challenging one. So far only two groups reported successful synthesis of free-standing $LiNbO_3$ (Grange et al., 2009; Santulli et al., 2010). In contrast multiple synthesis routes can be found for other alkaline materials as potassium niobate $KNbO_3$ (Li et al., 2009; Magrez et al., 2006; Nakayama et al., 2007; G. Z. Wang, Selbach et al., 2009; G. Z. Wang, Yu et al., 2009) and sodium niobate $NaNbO_3$ (Li et al., 2009; H. F. Shi et al., 2009; Wu et al., 2010) nanowires. We have investigated two chemical synthesis methods for the nanowires fabrication: hydrothermal synthesis and molten salt synthesis (MSS).
Hydrothermal synthesis is a technique used to crystallize substances under moderate temperatures (200-250°C) and high pressures. Thanks to the use of an autoclave, a thick-walled steel cylinder with an hermetic seal, the wanted crystalline materials can be obtained in one step (Fig. 1 a). The large amount of material that can be fabricated, the easiness (one step synthesis) and the speed of the synthesis make this approach very convenient. To optimize the synthesis parameters one can adjust the temperature, the time, the pressure (by an external pressure or the degree of the autoclave filling), the caustic soda concentration, the solid-liquid ratio and additives to control the properties of the end product. Thus, hydrothermal synthesis is promising because many operation parameters can be modulated to control the particle size and morphology. Due to its simplicity the hydrothermal technique has been widely studied and employed in inorganic synthesis for many years.

Molten salt synthesis (MSS) is used to prepare complex oxides from their constituent oxide. It is a multi steps synthesis and it is way longer than hydrothermal synthesis approach but more adaptable to other alkaline materials than $KNbO_3$ as shown in the literature (Li et al., 2009; Santulli et al., 2010). Oxides corresponding to a perovskite compound are mixed with one or two kinds of salt sand then heated at a temperature above the melting point of the salt to form a flux of the salt composition (Fig. 1 b). At this temperature, the oxides are rearranged and then diffused rapidly in a liquid state of the salt. With further heating, particles of the perovskite phase are formed through the nucleation and growth processes.
A huge advantage of chemical synthesis in respect to chemical vapor deposition (CVD) or lithographic fabrication of nanowires is the ability to synthesize free-standing nanowires. Thus no further step is needed to isolate or detached the nanowires, because no substrate is involved in the synthesis.

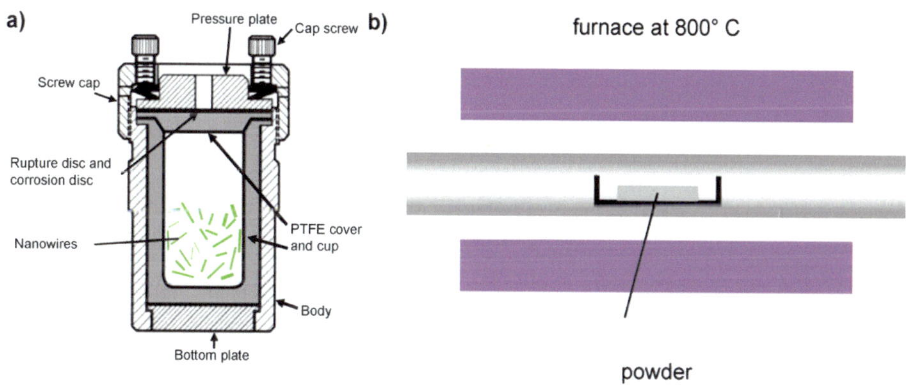

Fig. 1. a) Schematic diagram of a typical laboratory autoclave from Parr (Acid digestion bomb, 125ml, Model 4748). b) Schematic of the experimental setup for molten salt synthesis.

Fig. 2 shows typical results of $KNbO_3$ nanowires hydrothermally synthesized following (Magrez et al., 2006) recipe, $NaNbO_3$ nanowires hydrothermal synthesized following (Zhu et al., 2006) recipe and $LiNbO_3$ nanowires molten salt synthesized following (Santulli et al., 2010) recipe. Due to the low solubility of lithium hydroxide, molten salt synthesis is preferred to fabricate $LiNbO_3$ nanowires even if it is a multistep synthesis.

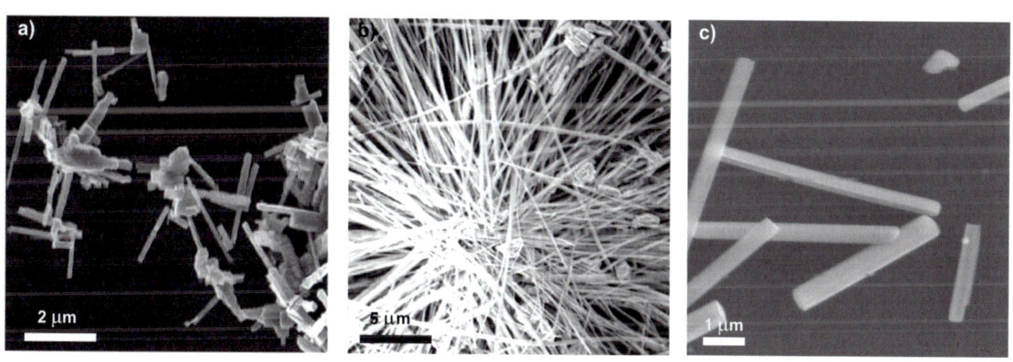

Fig. 2. SEM images of a) $KNbO_3$ nanowires with an aspect ratio up to 25. b) $NaNbO_3$ nanowires with an aspect ratio up to 50. c) $LiNbO_3$ nanowires with an aspect ratio up to 10.

3. Material properties and characterization

To investigate the structure of produced nanowires XRD characterization was employed, confirming the $KNbO_3$ orthorhombic structure in a percentage of 100%, $NaNbO_3$ orthorhombic phase in a percentage of 15% and $LiNbO_3$ trigonal phase in a percentage of 75% (see Fig. 3 a to c).

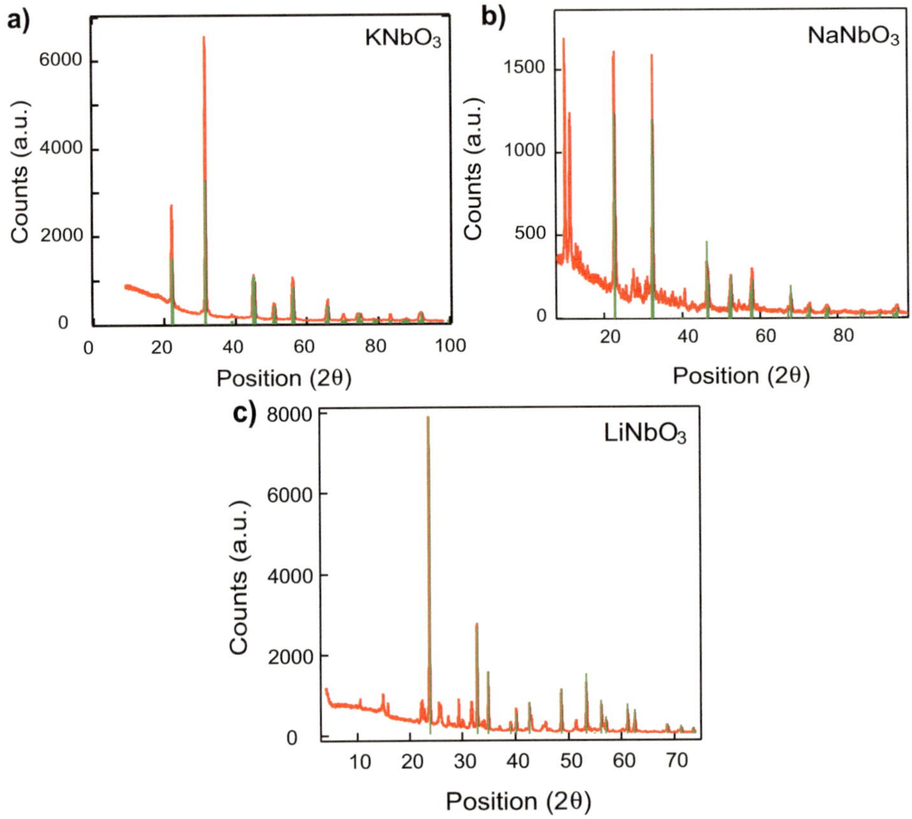

Fig. 3. X-ray diffraction (XRD) measurements (red) together with theoretical peaks (green) expected for desired phases of a), $KNbO_3$ b) $NaNbO_3$, c) $LiNbO_3$.

In the hydrothermal synthesis procedure, to increase the percentage of preferred nanowire crystal structure, a calcination step (annealing at high temperatures such as 550°C) can be performed. This calcination step increases the amount of ordered material phase in respect to other phases present in the sample, without changing the shape of the sample (Ke et al., 2008; H. Shi et al., 2009). For $NaNbO_3$ synthesis, this additional step resulted in the drastic increase in the amount of pure material switching from 15% to 64% of presence of $NaNbO_3$ ordered phase in the sample (Fig. 3 b).

Besides XRD, that will give results over a high number of nanowires, it is possible to perform Raman scattering measurements on single nanowires that allow to distinguish materials and reveal unexpected phenomena (Louis et al.). As a preliminary result, we performed Raman measurements on $KNbO_3$ nanowires and for comparison we show $LiNbO_3$ nanoflakes Raman measurements too (Fig. 4). Typical $LiNbO_3$ traces show peaks between 300 cm^{-1} and 480 cm^{-1}

(Santulli et al., 2010). $KNbO_3$ has one strong peak right below 300 cm^{-1} and nothing else between 300 and the silicon substrate peak (Louis et al.). The spectra are then different enough even between two alkaline niobates to easily distinguish the materials.

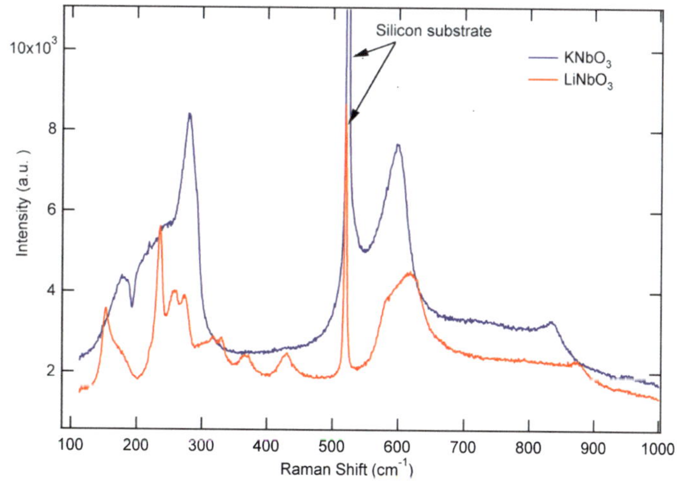

Fig. 4. Raman spectrum of $KNbO_3$ nanowires (blue) at 80K showing the typical shape known for this material and of $LiNbO_3$ nanoflakes (red).

4. Nonlinear optical characterization

The perovskite nanowires with their non centrosymmetric crystal structures exhibit second-order optical effects which open up a wide range of applications even at the nanoscale level. Indeed, this nonlinear effect scales with the square of the electric field, but it is a volume effect which is measureable down to a single nanoparticle level even in far field microscopy (Hsieh et al., 2010), contrary to weak surface SHG effect of centrosymmetric materials (Dadap, 2008; Dadap et al., 1999). In the nonlinear regime, the optical response is expressed by the polarization P as a power series in the electric field E as

$$\vec{P} = \varepsilon_0 \chi_1 \vec{E} + \varepsilon_0 \chi_2 \vec{E}^2 + \varepsilon_0 \chi_3 \vec{E}^3 + ... \quad (1)$$

where ε_0 is the permittivity of free space and χ_i is the i th-order nonlinear optical susceptibility tensor. Each χ_i represents a different optical effect that can be summarized as follows for a physical understanding of Eq. (1). χ_1, the linear susceptibility, is related to absorption and reflection of light. χ_2 encompasses sum and difference frequency generation such as SHG. χ_3 describes multiphoton absorption, third harmonic generation or coherent anti-Stokes Raman scattering.

Our applications are related to SHG as illustrated in Fig.5 a and b. When a nanocrystal of non centrosymmetric structure is optically excited at a fundamental frequency, it emits the optical signal at the exact doubled frequency. Only materials with crystalline structures lacking a centre of symmetry are capable of efficient SHG. Not only SHG will be scattered but the fundamental frequency too, thus efficient filters are needed to cut the fundamental.

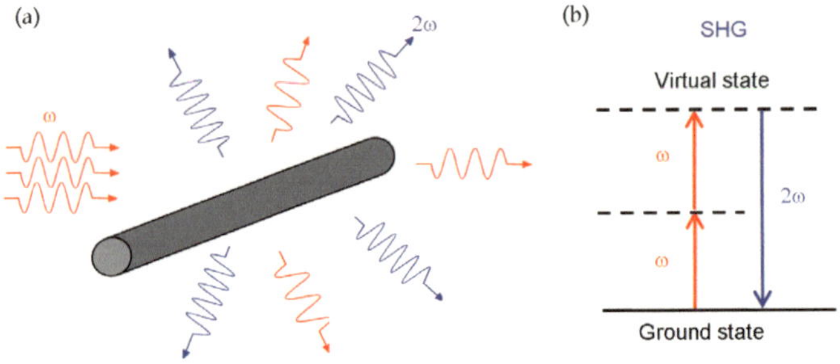

Fig. 5. (a) Schematic diagram of the SHG mechanism. (b) Energy diagram of the physical SHG mechanism.

The setup for the optical characterization of the SHG signal is shown in Fig. 6. A near infrared laser light (100 fs Ti:Sa oscillator) is focused onto a sample by lens L1 and the objective (OBJ) collects the signal imaged through a 4f configuration with lens L2. Filters are used to cut the fundamental frequency and detect only a narrow band around the SHG frequency onto an electron multiplying charges coupled device (EMCCD).

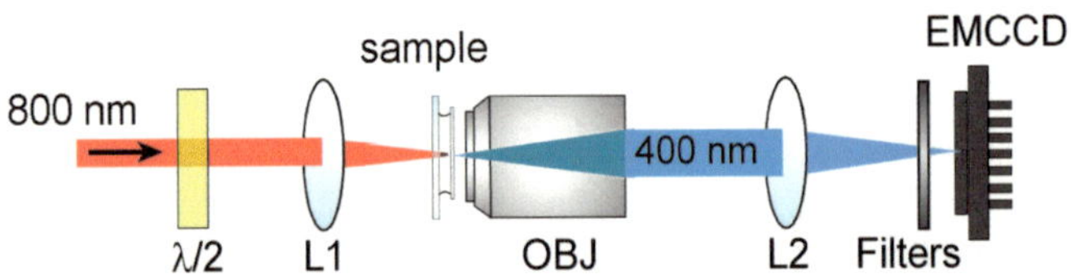

Fig. 6. Setup for measuring SHG from nanowires dried on a microscope glass slide. L1, L2, lenses; λ/2 half wave plate; OBJ, objective.

Typical SHG measurements at different incident polarization angle are displayed in Fig.7. When the polarization is parallel to the nanowire the signal is the strongest, which is expected for a nanowires with an optical axis along the nanowire. Full polar SHG characterization is then possible and a fit of the experimental data can confirm the crystal orientation of a nanowire, which is not a priori known for chemically bottom-up synthesized nanowires (Grange et al., 2009).

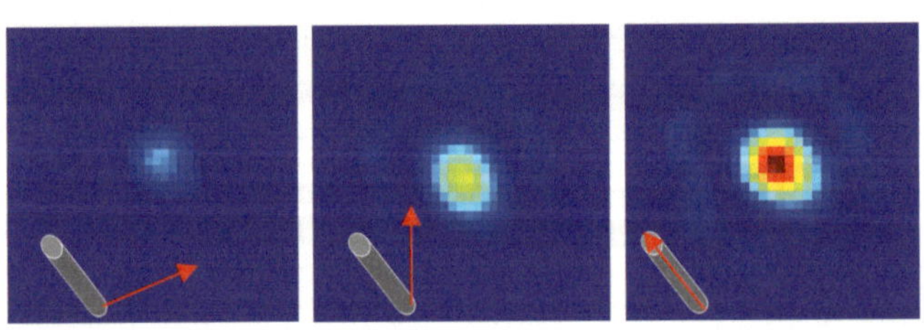

Fig. 7. EMCCD images of a single niobate nanowire under different incident light polarization. The insets sketched the wire position and the incident light polarization (red arrow). Images size 6x6 μm.

In addition to conventional measurements of SHG from niobate nanowires we have carried out SHG characterization by using Optical Tweezers (OT) (see Fig.8). An OT uses a focused laser beam to provide an attractive or repulsive force (typically on the order of pN), depending on the refractive index mismatch, to physically hold and move microscopic dielectric objects. Optical tweezers are routinely used tools in both physical and life sciences for manipulating objects from micron to the atomic scale and as force transducers in the pN range (Neuman & Block, 2004).

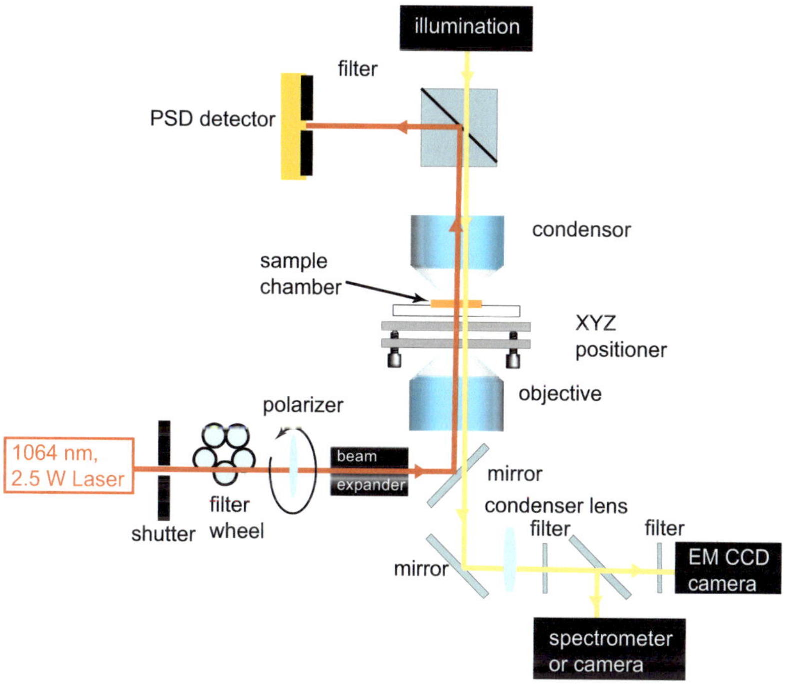

Fig. 8. Single beam optical tweezers setup for SHG characterization of the trapped nanowires. A Nd:YVO_4 CW 1064 laser is used simultaneously for nanowire trapping and pumping. As in the conventional SHG setup, the signal is recorded using an EMCCD camera. After passing through the condenser, the output IR beam passes through an aperture and it is detected by a position sensitive detector (PSD) positioned at the rear focal plane of the condenser. Any transverse motion of the trapped nanowire will cause the center of the output beam to shift sideways on the PSD.

The ability to control and monitor the position of a mesoscopic object with nanometric precision is important for the rapid progress of nanoscience and it is perfectly suited for the study of biological phenomena. Besides its application in biology, OT are also an appealing tool for semiconductor nanowire integration owing to their ability to act in situ in closed aqueous chambers, their potential applicability to a broad range of dielectric materials, their spatial positioning accuracy (<1 nm), and the degree to which their intensity, wavelength and polarization can be controlled using tuneable lasers. All trapping experiments and subsequent SHG characterization were carried out on a home-built single beam optical tweezers system.

As in the conventional setup, we can probe SHG from individual niobate nanowire under different incident light polarization. The SHG signal under different incident polarization is much less sensitive in this orientation due to the tweezing geometry (Fig. 9). However, it may still slightly differ from wire to wire due to the variability of crystalline phases of the bottom-up synthesized nanowires.

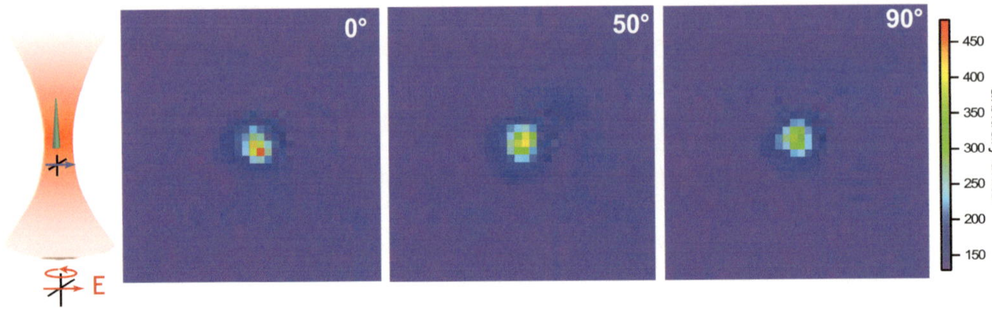

Fig. 9. EMCCD images of trapped single niobate nanowire under different incident light polarization. Images size 10x10μm.

3. Plasmonics nanoshells for enhanced SHG

We showed that non centrosymmetric nanowires exhibit SHG signal. However, nonlinear optical processes such as harmonic generation are generally inefficient at very small scale (the intensity goes down with the square of the particle volume). By developing nonlinear optical plasmonics core shell cavities, it was possible to strongly enhanced the SHG response of $BaTiO_3$ nanoparticles (Pu et al., 2010). Similarly, $KNbO_3$ nanowires are covered by a thin layer of gold to reach a near infrared plasmonics resonance and enhance the SHG signal for optimized used as imaging probes or localized light source.

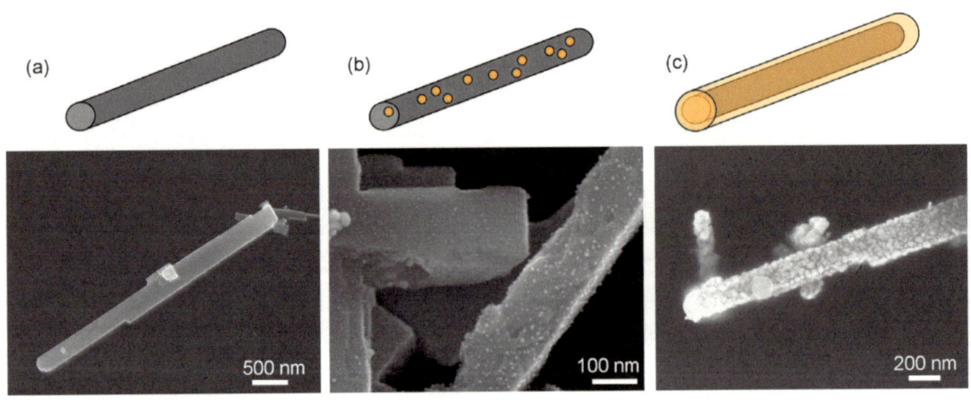

Fig. 10. Synthesis steps of $KNbO_3$/Au core-shell nanowires with their corresponding SEM images. (a) bare $KNbO_3$ nanowires (b) surface seeding with 2-3 nm colloidal gold particles (c) growth of a gold complete shell around the wires.

The gold coating process for $KNbO_3$/Au core-shell nanowires involved three main steps illustrated in Fig. 10. First of all, primary amines are coated on the surface of the $KNbO_3$ nanowires (Hsieh et al., 2009). The amino complex (Aminopropyltriethoxysilane) presents NH_2 complex at the surface of the wires. Then, 2-3 nm gold particles are adsorbed onto the surface of $KNbO_3$ wire thanks to the silane-amine functionalization (Au^{3+} cations are attracted by the primary amine NH_2 which become secondary amine) (Fig. 10 b). Finally, the Au shell is grown all around the particle from the seeded particles by a reaction of reduction of gold by hydroxylamine (Fig. 10 c).

The gold reduction process to coat the wire surface still needs some improvement for a better and more uniform coverage of the surface. However, inhomogeneities may even enhance the plasmonics effects as commonly used in photovoltaic devices based on plasmonics nanoparticles (Atwater & Polman, 2010). Further measurements will be performed to compare SHG signal from coated and uncoated wires.

5. Applications of optically tweezed nanowires

Recently, Nakayama et al. proved that several different types of nanowires can be stably trapped in optical traps; and that optical traps can thus be used to manipulate assembled three-dimensional nanowire heterostructures. Individual optically trapped nanowire can be placed and held in direct contact with living cells; in addition, one class of nanowires ($KNbO_3$ wires) exhibits efficient SHG and act as frequency converters, raising the possibility for a new type of scanning light microscopy (Nakayama et al., 2007) (Fig.11). Due to their small cross-section, nanowires represent ideal probes for mechanical and optical stimulation of cells and even organelles without overloading the sample with photons.

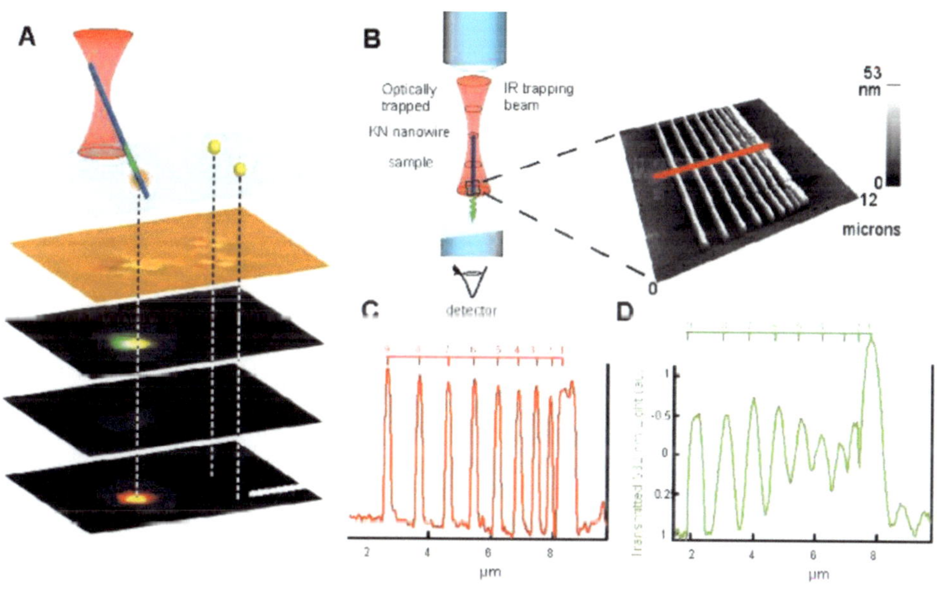

Fig. 11. Schematic of the experimental geometry where laser tweezers are used to accurately position the nanowire and excite fluorescence (A) Excitation of a fluorescent bead by waveguided SHG signal from an optically trapped $KNbO_3$ nanowire. (B) Schematic of inverted optical scanning configuration band AFM topographic image of thermally evaporated pattern of gold stripes on a glass coverslip. (C) AFM line scan from region indicated in B) (D) Optical transmission profile captured by scanning a single $KNbO_3$ nanowire over the metallic surface structure. The nanowire dimensions used to create the transmission line scan was measured by AFM: width=122 nm, length =1.4 μm and height=53 nm. From (Nakayama et al., 2007).

The nanowires can be also used as complementary biomarkers for long term cell tracking experiments. Two important nanowire features such as subwavelength waveguiding and frequency conversion capability with the ability to use optical tweezers to manipulate nanowires in realistic physiological environments can be used for various biological applications as local light and force sources (Fig. 12). Due to the subwavelength optical waveguiding nature of niobate nanowires, it is possible in combination with laser tweezers to create highly localized excitation source with the size which is determined by the nanowire tip diameter and to achieve the high spatial accuracy of nanowire position. This method is not only limited to the membrane proteins but it can be extended also to the imaging of intracellular compartments such as mitochondria. The success of intracellular imaging depends mostly on the nanowires functionalization with appropriate surface chemistries which will decrease the force required to introduce wires in the cell (Wallace & Sansom, 2008).

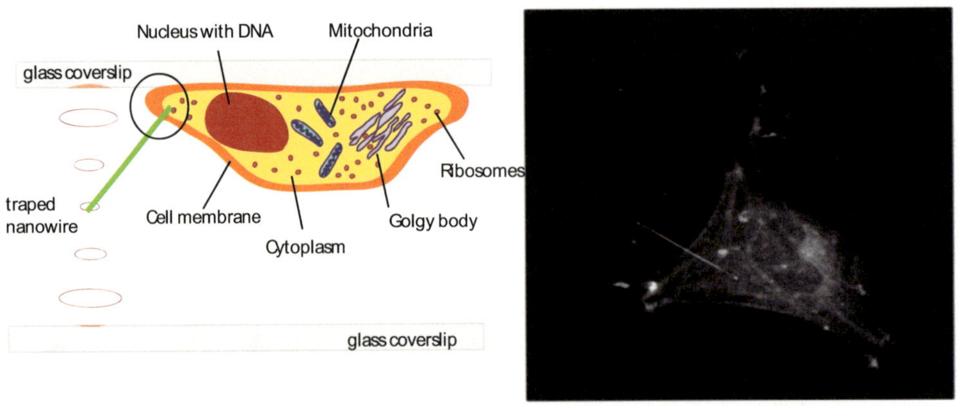

Fig. 12. Schematic of the experimental geometry where laser tweezers are used to accurately position the nanowire on the cell membrane while the nanowire through efficient SHG creates local excitation source. Fluorescence micrograph of a cultured mouse fibroblast (NIH 3T3) expressing YFP labeled ezrin and a nanowire.

6. Applications in optofluidics environment

In this section, we show applications of nanowires in an optofluidics environment. First we describe the devices fabrication, then a simulation of the device and finally experimental results with the nanowires.

The optofluidics device fabrication implies different process flows depending on the applications (Fig. 13). For dielectrophoresis applications, a 1 mm thick glass plate with a conductive electrode pattern in indium tin oxide (ITO) is used as described in (Choi et al., 2006). The sample is fabricated using contact photo lithography (Fig. 13 a). First, we obtain ITO coated glass plates rated at 30-60 Ω/square. Next, positive photoresist is spincoated onto the plates. A pattern is exposed using a UV mask aligner and developed. Finally, the ITO is etched to create the electrode pattern. In case of microfluidic channels that are made of polydimethylsiloxane (PDMS), the fabrication uses the replica molding technique (Fig. 13 b). A master mold is produced through UV lithography on a silicon wafer. After trimethylchlorosilane (TMCS) treatment of the master mold for 5 min, the PDMS is poured onto the mold with a 10 : 1 base-to-curing agent ratio. After curing in an oven at 80 °C for 1 h, the silicone is released from the mold.

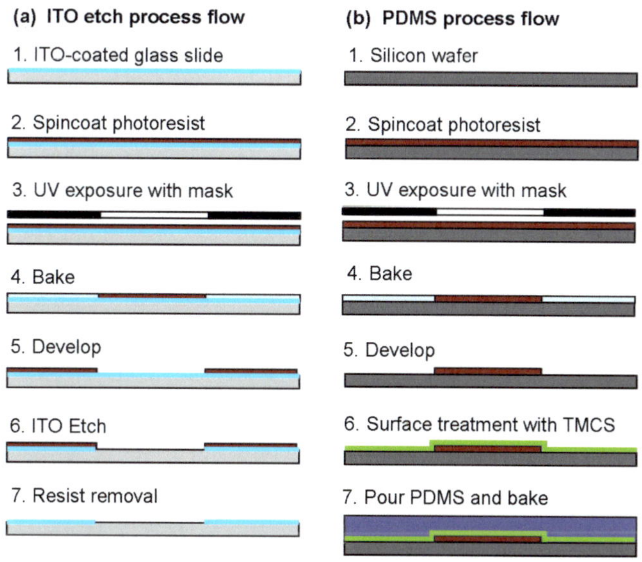

Fig. 13. (a) ITO process flow), (b) PDMS process flow.

For the first type of device, a floating dielectrophoresis electrode design was used to prevent shorting the nanowire after being trapped (Banerjee et al., 2007). This concept is best demonstrated by the COMSOL simulation (Fig. 14 a). A voltage is placed between the top and bottom electrodes. Due to the geometry of the floating electrodes, high electric fields are generated at the tips in the middle floating electrodes. The direction and magnitude of the electric field is indicated by the arrows. The high electric fields serve as dielectrophoretic traps and the direction of the electric field will align the nanowires due to electro-orientation. For the specific design, triangular ITO electrodes were designed with a 2 μm gap and the triangular ITO electrodes were separated by the contact electrodes with a 5 μm gap. The application of an external electric field induces an additional electro-orientation force caused by dielectrophoresis (DEP). DEP forces have been utilized to orient and manipulate several different types of dielectric nanowires (Burke, 2004).

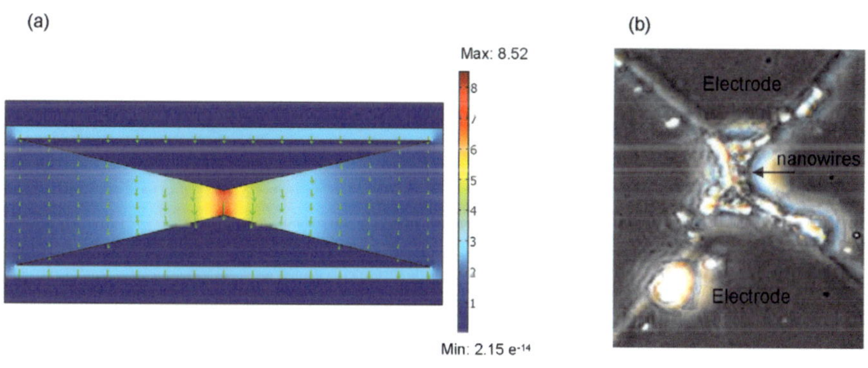

Fig. 14. (a) COMSOL simulation of the electrode design. Normalized surface electric field (V/m), arrow electric field (V/m).(b) White light image of the trapping of Lithium niobate nanowires at the tips of the floating electrodes.

A standard ITO etch with a 5:6:1 ratio of H_2O:HCl (40%):HNO_3 (60%) at room temperature was utilized. By changing the etching time, we were able to control the size of the gap between the triangular electrodes. With a 5 minutes etch, the electrodes were shorted and there was no gap. At 10 minutes, a 1 µm gap existed. By 15 minutes, a 2 µm gap was formed. Results utilized a 10 to 15 minutes electrode etch. A PDMS microfluidic channel with the dimensions of 100 µm wide, 10 µm deep and 1.5 cm long was placed to overlap with the electrode gap. The microfluidic channel can hold 15 nL of liquid. To obtain at least a single nanowire in the entire channel, a concentration of 6.7×10^4 particles/mL is required. Fig. 14 (b) shows the trapping of several nanowires with the floating electrodes design.
On top of the previous electrode device, we placed a PDMS chip with a single channel to orient the nanowires between the two electrodes (Fig. 15). The purpose of such design is to be able to detect electro-optic effects. Indeed, the bulk photovoltaic effect present in $LiNbO_3$ might be of interest to generate locally an electric field under laser illumination. As shown on Fig. 15 (b), it was possible to concentrate nanowires at the electrodes tip.

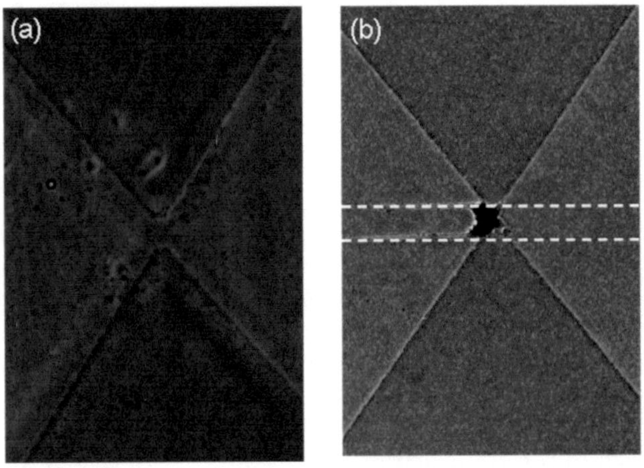

Fig. 15. (a) Empty electrodes (gap = 2 µm width)(b) Electrodes with PDMS channel (indicated with dashed lines) to concentrate the nanowires at the tip.

The last device we developed was to manipulate lithium niobate nanowires in a fluidic environment and monitor the second-harmonic generation (SHG) response using an optical trapping setup (Fig. 16 left). The conductive electrode pattern is interdigitated with specific spacing and width that were previously determined (Choi et al., 2006). The optical tweezer is generated with a specific polarization that is determined with a half wave plate and located outside of the fluidic region, close to either substrate. Therefore, the nanowire is not allowed to orient along the direction of propagation and it locates itself orthogonal to the beam. This polarization imparts a force to orient the nanowire. Furthermore, an external electric field is applied to the nanowires using the pattern of electrodes. This external electric field induces an additional electro-orientation force caused by the dielectrophoretic (DEP) response of the nanowires. Note that the torque on the nanowire due to the external electric field is almost ten times greater in magnitude than the torque on the nanowire due to the optical polarization of the laser for the particle orientation denoted above.

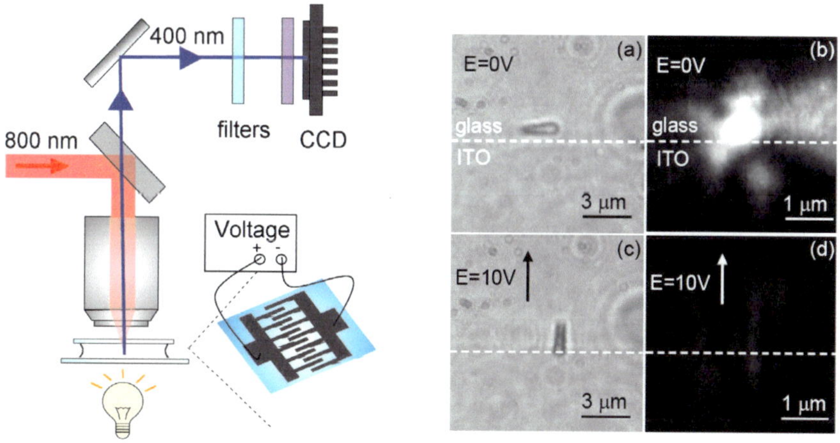

Fig. 16. Left: schematic of the setup for tweezing optically and electrically nanowires and detecting their SHG signal on a CCD after passing through filters to eliminate the incident light. Right: DEP response of a $LiNbO_3$ nanowire suspended in 170 µS/cm conductivity de-ionized water (a and b) without and (c and d) with an electric field of 10 V_{pp} at 150 kHz. (a) and (c) White light images; (b) and (d) SHG response (Reprinted with permission from (Grange et al., 2009). Copyright 2009, American Institute of Physics.)

Fig. 16 shows white light images (a and c) of a nanowire suspended in deionised water. At time t=0, no electric field ($E_{electrodes}$) is applied and the nanowire is only tweezed by the laser (E_{laser}) with a polarization parallel to the electrodes. Then a 10 V_{pp} electric field with a frequency of 150 kHz is applied and after 1 second, the wire is oriented along this electric field due to the stronger field than the one from the laser. Fig. 16 (b and d) shows the SHG response measured with and without the applied electric field.

The nanowires are aligned with the field due to positive DEP forces. Increasing the conductivity of the suspension as well as changing the frequency of the applied electric field can change the type of DEP force on the nanowire which will result in the nanowire to not be aligned with the field. We notice the crossover conductivity between negative and positive DEP when the conductivity of the solution is around 170 µS/cm. The combination provided by the applied external electric field and the polarization of the laser beam serves useful for several measurements in the nanometer scale. For instance, an estimation of the conductivity near the nanowire may be obtained. This may be useful if there is a conductivity gradient in the sample. Moreover, the polarization dependency of the nanowire allows for the detection of its position. The SHG signal can then be useful for detecting smaller wires close to the diffraction limit of a bright field microscope.

7. Conclusion and outlook

We described how most common types of alkaline niobates nanowires, $KNbO_3$, $NaNbO_3$ and $LiNbO_3$ can be synthesized via hydrothermal or molten salt synthesis. We performed materials characterization to determine the composition of the end product of the synthesis as well as electron imaging to determine the aspect ratio of the three types of nanowires. As a further characterization step, we provided nonlinear optical measurements on single niobates nanowires with different setup geometries: transmission setup or optical tweezers. A SHG signal was measured for all types of nanowires as it is well known in similar non

centrosymmetric bulk materials and it was possible to vary the SHG intensity by changing the polarization. Observed nonlinear properties of alkaline niobate nanowires suggest that all three types of nanowires could be used as frequency converters, mechano-optical probes or logic components at the nanoscale.
Then, plasmonics nanoshells were synthesized around $KNbO_3$ nanowires. The thin gold layer serves, first, as enhancing the SHG signal through the plasmon resonance in the near infrared wavelength range of the excitation laser light. And secondly, it makes niobates nanowires compatible with biological entities and it can be an excellent starting point for further DNA or protein functionalization.
Several applications are described and performed to best use the nanowires properties at the overlap between the micro- and the nanoscale. Nanowires are demonstrated as trapped SHG probes for localized illumination. Using the polarization dependency of the SHG signal allow us to determine the orientation of nanowires in optofluidics environment too.
To conclude, applications using the alkaline niobates nanowires are only at the beginning and they are not yet using all the physical properties known for these perovskite materials. Therefore, there is room to improve the synthesis, for instance aspect ratio, to better know the physical properties and to be able to measure small signal as electro-optic effects on single nanowire.

8. Acknowledgment

We like to thank Demetri Psaltis, Jae-Woo Choi, Chia-Lung Hsieh, Ye Pu, Grégoire Laporte, Yioannis Papadopoulos, Arnaud Magrez, Anna Fontcuberta i Morral and Bernt Ketterer for helpful discussions and measurements.

9. References

An, C. H., Tang, K. B., Wang, C. R., Shen, G. Z., Jin, Y., & Qian, Y. T. (2002). Characterization of $LiNbO_3$ nanocrystals prepared via a convenient hydrothermal route. *Materials Research Bulletin,* Vol. 37, No. 11, pp. 1791-1796.

Atwater, H. A., & Polman, A. (2010). Plasmonics for improved photovoltaic devices. *Nature Materials,* Vol. 9, No. 10, pp. 865-865.

Banerjee, S., White, B., Huang, L., Rego, B. J., O'Brien, S., & Herman, I. P. (2007). Precise positioning of carbon nanotubes by ac dielectrophoresis using floating posts. *Applied Physics a-Materials Science & Processing,* Vol. 86, No. 4, pp. 415-419.

Bhalla, A. S., Guo, R. Y., & Roy, R. (2000). The perovskite structure - a review of its role in ceramic science and technology. *Materials Research Innovations,* Vol. 4, No. 1, pp. 3-26.

Boyd, R. W. (2008). *Nonlinear Optics, Third Edition,* Academic press, ISBN 978-0-12-369470-6 Burlington, Mass.

Burke, P. J. (2004). Nano-dielectrophoresis: Electronic Nanotweezers. In *Encyclopedia of Nanoscience and Nanotechnology* H. S. Nalwa (Ed.),Vol. 6, pp. 623. ISBN 1-58883-001-2, Valencia, California.

Choi, J. W., Pu, A., & Psaltis, D. (2006). Optical detection of asymmetric bacteria utilizing electro orientation. *Optics Express,* Vol. 14, No. 21, pp. 9780-9785.

Dadap, J. I. (2008). Optical second-harmonic scattering from cylindrical particles. *Physical Review B,* Vol. 78, No. 20, pp. 1098-0121.

Dadap, J. I., Shan, J., Eisenthal, K. B., & Heinz, T. F. (1999). Second-harmonic Rayleigh scattering from a sphere of centrosymmetric material. *Physical Review Letters,* Vol. 83, No. 20, pp. 4045-4048.

Fluck, D., & Gunter, P. (2000). Second-harmonic generation in potassium niobate waveguides. *Ieee Journal of Selected Topics in Quantum Electronics,* Vol. 6, No. 1, pp. 122-131.

Grange, R., Choi , J.-W., Hsieh, C.-L., Pu, Y., Magrez, A., Smadja, R., Forro, L., & Psaltis, D. (2009). Lithium niobate nanowires synthesis, optical properties, and manipulation. *Applied Physics Letters,* Vol. 95, No. 14, pp. 143105-143108.

Guo, Y. P., Kakimoto, K., & Ohsato, H. (2004). Phase transitional behavior and piezoelectric properties of $(Na_{0.5}K_{0.5})NbO_3$-$LiNbO_3$ ceramics. *Applied Physics Letters,* Vol. 85, No. 18, pp. 4121-4123.

Hsieh, C.-L., Grange, R., Pu, Y., & Psaltis, D. (2009). Three-dimensional harmonic holographic microcopy using nanoparticles as probes for cell imaging. *Opt. Express,* Vol. 17, No. 4, pp. 2880-2891.

Hsieh, C.-L., Pu, Y., Grange, R., & Psaltis, D. (2010). Second harmonic generation from nanocrystals under linearly and circularly polarized excitations. *Opt. Express,* Vol. 18, No. 11, pp. 11917-11932.

Johnston, K. E., Tang, C. C., Parker, J. E., Knight, K. S., Lightfoot, P., & Ashbrook, S. E. (2010). The Polar Phase of $NaNbO_3$: A Combined Study by Powder Diffraction, Solid-State NMR, and First-Principles Calculations. *Journal of the American Chemical Society,* Vol. 132, No. 25, pp. 8732-8746.

Ke, T. Y., Chen, H. A., Sheu, H. S., Yeh, J. W., Lin, H. N., Lee, C. Y., & Chiu, H. T. (2008). Sodium niobate nanowire and its piezoelectricity. *Journal of Physical Chemistry C,* Vol. 112, No. 24, pp. 8827-8831.

Kong, L. B., Zhang, T. S., Ma, J., & Boey, F. (2008). Progress in synthesis of ferroelectric ceramic materials via high-energy mechanochemical technique. *Progress in Materials Science,* Vol. 53, No. 2, pp. 207-322.

Li, L. H., Deng, J. X., Chen, J., Sun, X. Y., Yu, R. B., Liu, G. R., & Xing, X. R. (2009). Wire Structure and Morphology Transformation of Niobium Oxide and Niobates by Molten Salt Synthesis. *Chemistry of Materials,* Vol. 21, No. 7, pp. 1207-1213.

Louis, L., Gemeiner, P., Ponomareva, I., Bellaiche, L., Geneste, G., Ma, W., Setter, N., & Dkhil, B. Low-Symmetry Phases in Ferroelectric Nanowires. *Nano Letters,* Vol. 10, No. 4, pp. 1177-1183.

Magrez, A., Vasco, E., Seo, J. W., Dieker, C., Setter, N., & Forro, L. (2006). Growth of single-crystalline $KNbO_3$ nanostructures. *Journal of Physical Chemistry B,* Vol. 110, No. 1, pp. 58-61.

Matthias, B. T., & Remeika, J. P. (1949). Ferroelectricity in the Ilmenite Structure. *Physical Review,* Vol. 76, No. 12, pp. 1886-1887.

Nakayama, Y., Pauzauskie, P. J., Radenovic, A., Onorato, R. M., Saykally, R. J., Liphardt, J., & Yang, P. D. (2007). Tunable nanowire nonlinear optical probe. *Nature,* Vol. 447, No. 7148, pp. 1098-1101.

Neuman, K. C., & Block, S. M. (2004). Optical trapping. *Rev. Sci. Instrum.,* Vol. 75, No. 9, pp. 2787-2809.

Niederberger, M., Pinna, N., Polleux, J., & Antonietti, M. (2004). A general soft-chemistry route to perovskites and related materials: Synthesis of $BaTiO_3$, $BaZrO_3$, and $LiNbO_3$ nanoparticles. *Angewandte Chemie-International Edition,* Vol. 43, No. 17, pp. 2270-2273.

Pribosic, I., Makovec, D., & Drofenik, M. (2005). Formation of nanoneedles and nanoplatelets of $KNbO_3$ perovskite during templated crystallization of the precursor gel. *Chemistry of Materials,* Vol. 17, No. 11, pp. 2953-2958.

Pu, Y., Grange, R., Hsieh, C.-L., & Psaltis, D. (2010). Nonlinear optical properties of core-shell nanocavities for enhanced second-harmonic generation. *Physical Review Letters,* Vol. 104, No. 20, pp. 207401-207405.

Saito, Y., Takao, H., Tani, T., Nonoyama, T., Takatori, K., Homma, T., Nagaya, T., & Nakamura, M. (2004). Lead-free piezoceramics. *Nature,* Vol. 432, No. 7013, pp. 84-87.

Santulli, A. C., Zhou, H., Berweger, S., Raschke, M. B., Sutter, E., & Wong, S. S. (2010). Synthesis of single-crystalline one-dimensional $LiNbO_3$ nanowires. *Crystengcomm,* Vol. 12, No. 10, pp. 2675-2678.

Schwesyg, J. R., Eggert, H. A., Buse, K., Sliwinska, E., Khalil, S., Kaiser, M., & Meerholz, K. (2007). Fabrication and optical characterization of stable suspensions of iron- or copper-doped lithium niobate nanocrystals in heptane. *Applied Physics B-Lasers and Optics,* Vol. 89, No. 1, pp. 15-17.

Shi, H., Li, X., Wang, D., Yuan, Y., Zou, Z., & Ye, J. (2009). $NaNbO_3$ Nanostructures: Facile Synthesis, Characterization, and Their Photocatalytic Properties. *Catalysis Letters,* Vol. 132, No. 1, pp. 205-212.

Shi, H. F., Li, X. K., Wang, D. F., Yuan, Y. P., Zou, Z. G., & Ye, J. H. (2009). $NaNbO_3$ Nanostructures: Facile Synthesis, Characterization, and Their Photocatalytic Properties. *Catalysis Letters,* Vol. 132, No. 1-2, pp. 205-212.

Shoji, I., Kondo, T., & Ito, R. (2002). Second-order nonlinear susceptibilities of various dielectric and semiconductor materials. *Optical and Quantum Electronics,* Vol. 34, No. 8, pp. 797-833.

Sutherland, R. L. (2003). *Handbook of Nonlinear Optics* Hardcover Marcel Dekker, ISBN 0824794265, New York.

Wallace, E. J., & Sansom, M. S. P. (2008). Blocking of carbon nanotube based nanoinjectors by lipids: A simulation study. *Nano Letters,* Vol. 8, No. 9, pp. 2751-2756.

Wang, G. Z., Selbach, S. M., Yu, Y. D., Zhang, X. T., Grande, T., & Einarsrud, M. A. (2009). Hydrothermal synthesis and characterization of $KNbO_3$ nanorods. *Crystengcomm,* Vol. 11, No. 9, pp. 1958-1963.

Wang, G. Z., Yu, Y. D., Grande, T., & Einarsrud, M. A. (2009). Synthesis of $KNbO_3$ Nanorods by Hydrothermal Method. *Journal of Nanoscience and Nanotechnology,* Vol. 9, No. 2, pp. 1465-1469.

Wang, L. H., Yuan, D. R., Duan, X. L., Wang, X. Q., & Yu, F. P. (2007). Synthesis and characterization of fine lithium niobate powders by sol-gel method. *Crystal Research and Technology,* Vol. 42, No. 4, pp. 321-324.

Weber, M. J. (2003). *Handbook of optical materials,* CRC Press, ISBN 0849335124, New York.

Wood, B. D., Mocanu, V., & Gates, B. D. (2008). Solution-Phase Synthesis of Crystalline Lithium Niobate Nanostructures. *Advanced Materials,* Vol. 20, No. 23, pp. 4552-4556.

Wooten, E. L., Kissa, K. M., Yi-Yan, A., Murphy, E. J., Lafaw, D. A., Hallemeier, P. F., Maack, D., Attanasio, D. V., Fritz, D. J., McBrien, G. J., & Bossi, D. E. (2000). A review of lithium niobate modulators for fiber-optic communications systems. *Ieee Journal of Selected Topics in Quantum Electronics,* Vol. 6, No. 1, pp. 69-82.

Wu, S. Y., Liu, X. Q., & Chen, X. M. (2010). Hydrothermal synthesis of $NaNbO_3$ with low NaOH concentration. *Ceramics International,* Vol. 36, No. 3, pp. 871-877.

Zhao, L. L., Steinhart, M., Yosef, M., Lee, S. K., & Schlecht, S. (2005). Large-scale template-assisted growth of $LiNbO_3$ one-dimensional nanostructures for nano-sensors. *Sensors and Actuators B-Chemical,* Vol. 109, No. 1, pp. 86-90.

Zhu, H. Y., Zheng, Z. F., Gao, X. P., Huang, Y. N., Yan, Z. M., Zou, J., Yin, H. M., Zou, Q. D., Kable, S. H., Zhao, J. C., Xi, Y. F., Martens, W. N., & Frost, R. L. (2006). Structural evolution in a hydrothermal reaction between Nb_2O_5 and NaOH solution: From Nb_2O_5 grains to microporous $Na_2Nb_2O_6$ $^2/_3$ H_2O fibers and $NaNbO_3$ cubes. *Journal of the American Chemical Society,* Vol. 128, No. 7, pp. 2373-2384.

24

Field Emission from Nanowires

Dilip S. Joag[1], Mahendra A. More[1] and Farid Jamali Sheini[2]
[1]*Centre for Advanced Studies in Materials Science and Condensed Matter Physics, Department of Physics, University of Pune, Pune,*
[2]*Department of Physics, Islamic Azad University, Ahwaz Branch, Ahwaz,*
[1]*India*
[2]*Iran*

1. Introduction

Nanowires of a variety of materials have gained importance in the past decade owing to their potential for reliable integration into electronic devices. In particular, this importance is due to their application in flat panel display technology as well as in the cold cathode technology for electron tube devices such as microwave tubes. This is because the conventional thermionic emission devices are accompanied with high power dissipation due to high cathode temperature. With the emergence of nanomaterials, there has been an upsurge in the research publications reporting on their field emission behaviour. This is because of the advantage of nanometric dimensions, making them suitable for field emission. Several metallic and semiconducting nanomaterials are found to operate delivering high current density at a lower applied potential as compared to the conventional field emitter counterparts. The nanowire form of various materials is suitable for the observed field emission properties. Additionally, the materials to be investigated must possess vacuum compatibility (low vapour pressure), high mechanical strength (should be able to withstand mechanical stress due to high electric field), and resistance to ion bombardment (due to residual gases in the vacuum chamber) and low work function. The emission current density is usually in the range of 10-100 mA/cm^2 for the applied field of 1-20 V/μm. It should be possible to deposit the nanomaterials on the flat substrate such as Si, with uniform density and with relative ease, to be used as a field emitter cathode.
The materials of interest from field emission point of view can be classified into the following broad classes: Carbon nanotubes, graphene, carbon fibre; Nanoforms of wide band gap materials such as SiC, GaN, AlN, BN, ZnO; other semiconductors such as Si, CdTe, CdS; and nanostructures and nanowires of established electron emitters such as LaB_6. In this chapter, emphasis will be given on the field emission phenomenon, an account of development in theory of field emission, basic methodology; and a brief review of the field emission work carried out on nanowires.

2. The field emission phenomenon

Field electron emission from metals and semiconductors has attracted the attention of researchers for the past more than sixty years. In the phenomenon called field emission, electrons tunnel out from metal or semiconductor surface into vacuum under the action of high electrostatic field ~ 10^6-10^7 V/cm. A systematic study of field emission under ultrahigh vacuum conditions has led to the development of the technique Field Emission Microscopy (FEM), the detailed treatment of which can be found in the literature (Gomer, 1961). The quantum mechanical tunneling emission current – field strength relation was derived by Fowler and Nordheim in the year 1928 (Fowler & Nordheim, 1928). A refined account of field emission and field emission spectroscopy has also been taken (Modinos, 1984).

The traditional and most useful application of field emission to surface science is in the study of surface diffusion of adsorbates over clean metal surfaces. In this application of the original FEM technique, an emitter is uniformly covered with an adsorbate and the current emanating from a very small region of the surface of the emitter is measured. This technique is called the probe hole technique (Oostrom, 1966), in which various single crystal planes of an emitter can be accessed in a single experiment.

Apart from the usefulness of field emission in surface science, there are technological applications. Field emission sources are routinely used in scanning electron microscopy and e- beam lithography because of their small optical size. In fact, a high density two dimensional array of emitters can be fabricated by a variety of techniques resulting into emitters suitable for flat panel displays and even for electron tube devices (Spindt, 1991). Early work on field emission devices consisted of micro-fabricated metallic tip array (commonly from Mo), but these devices suffer from several factors degrading the operating performance. One of the crucial factors limiting the operational stability and lifetime of these metallic emitters is tip degradation due to Joule heating at high emission currents and ion bombardment causing erosion and blunting of the tips. These Spindt cathodes currently find use in ion sources, mass spectrometers. Huge efforts have been invested in developing displays based on Spindt cathodes with some success (Chalamala, 1998). Obtaining sharp and robust materials capable of delivering high emission current densities at lower operating voltages is an ongoing effort.

2.1 The Fowler-Nordheim equation

The basic formulation of quantum mechanical tunneling of conduction electrons from a plane metal surface into vacuum was conceived by Fowler and Nordheim (Fowler &

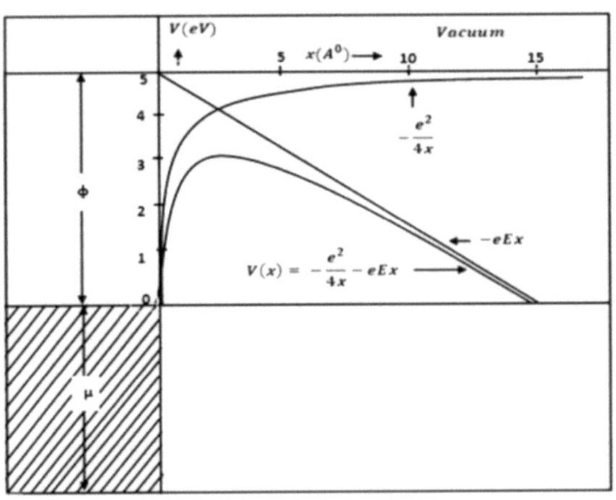

Fig. 1. Potential energy of an electron as a function of its distance x to the surface (schematic).

Nordheim, 1928) resulting into the Fowler-Nordheim (F-N) equation. The equation representing tunneling from an exact triangular barrier (Fig. 1) without any correction factors is given by

$$J = 6.2 \times 10^6 \frac{(\mu/\varphi)^{1/2}}{(\mu+\varphi)} E^2 \, exp \, [-6.8 \times 10^7 \frac{\varphi^{3/2}}{E}] \, amp/cm,^2 \quad (1)$$

for all energies expressed in electron volts and the electric field E in volts per centimetre. Here, φ is the work function of the metal, μ is the Fermi energy and J the field emission current density in amps/cm^2.

The above derivation is based on the exact triangular surface potential barrier. However, the electronic charge density at the surface does not end abruptly at the terminating metal surface and the barrier, therefore, cannot be expected to be straight up. Also, the image force potential reduces the surface potential barrier by a corresponding image term V_x.

$$V_x = \frac{-e^2}{4x}, \quad (2)$$

where e is the electronic charge and x is the distance from the metal surface. With this correction in the potential, the F-N equation acquires the form

$$J = 6.2 \times 10^6 \frac{(\mu/\varphi)^{1/2}}{\alpha^2(\mu+\varphi)} E^2 \exp \, [-6.8 \times 10^7 \frac{\varphi^{3/2}\alpha}{E}], \quad (3)$$

α is the image correction term

$$\alpha = (1-y)^{1/2} \quad (4)$$

and

$$y = 3.8 \times 10^{-4} E^{1/2} / \varphi \quad (5)$$

In the above, the zero point of the potential energy is taken to be at $\varphi + \mu$ above the bottom of the conduction band. Detailed treatment of the F-N equation for electron tunneling from metal surface can be found in the literature (Gomer, 1961; Fowler & Nordheim, 1928; Modinos, 1984).

In case of the semiconductors, the field electron emission theory is complex. Various aspects such as the contribution of the effective electron mass, field penetration, surface states, doping with n type or p type impurities and band structure have to be taken into account. Field emission from semiconductors has been dealt with by Stratton (Stratton, 1962) who considered the effect of the field penetration and surface states on electron emission. He pointed out that the field emission characteristic dominated by the surface states should be strongly temperature dependent while that dominated by field penetration is not. Baskin et al (Baskin, 1971) have analysed field electron emission behaviour of an n-type semiconductor based on the field emission from conduction band. A detailed account of the theory field emission from semiconductors is available (Modinos, 1984).

Recently, there have been attempts to revisit the theory of Fowler-Nordheim tunneling. The standard theory has been reformulated by Forbes and Deane (Forbes & Deane, 2007) leading to an insight into linearity of F-N plot. A revised mathematical form for empirical field emission current –voltage characteristics has also been obtained by Forbes (Forbes, 2008).

2.2 Application of F-N theory to the field emission from nanomaterials

The F-N theory primarily considers tunneling of electrons from flat planar surface of metals. The image potential corrected F-N equation (Eq.3) can be written as

$$J = a\varphi^{-1} P_f E^2 \exp\left(-\frac{b\varphi^{\frac{3}{2}}}{E}\right), \tag{6}$$

Where P_f is the prefactor given by

$$P_f = 4\,\varphi^{(3/2)}\,(\mu)^{1/2} / (\varphi + \mu), \tag{7}$$

μ is the Fermi energy and φ is the work function. P_f is of the order of unity and can be set as equal to unity.
Thus,

$$J = a\varphi^{-1} E^2 \exp\left(-\frac{b\varphi^{\frac{3}{2}}}{E}\right), \tag{8}$$

Here, a and b are constant (a= 1.54 x 10^{-6}A eV V^{-2}, b = 6.83 $V^{-3/2}$ Vnm^{-1})
The electric field E is the local electric field (surface field).
In case of the nanoparticles deposited on the flat cathode surface in the diode geometry, the electric field (applied or macroscopic field) between the cathode and the anode separated by a distance d with a potential difference V is given by,

$$\mathbf{E} = V/d \tag{9}$$

However, local electric field E at the surface of a nanoparticle can be much higher than **E** by a factor β,

$$E = \beta\,\mathbf{E} \tag{10}$$

Here, β is the field enhancement factor.
From eq (9) and (10),

$$E = \beta\,(V/d) = (\beta/d)\,V \tag{11}$$

The factor (β/d) is the voltage to local field conversion factor.
In terms of the macroscopic field **E**, the F-N equation (8) becomes,

$$J = a\varphi^{-1}\beta^2 \mathbf{E}^2 \exp\left(-\frac{b\varphi^{\frac{3}{2}}}{\beta\mathbf{E}}\right), \tag{12}$$

$$I = S \times J, \tag{13}$$

where I is the emission current (A) and S the emitting area, β is the field enhancement factor. The factor β is related to the emitter geometry. In certain cases, the field enhancement factor β can be expressed as β = h/r where h is the height and r is the radius of curvature of the tip of the emitting structure. The local electric field at the tip surface depends upon actual tip geometry (shape). For the elongated, cylindrical emitters the height to diameter (thickness) ratio is called the 'aspect ratio'. The emission performance of an emitter can be enhanced by increasing the 'aspect ratio'. Usually calculation of the factor β is complex. There exist several models for the calculation of β. These models are not uniformly applicable and may

vary for different types of materials as well as geometry. The value of β can be calculated from the slope of the F-N plot, but the value obtained may be physically unrealistic.
Many of the field emission studies on nanomaterials have been motivated from the electron sources point of view. Therefore, the current stability for a long duration operation, the effect of residual gas pressure, effect of adsorption and desorption causing local work function variations become important parameters.
The areal density of emitters on a cathode surface plays an important role in controlling the local electric field at the surface of an emitter. Higher density leads to a screening effect and lower density reduces this screening and causes rise in the local field at each emitter.

2.3 Recent developments in the theory of field electron emission from nanomaterials

As seen from the literature Eq. (12) has been widely used to fit the experimental data and calculate from the slope of the ln (J/E^2) versus $1/E$ (F-N) Plot, the field enhancement factor β (dimensionless) even for multiple emitters. Strictly, the F-N theory is valid for a single, metallic, planar emitter. Even for semiconducting emitters, the same equation is seen to be assumed. For semiconductors, the effect of band structure, surface states and field penetration have to be accounted for, and the theory is not simple to apply to practical emitters. The theory of field electron emission from carbon based emitters has been recently addressed to. An excellent account of early work on carbon based film emitters and other nanostructured heterogeneous materials and also hypotheses about emission mechanism have been reported by Forbes (Forbes, 2001). The F-N equation (Eq.12) has also been used by researchers to describe field emission from carbon nanotubes (CNT's) against many theoretical difficulties (Forbes, 2010). Recently a detailed analytical treatment of field emission from nanowall emitter has been attempted by Qin et. al. (Qin, 2011). The theory is very complex. The authors have shown that the F-N equation will not describe the electron emission from a nanowall of very small width (less than ≈ 2nm).
The F-N behaviour of nanomaterials is important from the basic point of view. However, field emission from nanomaterials has been very attractive field of research due to its technological importance. The current density obtainable for lower and lower turn-on voltage, stability of the current, and life are the main considerations of the study. Ease of synthesis, size and shape, electrical conductivity are also important consideration for a material to be a candidate as an emitter. Finally, ease of fabrication of array of emitters on lithographically patterned substrate decides the applicability of such sources in electronic devices.

3. Nanowires as field emitters

Nanowires of different materials are of immense interest due to their one dimensional structures. Such structures are known to show quantum confinement effects. Nanowires also possess high surface to volume ratio. In particular, nanowires with low threshold and low operating voltage have become potential candidates for field emission display (FED) application. Aligned nanowires with a high packing density can significantly enhance the field emission behaviour. Nanowires possess axial geometry and such nanorods like structures can be grown vertically on flat substrate with desirable density of wires per unit area. Final aim is to form regular array of nanowires with a good uniformity in their aspect ratio (length to radius). This is usually done by growing nanowires on a lithographically patterned substrate. Field emission investigations are carried out on the nanowires grown on the substrate.

3.1 Basic methodology

Most of the nanostructured materials are deposited/coated on flat substrate (nominally 1 cm^2) forming broad area electron emitter. The substrate is an electrical conductor/semiconductor. The method of deposition depends on the properties of the material to be deposited. Various methods such as chemical vapour deposition, Arc plasma deposition, pulsed Laser Deposition, electrodeposition, electrochemical deposition, spin coating etc. have been used by researchers. Amongst the various substrates Si is most common as it is suitable for integrating with electronic device technology. This forms the cathode. The anode is indium tin oxide (ITO) coated conducting glass plate with or without phosphor coating. The anode-cathode separation is in the range ~ usually a few microns to a few mm. The anode and the cathode are held separated with the quartz or alumina spacer. This is called the diode configuration. A typical anode cathode system is shown in Fig.2. A holder carrying the anode and the cathode is then mounted in an ultra high vacuum chamber which is subsequently evacuated to a pressure ~ 10^{-7}-10^{-8} mbar. The high voltage (0-10 kV) from stable DC supply is applied across the anode and the cathode. The current is measured by an electrometer amplifier (10^{-9}-10^{-3}A). The current voltage characteristics are measured over a wide range of voltages. Current stability studies are performed by recording the field emission current (at the pre-set value) at regular time intervals. The field emission pattern is observed on the anode screen and recorded if necessary.

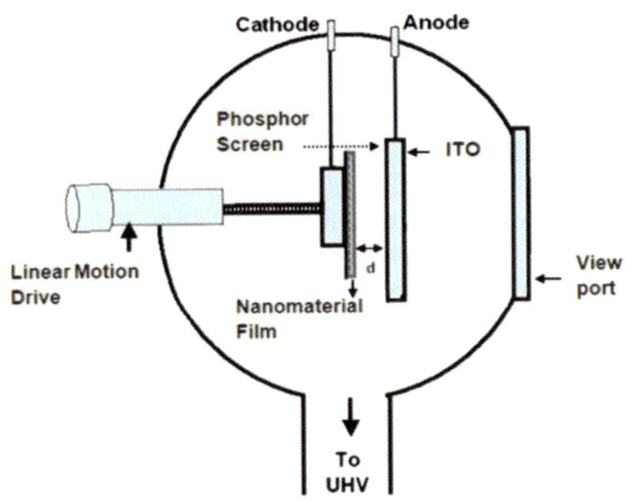

Fig. 2. Schematic of the planar diode configuration arrangement in a vacuum chamber.

3.2 Metallic nanowires

Traditional, Spindt type cathodes consisted of metallic emitters such as W, Mo, and also Si etc but will not be discussed here. Metallic nanowires of Mo, Cu, Co have been used as nanotips for field emission application (Hwang, 2005; Laurent, 2005; Liao, 2005). Field emission characteristics of electrochemically synthesized nickel nanowires on alumina nanoporous template, with oxygen plasma post treatment, have been reported (Joo, 2006). Nickel nanorods arrays have been vertically grown on a Ta-coated Si substrate by electrochemical process (Banerjee, 2011). The authors have reported the experimentally obtained geometrical field enhancement factor from F-N plot to be 4000, agreeing with existing models.

3.3 Inorganic semiconducting nanowires

Wide bandgap semiconductors such as SiC, GaN, ZnO, AlN have attractive physico chemical properties suitable to operate as efficient field emitters in their nanoforms. Field emission characteristics of SiC nanowires and nanorods (Wong, 1999) and GaN nanorods (Kim, 2003) have been investigated. Field emission from SnO_2 nanowhiskers, nanograss has been studied (Luo, 2004; Wang, 2005). The 1D SnO_2 structures are potentially important in application such as optoelectronics devices and sensors. ZnO is fore-runner amongst the wide band gap materials, so far as the field emission studies are concerned.

The field emission properties of ZnO nanowires were first studied by Lee's group in 2002 (Lee, 2002). In their work, well aligned ZnO nanowires were grown on Si substrate using a metal vapour deposition method. Before growth, nanoscale Co particles (average diameter ~ 6–8 nm) were distributed onto Si substrate as the catalyst. The average length and typical diameter of the ZnO nanowires grown at 550 °C were 13 μm and 50 nm, respectively. The turn-on voltage for the ZnO nanowires was found to be about ~ 6.0 V/μm at a current density of ~ 0.1mA/cm^2. The emission current density from the ZnO nanowires reached 1 mA/cm^2 at a bias field of 11.0 V/μm.

The ZnO nanoneedle arrays with sharp tips show improved field emission performance (Zhu, 2003). The nanoneedles were grown via selenium-assisted vapour phase deposition. The height and diameter of the nanoneedles were about 3 μm and 100 nm, respectively. The turn-on field was found to be ~ 2.4 V/μm. The emission current density can reach ~ 2.4 mA/cm^2 at an applied electric field of ~ 7.0 V/μm. The field emission property of single nanowire was also reported (Dong, 2003). The ZnO nanowires were formed on tungsten substrates using a vapour transport method. In this synthesis gold served as the catalyst during the growth. The lengths of the nanowires were several tens of microns, and their diameters ranging from 25 to 200 nm. Measurement of the I-V current for an individual ZnO nanowire was carried out using a movable Faraday cup. In this research the emission current fluctuations have been studied at different temperatures. ZnO films, nanowires array and nanosheets via seed-layer assisted electrochemical deposition route have been fabricated (Cao, 2007). They found that the ordered ZnO nanowire arrays with high density only show the field emission property a little better than that of the ZnO film because of the local electric field screening effect. The hierarchical ZnO nanowire arrays show good field emission properties due to their high aspect ratio, small radius of curvature, and proper density. Field emission characteristics of ZnO nanowires grown on carbon cloth have been reported (Jo, 2004). ZnO nanowires were synthesized on carbon cloth by vaporization and condensation method. Aligned nanowires carry a lot of importance in their functioning as field emitters. Jamali Sheini et al. (Sheini, 2010) have synthesized aligned ZnO nanowires by annealing gold deposited zinc foil in air at 400°C. Fig. 3 a shows the SEM image of the aligned nanowires. From the field emission studies, the value of the turn-on field corresponding to an emission current density of ~ 0.1 μA/cm^2 is found to be ~ 2.4 V/μm and a current density of 100 μA/cm^2 is obtained at an applied field of ~ 3.4 V/μm. The field emission current stability investigated over a period of 3 hours at the preset value of 1 μA is found to be good. Attempt has been made to synthesize ZnO nanowires by cathodic electrodeposition of nanostructured ZnO thin films on zinc substrate, followed by annealing in air (Sheini, 2009). The field emission studies showed the value of 1.36 V/μm for drawing the emission current density of ~ 1 μA/cm^2 for the best specimen. The field emission was measured in a planar diode configuration. The photoenhanced field emission from Sn doped ZnO nanowires has also been reported by Jamali Sheini et al. (Sheini, 2010).

Fig. 3. (a) An SEM image of aligned ZnO nanowires grown on Au coated Zn Substrate and annealed at 400^0 C for 4 hours. Reproduced from Materials Chemistry and Physics (Jamali Sheini, 2010) Copyright © 2010, *Elsevier.*

Fig. 3. (b) An SEM image of as-synthesized ZnO nanowires grown on Zn substrate. Reproduced from Ultramicroscopy (Jamali sheini, 2009) Copyright © 2009. *Elsevier.*

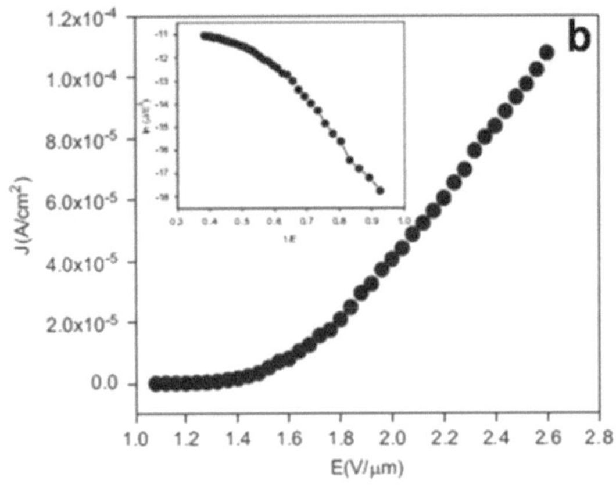

Fig. 3. (c) A field emission current density-applied field (J–**E**) plot of the ZnO nanowires and inset showing the corresponding Fowler-Nordheim (F-N) plot. Reproduced from Ultramicroscopy (Jamali sheini, 2009) Copyright © 2009. *Elsevier.*

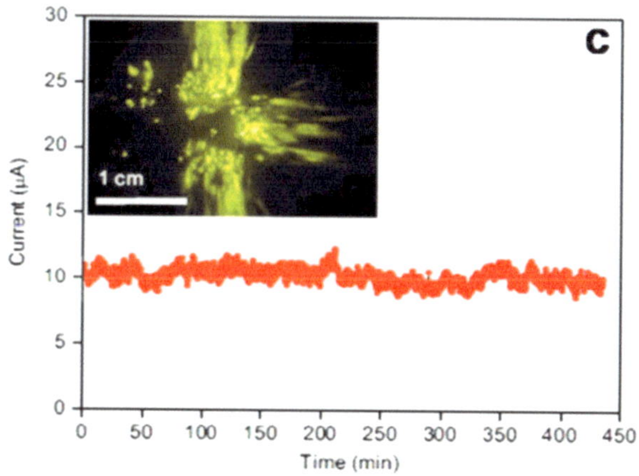

Fig. 3. (d) Current versus time (I–t) plot of the ZnO nanowires and inset showing the field emission image. Reproduced from Ultramicroscopy (Jamali sheini, 2009) Copyright © 2009. *Elsevier.*

Fig. 3 b shows the scanning electron micrograph (SEM) of ZnO nanowires grown on Zn foil by electrodeposition followed by annealing at 400°C for 4 h. The field emission J-E plot and F-N plot are depicted in Fig 3 c and the inset, respectively. Fig. 3 d shows the I-t stability plot and the field emission pattern.

The non-linearity of the F-N plot has been attributed to the semiconducting properties of ZnO. Excellent current stability has been observed with fluctuations within ± 5% of the average value. A good account of electrochemical deposition of ZnO nanowires array has been taken by Zeng et al. (Zeng, 2010). These authors have discussed the effect of doping on the field emission properties. The narrow band gap semiconductors and other semiconducting nanomaterials such as copper sulfide (Cu_2S) nanowires arrays have been studied (Chen, 2002). The CuO nanobelts, nanorods, and nanofibres have also been studied by Hsieh et al. (Hsieh, 2003). The MoO_3 nanowires and nanobelt structures have been studied by Zhou et al. (Zhou, 2003).

One dimensional chalcogenides nanowires have proved to be significant for their novel physical properties and in particular, field emission properties (Datta, 2009). Fig. 4a shows the SEM image of the CdS nanowires synthesized by simple thermal evaporation route on Si substrate.

Fig. 4. (a) An SEM image of as-synthesized CdS nanowires grown on Si substrate by simple thermal evaporation technique.

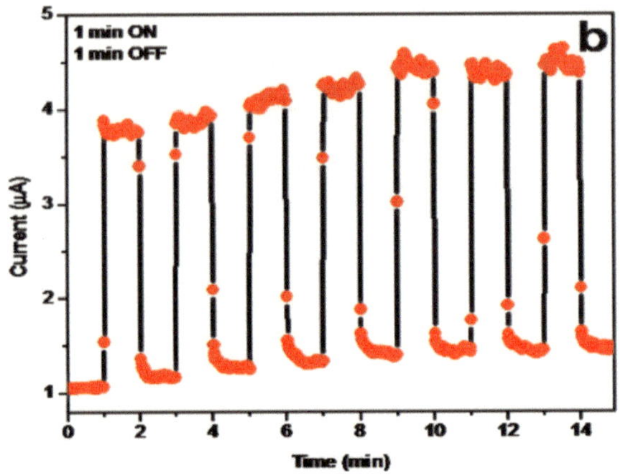

Fig. 4. (b) Photo-switching properties of the CdS nanowires in current – time (I-t) mode.

The turn on field for the emission current density of 0.1 μA/cm² is found to be 1.4 V/μm, lower than the reported value for CdS nanostructures. A current density of 27 μA/cm² was drawn at an applied field of ~ 2.4 V/μm. Upon illumination with visible light, the CdS nanowires switch to the enhanced emission state and return to the original field emission current after the light source is turned off. Thus it is demonstrated that the CdS nanowires act as photo-field emitter switch. A sequence of current pulses generated due to repetitively switching the light source, on and off, is shown in Fig 4b. The effect has been attributed to the photoconductive property of CdS (band gap 2.4 eV) (Chavan, 2011). Fig. 4c shows the J-E characteristics without and with illumination. Fig 5 shows field emission current stability continuously recorded for 24 hrs at the preset current of 1 μA. The current is seen to be highly stable. The inset shows field emission image showing the emission spots due to nanowires.

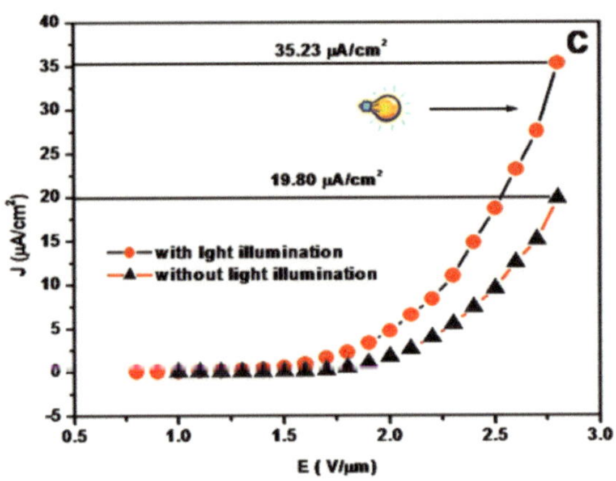

Fig. 4. (c) J–**E** plots of CdS nanowires with and without light illumination.

The high sensitivity and fast response of CdS nanowires points towards possible application of the CdS nanowire field emitters in devices and in pulsed electron beam technology.

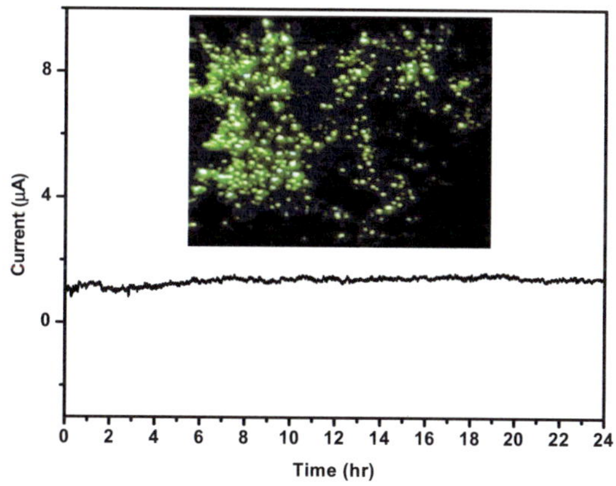

Fig. 5. Current versus time (I–t) plot of the CdS nanowires and inset showing the field emission image.

The number of research articles in the area of the field emission from nanowires and nanomaterials in general is too large to review here. A good account of inorganic nanostructures and their field emission applications has been taken by Fang et al (Fang, 2008). Lanthanum hexaboride (LaB_6) is a very well known thermionic electron emitter. Its potential as a field emitter has been recognised by researchers world over. LaB_6 nanowires are ideal for applications as electron sources due to their inherent characteristics such as low work function (2.6 eV). Recently it is reported that excellent field electron emission is obtained from single LaB_6 nanowire emitter with stability better than the LaB_6 needle type emitter or tungsten field emitter (Zhang, 2010). Extremely high emission current density from LaB_6 arrays is achieved by Qi et al (Qi, 2008). Field emitter arrays of LaB_6 have been fabricated by employing the transfer mold technique (Nakamoto, 2002). LaB_6 nanowires have been synthesized using arc plasma synthesis (Late, 2010). More efforts are required to fabricate LaB_6 nanowires arrays in a reliable and efficient manner.

4. Future outlook

Nanowire field emitter arrays of various materials will be the area of research interests for some more time. More efforts are required to deliver improved field emitter nanowire arrays as sources in electron tube devices and electron beam accelerators.

5. Acknowledgments

The authors thank Padmakar G. Chavan for discussion and help in preparing this manuscript. DSJ wishes to thank CSIR New Delhi for Emeritus Scientist scheme.

6. References

Banerjee, A. N.; Qian, S. & Joo, S. W. (2011). Large field enhancement at electrochemically grown quasi-1D Ni nanostructures with low-threshold cold-field electron emission *Nanotechnology,* Vol.22, No.3, (January 2011), pp. 035702-035709, ISSN 1530-6992

Baskin, L. M. Lvov, O. L. & Fursey, G. N. (1971). General features of field emission from semiconductors. *Phys. Stat. Sol (b),* Vol.47, No.1, (Sept 1971), pp. 49-62, ISSN 1521-3951

Cao, B.; Teng, X.; Heo, S. H.; Li, Y.; Cho, S. O.; Li, G. & Cai, W. (2007). Different ZnO Nanostructures Fabricated by a Seed-Layer Assisted Electrochemical Route and Their Photoluminescence and Field Emission Properties. *J. Phys. Chem. C* Vol.111, No.6, (Feb 2007), pp. 2470-2476, ISSN 1932-7455.

Chalamala, B. R.; Wei, Y. & Gnade, B. E. (1998). FED up with fat tubes. *IEEE Spectrum,* Vol.35 No.4, (April 1998),pp. 42-51, ISSN 0018-9235

Chavan, P. G.; Badadhe, S. S.; Mulla, I. S.; More, M. A. & Joag, D. S. (2011). Synthesis of single crystalline CdS nanocombs and their application in photo-sensitive field emission switches *Nanoscale,* Vol.3, No.3, (March 2011), pp. 1078-1083, ISSN 2040-3372

Chen, J.; Deng, S.Z.; Xu, N. S.; Wang, S.; Wen, X.; Yang, S.; Yang, C.; Wang, J. & Ge, W. (2002). Field emission from crystalline copper sulphide nanowire arrays *Appl. Phys. Lett., Vol.*80, No.19, (May 2002), pp. 3620-3622, ISSN 1077-3118

Datta, A.; Chavan, P.G.; Sheini, F.J.; More, M. A.; Joag, D. S. & Patra, A. (2009). Growth, Optical, and Field Emission Properties of Aligned CdS Nanowires *Cryst. Growth Des.,* Vol.9, No.9, (Sept 2009), pp. 4157-4162, ISSN 1528-7505

Dong, L.; Jiao, J.; Tuggle, D. W.; Petty, J. M.; Elliff, S. A. & Coulter, M. (2003). ZnO nanowires formed on tungsten substrates and their electron field emission properties *Appl. Phys. Lett.,* Vol.82, No.7, (Feb 2003), pp. 1096-1098, ISSN 1077-3118

Fang, X.; Bando, Y.; Gautam, U. K.; Ye, C. & Golberg, D. (2008). Inorganic semiconductor nanostructures and their field-emission applications *J. Mater. Chem.,* Vol.18, No.5, (Jan 2008), pp. 509-522, ISSN 1364-5501.

Forbes, R. G. (2001). Low-macroscopic-field electron emission from carbon films and other electrically nanostructured heterogeneous materials: hypotheses about emission mechanism. *Solid-State Electronics,* Vol.45, No.6, (June 2001), pp. 779-808, ISSN 0038-1101

Forbes, R. G. (2008). Call for experimental test of a revised mathematical form for empirical field emission current-voltage characteristics. *Appl. Phys. letts.,* Vol.92, No. 19, (May 2008), pp. 193105-193107, ISSN 1077-3118

Forbes R. G. (2010). Thin-slab model for field electron emission *J. Vac. Sci. Technol. B,* Vol.28, No. 2, (March 2010), pp.C2A43-C2A49, ISSN 1520-8567.

Forbes, R. G. & Deane, J. H. B. (2007). Reformulation of the standard theory of Fowler-Nordheim tunneling and cold field electron emission. *Proc. Royal Soc. A,* Vol.463, No.2087, (Nov 2007), pp.2907-2927, ISSN 1471-2946

Fowler, R. H.; & Nordheim, L. (1928). Electron Emission in Intense Electric Fields *Proc. Royal. Soc. London,* Vol.A119, No.781, (May 1928), pp. 173-181, ISSN 1471-2946

Gomer, R. (1961). *Field emission and field ionisation, Harward University press,* ISBN 1-56396-124-5, Cambridge, MA,

Hsieh, C.T.; Chen, J.M.; Lin, H. H. & Shih, H.C. (2003). Field emission from various CuO nanostructures *Appl. Phys. Lett.,* Vol.83, No. 16, (Oct 2003), pp. 3383-3385, ISSN 1077-3118

Hwang, S. K.; Lee, J. H.; Jeong, S. H.; Lee, P. S. & Lee, K. H. (2005). Fabrication of carbon nanotube emitters in an anodic aluminium oxide nanotemplate on a Si wafer by multi-step anodization *Nanotechnology*, Vol.16, No.6, (June 2005), pp. 850-858, ISSN 1361-6528

Jamali Sheini, F.; Joag, D. S. & More, M. A. (2009). Field emission studies on electrochemically synthesized ZnO nanowires *Ultramicroscopy*, Vol.109, No.5, (April 2009), pp. 418-422, ISSN 0304-3991

Jamali Sheini, F.; Joag, D. S.; More, M. A.; Singh, J. & Srivastava, O. N. (2010). Low temperature growth of aligned ZnO nanowires and their application as field emission cathodes. *Materials Chemistry and Physics* Vol.120, No.2-3, (Apr 2010), pp. 691-696, ISSN 0254-0584

Jamali Sheini, F.; More, M. A.; Jadkar, S. R.; Patil, K. R.; Pillai, V. K. & Joag, D. S. (2010). Observation of Photoconductivity in Sn-Doped ZnO Nanowires and Their Photoenhanced Field Emission Behavior *J. Phys. Chem. C*, Vol.114, No.9, (March 2010), pp. 3843-3849, ISSN 1932-7455

Jo, S. H.; Banerjee, D. & Ren, Z. F. (2004). Field emission of zinc oxide nanowires grown on carbon cloth. *Appl. Phys. Lett.*, Vol.85, No.8, (Aug 2004), pp. 1407-1409, ISSN 1077-3118

Joo, J.; Lee, S. J.; Park, D. H.; Kim, Y. S.; Lee, Y.; lee, C. J. & Lee S-R. (2006). Field emission characteristics of electrochemically synthesized nickel nanowires with oxygen plasma post-treatment *Nanotechnology*, Vol.17, No.14, (July 2006), pp. 3506-3511, ISSN 1530-6992

Kim, H. M.; Kang, T.W.; Chung, K. S.; Hong, J. P. & Choi, W. B. (2003), Field emission displays of wide-bandgap gallium nitride nanorod arrays grown by hydride vapor phase epitaxy *Chem. Phys. Lett.*, Vol.377, No.5-6, (August 2003), pp. 491-494, ISSN 0009-2614

Late, D. J.; Karmakar, S.; More, M. A.; Bhoraskar S. V. & Joag, D. S. (2010). Arc plasma synthesized LaB6 nanocrystallite film on various substrates as a field emitter *J Nanopart Res*, Vol.12, No.7, (Sept 2010), pp. 2393-2403 ISSN 1572-896X

Lee, C. J.; Lee, T. J.; Lyu, S. C.; Zhang, Y.; Ruh, H. & Lee, H. J. (2002). Field emission from well-aligned zinc oxide nanowires grown at low temperature *Appl. Phys. Lett.*, Vol.81, No.19, (Nov 2002), pp. 3648-3650, ISSN 1077-3118

Liao, L.; Li, J. C.; Wang, D. F.; Liu, C.; Liu, C. S.; Fu, Q. & Fan, L. X. (2005). Field emission property improvement of ZnO nanowires coated with amorphous carbon and carbon nitride films *Nanotechnology*, Vol.16, No.6, (June 2005), pp. 985-989, ISSN 1361-6528

Luo, S. H.; Wan, Q.; Liu, W. L.; Zhang, M.; Di, Z. F.; Wang, S.Y.; Song, Z.T.; Lin, C. L. & Dai, J.Y. (2004). Vacuum electron field emission from SnO_2 nanowhiskers synthesized by thermal evaporation. *Nanotechnology*, Vol.15, No.11, (November 2004), pp. 1424 1427, ISSN 1530-6992

Modinos, A. (1984). *Field, Thermionic and secondary Electron Emission Spectroscopy* (Feb 1984), (*Plenum Press*), ISBN 0306413213, New York

Nakamoto, M. & Fukuda, K. (2002). Field electron emission from LaB_6 and TiN emitter arrays fabricated by transfer mold technique *Appl. Surf. Sci,* Vol.202, No.3-4, (Dec 2002), pp. 289-294, ISSN 0169-4332

Oostrom, A. G. J. Van *Philips Res. Rep.* 1, 11 (1966), ISSN 0031-7918.

Qi, K. C.; Lin, Z. L.; Chen, W. B.; Cao, G. C.; Cheng J. B. & Sun, X. W. (2008). Formation of extremely high current density LaB_6 field emission arrays via e-beam deposition *Appl. Phys. Lett,*Vol.93, No.9, (Sept 2008), pp. 093503-093505 ISSN 1077-3118

Qin, X-Z.; Wang, W-L.; Xu, N-S.; Bing, Z. & Forbes, R. G. (2011). Analytical treatment of cold field electron emission from a nanowall emitter, including quantum confinement effects *Proc. R. Soc. A,* Vol.467, No.2128, (April 2011), pp. 1029-1051, ISSN 1471-2946

Spindt C.; Holland C. E.; Rosengreen A. & Brodie I. (1991). Field emitter array development for gigahertz operation, *Proceeding of IEEE Transation on Electron Devices ED- 38,* ISBN 0018-9383, San Francisco, CA, May, 1991

Stratton, R. (1962). Theory of Field Emission from Semiconductors. *Phys. Rev.,* Vol.125, No.1, (Jan 2062), pp. 67-82, ISSN 1943-2879

Vila, L.; Vincent, P.; Pra, L. D.-D.; Pirio, G.; Minoux, E.; Gangloff, L.; Demoustier-Champagne S.; Sarazin, N.; Ferain, E.; Legras, Piraux, R. L. & P. Legagneux (2004). Growth and Field-Emission Properties of Vertically Aligned Cobalt Nanowire Arrays *Nano Lett.,* Vol.4, No.4, (March 2004), pp. 521-524, ISSN 1530-6992

Wang,, B.; Yang, Y .H.; Wang, C. X.; Xu, N. S. & Yang, G. W. (2005). Field emission and photoluminescence of SnO_2 nanograss. *J. Appl. Phys.,* Vol.98, No.12, (Dec 2005), pp. 124303-124306, ISSN 0021-8979

Wong, K. W.; Zhou, X. T.; Frederick, C. K.; Au, C. K.; Lai, H. L.; Lee, C. S. & Lee, S. T. (1999). Field-emission characteristics of SiC nanowires prepared by chemical-vapor deposition *Appl. Phys. Lett,* Vol.75, No.19, (Nov 1999), pp. 2918-2920, ISSN 1077-3118

Zeng, H.; Cui, J.; Cao, B.; Gibson, U.; Bando, Y. & Golberg, D. (2010). Electrochemical Deposition of ZnO Nanowire Arrays: Organization, Doping, and Properties *Science of Advanced Materials* Vol.2, No.3, (Sept 2010), pp. 336-358, ISSN 1947-2943

Zhang, H.; Tang, J.; Yuan, J.; Ma, J.; Shinya, N.; Nakajima, K.; Murakami, H.; Ohkubo, T. & Qin, L.-C. (2010). Nanostructured LaB_6 Field Emitter with Lowest Apical Work Function *Nano Lett.* Vol.10, No.9, (Sept 2010), pp. 3539-3544, ISSN 1530-6992

Zhou, J.; Deng, S. Z.; Xu, N. S.; Chen, J. & She, J. C. (2003). Synthesis and field-emission properties of aligned MoO_3 nanowires *Appl. Phys. Lett.,* Vol.83, No.13, (Sept 2003), pp. 2653-2655, ISSN 1077-3118

Zhu, Y. W.; Zhang, H. Z.; Sun, X. C.; Feng, S. Q.; Xu, J.; Zhao, Q.; Xiang, B.; Wang, R. M. & Yu, D. P. (2003). Efficient field emission from ZnO nanoneedle arrays *Appl. Phys. Lett.,* Vol.83, No.1, (July 2003), pp. 144-146, ISSN 1077-3118